VISUAL DIFFERENTIAL GEOMETRY *and* FORMS

VISUAL DIFFERENTIAL GEOMETRY *and* FORMS

A mathematical drama in five acts

TRISTAN NEEDHAM

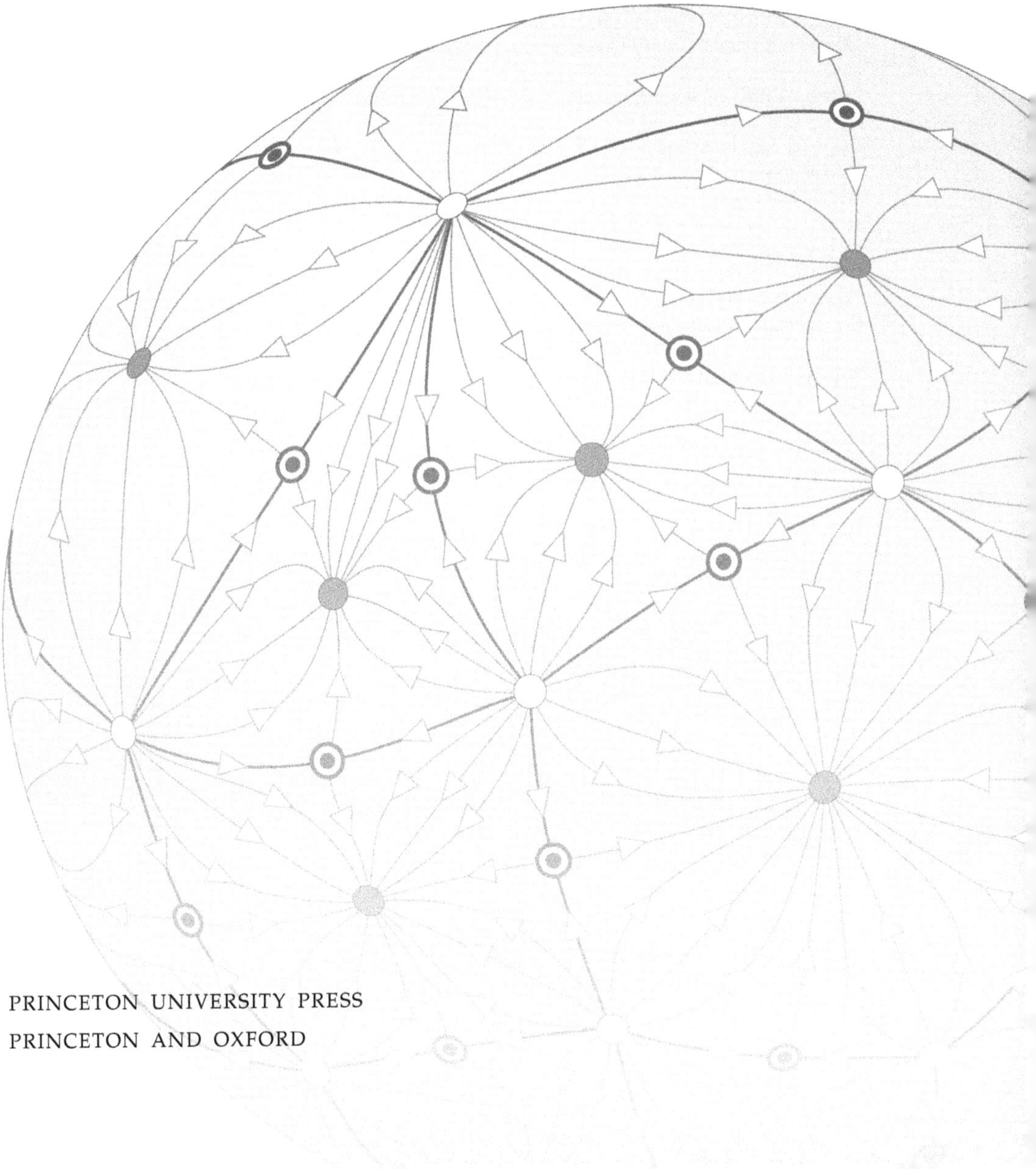

PRINCETON UNIVERSITY PRESS
PRINCETON AND OXFORD

Published by Princeton University Press
41 William Street, Princeton, New Jersey 08540
6 Oxford Street, Woodstock, Oxfordshire OX20 1TR

press.princeton.edu

ISBN 9780691203690
ISBN (pbk.) 9780691203706
ISBN (ebook) 9780691219899

Library of Congress Control Number: 2021934723

British Library Cataloging-in-Publication Data is available

Editorial: Susannah Shoemaker, Kristen Hop
Production Editorial: Terri O'Prey
Text Design: Wanda España
Jacket/Cover Design: Wanda España
Production: Brigid Ackerman
Publicity: Matthew Taylor, Amy Stewart
Copyeditor: Gregory W. Zelchenko
Cover image: Stiefel vector field (figure [19.12])

This book has been composed in Palatino (text) and Euler (mathematics)

10 9 8 7 6 5 4 3 2 1

For Roger Penrose

Contents

Prologue *xvii*
Acknowledgements *xxv*

ACT I
The Nature of Space

1 Euclidean and Non-Euclidean Geometry **3**

1.1 Euclidean and Hyperbolic Geometry 3
1.2 Spherical Geometry 6
1.3 The Angular Excess of a Spherical Triangle 8
1.4 Intrinsic and Extrinsic Geometry of Curved Surfaces 9
1.5 Constructing Geodesics via Their Straightness 11
1.6 The Nature of Space 14

2 Gaussian Curvature **17**

2.1 Introduction 17
2.2 The Circumference and Area of a Circle 19
2.3 The Local Gauss–Bonnet Theorem 22

3 Exercises for Prologue and Act I **24**

ACT II
The Metric

4 Mapping Surfaces: The Metric **31**

4.1 Introduction 31
4.2 The Projective Map of the Sphere 32
4.3 The Metric of a General Surface 34
4.4 The Metric Curvature Formula 37
4.5 Conformal Maps 38
4.6 Some Visual Complex Analysis 41
4.7 The Conformal Stereographic Map of the Sphere 44
4.8 Stereographic Formulas 47
4.9 Stereographic Preservation of Circles 49

5 The Pseudosphere and the Hyperbolic Plane **51**

5.1 Beltrami's Insight 51
5.2 The Tractrix and the Pseudosphere 52
5.3 A Conformal Map of the Pseudosphere 54
5.4 The Beltrami–Poincaré Half-Plane 56

5.5 Using Optics to Find the Geodesics 58
5.6 The Angle of Parallelism 60
5.7 The Beltrami–Poincaré Disc 62

6 Isometries and Complex Numbers 65

6.1 Introduction 65
6.2 Möbius Transformations 67
6.3 The Main Result 72
6.4 Einstein's Spacetime Geometry 74
6.5 Three-Dimensional Hyperbolic Geometry 79

7 Exercises for Act II 83

ACT III
Curvature

8 Curvature of Plane Curves 97

8.1 Introduction 97
8.2 The Circle of Curvature 98
8.3 Newton's Curvature Formula 100
8.4 Curvature as Rate of Turning 101
8.5 Example: Newton's *Tractrix* 104

9 Curves in 3-Space 106

10 The Principal Curvatures of a Surface 109

10.1 Euler's Curvature Formula 109
10.2 Proof of Euler's Curvature Formula 110
10.3 Surfaces of Revolution 112

11 Geodesics and Geodesic Curvature 115

11.1 Geodesic Curvature and Normal Curvature 115
11.2 Meusnier's Theorem 117
11.3 Geodesics are "Straight" 118
11.4 Intrinsic Measurement of Geodesic Curvature 119
11.5 A Simple Extrinsic Way to Measure Geodesic Curvature 120
11.6 A New Explanation of the Sticky-Tape Construction of Geodesics 120
11.7 Geodesics on Surfaces of Revolution 121
11.7.1 Clairaut's Theorem on the Sphere 121
11.7.2 Kepler's Second Law 123
11.7.3 Newton's Geometrical Demonstration of Kepler's Second Law 124
11.7.4 Dynamical Proof of Clairaut's Theorem 126
11.7.5 Application: Geodesics in the Hyperbolic Plane (Revisited) 128

12 **The Extrinsic Curvature of a Surface** 130

12.1 Introduction 130
12.2 The Spherical Map 130
12.3 Extrinsic Curvature of Surfaces 131
12.4 What Shapes Are Possible? 135

13 **Gauss's *Theorema Egregium*** 138

13.1 Introduction 138
13.2 Gauss's *Beautiful Theorem* (1816) 138
13.3 Gauss's *Theorema Egregium* (1827) 140

14 **The Curvature of a Spike** 143

14.1 Introduction 143
14.2 Curvature of a Conical Spike 143
14.3 The Intrinsic and Extrinsic Curvature of a Polyhedral Spike 145
14.4 *The Polyhedral Theorema Egregium* 147

15 **The Shape Operator** 149

15.1 Directional Derivatives 149
15.2 The Shape Operator S 151
15.3 The Geometric Effect of S 152
15.4 DETOUR: The *Geometry* of the Singular Value Decomposition and of the Transpose 154
15.5 The General Matrix of S 158
15.6 Geometric Interpretation of S and Simplification of [S] 159
15.7 [S] Is Completely Determined by Three Curvatures 161
15.8 Asymptotic Directions 162
15.9 Classical Terminology and Notation: The Three *Fundamental Forms* 164

16 **Introduction to the Global Gauss–Bonnet Theorem** 165

16.1 Some Topology and the Statement of the Result 165
16.2 Total Curvature of the Sphere and of the Torus 168
16.2.1 Total Curvature of the Sphere 168
16.2.2 Total Curvature of the Torus 169
16.3 Seeing $\mathcal{K}(\mathcal{S}_g)$ via a Thick Pancake 170
16.4 Seeing $\mathcal{K}(\mathcal{S}_g)$ via Bagels and Bridges 171
16.5 The Topological Degree of the Spherical Map 172
16.6 Historical Note 174

17 **First (Heuristic) Proof of the Global Gauss–Bonnet Theorem** 175

17.1 Total Curvature of a Plane Loop: Hopf's *Umlaufsatz* 175
17.2 Total Curvature of a Deformed Circle 178
17.3 Heuristic Proof of Hopf's *Umlaufsatz* 179

17.4 Total Curvature of a Deformed Sphere 180
17.5 Heuristic Proof of the Global Gauss–Bonnet Theorem 181

18 Second (Angular Excess) Proof of the Global Gauss–Bonnet Theorem 183

18.1 The Euler Characteristic 183
18.2 Euler's (Empirical) Polyhedral Formula 183
18.3 Cauchy's Proof of Euler's Polyhedral Formula 186
18.3.1 Flattening Polyhedra 186
18.3.2 The Euler Characteristic of a Polygonal Net 187
18.4 Legendre's Proof of Euler's Polyhedral Formula 188
18.5 Adding Handles to a Surface to Increase Its Genus 190
18.6 Angular Excess Proof of the Global Gauss–Bonnet Theorem 193

19 Third (Vector Field) Proof of the Global Gauss–Bonnet Theorem 195

19.1 Introduction 195
19.2 Vector Fields in the Plane 195
19.3 The Index of a Singular Point 196
19.4 The Archetypal Singular Points: Complex Powers 198
19.5 Vector Fields on Surfaces 201
19.5.1 The Honey-Flow Vector Field 201
19.5.2 Relation of the Honey-Flow to the Topographic Map 203
19.5.3 Defining the Index on a Surface 204
19.6 The Poincaré–Hopf Theorem 206
19.6.1 Example: The Topological Sphere 206
19.6.2 Proof of the Poincaré–Hopf Theorem 207
19.6.3 Application: Proof of the Euler–L'Huilier Formula 208
19.6.4 Poincaré's Differential Equations Versus Hopf's *Line Fields* 210
19.7 Vector Field Proof of the Global Gauss–Bonnet Theorem 214
19.8 The Road Ahead 218

20 Exercises for Act III 219

ACT IV

Parallel Transport

21 An Historical Puzzle 231

22 Extrinsic Constructions 233

22.1 Project into the Surface as You Go! 233
22.2 Geodesics and Parallel Transport 235
22.3 Potato-Peeler Transport 236

23 Intrinsic Constructions 240

23.1 Parallel Transport via Geodesics 240
23.2 The Intrinsic (aka, "Covariant") Derivative 241

24 Holonomy **245**

24.1 Example: The Sphere 245
24.2 Holonomy of a General Geodesic Triangle 246
24.3 Holonomy Is Additive 248
24.4 Example: The Hyperbolic Plane 248

25 An Intuitive Geometric Proof of the *Theorema Egregium* **252**

25.1 Introduction 252
25.2 Some Notation and Reminders of Definitions 253
25.3 The Story So Far 253
25.4 The Spherical Map Preserves Parallel Transport 254
25.5 The Beautiful Theorem and *Theorema Egregium* Explained 256

26 Fourth (Holonomy) Proof of the Global Gauss–Bonnet Theorem **257**

26.1 Introduction 257
26.2 Holonomy Along an *Open* Curve? 257
26.3 Hopf's Intrinsic Proof of the Global Gauss–Bonnet Theorem 258

27 Geometric Proof of the Metric Curvature Formula **261**

27.1 Introduction 261
27.2 The Circulation of a Vector Field Around a Loop 262
27.3 Dry Run: Holonomy in the Flat Plane 264
27.4 Holonomy as the Circulation of a Metric-Induced Vector Field in the Map 266
27.5 Geometric Proof of the Metric Curvature Formula 268

28 Curvature as a Force between Neighbouring Geodesics **269**

28.1 Introduction to the Jacobi Equation 269
28.1.1 Zero Curvature: The Plane 269
28.1.2 Positive Curvature: The Sphere 270
28.1.3 Negative Curvature: The Pseudosphere 272
28.2 Two Proofs of the Jacobi Equation 274
28.2.1 Geodesic Polar Coordinates 274
28.2.2 Relative Acceleration = Holonomy of Velocity 276
28.3 The Circumference and Area of a Small Geodesic Circle 278

29 Riemann's Curvature **280**

29.1 Introduction and Summary 280
29.2 Angular Excess in an n-Manifold 281
29.3 Parallel Transport: Three Constructions 282
29.3.1 Closest Vector on Constant-Angle Cone 282
29.3.2 Constant Angle within a Parallel-Transported Plane 283
29.3.3 *Schild's Ladder* 284

29.4 The Intrinsic (aka "Covariant") Derivative $\nabla_{\mathbf{v}}$ 284
29.5 The Riemann Curvature Tensor 286
29.5.1 Parallel Transport Around a Small "Parallelogram" 286
29.5.2 Closing the "Parallelogram" with the Vector Commutator 287
29.5.3 The General Riemann Curvature Formula 288
29.5.4 Riemann's Curvature Is a *Tensor* 291
29.5.5 Components of the Riemann Tensor 292
29.5.6 For a Given $\mathbf{w}_o$, the Vector Holonomy *Only* Depends on the *Plane* of the Loop and Its *Area* 293
29.5.7 Symmetries of the Riemann Tensor 294
29.5.8 Sectional Curvatures 296
29.5.9 Historical Notes on the Origin of the Riemann Tensor 297
29.6 The Jacobi Equation in an n-Manifold 299
29.6.1 Geometrical Proof of the Sectional Jacobi Equation 299
29.6.2 Geometrical Implications of the Sectional Jacobi Equation 300
29.6.3 Computational Proofs of the Jacobi Equation and the Sectional Jacobi Equation 301
29.7 The Ricci Tensor 302
29.7.1 Acceleration of the Area Enclosed by a Bundle of Geodesics 302
29.7.2 Definition and Geometrical Meaning of the Ricci Tensor 304
29.8 Coda 306

30 Einstein's Curved Spacetime 307

30.1 Introduction: *"The Happiest Thought of My Life."* 307
30.2 Gravitational Tidal Forces 308
30.3 Newton's Gravitational Law in Geometrical Form 312
30.4 The Spacetime Metric 314
30.5 Spacetime Diagrams 315
30.6 Einstein's Vacuum Field Equation in Geometrical Form 317
30.7 The Schwarzschild Solution and the First Tests of the Theory 319
30.8 Gravitational Waves 323
30.9 The Einstein Field Equation (with Matter) in Geometrical Form 326
30.10 Gravitational Collapse to a Black Hole 329
30.11 The Cosmological Constant: *"The Greatest Blunder of My Life."* 331
30.12 The End 333

31 Exercises for Act IV 334

ACT V
Forms

32 1-Forms **345**

32.1 Introduction 345
32.2 Definition of a 1-Form 346
32.3 Examples of 1-Forms 347
32.3.1 Gravitational Work 347
32.3.2 Visualizing the Gravitational Work 1-Form 348
32.3.3 Topographic Maps and the Gradient 1-Form 349
32.3.4 Row Vectors 352
32.3.5 Dirac's Bras 352
32.4 Basis 1-Forms 352
32.5 Components of a 1-Form 354
32.6 The Gradient as a 1-Form: $\mathbf{d}f$ 354
32.6.1 Review of the Gradient as a Vector: $\boldsymbol{\nabla} f$ 354
32.6.2 The Gradient as a 1-Form: $\mathbf{d}f$ 355
32.6.3 The Cartesian 1-Form Basis: $\{\mathbf{d}x^j\}$ 356
32.6.4 The 1-Form Interpretation of $df = (\partial_x f)\, dx + (\partial_y f)\, dy$ 357
32.7 Adding 1-Forms Geometrically 357

33 Tensors **360**

33.1 Definition of a Tensor: Valence 360
33.2 Example: Linear Algebra 361
33.3 New Tensors from Old 361
33.3.1 Addition 361
33.3.2 Multiplication: The Tensor Product 361
33.4 Components 362
33.5 Relation of the Metric Tensor to the Classical Line Element 363
33.6 Example: Linear Algebra (Again) 364
33.7 Contraction 365
33.8 Changing Valence with the Metric Tensor 366
33.9 Symmetry and Antisymmetry 368

34 2-Forms **370**

34.1 Definition of a 2-Form and of a p-Form 370
34.2 Example: The Area 2-Form 371
34.3 The Wedge Product of Two 1-Forms 372
34.4 The Area 2-Form in Polar Coordinates 374
34.5 Basis 2-Forms and Projections 375
34.6 Associating 2-Forms with Vectors in $\mathbb{R}^3$: Flux 376
34.7 Relation of the Vector and Wedge Products in $\mathbb{R}^3$ 379
34.8 The Faraday and Maxwell Electromagnetic 2-Forms 381

35 3-Forms **386**

35.1 A 3-Form Requires Three Dimensions 386
35.2 The Wedge Product of a 2-Form and 1-Form 386
35.3 The Volume 3-Form 387
35.4 The Volume 3-Form in Spherical Polar Coordinates 388
35.5 The Wedge Product of Three 1-Forms and of p 1-Forms 389
35.6 Basis 3-Forms 390
35.7 Is $\boldsymbol{\Psi} \wedge \boldsymbol{\Psi} \neq 0$ Possible? 391

36 Differentiation **392**

36.1 The Exterior Derivative of a 1-Form 392
36.2 The Exterior Derivative of a 2-Form and of a p-Form 394
36.3 The Leibniz Rule for Forms 394
36.4 Closed and Exact Forms 395
36.4.1 A Fundamental Result: $\mathbf{d}^2 = 0$ 395
36.4.2 Closed and Exact Forms 396
36.4.3 Complex Analysis: Cauchy–Riemann Equations 397
36.5 Vector Calculus via Forms 398
36.6 Maxwell's Equations 401

37 Integration **404**

37.1 The Line Integral of a 1-Form 404
37.1.1 Circulation and Work 404
37.1.2 Path-Independence $\Longleftrightarrow$ Vanishing Loop Integrals 405
37.1.3 The Integral of an Exact Form: $\boldsymbol{\varphi} = \mathbf{d}f$ 406
37.2 The Exterior Derivative as an Integral 406
37.2.1 **d**(1-Form) 406
37.2.2 **d**(2-Form) 409
37.3 Fundamental Theorem of Exterior Calculus (Generalized Stokes's Theorem) 411
37.3.1 Fundamental Theorem of Exterior Calculus 411
37.3.2 Historical Aside 411
37.3.3 Example: Area 412
37.4 The Boundary of a Boundary Is Zero! 412
37.5 The Classical Integral Theorems of Vector Calculus 413
37.5.1 $\boldsymbol{\Phi} = 0$-Form 413
37.5.2 $\boldsymbol{\Phi} = 1$-Form 414
37.5.3 $\boldsymbol{\Phi} = 2$-Form 415
37.6 Proof of the Fundamental Theorem of Exterior Calculus 415
37.7 Cauchy's Theorem 417
37.8 The Poincaré Lemma for 1-Forms 418
37.9 A Primer on de Rham Cohomology 419
37.9.1 Introduction 419

37.9.2 A Special 2-Dimensional Vortex Vector Field 419
37.9.3 The Vortex 1-Form Is Closed 420
37.9.4 Geometrical Meaning of the Vortex 1-Form 420
37.9.5 The Topological Stability of the Circulation of a Closed 1-Form 421
37.9.6 The First de Rham Cohomology Group 423
37.9.7 The Inverse-Square Point Source in $\mathbb{R}^3$ 424
37.9.8 The Second de Rham Cohomology Group 426
37.9.9 The First de Rham Cohomology Group of the Torus 428

38 Differential Geometry via Forms 430

38.1 Introduction: Cartan's Method of Moving Frames 430
38.2 Connection 1-Forms 432
38.2.1 Notational Conventions and Two Definitions 432
38.2.2 Connection 1-Forms 432
38.2.3 WARNING: Notational Hazing Rituals Ahead! 434
38.3 The Attitude Matrix 435
38.3.1 The Connection Forms via the Attitude Matrix 435
38.3.2 Example: The Cylindrical Frame Field 436
38.4 Cartan's Two Structural Equations 438
38.4.1 The Duals $\boldsymbol{\theta}^i$ of $\mathbf{m}_i$ in Terms of the Duals $\mathbf{d}x^j$ of $\mathbf{e}_j$ 438
38.4.2 Cartan's First Structural Equation 439
38.4.3 Cartan's Second Structural Equation 440
38.4.4 Example: The Spherical Frame Field 441
38.5 The Six Fundamental Form Equations of a Surface 446
38.5.1 Adapting Cartan's Moving Frame to a Surface: The Shape Operator and the Extrinsic Curvature 446
38.5.2 Example: The Sphere 447
38.5.3 Uniqueness of Basis Decompositions 447
38.5.4 The Six Fundamental Form Equations of a Surface 448
38.6 Geometrical Meanings of the Symmetry Equation and the Peterson–Mainardi–Codazzi Equations 449
38.7 Geometrical Form of the Gauss Equation 450
38.8 Proof of the Metric Curvature Formula and the *Theorema Egregium* 451
38.8.1 Lemma: Uniqueness of $\boldsymbol{\omega}_{12}$ 451
38.8.2 Proof of the Metric Curvature Formula 451
38.9 A New Curvature Formula 452
38.10 Hilbert's Lemma 453
38.11 Liebmann's Rigid Sphere Theorem 454
38.12 The Curvature 2-Forms of an n-Manifold 455
38.12.1 Introduction and Summary 455
38.12.2 The Generalized Exterior Derivative 457

38.12.3 Extracting the Riemann Tensor from the Curvature 2-Forms 459
38.12.4 The Bianchi Identities Revisited 459
38.13 The Curvature of the Schwarzschild Black Hole 460

39 Exercises for Act V **465**

Further Reading *475*
Bibliography *485*
Index *491*

Prologue

The Faustian Offer

> Algebra is the offer made by the devil to the mathematician. The devil says: *"I will give you this powerful machine, and it will answer any question you like. All you need to do is give me your soul: give up geometry and you will have this marvellous machine."* ... the danger to our soul is there, because when you pass over into algebraic calculation, essentially you stop thinking: you stop thinking geometrically, you stop thinking about the meaning.
>
> Sir Michael Atiyah[1]

"Differential Geometry" contains the word *"Geometry."*

A tautology? Well, the undergraduate who first opens up the assigned textbook on the subject may care to disagree! In place of geometry, our hapless student is instead confronted with a profusion of *formulas*, and their proofs consist of lengthy and opaque *computations*. Adding insult to injury, these computations are frequently *ugly*, involving a "debauch of indices"[2]—a phrase coined by Élie Cartan (one of the principal heroes of our drama) in 1928. If the student is honest and brave, the professor may be forced to confront an embarrassingly blunt question: "Where has the *geometry* gone?!"

Now, truth be told, most modern texts *do* in fact contain many *pictures*, usually of computer-generated curves and surfaces. But, with few exceptions, these pictures are of specific, concrete examples, which merely *illustrate* theorems whose proofs rest entirely upon symbolic manipulation. In and of themselves, these pictures *explain nothing!*

The present book has *two* distinct and equally ambitious objectives, the first of which is the subject of the first four Acts—to put the "Geometry" back into introductory "Differential Geometry." The 235 hand-drawn diagrams contained in the pages that follow are qualitatively and fundamentally of a different character than mere computer-generated examples. They are the conceptual fruits of many years of intermittent but intense effort—they are the visual embodiment of *intuitive geometric explanations of stunning geometric facts.*

The words I wrote in the Preface to VCA[3] apply equally well now: "A significant proportion of the geometric observations and arguments contained in this book are, to the best of my knowledge, new. I have not drawn attention to this in the text itself as this would have served no useful purpose: students don't need to know, and experts will know without being told. However, in cases where an idea is clearly unusual but I am aware of it having been published by someone else, I have tried to give credit where credit is due." In addition, I have attributed *exercises* that appear to be original, but that are not of my making.

On a personal note (but with a serious mathematical point to follow), the roots of the present endeavour can be traced back decades, to my youth. The story amounts to a tale of two books.

---

[1] *Mathematics in the 20th Century* (Shenitzer & Stillwell, 2002, p. 6)

[2] The full quotation begins to reveal Cartan's heroic stature: "The utility of the absolute differential calculus of Ricci and Levi-Civita must be tempered by an avoidance of excessively formal calculations, where the debauch of indices disguises an often very simple geometric reality. It is this reality that I have sought to reveal." (From the preface to Cartan 1928.)

[3] Given the frequency with which I shall have occasion to refer back to my first book, *Visual Complex Analysis: 25th Anniversary Edition* (Needham 2023), I shall adopt the compact conceit of referring to it simply as VCA.

The *first book* ignited my profound fascination with Differential Geometry and with Einstein's General Theory of Relativity. Perhaps the experience was so intense because it was my *first* love; I was 19 years old. One day, at the end of my first year of studying physics at Merton College, Oxford, I stumbled upon a colossal black book in the bowels of Blackwell's bookshop. Though I did not know it then, the 1,217-page tome was euphemistically referred to by relativity theorists as "The Bible." Perhaps it is appropriate, then, that this remarkable work altered the entire course of my life. Had I not read *Gravitation* (Misner, Thorne, and Wheeler 1973), I would never have had the opportunity[4] to study under (and become lifelong friends with) Roger Penrose, who in turn fundamentally transformed my understanding of mathematics and of physics.

In the summer of 1982, having been intrigued by the mathematical glimpses contained in Westfall's (1980) excellent biography of Newton, I made an intense study of Newton's (1687) masterpiece, *Philosophæ Naturalis Principia Mathematica*, usually referred to simply as the *Principia*. This was the *second book* that fundamentally altered my life. While V. I. Arnol'd[5] and S. Chandrasekhar (1995) sought to lay bare the remarkable nature of Newton's *results* in the *Principia*, the present book instead arose out of a fascination with Newton's *methods*.

As we have discussed elsewhere,[6] Newtonian scholars have painstakingly dismantled the pernicious myth[7] that the results in the 1687 *Principia* were first derived by Newton using his original 1665 version of the calculus, and only later recast into the geometrical form that we find in the finished work.

Instead, it is now understood that by the mid-1670s, having studied Apollonius, Pappus, and Huygens, in particular, the mature Newton became disenchanted with the form in which he had originally discovered the calculus in his youth—which is different again from the Leibnizian form we all learn in college today—and had instead embraced purely geometrical methods.

Thus it came to pass that by the 1680s Newton's algebraic infatuation with power series gave way to a new form of calculus—what he called the "synthetic method of fluxions"[8]—in which the geometry of the Ancients was transmogrified and reanimated by its application to shrinking geometrical figures in their moment of vanishing. *This* is the potent but nonalgorithmic form of calculus that we find in full flower in his great *Principia* of 1687.

Just as I did in VCA, I now wish to take full advantage of Newton's approach throughout this book. Let me therefore immediately spell it out, and in significantly greater detail than I did in VCA, in the vain hope that this second book may inspire more mathematicians and physicists to adopt Newton's intuitive (yet rigorous[9]) methods than did my first.

If two quantities A and B depend on a small quantity ϵ, and their ratio approaches unity as ϵ approaches zero, then we shall avoid the more cumbersome language of limits by following Newton's precedent in the *Principia*, saying simply that, "A is ultimately equal to B." Also, as we did in earlier works (Needham 1993, 2014), we shall employ the symbol $\asymp$ to denote this concept of ultimate equality.[10] In short,

$$\text{"A is ultimately equal to B"} \quad \Longleftrightarrow \quad A \asymp B \quad \Longleftrightarrow \quad \lim_{\epsilon\to 0} \frac{A}{B} = 1.$$

---

[4]Years later I was privileged to meet with Wheeler several times, and to correspond with him, so I was finally able to thank him directly for the impact that his *Gravitation* had had upon my life.

[5]See Arnol'd and Vasil'ev (1991); Arnol'd (1990).

[6]See Needham (1993), the Preface to VCA, and Needham (2014).

[7]Sadly, this myth originated with Newton himself, in the heat of his bitter priority battle with Leibniz over the discovery of the calculus. See Arnol'd (1990), Bloye and Huggett (2011), de Gandt (1995), Guicciardini (1999), Newton (1687, p. 123), and Westfall (1980).

[8]See Guicciardini (2009, Ch. 9).

[9]Fine print to follow!

[10]This notation was subsequently adopted by the Nobel physicist, Subrahmanyan Chandrasekhar (see Chandrasekhar 1995, p. 44).

It follows [exercise] from the theorems on limits that ultimate equality is an equivalence relation, and that it also inherits additional properties of ordinary equality, e.g., $X \asymp Y \;\&\; P \asymp Q \Rightarrow X \cdot P \asymp Y \cdot Q$, and $A \asymp B \cdot C \Leftrightarrow (A/B) \asymp C$.

Before we begin to apply this idea in earnest, we also note that the jurisdiction of ultimate equality can be extended naturally to things other than numbers, enabling one to say, for example, that two triangles are "ultimately similar," meaning that their angles are ultimately equal.

Having grasped Newton's method, I immediately tried my own hand at using it to simplify my teaching of introductory calculus, only later realizing how I might apply it to Complex Analysis (in VCA), and now to Differential Geometry. Though I might choose any number of simple, illustrative examples (*see* Needham 1993 for more), I will reuse the specific one I gave in the preface to VCA, and for one simple reason: *this* time I will use the "$\asymp$"–notation to *present* the argument rigorously, whereas in VCA I did not. Indeed, this example may be viewed as a recipe for transforming most of VCA's "explanations" into "proofs,"[11] merely by sprinkling on the requisite $\asymp$ s.

Let us show that if $T = \tan\theta$, then $\frac{dT}{d\theta} = 1 + T^2$. See figure below. If we increase θ by a small (ultimately vanishing) amount $\delta\theta$, then T will increase by the length of the vertical hypotenuse δT of the small triangle, in which the other two sides of this triangle have been constructed to lie in the directions $(\theta + \delta\theta)$ and $(\theta + \frac{\pi}{2})$, as illustrated. To obtain the result, we first observe that in the limit that $\delta\theta$ vanishes, the small triangle with hypotenuse δT is ultimately similar to the large triangle with hypotenuse L, because $\psi \asymp \frac{\pi}{2}$. Next, as we see in the magnifying glass, the side δs adjacent to θ in the small triangle is ultimately equal to the illustrated arc of the circle with radius L, so $\delta s \asymp L\,\delta\theta$. Thus,

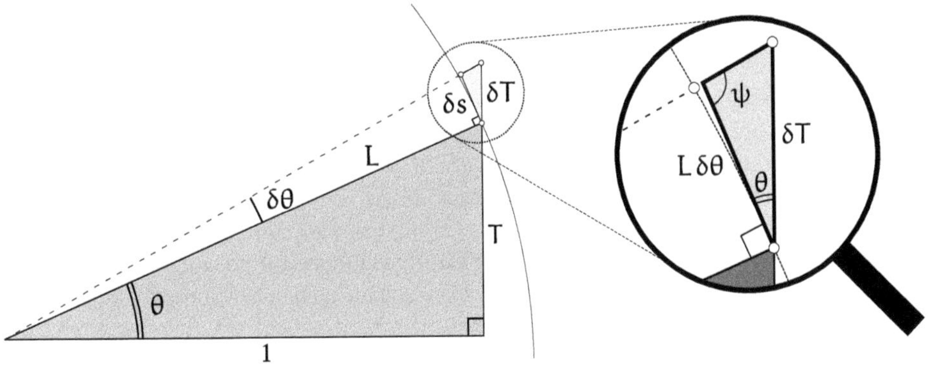

$$\frac{dT}{L\,d\theta} \asymp \frac{\delta T}{L\,\delta\theta} \asymp \frac{\delta T}{\delta s} \asymp \frac{L}{1} \quad \Longrightarrow \quad \frac{dT}{d\theta} = L^2 = 1 + T^2.$$

So far as I know, Newton never wrote down this specific example, but compare the illuminating directness of his *style*[12] of geometrical reasoning with the unilluminating computations we teach our students today, more than three centuries later! As Newton himself put it,[13] the geometric method is to be preferred by virtue of the "clarity and brevity of the reasoning involved and because of the simplicity of the conclusions and the illustrations required." Indeed, Newton went even further, resolving that *only* the synthetic method was "worthy of public utterance."

---

[11] I was already using the $\asymp$ notation (both privately and in print) at the time of writing VCA, and, in hindsight, it was a mistake that I did not employ it in that work; this led some to suppose that the arguments presented in VCA were less rigorous than they actually were (and remain).

[12] The best ambassador for Newton's approach will be you yourself. We therefore suggest that you *immediately* try your *own* hand at Newtonian reasoning, by doing Exercises 1, 2, 3, and 4, on page 24.

[13] See Guicciardini (2009, p. 231)

Newton himself did not employ *any* symbol to represent his concept of "ultimate equality." Instead, his devotion to the geometrical *method* of the Ancients spilled over into emulating their *mode* of expression, causing him to write out the words "ultimately have the ratio of equality," every single time the concept was invoked in a proof. As Newton (1687, p. 124) explained, the *Principia* is "written in words at length in the manner of the Ancients." Even when Newton claimed that two ratios were ultimately equal, he insisted on expressing *each ratio* in words. As a result, I myself was quite unable to follow Newton's reasoning without first transcribing and summarizing each of his paragraphs into "modern" form (which was in fact already quite common in 1687). Indeed, back in 1982, this was the catalyst for my private introduction and use of the symbol, $\asymp$.

It is my view that Newton's choice *not* to introduce a symbol for "ultimate equality" was a tragically consequential error for the development of mathematics. As Leibniz's symbolic calculus swept the world, Newton's more penetrating geometrical method fell by the wayside. In the intervening centuries only a handful of people ever sought to repair this damage and revive Newton's approach, the most notable and distinguished recent champion having been V. I. Arnol'd[14] (1937–2010).

Had Newton shed the trappings of this ancient mode of exposition and instead employed some symbol (*any* symbol!) in place of the words "ultimately equal," his dense, paragraph-length proofs in the *Principia* might have been reduced to a few succinct lines, and his mode of thought might still be widely employed today. Both VCA and this book are attempts to demonstrate, very concretely, the continuing relevance and vitality of Newton's geometrical approach, in areas of mathematics whose discovery lay a century in the future at the time of his death in 1727.

Allow me to insert some fine print concerning my use of the words "rigour" and "proof." Yes, my explicit use of Newtonian ultimate equalities in this work represents a quantum jump in rigour, as compared to my exposition in VCA, but there will be some mathematicians who will object (with justification!) that even this increase in rigour is insufficient, and that *none* of the "proofs" in this work are worthy of that title, including the one just given: I did not actually prove that the side of the triangle is ultimately equal to the arc of the circle.

I can offer no *logical* defence, but will merely repeat the words I wrote in the Preface of VCA, more than two decades ago: "My book will no doubt be flawed in many ways of which I am not yet aware, but there is one 'sin' that I have intentionally committed, and for which I shall not repent: many of the arguments are not rigorous, at least as they stand. This is a serious crime if one believes that our mathematical theories are merely elaborate mental constructs, precariously hoisted aloft. Then rigour becomes the nerve-racking balancing act that prevents the entire structure from crashing down around us. But suppose one believes, as I do, that our mathematical theories are attempting to capture aspects of a robust Platonic world that is not of our making. I would then contend that an initial lack of rigour is a small price to pay if it allows the reader to see into this world more directly and pleasurably than would otherwise be possible." So, to preemptively address my critics, let me therefore concede, from the outset, that when I claim that an assertion is "proved," it may be read as, *"proved beyond a reasonable doubt"!*[15]

Separate and apart from the issue of rigour is the sad fact that in rethinking so much classical mathematics I have almost certainly made mistakes: The blame for all such errors is mine, and mine alone. But please do not blame my geometrical tools for such poor craftsmanship—I am *equally* capable of making mistakes when performing symbolic computations! Corrections will be received with gratitude at VDGF.correction@gmail.com.

The book can be fully understood without giving a second thought to the complete arc of the unfolding drama, told as it is in five Acts. That said, I think that plot matters, and that the book's unorthodox structure and title are fitting, for the following reasons. First, I have sought

[14]See, for example, Arnol'd (1990).

[15]Upon reading these words, a strongly supportive member of the Editorial Board of Princeton University Press suggested to my editor that in place of "Q.E.D.," I conclude each of my proofs with the letters, "P.B.R.D."!

to present the ideas as dramatically as I myself see them, not only in terms of their historical development,[16] but also (more importantly) in terms of the cascading, interconnected flow of the ideas themselves, and their startling implications for the rest of mathematics and for physics. Second, more by instinct than design, the role of each of the five Acts does indeed follow (more or less) the classical structure of a Shakespearean drama; in particular, the anticipated "Climax" is indeed Act III: "Curvature." It was in fact years after I had begun work on the book that one day it suddenly became clear to me that what I had been composing all along had been *a mathematical drama in five acts.* That very day I "corrected" the title of the work, and correspondingly changed its five former "Parts" into "Acts":

- Act I: The Nature of Space
- Act II: The Metric
- Act III: Curvature
- Act IV: Parallel Transport
- Act V: Forms

The first four Acts fulfill the promise of a self-contained, *geometrical* introduction to Differential Geometry. Act IV is the true mathematical powerhouse that finally makes it possible to provide *geometric proofs* of many of the assertions made in the first three Acts.

Several aspects of the *subject matter* are as unorthodox (in a first course) as the geometrical methods by which they are treated. Here we shall describe only the three most important examples.

First, the climax *within* the climax of Act III is the *Global Gauss–Bonnet Theorem*—a remarkable link between local geometry and global topology. While the inclusion of this topic is standard, our treatment of it is not. Indeed, we celebrate its centrality and fundamental importance with an extravagant display of mathematical fireworks: we devote *five* chapters to it, offering up *four* quite distinct proofs, each one shedding new light on the result, and on the nature of Differential Geometry itself.

Second, the transition (usually in graduate school) from 2-dimensional surfaces to n-dimensional spaces (called "manifolds") is often confusing and intimidating for students. Chapter 29—the second longest chapter of the book—seeks to bridge this gap by focusing (initially) on the curvature of **3**-dimensional manifolds, which can be *visualized*; yet we frame the discussion so as to apply to *any* number of dimensions. We use this approach to provide an intuitive, geometrical, yet technically complete, introduction to the famous *Riemann tensor*, which measures the curvature of an n-dimensional manifold.

Third, having committed to a full treatment of the Riemann tensor, we felt it would have been *immoral* to have hidden from the reader its single greatest triumph in the arena of the natural world. We therefore conclude Act IV with a prolonged, *geometrical* introduction to Einstein's glorious *General Theory of Relativity*, which explains gravity as the curvature impressed upon 4-dimensional spacetime by matter and energy. This is the third longest chapter of the book. Not only does it treat (in complete geometrical detail) the famous *Gravitational Field Equation* (which Einstein discovered in 1915) but it also explains some of the most recent and exciting discoveries regarding its implications for black holes, gravitational waves, and cosmology!

Now let us turn to Act V, which is quite different in character from the four Acts that precede it, for it seeks to accomplish a *second* objective of the work, one that is quite distinct from the first, but no less ambitious.

Even the most rabid geometrical zealot must concede that Atiyah's diabolical machine (described in the opening quotation) is a *necessary* evil; but if we *must* calculate, let us at least

[16] As I did in VCA, I *strongly* recommend Stillwell's (2010) masterpiece, *Mathematics and Its History*, as a companion to this book, for it provides deeply insightful and detailed analysis of many historical developments that we can only touch on here.

do so gracefully! Fortunately, starting in 1900, Élie Cartan developed a powerful and elegant new method of *computation*, initially to investigate Lie Groups, but later to provide a new approach to Differential Geometry.

Cartan's discovery is called the "Exterior Calculus," and the objects it studies and differentiates and integrates are called "Differential Forms," here abbreviated simply to *Forms*. We shall ultimately follow Cartan's lead, illustrating his method's power and elegance in the final chapter of Act V—the longest chapter of the book—*reproving symbolically results that were proven geometrically in the first four Acts.* But Forms will carry us *beyond* what was possible in the first four Acts: in particular, they will provide a beautifully efficient method of calculating the Riemann tensor of an n-manifold, via its *curvature 2-forms*.

First, however, we shall fully develop Cartan's ideas in their own right, providing a self-contained introduction to Forms that is *completely independent* of the first four Acts. Lest there be any confusion, we repeat, *the first six chapters*—out of seven—*of Act V make no reference whatsoever to Differential Geometry!* We have done this because Forms find fruitful applications across diverse areas of mathematics, physics, and other disciplines. *Our aim is to make Forms accessible to the widest possible range of readers, even if their primary interest is* not *Differential Geometry.*

To that end, we have sought to treat Forms much more intuitively and *geometrically* than is customary. That said, the reader should be under no illusions: the principal purpose of Act V is to construct, at the *undergraduate* level, the "Devil's machine"—a remarkably powerful method of *computation*.

The immense power of these Forms is reminiscent of the complex numbers: a tiny drop goes in, and an ocean pours out—Cartan's Forms explain vastly more than was asked of them by their discoverer, a sure sign that he had hit upon *Platonic* Forms!

To give just one example, Forms unify and clarify *all* of Vector Calculus, in a way that would be a *revelation* to undergraduates, if only they were permitted to see it. Indeed, Green's Theorem, Gauss's Theorem, and Stokes's Theorem are merely different manifestations of a *single* theorem about Forms that is simpler than any of these special cases! Despite the indisputable importance of Differential Forms across mathematics and physics, most *undergraduates* will leave college without ever having seen them, and I have long considered this a scandal. Only a precious handful[17] of undergraduate textbooks (on either Vector Calculus *or* Differential Geometry) even mention their existence, and they are instead relegated to graduate school.

This lamentable state of affairs is now well into its second century, and I see no signs of an impending sea change. In response, Act V seeks not to curse the dark, but rather to light a candle,[18] striving to convince the reader that Cartan's Forms (and their underlying "tensors") are as *simple* as they are beautiful, and that they (and the name Cartan!) deserve to become a standard part of the *undergraduate* curriculum. *This* is the brazenly ambitious goal of Act V. After drowning the reader in *pure* Geometry for the first four Acts, we hope that the computational aspect of this final Act may serve as a suitably cathartic dénouement!

Before we close, let us simply list some housekeeping details:

- First, I have made no attempt to write this book as a classroom textbook. While I hope that some brave souls may nevertheless choose to use it for that purpose—as some previously did with VCA—my primary goal has been to communicate a majestic and powerful subject to the reader as honestly and as lucidly as I am able, regardless of whether that reader is a tender neophyte, or a hardened expert.

---

[17]See *Further Reading*, at the end of this book.

[18]Ours is certainly not the first such candle to be lit. Indeed, just as our work was nearing completion, Fortney (2018) published an entire book devoted to this same goal. However, Fortney's work does not include any discussion of Differential Geometry, and, at 461 pages, Fortney's book is considerably longer than the 100-page introduction to Forms contained in Act V of this book.

- My selection of topics may seem eclectic at times: for example, why is no attention paid to the fascinating and important topic of minimal surfaces? Frequently, as in this case, it is for one (or both) of the following two reasons: (1) our focus is on intrinsic geometry, *not* extrinsic[19] geometry; (2) an excellent literature already exists on the subject; in such cases, I have tried to provide useful pointers in the *Further Reading* section at the end of the book.
- *Equations* are numbered with (ROUND) brackets, while *figures* are numbered with [SQUARE] brackets.
- Bold italics are used to highlight the ***definition of a new term***.
- For ease of reference when flipping through the book, noteworthy results are `framed`, while doubly remarkable facts are `double – framed`. In the entire work, only a handful of results are *so* fundamental that they are *triple*-framed; we hope the reader will enjoy finding them, like Easter eggs.
- I have tried to make you, the reader, into an active participant in developing the ideas. For example, as an argument progresses, I have frequently and deliberately placed a pair of logical stepping stones sufficiently far apart that you may need to pause and stretch slightly to pass from one to the next. Such places are marked "[exercise]"; they often require nothing more than a simple calculation or a moment of reflection.
- Last, we encourage the reader to take full advantage of the *Index*; its creation was a painful labour of love!

We bring this Prologue to a close with a broader philosophical objective of the work, one that transcends the specific mathematics we shall seek to explain.

One of the rights [*sic*] of passage from mathematical adolescence to adulthood is the ability to distinguish *true miracles* from *false miracles*. Mathematics itself is replete with the former, but examples of the latter also abound: "I can't believe all those ugly terms cancelled and left me such a beautifully simple answer!"; or, "I can't believe that this complicated expression has such a simple meaning!"

Rather than congratulating oneself in such a circumstance, one should instead hang one's head in shame. For if all those ugly terms cancelled, *they should never have been there in the first place!* And if that complicated expression has a wonderfully simple meaning, *it should never have been that complicated in the first place!*

In my own case, I am embarrassed to confess that mathematical puberty lasted well into my 20s, and I only *started* to grow up once I became a graduate student, thanks to the marvellous twin influences of Penrose and of my close friend George Burnett-Stuart, a fellow advisee of Penrose.

The Platonic Forms of mathematical reality are always perfectly beautiful and they are always perfectly simple; transient impressions to the contrary are manifestations of our own imperfection. My hope is that this book may help nudge the reader towards humility in the face of this perfection, just as my two friends first nudged me down this path, so many years ago amidst the surreal, Escher-like spires of Oxford.

T. N.

Mill Valley, California
Newtonmas, 2019

---

[19]The meanings of "intrinsic" and "extrinsic" are explained in Section 1.4.

Acknowledgements

Roger Penrose transformed my understanding of mathematics and of physics. From the very first paper of his I ever read, when I was 20 years old, the *perfection*, beauty, and almost musical counterpoint of his ideas elicited in me a profound aesthetic exhilaration that I can only liken to the experience of listening to the opening of Bach's Cantata 101 or Beethoven's *Grosse Fuge*.

From the time that I was his student, Roger's ability to unravel the deepest of mysteries through *geometry* left an indelible mark—it instilled in me a lifelong, unshakable *faith* that a geometric explanation must always exist. (My study of Newton's *Principia* later served to *deepen* my belief in the universality of the geometrical approach.) Without that faith, this work could not exist, for it sometimes took me many *years* of groping before I discovered the geometric explanation of a particular mathematical phenomenon.

To be able to count myself amongst Roger's friends has been a great joy and a high honour for 40 years, and my dedication of this imperfect work to Roger can scarcely repay the intellectual debt that I owe to him, but it is the best that I can do.

In order to properly introduce the next person to whom I owe thanks, I am forced to reveal a somewhat shabby detail about myself: when I first came to America from England in 1989, I smoked two packs of cigarettes per day! In 1995 I was finally able to quit: it was the hardest thing I had ever done, and I would likely have failed, had it not been for the invention of the nicotine patch.

Perhaps five years later, in response to the 1997 publication of my *Visual Complex Analysis* (VCA), I received a "fan letter" from a *medical* researcher at Duke University; he planned to visit the Bay Area and asked if we could meet. With some trepidation, I agreed. My visitor turned out to be Professor Jed Rose, *inventor* of my (saviour) nicotine patch! Jed had started out in mathematics and physics, and had never lost his love of those disciplines, but shrewdly calculated that he could have a greater impact if he directed his energies to medical research, instead. I'm so glad that he did!

Once I began work on VDGF (*Visual Differential Geometry and Forms*, this book) in 2011, Jed became my most enthusiastic supporter, demonstrating great generosity in using funds from his medical inventions to buy out some of my teaching, thereby greatly assisting my research for VDGF. Every time Jed visited me and my family in California during the nine years of work on the book, my spirits were lifted by Jed's relentlessly upbeat personality and his belief in the importance of what I was trying to accomplish. And, as the manuscript slowly evolved, Jed offered a remarkably large number of detailed and helpful suggestions and corrections; the finished work is significantly better as a result of his helpful observations. Thus, as you see, Jed helped me in three linearly independent directions, and I cannot thank him enough. And as if all this were not enough, what started out as a purely intellectual relationship, subsequently blossomed into a very warm and close friendship between our two families.

The next key person I would like to thank is Professor Thomas Banchoff, the distinguished geometer of Brown University. During the writing of this book, I managed to arrange for Tom to come to USF as a visiting scholar in two separate years, for one semester each. Tom was extremely generous to me during both of those semesters, offering to read my evolving manuscript and giving me extremely valuable feedback. Each week he would read the latest installment of the manuscript, hot off the presses, and then he and I would meet in his office each Friday afternoon, and go over his detailed corrections and suggestions, written in red pen in the margins, line by line. Although this partnership sadly ended when the book was only half done, I have adopted essentially all of his helpful suggestions and corrections, and I am immensely grateful to him for sharing his deep geometrical wisdom and expertise with me.

I wish to express my sincere gratitude to Dr. Wei Liu[20] (a physicist specializing in optics research, whom I hope to eventually meet, some fine day). In 2019 he wrote to me to express his appreciation of VCA, and he enclosed a research paper of his[21] that cited my treatment of the Poincaré–Hopf Theorem. This paper totally opened my eyes to how physicists continue to make wonderful use of a beautiful result of Hopf that seems to have completely evaporated from all *mathematical* textbooks. The result is the subject of Section 19.6.4, and it says this: The Poincaré–Hopf Theorem not only applies to vector fields, but also to Hopf's *line fields*,[22] which greatly generalize vector fields and which can have singular points with *fractional* indices. Witness the examples shown in [19.14] on page 212. At my request, Dr. Liu then further assisted me by pointing out many other applications that physicists have made of Hopf's ideas. I have in turn shared his kind guidance with you, dear reader, in the *Further Reading* section at the end of the book.

In addition to the principal players above, I have received all manner of advice, support, and suggestions from colleagues and friends, near and far.

My beloved brother Guy is an anchor whose love and faith in me is too often taken for granted, but it should not be!

Stanley Nel and Paul Zeitz, my friends of 30 years, have always believed in me more than I have believed in myself, and their encouragement has meant a great deal to me over the many years it has taken to create this book, first struggling to discover the needed geometrical insights, then writing and *drawing* the book.

Douglas Hofstadter—whose *Gödel, Escher, Bach* transfixed me (and millions more) as an undergraduate—has honoured me with his support for more than 20 years. First, he has repeatedly and forcefully promoted VCA, both in print and in interviews. Second, he read and provided very valuable feedback on Needham (2014), which was ultimately incorporated into this book.

Dr. Ed Catmull—co-founder with Steve Jobs of Pixar, and later president of both Pixar and Walt Disney Animation Studios—wrote me a very flattering email about VCA back in 1999. At first I was convinced that one of my the University of San Francisco (USF) maths-pals was playing a practical joke on me, but the email was real. Ed invited me to visit the Pixar campus (still in Point Richmond back in those days), gave me a guided tour of the studios, took me out to lunch, and offered me a job! (I will leave it to the reader to assign a numerical value to my stupidity in turning that offer down.) Though my contact with Ed has been sporadic, he has been a faithful bolsterer of VCA (praising it in interviews he has done), he also wrote me a letter of recommendation for a grant, and he has been very supportive of my effort to create VDGF. I deeply appreciate Ed's encouragement over the years, and I thank him for the extremely kind words on the back of this book.

Professor Frank Morgan (whom I know by reputation only) was originally approached by Princeton University Press to provide an *anonymous* review of the manuscript of VDGF. But when he submitted his review to my editor, he *also* sent it to me directly, under his own name. I am very grateful that he chose to do this, as it now allows me to thank him publicly for his concrete suggestions and corrections. Furthermore, I was especially grateful for the tremendous boost his report gave to my *morale* at the time. Finally, I offer him my sincerest thanks for being willing to share his remarkably generous assessment of VDGF on the back of this book.

---

[20]College for Advanced Interdisciplinary Studies, National University of Defense Technology, Changsha, Hunan, P. R. China.

[21]Chen et al. (2020b).

[22]This is the modern terminology; Hopf (1956) originally called them *fields of line elements*.

I am likewise also grateful for the constructive criticism, suggestions, and corrections I have received from all of the truly anonymous reviewers—I have tried to incorporate all of their improvements, and I'm sorry I cannot thank each of them by name.

I thank The M. C. Escher Company for permission to reproduce two modifications of *Circle Limit I*: [5.11] and [5.12], the latter being an explicit mathematical transformation, carried out by John Stillwell, and used with his kind permission. Note that M. C. Escher's *Circle Limit I* is © 2020, The M. C. Escher Company, The Netherlands. All rights reserved. www.mcescher.com.

Finally, I am very grateful to Professor Henry Segerman for supplying me with the image of his *Topology Joke* [18.8], page 191, and for granting me permission to reproduce it here.

This is my second book, and it is also my last book. I therefore wish to not only thank all those who directly helped me create VDGF, listed above, but also those who influenced and supported me much earlier in my life. Some of these people were so seamlessly integrated into the fabric of my existence that they became invisible, and, shamefully, I failed to thank them properly in VCA; now is my last chance to put things right.

First amongst these is Anthony Levy, my oldest friend, from our undergraduate days together at Merton College, Oxford. Inexplicably, Anthony (or Tony, as I knew him then) believed in me long before there was any evidence to support such a belief, and that continued belief in me has buoyed me up repeatedly during periods of mathematical self-doubt over the decades. And, beyond the world of pure intellect, Anthony's sage advice and love have helped me navigate some of the most fraught episodes of my life.

Also from those undergraduate Merton days, I will always be grateful to my two physics tutors, Dr. Michael Baker (1930–2017) and Dr. Michael G. Bowler, who not only taught me a great deal of physics themselves, but who also went out of their way to arrange for me to have more advanced, individual tuition on General Relativity and on spinors, from two remarkable Fellows of Merton, Dr. Brian D. Bramson and Dr. Richard S. Ward. In particular, Dr. Bramson's enthusiasm for science was utterly contagious, and it was he who gave me my very first exposure to the (revelatory!) work of Penrose, and who pushed me to apply to undertake a DPhil in Penrose's "Relativity Group."

Moving forward to my graduate student days studying under Penrose, I wish to thank, once again, my friend George Burnett-Stuart, a fellow advisee of Penrose. George and I shared a small house together on Great Clarendon St. for several years as we carried out our doctoral work, and in the course of our endless discussions of music, physics, and mathematics, George helped me to refine both my conception of the nature of mathematics, and of what constitutes an acceptable explanation within that subject. For better or for worse, George bears great responsibility for the mathematician that I am today.

Moving forward again, to my life at USF, I am grateful to John Stillwell—who just retired—and his wonderful wife, Elaine, for 20 happy and productive years as colleagues and friends. While I have picked his brains many times during the course of writing this book, my greatest debt is to his many *writings*. Indeed, our relationship began with a "fan letter" I wrote to him in reaction to the first edition of his magnificent magnum opus, *Mathematics and Its History*. A few years later, while serving as Associate Dean for Sciences, I was able to lure John away from Australia, creating a professorship for him at USF. Both VCA and VDGF owe a great deal to John's holistic grasp of the entire expanse of mathematics, and his ability to use this perspective to give back *meaning* to mathematical ideas, all of which he has so generously shared with the world through his many wonderful books.

In 1996, I concluded the Acknowledgements of VCA with these words: "Lastly, I thank my dearest wife Mary. During the writing of this book she allowed me to pretend that science was the most important thing in life; now that the book is over, she is my daily proof that there is something even more important." Today, more than two decades later, my love for Mary has only grown more profound, but I now have two *more* daily proofs than I had before!

In 1999, Mary and I were blessed with twins: Faith and Hope have been our dazzling beacons of pride and joy ever since. I'm sorry that VDGF has hung like a dark cloud over the life of my family for almost half of my daughters's lives, and that it has robbed us of so much time together. Yet it is the love of these three souls that has given my life meaning and purpose, and has sustained me throughout the long struggle that created this work.

ACT I

The Nature of Space

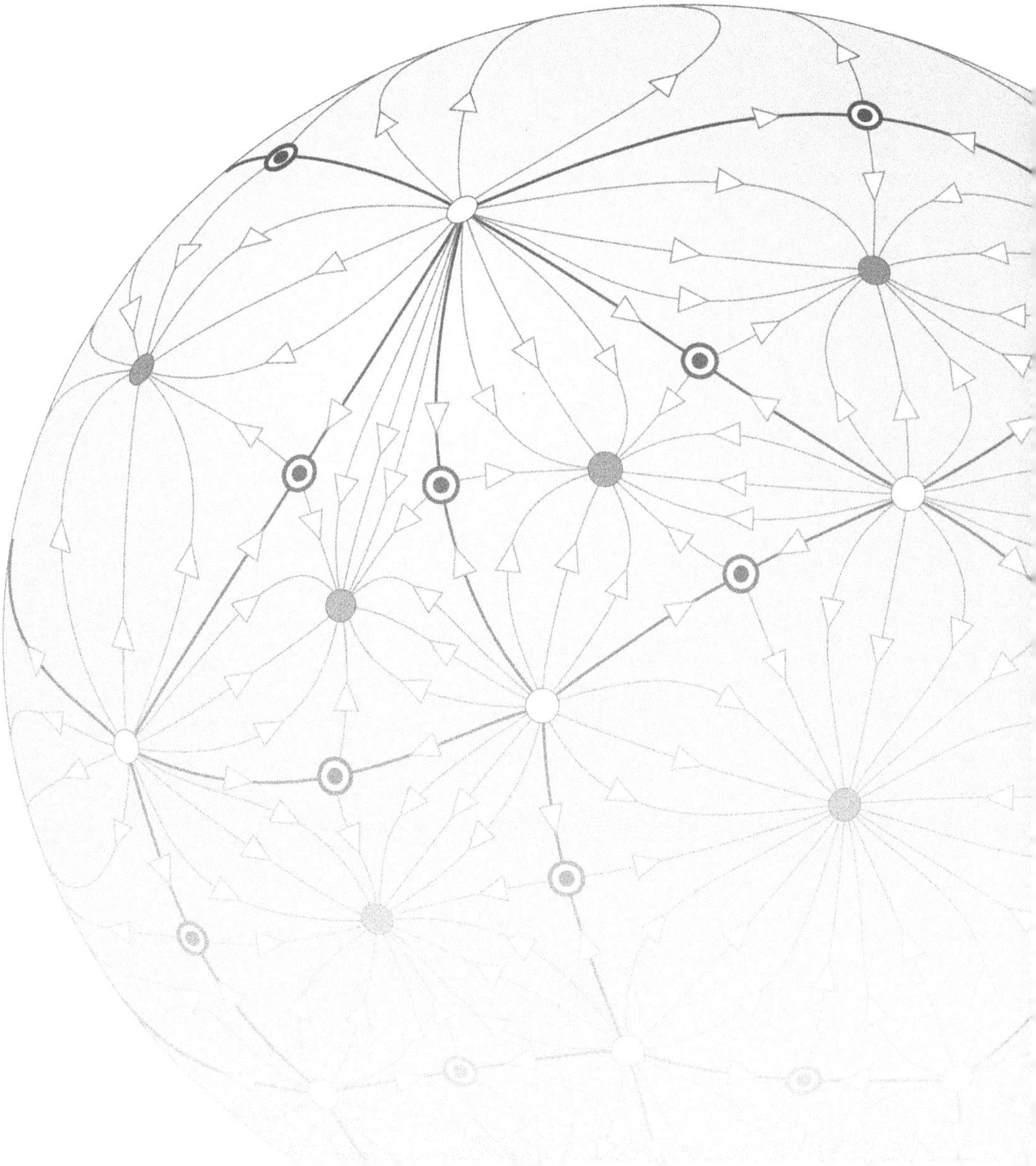

Chapter 1

Euclidean and Non-Euclidean Geometry

1.1 Euclidean and Hyperbolic Geometry

Differential Geometry is the application of calculus to the geometry of space that is *curved*. But to understand space that is curved we shall first try to understand space that is *flat*.

We inhabit a natural world pervaded by curved objects, and if a child asks us the meaning of the word "flat," we are most likely to answer in terms of the *absence* of curvature: a smooth surface *without* any bumps or hollows. Nevertheless, the very earliest mathematicians seem to have been drawn to the singular simplicity and uniformity of the flat plane, and they were rewarded with the discovery of startlingly beautiful facts about geometric figures constructed within it. With the benefit of enormous hindsight, some of these facts can be seen to *characterize* the plane's flatness.

One of the earliest and most profound such facts to be discovered was Pythagoras's Theorem. Surely the ancients must have been awed, as any sensitive person must remain today, that a seemingly unalloyed fact about *numbers*,

$$3^2 + 4^2 = 5^2,$$

in fact has *geometrical* meaning, as seen in [1.1].[1]

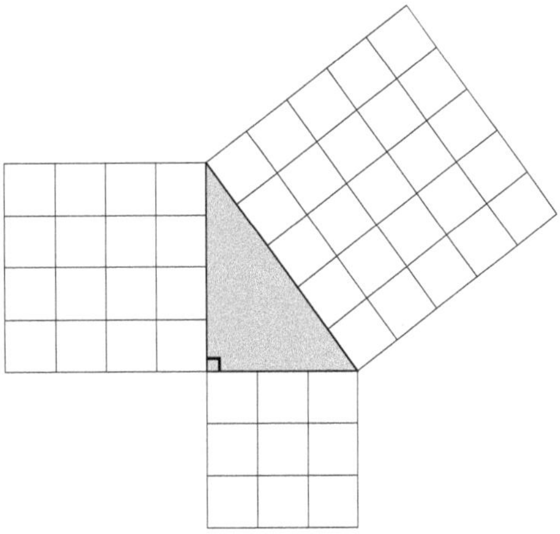

[1.1] *Pythagoras's Theorem: the geometry of* $3^2 + 4^2 = 5^2$.

While Pythagoras himself lived in Greece around 500 BCE, the theorem bearing his name was discovered much earlier, in various places around the world. The earliest known example of such knowledge is recorded in the Babylonian clay tablet (catalogued as "Plimpton 322") shown in [1.2], which was unearthed in what is now Iraq, and which dates from about 1800 BCE.

The tablet lists ***Pythagorean triples***:[2] integers (a, b, h) such that h is the hypotenuse of a right triangle with sides a and b, and therefore $a^2 + b^2 = h^2$. Some of these ancient examples are impressively large, and it seems clear that they did not stumble upon them, but rather possessed a mathematical process for generating solutions. For example, the fourth row of the tablet records the fact that $13500^2 + 12709^2 = 18541^2$.

The deeper knowledge that underlay these ancient results is not known,[3] but to find the first evidence of the "modern," logical, deductive approach to mathematics we must jump 1200 years into the future of the clay tablet. Scholars believe that it was Thales of Miletus (around 600 BCE)

---

[1] We repeat what was said in the Prologue: *equations* are labelled with parentheses (*round* brackets)—(...), while *figures* are labelled with *square* brackets—[...].

[2] In fact the tablet only records *two* members (a, h) of the triples (a, b, h).

[3] In the seventeenth century, Fermat and Newton reconstructed and generalized a *geometrical* method of generating the general solution, due to Diophantus. See Exercise 5.

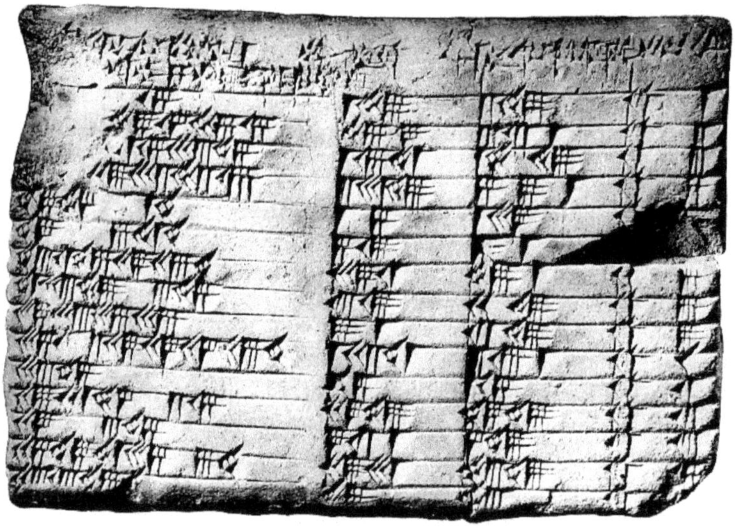

[1.2] *Plimpton 322: A clay tablet of Pythagorean triples from 1800 BCE.*

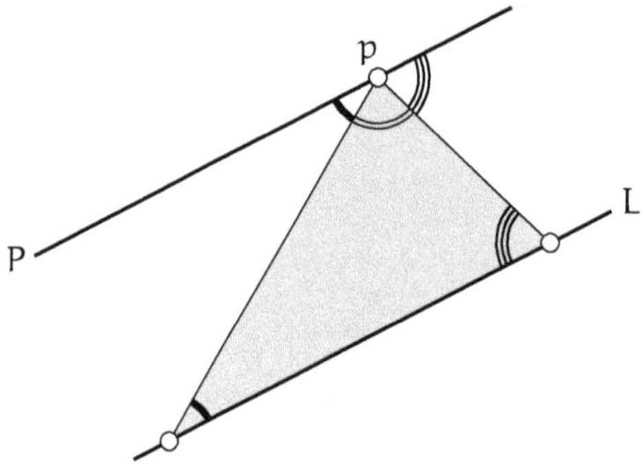

[1.3] *Euclid's Parallel Axiom:* P *is the* unique *parallel to* L *through* p*, and the angle sum of a triangle is* π.

who first pioneered the idea of deducing new results from previously established ones, the logical chain beginning at a handful of clearly articulated assumptions, or ***axioms***.

Leaping forward again, 300 years beyond Thales, we find one of the most perfect exemplars of this new approach in Euclid's *Elements*, dating from 300 BCE. This work sought to bring order, clarity, and rigour to geometry by deducing everything from just five simple axioms, the fifth and last of which dealt with parallel lines.

Defining two lines to be ***parallel*** if they do not meet, Euclid's Fifth Axiom[4] is illustrated in [1.3]:

> ***Parallel Axiom.*** *Through any point* p *not on the line* L *there exists precisely one line* P *that is parallel to* L.

But the character of this axiom was more complex and less immediate than that of the first four, and mathematicians began a long struggle to dispense with it as an assumption, instead seeking to show that it must be a logical *consequence* of the first four axioms.

This tension went unresolved for the next *2000 years*. As the centuries passed, many attempts were made to prove the Parallel Axiom, and the number and intensity of these efforts reached a crescendo in the 1700s, but all met with failure.

Yet along the way useful *equivalents* of the axiom emerged. For example: *There exist similar triangles of different sizes* (Wallis in 1663; see Stillwell (2010)). But the very first equivalent was already present in Euclid, and it is the one still taught to every school child: *the angles in a triangle add up to two right angles.* See [1.3].

The explanation of these failures only emerged around 1830. Completing a journey that had begun 4000 years earlier, Nikolai Lobachevsky and János Bolyai independently announced the

[4]Euclid did not state his axiom in this form, but it is logically equivalent.

discovery of an entirely new form of geometry (now called ***Hyperbolic Geometry***) taking place in a new kind of plane (now called the ***hyperbolic plane***). In this Geometry the first four Euclidean axioms hold, but the parallel axiom does *not*. Instead, the following is true:

Hyperbolic Axiom. *There are at least two parallel lines through* p *that do not meet* L. (1.1)

These pioneers explored the logical consequences of this axiom, and by purely abstract reasoning were led to a host of fascinating results within a rich new geometry that was bizarrely different from that of Euclid.

[1.4] *Johann Heinrich Lambert (1728–1777).*

Many others before them, perhaps most notably Saccheri (in 1733; see Stillwell 2010) and Lambert (in 1766; see Stillwell 2010), had discovered some of these consequences of (1.1), but their aim in exploring these consequences had been to find a *contradiction*, which they believed would finally prove that Euclidean Geometry to be the One True Geometry.

Certainly Saccheri believed he had found a clear contradiction when he published "Euclid Freed of Every Flaw." But Lambert is a much more perplexing case, and he is perhaps an unsung hero in this story. His results penetrated so deeply into this new geometry that it seems impossible that he did not at times believe in the reality of what he was doing. Regardless of his motivation and beliefs[5], Lambert (shown in [1.4]) was certainly the first to discover a remarkable fact[6] about the angle sum of a triangle under axiom (1.1), and his result will be central to much that follows in Act II.

Nevertheless, Lobachevsky and Bolyai richly deserve their fame for having been the first to recognize and fully embrace the idea that they had discovered an entirely new, consistent, non-Euclidean Geometry. But what this new geometry really *meant*, and what it might be useful for, even they could not say.[7]

Remarkably and surprisingly, it was the *Differential Geometry of curved surfaces* that ultimately resolved these questions. As we shall explain, in 1868 the Italian mathematician Eugenio Beltrami finally succeeded in giving Hyperbolic Geometry a concrete interpretation, setting it upon a firm and intuitive foundation from which it has since grown and flourished. Sadly, neither Lobachevsky nor Bolyai lived to see this: they died in 1856 and 1860, respectively.

This non-Euclidean Geometry had in fact already manifested itself in various branches of mathematics throughout history, but always in disguise. Henri Poincaré (beginning around 1882) was the first not only to strip it of its camouflage, but also to recognize and exploit its power

[5]I thank Roger Penrose for making me see that Lambert deserves greater credit than he is usually granted. Penrose did so by means of the following analogy: "Should we not give credit to Einstein for the cosmological constant, even if he introduced it for the wrong reasons? And should we blame him for later retracting it, calling it the "greatest blunder of my life"? Or what about General Relativity itself, which Einstein seemed to become less and less convinced was the right theory (needing to be replaced by some kind of non-singular unified field theory) as time went on?" [Private communication.]

[6]If you cannot wait, it's (1.8).

[7]Lobachevsky did in fact put this geometry to use to evaluate previously unknown integrals, but (at least in hindsight) this particular application must be viewed as relatively minor.

in such diverse areas as Complex Analysis, Differential Equations, Number Theory, and Topology. Its continued vitality and centrality in the mathematics of the 20th and twenty-first centuries is demonstrated by Thurston's work on 3-manifolds, Wiles's proof of ***Fermat's Last Theorem***, and Perelman's proof of the ***Poincaré Conjecture*** (as a special case of Thurston's ***Geometrization Conjecture***), to name but three examples.

In Act II we shall describe Beltrami's breakthrough, as well as the nature of Hyperbolic Geometry, but for now we wish to explore a different, simpler kind of non-Euclidean Geometry, one that was already known to the Ancients.

1.2 Spherical Geometry

To construct a non-Euclidean Geometry we must deny the existence of a unique parallel. The Hyperbolic Axiom assumes two or more parallels, but there is one other logical possibility—*no* parallels:

> ***Spherical Axiom.*** *There are* no *lines through* p *that are parallel to* L : every *line meets* L. (1.2)

Thus there are actually *two* non-Euclidean[8] geometries: spherical and hyperbolic.

As the name suggests, Spherical Geometry can be realized on the surface of a sphere—denoted $\mathbb{S}^2$ in the case of the *unit* sphere—which we may picture as the surface of the Earth. On this sphere, what should be the analogue of a "straight line" connecting two points on the surface? Answer: the shortest route between them! But if you wish to sail or fly from London to New York, for example, what *is* the shortest route?

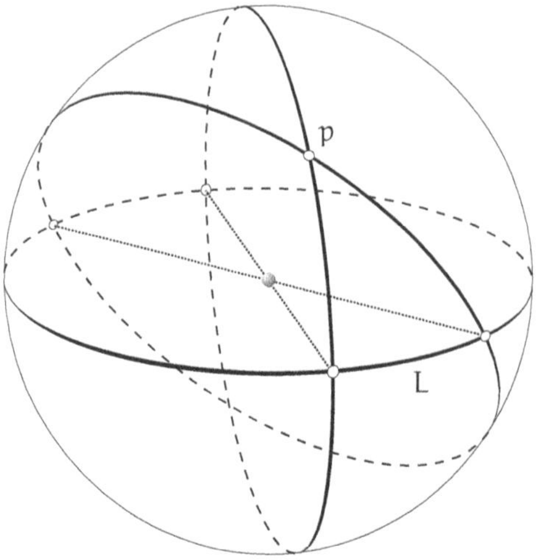

[1.5] *The great circles of $\mathbb{S}^2$ intersect in pairs of antipodal points.*

The answer, already known to the ancient mariners, is that the shortest route is an arc of a ***great circle***, such as the equator, obtained by cutting the sphere with a plane passing through its centre. In [1.5] we have chosen L to be the equator, and it is clear that (1.2) is satisfied: every line through p meets L in a pair of ***antipodal*** (i.e., diametrically opposite) points.

In the plane, the shortest route is also the *straightest* route, and in fact the same is true on the sphere: in a precise sense to be discussed later, the great circle trajectory bends neither to the right nor to the left as it traverses the spherical surface.

There are other ways of constructing the great circles on the Earth that do not require thinking about planes passing through the completely inaccessible centre of the Earth. For example, on a globe you may map out your great circle journey by holding down one end of a piece of string on London and pulling the string tightly over the surface so that the other end is on New York. The taut string has

[8] Nevertheless, the reader should be aware that in modern usage "non-Euclidean Geometry" is usually synonymous with "Hyperbolic Geometry."

automatically found the shortest, straightest route—the shorter[9] of the two arcs into which the great circle through the two cities is divided by those cities.

With the analogue of straight lines now found, we can "do geometry" within this spherical surface. For example, given three points on the surface of the Earth, we can connect them together with arcs of great circles to obtain a "triangle." Figure [1.6] illustrates this in the case where one vertex is located at the north pole, and the other two are on the equator.

But if this non-Euclidean Spherical Geometry was already used by ancient mariners to navigate the oceans, and by astronomers to map the spherical night sky, what then was so shocking and new about the non-Euclidean geometry of Lobachevsky and Bolyai?

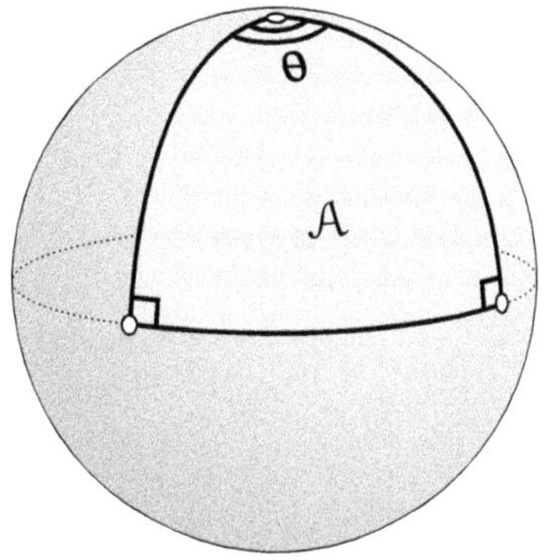

[1.6] *A particularly simple "triangle" on the sphere.*

The answer is that this Spherical Geometry was merely considered to be inherited from the *Euclidean* Geometry of the 3-dimensional space in which the sphere resides. No thought was given in those times to the sphere's internal 2-dimensional geometry as representing an alternative to Euclid's plane. Not only did it violate Euclid's fifth axiom, it also violated a much more basic one (Euclid's first axiom) that we can always draw a unique straight line connecting two points, for this fails when the points are antipodal.

On the other hand, the Hyperbolic Geometry of Lobachevsky and Bolyai was a much more serious affront to Euclidean Geometry, containing familiar lines of infinite length, yet flaunting multiple parallels, ludicrous angle sums, and many other seemingly nonsensical results. Yet the 21-year-old Bolyai was confident and exuberant in his discoveries, writing to his father, *"From nothing I have created another entirely new world."*

We end with a tale of tragedy. Bolyai's father was a friend of Gauss, and sent him what János had achieved. By this time Gauss had himself made some important discoveries in this area, but had kept them secret. In any case, János had seen further than Gauss. A kind word in public from Gauss, the most famous mathematician in the world, would have assured the young mathematician a bright future. But Nature and nurture sometimes conspire to pour extraordinary mathematical gifts into a vessel marred by very ordinary human flaws, and Gauss's reaction to Bolyai's marvellous discoveries was mean-spirited and self-serving in the extreme.

First, Gauss kept Bolyai in suspense for six months, then he replied as follows:

> Now something about the work of your son. You will probably be shocked for a moment when I begin by saying that *I cannot praise it*, but I cannot do anything else, since to praise it would be to praise myself. The whole content of the paper, the path that your son has taken, and the results to which he has been led, agree almost everywhere with my own meditations, which have occupied me in part for 30–35 years.

Gauss did however "thank" Bolyai's son for having "saved him the trouble"[10] of having to write down theorems he had known for decades.

János Bolyai never recovered from the surgical blow delivered by Gauss, and he abandoned mathematics for the rest of his life.[11]

---

[9]If the two points are antipodal, such as the north and south poles, then the two arcs are the *same* length. Furthermore, the great circle itself is no longer unique: *every* meridian is a great circle connecting the poles.

[10]Gauss had previously denigrated Abel's discovery of elliptic functions in precisely the same manner; see Stillwell (2010, p. 236).

[11]If this depresses you, turn your thoughts to the uplifting counterweight of Leonhard Euler. An intellectual volcano erupting with wildly original thoughts (some of which we shall meet later) he was also a kind and generous spirit. We cite one, parallel

1.3 The Angular Excess of a Spherical Triangle

As we have said, the parallel axiom is equivalent to the fact that the angles in a triangle sum to π. It follows that both the spherical axiom and the hyperbolic axiom must lead to geometries in which the angles do *not* sum to π. To quantify this departure from Euclidean Geometry, we introduce the ***angular excess***, defined to be the amount $\mathcal{E}$ by which the angle sum exceeds π:

$$\mathcal{E} \equiv (\text{angle sum of triangle}) - \pi.$$

For example, for the triangle shown in [1.6], $\mathcal{E} = (\theta + \frac{\pi}{2} + \frac{\pi}{2}) - \pi = \theta$.

A crucial insight now arises if we compare the triangle's angular excess with its area $\mathcal{A}$. Let the radius of the sphere be R. Since the triangle occupies a fraction $(\theta/2\pi)$ of the northern hemisphere, $\mathcal{A} = (\theta/2\pi)\, 2\pi R^2 = \theta\, R^2$. Thus,

$$\mathcal{E} = \frac{1}{R^2}\mathcal{A}. \tag{1.3}$$

[1.7] *Thomas Harriot (1560–1621).*

In 1603 the English mathematician Thomas Harriot (see [1.7]) made the remarkable discovery[12] that this relationship holds for *any* triangle Δ on the sphere; see [1.8a]. Harriot's elementary but ingenious argument[13] goes as follows.

Prolonging the great-circle sides of Δ divides the surface of the sphere into eight triangles, the four triangles labelled $\Delta, \Delta_\alpha, \Delta_\beta, \Delta_\gamma$ each being paired with a congruent antipodal triangle. This is clearer in [1.8b]. Since the area of the sphere is $4\pi R^2$, we deduce that

$$\mathcal{A}(\Delta) + \mathcal{A}(\Delta_\alpha) + \mathcal{A}(\Delta_\beta) + \mathcal{A}(\Delta_\gamma) = 2\pi R^2. \tag{1.4}$$

On the other hand, it is clear in [1.8b] that Δ and Δ_α together form a wedge whose area is a fraction $(\alpha/2\pi)$ of the area of the sphere:

$$\mathcal{A}(\Delta) + \mathcal{A}(\Delta_\alpha) = 2\alpha R^2.$$

Similarly,

$$\mathcal{A}(\Delta) + \mathcal{A}(\Delta_\beta) = 2\beta R^2,$$
$$\mathcal{A}(\Delta) + \mathcal{A}(\Delta_\gamma) = 2\gamma R^2.$$

---

example. When the then-unknown 19-year-old Lagrange wrote to him with overlapping discoveries in the calculus of variations, Euler wrote back: "... I deduced this myself. However, I decided to conceal this until you publish your results, since in no way do I want to take away from you any of the glory that you deserve." See Gindikin (2007, p. 216). Incidentally, Euler also personally intervened to rescue Lambert's career!

[12]This discovery is most often attributed to Girard, who rediscovered it about 25 years later.

[13]This argument was later rediscovered by Euler in 1781.

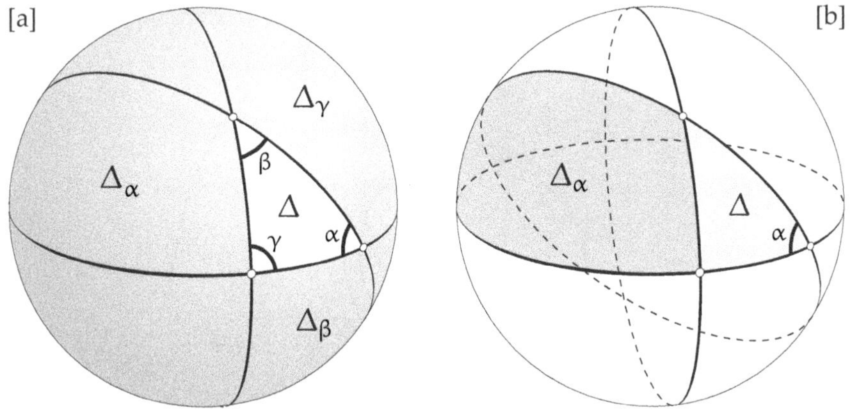

[1.8] *Harriot's Theorem (1603):* $\mathcal{E}(\Delta) = \mathcal{A}(\Delta)/R^2$.

Adding these last three equations, we find that

$$3\mathcal{A}(\Delta) + \mathcal{A}(\Delta_\alpha) + \mathcal{A}(\Delta_\beta) + \mathcal{A}(\Delta_\gamma) = 2(\alpha + \beta + \gamma)R^2. \tag{1.5}$$

Finally, subtracting (1.4) from (1.5), we find that

$$\mathcal{A}(\Delta) = R^2(\alpha + \beta + \gamma - \pi) = R^2\,\mathcal{E}(\Delta),$$

thereby proving (1.3).

1.4 Intrinsic and Extrinsic Geometry of Curved Surfaces

The mathematics associated with this stretched-string construction of a "straight line" will be explored in depth later in the book. For now we merely observe that the construction can be applied equally well to a *non*spherical surface, such as the crookneck squash shown in [1.9].

Just as on the sphere, we stretch a string over the surface, thereby finding the shortest, straightest route between two points, such as a and b. Provided that the string can slide around on the surface easily, the tension in the string ensures that the resulting path is as short as possible. Note that in the case of cd, we must imagine that the string runs over the *inside* of the surface.

In order to deal with all possible pairs of points in a uniform way, it is therefore best to imagine the surface as made up of two thinly separated layers, with the string trapped between them. On the other hand, this is only useful for thought experiments, not actual experiments. We shall overcome this obstacle shortly by providing a *practical* method of constructing these straightest curves on the surface of a physical object, even if the patch of surface bends the wrong way for a string to be stretched tightly over the outside of the object.

These shortest paths on a curved surface are the equivalent of straight lines in the plane, and they will play a crucial role throughout this book—they are called ***geodesics***. Thus, to use this new word, we may say that geodesics in the plane are straight lines, and geodesics on the sphere are great circles.

But even on the sphere the *length-minimizing* definition of "geodesic" is provisional, because we see that nonantipodal points are connected by *two* arcs of the great circle passing through them: the short one (which *is* the shortest route) and the long one. Yet the long arc is every bit as much a geodesic as the short one. There is the additional complication on the sphere that antipodal points

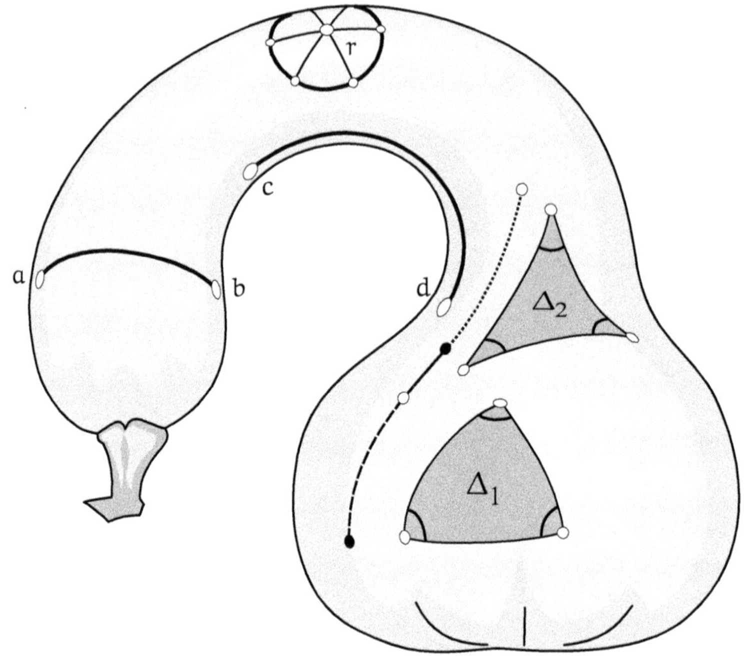

[1.9] *The* **intrinsic geometry** *of the surface of a crookneck squash:* **geodesics** *are the equivalents of straight lines, and triangles formed out of them may possess an angular excess of either sign, depending on how the surface bends:* $\mathcal{E}(\Delta_1) > 0$ *and* $\mathcal{E}(\Delta_2) < 0$.

are connected by *multiple* geodesics, and this *nonuniqueness* occurs on more general surfaces, too. What *is* true is that any two points that are *sufficiently close together* can be joined by a unique geodesic segment that is the shortest route between them.

Just as a line segment in the plane can be extended indefinitely in either direction by laying down overlapping segments, so too can a geodesic segment be extended on a curved surface, and this extension is unique. For example, in [1.9] we have extended the dashed geodesic segment connecting the black dots, by laying down the overlapping dotted segment between the white points.

Because of the subtleties associated with the length-minimizing characterization of geodesics, before long we will provide an alternative, purely *local* characterization of geodesics, based on their straightness.

With these caveats in place, it is now clear how we should define distance within a surface such as [1.9]: the distance between two sufficiently close points a and b is the length of the geodesic segment connecting them.

Figure [1.9] shows how we may then define, for example, a "circle of radius r and centre c" as the locus of points at distance r from c. To construct this ***geodesic circle*** we may take a piece of string of length r, hold one end fixed at c, then (keeping the string taut) drag the other end round on the surface. But just as the angles in a triangle no longer sum to π, so now the circumference of a circle no longer is equal to $2\pi r$. In fact you should be able to convince yourself that for the illustrated circle the circumference is *less* than $2\pi r$.

Given three points on the surface, we may join them with geodesics to form a ***geodesic triangle***; [1.9] shows two such triangles, Δ_1 and Δ_2:

- Looking at the angles in Δ_1, it seems clear that they sum to *more* than π, so $\mathcal{E}(\Delta_1) > 0$, like a triangle in Spherical Geometry.

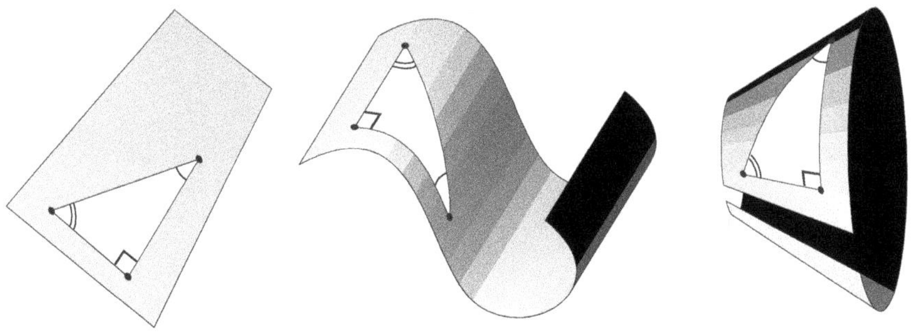

[1.10] *Bending a piece of paper changes the extrinsic geometry, but not the intrinsic geometry.*

- On the other hand, it is equally clear that the angles of Δ_2 sum to *less* than π: $\mathcal{E}(\Delta_2) < 0$, and (as we shall explain) this opposite behaviour is in fact exhibited by triangles in *Hyperbolic Geometry*. Note also that if we construct a circle in this saddle-shaped part of the surface, the circumference is now *greater* than $2\pi r$.

The concept of a geodesic belongs to the so-called ***intrinsic geometry*** of the surface—a fundamentally new view of geometry, introduced by Gauss (1827). It means the geometry that is knowable to tiny, ant-like, intelligent (but 2-dimensional!) creatures living *within* the surface. As we have discussed, these creatures can, for example, define a geodesic "straight line" connecting two nearby points as the shortest route within their world (the surface) connecting the two points. From there they can go on to define triangles, and so on. Defined in this way, it is clear that the intrinsic geometry is unaltered when the surface is bent (as a piece of paper can be) into quite different shapes in space, as long as distances *within* the surface are not stretched or distorted in any way. To the ant-like creatures within the surface, such changes are utterly undetectable.

Under such a bending, the so-called ***extrinsic geometry*** (how the surface sits in space) most certainly does change. See [1.10]. On the left is a flat piece of paper on which we have drawn a triangle Δ with angles $(\pi/2)$, $(\pi/6)$, and $(\pi/3)$. Of course $\mathcal{E}(\Delta) = 0$. Clearly we can bend such a flat piece of paper into either of the two (extrinsically) curved surfaces on the right.[14] However, *intrinsically* these surfaces have undergone no change at all—they are both as flat as a pancake! The illustrated triangles on these surfaces (into which Δ is carried by our stretch-free bending of the paper) are identical to the ones that intelligent ants would construct using geodesics, and in both cases $\mathcal{E} = 0$: geometry on these surfaces is Euclidean.

Even if we take a patch of a surface that is intrinsically *curved*, so that a triangle within it has $\mathcal{E} \neq 0$, it too can generally be bent somewhat without stretching or tearing it, thereby altering its extrinsic geometry while leaving its intrinsic geometry unaltered. For example, cut a ping pong ball in half and gently squeeze the rim of one of the hemispheres, distorting that circular rim into an oval (but not an oval lying in a single plane).

1.5 Constructing Geodesics via Their Straightness

We have already alluded to the fact that geodesics on a surface have at least *two* characteristics in common with lines in the plane: (1) they provide the *shortest* route between two points that are not too far apart *and* (2) they provide the "straightest" route between these points. In this section we seek to clarify what we mean by "straightness," leading to a very simple and *practical* method of constructing geodesics on a physical surface.

---

[14]But note that we must first trim the edges of the rectangle to bend it into the shape on the far right.

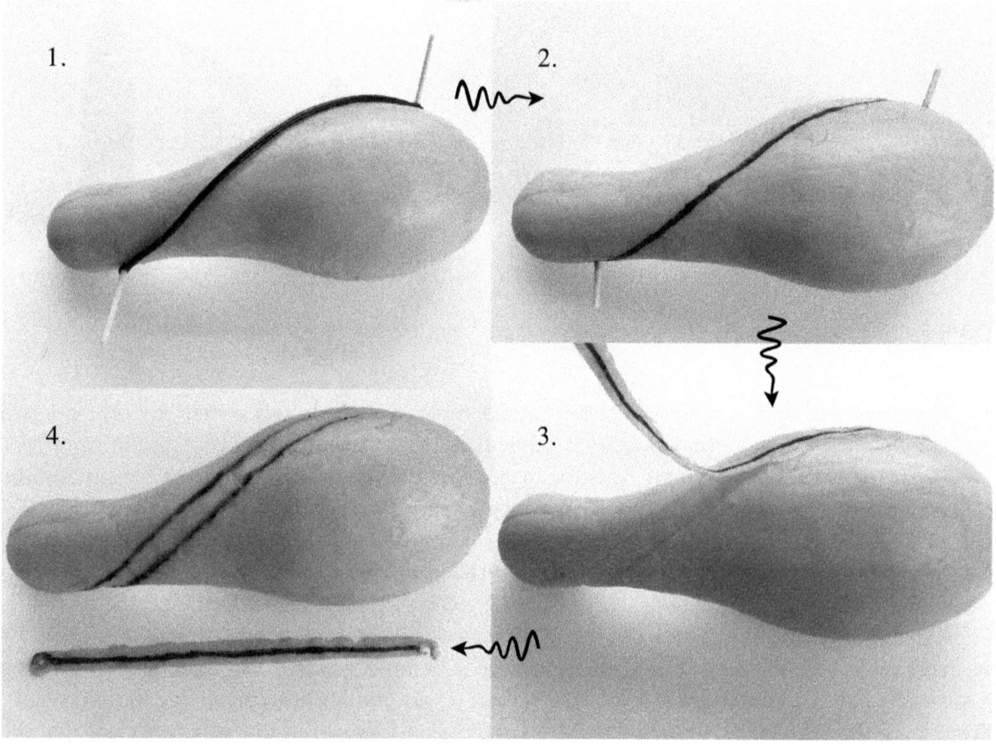

[1.11] *On the curved surface of a fruit or vegetable, peel a narrow strip surrounding a geodesic, then lay it flat on the table. You will obtain a* straight line *in the plane!*

Most texts on Differential Geometry pay scant attention to such practical matters, and it is perhaps for this reason that the construction we shall describe is surprisingly little known in the literature.[15] In sharp contrast, in this book we *urge* you to explore the ideas by all means possible: theoretical contemplation, drawing, computer experiments, and (especially!) physical experiments with actual surfaces. Your local fruit and vegetable shop can supply your laboratory with many interesting shapes, such as the yellow summer squash shown in [1.11].

We can now use this vegetable to reveal the hidden straightness of geodesics via an experiment that we hope you will repeat for yourself:

1. On a fruit or vegetable, construct a geodesic by stretching a string over its curved surface.
2. Use a pen to trace the path of the string, then remove the string.
3. Make shallow incisions on either side of (and close to) the inked path, then use a vegetable peeler or small knife to remove the narrow strip of peel between the two cuts.
4. Lay the strip of peel flat on the table, and witness the marvellous fact that the geodesic within the peeled strip has become a *straight line* in the plane!

But why?!

To understand this, first let us be clear that although the strip is free to bend in the direction perpendicular to the surface (i.e., perpendicular to itself), it is *rigid* if we try to bend it sideways, tangent to the surface. Now let us employ proof by contradiction, and imagine what would happen if such a peeled geodesic did *not* yield a straight line when laid flat on the table. It is both a

[15] One of the rare exceptions is Henderson (2013), which we strongly recommend to you; for more details, see the *Further Reading* section at the end of this book.

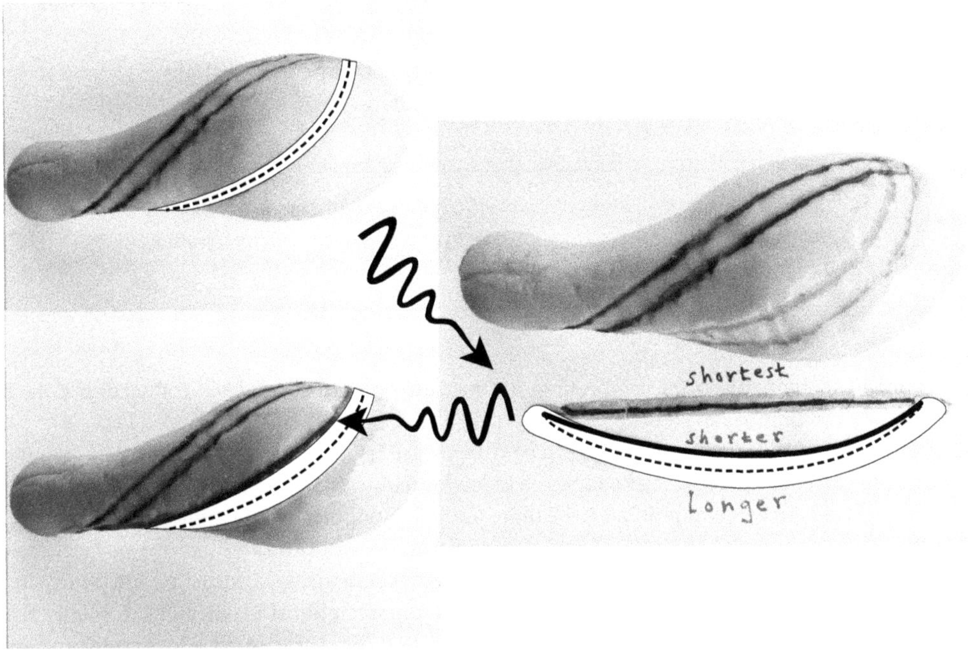

[1.12] *Suppose that the illustrated dotted path is a geodesic such that a narrow (white) strip surrounding it does* not *become a straight line when laid flat in the plane. But in that case we can shrink the dotted path in the plane (towards the shortest, straight-line route in the plane) thereby producing the solid path. But if we then reattach the strip to the surface, this solid path is still shorter than the original dotted path, which was supposed to be the shortest path within the surface—a contradiction!*

drawback and an advantage of conducting such physical experiments that they will simply not *permit* us to construct something that is impossible, as is required in our desired mathematical proof by contradiction. Nevertheless, let us *suppose* that there exists a geodesic path, such as the dotted one shown on top left of [1.12], that when peeled and laid flat on the table (on the right) does *not* become a straight line.

The shortest route between the ends of this dotted (nonstraight) plane curve is the straight line connecting them. (As illustrated, this is the path of the *true* geodesic we already found using the string—but pretend you don't know that for now!) Thus we may shorten the dotted curve by deforming it slightly towards this straight, shortest route, yielding the solid path along the edge of the peeled strip. Therefore, after reattaching the strip to the surface (bottom left) the solid curve provides a shorter route over the surface than the dotted one, which we had supposed to be the shortest: a contradiction! Thus we have proved our previous assertion:

> *If a narrow strip surrounding a segment* G *of a geodesic is cut out of a surface and laid flat in the plane, then* G *becomes a segment of a* straight line. (1.6)

We are now very close to the promised simple and practical construction of geodesics. Look again at step 3 of [1.11], where we peeled off the strip of surface. But imagine now that we are *reattaching* the strip to the surface, instead. Ignore the history of how we got to this point: what are we actually doing right now in this reattachment process? We have picked up a narrow straight strip (of three-dimensional peel—but mathematically idealized as a two-dimensional strip) and we have unrolled it back onto the surface into the shallow channel from which we cut it. But here is the crucial observation: this shallow channel need not exist—the *surface* decides where the strip must go as we unroll it!

Thus, as a kind of time-reversed converse of (1.6), we obtain a remarkably simple and practical method[16] of constructing geodesics on a physical surface:

> *To construct a geodesic on a surface, emanating from a point* p *in direction* **v**, *stick one end of a length of narrow sticky tape down at* p *and unroll it onto the surface, starting in the direction* **v**. (1.7)

(Note, however, that this does *not* provide a construction of the geodesic connecting p to a specified *target* point q.)

If this construction seems too simple to be true, please try it on any curved surface you have to hand. You can check that the sticky tape[17] is indeed tracing out a geodesic by stretching a string over the surface between two points on the tape: the string will follow the same path as the tape. But note that, as a promised bonus, this new tape construction works on *any* part of a surface, even where the surface is concave towards you, so that the stretched-string construction breaks down.

Of course all of this is a concrete manifestation of a mathematical idealization. A totally flat narrow strip of tape of nonzero width *cannot*[18] be made to fit perfectly on a genuinely curved surface, but its centre line *can* be made to rest on the surface, while the rest of the tape is tangent to the surface.

1.6 The Nature of Space

Let us return to the history of the discovery of non-Euclidean Geometry and take our first look at how these two new geometries differ from Euclid's.

As we have said, Euclidean Geometry is characterized by the vanishing of $\mathcal{E}(\Delta)$. Note that, unlike the original formulation of the parallel axiom, *this statement can be checked against experiment*: construct a triangle, measure its angles, and see if they add up to π. Gauss may have been the first person to ever conceive of the possibility that physical space might not be Euclidean, and he even attempted the above experiment, using three mountain tops as the vertices of his triangle, and using light rays for its edges.

Within the accuracy permitted by his equipment, he found $\mathcal{E} = 0$. Quite correctly, Gauss did not conclude that physical space is definitely Euclidean in structure, but rather that if it is *not* Euclidean then its deviation from Euclidean Geometry is extremely small. But he did go so far as to say (see Rosenfeld 1988, p. 215) that he wished that this non-Euclidean Geometry might apply to the real world. In Act IV we shall see that this was a prophetic statement.

---

[16]This important fact is surprisingly hard to find in the literature. After we (re)discovered it, more than 30 years ago, we began searching, and the earliest mention of the underlying idea we could find at that time was in Aleksandrov (1969, p. 99), albeit in a less practical form: he imagined pressing a flexible metal ruler down onto the surface. Later, the basic idea also appeared in Koenderink (1990), Casey (1996), and Henderson (2013). *However*, we have since learned that the essential idea (though not in our current, *practical* form) goes all the way back to Levi-Civita, more than a century ago! See the footnote on page 236.

[17]We recommend using masking tape (aka painter's tape) because it comes in bright colours, and once a strip has been created, it can be detached and reattached repeatedly, with ease. A simple way to create narrow strips (from the usually wide roll of tape) is to stick a length of tape down onto a kitchen cutting board, then use a sharp knife to cut down its length, creating strips as narrow as you please.

[18]This is a consequence of a fundamental theorem we shall meet later, called the *Theorema Egregium*.

Although Gauss had bragged to friends that he had anticipated the Hyperbolic Geometry of Lobachevsky and Bolyai by decades, even he had unknowingly been scooped on some of its central results.

In 1766 (eleven years before Gauss was born) Lambert rediscovered Harriot's result on the sphere and then broke totally new ground in pursuing the analogous consequences of the Hyperbolic Axiom (1.1). First, he found that a triangle in Hyperbolic Geometry (if such a thing even existed) would behave *oppositely* to one in Spherical Geometry:

- In Spherical Geometry the angle sum of a triangle is greater than π: $\mathcal{E} > 0$.
- In Hyperbolic Geometry the angle sum of a triangle is less than π: $\mathcal{E} < 0$.

Thus a hyperbolic triangle behaves like a triangle drawn on a saddle-shaped piece of surface, like Δ_2 in [1.9]. Later we shall see that this is no accident.

Furthermore, Lambert discovered the crucial fact that $\mathcal{E}(\Delta)$ is again simply proportional to $\mathcal{A}(\Delta)$:

> *In both Spherical and Hyperbolic Geometry,*
>
> $$\mathcal{E}(\Delta) = \mathcal{K}\,\mathcal{A}(\Delta), \qquad (1.8)$$
>
> *where $\mathcal{K}$ is a constant that is positive in Spherical Geometry, and negative in Hyperbolic Geometry.*

Several interesting observations can be made in connection with this result:

- Although there are no qualitative differences between them, there are nevertheless *infinitely many* different Spherical Geometries, depending on the value of the positive constant $\mathcal{K}$. Likewise, each negative value of $\mathcal{K}$ yields a different Hyperbolic Geometry.
- Since the angle sum of a triangle cannot be negative, $\mathcal{E} \geqslant -\pi$. Thus in Hyperbolic Geometry ($\mathcal{K} < 0$) we have the strange and surprising result that *no triangle can have an area greater than* $|\pi/\mathcal{K}|$.
- From (1.8) we deduce that two triangles of different size cannot have the same angles. In other words, in non-Euclidean Geometry, *similar triangles of different size do not exist!* (This accords with Wallis's 1663 discovery that the existence of similar triangles is equivalent to the Parallel Axiom.)
- Closely related to the previous point is the fact that in non-Euclidean Geometry there exists an *absolute unit of length*. (Gauss himself found it to be an exciting possibility that this purely mathematical fact might be realized in the physical world.) For example, in Spherical Geometry we could define this absolute unit of length to be the side of *the* equilateral triangle having, for instance, angle sum 1.01π. Similarly, in Hyperbolic Geometry we could define it to be the side of *the* equilateral triangle having angle sum 0.99π.
- A somewhat more natural way of defining the absolute unit of length is in terms of the constant $\mathcal{K}$. Since the radian measure of angle is defined as a ratio of lengths, $\mathcal{E}$ is a pure number. On the other hand, the area $\mathcal{A}$ has units of (length)^2. It follows that $\mathcal{K}$ must have units of $1/(\text{length})^2$, and so there exists a length R such that $\mathcal{K}$ can be written as follows: $\mathcal{K} = +(1/R^2)$ in Spherical Geometry; $\mathcal{K} = -(1/R^2)$ in Hyperbolic Geometry. Of course in Spherical Geometry we already know that the length R occurring in the formula $\mathcal{K} = +(1/R^2)$ is simply the

radius of the sphere. Later we will see that this length R occurring in the formula $\mathcal{K} = -(1/R^2)$ can be given an equally intuitive and *concrete* interpretation in Hyperbolic Geometry.

- The smaller the triangle, the harder it is to distinguish it from a Euclidean triangle: only when the linear dimensions are a significant fraction of R will the differences become discernable. For example, we humans are small compared to the radius of the Earth, so if we find ourselves in a boat in the middle of a lake, its surface appears to be a Euclidean plane, whereas in reality it is part of a sphere. This Euclidean illusion for small figures is the reason that Gauss chose the largest possible triangle to conduct his light-ray experiment, thereby increasing his chances of detecting any small curvature that might be present in the space through which the light rays travelled.

Chapter 2

Gaussian Curvature

2.1 Introduction

The proportionality constant,

$$\mathcal{K} = +\frac{1}{R^2},$$

that enters into Spherical Geometry via Harriot's result (1.3), is called the ***Gaussian curvature***[1] of the sphere. The smaller the radius R, the more tightly curved is the surface of the sphere, and the greater the value of the Gaussian curvature $\mathcal{K}$.

Likewise, in Hyperbolic Geometry the negative constant

$$\mathcal{K} = -\frac{1}{R^2}$$

occurring in (1.8) is *again* called the Gaussian curvature, for reasons we shall explain shortly.

This intrinsic[2] concept $\mathcal{K}$ was announced by Gauss (after a decade of private investigation) in his revolutionary "General Investigations of Curved Surfaces,"[3] published in 1827.

[2.1] *Carl Friedrich Gauss (1777–1855).*

As we now explain, Gauss introduced this concept to measure the *curvature at each point* of a general, irregular surface such as that depicted in [1.9]. This one idea of curvature will dominate all that is to come. According to Harriot's and Lambert's results (1.8),

$$\mathcal{K} = \frac{\mathcal{E}(\Delta)}{\mathcal{A}(\Delta)} = \textit{angular excess per unit area.}$$

In both Spherical and Hyperbolic Geometry this interpretation holds for a triangle Δ of any size and any location. But on a more general surface such as in [1.9] this definition makes no sense, for even the *sign* of $\mathcal{E}$ varies between triangles, such as Δ_1 and Δ_2, that reside in different parts of the surface.

---

[1] Also called the ***Gauss curvature***, ***intrinsic curvature***, ***total curvature***, or just plain ***curvature***.

[2] As we shall discuss later, Olinde Rodrigues had arrived at and published the same concept as early as 1815, but from an *extrinsic* point of view. Gauss was not aware that he had been anticipated in this way.

[3] Gauss (1827).

[2.2] *The Gaussian curvature $\mathcal{K}(p)$ at a point is the angular excess per unit area as a geodesic triangle shrinks to that point. In this example, $\mathcal{K}(p) > 0$ and $\mathcal{K}(q) < 0$.*

To define the Gaussian curvature *at* a point p on such a surface we now imagine a small geodesic triangle Δ_p containing p, and then allow the triangle to shrink down towards p.

Using the construction of geodesics discovered in the previous section, [2.2] depicts such a sequence of shrinking triangles, converging towards a point on the surface of an inflatable swimming pool ring, the mathematical name for which is a ***torus***. We now define the Gaussian curvature $\mathcal{K}(p)$ at p to be the limit as this triangle shrinks down towards p:

$$\mathcal{K}(p) = \lim_{\Delta_p \to p} \frac{\mathcal{E}(\Delta_p)}{\mathcal{A}(\Delta_p)} = \textit{angular excess per unit area at } p. \tag{2.1}$$

At this stage it is *not* meant to be obvious to you that this limit exists, independently of the shape of the triangle and the precise manner in which it shrinks; this will be proved later. As our drama unfolds we shall discover several other ways[4] of interpreting the Gaussian curvature and of calculating its value in concrete cases.

[4]For a mathematical concept to be truly fundamental it must lie at the intersection of different branches of mathematics. Thus it is to be expected that each of these branches will provide a seemingly distinct yet equally natural way of looking at one and the same concept.

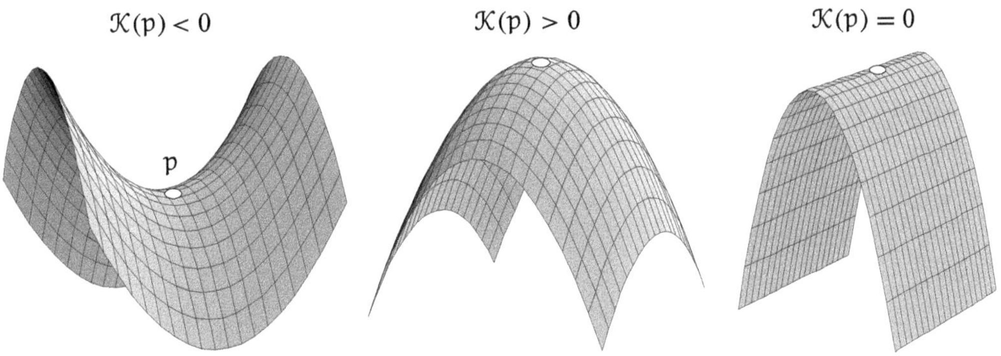

[2.3] *The Gaussian curvature $\mathcal{K}$ is the local angular excess per unit area: its sign is negative if the surface looks like a saddle, positive if it's like a hill, and it vanishes if it's like a curled piece of paper.*

The definition in (2.1) extends beyond triangles. If we replace Δ_p with a small n-gon then (see Ex. 10),

$$\boxed{\mathcal{E}(\text{n-gon}) \equiv [\text{angle sum}] - (n-2)\pi,} \tag{2.2}$$

and the interpretation of curvature in (2.1) as *angular excess per unit area* applies without change.

Inspection of the inflatable pool ring in [2.2] should make it clear that $\mathcal{K}(p) > 0$ at every point p on the outer half, where the immediate neighbourhood of p resembles a hill, whereas $\mathcal{K}(q) < 0$ at every point q on the inner half, where the immediate neighbourhood of q resembles a saddle. Figure [2.3] summarizes this.

2.2 The Circumference and Area of a Circle

But why is $\mathcal{K}(p)$ so important? Yes, clearly it controls small triangles to some extent, but there is so much more to geometry than just triangles! The answer is that while we may have chosen to define $\mathcal{K}(p)$ (for the moment) in terms of small triangles, we will gradually discover that the curvature has an iron grip over *every* aspect of geometry within the surface. Let us give just two examples for now.

In [1.9] we indicated how a "circle of radius r" centred at c could be defined by taking the end p of a geodesic segment cp of fixed length r and swinging it fully around c. Let us calculate the circumference $C(r)$ of such a circle constructed on the sphere of radius R.

Referring to [2.4], we see that

$$\rho = R\sin\phi \quad \text{and} \quad \phi = \frac{r}{R} \quad \Longrightarrow \quad C(r) = 2\pi R\sin(r/R). \tag{2.3}$$

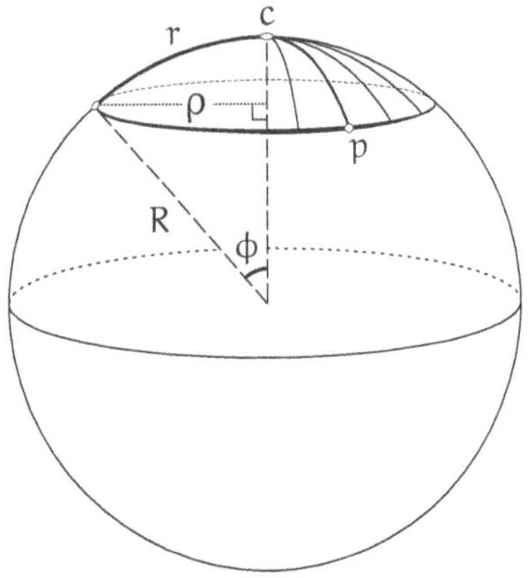

[2.4] *A circle of radius r on a sphere of radius R has circumference $C(r)$, given by $C(r) = 2\pi R\sin(r/R)$.*

Just as the curvature governs the departure of the angle sum of a triangle from the Euclidean prediction of π, so too does it govern the departure of $C(r)$ from the Euclidean prediction of $2\pi r$. To see this, recall the power series for sine:

$$\sin\phi = \phi - \frac{1}{3!}\phi^3 + \frac{1}{5!}\phi^5 + \cdots.$$

Thus, as ϕ vanishes,

$$\phi - \sin\phi \asymp \frac{1}{6}\phi^3.$$

(We remind the reader that here, $\asymp$ denotes Newton's concept of ***ultimate equality***, as introduced in the Prologue.) It follows from (2.3) that as r shrinks to zero,

$$2\pi r - C(r) = 2\pi R[(r/R) - \sin(r/R)] \asymp \frac{\pi r^3}{3R^2}.$$

In other words, the inhabitants of S^2 can now determine the curvature of their world by examining the circumference of a small circle, just as easily as they previously could by examining the angles of a small triangle:

$$\mathcal{K} \asymp \frac{3}{\pi}\left[\frac{2\pi r - C(r)}{r^3}\right]. \tag{2.4}$$

Remarkably, as we will be able to show much later, in Act IV, this formula continues to correctly measure the Gaussian curvature on a *general* surface! (Note that the power of r in the denominator could have been anticipated: we know that $\mathcal{K}$ has dimensions of $1/(\text{length})^2$, and circumference is a length, so we require $(\text{length})^3$ in the denominator.)

Continuing with this example, let us instead examine the *area* $\mathcal{A}(r)$ of the polar cap bounded by this circle. Again it is the curvature that governs how the area departs from the Euclidean prediction of πr^2. With the assistance of the formula for the polar cap (see Ex. 10, p. 85) it is not hard to verify [exercise] that, in fact,

$$\mathcal{K} \asymp \frac{12}{\pi}\left[\frac{\pi r^2 - \mathcal{A}(r)}{r^4}\right]. \tag{2.5}$$

And *again* this formula turns out to be universal! (Again, the same reasoning as above explains the fourth power in the denominator.)

While we are not yet in a position to prove the universality of (2.4) and (2.5), we can at least see that they do indeed yield the correct *sign* at each point of a variably curved surface, such as that shown in [1.9]. For if the immediate vicinity of a point on such a surface is positively curved, then it is hill-shaped near that point (as it is everywhere in the region of [1.9] containing Δ_1). Thus both the circumference and area of a small circle centred there will indeed be squeezed by the curvature and be *less* than they would have been in a flat Euclidean plane. Thus, both the preceding formulas yield $\mathcal{K} > 0$, as they should.

On the other hand, if the surface is saddle-shaped near the point, the opposite happens. Recall that we pointed out in [1.9] that a circle drawn in the saddle-shaped part of the surface (where Δ_2 is located) will have $C(r) > 2\pi r$. To grasp this, stand up and hold one arm out at right angles to your body. If you spin around on your heels, the tip of your hand will trace out a horizontal circle. Now repeat this pirouette, but this time wave your arm up and down as you turn; clearly the tip of hand has travelled *further* than before. But this waving up and down is just what happens when we trace out a circle on a saddle-shaped surface, and therefore both of the preceding formulas yield $\mathcal{K} < 0$, as they should.

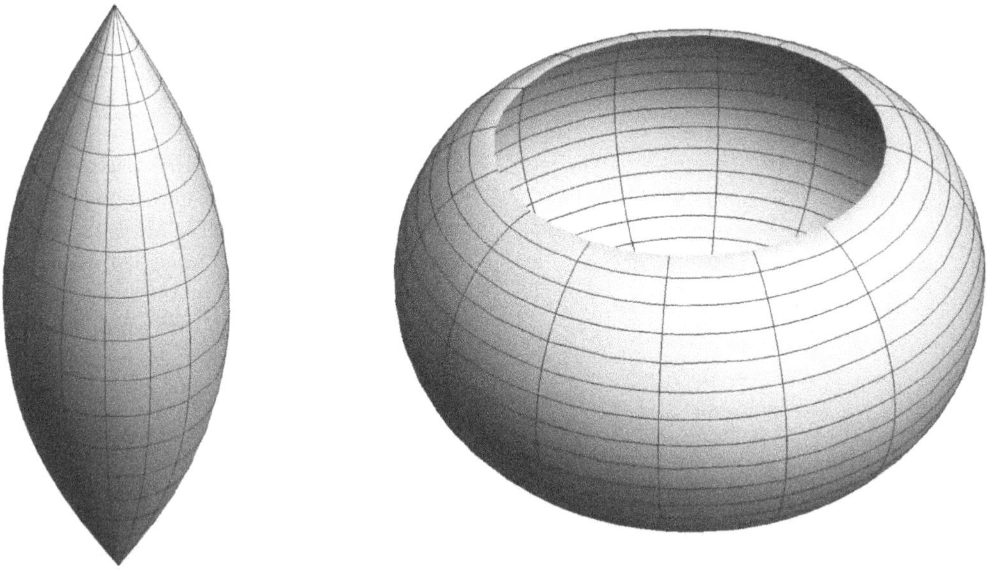

[2.5] *Nonspherical surfaces of revolution exist that possess constant positive curvature, but these necessarily have either spikes or edges.*

We have said that curvature has an "iron grip" on geometry, but just how absolute is this control? For example, if we know that a patch of surface has constant positive curvature $\mathcal{K} = (1/R^2)$, must it in fact be a portion of a sphere of radius R? Well, take a ping pong ball and cut it in half—now flex one of the hemispheres slightly. Clearly we have obtained a new *non*spherical patch of surface, but since we have not stretched distances within the surface, geodesics and angles are unchanged, and therefore according to the definition (2.1) the curvature $\mathcal{K}$ has not changed. Thus we certainly can obtain at least patches of surface of constant curvature that are not extrinsically spherical, although they all have identical intrinsic geometry.

Figure [2.5] illustrates that even if we restrict attention just to surfaces of revolution, the sphere is not the only one of constant positive curvature. In fact there is an entire family of such surfaces, with the sphere representing the boundary case between the two illustrated types (see Ex. 22, p. 89). Though they hardly look like spheres, an intelligent ant living on either of these surfaces would never know that she wasn't living on a sphere. Well, that's almost true: eventually she might discover sharp creases or spikes at which the surface is not smooth, or else she might run into an edge where the surface ends. In 1899 Heinrich Liebmann proved[5] that if a surface of constant positive curvature does not suffer from these defects then it can *only* be a sphere.

Ignoring such superficial extrinsic differences, can two surfaces have *constant* positive curvature $\mathcal{K} = (1/R^2)$ and yet have genuinely different *intrinsic* geometries? More explicitly, if our intelligent ant were suddenly transported from one surface to the other, could she devise an experiment to discover that her world had changed? In 1839 Minding (one of Gauss's few students) discovered the answer: "no!" In other words, Minding found[6] that if two surfaces have *constant* positive curvature $\mathcal{K} = (1/R^2)$ then their intrinsic geometries are locally *identical* to that of the sphere of radius R.

We have discussed the fact that the inner rim of the pool ring in [2.2] has negative curvature, but it is not *constant* negative curvature. Indeed, if C is the circle of contact between the ring and the ground, separating the inner and outer halves, then it's clear that the negative curvature $\mathcal{K}(q)$

[5] The proof will have to wait till Section 38.11.

[6] The proof will have to wait till Act IV (Exercise 7, p. 336).

must die away to zero as q approaches C, switching to positive if q crosses over C to the outer half. (This is investigated in Act II: Ex. 23, page 89.)

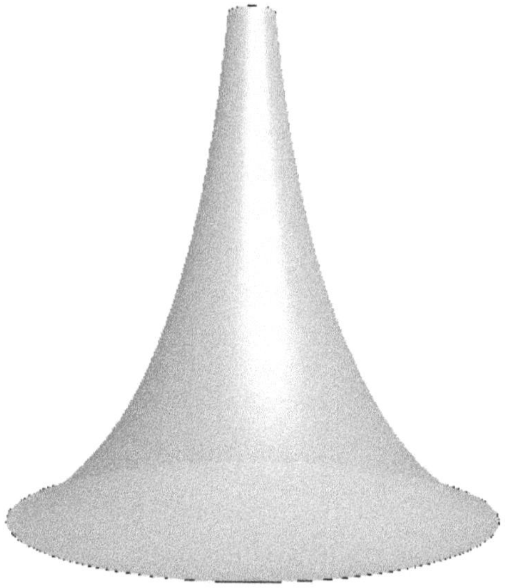

[2.6] *The* **pseudosphere** *of base radius* R *has constant negative curvature* $\mathcal{K} = -(1/R^2)$.

In fact there do exist surfaces that have constant negative curvature. Eugenio Beltrami (whom we shall meet properly shortly) called all such surfaces ***pseudospherical***, in honour of the simplest exemplar, the ***pseudosphere***,[7] shown in [2.6]. (The precise construction of this surface will be explained in Act II, but it is the surface of revolution generated by a curve called the ***tractrix***, which was first investigated by Newton, in 1676.) If R is the radius of the circular base of the pseudosphere then (as we shall prove later) the constant negative value of the curvature over the entire surface is $\mathcal{K} = -(1/R^2)$.

While the name of this surface is too established to be changed, it is perhaps unfortunate. As you see, it is certainly not closed like the sphere. In fact we shall prove later in the book that no closed pseudospherical surfaces can exist. Further, we see that while the pseudosphere extends upwards indefinitely, it has a circular edge at its base. It turns out to be impossible to extend the pseudosphere beyond this edge while maintaining its constant negative curvature. This is no accident: in 1901 David Hilbert proved that if a surface of constant negative curvature is to be embedded within ordinary Euclidean 3-space then it must *necessarily* have an edge beyond which it cannot be extended.

Minding's result applies here too: if two patches of surfaces have constant negative curvature $\mathcal{K} = -(1/R^2)$ then their intrinsic geometry is *identical* to that of the pseudosphere of radius R.

To sum up, if a surface has constant curvature $\mathcal{K}$ (positive or negative) then this single number determines the intrinsic geometry of the surface *completely*.

But what of more typical surfaces, within which the curvature varies? While the control of the curvature remains very great, it is no longer absolute: it is possible for two surfaces to have different intrinsic geometries while still having identical curvatures at corresponding points. (A concrete example is provided in Ex. 19, p. 224.)

2.3 The Local Gauss–Bonnet Theorem

Recall Harriot's 1603 result (1.3) on the sphere: *the angular excess of a triangle is the curvature times the area*, which we may think of as the total amount of curvature residing within the triangle.

The ***Local***[8] ***Gauss–Bonnet Theorem***, as originally stated by Gauss in the *Disquisitiones Generales* of 1827, is a stunning generalization of this result to a geodesic[9] triangle Δ on a *general* curved

[7] This surface, also known as the ***tractroid***, was first investigated by Huygens in 1693; see Stillwell (2010, p. 345).

[8] Here the word "local" does not signify infinitesimal, but instead distinguishes this result from a subsequent "global" version that applies to an *entire*, closed surface.

[9] In 1865 Bonnet generalized the formula to nongeodesic triangles, hence the name of the theorem. The most general version of the theorem will not be proved until the end of Act IV (Ex. 6, p. 336).

surface of variable curvature, illustrated in [2.7]. It says that the angular excess of such a triangle is simply the *total curvature* inside it:

$$\mathcal{E}(\Delta) = \alpha + \beta + \gamma - \pi = \iint_{\Delta} \mathcal{K}\, d\mathcal{A}. \tag{2.6}$$

In the case of the sphere, $\mathcal{K} = 1/R^2$, so (2.6) yields Harriot's Formula (1.3) as a very special case.

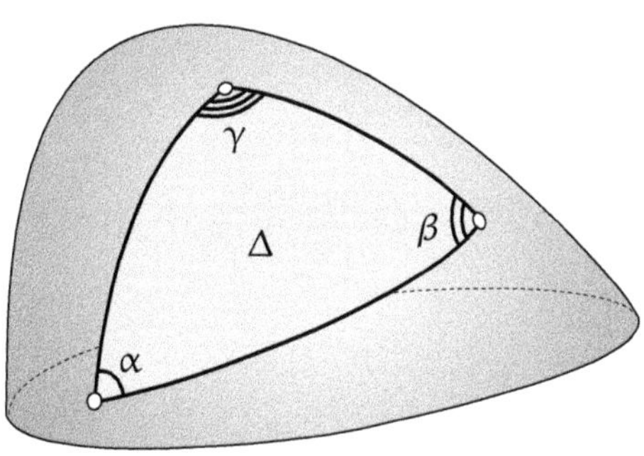

[2.7] *A general geodesic triangle on a general surface.*

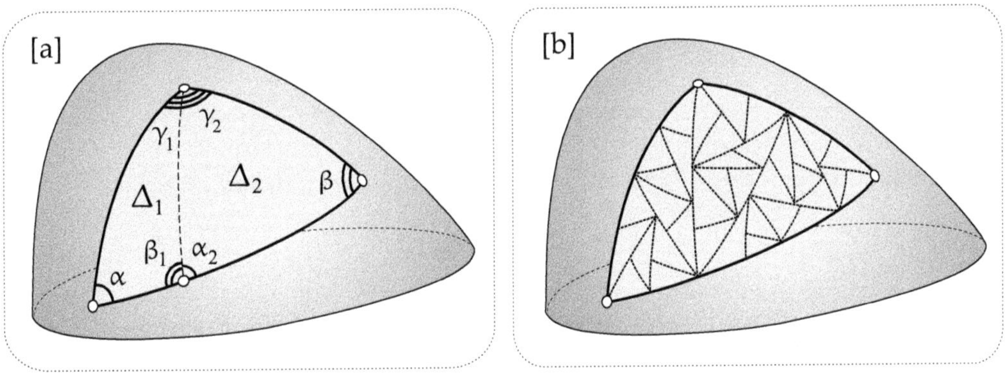

[2.8] **[a]** *Angular excess is additive:* $\mathcal{E}(\Delta) = \mathcal{E}(\Delta_1) + \mathcal{E}(\Delta_2)$. **[b]** *This persists if we continue subdividing triangles:* $\mathcal{E}(\Delta) = \sum \mathcal{E}(\Delta_i)$.

To see why this is so, first recall our original definition of curvature, (2.1). As the triangle Δ_p shrinks towards a point p on $\mathcal{S}$,

$$\mathcal{E}(\Delta_p) \asymp \mathcal{K}(p)\, \mathcal{A}(\Delta_p). \tag{2.7}$$

The key fact is that the angular excess is ***additive***.

In [2.8a] a geodesic segment (dashed) has been drawn from one vertex of Δ to an arbitrary point on the opposite edge, thereby splitting Δ into two geodesic subtriangles, Δ_1 and Δ_2. Observing that $\beta_1 + \alpha_2 = \pi$, we find that

$$\mathcal{E}(\Delta_1) + \mathcal{E}(\Delta_2) = [\alpha + \beta_1 + \gamma_1 - \pi] + [\alpha_2 + \beta + \gamma_2 - \pi] = \alpha + \beta + \gamma_1 + \gamma_2 - \pi,$$

and therefore

$$\mathcal{E}(\Delta) = \mathcal{E}(\Delta_1) + \mathcal{E}(\Delta_2).$$

These subtriangles may then be subdivided in their turn, and so on and so forth, yielding [2.8b], and the additive property ensures that $\mathcal{E}(\Delta) = \sum \mathcal{E}(\Delta_i)$. As the subdivision becomes finer and finer, the curvature varies less and less within each Δ_i, approaching the constant value $\mathcal{K}_i$, and in this limit (2.7) yields $\mathcal{E}(\Delta) \asymp \sum \mathcal{K}_i \mathcal{A}_i$, and so we arrive at the Local Gauss–Bonnet Theorem, (2.6).

Chapter 3

Exercises for Prologue and Act I

Prologue: Newtonian Ultimate Equality ($\asymp$)

1. (This is a model of how an "ultimate equality"—see Prologue—becomes an equality.) Sketch a cube of side x, hence of volume $\mathcal{V} = x^3$. In the same picture, keeping one vertex fixed, sketch a slightly larger cube with side $x + \delta x$. If $\delta\mathcal{V}$ is the resulting increase in volume, use your sketch to deduce that as δx vanishes,

$$\delta\mathcal{V} \asymp 3x^2\,\delta x \quad\Longrightarrow\quad \frac{d\mathcal{V}}{dx} \asymp \frac{\delta\mathcal{V}}{\delta x} \asymp 3x^2 \quad\Longrightarrow\quad \frac{d\mathcal{V}}{dx} \asymp 3x^2.$$

But *the quantities in the final ultimate equality are independent of* δx, so they are equal:

$$(x^3)' = \frac{d\mathcal{V}}{dx} = 3x^2.$$

2. (This example is taken from Needham 1993.) Let $c = \cos\theta$ and $s = \sin\theta$. In the first quadrant of $\mathbb{R}^2$, draw a point $p = (c, s)$ on the unit circle. Now let p rotate by a small (ultimately vanishing) angle $\delta\theta$. With one vertex at p, draw the small triangle whose sides are δc and δs. By emulating the Newtonian geometrical argument in the Prologue, *instantly and simultaneously* deduce that

$$\frac{ds}{d\theta} = c \qquad \textit{and} \qquad \frac{dc}{d\theta} = -s.$$

3. (This example is taken from Needham 1993.) Let L be a general line through the point (a, b) in the first quadrant of $\mathbb{R}^2$, and let $\mathcal{A}$ be the area of the triangle bounded by the x-axis, the y-axis, and L.

(i) Use ordinary calculus to find the position of L that minimizes $\mathcal{A}$, and show that $\mathcal{A}_{min} = 2ab$.

(ii) Use Newtonian reasoning to solve the problem *instantly*, without calculation! (*Hints:* Let $\delta\mathcal{A}$ be the change in the area resulting from a small (ultimately vanishing) rotation $\delta\theta$ of L. By *drawing* $\delta\mathcal{A}$ in the form of two triangles, and observing that each triangle is ultimately equal to a sector of a circle, write down an ultimate equality for $\delta\mathcal{A}$ in terms of $\delta\theta$. Now set $\delta\mathcal{A} = 0$.)

4. The following problem is taken from Arnol'd (1990, p. 28), which also contains the solution. Evaluate the limit

$$\lim_{x\to 0} \frac{\sin\tan x - \tan\sin x}{\sin^{-1}\tan^{-1} x - \tan^{-1}\sin^{-1} x},$$

(i) using *any* traditional method you can think of. (We would be remiss if we did not wish you the best of British luck in this endeavour! Arnol'd notes that the *only* mathematician who was ever able to solve this problem quickly was the Fields Medalist, Gerd Faltings.)

(ii) using Newtonian geometrical reasoning.

Euclidean and Non-Euclidean Geometry

5. It is not known with certainty how the Babylonians generated the Pythagorean triples in [1.2], but, 1500 years later, Euclid (around 300 BCE) was the first to state and prove the most general formulas for such triples. Half a millennium after that, Diophantus[1] (around 250 CE) was the first to use a *geometrical* construction[2] to systematically generate ***rational points*** on the unit circle, meaning points with rational coordinates. These rational points can then be used to generate Pythagorean triples, as follows.

(i) Let L be the line $y = m(x+1)$ through the point $(-1, 0)$ of the unit circle $\mathcal{C}$, and let (X, Y) be the second intersection point of L and $\mathcal{C}$. Prove that

$$X = \frac{1-m^2}{1+m^2} \quad \text{and} \quad Y = \frac{2m}{1+m^2}.$$

(ii) Deduce that *if the slope* $m = (q/p)$ *is* rational, *then so are* X *and* Y:

$$X = \frac{p^2-q^2}{p^2+q^2} \quad \text{and} \quad Y = \frac{2pq}{p^2+q^2}.$$

(iii) Deduce that if (a, b, h) is an arbitrary Pythagorean triple, then

$$\frac{a}{h} = \frac{p^2-q^2}{p^2+q^2} \quad \text{and} \quad \frac{b}{h} = \frac{2pq}{p^2+q^2},$$

for some integers p and q.

(iv) Deduce that the most general Pythagorean triple is given by the following formulas, first stated by Euclid:

$$a = (p^2-q^2)r \quad \text{and} \quad b = 2pqr \quad \text{and} \quad h = (p^2+q^2)r,$$

where p, q, r are arbitrary integers.

6. Use (1.3) to deduce that as the size of a triangle on the sphere shrinks to nothing, it ultimately appears to the inhabitants of the sphere to be *Euclidean*, i.e., with angle sum equal to π.

7. Let p and q be distinct, nonantipodal points on the sphere, and consider the unique great circle $\mathcal{C}$ through them. Let m_1 and m_2 be the midpoints of the two arcs into which $\mathcal{C}$ is split by p and q. Show that the locus of points that are equidistant from p and q is the great circle through m_1 and m_2 that cuts $\mathcal{C}$ orthogonally: this is the generalized "perpendicular bisector" of pq. (*Hint:* while it is mathematically immaterial, it can be *psychologically* helpful to imagine that the sphere has been rotated so that p and q lie on the equator.)

8. If the sides of a triangle on the unit sphere $\mathbb{S}^2$ are each less than π, show that the triangle is contained within a hemisphere. (*Hint:* while it is mathematically immaterial, it can be *psychologically* helpful to imagine that the sphere has been rotated so that one vertex is at the north pole.)

9. One of the characteristics of the Euclidean plane is that it possesses ***regular tessellations***: it can be ***tiled*** (completely filled, without any gaps) by regular polygons. There are precisely three such regular tessellations of the plane, using equilateral triangles, squares, or regular

---

[1] For the little that is known of his life, see Stillwell (2010, §3.6).

[2] This is the prototype of Newton's method of generating rational points on *cubic* curves, using chords and tangents. See Stillwell (2010, §3.5).

[3.1] Icosahedral tessellation of the sphere. *Here the image of each triangular face of the icosahedron has been further divided into six congruent triangles, by joining its centre to the midpoints of its edges. (This lovely drawing is taken from Fricke (1926).)*

hexagons. The sphere *also* admits regular tessellations. Imagine a wire-frame icosahedron inscribed in a sphere of radius R. Now imagine a point source of light at the centre of the sphere. The shadow on the sphere is shown in [3.1]. Here the image of each triangular face of the icosahedron has been further divided into six congruent triangles, by joining its centre to the midpoints of its edges.

(i) Explain why the shadows of the icosahedral edges are arcs of *great circles* on the sphere, thereby creating genuine spherical triangles.

(ii) Verify that if we instead inscribe a dodecahedron, centrally project its pentagons onto the sphere, then join their centres to the midpoints of the edges, we obtain the *same* tessellation.

(iii) Into how many congruent triangles has the sphere (of area $4\pi R^2$) been divided? So what is the area $\mathcal{A}$ of each triangle?

(iv) By observing how many like angles come together at each vertex, deduce that the angles of the triangles are $(\pi/2)$, $(\pi/3)$, and $(\pi/5)$. Hence calculate the angular excess $\mathcal{E}$ of each triangle.

(v) Verify that the previous two answers are in accord with Harriot's Theorem, (1.3).

10. (i) Prove that in Euclidean Geometry the angle sum of a quadrilateral is 2π.

(ii) If $\mathcal{Q}$ is a geodesic quadrilateral on the sphere of radius R, its angular excess is therefore

$$\mathcal{E}(\mathcal{Q}) = (\text{angle sum of } \mathcal{Q}) - 2\pi.$$

By drawing in a diagonal of $\mathcal{Q}$, thereby splitting it into two geodesic triangles, deduce that (1.3) generalizes to

$$\mathcal{E}(\mathcal{Q}) = \frac{1}{R^2}\mathcal{A}(\mathcal{Q}).$$

(iii) Prove (2.2), and hence generalize (ii) to geodesic n-gons on the sphere.

11. Using the technique described in the footnote on page 14, or otherwise, manufacture narrow strips of sticky tape, ideally coloured masking tape. Then use (1.7) to conduct the following experiments on the surface of a vase of the approximate form shown in [11.7], on page 127. If you do not own such a vase, we suggest you borrow one—this exercise is too interesting to be missed!

(i) Starting at a point on the horizontal circle of greatest radius, ρ_{max}, launch a geodesic straight up the vase, creating a ***meridian*** of the surface of revolution.

(ii) Starting at the same point, launch several more geodesics in directions that make ever-larger angles ψ with the meridian.

(iii) Note that beyond some critical angle ψ_c, the geodesics initially make their way up the vase, but then turn back and come down the vase!

(iv) As best you can, try to find the critical geodesic—the one that separates those that turn back from those that don't. By pressing a protractor against the surface at its launch point, measure its angle, ψ_c. NOTE: To find the critical geodesic, you will need to construct extra-long geodesic segments, which you may do by extending an existing segment, overlapping the new with the old, as illustrated in [1.9].

(v) Let ρ_{max} be the maximum radius of the vase (at the launch point), and let ρ_{min} be the minimum radius, at the throat of the vase. By measuring diameters and dividing by 2, find these two radii as accurately as you can. Now verify that, within experimental error,

$$\psi_c = \sin^{-1}\left[\rho_{min}/\rho_{max}\right]!$$

(This is a physical instantiation of ***Clairaut's Theorem***, which we will prove much later, in Section 11.7.4.)

Gaussian Curvature

12. ***Zero*** **Curvature.** Using the technique described in the footnote on page 14, or otherwise, manufacture narrow strips of sticky tape, ideally coloured masking tape. Then use (1.7) to conduct the following experiments.

(i) Roll up a piece of paper into a cone, and tape it so that it does not unfurl. *Start* to create a long geodesic segment originating at a point on the rim, but try to guess its form *before* you stick your strip down onto the surface. Starting at the same point on the rim, repeat this construction, launching new geodesics in different directions.

(ii) Next, construct a geodesic triangle, and use a protractor to verify that $\mathcal{E} = 0$. (This is true of *all* geodesic triangles on this surface, proving that $\mathcal{K} = 0$.)

(iii) Finally, cut open the cone (along a generator) and press the paper flat again, and observe the Euclidean form of your tape constructions.

13. ***Positive*** **Curvature.** Stick a toothpick into the north pole of any approximately spherical fruit (of radius R), such as a melon. Tie one end of a piece of string or dental floss to the toothpick and stretch the string tightly over the surface to a point about halfway to the equator. Holding a pen at the end of this geodesic segment of length r, drag the pen around (keeping the string taut) to create a circle of latitude of length $C(r)$. Stick toothpicks into this circle at 16 evenly spaced points. Now wrap a piece of string or dental floss around the outside of the approximately circular ring of toothpicks, and gently pull it tight, so that it follows the circle of latitude *on* the surface of the fruit. With a pen, carefully mark the string at the start and end of the circle. Unwrap it and stretch it along a ruler to measure the length $C(r)$ between the two marks.

(i) Treating (2.4) as an approximate equality (instead of an exact ultimate equality), estimate $\mathcal{K}$. From these *intrinsic* measurements, estimate the *extrinsic* radius R of the fruit. Compare your answer with a direct measurement of R.

(ii) Continuing from (i), suppose that your measurements of r and C(r) are perfect. Use the third term of the (decreasing and alternating) Taylor expansion of sin(r/R) to show that an upper bound on the percentage error in $\mathcal{K}$ that results from applying (2.4) in the manner of part (i)—i.e., *without* taking the limit implied by the ultimate equality—is given by

$$\left|\frac{\Delta\mathcal{K}}{\mathcal{K}}\right| < 5\left[\frac{r}{R}\right]^2 \%$$

Deduce that even for a circle as large as the one you constructed, the error cannot be larger than approximately 3%!

(iii) Use the result of (ii) to deduce a formula for the upper bound of the percentage error in R.

14. ***Negative* Curvature.** Using the technique described in the footnote on page 14, or otherwise, manufacture narrow strips of sticky tape, ideally coloured masking tape. Then use (1.7) to conduct the following experiments.

(i) By following the instructions that accompany [5.3], page 53, construct your own, personal pseudosphere out of discs of radius R—the more cones, the better; the bigger, the better!

(ii) Starting at a point on the circular base of radius R, launch geodesics in various directions, and try to predict their course *before* you lay the tape down onto the surface. When a strip of tape runs out, continue the geodesic by simply *overlapping* a new strip with the old one, as illustrated in [1.9]. With the sole exception of the meridian geodesic that heads straight up the surface—tracing a ***tractrix*** generator of the surface of revolution—note that every geodesic initially heads up the pseudosphere but then turns around and comes back down the pseudosphere, ultimately returning to the base circle.

(iii) Construct a right-angled geodesic triangle, Δ, measure its angles, and hence estimate its angular excess, $\mathcal{E}(\Delta)$. Estimate (as best you can) its areas, $\mathcal{A}(\Delta)$. Hence estimate the (constant) curvature $\mathcal{K}$ of your pseudosphere, using

$$\mathcal{K} = \frac{\mathcal{E}(\Delta)}{\mathcal{A}(\Delta)}.$$

(iv) The larger the triangle, the larger (i.e., more negative) the value of $\mathcal{E}(\Delta)$, making its measurement easier and more accurate. But the tradeoff is that it becomes harder to accurately estimate the area $\mathcal{A}(\Delta)$. To overcome this difficulty, do the following. Make narrow strips of the same kind that you use to generate geodesics, but create them all with the *same* (accurately measured) width, W, perhaps 1/4 inch. Now fill your Δ with these strips, cutting them off when they hit an edge. Remove the strips and lay them end-to-end on a flat surface, and measure the total length, L. Then $\mathcal{A}(\Delta) \asymp LW$.

(v) Repeat (iii) with several more triangles, but no longer restrict them to be right-angled, because (iv) now makes it easy to measure $\mathcal{A}(\Delta)$ for any shape of triangle. Verify that (within experimental error) all triangles yield the *same* value of $\mathcal{K}$.

(vi) Assuming that

$$\mathcal{K} = -\frac{1}{R^2},$$

estimate R, and compare this to the actual radius of the discs you used to construct your pseudosphere.

ACT II

The Metric

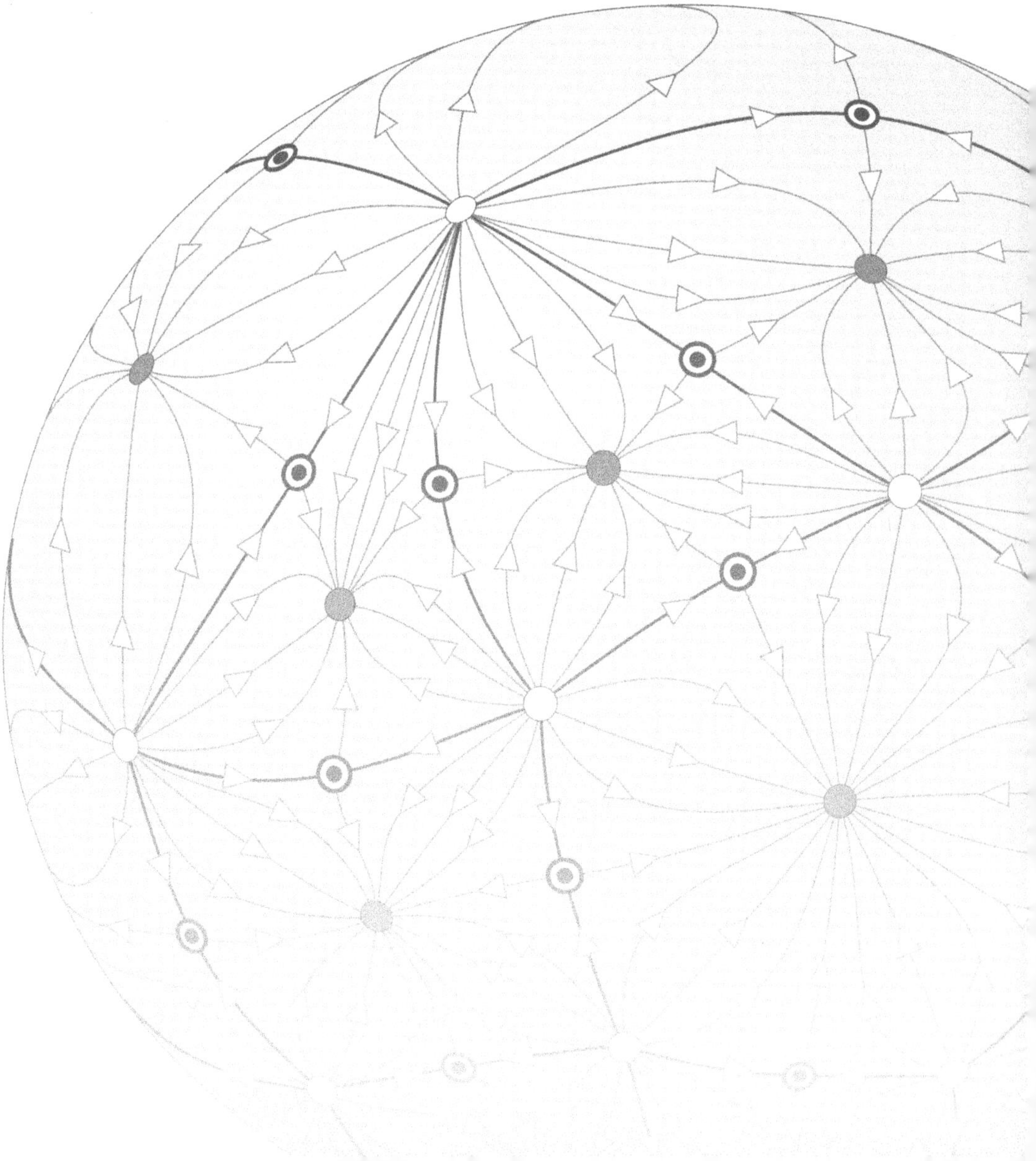

Chapter 4

Mapping Surfaces: The Metric

4.1 Introduction

The perfect extrinsic symmetry of the sphere has the advantage of making it obvious that its intrinsic geometry is likewise uniform. By contrast, in [2.5] it's certainly not clear that a flexible but unstretchable shape fitted to the surface can be freely slid about and rotated on the surface, all the while remaining fully in contact with the surface. But in fact this must be so, for the above discussion shows that the actual shape of a surface in space is a distraction: these surfaces are intrinsically indistinguishable from the sphere (at least locally).

From this point of view it would be better to have a more abstract model that captured the essence of all possible surfaces having the same intrinsic geometry. By the "essence" we mean knowledge of the distance between any two points, for this and this alone determines the intrinsic geometry. In fact—and this was Gauss's fundamental insight into Differential Geometry—it is sufficient to have a rule for the *infinitesimal* distance between neighbouring points. This rule is called the ***metric***.[1] Given this metric, we may determine the length of any curve as an infinite sum (i.e., integral) of the infinitesimal segments into which it may be divided. Consequently, we may also identify the geodesics of the geometry as the shortest routes from one point to another, and we can likewise determine angles.

This leads to the following strategy for capturing the essence of any curved surface $\mathcal{S}$ (not necessarily one of constant curvature). To avoid the distraction of the shape of the surface in space, we draw a (cartographic) ***map*** of $\mathcal{S}$ on a flat piece of paper. That is, we set up a one-to-one correspondence between points $\widehat{z}$ on $\mathcal{S}$ and points z on the plane. Of course in the case of the spherical Earth and the spherical night sky, mariners and astronomers have been devising such geographical and celestial maps for thousands of years.

IMPORTANT NOTE ON TERMINOLOGY: In most of mathematics, "map" is synonymous with "mapping" or "function." However, when used as a noun, we shall always use "map" in the current sense of a *cartographic* map. When we mean "mapping" or "function," we shall *say* "mapping" or "function"! That said, we will retain some of the traditional meaning when it comes to the *verb*: e.g., the coordinate functions map this curve on the surface to that curve in the map.

In general, such a map of a curved surface cannot be created without introducing some kind of distortion: if you peel part of an orange and try to press the patch of peel flat on the table, it will tear. Euler was the first to prove (in 1775) the mathematical impossibility of a perfect map of the Earth, in which all "straight lines" (geodesics) on the surface become straight lines in the maps, and all terrestrial distances are represented by proportional distances on the map.

The preceding discussion explains this impossibility of a perfect map. A triangle on the Earth's surface has $\mathcal{E} \neq 0$, but if this triangle could be pressed flat without altering distances then its image in the map would be a Euclidean triangle with $\mathcal{E} = 0$: a contradiction.

We will eventually discover that there exist deep and mysterious connections between non-Euclidean Geometry and *complex numbers*. Let us therefore, from the very outset, imagine that the flat piece of paper on which we draw our map is in fact the *complex plane*, $\mathbb{C}$.

---

[1] Another common name, particularly in older works, is the ***First Fundamental Form***.

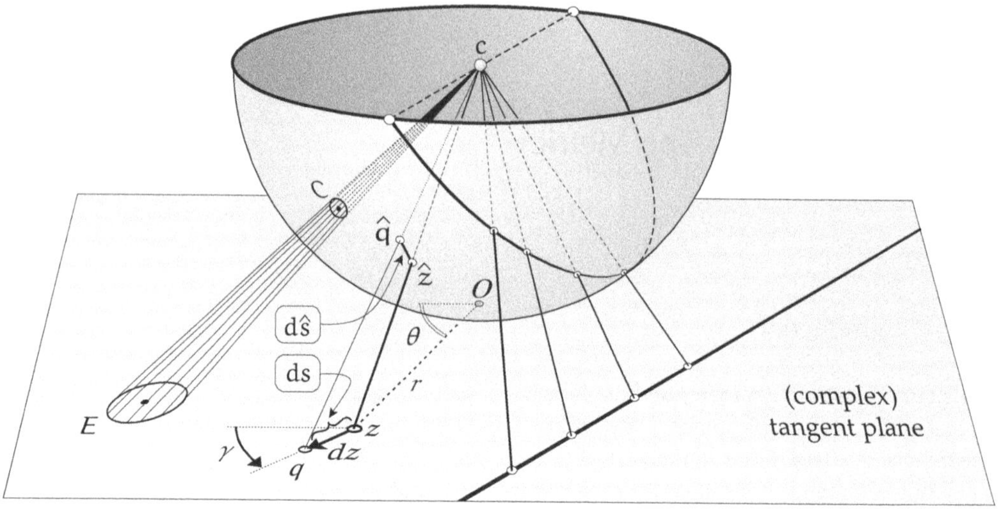

[4.1] *The central projection of the sphere maps geodesics to straight lines and circles to ellipses. The* metric *tells us the local scale factor relating map distance* $\mathrm{d}s$ *and surface distance* $\mathrm{d}\widehat{s}$.

Now consider the distance $\delta\widehat{s}$ separating two neighbouring points $\widehat{z}$ and $\widehat{q}$ on $\mathcal{S}$. The points $\widehat{z}$ and $\widehat{q}$ will then be represented by the complex numbers $z = r\,e^{i\theta}$ and $q = z + \delta z$ in this complex plane, separated by (Euclidean) distance $\delta s = |\delta z|$. For a concrete example of such a map (to be explained in a moment) look ahead to [4.1]. Once we have a rule for calculating the actual separation $\delta\widehat{s}$ on $\mathcal{S}$ from the apparent separation δs in the map, then (in principle) we know everything there is to know about the intrinsic geometry of $\mathcal{S}$.

The rule giving $\delta\widehat{s}$ in terms of δz is called the ***metric***. In general, $\delta\widehat{s}$ depends on the direction of δz as well as its length δs: writing $\delta z = e^{i\gamma}\,\delta s$, so $\delta\widehat{s} \asymp \Lambda(z,\gamma)\,\delta s$, in which we remind the reader that $\asymp$ denotes Newton's concept of ***ultimate equality***, which was introduced in the Prologue. This relation is more traditionally expressed using the notation of infinitesimals, as,

$$\mathrm{d}\widehat{s} = \Lambda(z,\gamma)\,\mathrm{d}s. \tag{4.1}$$

According to this formula, $\Lambda(z,\gamma)$ is the amount by which we must expand the apparent separation $\mathrm{d}s$ in the map—located at z, and in the direction γ—to obtain the true separation $\mathrm{d}\widehat{s}$ on the surface $\mathcal{S}$.

4.2 The Projective Map of the Sphere

Figure [4.1] illustrates the meaning of (4.1) for one particular method of drawing a map of the southern hemisphere, called ***central projection***. Imagine the hemisphere to be a glass bowl resting on the complex plane at the origin 0, and imagine a light source at the centre of the sphere, c. Then a light ray passing through $\widehat{z}$ on the hemisphere goes on to hit a point z in $\mathbb{C}$. The resulting plane map is the so-called ***projective map*** (or ***projective model***) of the southern hemisphere.

If we draw a circle C on the hemisphere then the rays passing through it form a cone in space, and hence they hit $\mathbb{C}$ in a perfect ellipse E. This is a very special and unusual property of this particular method of drawing a map. However, it turns out[2] that if C is an intrinsically defined circle on a *general* surface $\mathcal{S}$ (such as in [1.9]) then as its radius shrinks its image E in a *general* map

[2]Because if the mapping is differentiable then its local effect is a linear transformation.

will ultimately be an ellipse, also. Returning to the perfect ellipses of the central projection, it's clear that E's major axis will be radial, pointed directly away from the bowl's point of contact with the plane. In other words, if you imagine dz rotating about z, then $\Lambda(z, \gamma)$ achieves its minimum and maximum values at $\gamma = \theta$ and $\gamma = \theta + (\pi/2)$, respectively.

How we choose to draw a map of a surface depends on which features we wish to accurately or ***faithfully*** represent. For example, [4.1] illustrates that the projective map faithfully represents lines: a straight line in the map represents a great-circle geodesic on the sphere. But, the price that we pay for preserving the concept of lines is that angles are not faithfully represented: the angle at which two curves meet on the sphere is not (in general) the angle at which they meet on the map.

That said, there are in fact two orthogonal families of curves on the sphere that map to orthogonal curves in the plane: they are the circles of longitude and latitude. A circle of latitude (i.e., a horizontal cross section of the hemisphere) maps to an origin-centred circle in the plane, and a (semi-)circle of longitude (i.e., a vertical cross section of the hemisphere through its centre) maps to a line through the origin. As claimed, these circles and lines do indeed meet at right angles. We now use this fact to derive a formula for the metric of the sphere in terms of polar coordinates (r, θ) in the projective map. It is not hard to accomplish this by calculation [exercise], but we shall instead use the Newtonian form of geometric reasoning that was introduced in the Prologue, and which we shall employ throughout this work.

See [4.2]. A small rotation of $\delta\theta$ in the plane rotates z a distance $r\,\delta\theta$ round the circle of radius r and rotates $\hat{z}$ by $\delta\hat{s}_1$ round a horizontal circle of latitude. If z then moves radially outward by δr then $\hat{z}$ moves north by $\delta\hat{s}_2$ along a vertical circle of longitude. By Pythagoras's Theorem, $\delta\,\hat{s}^2 \asymp \delta\,\hat{s}_1{}^2 + \delta\,\hat{s}_2{}^2$, and we shall now find each of these terms separately. (Recall from the Prologue that $\asymp$ is our symbol for Newton's concept of "ultimate equality.")

Let $H = cz$ denote the distance in $\mathbb{R}^3$ from the centre c of the sphere to the complex number z, as illustrated in [4.2]. Since $\hat{z}$ is at distance R from c, and since the rotation causes $\hat{z}$ and z to move in proportion to their distances from the centre of the sphere:

$$\frac{d\hat{s}_1}{r\,d\theta} \asymp \frac{\delta\hat{s}_1}{r\,\delta\theta} = \frac{R}{H}.$$

Next, imagine H to be a string of fixed length, attached to c. If we swing its free end at z upward a distance ϵ in the plane perpendicular to the complex plane, then the northward motion $\delta\hat{s}_2$ of $\hat{z}$ will be in the same proportion as before. Furthermore, we see that the small vertical grey triangle with edges ϵ and δr is ultimately similar to the large triangle 0cz with sides R and H. Thus,

$$\frac{\delta\hat{s}_2}{\epsilon} \asymp \frac{R}{H} \quad \text{and} \quad \frac{\epsilon}{\delta r} \asymp \frac{R}{H} \quad \Longrightarrow \quad \frac{d\hat{s}_2}{dr} = \frac{R^2}{H^2}$$

Finally, by Pythagoras's Theorem again, $H^2 = R^2 + r^2$. So, reverting to the industry-standard, infinitesimal-based notation, the metric describing the true distance on the sphere in terms of the coordinates in the projective map is given by

$$d\hat{s}^2 = \frac{1}{1+(r/R)^2}\left[\frac{dr^2}{1+(r/R)^2} + r^2\,d\theta^2\right]. \tag{4.2}$$

In geography, the angle of latitude ϕ is usually defined to be zero at the equator, but we choose to instead measure it from a pole (as shown in both [4.2] and [2.4]). Returning to [4.1], we can now quantify the amount of distortion induced by the map, by looking at the shape of the small ellipse E. As C shrinks, you should now be able to confirm [exercise] that

$$\left(\frac{\text{major axis of E}}{\text{minor axis of E}}\right) \asymp \sec\phi, \tag{4.3}$$

so as C moves north towards the rim ($\phi = \frac{\pi}{2}$), the distortion of shapes increases without limit.

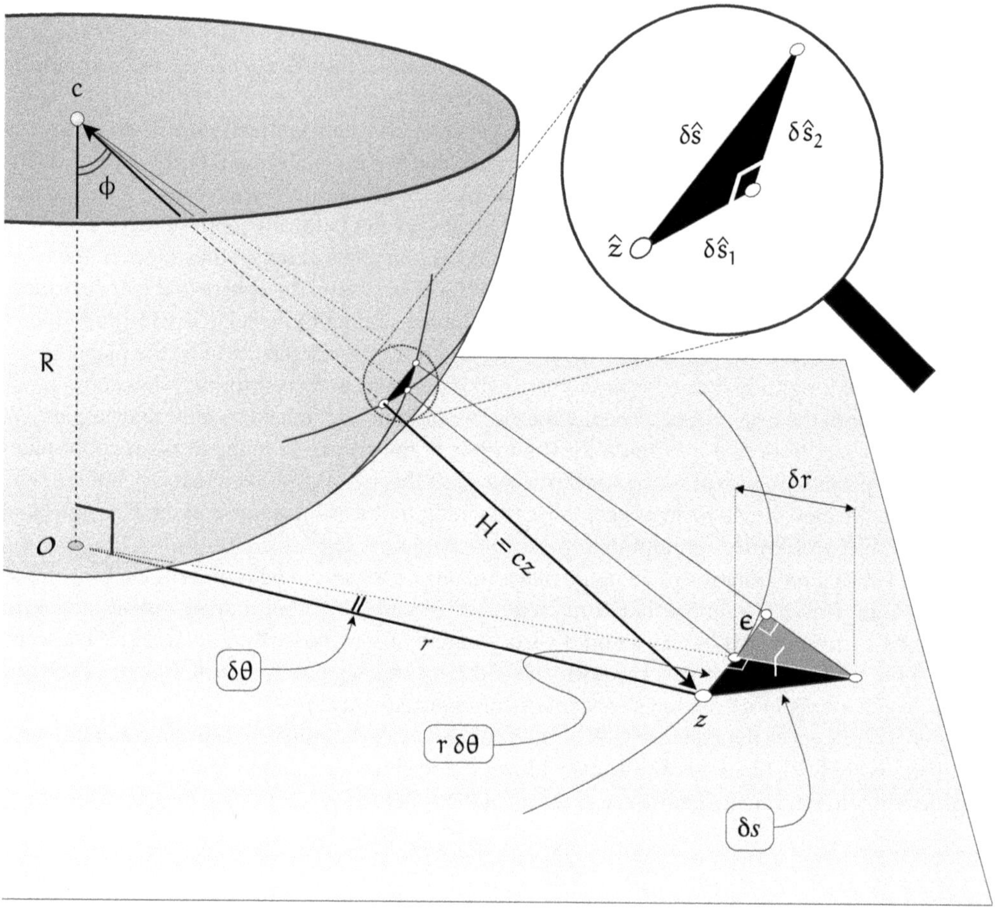

[4.2] *A small rotation of* $\delta\theta$ *in the plane moves* z *by* $r\,\delta\theta$ *and it moves* $\hat{z}$ *by* $\delta\widehat{s_1}$ *round a horizontal circle of latitude. If* z *then moves radially outward by* δr *then* $\hat{z}$ *moves north by* $\delta\widehat{s_2}$ *along a vertical circle of longitude.*

4.3 The Metric of a General Surface

Given the critical importance of maps in navigation, mathematicians have for centuries explored many different methods of drawing maps of the Earth, and we shall meet an especially important one shortly; other maps are explored in the exercises at the end of this Act. For now we wish merely to point out that each such map has a different metric formula associated with it, despite the fact that they all describe the *same* intrinsic geometry.

For example, the most common way of describing the location of a place on Earth is to supply its longitude θ and latitude ϕ. Suppose we use these two angles to draw a very straightforward kind of map in the plane, with θ along the horizontal axis, and ϕ along the vertical axis. That is, if a particular house has longitude θ and latitude ϕ then in the flat map we represent it by the point with Cartesian coordinates (θ, ϕ). With the assistance of [2.4], you should now easily find [exercise] that the metric formula telling us the true distance on the sphere corresponding to neighbouring points in the map is

$$d\hat{s}^2 = R^2\left[\sin^2\phi\, d\theta^2 + d\phi^2\right]. \tag{4.4}$$

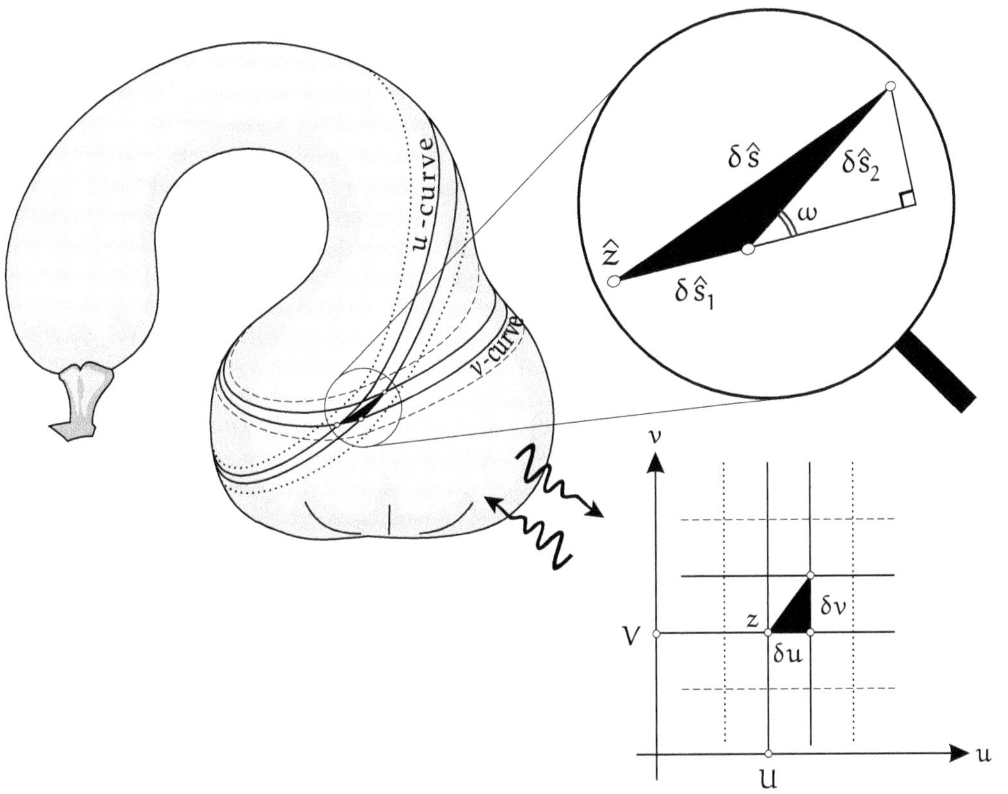

[4.3] *On a general surface, draw two parameterized families of curves: "u-curves" (*$u = \text{const.}$*) and "v-curves." A point is then specified by the particular pair of curves* (U, V) *that intersect there, yielding a point* $z = U + iV$ *in the map. A small horizontal movement* δu *in the map ultimately produces a proportional movement* $\delta\hat{s}_1 \asymp A\,\delta u$ *on the surface, along a v-curve.*

This certainly *looks* very different from our previous formula, (4.2), but we know that it in fact describes exactly the same intrinsic geometry.

Even if we simply take our surface to be a flat plane, an infinite variety of metric formulas are possible. For example, if we use regular Cartesian coordinates then $d\hat{s}^2 = dx^2 + dy^2$, whereas if we use polar coordinates then $d\hat{s}^2 = dr^2 + r^2 d\theta^2$.

Now let us investigate the form and meaning of the most general metric formula for a general surface. Figure [4.3] illustrates how we may draw a general map of a patch of such a surface $\mathcal{S}$. For each point $\hat{z}$ in this patch, our first aim is to assign a pair of coordinates (u, v) so that it can be represented by the complex number $z = u + iv$ in the flat map.

To get started, we simply draw, quite arbitrarily, a family of curves covering the patch, so that one (and only one) curve from our family passes though each point $\hat{z}$ on the patch. We now label each of these curves with a unique number u, and agree to call these curves, with their assigned u-values, the u-***curves***. The labelling can be done quite arbitrarily, *except* that we shall insist that the numbering vary *smoothly*, in the sense that as we move over the surface across the u-curves, the u-values should change at a definite *rate*, i.e., differentiably—more on this in a moment.

To complete the coordinate system, we now draw a second family of curves, again quite arbitrarily, except that they should cross the u-curves and never coincide with them. Now label these new curves with v-values, and, again, the v-values should vary differentiably, at a definite *rate* (in the same manner as before); call these the v-***curves***. Thus, as illustrated, the point $\hat{z}$ can be labelled by the unique u-curve (say $u = U$) and v-curve (say $v = V$) that intersect there. So, in the map, $\widehat{z}$ can now be represented as $z = U + iV$. Thus, as illustrated, u-curves are represented in the map by vertical lines, while v-curves are mapped to horizontal lines.

Now that we have the coordinates on at least some patch of the surface, the task is to find the metric formula that gives the distance between neighbouring points. Suppose that in the map we make a small movement away from z along $\delta z = \delta u + i\,\delta v$. By virtue of the differentiable assignment of u-values to the u-curves, the small change δu produces—and we now come to the *definition* of ***differentiable***—a small movement $\delta\widehat{s}_1$ on the surface (along a v-curve) that is ultimately *proportional* to δu, and we define A to be this proportionality constant at the point:

$$A \equiv \frac{\partial \widehat{s}_1}{\partial u} \asymp \frac{\delta \widehat{s}_1}{\delta u}.$$

This is important, so we reiterate: A *is the local scale factor in the horizontal direction of the* (u, v)*-map,* the factor by which we must stretch a small horizontal distance in the map to obtain the true distance on the surface.

There is also a way to visualize this without even looking at the map. If in [4.3] we imagine that the u-curves have been drawn at small fixed increments of ϵ ($u = U, u = U + \epsilon, u = U + 2\epsilon, \ldots$) then A can also be visualized as being inversely proportional to the crowding or density of the u-curves. For the greater the crowding, the greater the change δu resulting from a given movement $\delta\widehat{s}_1$ on the surface.

In exactly the same way, the small change δv (while keeping u constant) produces a movement $\delta\widehat{s}_2$ on the surface that is ultimately proportional to it, enabling us to define B to be the local scale factor in the vertical direction of the map:

$$B \equiv \frac{\partial \widehat{s}_2}{\partial v} \asymp \frac{\delta \widehat{s}_2}{\delta v}.$$

Finally, as illustrated, we define ω to be the angle between the u-curves and the v-curves. Obviously this angle is not a constant: just as with the scale factors A and B, the angle ω is a function of position.

We can now apply Pythagoras's Theorem to the right triangle shown in the magnifying glass of [4.3]:

$$\delta\widehat{s}^2 \asymp (\delta\widehat{s}_1 + \delta\widehat{s}_2 \cos\omega)^2 + (\delta\widehat{s}_2 \sin\omega)^2 \tag{4.5}$$

$$\asymp (A\,\delta u + B\,\delta v \cos\omega)^2 + (B\,\delta v \sin\omega)^2. \tag{4.6}$$

After simplifying, and reverting to more standard, infinitesimal-based notation, we find that

The general metric formula for a general surface is

$$d\widehat{s}^2 = A^2\,du^2 + B^2\,dv^2 + 2F\,du\,dv, \quad \textit{where} \quad F = AB\cos\omega. \tag{4.7}$$

Assuming that you will eventually look at other books, we should immediately issue a NOTATIONAL WARNING. In his original masterpiece of 1827, Gauss made the decision to write[3] the metric as

$$d\widehat{s}^2 = E\,du^2 + G\,dv^2 + 2F\,du\,dv,$$

[3]In fact Gauss wrote p and q in place of u and v, which later became the standard notation. We have chosen not to mind our historical p's and q's.

In the subsequent centuries, almost every[4] research paper and textbook on Differential Geometry has slavishly perpetuated this E, F, G-notation. And yet we have seen that it is $\sqrt{E} = A$ and $\sqrt{G} = B$ that have the simple geometric interpretation given above, and thus it should come as no surprise that it is they (*not* E and G) that manifest themselves in many important formulas. The consequence has been that a literature has arisen that is needlessly cluttered with square roots. We shall therefore continue to employ the notation A and B (in place of $\sqrt{E}$ and $\sqrt{G}$) throughout the book, so we stress that when you look elsewhere you must translate using this

$$\text{NOTATIONAL DICTIONARY:} \quad E \equiv A^2, \quad G \equiv B^2, \quad F \equiv AB \cos\omega. \tag{4.8}$$

Next, the general metric formula can be simplified as follows. It should be clear that once we have drawn the family of u-curves we may then draw the family of *orthogonal trajectories*, and then *choose* these as the v-curves. Since this construction insists that $\omega = (\pi/2)$, it follows that $F = 0$, and hence,

Locally, we may always construct an **orthogonal** (u, v)*-coordinate system on a general surface; the metric then takes the form,*

$$d\hat{s}^2 = A^2\, du^2 + B^2\, dv^2. \tag{4.9}$$

Note, however, that it is generally *not* possible to cover the *entire* surface with a single (u, v)-coordinate system, even if we do *not* insist on orthogonal coordinates. The problem that arises is that two u-curves (and/or v-curves) may be forced to intersect, in which case the intersection point would have to be assigned two *different* u-values. In fact, as will eventually see in Chapter 19, such problems are in fact *inevitable* on *every closed surface*, other than the doughnut.

For example, on the surface of the Earth, suppose we choose our u-curves to be circles of latitude (though not necessarily with u = latitude). Then the orthogonal trajectories (i.e., the v-curves) are *necessarily* the circles of longitude: great circles intersecting at the north and south poles. Thus each pole must be assigned infinitely many v-values.

4.4 The Metric Curvature Formula

Now suppose we are simply handed a metric *formula* (4.9), without any direct geometric knowledge of the surface $\mathcal{S}$ it describes, nor the geometric meaning of the coordinates u and v, themselves. What does this surface actually look like? As far as the intrinsic geometry of $\mathcal{S}$ is concerned, this formula tells us *everything*, but only in principle. How can we actually extract useful information from the formula?

If we knew the curvature $\mathcal{K}$ at each point then (via [2.3]) we would possess a clear understanding of the nature and shape of $\mathcal{S}$. But since the metric knows everything about the intrinsic geometry, it must (in particular) contain this information about the curvature. Thus the existence of a formula for $\mathcal{K}$ is assured, and, by virtue of the symmetry of the metric formula, it is also clear that the formula for $\mathcal{K}$ must be symmetrical under the simultaneous exchanges $u \leftrightarrow v$ and $A \leftrightarrow B$.

---

[4]There are only a few exceptions, but their high pedigree perhaps lends a modicum of respectability to our choice of notation. For example, Hopf (1956, p. 92) occasionally wrote $E = e^2$ and $G = g^2$, while Blaschke (1929, p. 162) occasionally employed precisely our A, B-notation.

Nevertheless, it is remarkable how *beautiful*[5] the formula for $\mathcal{K}$ turns out to be:

$$\boxed{\mathcal{K} = -\frac{1}{AB}\left(\partial_v\left[\frac{\partial_v A}{B}\right] + \partial_u\left[\frac{\partial_u B}{A}\right]\right).} \tag{4.10}$$

The road will be long, but we shall ultimately derive this formula with the simplicity and grace it deserves, first geometrically in Act IV (Chapter 27), and finally by calculation (using Cartan's Differential Forms) in Act V (Section 38.8.2). For now, though, we treat it as hyper-advanced technology delivered to us from the future, like a phaser weapon from *Star Trek*'s twenty-third century: we may fire it at targets without beginning to understand (for now) how it works.[6]

For example, the Euclidean metric $d\widehat{s}^2 = dr^2 + r^2 d\theta^2$ has $u = r$, $v = \theta$, $A = 1$, and $B = r$, so

$$\mathcal{K} = -\frac{1}{r}\left(\partial_\theta\left[\frac{\partial_\theta 1}{r}\right] + \partial_r\left[\frac{\partial_r r}{1}\right]\right) = -\frac{1}{r}(\partial_r 1) = 0,$$

as it should.

Next, let's aim at the spherical metric (4.4). In this case, $u = \theta$, $v = \phi$, $A = R\sin\phi$, and $B = R$, so

$$\mathcal{K} = -\frac{1}{R^2\sin\phi}\left(\partial_\phi\left[\frac{\partial_\phi R\sin\phi}{R}\right] + \partial_\theta\left[\frac{\partial_\theta R}{R\sin\phi}\right]\right) = -\frac{\partial_\phi\cos\phi}{R^2\sin\phi} = +\frac{1}{R^2},$$

as it should.

Although the calculation is longer, we encourage you to try your own hand at applying this formula to the projective metric formula (4.2) for the sphere; this too should yield $\mathcal{K} = +(1/R^2)$.

Before moving on, we note another result we shall need later. The metric tells us how to convert small distances in the map into distances on the surface. But how should we convert *areas*? Well, in [4.3] a small rectangle in the map has area $\delta u\,\delta v$, and on the surface the corresponding parallelogram has an area that is ultimately equal to $(A\,\delta u)(B\,\delta v\sin\omega)$. Thus, using (4.7), the formula for the infinitesimal element of area $d\mathcal{A}$ on the surface is

$$\boxed{d\mathcal{A} = \sqrt{(AB)^2 - F^2}\,du\,dv.} \tag{4.11}$$

If we specialize (as we usually shall) to an *orthogonal coordinate system*, so that $F = 0$, then this formula simplifies to

$$\boxed{d\mathcal{A} = AB\,du\,dv.} \tag{4.12}$$

4.5 Conformal Maps

While the projective map of the sphere enjoys the advantage of preserving straight lines, for almost all purposes it is much better to sacrifice straight lines in favour of preserving *angles*. A map that

---

[5]Again, most texts retain Gauss's original E, G-notation, so the beauty of this remarkable formula is traditionally marred by the appearance of *five* distracting and unnecessary square roots!

[6]This approach is not without risk: witness the tragic twentieth-century use of Dr. McCoy's stolen phaser in *The City on the Edge of Forever*!

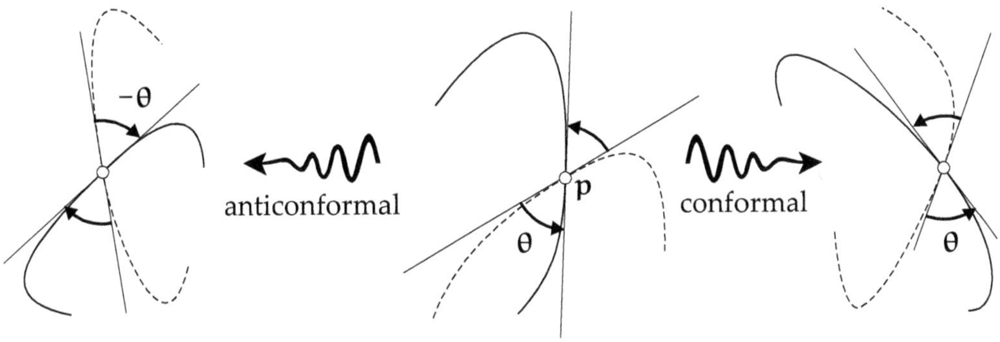

[4.4] *In the centre, two curves intersect at* p*, and the angle between them (from dashed to solid) is defined to be the angle* θ *between their tangents. A* conformal *mapping (right) preserves both the magnitude and sense of the angle, whereas an* anticonformal *mapping (left) preserves the magnitude but reverses the sense.*

preserves the magnitude and ***sense***[7] of angles is called ***conformal***; if it preserves the magnitude but reverses the sense, it is called ***anticonformal***.

When we speak of the angle between two curves, we mean the angle between their tangents. See [4.4].

In terms of the metric formula (4.1), a map is conformal if and only if the expansion factor Λ does not depend on the direction γ of the infinitesimal vector dz emanating from z:

$$\textbf{\textit{Conformal map}} \iff d\widehat{s} = \Lambda(z)\, ds. \tag{4.13}$$

The great advantage of such a map is that

> *An infinitesimal shape on the surface* $\mathcal{S}$ *is represented in a* **conformal map** *by a* similar *shape that differs from the original only in size: the linear dimensions of the shape on* $\mathcal{S}$ *are just* Λ *times bigger than the linear dimensions of the shape in the map.*

Indeed, eighteenth-century mathematicians used the expression ***similar in the small***, in place of the modern term ***conformal***.

It is clear that (4.13) implies conformality, and to see that the converse is true (as claimed) consider [4.5]. On the left we see a triangle shrinking down towards a point in a *conformal* map of a surface. As it does so, the corresponding curvilinear triangle on the surface shrinks towards a rectilinear triangle that—to use the language introduced in the Prologue—is ***ultimately similar***:

$$\frac{\delta\widehat{s_1}}{\delta s_1} \asymp \frac{\delta\widehat{s_2}}{\delta s_2} \asymp \Lambda,$$

for some Λ, independent of the directions of the triangle sides δs_1 and δs_2. Thus, we see that conformality implies (4.13).

In discussing the general metric formula (4.7), it was clear that we could always specialize to an orthogonal coordinate system of u-curves and v-curves on the surface, corresponding to orthogonal vertical and horizontal lines in the map. But at this stage the expansion factors A and B

[7] Anticlockwise $(+)$ or clockwise $(-)$.

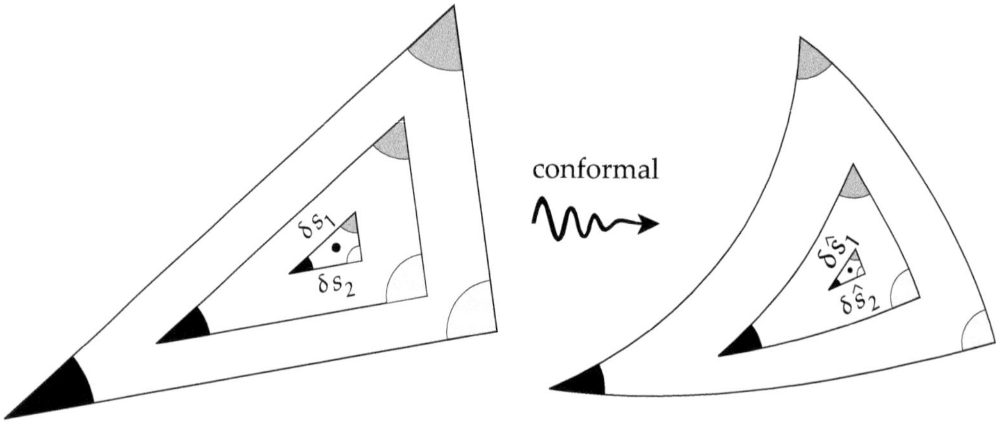

[4.5] *As a triangle shrinks in the map (left) its conformal image on the surface (right) is* **ultimately similar**: $\frac{\delta \widehat{s_1}}{\delta s_1} \asymp \frac{\delta \widehat{s_2}}{\delta s_2}$.

in these directions were different, so an infinitesimal circle in the map would be stretched into an ellipse on the surface, and angles would be changed, in general.

We are now contemplating an even greater specialization in which the expansion factor is the *same* in all directions, so that $A = B = \Lambda$, and infinitesimal circles are mapped to infinitesimal *circles*, and *angles are preserved*. In this case, the (u, v)-coordinates are called ***conformal coordinates*** (or ***isothermal coordinates***), and (4.9) reduces to being a simple multiple of the Euclidean metric:

$$d\widehat{s}^2 = \Lambda^2 \, [du^2 + dv^2]. \tag{4.14}$$

This is such a strong restriction that one might fear that such maps might not even exist. But as Gauss discovered in 1822, it is *always* possible to draw such a map of a general surface, at least locally. Remarkably, the proof (see Ex. 8) depends on *complex numbers*—indeed, the deep connection between Complex Analysis and conformal maps is the subject of the next section.

The already elegant curvature formula (4.10) now becomes even *more* elegant.[8] Recall that the second-order ***Laplacian***[9] differential operator ∇^2 is defined by

$$\nabla^2 \equiv \partial_u^2 + \partial_v^2. \tag{4.15}$$

[8] According to Dombrowski (1979, p. 128), this was the very *first* curvature formula that Gauss discovered, recorded in his private notes, dated 13th of December, 1822; see Gauß (1973, p. 381). Only in 1825 did he find a formula in more general (nonconformal) orthogonal coordinates, but, again, he kept this secret. Finally, in 1827, he found the completely general formula in a nonorthogonal coordinate system, and *only* this horrendously complicated and ugly formula appears in print in the final masterpiece (Gauss 1827). We note that Gauss was positively *proud* of his deliberate and utterly *deplorable* (in our view) habit of concealing his motivations and his path of discovery, declaring, "No architect leaves the scaffolding in place after completing the building."

[9] While physicists tend to denote this operator ∇^2, mathematicians instead tend to write it as Δ. We have avoided the latter notation because we often use Δ to denote a triangle.

Since now $A = B = \Lambda$, we easily find [exercise] that (4.10) simplifies to

$$\boxed{\mathcal{K} = -\frac{\nabla^2 \ln \Lambda}{\Lambda^2}.} \tag{4.16}$$

4.6 Some Visual Complex Analysis

Even if we take the surface $\mathcal{S}$ to be the plane, so that we are dealing with conformal mappings of the plane to itself, this turns out to be a deep and rich field of study. Remarkably, such conformal mappings are inextricably intertwined with the complex numbers, as we shall briefly explain in this section. (Additional concrete examples will appear later in this Act.) For a *full* exploration of how the spectacular results of Complex Analysis emerge from this geometric foundation, we shall immodestly recommend our first book, *Visual Complex Analysis* (VCA).

Not only does every surface $\mathcal{S}$ possess a conformal map, it possesses an *infinite variety* of conformal maps! We begin by noting that there is nothing unique about the specific u-curves and orthogonal v-curves that yield a conformal (u, v)-coordinate system with metric (4.14):

$$d\widehat{s} = \Lambda(u, v)\, ds.$$

The real magic resides within the conformal *mapping* $F : \mathbb{C} \to \mathcal{S}$ itself. Given such a conformal mapping F, we may create an infinite variety of conformal $(\widetilde{u}, \widetilde{v})$-coordinates on $\mathcal{S}$ simply by rotating, expanding, and translating our (u, v)-coordinate grid in $\mathbb{C}$, yielding (by application of F) completely new $\widetilde{u}$-curves and orthogonal $\widetilde{v}$-curves on $\mathcal{S}$. And this new, orthogonal $(\widetilde{u}, \widetilde{v})$-coordinate system on $\mathcal{S}$ is just as conformal as the original one was.

Let us introduce some notation to explain this fully. As is customary in Complex Analysis, we shall think of $z = u + iv$ as living in one copy of $\mathbb{C}$, and its image $\widetilde{z} = \widetilde{u} + i\widetilde{v}$ under a complex function $z \mapsto \widetilde{z} = f(z)$ as living in a separate copy of $\mathbb{C}$:

$$z \mapsto \widetilde{z} = \widetilde{u} + i\widetilde{v} = f(z) = f(u + iv).$$

In the example just discussed, we have $f(z) = a e^{i\tau} z + w$, which does an expansion by the (real) factor a, a rotation through angle τ, followed by a translation by the (complex) number w.

If we compose this mapping f with the mapping F then we obtain the new conformal mapping $\widetilde{F} \equiv F \circ f$ from $\mathbb{C}$ to $\mathcal{S}$. If z moves along a small complex number δz then its image under the first mapping $z \mapsto \widetilde{z} = f(z)$ will move along what we shall call δz's "image" $\delta\widetilde{z}$, emanating from $\widetilde{z}$, and clearly

$$\delta\widetilde{z} = a e^{i\tau} \delta z. \tag{4.17}$$

Thus the length $\delta s = |\delta z|$ is stretched by factor a, so that $\delta\widetilde{s} = |\delta\widetilde{z}| = a\, \delta s$. Next, under the second mapping F (up to the surface $\mathcal{S}$), the length of $\delta\widetilde{s}$ will be stretched by the conformal metric factor $\Lambda(\widetilde{z})$. Thus the net magnification of δs under the new mapping $\widetilde{F}$ is given by the *product* of these two expansion factors:

$$d\widehat{s} = \widetilde{\Lambda}(z)\, ds, \qquad \text{where} \qquad \widetilde{\Lambda}(z) = a\, \Lambda(\widetilde{z}).$$

But even this freedom barely scratches the surface, and to explain why we must simply quote one fundamental fact from Complex Analysis; see VCA for details. *Every* familiar, useful real function $f(x)$ you have ever studied (such as x^m, e^x, and $\sin x$) has a unique generalization $f(z)$ to complex number inputs z. Picturing this, as before, as a mapping from one $\mathbb{C}$ to another $\mathbb{C}$, the miracle of Complex Analysis is that *all* of these naturally occurring mappings $f(z)$ are automagically *conformal*!

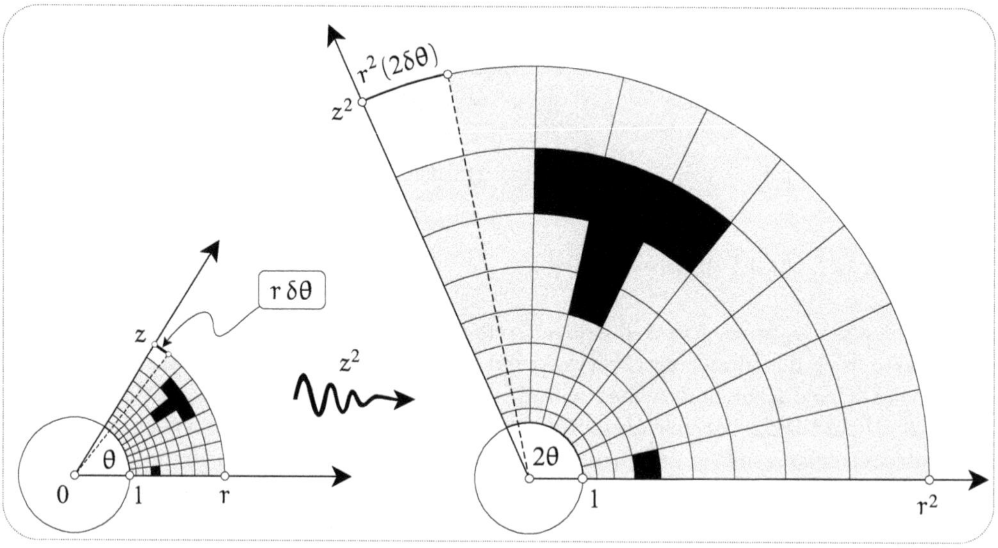

[4.6] *The mapping $z \mapsto f(z) = z^2$ (in common with* all *powers of z) is* conformal, *so the fine grid of "squares" on the left is mapped to approximate squares on the right, and these ultimately become perfect squares in the limit that the size of the squares shrinks to zero.*

For example, [4.6] illustrates the effect of

$$z \mapsto \widetilde{z} = f(z) = z^2 = \left[r\,e^{i\theta}\right]^2 = r^2\,e^{i2\theta},$$

which squares the length, and doubles the angle, of every point. As we see, the small "squares" on the left are sent to *similar* "squares" on the right. Of course both sets of "squares" only become *actual* squares in the limit that they shrink to zero. Likewise, the smaller the black T-shape on the left, the more perfectly similar is the image T-shape on the right.

To see how remarkable this is, suppose you were to randomly write down two real functions $\widetilde{u}$ and $\widetilde{v}$ (of the real variables u and v), and then perform a shotgun wedding between these two functions to amalgamate them into a single complex mapping, $f(u,v) = \widetilde{u} + i\widetilde{v}$. Then there is essentially *no chance* that f will be conformal, and, as we shall now see, this also means that there is no chance that $f'(z)$ exists!

Let us begin to redo the earlier analysis, but now replacing the linear function $f(z)$ with a *general* mapping *such that the derivative* $f'(z)$ *exists*. Such functions are called ***analytic***. As we have just indicated, such analytic mappings are extraordinarily rare, and *yet* they include essentially *all* the useful functions that naturally arise out of mathematics and physics.

The analogue of (4.17) is now

$$\delta\widetilde{z} \asymp f'(z)\,\delta z = a e^{i\tau}\,\delta z. \tag{4.18}$$

The main difference is that now both the expansion factor $a(z)$ and the angle of rotation $\tau(z)$ both depend on the location z, rather than being constant throughout $\mathbb{C}$. For example, in [4.6] we can see that the black square adjacent to the real axis undergoes a smaller expansion than does the white square in the corner of the grid, so $a(z)$ is smaller at the black square than it is at the white square. Likewise, it is clear that the black square undergoes *no* rotation, so $\tau = 0$ there, but the white square clearly *does* need to be rotated to obtain its image.

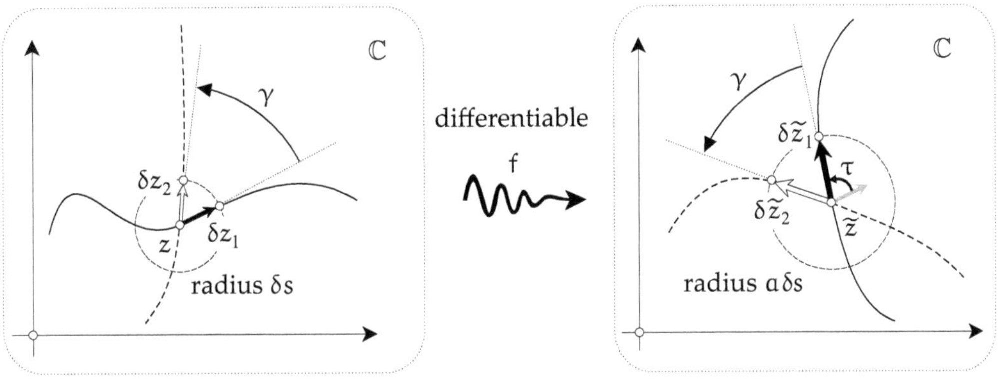

[4.7] The Amplitwist. *The local effect of a differentiable complex mapping* $f(z)$ *is an* **amplitwist** *(***amplification** *and* **twist***) described by the complex number* $f'(z) =$ amplitwist *of* $f(z) =$ (amplification) $e^{i\,(\text{twist})} = a e^{i\tau}$. *Any pair of tiny complex numbers* $\delta z_{1,2}$ *emanating from* z *are* **amplitwisted** equally *to produce a pair* $\delta \widetilde{z}_{1,2} \asymp a e^{i\tau}\, \delta z_{1,2}$ *emanating from* $\widetilde{z} = f(z)$, *and thus the angle* γ *between curves is preserved:* f *is* conformal!

Provided that $f'(z) \neq 0$, equation (4.18) says that every tiny complex arrow δz emanating from z undergoes the *same* expansion a and rotation τ to obtain its image arrow $\delta \widetilde{z}$ emanating from $\widetilde{z} = f(z)$. See [4.7]. As illustrated, the angle between a pair of δz emanating from z will be the same as the angle between the image pair of $\delta \widetilde{z}$ at $\widetilde{z} = f(z)$, so it follows that differentiable complex mappings are automatically *conformal!*

We see that the derivative $f'(z)$ describes the local behaviour of the conformal mapping, and in VCA we introduced (*nonstandard*) terminology to describe this geometrically. We call the local expansion factor a the ***amplification***; we call the local angle of rotation τ the ***twist***; and we call the combined amplification and twist (needed to transform the original shape into its image) the ***amplitwist***. In summary,

$$f'(z) = \text{amplitwist of } f \text{ at } z = (\text{amplification})\, e^{i[\text{twist}]} = a\, e^{i\tau}. \tag{4.19}$$

Before we turn to conformal metric formulas for surfaces in space, let us return to $f(z) = z^2$ and show how we may use [4.6] to geometrically deduce its amplitwist.

Focus attention on the illustrated white square, with one corner at $z = re^{i\theta}$. Since the radial edge through z at angle θ is mapped to a radial edge through z^2 at angle 2θ, this edge has undergone a rotation of θ, so

$$\tau = \text{twist} = \theta.$$

To find the amplification a, consider the highlighted outer edge of the square (ultimately equal to the arc of the circle through z, connecting the white dots). If this subtends angle $\delta\theta$ at the origin, then its length is ultimately $r\,\delta\theta$. Since angles are doubled by the mapping, the image arc will subtend angle $2\delta\theta$, and since it now lies on a circle of radius r^2, the length of this image edge is ultimately $r^2(2\delta\theta)$. Thus,

$$(\text{image edge}) \asymp 2r\,(\text{original edge}) \implies a = \text{amplification} = 2r.$$

We conclude that,

$$(z^2)' = \text{amplitwist of } z^2 = (\text{amplification})\, e^{i[\text{twist}]} = 2r\, e^{i\theta} = 2z,$$

a result that looks formally identical to the real result, $(x^2)' = 2x$, but which now means so much more.

It is straightforward [exercise] to generalize this geometric argument to deduce that $(z^m)' = m\,z^{m-1}$. The amplitwists of other important mappings can likewise be deduced purely geometrically; see VCA for details.

Now let us return to our main interest: conformal coordinates on a surface. We can now replace our simple linear function with our enormously richer class of differentiable (i.e., conformal) mappings, $f(z)$. Once again defining $\widetilde{F} \equiv F \circ f$ from $\mathbb{C}$ to $\mathcal{S}$, the new metric formula is

$$d\widehat{s} = \widetilde{\Lambda}(z)\,ds, \qquad \text{where} \qquad \widetilde{\Lambda}(z) = (\text{amplification})\,\Lambda(\widetilde{z}) = |f'(z)|\,\Lambda(\widetilde{z}).$$

The existence of a conformal mapping $F: \mathbb{C} \mapsto \mathcal{S}$ allows us to take any analytic mapping $f: \mathbb{C} \mapsto \mathbb{C}$, and *transfer* it to $\mathcal{S}$, so that it becomes a conformal mapping of $\mathcal{S}$ to itself. To see this, consider the effect of

$$\widehat{f} \equiv F \circ f \circ F^{-1}, \tag{4.20}$$

acting on $\mathcal{S}$. First F^{-1} conformally maps $\mathcal{S}$ to $\mathbb{C}$, then f conformally maps $\mathbb{C}$ to $\mathbb{C}$, and finally F conformally maps $\mathbb{C}$ back to $\mathcal{S}$. Since each of the three mappings preserves angles, so too does their composition, so $\widehat{f}: \mathcal{S} \mapsto \mathcal{S}$ is indeed conformal.

In the next section we shall meet an extremely important example of a conformal mapping F in the case where $\mathcal{S} = S^2$. Later, we will use this F, via (4.20), to express the rotations of S^2 as complex functions (given by (6.10)). These rotations are not merely conformal, they are the (orientation-preserving) ***isometries*** of the sphere.

4.7 The Conformal Stereographic Map of the Sphere

Hipparchus (c. 150 BCE) may have been the first to construct a conformal[10] map of the sphere using the method illustrated in [4.8], which is called ***stereographic projection***. Certainly, by 125 CE Ptolemy (who is instead usually credited as the discoverer) was using it to plot the positions of heavenly bodies on the ***celestial sphere***.

The construction is similar to central projection, except now we imagine a point source of light at the north pole N, and we project onto a plane that passes through the equator, instead of touching the south pole.[11] Then a light ray passing through $\widehat{z}$ on Σ goes on to hit a point z in $\mathbb{C}$, which we call the ***stereographic image*** of $\widehat{z}$. Since this gives us a one-to-one correspondence between points in $\mathbb{C}$ and points on Σ, let us also say that $\widehat{z}$ is the stereographic image of z. No confusion should arise from this, the context making it clear whether we are mapping $\mathbb{C}$ to Σ, or vice versa.

Note the following immediate facts: (i) the southern hemisphere of Σ is mapped to the interior in $\mathbb{C}$ of the circle $|z| = R$, and in particular the south pole S is mapped to the origin 0 of $\mathbb{C}$; (ii) each point on the equator is mapped to itself; (iii) in $\mathbb{C}$, the exterior of the circle $|z| = R$ is mapped to the northern hemisphere of Σ, *except* that N is not the image of any finite point in the plane. However, it is clear that as z moves further and further away from the origin (in any direction), $\widehat{z}$ moves closer and closer to N. In Complex Analysis, and later in this Act, one takes Σ to be the *unit* sphere, and once its points are stereographically labelled with complex numbers, it is called the ***Riemann sphere***. The point N is then the concrete manifestation of the ***point at infinity*** of the so-called ***extended complex plane***.

Figure [4.8] illustrates the fact that

---

[10]The conformality of the construction in [4.8] is *not* meant to be immediately obvious; it will be explained shortly, in [4.9].

[11]In some older texts the plane *is* in fact taken to touch the south pole; this [exercise] alters the map only by a constant scale factor of 2.

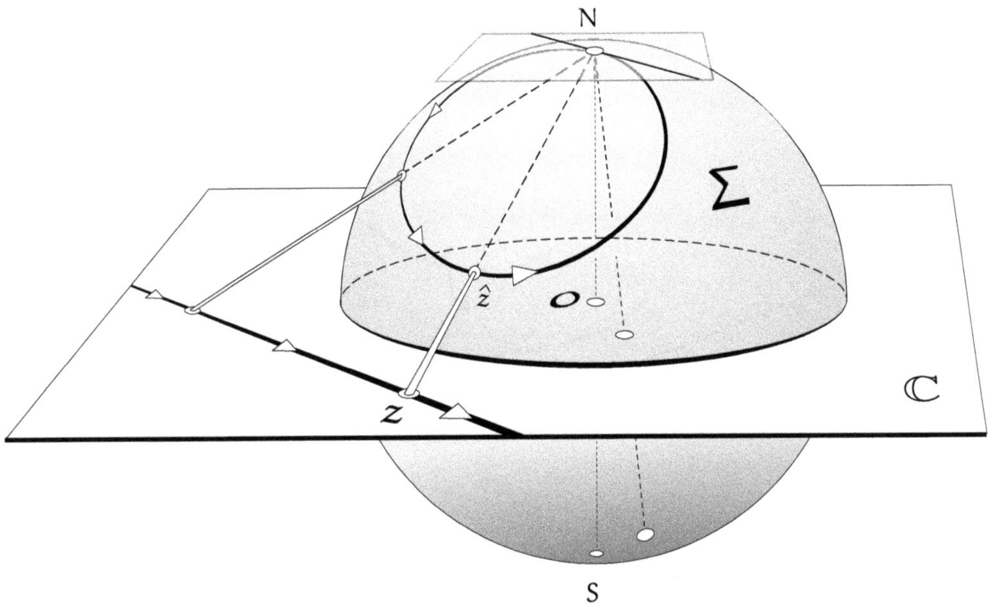

[4.8] Stereographic Projection: *light from N shines through the glass sphere Σ, casting shadows onto ℂ. The shadow of a circle through N is a line in ℂ that is parallel to the circle's tangent at N.*

> *The stereographic image of a line in the plane is a circle on Σ that passes through N in a direction parallel to the original line.* (4.21)

To see this, first observe that as z moves along the line shown in [4.8], the line connecting N to z sweeps out part of a plane through N. Thus $\widehat{z}$ moves along the intersection of this plane with Σ, which is a circle passing through N. Next, note that the tangent plane to Σ at N is parallel to ℂ. But if we slice through two parallel planes with a third plane, we obtain two parallel lines of intersection. Therefore, the tangent at N to the circle is parallel to the original line, as claimed.

From this last fact it follows that *stereographic projection preserves angles.* Consider [4.9], which shows two lines intersecting at z, together with their circular, stereographic projections, passing through N.

First note that the magnitude of the angle of intersection between the circles is the *same* at their two intersection points, $\widehat{z}$ and N. This is because the figure has *mirror symmetry* in the plane passing through the centre of the sphere and the centres of the two circles.[12] Since the tangents to the circles at N are parallel to the original lines in the plane, it follows that the illustrated angles at z and $\widehat{z}$ are of equal *magnitude*. But before we can say that stereographic projection is "conformal," we must assign a ***sense*** to the angle on the sphere.

According to our convention, the illustrated angle at z (from the black curve to the white one) is *positive*; i.e., it is counterclockwise when viewed from above the plane. From the perspective from which we have drawn [4.9], the angle at $\widehat{z}$ is negative (clockwise). However, if we were looking at this angle from *inside* the sphere then it would be positive. Thus,

> *If we define the sense of an angle on Σ by its appearance to an observer inside Σ, then stereographic projection is* conformal.

[12]This will become crystal clear if you draw for yourself any two intersecting circles on an orange.

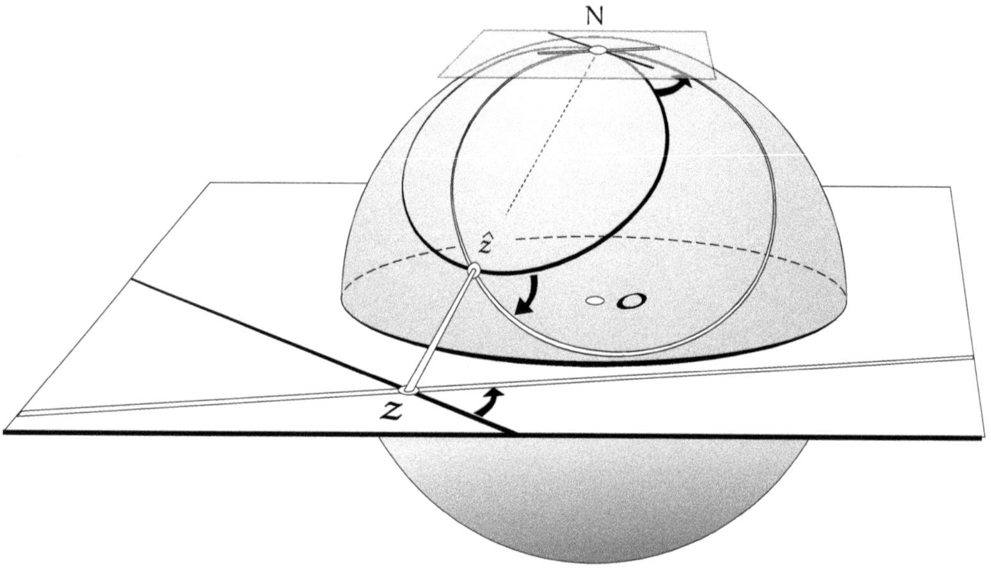

[4.9] Stereographic Projection Is Conformal. *As the line rotates around z, the tangent at N of its circular image rotates with it, so the angles at z and N are equal. But, by symmetry, the angle at $\hat{z}$ equals the angle at N, so this equals the angle at z, proving that stereographic projection is conformal.*

HISTORICAL NOTE: It is remarkable that although stereographic projection had been well known since Ptolemy first put it to practical use around 125 CE, its beautiful and fundamentally important *conformality* was not discovered for another 1500 years! This was first done around 1590 by Thomas Harriot[13]—yes, the *same* Thomas Harriot who in 1603 discovered the fundamental formula (1.3) linking angular excess and area on the sphere!

It follows from conformality that the metric takes the form (4.13). Thus a very small circle of radius $\delta\hat{s}$ on the sphere is ultimately mapped to a circle of radius δs in the plane, where

$$\delta\hat{s} \asymp \Lambda\,\delta s.$$

See [4.10]. Our task now is to find Λ.

Since Λ is independent of the direction of radius $\delta\hat{s}$ emanating from $\hat{z}$, we are free to choose this direction so as to make the analysis of the geometry as simple as possible. Thus, as illustrated, let us choose the direction to be horizontal, along a circle of latitude.

Under the stereographic projection, this circle of latitude undergoes a uniform expansion to yield an origin-centred circle through z, with δs pointing along it. As $\hat{z}$ rotates around its circle of latitude, z rotates with it, and their motions are in proportion to their distances from N. Thus,

$$\frac{\delta\hat{s}}{\delta s} = \frac{N\hat{z}}{Nz}.$$

Now consider [4.11], which shows the vertical cross section of [4.10] taken through N, $\hat{z}$, z. The triangle N$\hat{z}$S is similar to N0z, so

$$\frac{N\hat{z}}{2R} = \frac{R}{Nz}.$$

Combining this with the previous equation, we deduce that

$$\frac{\delta\hat{s}}{\delta s} = \frac{2R^2}{[Nz]^2}.$$

[13]See Stillwell (2010, §16.2). For a short sketch of Harriot's life, see Stillwell (2010, §17.7). For a full biography of Harriot, see Arianrhod (2019).

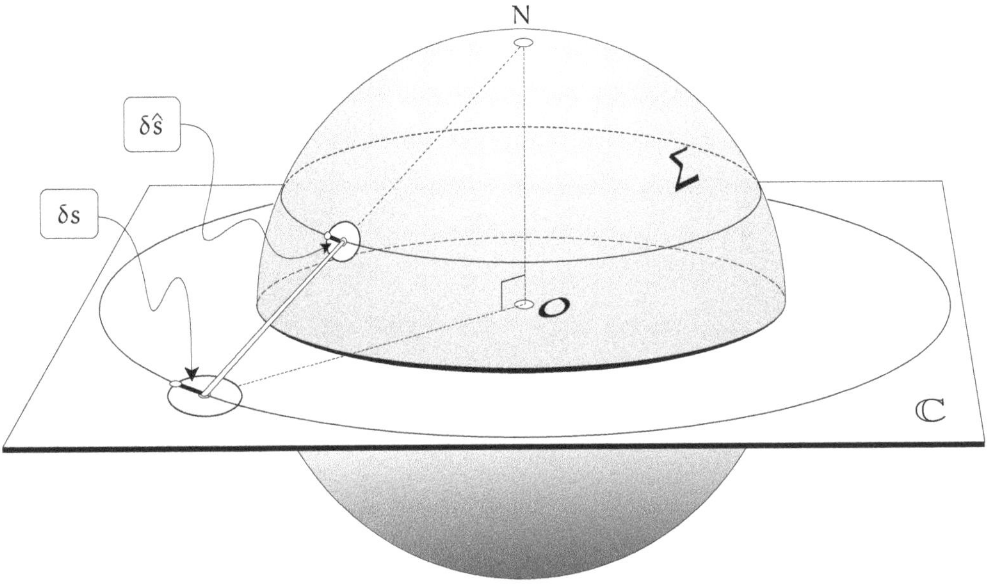

[4.10] *Conformality implies that a very small circle of radius δŝ on the sphere is ultimately mapped to a circle of radius δs in the plane. To find the metric, we must find the ratio of their radii, which we do by choosing δŝ along the illustrated circle of latitude.*

Finally, writing $r = |z|$, and applying Pythagoras's Theorem to the triangle N0z, we obtain $[Nz]^2 = R^2 + r^2$, so the conformal metric for the stereographic map is

$$d\hat{s} = \frac{2}{1+(r/R)^2}\, ds. \tag{4.22}$$

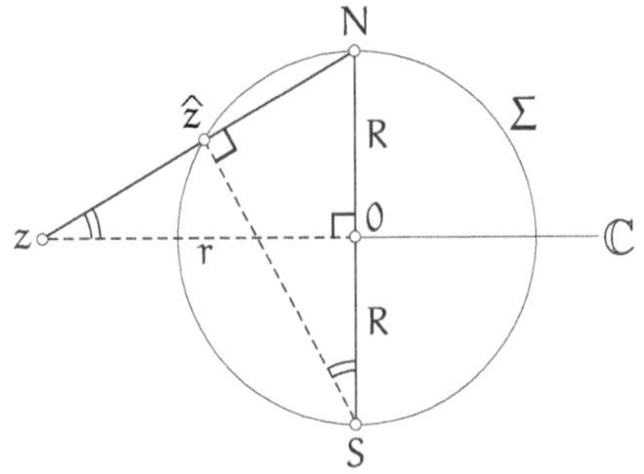

[4.11] *The triangle* $N\hat{z}S$ *is similar to* $N0z$.

Here is our first opportunity (there will be others) to try out the conformal curvature formula (4.16); of course we should find that $\mathcal{K} = (1/R^2)$. We suggest you confirm this in two ways. First, write $r^2 = x^2 + y^2$ and use our original Cartesian form of the Laplacian operator, (4.15). Second, use the fact that in polar coordinates the Laplacian instead takes the form

$$\nabla^2 = \partial_r^2 + \frac{1}{r}\partial_r + \frac{1}{r^2}\partial_\theta^2. \tag{4.23}$$

4.8 Stereographic Formulas

In this section we derive explicit formulas connecting the coordinates of a point $\hat{z}$ on Σ and its stereographic projection z in $\mathbb{C}$. To simplify matters, *we now restrict ourselves to the standard case* $R = 1$.

To begin with, let us describe z with Cartesian coordinates: $z = x + iy$. Similarly, let (X, Y, Z) be the Cartesian coordinates of $\hat{z}$ on Σ; here the X- and Y-axes are chosen to coincide with the x- and

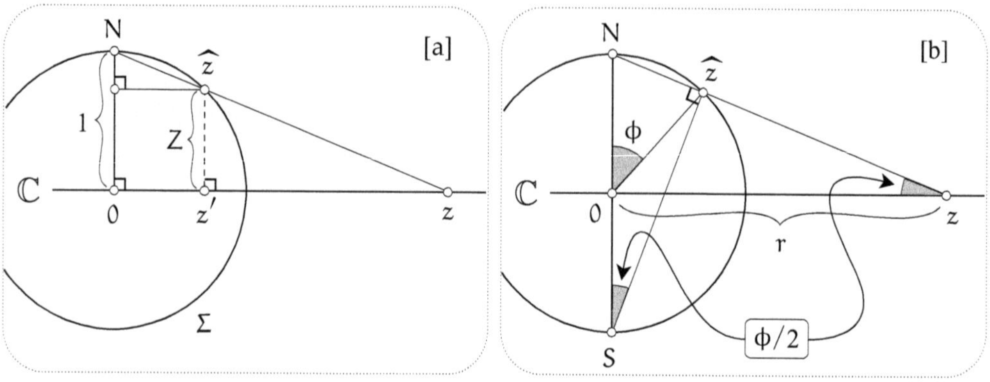

[4.12] [a] *By similar triangles,* $|z|/|z'| = 1/(1-Z)$. **[b]** $r = \cot(\phi/2)$.

y-axes of $\mathbb{C}$, so that the positive Z-axis passes through N. To make yourself comfortable with these coordinates, check the following facts: the equation of Σ is $X^2+Y^2+Z^2=1$, the coordinates of N are $(0,0,1)$, and similarly $S=(0,0,-1)$, $1=(1,0,0)$, and $i=(0,1,0)$.

Now let us find the formula for the stereographic projection $z = x+iy$ of the point $\hat{z}$ on Σ in terms of the coordinates (X,Y,Z) of $\hat{z}$. Let $z' = X+iY$ be the foot of the perpendicular from $\hat{z}$ to $\mathbb{C}$. Clearly, the desired point z is in the same direction as z', so

$$z = \frac{|z|}{|z'|}\, z'.$$

Now look at [4.12a], which shows the vertical cross section of Σ and $\mathbb{C}$ taken through N and $\hat{z}$; note that this vertical plane necessarily also contains z' and z. From the similarity of the illustrated right triangles with hypotenuses $N\hat{z}$ and Nz, we immediately deduce [exercise] that

$$\frac{|z|}{|z'|} = \frac{1}{1-Z},$$

and so we obtain our first stereographic formula:

$$x+iy = \frac{X+iY}{1-Z}. \tag{4.24}$$

For later use, let us now also invert this formula to find the coordinates of $\hat{z}$ in terms of those of z. Since [exercise]

$$|z|^2 = \frac{1+Z}{1-Z},$$

we obtain [exercise]

$$X+iY = \frac{2z}{1+|z|^2} = \frac{2x+i2y}{1+x^2+y^2}, \quad \text{and} \quad Z = \frac{|z|^2-1}{|z|^2+1}. \tag{4.25}$$

Although it is often useful to describe the points of Σ with the three coordinates (X,Y,Z), this is certainly unnatural, for the sphere is intrinsically 2-dimensional. If we instead describe $\hat{z}$ with the more natural (2-dimensional) spherical polar coordinates (ϕ, θ) then we obtain a particularly neat stereographic formula.

First recall that θ measures the angle around the Z-axis, with $\theta = 0$ being assigned to the vertical half-plane through the positive X-axis. Thus for a point z in $\mathbb{C}$, the angle θ is simply the usual angle from the positive real axis to z. The definition[14] of ϕ is illustrated in [4.12b]—it is the angle subtended at the centre of Σ by the points N and $\widehat{z}$; for example, the equator corresponds to $\phi = (\pi/2)$. By convention, $0 \leqslant \phi \leqslant \pi$.

If z is the stereographic projection of the point $\widehat{z}$ having coordinates (ϕ, θ), then clearly $z = r\, e^{i\theta}$, and so it only remains to find r as a function of ϕ. From [4.12b] it is clear [exercise] that the triangles $N\widehat{z}S$ and $N0z$ are similar, and because the angle $\angle NS\widehat{z} = (\phi/2)$, it follows [exercise] that $r = \cot(\phi/2)$. Thus our new stereographic formula is

$$\boxed{z = \cot(\phi/2)\, e^{i\theta}.} \tag{4.26}$$

In Exercise 33 we show how this formula may be used to establish a beautiful alternative interpretation of stereographic projection, due to Sir Roger Penrose.

We will now illustrate this formula with an application that we shall need shortly: the relationship between the complex numbers representing ***antipodal points***. Recall that this means two points on a sphere that are diametrically opposite each other, such as the north and south poles. Let us show that

If $\widehat{p}$ and $\widehat{q}$ are antipodal points of Σ, then their stereographic projections p and q are related by the following formula:

$$q = -(1/\overline{p}). \tag{4.27}$$

Note that the relationship between p and q is actually symmetrical (as clearly it should be): $p = -(1/\overline{q})$. To verify (4.27), first check for yourself that if $\widehat{p}$ has coordinates (ϕ, θ) then $\widehat{q}$ has coordinates $(\pi - \phi, \pi + \theta)$. Thus,

$$q = \cot\left[\frac{\pi}{2} - \frac{\phi}{2}\right] e^{i(\pi+\theta)} = -\frac{1}{\cot(\phi/2)}\, e^{i\theta} = -\frac{1}{\cot(\phi/2)\, e^{-i\theta}} = -\frac{1}{\overline{p}},$$

as was to be shown.

For an elementary geometric proof of (4.27), see Exercise 7.

4.9 Stereographic Preservation of Circles

This section is devoted to proving a single fact that is beautiful, surprising, and critically important:

Stereographic projection preserves circles! (4.28)

That is, not only does it send infinitesimal circles to infinitesimal circles (as all conformal maps must do) but it sends a finite circle of any size and location on the sphere to a perfect circle in

[14]This is the American convention; in my native England the roles of θ and ϕ are the reverse of those stated here. For a lovely example of the same diagram, but labelled per British convention, see Penrose and Rindler (1984, Vol. 1, p. 12).

the plane, although the centre on the sphere is *not* mapped to the centre in the plane. Note that if the circle passes close to N then its image will be extremely large, and in the limit that it passes through N this very large circle becomes the line shown in [4.8].

There exists a beautiful, completely conceptual, geometric explanation of (4.28)—see VCA, page 161—but in our current haste we must instead settle for a calculation.

Every circle on the unit sphere Σ is the intersection of Σ with a plane whose distance from O is less than one:

$$lX + mY + nZ = k, \qquad \text{where} \qquad l^2 + m^2 + n^2 \geqslant k^2.$$

Substituting (4.25) into the equation of this plane we find [exercise] that the circle of intersection on Σ stereographically projects to a curve in $\mathbb{C}$ whose equation is

$$2lx + 2my + n(x^2 + y^2 - 1) = k(x^2 + y^2 + 1).$$

If $k = n$ then [exercise] the circle on Σ passes through N and its image is a line (as it should be!) with equation $lx + my = n$. If $k \neq n$ then we may complete squares [exercise] to rewrite the equation as

$$\left[x - \frac{l}{k-n}\right]^2 + \left[y - \frac{m}{k-n}\right]^2 = \frac{l^2 + m^2 + n^2 - k^2}{(k-n)^2},$$

which is indeed a circle with

$$\text{centre} = \left(\frac{l}{k-n}, \frac{m}{k-n}\right) \qquad \text{and} \qquad \text{radius} = \frac{\sqrt{l^2 + m^2 + n^2 - k^2}}{|k-n|}.$$

(Comment: This is an excellent example of the seductive but corrupting power of calculation. We invoked the "Devil's machine"—see Prologue—and in just a couple of lines its work was complete and the result was proved. Yet here we stand, bereft of any understanding of *why* the result is true!)

From the preservation of circles we readily deduce from [1.5] that geodesics on the sphere (the great circles) appear in the map as circles that intersect the equatorial circle at opposite ends of a diameter.

Chapter 5

The Pseudosphere and the Hyperbolic Plane

5.1 Beltrami's Insight

While the long investigation of parallel lines reached a climax around 1830 with the discovery of Hyperbolic Geometry by Lobachevsky and Bolyai, a quite different parallel line—pun intended!—of enquiry into Differential Geometry also reached a climax at almost the same time, with Gauss's differential-geometric discoveries of 1827.

Just as lines on the sphere that are initially parallel[1] must eventually intersect, so too did these two parallel lines of thought collide, and in a powerful and fruitful way.

In 1868 the Italian geometer Eugenio Beltrami (see [5.1]) recognized that a connection might exist between two results from these seemingly unrelated realms of thought. On the one hand, he knew of Lambert's result (1.8)—later rediscovered by Gauss, Lobachevsky, and Bolyai—that in Hyperbolic Geometry the angular excess of a triangle is a fixed *negative* multiple of its area. On the other hand, he also knew of the Local Gauss–Bonnet Theorem.

Beltrami had the insight that if one could find a surface of *constant negative curvature* $\mathcal{K} = -(1/R^2)$, then, by virtue of (2.6), geodesic triangles constructed within it would automatically obey the central law of Hyperbolic Geometry:

$$\mathcal{E}(\Delta) = -\frac{1}{R^2}\,\mathcal{A}(\Delta).$$

[5.1] *Eugenio Beltrami (1835–1900).*

Up to this point, the bizarre Hyperbolic Geometry of Lobachevsky and Bolyai had languished in obscurity for close to 40 years, vilified by some, but ignored by most. Now, at last, Beltrami had an idea that could put it on a secure and intuitive foundation. Perhaps Hyperbolic Geometry simply *was* the intrinsic geometry of a surface of constant negative curvature! A 2000-year-old struggle was about to come to an end.

---

[1] Think of two neighbouring points on the equator, and the meridians passing through them.

5.2 The Tractrix and the Pseudosphere

Beltrami already knew that the pseudosphere shown in [2.6] was indeed a surface of constant negative curvature $\mathcal{K} = -(1/R^2)$, where R is the radius of its circular bottom rim. (We shall prove this fact within this section.) Thus, more concretely, his insight was that the local geometry within this surface obeys the laws of the abstract non-Euclidean Geometry of Lobachevsky and Bolyai. But this abstract Hyperbolic Geometry is understood to take place in an infinite ***hyperbolic plane*** that is exactly like the Euclidean plane, obeying the first four of Euclid's axioms, *but* with lines that obey the Hyperbolic Axiom (1.1), instead of Euclid's Parallel Axiom.

While the constant negative curvature of the pseudosphere ensures that it faithfully embodies local consequences of this axiom, the pseudosphere will not do as a model of the *entire* hyperbolic plane, because it departs from the Euclidean plane in two unacceptable ways: (1) the pseudosphere is akin to a cylinder instead of a plane and (2) a line segment cannot be extended indefinitely in both directions: we hit the rim. (As we noted earlier, Hilbert discovered in 1901 that such a rim is an *essential* feature of *all* surfaces of constant negative curvature—not intrinsically, but by virtue of trying to force the surfaces to fit inside ordinary Euclidean 3-space.)

Beltrami recognized these obstacles, and in the next section we will see how he overcame them both, in one fell swoop, by constructing a conformal map of the pseudosphere. For now, though, we turn to the construction of the pseudosphere itself.

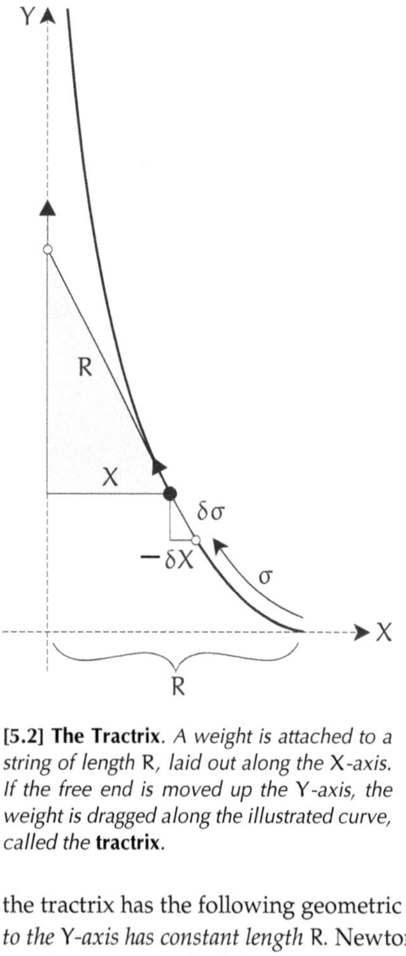

[5.2] The Tractrix. *A weight is attached to a string of length R, laid out along the X-axis. If the free end is moved up the Y-axis, the weight is dragged along the illustrated curve, called the* **tractrix.**

Try the following experiment. Take a small heavy object, such as a paperweight, and attach a length of string to it. Now place the object on a table and drag it by moving the free end of the string along the edge of the table. You will see that the object moves along a curve like that in [5.2], where the Y-axis represents the edge of the table. This curve is called the ***tractrix***, and the Y-axis (which the curve approaches asymptotically) is called the ***axis***. The tractrix was first investigated by Newton, in 1676.

If the length of the string is R, then it follows that the tractrix has the following geometric property: *the segment of the tangent from the point of contact to the Y-axis has constant length* R. Newton identified this as the defining property of the tractrix.

Returning to [5.2], let σ represent arc length along the tractrix, with $\sigma = 0$ corresponding to the starting position $X = R$ of the object we are dragging. Just as the object is about to pass through (X, Y), let δX denote the very small change in X that occurs while the object moves a distance $\delta\sigma$ along the tractrix. From the ultimate similarity of the illustrated triangles, we deduce that

$$\frac{-dX}{d\sigma} = \frac{X}{R},$$

and therefore

$$\boxed{X = R\,e^{-\sigma/R}.} \tag{5.1}$$

The ***pseudosphere*** of radius R, [2.6], may now be simultaneously defined and constructed as the surface obtained by rotating the tractrix about its axis. Remarkably, this surface was investigated as early as 1693 (by Christiaan Huygens), two centuries prior to its catalytic role in the acceptance of Hyperbolic Geometry, and the constancy of its curvature was already known to Gauss's student Minding as early as 1839.

To demonstrate this constancy of the curvature, let us find the metric of the pseudosphere. We choose very natural orthogonal coordinate curves on the surface, as follows. See [5.4a] (but please ignore [5.4b] for now). To specify a point on the pseudosphere we need only say (i) which tractrix generator it lies on and (ii) how far along this tractrix it lies. To answer (i) we specify the angle x around the axis of the pseudosphere, and to answer (ii) we specify the distance σ travelled up the tractrix, starting at the base.

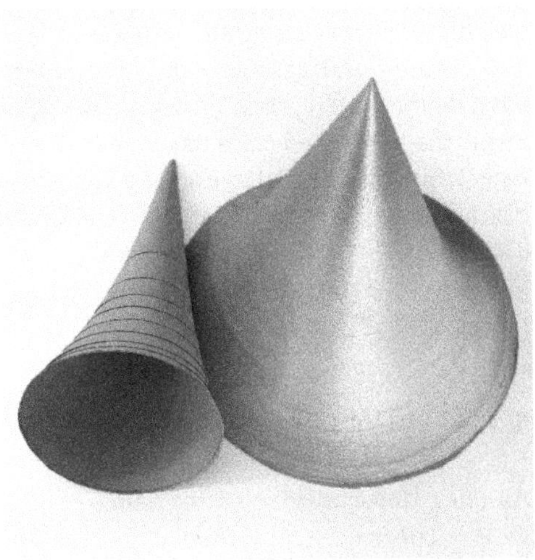

[5.3] *The author's personal pseudophere, built out of cones, themselves built out of discs of radius R. The top half has been removed to make the construction easier to see.*

Thus the curves $x = \textit{const.}$ are the tractrix generators of the pseudosphere (note that these are clearly geodesics[2]), and the curves $\sigma = \textit{const.}$ are circular cross sections of the pseudosphere (note that these are clearly *not* geodesics). Since the radius of such a circle is the same thing as the X-coordinate in [5.2], it follows from (5.1) that

> *The radius* X *of the circle* $\sigma = \text{const.}$ *passing through the point* (x, σ) *on the pseudosphere is given by* $X = R\,e^{-\sigma/R}$.

An increase dx therefore rotates the point through the arc $X\,dx$, as illustrated. Thus, the metric is

$$d\hat{s}^2 = X^2\,dx^2 + d\sigma^2 = (R\,e^{-\sigma/R})^2\,dx^2 + d\sigma^2. \tag{5.2}$$

Finally, we may now apply the curvature formula (4.10) to this metric to obtain

$$\mathcal{K} = -\frac{1}{R\,e^{-\sigma/R}}\,\partial_\sigma\left[\frac{\partial_\sigma(R\,e^{-\sigma/R})}{1}\right] = -\frac{1}{R^2},$$

thereby confirming the crucial fact that Beltrami needed to interpret Hyperbolic Geometry:

> *The pseudosphere has constant negative curvature* $\mathcal{K} = -(1/R^2)$, *where* R *is its base radius.* (5.3)

This proposition carries such great mathematical and historical interest that in the course of this book we shall attempt to understand it as directly and geometrically as possible. Indeed, in

[2]By symmetry, the meridians of *any* surface of revolution are geodesics.

later Acts we shall offer *two* geometric proofs: one using extrinsic geometry (Act III), and one using intrinsic geometry (Act IV).

Before moving on, we can think of no better way to develop a feel (literally!) for the geometry of the pseudosphere than to *build your own!* To see the idea behind the construction, imagine rotating [2.6] to create the pseudosphere. In this process, no matter the position of the dragged weight, *the rotating string always traces out a cone (tangent to the pseudosphere) of fixed slant length* R.

Therefore, take as many sheets of paper as you can cut through with your scissors, and staple them together along three edges, a few staples per edge. Find the largest bowl or plate that will fit inside the paper, and trace its circular edge. Now cut along this circle to produce identical discs; repeat this step till you have at least 20 discs—the more, the better! Cut a small wedge from the first disc and tape the edges together to create a very shallow cone. Take the next disc and repeat with a slightly larger wedge[3] to create a slightly taller cone, but still with the same slant length. Place this new cone on top the previous cone and repeat, and repeat, ... Behold your own personal pseudosphere!

5.3 A Conformal Map of the Pseudosphere

As a first step towards creating a map of an infinite hyperbolic plane akin to the Euclidean plane, we now construct a conformal map of the pseudosphere in $\mathbb{C}$. In our map, let us choose the angle x as our horizontal axis, so that the tractrix generators of the pseudosphere are represented by vertical lines. See [5.4b]. Thus a point on the pseudosphere with coordinates (x, σ) will be represented in the map by a point with Cartesian coordinates (x, y), which we will think of as the complex number $z = x + iy$.

If our map were not required to be special in any way, then we could simply choose $y = y(x, \sigma)$ to be an arbitrary function of x and σ. But now let us insist that our map be *conformal*. Thus an infinitesimal[4] triangle on the pseudosphere is mapped to a *similar* infinitesimal triangle in the map, and more generally it follows that any small shape on the pseudosphere looks the same (only bigger or smaller) in the map. Having decided upon such a conformal map, it turns out there is (virtually) no freedom in the choice of the y-coordinate. Let us see why. First, the tractrix generators $x = const.$ are orthogonal to the circular cross sections $\sigma = const.$, so the same must be true of their images in our conformal map. Thus the image of $\sigma = const.$ must be represented by a horizontal line $y = const.$, and from this we deduce that $y = y(\sigma)$ must be a function solely of σ.

Second, on the pseudosphere consider the arc of the circle $\sigma = const.$ (of radius X) connecting the points (x, σ) and $(x + dx, \sigma)$. By the definition of x, these points subtend angle dx at the centre of the circle, so their separation on the pseudosphere is $X\,dx$, as illustrated. In the map, these two points have the same height and are separated by distance dx. Thus in passing from the pseudosphere to the map, this particular line segment is shrunk by factor X.

However, since the map is conformal, an infinitesimal line segment emanating from (x, σ) in *any* direction must be multiplied by the *same* factor $(1/X) = \frac{1}{R} e^{\sigma/R}$. In other words, the metric is

$$\boxed{d\widehat{s} = X\, ds.}$$

Third, consider the uppermost black disc on the pseudosphere shown in [5.4a]. Think of this disc as infinitesimal, say of diameter ϵ. In the map, it will be represented by *another disc*, whose diameter (ϵ/X) may be interpreted more vividly as the angular width of the original disc as seen by an observer at the same height on the pseudosphere's axis. Now suppose we repeatedly

[3] In practice you may find it easier simply to cut a radial slit, then *overlap* the paper to create the cone.

[4] Here, for once, we shall prefer the intuitive abbreviation "infinitesimal," instead of spelling out the rigorous "ultimate equalities."

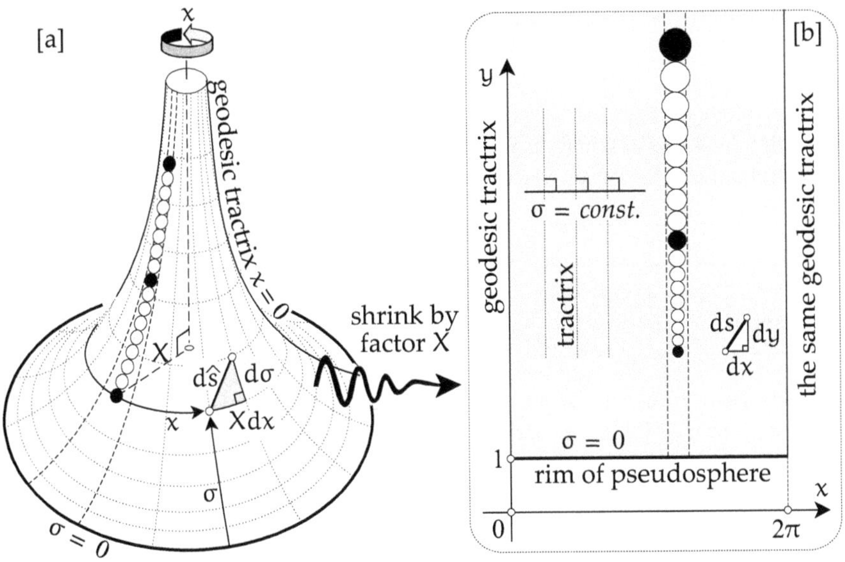

[5.4] *A conformal map of the pseudosphere. First, choose the* x*-coordinate to be the angle round the axis. Then conformality dictates the* y*-coordinate via the similarity of the illustrated infinitesimal triangles:* $\frac{dy}{d\sigma} = \frac{1}{X}$.

translate the original disc towards the pseudosphere's rim, moving it a distance ϵ each time. Figure [5.4a] illustrates the resulting chain of touching, congruent discs. As the disc moves down the pseudosphere, it recedes from the axis, and its angular width as seen from the axis therefore diminishes. Thus the image disc in the map appears to gradually shrink as it moves downward, and the equal distances 8ϵ between the successive black discs certainly do not appear equal in the map.

Having developed a feel for how the map works, let's actually calculate the y-coordinate corresponding to the point (x, σ) on the pseudosphere. From the above observations (or directly from the requirement that the illustrated triangles be similar) we deduce that

$$\frac{dy}{d\sigma} = \frac{1}{X} = \frac{1}{R} e^{\sigma/R} \quad \Longrightarrow \quad y = e^{\sigma/R} + const.$$

The standard choice of this constant is 0, so that

$$y = e^{\sigma/R} = (R/X). \tag{5.4}$$

Thus the entire pseudosphere is represented in the map by the shaded region lying above the line $y = 1$ (which itself represents the pseudosphere's rim), and the metric associated with the map is

$$d\hat{s} = \frac{R\,ds}{y} = \frac{R\sqrt{dx^2 + dy^2}}{y}. \tag{5.5}$$

For future use, also note that an infinitesimal rectangle in the map with sides dx and dy represents a similar infinitesimal rectangle on the pseudosphere with sides $(R\,dx/y)$ and $(R\,dy/y)$. Thus the apparent area dx dy in the map is related to the true area $d\mathcal{A}$ on the pseudosphere by

$$d\mathcal{A} = \frac{R^2\,dx\,dy}{y^2}. \tag{5.6}$$

(Of course this is just a special case of (4.12), with $A = B = (R/y)$.)

5.4 The Beltrami–Poincaré Half-Plane

We now have a conformal map of the *cylinder-like, rimmed,* pseudosphere: $\{(x, y): 0 \leqslant x < 2\pi,\ y \geqslant 1\}$. To instead create a map of an infinite hyperbolic plane, Beltrami knew that he must remove both of these adjectives. Note that while different choices of R yield quantitatively different geometries, they are all qualitatively the same, so there is no harm in making a specific choice:

> *In essentially all books and papers on Hyperbolic Geometry, the specific choice* $R = 1$ *is made, so that* $\mathcal{K} = -1$. *In this section we too shall make this conventional choice.*

If one wishes to return from this specific case to the general case, one need only insert the appropriate power of R into the special case ($R = 1$) formulas. For example, if you are dealing with area, then you must multiply by R^2.

To remove the "cylinder-like" adjective, imagine painting a wall with a standard cylindrical paint roller (of unit radius). After one revolution you have painted a strip of wall of width 2π, and every point on the surface of the roller has been mapped to a unique point within this strip of flat wall. To paint the entire wall, you can simply keep on rolling! Now imagine that our paint roller instead takes the form of a pseudosphere. To make it fit onto the flat wall you must first stretch out its surface, according to the metric (5.5), but then, just as before, you can keep on rolling (let's say horizontally). If a particle moves along a horizontal line on the wall, for example, the corresponding particle on the pseudosphere goes round and round the horizontal circle $\sigma = const.$ The "cylinder-like" adjective has been successfully removed[5] and we now have map $\{(x, y): -\infty < x < \infty,\ y \geqslant 1\}$.

Our second problem, the pseudosphere's rim, is solved by the conformal map with equal ease. On the left of [5.5] is the image of a particle moving down the pseudosphere along a tractrix. Of course on the pseudosphere the journey is rudely interrupted at some point $\widehat{p}$ on the rim ($\sigma = 0$), corresponding to a point p on the line $y = 1$. But in the map this point p is just like any other, and there is absolutely nothing preventing us from continuing all the way down to the point q on $y = 0$, with the true distances $d\widehat{s}$ continuing to be given by the ***standardized hyperbolic metric,***

$$d\widehat{s} = \frac{ds}{y}. \tag{5.7}$$

Why stop at q? The answer is that the particle will never even get that far, because q is infinitely far from p! Consider the small disc D of diameter ds on the line $y = 2$ shown on the left of [5.5]. Its true size on the pseudosphere is $d\widehat{s} = \frac{ds}{y}$, and this is ultimately equal [exercise] to the illustrated angle it subtends at the point h directly below it on the line $y = 0$. Now imagine D moving down the pseudosphere at steady speed. Its apparent size in the map must shrink so that its subtends a constant angle at h. In the map it hits $y = 1$ ... and keeps on going!

Assuming it took one unit of time to go from $y = 2$ to $y = 1$, then in the next unit of time it will reach $y = (1/2)$, then $y = (1/4), \ldots,$ for these points are all separated by the same hyperbolic distance:

$$\ln 2 = \int_1^2 \frac{dy}{y} = \int_{1/2}^1 \frac{dy}{y} = \int_{1/4}^{1/2} \frac{dy}{y} = \cdots.$$

---

[5]Stillwell (1996) notes that this was perhaps the first appearance in mathematics of what topologists now call a ***universal cover.***

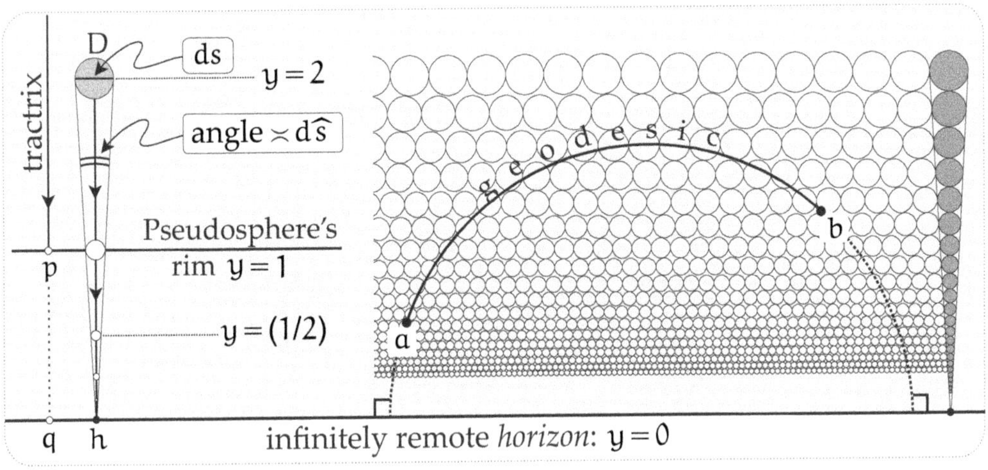

[5.5] *The hyperbolic diameter of the disc* D *is the (Euclidean) angle it subtends on the horizon. Thus, as it moves downward, its image in the map shrinks. The discs on the right are all the same size, and a geodesic* ab *therefore passes through the smallest number of them.*

Thus, viewed within the map, the motion *slows down* and each successive unit of time only halves the distance from $y = 0$, and therefore D will never reach it. (An appropriate name for this phenomenon might be "Zeno's Revenge"!)

At last, we now possess a concrete model of

> ***The Hyperbolic Plane*** $\mathbb{H}^2$: *the entire shaded half-plane* $y > 0$, *with metric* $d\widehat{s} = \frac{ds}{y}$. (5.8)

The points on the real axis $y = 0$ are infinitely far from ordinary points and are not (strictly speaking) considered part of the hyperbolic plane. They are called ***ideal points***, or ***points at infinity***. The complete line $y = 0$ of points at infinity is called the ***horizon***.

Although Beltrami discovered this map in 1868 (anticipating Poincaré by 14 years) it is now universally known as the ***Poincaré half-plane***. But in an attempt to restore some semblance of historical balance, we shall doggedly refer to this map as the ***Beltrami–Poincaré half-plane***.

Let us attempt to make the metric of this map more vivid. On the far right of [5.5] is a vertical string of touching circles of equal hyperbolic size ϵ, as in [5.4]. To the left of this we have filled part of the hyperbolic plane with such circles, *all of equal hyperbolic diameter* ϵ. Thus the hyperbolic length of any curve is ultimately equal to the number of circles it intercepts, multiplied by ϵ. This makes it clear that the shortest route from a to b is the one that intercepts the smallest number of circles, and therefore has the approximate shape shown.

If you followed our earlier advice and built your own model pseudosphere, you can also check the shape of a geodesic by stretching a string over its surface between two points at similar heights. To investigate geodesics over regions where you cannot stretch a string, such as along the tractrix generators themselves, you may instead employ the tape construction—(1.7), page 14—which works *everywhere*.

Our next task is to confirm the lovely fact illustrated in [5.5]: the exact form of such a geodesic is a perfect *semicircle that meets the horizon at right angles*. The only geodesics that are not of this form are the vertical half-lines (extending the tractrix generators), but even these may be viewed as a limiting case in which the radius of the semicircle tends to infinity.

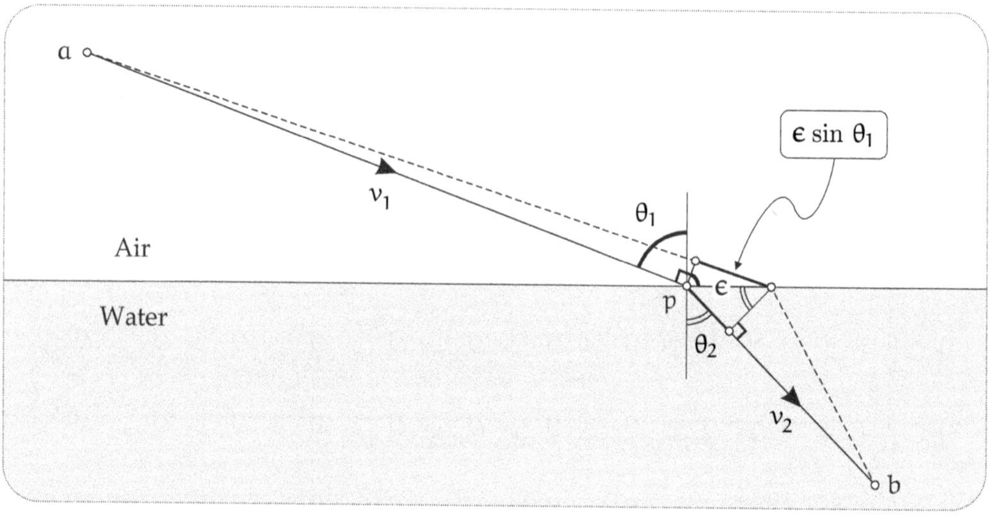

[5.6] *Snell's Law: In order for light to take equal time to travel along the two neighbouring routes, the extra time in the air must be exactly compensated by the reduced time in the water:* $\sin \theta_1 / v_1 = \sin \theta_2 / v_2$.

5.5 Using Optics to Find the Geodesics

In order to explain the semicircular form of geodesics in the Beltrami–Poincaré half-plane model of Hyperbolic Geometry, we shall draw on ideas from physics, specifically from *optics*.[6] Our inspiration derives from a law called ***Fermat's Principle***,[7] discovered in 1662:

Light travels from one place to another in the least amount of time. (5.9)

We shall begin our brief excursion into physics by using Newtonian reasoning[8] to see how Fermat's Principle allows us to determine geometrically the abrupt bending of light (called ***refraction***) that occurs when it passes from air into water (for example). This is why [exercise] your spoon appears to bend when you stick it into a cup of tea.

In [5.6], a ray of light heads out from a in direction θ_1 to the vertical, travelling at speed v_1 through the air, hitting the water at p. It is then refracted into direction θ_2, travelling through the water at reduced speed v_2, finally arriving at b. As early as 130 CE, Ptolemy conducted such experiments and compiled a fairly accurate table of the pair of angles, θ_1 and θ_2. But the precise mathematical relationship between the two angles eluded Ptolemy, as it would continue to elude scientists for centuries to come.

Finally, in 1621 the Dutch mathematician Willebrord Snell van Royen (1591–1626) discovered the correct law, now universally known as ***Snell's Law***:[9]

---

[6]This approach does not seem to be widely known. I thank Sergei Tabachnikov for pointing out to me that it was previously published by Gindikin (2007, p. 324). The essential idea of using Fermat's principle to find a path that minimizes some quantity goes back to Johann Bernoulli's solution of the Brachistochrone Problem in 1697.

[7]First, this is the same Pierre de Fermat (1601–1665) famous for discoveries in number theory, including Fermat's Last Theorem. Second, Feynman discovered that there is a beautiful quantum-mechanical explanation of this principle, a masterful account of which can be found in Feynman (1985).

[8]We were honoured to realize that here we merely retraced the path taken by Feynman: see Feynman et al. (1963, Vol. 1, 26-3). Fermat himself first gave an analytic proof and later a geometric one, but neither is as elegant as the present Newtonian argument; both of Fermat's proofs can be found in Mahoney (1994, pp. 399–401).

[9]As usual, the history is far more complex than the name suggests. The same Thomas Harriot that we met earlier also discovered "Snell's Law," 20 years before Snell, but as with most of his discoveries Harriot was secretive, though this result he did communicate to Kepler. But even Harriot had been beaten to the result—by 600 years! The Islamic mathematician and physicist Ibn Sahl published it in 984 CE, and even used it to design sophisticated anaclastic lenses.

$$\boxed{\sin\theta_1 = n\sin\theta_2, \quad \text{where } n = const.} \tag{5.10}$$

The value of n (called the ***index of refraction***) depends on the materials on either side of the interface, but for air/water $n \approx 1.33$.

It should be clear, at least qualitatively, that Fermat's Principle requires the light to bend. For if it were to travel in a straight line from a to b then it would waste valuable time travelling relatively slowly through the water, instead of travelling quickly through the air. Quantitatively, the amount of bending needed to minimize the flight time will occur when the derivative of the time (with respect to the position of p) is equal to zero.

Put geometrically, if the position of p minimizes the time, then an infinitesimal displacement ϵ of p should produce no change (to first order in ϵ) in the time. But, as we see in [5.6], this displacement causes the light to travel an additional distance through the air that is ultimately equal to $\epsilon \sin\theta_1$, increasing the flight time by an amount that is ultimately equal to $(\epsilon\sin\theta_1)/v_1$. On the other hand, by the same reasoning, the time *saved* in the water is ultimately equal to $(\epsilon\sin\theta_2)/v_2$. For the net time change to vanish, these two individual time changes must be equal. Thus, by cancelling ϵ,

$$\frac{\sin\theta_1}{v_1} = \frac{\sin\theta_2}{v_2}. \tag{5.11}$$

This not only proves Snell's Law (5.10), but it also makes a physical prediction, confirmed by direct experiment: the index of refraction is the ratio of the speed of light in the two materials, $n = (v_1/v_2)$.

Now suppose that our water sits at the bottom of a drinking glass. See [5.7]. How will the light bend when it gets to the bottom of the water and enters the glass base, where its speed is v_3? Applying the same law as before,

$$\frac{\sin\theta_1}{v_1} = \frac{\sin\theta_2}{v_2} = \frac{\sin\theta_3}{v_3}.$$

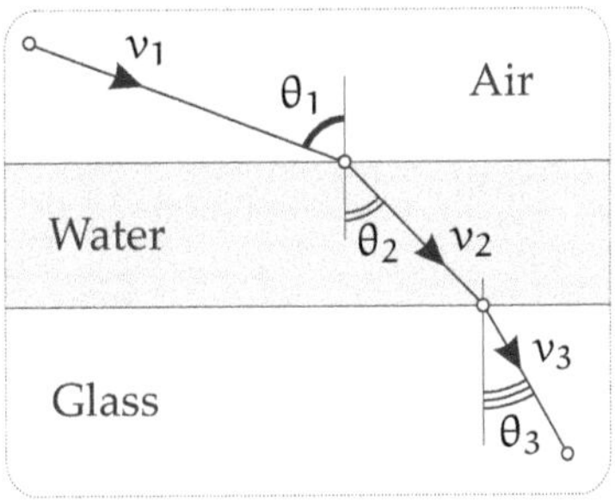

[5.7] *Snell's Law applied to multiple layers of material.*

Thus, more generally, if we have many thin horizontal strips of material, the speed of light being v_i within each, then the entire journey of the light through the layers is governed by the law

$$\frac{\sin\theta_i}{v_i} = const. = k.$$

Finally, let us go one step further and imagine the passage of light through a block of nonuniform material whose density is the same on each horizontal slice $y = const.$, but which varies *continuously* as y varies, the speed of light at height y being $v(y)$, and the angle of the light there being $\theta(y)$. Then the Generalized Snell's Law is

$$\boxed{\frac{\sin\theta(y)}{v(y)} = k.} \tag{5.12}$$

All very interesting, you say, but what has this to do with proving the semicircular form of geodesics in the hyperbolic plane?! Well, suppose points $\widehat{a}$ and $\widehat{b}$ on the pseudosphere correspond to the points a and b in $\mathbb{H}^2$, [5.5]. Imagine a particle that travels along different routes over the surface of the pseudosphere from $\widehat{a}$ and $\widehat{b}$, but always at the same constant speed, say, 1. As the particle follows one of these paths, its image in the hyperbolic plane will follow a corresponding path from a to b, but *not* with uniform speed: as we saw in [5.5], if a particle moves down the pseudosphere at constant speed, its image in the hyperbolic plane *slows down*.

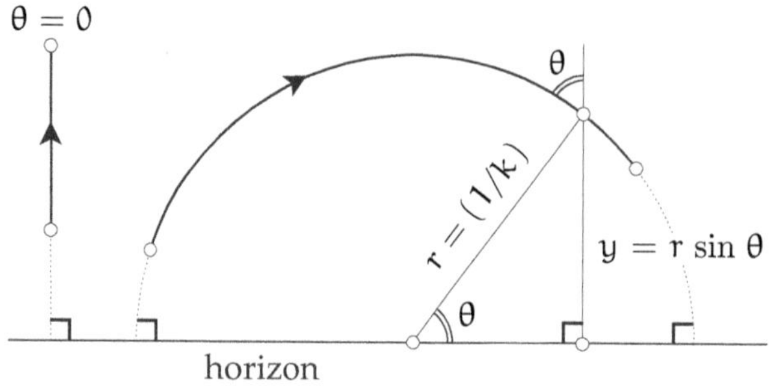

[5.8] *The hyperbolic geodesics must satisfy the Generalized Snell's Law* $(\sin\theta/y) = k$*, and are therefore semicircles and half-lines orthogonal to the horizon.*

The next crucial observation is that because the map is *conformal*, the slowing down only depends on the *location* of the particle, not on the direction in which it moves. For suppose we simultaneously launch from $\widehat{a}$ a multitude of unit-speed particles in all directions. After infinitesimal time ϵ they will form a circle of radius ϵ centred at $\widehat{a}$ on the pseudosphere, while in the hyperbolic plane (5.8) tells us they will also form a circle (centred at a) but of radius ϵy, where y is the height of a. In other words, the speed of the particles emanating from a is $v(y) = y$: the closer to the horizon we are, the slower the particles move.

Of course the time taken for each journey $\widehat{a}\widehat{b}$ on the pseudosphere will be the same as the time taken along ab in the hyperbolic plane. But on the pseudosphere, the geodesic route, being the shortest, will also be the path of least time, and therefore the geodesic in the hyperbolic plane is *also* the quickest route from a to b: *geodesic motion in the hyperbolic plane automatically obeys Fermat's Principle*, and its shape is therefore dictated by the Generalized Snell's Law! Substituting $v(y) = y$ into (5.12), suddenly the answer becomes clear in [5.8]:

> *Geodesics in the Beltrami–Poincaré half-plane model of* $\mathbb{H}^2$ *satisfy* $(\sin\theta/y) = k$. *If* $k \neq 0$ *this is a semicircle of radius* $r = (1/k)$ *centred on the horizon. If* $k = 0$ *then it is a vertical half-line* $\theta = 0$. (5.13)

Later, in Section 11.7.5, we will provide a *second* physical explanation of this important fact, based on *angular momentum*!

5.6 The Angle of Parallelism

Now that we grasp the form of geodesics in the hyperbolic plane, we can return to our starting point and visually confirm in [5.9] the truth of the Hyperbolic Axiom (1.1). There are indeed infinitely many lines (shown dashed) through p that do not meet the line L. Such lines are said to be ***ultra-parallel*** to L.

Separating the ultra-parallels from the lines that do intersect L, we see that there are precisely two lines that fail to meet L anywhere within the hyperbolic plane proper, but that do meet it on the horizon. These two lines are called ***asymptotic***.[10]

As in Euclidean Geometry, the figure makes it clear that there is precisely one line M (dotted) passing through p that cuts L at right angles (say at q). The existence of M makes it possible to define the distance of a point p from a line L in the usual way, namely, as the (hyperbolic, $d\widehat{s}$) length D of the segment pq of M.

In fact, as illustrated, M bisects the angle at p contained by the asymptotic lines, though for the moment this is far from obvious. The angle between M and either asymptotic line is called the ***angle of parallelism***, and is usually denoted Π. As one rotates the line M about p, its intersection

[10] Another commonly used name is ***parallel***.

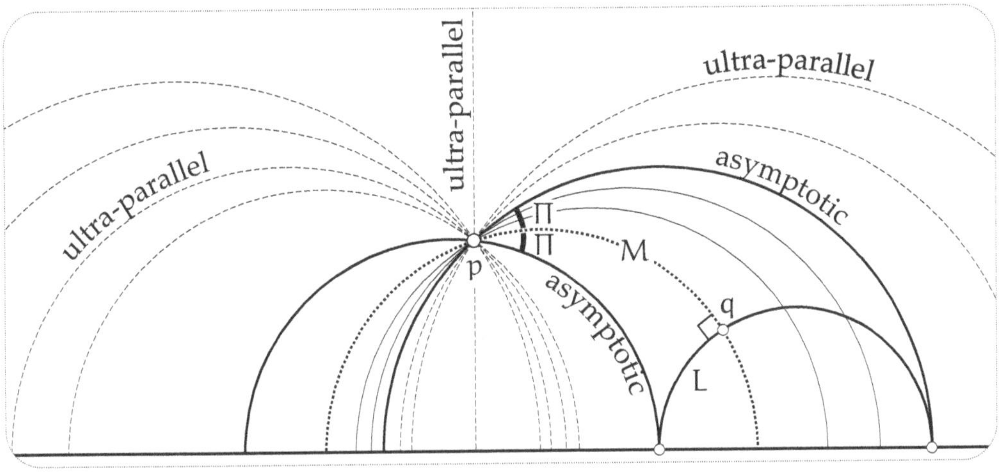

[5.9] *Explicit confirmation of the Hyperbolic Axiom: there are infinitely many lines* (ultra-parallels) *(shown dashed) through* p *that fail to meet the given line* L.

point on L moves off towards infinity, and Π tells you how far one can rotate M before it starts missing L entirely.

As Lobachevsky and Bolyai both discovered, there exists a remarkable relationship between this angle Π and the hyperbolic distance D of p from L:

$$\boxed{\tan(\Pi/2) = e^{-D}.} \qquad (5.14)$$

This is usually called the ***Bolyai–Lobachevsky Formula***.

So, if p is close to L then $\Pi \approx (\pi/2)$ and the Euclidean result almost holds: about half the rays (half-lines) emanating from p eventually hit L. Of course in Euclidean Geometry exactly half the rays hit L, no matter how far p is from L. But in the hyperbolic plane we see—qualitatively in [5.9] and quantitatively in (5.14)—that as p recedes from L, the proportion of rays emanating from p that hit L shrinks to zero!

It is essential to realize that from the point of view of microscopic inhabitants of the pseudosphere, or the true hyperbolic plane, there is no way to distinguish between geodesics—every straight line is like every other. Thus, intrinsically, geodesics that are represented in our map by vertical half-lines are completely indistinguishable from geodesics represented by semicircles.

But what about the fact that the semicircles have two ends on the horizon, whereas the vertical lines appear to only have one? The answer is that, in addition to the points on the horizon, there is one more point at infinity, and all the vertical lines meet there. According to (5.8), as we move upward along two neighbouring, vertical lines, the distance between them dies away as $(1/y)$, and they converge to a single point at infinity; this is particularly vivid on the pseudosphere.

Having stressed that the two manifestations of lines in the map are mathematically identical, we now do an about-face and also stress that *psychologically* they are *not* identical. That is, standing outside this non-Euclidean world and looking in via our map, we may find it easier to *see* that some mathematical relation holds if we look at the simpler (but less typical) case where a line is vertical rather than semicircular.

What gives this idea real power is the existence of rigid motions of $\mathbb{H}^2$, such as rotations about a point. Such distance-preserving motions are called ***isometries***, and they are the subject of the next chapter. For now, we note that the existence of isometries makes it possible to rigidly

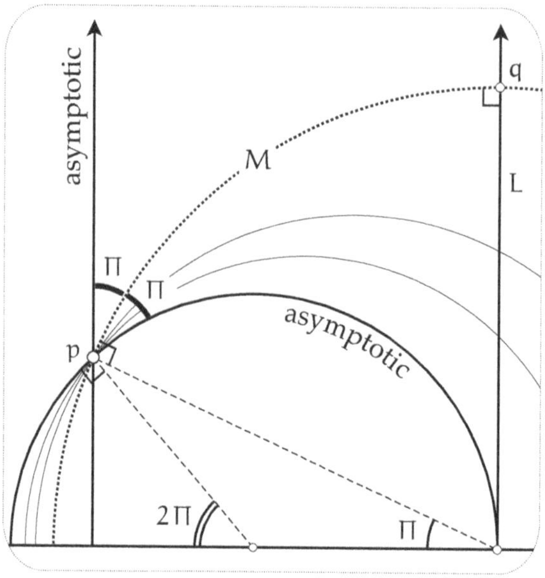

[5.10] *The same geometry as [5.9], but shown after rotating the hyperbolic plane about* p *until* L *appears vertical. This simplified picture can be used to derive the Bolyai–Lobachevsky Formula, (5.14).*

move any hyperbolic figure involving a semicircular line so that it appears in the map as a vertical line.

For example, returning to [5.9], we may rotate $\mathbb{H}^2$ about p until L becomes vertical, in which case the diagram now takes on the simpler form shown in [5.10]. An immediate payoff is that now we *can* see that M bisects the angle between the asymptotics at p: simply verify all the marked angles [exercise]. But this means that M must indeed have bisected the angle when L was in general position in [5.9], *before* we did the rotation.

By the same token, the simpler geometry of this new picture also makes it much easier to confirm the truth of the Bolyai–Lobachevsky Formula, (5.14); for details, see VCA, p. 348.

5.7 The Beltrami–Poincaré Disc

The Beltrami–Poincaré upper half-plane with metric $d\widehat{s} = ds/y$ is merely one way of depicting the abstract hyperbolic plane $\mathbb{H}^2$. Several alternative models exist.[11] While these different models are, by definition, all intrinsically identical, they are not *psychologically* identical: a particular fact or formula can be very hard to see in one model, yet be transparent in another. Facility in switching between models is therefore a valuable skill when trying to come to grips with the wonders of Hyperbolic Geometry.

For now we wish only to illustrate one particularly useful and famous model. The new model is drawn in the unit disc; see [5.11]. Like the upper half-plane, this is a conformal model, and geodesics are again represented as arcs of circles that meet the horizon at right angles, but now the infinitely distant horizon is the boundary of this disc, the unit circle. If r is distance from the centre of the disc, the new metric formula is (see Ex. 25)

$$d\widehat{s} = \frac{2}{1 - r^2}\, ds. \tag{5.15}$$

For further details we refer you to VCA or to Stillwell (2010). You can at least confirm that this is indeed $\mathbb{H}^2$ by using the conformal curvature formula (4.16) to verify [exercise] that this surface has constant negative curvature $\mathcal{K} = -1$.

It was Beltrami who first discovered this model, announcing it in the same 1868 paper as the half-plane model; see Stillwell (1996). Again, Poincaré rediscovered this model 14 years later, and again it became universally known as the "Poincaré disc." However, as before, we shall steadfastly

[11] See VCA or Stillwell (2010) for a full account.

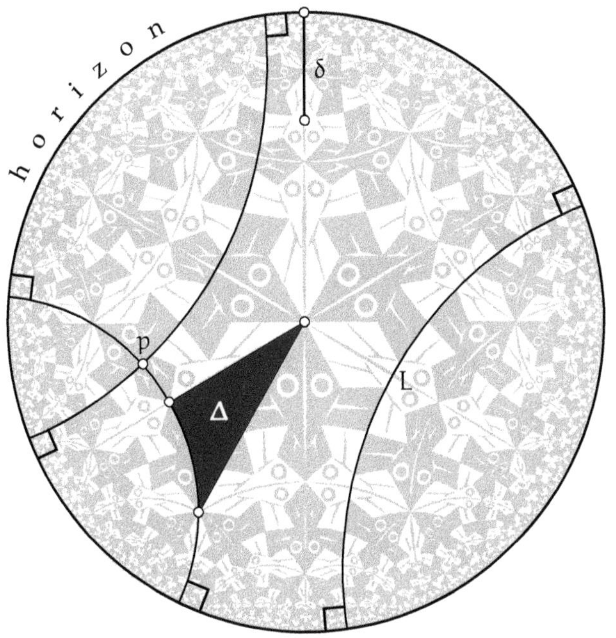

[5.11] *The Beltrami–Poincaré disc model of the hyperbolic plane. The background is Escher's* Circle Limit I; *superimposed are hyperbolic lines, which are diameters and circular arcs orthogonal to the infinitely distant boundary circle (the horizon). Clearly the Hyperbolic Axiom (1.1) is satisfied, and* $\mathcal{E}(\Delta) < 0$. *M. C. Escher's* Circle Limit I *© 2020 The M. C. Escher Company-The Netherlands. All rights reserved. www.mcescher.com.*

insist upon giving credit to both by calling this the ***Beltrami–Poincaré disc***; less contentiously, it is also called the ***conformal disc model***.

In 1958 the famous British geometer H.S.M. Coxeter (1907–2003) introduced the Dutch artist M. C. Escher (1898–1972) to the conformal disc model of $\mathbb{H}^2$. This led Escher to create his famous *Circle Limit* series of woodcuts,[12] the first of which is reproduced (deliberately faintly) in [5.11], but with hyperbolic lines prominently superimposed. (The idea for this diagram came directly from Penrose (2005, Fig. 2.12).) We see, for example, that there are indeed infinitely many lines through p that fail to meet L, in accord with the Hyperbolic Axiom (1.1). Diameters of the circle are also hyperbolic lines, so the illustrated triangle Δ is a genuine hyperbolic triangle. Note that it is visually apparent (and easily provable) that $\mathcal{E}(\Delta) < 0$, as it should be.

As you stare at this, try to imagine yourself as one of the fish. You are exactly the same size and shape as every other fish, and you can swim in a straight line forever without ever seeing any change in your surroundings or in your fellow fish. But looking into the map from outside, the compression of distances makes you appear to be swerving along a circular path and to be shrinking as you go. In fact if $\delta \equiv (1 - r)$ is the illustrated *Euclidean* distance of fish from the horizon, and we look at a point close to the edge of the map, we find that (5.15) implies [exercise] that the *(apparent size of fish)* $\propto \delta$.

In Exercise 25 you will see that there is in fact a simple *conformal* transformation that relates this new disc model to our previous conformal half-plane model, thereby explaining the conformality of the new disc model. Computer application of this transformation to Escher's original art [5.11] yields [5.12], a picture that Escher himself never created, but which he surely would have appreciated. (This image is reproduced, with permission, from Stillwell (2005, p. 195).)

[12]One can find online a short video of Coxeter himself discussing the mathematics of these Escher constructions.

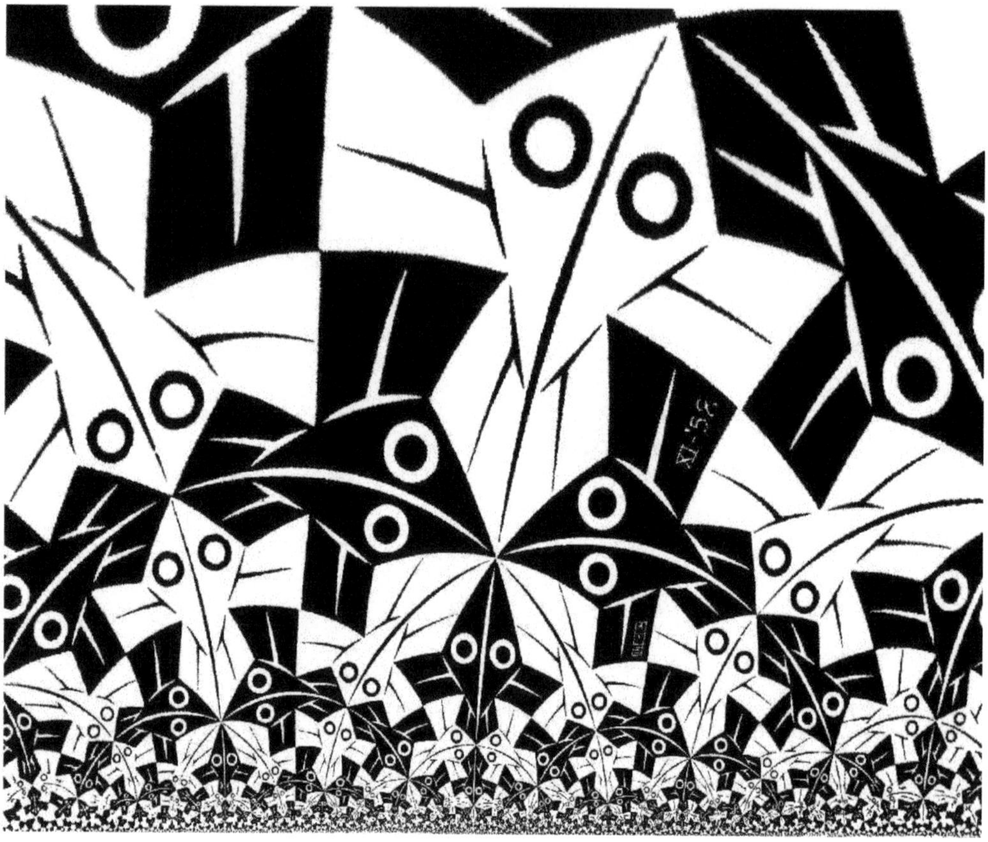

[5.12] *Escher's* Circle Limit I *transformed (by Professor John Stillwell) from its original conformal disc model [5.11] to the conformal half-plane model. M. C. Escher's* Circle Limit I *© 2020 The M. C. Escher Company-The Netherlands. All rights reserved. www.mcescher.com.*

Let us pause, catch our breath, and look back at how far we have come. The story[13] with which we began this book has arrived at a happy conclusion, an ending of sorts. For 2000 years Euclidean Geometry had been vexed by confusion and doubt surrounding the Parallel Axiom. Now Beltrami's concrete vindication of Hyperbolic Geometry as a legitimate alternative had at last brought about clarity and cathartic release from two millennia of mathematical tension. What a splendid place to end Act II!

Or not . . .

[13]For the complete saga, see Gray (1989).

Chapter 6

Isometries and Complex Numbers

6.1 Introduction

Great mathematical ideas not only put past mysteries to rest, they also reveal *new* ones: tunnels at the end of the light! It is to such glimpses of strange new connections with other areas of mathematics and physics that we now turn.

The first of these new mysterious connections is between the three geometries of constant curvature (Euclidean, Spherical, and Hyperbolic) and the *complex numbers.*

We have already asked you to imagine our maps of curved surfaces as being drawn in the complex plane. However, the astute reader will have noticed that, up to this point, we have made but slight use of the essential structure of the complex numbers. That is about to change. First, however, we need to make some more general observations about the very concept of an isometry.

Isometries necessarily preserve the magnitude of every angle, and the ones that also preserve the *sense* of the angle (clockwise versus counterclockwise) are called ***direct***, while those that reverse the sense are called ***opposite***. Thus a direct isometry is a very special kind of conformal mapping, while an opposite isometry is a very special kind of anticonformal mapping. For example, in the plane, a rotation is a direct isometry, whereas a reflection in a line is an opposite isometry.

Next we observe that, under the operation of composition,

> *The complete set of isometries (both direct and opposite) of a given surface $\mathcal{S}$ automatically has the structure of a* **group**, $\mathcal{G}(\mathcal{S})$.

To confirm this, let $e = (\text{do nothing})$ and let a, b, c be any three isometries of $\mathcal{S}$. Then the ***group axioms***[1] are satisfied:

- Since e clearly preserves distances, $e \in \mathcal{G}(\mathcal{S})$, and since $a \circ e = a = e \circ a$, we deduce that e is the ***group identity***.
- If we apply a and then apply b (both of which preserve distances) then the net transformation also preserves distances: $b \circ a \in \mathcal{G}(\mathcal{S})$.
- Since the transformation a preserves distances, its inverse does too: $a^{-1} \in \mathcal{G}(\mathcal{S})$.
- Composition of transformations (not necessarily isometries) is associative: $(a \circ b) \circ c = a \circ (b \circ c)$.

Note that (under composition) direct and opposite isometries behave like $(+)$ and $(-)$ under multiplication: $(+)(+) = (+)$, $(+)(-) = (-)$, and $(-)(-) = (+)$. It follows that

> *The direct isometries form a* **subgroup** $\mathcal{G}_+(\mathcal{S})$ *of the full group* $\mathcal{G}(\mathcal{S})$.

[1]These define the mathematical concept of a ***group***. Even if you have not met this concept before, you can still follow along by simply accepting the following axioms as the definition.

On the other hand, the opposite isometries do not form a group at all. But they do belong to the full group $\mathcal{G}(\mathcal{S})$, so how are they related to $\mathcal{G}_+(\mathcal{S})$?

Let ξ be a specific, fixed opposite isometry; then ξ^{-1} is too. Let ζ be a general opposite isometry—think of this varying over all possible opposite isometries. Then $\xi^{-1} \circ \zeta \in \mathcal{G}_+(\mathcal{S}) \Rightarrow \zeta \in \xi \circ \mathcal{G}_+(\mathcal{S})$. By the same token, $\zeta \in \mathcal{G}_+(\mathcal{S}) \circ \xi$. Thus,

> *If ξ is any opposite isometry, the complete set of opposite isometries is $\xi \circ \mathcal{G}_+(\mathcal{S}) = \mathcal{G}_+(\mathcal{S}) \circ \xi$, and the full symmetry group is therefore*
>
> $$\mathcal{G}(\mathcal{S}) = \mathcal{G}_+(\mathcal{S}) \cup [\xi \circ \mathcal{G}_+(\mathcal{S})] = \mathcal{G}_+(\mathcal{S}) \cup [\mathcal{G}_+(\mathcal{S}) \circ \xi]. \tag{6.1}$$

Does every surface $\mathcal{S}$ possess a nontrivial group $\mathcal{G}(\mathcal{S})$ of isometries? No, for an isometry must also preserve the *curvature*. Suppose an isometry carries a very small (ultimately vanishing) triangle Δ at p to a congruent triangle Δ' at p'. Then

$$\mathcal{K}(p) \asymp \frac{\mathcal{E}(\Delta)}{\mathcal{A}(\Delta)} = \frac{\mathcal{E}(\Delta')}{\mathcal{A}(\Delta')} \asymp \mathcal{K}(p').$$

Thus an irregular surface like the squash depicted in [1.9] will not have any isometries.[2]

On the other hand, it is not necessary for the curvature to be constant over $\mathcal{S}$ for isometries to exist.[3] For example, consider any surface of revolution. By its very construction, this surface of (typically) nonconstant curvature does admit a group of isometries. Rotations about the surface's axis are direct isometries, and reflections in planes passing through this axis are opposite isometries; additional isometries may exist as well.

The greater the symmetry of $\mathcal{S}$, the bigger the group of isometries, and the greatest symmetry occurs in the three cases of constant curvature: $\mathcal{K}=0$, $\mathcal{K}>0$, and $\mathcal{K}<0$. Extrinsically, the archetypal surfaces possessing these geometries are the Euclidean plane, the sphere, and the pseudosphere. However, the concept of an isometry belongs to *intrinsic* geometry; so, for example, the Beltrami–Poincaré half-plane map of $\mathbb{H}^2$ is in fact a much better depiction of Hyperbolic Geometry than is the pseudosphere, and it is this map (or the conformal disc model) that will be relevant in what follows.

We will refine the statement shortly, but here in brief is the startling connection between these three, maximally symmetric geometries and the complex numbers—the **Main Result**:

> *All three geometries of constant curvature have symmetry groups $\mathcal{G}_+(\mathcal{S})$ (of direct isometries) that are subgroups of the group of Möbius transformations of the complex plane, $z \mapsto M(z) = \frac{az+b}{cz+d}$, where a, b, c, d are complex constants.* (6.2)

[2] I have not thought through precisely how to quantify the degree of irregularity needed to preclude the existence of isometries. A starting point might be to observe that if the value of $\mathcal{K}(p)$ is *uniquely* attained at p, then p must be a *fixed point* of any putative isometry—it has nowhere to go! Three such points would result in three fixed points, and I would guess this would preclude nontrivial isometries.

[3] We will be concerned with infinite, continuous sets of isometries, but it is also possible to have a *finite* set of isometries, even for a smooth surface. For example, consider the symmetries of a gaming die (a cube with its edges and corners rounded off).

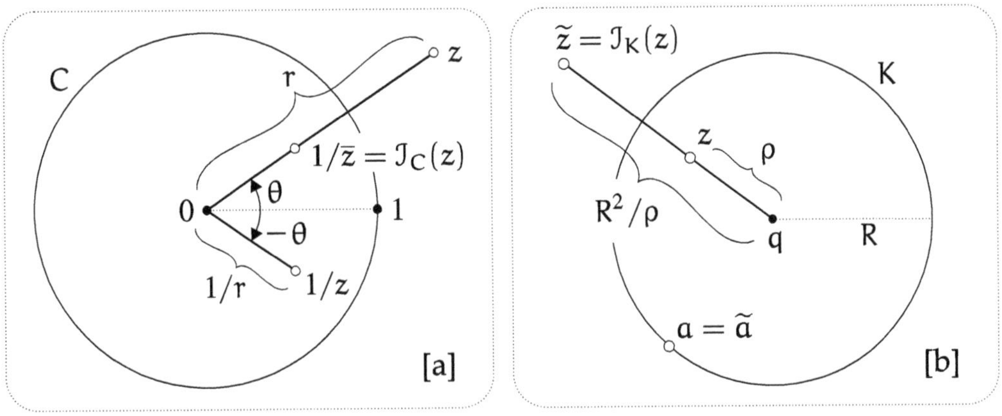

[6.1] [a] Complex inversion *is the composition of geometric inversion and conjugation.* **[b] Geometric inversion** *in a general circle.*

6.2 Möbius Transformations

As the above result alone may convince you, Möbius[4] transformations are extraordinarily important in modern mathematics (and, as we shall see, in physics too). We now summarize[5] some facts that we shall need concerning these transformations.

- **Decomposition into Simpler Transformations.** Let us decompose [exercise] $z \mapsto M(z) = \frac{az+b}{cz+d}$ into the following sequence of transformations:

$$\left.\begin{array}{ll} \text{(i)} & z \mapsto z + \frac{d}{c}, \text{ which is a translation;} \\ \text{(ii)} & z \mapsto (1/z), \text{ which we shall call } \textit{complex inversion;} \\ \text{(iii)} & z \mapsto -\frac{(ad-bc)}{c^2} z, \text{ which is an expansion and a rotation; and} \\ \text{(iv)} & z \mapsto z + \frac{a}{c}, \text{ which is another translation.} \end{array}\right\} \qquad (6.3)$$

  Note that if $(ad - bc) = 0$ then $M(z)$ crushes the entire complex plane down to a single image point (a/c); in this exceptional case $M(z)$ cannot be undone, and is called ***singular***. In discussing Möbius transformations we shall therefore always assume that $M(z)$ is ***nonsingular***, meaning that it is invertible (i.e., $(ad - bc) \neq 0$).

  Of the four transformations above, only the second one (the reciprocal mapping) is unfamiliar and requires further investigation.

- **Inversion in a Circle.** This mapping $z \mapsto (1/z)$ holds the key to understanding the Möbius transformations. As we did in VCA, we shall call this reciprocal mapping ***complex inversion***. In polar coordinates, the image of $z = r\,e^{i\theta}$ under complex inversion is $1/(r\,e^{i\theta}) = (1/r)\,e^{-i\theta}$: the new length is the reciprocal of the original, and the new angle is the negative of the original. See [6.1a]. Note how a point outside the unit circle C is mapped to a point inside C, and vice versa. Figure [6.1a] also illustrates a particularly fruitful way of decomposing complex inversion into a two-stage process:

  1. Send $z = r\,e^{i\theta}$ to the point that is in the same direction as z but that has reciprocal length, namely the point $(1/r)\,e^{i\theta} = (1/\bar{z})$.

---

[4]Also called ***fractional linear*** or ***bilinear*** transformations.

[5]For greater depth, see Chapters 3 and 6 of VCA.

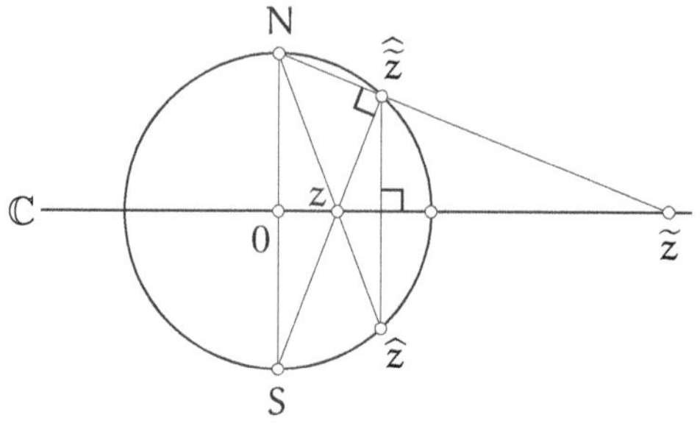

[6.2] A vertical cross section of the Riemann sphere. Reflecting the sphere in its equatorial plane $\mathbb{C}$ sends $\widehat{z} \mapsto \widehat{\widehat{z}}$ and $z \mapsto \widetilde{z} = \mathcal{J}_{\mathbb{C}}(z)$.

2. Apply complex conjugation (i.e., reflection in the real axis), which sends $(1/\overline{z})$ to $\overline{(1/\overline{z})} = (1/z)$.

Check for yourself that the order in which we apply these two mappings is immaterial. (This is atypical: in (6.3), for example, the order certainly does matter.)

While stage (2) is geometrically trivial, we shall see that the mapping in stage (1) is filled with surprises; it is called[6] ***geometric inversion***, or simply ***inversion***. Clearly, the unit circle C plays a special role for this mapping: the inversion interchanges the interior and exterior of C, while each point *on* C remains fixed (i.e., is mapped to itself). For this reason we write the mapping as $z \mapsto \mathcal{J}_C(z) = (1/\overline{z})$, and we call $\mathcal{J}_C$ (a little more precisely than before) "inversion in C."

This added precision in terminology is important because, as illustrated in [6.1b], there is a natural way of generalizing $\mathcal{J}_C$ to inversion in an *arbitrary* circle K (say with centre q and radius R). Clearly, this "inversion in K," written $z \mapsto \widetilde{z} = \mathcal{J}_K(z)$, should be such that the interior and exterior of K are interchanged, while each point on K remains fixed. If ρ is the distance from q to z, then *we define $\widetilde{z} = \mathcal{J}_K(z)$ to be the point in the same direction from q as z, and at distance (R^2/ρ) from q*. Check for yourself that this definition is forced on us if we imagine this figure to be [6.1a] expanded by factor R.

- **Inversion Is Equatorial Reflection of the Riemann Sphere.** If we stereographically suck the complex numbers off the plane and onto the unit sphere, thereby creating the Riemann sphere, then the effect of inversion becomes startlingly simple:

> *Inversion of $\mathbb{C}$ in the unit circle induces a reflection of the Riemann sphere in its equatorial plane, $\mathbb{C}$.* (6.4)

To confirm (6.4), see [6.2], which shows a vertical cross section of the Riemann sphere. The point z is stereographically projected to $\widehat{z}$, reflected across the equatorial plane to $\widehat{\widehat{z}}$, and finally projected to $\widetilde{z}$. We see [exercise] that the triangles $N0\widetilde{z}$ and $z0N$ are similar, and so $|\widetilde{z}|/1 = 1/|z|$. Thus, $\widetilde{z} = \mathcal{J}_C(z)$, as claimed.

Note that this figure also shows that inversion in the unit circle is equivalent to stereographic projection from the south pole, followed by stereographic projection from the north pole (or vice versa).

---

[6] Another common and appropriate name is ***reflection*** in a circle (for reasons that will become apparent shortly). In older works it is often called "transformation by reciprocal radii."

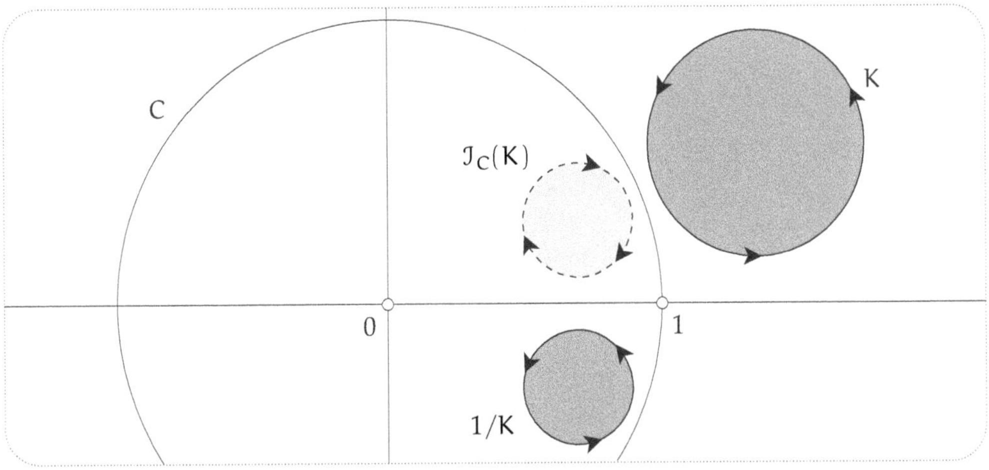

[6.3] *Complex inversion in the unit circle C sends the oriented circle K to $(1/K)$, and the shaded region to the left of the direction of travel along K is mapped to the shaded region to the left of the direction of travel along $(1/K)$.*

- **Inversion Preserves Circles.** By virtue of (4.28), a circle K in $\mathbb{C}$ projects to a circle on Σ, and the anticonformal reflection of the sphere in its equatorial plane (inversion) maps this to another circle on Σ, which finally projects back to a circle $\mathcal{I}_C(K)$ in $\mathbb{C}$. See [6.3].

  If we instead begin with a line, we get a circle on Σ passing through N (recall (4.21)) which reflects to a circle through S, which projects back to a circle in $\mathbb{C}$ passing through 0. Conversely, because inversion *swaps* points, a circle K through 0 is sent to a line $\mathcal{I}_C(K)$. See [6.4].

  We may think of the second result as a limiting case of the first, a line being a limiting form of a circle. Indeed, on the Riemann sphere a line is literally a circle that happens to pass through the north pole. With this unified language in place, we may summarize by saying,

Inversion is anticonformal and maps circles to circles. (6.5)

- **Complex Inversion is a Rotation of the Riemann Sphere.** If we now follow this inversion by conjugation, $z \mapsto \bar{z}$, then the net effect is complex inversion. But the effect of conjugation on the Riemann sphere is another reflection, this time in the vertical plane passing through the real axis. As you may easily check (perhaps with the assistance of an orange) the net effect of these two reflections in perpendicular planes passing through the real axis is a *rotation* about that axis:

Complex inversion, $z \mapsto (1/z)$, is a rotation of the Riemann sphere, through angle π about the real axis. It is therefore conformal and maps circles to circles. (6.6)

We can also provide a second proof using the stereographic formula, (4.26). First check for yourself that if $\hat{z}$ has coordinates (ϕ, θ) then rotating it by π about the real axis carries it to a point with coordinates $(\pi - \phi, -\theta)$. So this rotated point corresponds to the complex number

$$\cot\left[\frac{\pi}{2} - \frac{\phi}{2}\right] e^{i(-\theta)} = \frac{1}{\cot(\phi/2)}\, e^{-i\theta} = \frac{1}{\cot(\phi/2)\, e^{i\theta}} = \frac{1}{z},$$

as was to be shown.

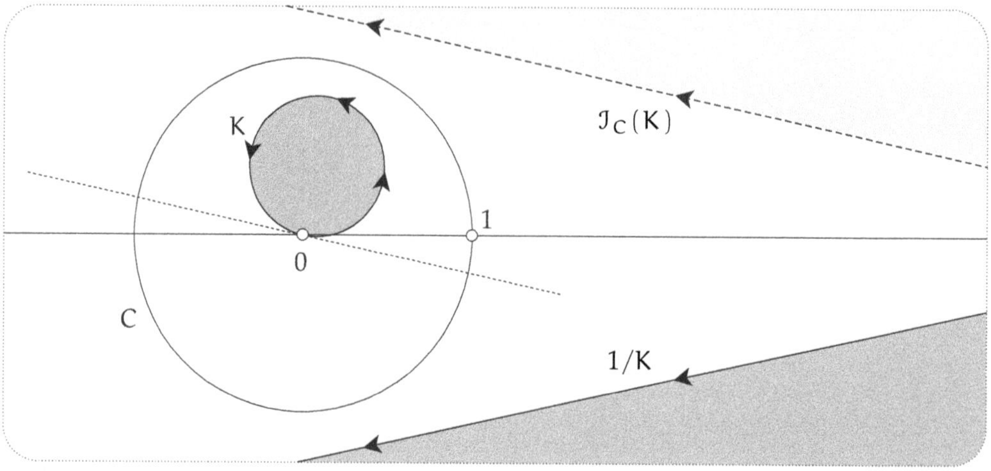

[6.4] *Complex inversion in the unit circle* C *sends the oriented circle* K *to* (1/K), *and the shaded region to the left of the direction of travel along* K *is mapped to the shaded region to the left of the direction of travel along* (1/K).

Figure [6.3] illustrates the effect of both inversion and complex inversion on an oriented circle K. Note how the darkly shaded region to the left of the direction of travel along K (the interior of K) is mapped to the region to the left of the direction of travel along (1/K). If K contains 0 then the orientation of (1/K) is *reversed* by the mapping, but the rule "left $\mapsto$ left" remains in force. Check all this for yourself, both directly and by using (6.6).

Figure [6.4] illustrates the same phenomenon in the case where K passes through 0 (the south pole) and (1/K) is therefore a line (a circle through the north pole).

Observe that this provides a conformal mapping between a half-plane and the interior of a disc, and this is in fact precisely how the conformal disc model [5.11] of the hyperbolic plane is constructed from the conformal half-plane model [5.12]. (The explicit Möbius transformation is given in Exercise 25, and the full explanation is given in VCA, Section 6.3.10.)

- **Preservation of Angles and Circles.** As illustrated in [6.3] and [6.4], it follows immediately from (6.3) and (6.6) that, using the new generalized sense of "circle,"

> *Möbius transformations are conformal and map each oriented circle* K *to an oriented circle* $\widetilde{K}$ *in such a way that the region to the left of* K *is mapped to the region to the left of* $\widetilde{K}$. (6.7)

- **Matrix Representation.** Geometrically, the Riemann sphere allows us to think of ∞ like any other point—it's simply the north pole. From (6.6), we see that complex inversion induces a rotation that swaps the north and south poles, so the statements $0 = 1/\infty$ and $\infty = 1/0$ are literally true.

  It would be nice if we could likewise do away with the exceptional role of infinity at the *algebraic* level. To do so, we adapt an idea from Projective Geometry, describing each point on the Riemann sphere as the ratio of a *pair* of complex numbers $(\mathfrak{z}_1, \mathfrak{z}_2)$ living in $\mathbb{C}^2$: $z = (\mathfrak{z}_1/\mathfrak{z}_2)$.

  The ordered pair of complex numbers[7] $[\mathfrak{z}_1, \mathfrak{z}_2]$ are called ***projective coordinates*** or ***homogeneous coordinates*** of z. In order that this ratio be well defined we demand that $[\mathfrak{z}_1, \mathfrak{z}_2] \neq [0, 0]$.

[7] We note, sadly only in passing, that $[\mathfrak{z}_1, \mathfrak{z}_2]$ may also be viewed as the coordinate representation of a ***2-spinor***. This concept lies at the heart of huge body of fundamental, pioneering work by Sir Roger Penrose—see Penrose and Rindler (1984) for

To each ordered pair [$\mathfrak{z}_1$ arbitrary, $\mathfrak{z}_2 \neq 0$] there corresponds precisely one ordinary, finite point $z = (\mathfrak{z}_1/\mathfrak{z}_2)$, but to each point z there corresponds an infinite set of projective coordinates, $[k\mathfrak{z}_1, k\mathfrak{z}_2] = k[\mathfrak{z}_1, \mathfrak{z}_2]$, where k is an arbitrary nonzero complex number. For example, i can be represented as $[1+i, 1-i]$, as $[-1, i]$, or as $[3+2i, 2-3i]$, to give just three examples out of infinitely many.

The *point at infinity*, on the other hand, can now be represented as an unremarkable pair of the form $[\mathfrak{z}_1, 0]$.

Just as a linear transformation of $\mathbb{R}^2$ is represented by a real 2×2 matrix, so a linear transformation of $\mathbb{C}^2$ is represented by a complex 2×2 matrix:

$$\begin{bmatrix} \mathfrak{z}_1 \\ \mathfrak{z}_2 \end{bmatrix} \longmapsto \begin{bmatrix} \mathfrak{w}_1 \\ \mathfrak{w}_2 \end{bmatrix} = \begin{bmatrix} a & b \\ c & d \end{bmatrix} \begin{bmatrix} \mathfrak{z}_1 \\ \mathfrak{z}_2 \end{bmatrix} = \begin{bmatrix} a\mathfrak{z}_1 + b\mathfrak{z}_2 \\ c\mathfrak{z}_1 + d\mathfrak{z}_2 \end{bmatrix}.$$

But if $[\mathfrak{z}_1, \mathfrak{z}_2]$ and $[\mathfrak{w}_1, \mathfrak{w}_2]$ are thought of as the projective coordinates in $\mathbb{C}^2$ of the point $z = (\mathfrak{z}_1/\mathfrak{z}_2)$ in $\mathbb{C}$ and its image point $w = (\mathfrak{w}_1/\mathfrak{w}_2)$, then the above linear transformation of $\mathbb{C}^2$ induces the following (nonlinear) transformation of $\mathbb{C}$:

$$z = \frac{\mathfrak{z}_1}{\mathfrak{z}_2} \quad \longmapsto \quad w = \frac{\mathfrak{w}_1}{\mathfrak{w}_2} = \frac{a\mathfrak{z}_1 + b\mathfrak{z}_2}{c\mathfrak{z}_1 + d\mathfrak{z}_2} = \frac{a(\mathfrak{z}_1/\mathfrak{z}_2) + b}{c(\mathfrak{z}_1/\mathfrak{z}_2) + d} = \frac{az+b}{cz+d}.$$

This is none other than the most general Möbius transformation!

Thus to every Möbius transformation $M(z)$ there corresponds a 2×2 matrix $[M]$,

$$M(z) = \frac{az+b}{cz+d} \quad \longleftrightarrow \quad [M] = \begin{bmatrix} a & b \\ c & d \end{bmatrix},$$

and we deduce that the matrix of the composition of two Möbius transformations is the product of the corresponding matrices:

$$[M_2 \circ M_1] = [M_2]\,[M_1]. \tag{6.8}$$

Likewise, the inverse of a Möbius transformation can be found by taking the inverse of the corresponding matrix:

$$[M^{-1}] = [M]^{-1}. \tag{6.9}$$

It follows easily [exercise] that the nonsingular Möbius transformations do indeed form a group (as was implicitly claimed earlier).

Since the coefficients of the Möbius transformation are not unique, neither is the corresponding matrix: if k is any nonzero constant, then the matrix $k[M]$ corresponds to the same Möbius transformation as $[M]$. However, if $[M]$ is ***normalized*** by imposing $(ad - bc) = 1$, then there are just two possible matrices associated with a given Möbius transformation: if one is called $[M]$, the other is $-[M]$; in other words, the matrix is determined "uniquely up to sign." This apparently trivial fact turns out to have deep significance in both mathematics and physics; see Penrose and Rindler (1984, Ch. 1) and Penrose (2005, §11.3).

---

technical details—having to do with "spinorial objects," that do not return to their original state after being rotated by 2π! For an intuitive explanation of the existence of such seemingly impossible objects, see Penrose (2005, §11.3).

6.3 The Main Result

We can now state and prove a more detailed version of the Main Result, (6.2):

The symmetry groups $\mathcal{G}_+(\mathcal{S})$ of all three geometries of constant curvature are subgroups of the group of Möbius transformations.

1. *Euclidean Geometry ($\mathcal{K}=0$):*

$$E(z) = e^{i\theta}z + k.$$

Note that $z \mapsto \bar{z}$ is an opposite isometry, being reflection in the real axis. Thus (6.1) tells us that the full group of isometries $\mathcal{G} = \{E(z)\} \cup \{E(\bar{z})\}$.

2. *Spherical Geometry ($\mathcal{K}=+1$) in the stereographic map:*

$$S(z) = \frac{az+b}{-\bar{b}z+\bar{a}}, \quad \textit{where} \quad |a|^2+|b|^2=1. \tag{6.10}$$

Note that $z \mapsto \bar{z}$ is an opposite isometry, being reflection of the sphere in the vertical plane through the real axis. Thus (6.1) tells us that the full group of isometries $\mathcal{G} = \{S(z)\} \cup \{S(\bar{z})\}$.

3. *Hyperbolic Geometry ($\mathcal{K}=-1$) in the Beltrami–Poincaré half-plane map:*

$$H(z) = \frac{az+b}{cz+d}, \quad \textit{where } a, b, c, d \textit{ are } \text{real}, \textit{ and } (ad-bc)=1. \tag{6.11}$$

Note that $z \mapsto -\bar{z}$ is an opposite isometry, being reflection in the imaginary axis. Thus, again by (6.1), the full group of isometries $\mathcal{G} = \{H(z)\} \cup \{H(-\bar{z})\}$.

The matrix results (6.8) and (6.9) make it a simple matter [exercise] to verify that each of these three sets is indeed a group. We also note that in the spherical case the matrices are of a special kind that plays an important role in physics. They are called ***unitary***, meaning that if you take the conjugate and then transpose (the net operation being denoted $*$) then [exercise] you obtain the matrix of the inverse transformation:

$$\textbf{\textit{unitary}} \text{ means} \quad [S][S]^* = \text{the identity matrix.}$$

The Euclidean transformation $E(z)$ is a rotation of θ followed by a translation of k, and it seems clear that this is indeed the most general rigid motion of the plane. We therefore rush past this and reserve our energy for the challenge of proving the surprising results claimed for $\mathcal{K}=\pm1$. In this work we cannot pause to do geometric justice to these lovely results; instead, we refer you to VCA. For now, let us show off the *computational* power of the metric machinery we have developed.

To do so, we need to pause and introduce a pinch of Complex Analysis. As we explained in Section 4.6, the derivative $f'(z)$ of a complex mapping $z \mapsto w = f(z)$ is defined exactly as in first-year calculus, and all the usual rules of differentiation apply without change. Thus, applying the quotient rule to the normalized (i.e., $(ad-bc)=1$) Möbius transformation, we find that

$$M(z) = \frac{az+b}{cz+d} \quad \Longrightarrow \quad M'(z) = \frac{1}{(cz+d)^2}. \tag{6.12}$$

But recall from (4.19) that the complex numbers imbue this derivative with the "amplitwist" interpretation, which is magically richer than in ordinary calculus.

We can now return to the Main Result. Euler was the first to prove (in 1775) that the rigid motions of the sphere are simply rotations, and it was Gauss (around 1819) who first recognized that they could be represented as Möbius transformations of the form (6.10).

Taking the sphere to have unit radius, we may rewrite the stereographic metric formula (4.22) in terms of the complex number z:

$$d\widehat{s} = \frac{2}{1+|z|^2}\,|dz|.$$

To prove that the mapping $z \mapsto w = S(z) = \frac{az+b}{-\overline{b}z+\overline{a}}$ is a direct isometry, we must therefore show that

$$\frac{2}{1+|w|^2}\,|dw| = \frac{2}{1+|z|^2}\,|dz|. \tag{6.13}$$

Direct calculation shows [exercise] that

$$1+|w|^2 = \frac{1+|z|^2}{|-\overline{b}z+\overline{a}|^2}.$$

It follows from (6.12) that

$$\left|\frac{dw}{dz}\right| = |S'(z)| = \frac{1}{|-\overline{b}z+\overline{a}|^2} = \frac{1+|w|^2}{1+|z|^2},$$

thereby confirming (6.13), and with it part 2 of the Main Result. (Well, this does not prove that these are the *only* direct isometries. This will be addressed by our deferred geometric analysis.)

But how could any mortal (even Gauss) have guessed the form of these Möbius transformations?! Here is a simple argument, based only on what we know so far: *If the rotation carries the point $\widehat{z}$ on the sphere to $\widehat{M(z)}$, then it must also carry the point antipodal to $\widehat{z}$ to the point antipodal to* $\widehat{M(z)}$. But we know that the stereographic images of antipodal points are related by (4.27), page 49, and so

$$M\left(-\frac{1}{\overline{z}}\right) = -\frac{1}{\overline{M(z)}}.$$

Thus,

$$\frac{a\left[-\frac{1}{\overline{z}}\right]+b}{c\left[-\frac{1}{\overline{z}}\right]+d} = -\frac{\overline{cz+d}}{\overline{az+b}} \quad\Longrightarrow\quad \frac{-b\,\overline{z}+a}{d\,\overline{z}-c} = \frac{\overline{c}\,\overline{z}+\overline{d}}{\overline{a}\,\overline{z}+\overline{b}},$$

which is clearly satisfied if $c = -\overline{b}$ and $d = \overline{a}$, which is precisely the form of (6.10)!

In Exercise 27 we give concrete examples of how to represent rotations as Möbius transformations, and we also provide a constructive proof that every rotation must be a Möbius transformation of the form (6.10).

Lastly, consider the hyperbolic plane. The metric formula (5.7) tells us that we must show that $z \mapsto w = H(z) = \frac{az+b}{cz+d}$ satisfies

$$\frac{|dw|}{\operatorname{Im} w} = \frac{|dz|}{\operatorname{Im} z}. \tag{6.14}$$

Direct calculation shows [exercise, recalling that $\overline{a} = a, \ldots$] that

$$\operatorname{Im} w = \frac{w-\overline{w}}{2i} = \frac{\operatorname{Im} z}{|cz+d|^2}.$$

It follows from (6.12) that

$$\left|\frac{dw}{dz}\right| = \left|H'(z)\right| = \frac{1}{|cz+d|^2} = \frac{\operatorname{Im} w}{\operatorname{Im} z},$$

thereby confirming (6.14), and with it the final part of the Main Result.

These hyperbolic isometries not only contain analogues of ordinary rotations and translations, but they also include a *third* type of rigid motion, called a ***limit rotation***, which has no counterpart in ordinary Euclidean Geometry. It is the limit of an ordinary rotation of $\mathbb{H}^2$ as the centre of rotation moves off to infinity, ultimately becoming a point on the horizon, $y = 0$. For details, see VCA, Section 6.3.7.

6.4 Einstein's Spacetime Geometry

If mere *sub*groups of the group of Möbius transformations have such deep significance for geometry, it is natural to ask if even more remarkable powers reside within the *full* group that gave rise to them.

Indeed, the full group plays a fundamental role in at least two seemingly unrelated spheres of knowledge: Relativity Theory and *3-dimensional* Hyperbolic Geometry. In fact this is *not* a coincidence. There is a deep connection between these two subjects, but we shall not be able to explore it here, and instead direct you to the wonderful exposition of Penrose (2005, §18.4).

The first role may be called "fundamental" without hyperbole:[8]

> *The Möbius group describes the symmetries of space and time, or, more precisely, of Einstein's unification of the two*—spacetime.

Clearly it would be neither appropriate nor feasible for us to explore Einstein's Special Theory of Relativity in detail here.[9] However, for readers who have not studied this theory previously, we shall try to say enough to make sense of this particular connection with the Möbius group.[10]

The starting point of Einstein's theory is an extraordinary and bizarre experimental fact of Nature:

> *The speed of light is the same for all observers in uniform relative motion.*

As Einstein first recognized in 1905, this can *only* be possible if such observers *disagree* about measurements of space and time!

To quantify this, let us combine the time T and the 3-dimensional Cartesian coordinates (X, Y, Z) of an ***event*** $\mathfrak{E}$ into a single ***4-vector*** (T, X, Y, Z) in 4-dimensional ***spacetime***.[11] These are the space and time coordinates of event $\mathfrak{E}$ for what we shall call our *first observer*.

Of course the spatial components of the first observer's vector have no absolute significance: if a *second observer* uses the same origin but has coordinate axes rotated relative to the first, then

---

[8]We encourage you not to peek, but, the mathematically precise statement of this result will be given in (6.20).

[9]For a wonderfully physical account of the theory, we recommend Taylor and Wheeler (1992). For more on the *geometry* of the theory, we strongly recommend Misner, Thorne, and Wheeler (1973) and Penrose (2005).

[10]According to Coxeter (1967, pp. 73–77), this connection was first recognized by H. Liebmann almost immediately, in 1905!

[11]In fact it was Hermann Minkowski who in 1908 recast Einstein's 1905 theory in these geometrical terms, and at first Einstein himself did not approve!

this second observer will ascribe different spatial coordinates $(\widetilde{X}, \widetilde{Y}, \widetilde{Z})$ to the same event $\mathfrak{E}$. Yet, if these two observers are *not in relative motion*, they will nevertheless agree on the value of $\widetilde{X}^2 + \widetilde{Y}^2 + \widetilde{Z}^2 = X^2 + Y^2 + Z^2$, for this represents the square of the distance to the point where the event happened.

In contrast to this, we are accustomed to thinking that the *time* component T *does* have an absolute significance. However, Einstein's theory—confirmed by innumerable experiments—tells us that this is wrong. *If our two (momentarily coincident) observers* are *in relative motion, they will disagree about the times at which events occur.* Furthermore, they will no longer agree about the value of $(X^2 + Y^2 + Z^2)$—this is the famous ***Lorentz contraction***.

Such effects are only discernible if the relative speed of the two observers is a significant fraction of the fantastic speed of light (which is roughly 186,000 miles per second). For example, even if the second observer shoots away from the first observer with the speed of a rifle bullet (2,000 miles per hour), then even over the period of their entire adult lifetimes (say 85 years), the total accumulated discrepancy between their two clocks would only amount to about one hundredth of a second! It is this accident of our snail-like existence relative to the speed of light that hid for thousands of years (and continues to hide, day-to-day) the truth of Einstein's discovery.

But if the second object *is* travelling at an appreciable fraction of the speed of light, the temporal and spatial distortion is dramatic, and this effect is witnessed daily in particle accelerators around the world. Under these strange circumstances, is there *any* aspect of spacetime that has absolute significance and upon which the two observers in uniform relative motion must agree?

Einstein's amazing answer is "Yes!": spacetime *does* possess an *absolute* structure that is independent of all observers. For this reason, Einstein personally disliked, and resisted for years, the name "Theory of Relativity"—it should be "Theory of the Absolute"!

Making a convenient choice of units in which the speed of light is equal to 1, Einstein[12] discovered that both observers will agree on the value of the ***spacetime interval*** $\gimel$ between the observer and the event. This interval[13] $\gimel$ is defined by means of its *square*:

$$\boxed{\gimel^2 \equiv T^2 - (X^2 + Y^2 + Z^2) = \widetilde{T}^2 - (\widetilde{X}^2 + \widetilde{Y}^2 + \widetilde{Z}^2).} \tag{6.15}$$

As Minkowski realized, this is the correct generalization of the concept of distance appropriate to spacetime, and the isometries/symmetries of spacetime must preserve it. But, quite unlike ordinary distance, here *the square of the interval between distinct events can be zero or even negative.*

Let us provide a more vivid interpretation of $\gimel$ in the case[14] $\gimel^2 > 0$. As we have said, we suppose our two observers (in uniform relative motions to one another) are momentarily coincident at one particular place and time, defining an origin event that we label $\mathfrak{O}$. (Of course for this event to be precisely defined we must imagine our observers to be pointlike, having a specific location.) Now suppose that (by accident or by design) the first observer arrives at the event $\mathfrak{E}$ just as it happens. As far as she is concerned, she has been sitting still the whole time, and both events $\mathfrak{O}$ and $\mathfrak{E}$ happened right where she sits $(X = Y = Z = 0)$, and therefore the spacetime interval between $\mathfrak{O}$ and $\mathfrak{E}$ (upon which *all* observers must agree) is simply $\gimel = T$:

---

[12]Just as we have previously asked Poincaré to step aside slightly, to make room for Beltrami in connection with $\mathbb{H}^2$, so now must we ask Einstein to step aside slightly for Poincaré: already in 1905 Poincaré, too, had discovered the invariance of the spacetime interval. Of course Hendrik Lorentz should be considered the third father of Special Relativity, but at least he is immortalized in the group of transformations that bear his name.

[13]Surprisingly, there is no standard symbol for this interval, but our use of ***gimel*** (the third letter of the Hebrew alphabet) seems appropriate: it resembles (visually) the English "I" of "Interval," and one connotation (culturally) of gimel is of a *bridge* connecting two points.

[14]If $\gimel^2 < 0$ there is a different but equally simple interpretation; see Taylor and Wheeler (1992, Ch. 1).

> *If an observer in uniform motion carries a wristwatch from one event to another, the invariant spacetime interval $\mathfrak{I}$ between the two events is simply the elapsed time on that watch.* (6.16)

Next, imagine that a spark occurs at the event $\mathfrak{O}$, so that photons (particles of light) travel out in all directions from this flash. If both observers focus their attention on a single photon, both will agree that every event along this photon's spacetime trajectory has $\mathfrak{I}=0$. Such a 4-vector of vanishing "length," pointing along a particular light ray, is called a ***null*** vector.

A ***Lorentz transformation*** $\mathfrak{L}$ is a linear transformation of spacetime (a 4×4 matrix) that maps one observer's description (T,X,Y,Z) of an event to another observer's description $(\widetilde{T},\widetilde{X},\widetilde{Y},\widetilde{Z})$ of the same event. Put differently, $\mathfrak{L}$ is a linear transformation that preserves the quantity $\mathfrak{I}^2$, upon which both observers must agree.

Now let's return to the imagined flash of light—an origin-centred sphere whose radius increases at the speed of light. After time $T=1$ these photons will form a sphere of unit radius, which we now choose to identify with the *Riemann sphere*. Thus this sphere is now thought of as made up of points that are simultaneously labelled with spacetime coordinates $(1,X,Y,Z)$ *and* with complex numbers, assigned stereographically via (4.25) on page 48.

Substituting the projective-coordinate description $z=(\mathfrak{z}_1/\mathfrak{z}_2)$ into the stereographic formulas (4.25), we obtain,

$$X=\frac{\mathfrak{z}_1\overline{\mathfrak{z}_2}+\mathfrak{z}_2\overline{\mathfrak{z}_1}}{|\mathfrak{z}_1|^2+|\mathfrak{z}_2|^2},\qquad Y=\frac{\mathfrak{z}_1\overline{\mathfrak{z}_2}-\mathfrak{z}_2\overline{\mathfrak{z}_1}}{i\,(|\mathfrak{z}_1|^2+|\mathfrak{z}_2|^2)},\qquad \text{and}\qquad Z=\frac{|\mathfrak{z}_1|^2-|\mathfrak{z}_2|^2}{|\mathfrak{z}_1|^2+|\mathfrak{z}_2|^2}.$$

But a light ray may be identified by *any* null vector along it, so instead of choosing $T=1$ let us now eliminate the denominators in the above expressions by scaling up our null vector by a factor of $(|\mathfrak{z}_1|^2+|\mathfrak{z}_2|^2)$ (i.e., by choosing $T=|\mathfrak{z}_1|^2+|\mathfrak{z}_2|^2$). The new null vector (T,X,Y,Z) (in the same spacetime direction as the original) is therefore given by

$$T=|\mathfrak{z}_1|^2+|\mathfrak{z}_2|^2,\qquad X=\mathfrak{z}_1\overline{\mathfrak{z}_2}+\mathfrak{z}_2\overline{\mathfrak{z}_1},$$

$$Y=-i\,(\mathfrak{z}_1\overline{\mathfrak{z}_2}-\mathfrak{z}_2\overline{\mathfrak{z}_1}),\qquad Z=|\mathfrak{z}_1|^2-|\mathfrak{z}_2|^2.$$

You may readily check [exercise] that these formulas may be inverted to yield

$$\begin{pmatrix} T+Z & X+iY \\ X-iY & T-Z \end{pmatrix}=2\begin{pmatrix} \mathfrak{z}_1\overline{\mathfrak{z}_1} & \mathfrak{z}_1\overline{\mathfrak{z}_2} \\ \mathfrak{z}_2\overline{\mathfrak{z}_1} & \mathfrak{z}_2\overline{\mathfrak{z}_2} \end{pmatrix}=2\begin{pmatrix} \mathfrak{z}_1 \\ \mathfrak{z}_2 \end{pmatrix}\begin{pmatrix} \overline{\mathfrak{z}_1} & \overline{\mathfrak{z}_2} \end{pmatrix},$$

or,

$$\boxed{\begin{pmatrix} T+Z & X+iY \\ X-iY & T-Z \end{pmatrix}=2\begin{pmatrix} \mathfrak{z}_1 \\ \mathfrak{z}_2 \end{pmatrix}\begin{pmatrix} \mathfrak{z}_1 \\ \mathfrak{z}_2 \end{pmatrix}^{*},} \tag{6.17}$$

where "$*$" again denotes the ***conjugate transpose***.

Note that the spacetime interval can now be expressed neatly as the determinant of this matrix:

$$\boxed{\mathfrak{I}^2=T^2-(X^2+Y^2+Z^2)=\det\begin{pmatrix} T+Z & X+iY \\ X-iY & T-Z \end{pmatrix}.} \tag{6.18}$$

This makes it easy to see [exercise] that this scaled-up spacetime vector is still null, as it should be.

Having now used stereographic projection to identify light rays with complex numbers,[15] via (6.17), let us find the effect on the flash of light effected by a Möbius[16] transformation $z \mapsto \widetilde{z} = M(z)$ of $\mathbb{C}$, or equivalently,

$$\begin{bmatrix} \mathfrak{z}_1 \\ \mathfrak{z}_2 \end{bmatrix} \longmapsto \begin{bmatrix} \widetilde{\mathfrak{z}_1} \\ \widetilde{\mathfrak{z}_2} \end{bmatrix} = [M] \begin{bmatrix} \mathfrak{z}_1 \\ \mathfrak{z}_2 \end{bmatrix}.$$

Substituting this into (6.17) we obtain the following *linear transformation* of the null vectors making up the flash of light:

$$\begin{pmatrix} T+Z & X+iY \\ X-iY & T-Z \end{pmatrix} \mapsto \begin{pmatrix} \widetilde{T}+\widetilde{Z} & \widetilde{X}+i\widetilde{Y} \\ \widetilde{X}-i\widetilde{Y} & \widetilde{T}-\widetilde{Z} \end{pmatrix} = [M] \begin{pmatrix} T+Z & X+iY \\ X-iY & T-Z \end{pmatrix} [M]^*. \tag{6.19}$$

Finally, imagine that this linear transformation is applied to *all* spacetime vectors (not just null ones). Because $\det[M] = 1 = \det[M]^*$, it follows from (6.18) that *this linear transformation preserves the spacetime interval*:

$$\widetilde{\mathfrak{I}}^2 = \det\left\{ [M] \begin{pmatrix} T+Z & X+iY \\ X-iY & T-Z \end{pmatrix} [M]^* \right\} = \mathfrak{I}^2.$$

Thus,

Every Möbius transformation of $\mathbb{C}$ yields a unique Lorentz transformation of spacetime. Conversely, it can be shown (Penrose and Rindler, 1984, Ch. 1) that to every Lorentz transformation there corresponds a unique (up to sign) Möbius transformation. (6.20)

Even amongst professional physicists, this beautiful "miracle" is not as well known as it should be.

It turns out that every Möbius or Lorentz transformation is fundamentally equivalent to one of just four archetypes. The essential idea is to focus on the so-called ***fixed points*** of the transformation, meaning points that remain fixed in the sense that they are mapped to themselves: $M(z) = z$.

Clearly [exercise] this leads to a quadratic equation with two solutions that may be distinct or coincide. If the fixed points are distinct, one can imagine[17] them to be the north and south poles; this yields the three archetypes shown in [6.5a–c]. If the two fixed points coincide, one can imagine them both to be the north pole; this yields the fourth archetype shown in [6.5d].

To make these four types of Lorentz transformation more vivid, imagine yourself in a spaceship in interstellar space, looking out in all directions at countless stars scattered over your ***celestial sphere***. This is an imagined unit sphere with you at its centre, a dot being marked on its surface wherever the starlight intercepts the sphere on its way to your eye. We suppose the sphere to be aligned with your spaceship, the north pole lying directly ahead of you.

[15] Exercise 33 explains Penrose's method of accomplishing this directly in spacetime.

[16] In the context of relativity theory, a Möbius transformation is called a ***spin transformation***, and the corresponding matrix $[M]$ is called a ***spin matrix***. See Penrose and Rindler (1984, Ch. 1) for details.

[17] For a detailed justification of this simplification in the case of Möbius transformations, see VCA, Ch. 3; in the case of Lorentz transformation, see Penrose and Rindler (1984, Ch. 1).

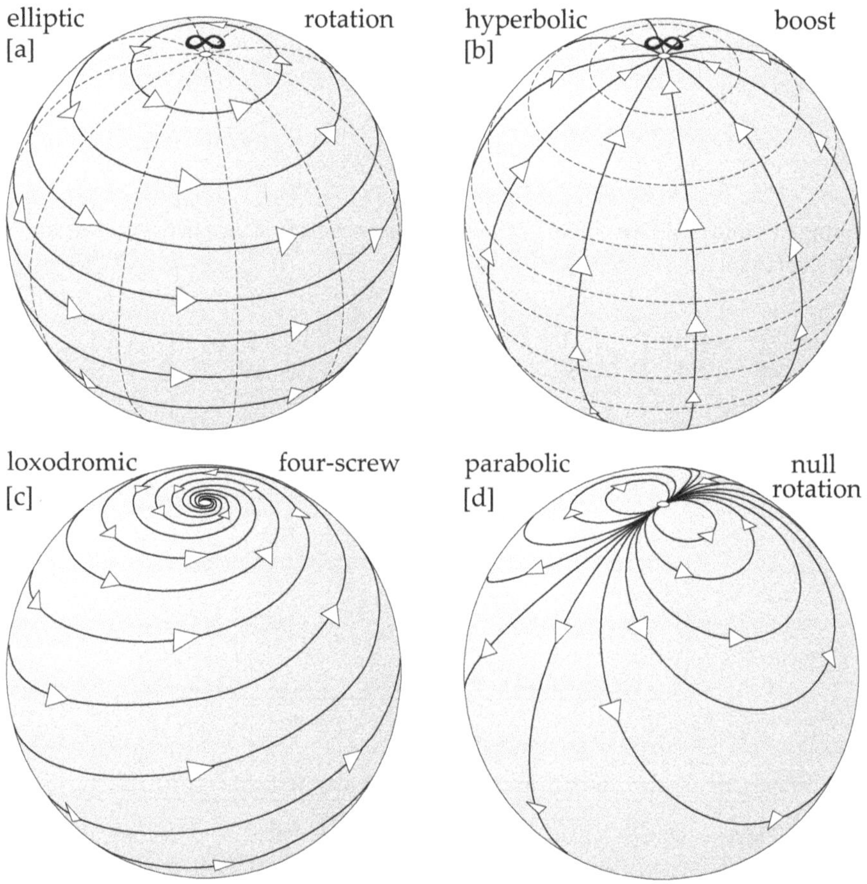

[6.5] Classification of Möbius and Lorentz Transformations. *Each of the four types of transformation has two names, depending on whether it is viewed as acting on* $\mathbb{C}$ *(name on the left) or on spacetime (name on the right).*

If you now fire your lateral thrusters so as to set your spaceship in a spin about its north–south axis, then the star field will rotate around you, as illustrated in [6.5a]. Unsurprisingly, this type of Lorentz transformation is called a ***rotation***. In the complex plane the corresponding Möbius transformation is also a rotation, $M(z) = e^{i\alpha}z$, and this type is called ***elliptic***.

Suppose you are no longer spinning and that you now fire your powerful main engines. Almost instantly this sends your craft hurtling at nearly light speed v directly ahead toward the north pole of your celestial sphere. As illustrated in [6.5b], you will in fact see the stars crowd *toward* your direction of travel—the exact opposite of the depiction in most science-fiction films! This type of Lorentz transformation is called a ***boost***. In the complex plane the corresponding Möbius transformation is a simple expansion by some real factor ρ, $M(z) = \rho z$, and this type is called ***hyperbolic***. In fact (see Ex. 34) the expansion factor ρ is related to the spacecraft velocity v by

$$\rho = \sqrt{\frac{1+v}{1-v}}. \tag{6.21}$$

If you fire your main engines *and* your lateral thrusters then the previous two effects combine to cause the star field to spiral toward your direction of motion, as shown in [6.5c]. This type of Lorentz transformation is called a ***four-screw***, and the corresponding Möbius transformation, $M(z) = \rho e^{i\alpha}z$, is called ***loxodromic***.

The last type of Lorentz transformation is shown in [6.5d]. It is called a ***null rotation*** and both of its fixed points are coincident at the north pole. While it is hard to give a vivid physical description of this spacetime transformation, the corresponding Möbius transformation is called ***parabolic*** and is a simple translation of $\mathbb{C}$: $M(z) = z + \tau$. This moves every complex number along a line parallel to τ. But, as we know from [4.8], page 45, these lines all stereographically project to circles through the north pole with a common tangent there that is parallel to τ, thereby explaining the form of [6.5d]. Note that movements on the sphere become smaller and smaller as the north pole is approached, leaving this as the only fixed point.

6.5 Three-Dimensional Hyperbolic Geometry

Remarkably, neither Lobachevsky nor Bolyai set out to develop a 2-dimensional non-Euclidean Geometry. Instead, from the outset, both men independently sought a hyperbolic alternative to *three*-dimensional[18] Euclidean space; hyperbolic planes then emerged naturally within this hyperbolic 3-space.

Recall that we created a conformal map of the hyperbolic plane in the upper-half-plane by taking the metric to be

$$d\widehat{s} = \frac{ds}{\text{Euclidean height above the boundary}} = \frac{\sqrt{dx^2 + dy^2}}{y}. \tag{6.22}$$

To generalize this to a *three*-dimensional hyperbolic space, $\mathbb{H}^3$, consider the half-space $Z > 0$ above the horizontal (X, Y)-plane of Euclidean 3-space (X, Y, Z). We may now create a conformal model of $\mathbb{H}^3$ by again taking the metric to be[19]

$$d\widehat{s} = \frac{ds}{\text{Euclidean height above the boundary}} = \frac{\sqrt{dX^2 + dY^2 + dZ^2}}{Z}. \tag{6.23}$$

Points on the (X, Y)-plane $(Z = 0)$ are therefore infinitely far away, and this bounding plane is the two-dimensional horizon, which we will now think of as the complex plane $\mathbb{C}$, with coordinates $X + iY$.

It follows immediately from this construction that every vertical half-plane is a copy of $\mathbb{H}^2$. See [6.6] and compare to [5.5]. (By extension, imagine this space filled with small spheres of equal hyperbolic size, shrinking in apparent Euclidean size as the horizon is approached.)

If we imagine light travelling within such a vertical plane, symmetry dictates that it *remain* in that plane, and our previous analysis of the path taken by the light therefore applies. Thus,

The geodesics of $\mathbb{H}^3$ are the vertical half-lines and the semicircles orthogonal to the horizon, $\mathbb{C}$.

In $\mathbb{H}^2$ the vertical half-lines are exceptional geodesics, the typical ones being the semicircles orthogonal to the horizon. Likewise, in $\mathbb{H}^3$ the vertical half-planes are the exceptional hyperbolic planes; the typical planes are in fact hemispheres orthogonal to the horizon $\mathbb{C}$, such as the one shown in [6.6].

---

[18] See Stillwell (2010, p. 365).

[19] This, too, was published by Beltrami in 1868, along with a generalization to n dimensions. See Stillwell (1996).

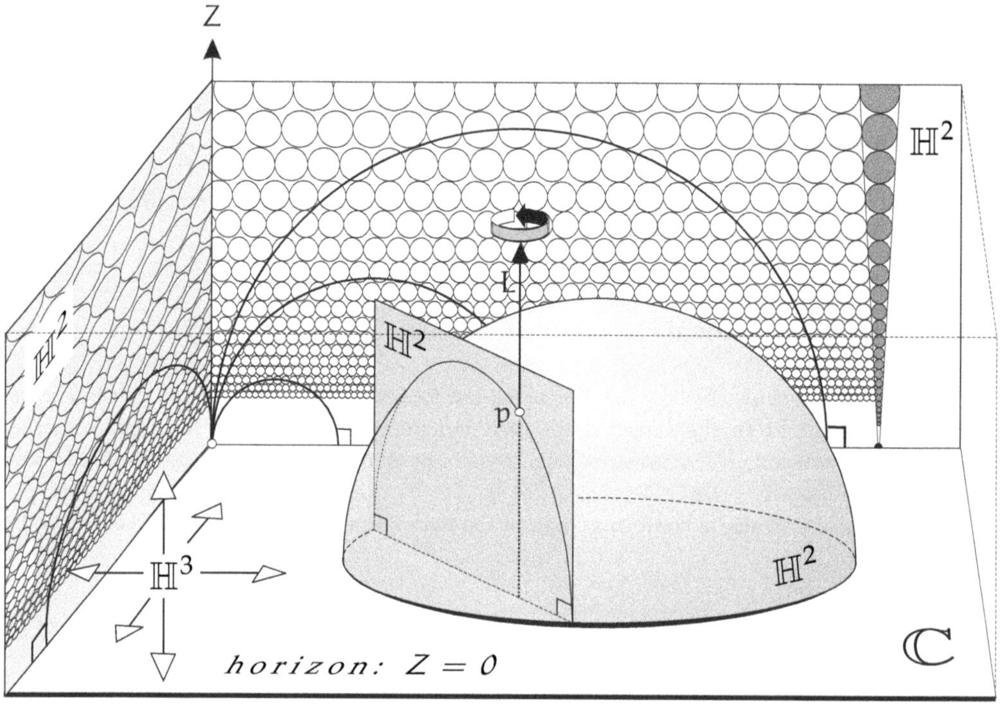

[6.6] Hyperbolic 3-Space, $\mathbb{H}^3$. *Hyperbolic planes appear as vertical half-planes and hemispheres orthogonal to* $\mathbb{C}$. *The intersection of two such planes is a hyperbolic line: a vertical half-line, or a semicircle orthogonal to* $\mathbb{C}$.

To begin to see this, imagine that you live in $\mathbb{H}^3$; what would you mean by the word "plane"? Suppose I hand you an infinitesimal disc D centred at p. To extend this disc (uniquely) into an infinite "plane" H, you presumably need to prolong its diameters into infinite hyperbolic lines, i.e., semicircles meeting $\mathbb{C}$ orthogonally. If the disc is vertical this clearly yields the vertical half-plane containing D. If the disc D is *not* vertical, we claim that the construction will instead generate a hemisphere H orthogonal to $\mathbb{C}$, with D coincident with the tangent plane to H at p. Incidentally, we may easily construct the centre of this hemisphere as the intersection with $\mathbb{C}$ of the (Euclidean) normal to D through p.

In [6.6] let us take the illustrated hemisphere to be this H. Consider the illustrated intersection of H with a vertical plane through p. Clearly this is a hyperbolic line extending a diameter of D, since we have *constructed* H to have its tangent plane coincident with D at p. Now let L be the illustrated vertical line through p. If we rotate the vertical plane about L, this hyperbolic line of intersection sweeps out all of H, so this is indeed the unique extension of D into a plane. Thus,

> *The hyperbolic planes of* $\mathbb{H}^3$ *are the vertical half-planes and the hemispheres orthogonal to the horizon,* $\mathbb{C}$.

It is clear that a vertical half-plane is a hyperbolic plane in two senses: it is what an inhabitant of $\mathbb{H}^3$ calls a plane, *and* its internal geometry is that of $\mathbb{H}^2$. Is the same true of the hemispherical planes? That is, if we measure distance within such a hemisphere using the hyperbolic metric (6.23) of the ambient space $\mathbb{H}^3$, do we obtain an $\mathbb{H}^2$?

The answer is "Yes": each hemisphere is intrinsically an $\mathbb{H}^2$. Indeed if you live in $\mathbb{H}^3$ then every direction is like every other direction, every line is like every other line, and every plane is like every other plane. By generalizing geometric inversion to 3-dimensional space, it is possible to provide a natural geometric explanation, by showing that there exist motions that carry a vertical hyperbolic plane into a hemispherical one, so their intrinsic geometries must be the same.[20] For now, though, we shall make do with a calculation.

If we use spherical polar coordinates (longitude and latitude) on the hemisphere of radius R, then $Z = R\cos\phi$, so (4.4) implies that the hyperbolic metric (6.23) reduces to

$$d\widehat{s}^2 = \frac{\sin^2\phi\, d\theta^2 + d\phi^2}{\cos^2\phi}. \tag{6.24}$$

To see that this really is $\mathbb{H}^2$, we can either introduce new coordinates that transform this metric into the standard form (6.22) (see Ex. 28) or else we can use the intrinsic metric curvature formula (4.10) to confirm [exercise] that $\mathcal{K} = -1$.

Let us describe a couple of simple isometries of $\mathbb{H}^3$. The metric (6.23) is clearly unaltered by a horizontal Euclidean translation,

$$X + iY \longmapsto X + iY + (\text{complex constant}) \quad \text{and} \quad Z \mapsto Z.$$

Likewise, hyperbolic distance is unaltered by an origin-centred Euclidean dilation (expansion) of space,

$$(X, Y, Z) \longmapsto k(X, Y, Z) = (kX, kY, kZ),$$

where $k > 0$ is the expansion factor, for this will scale up both ds and Z by the same factor k, leaving $d\widehat{s}$ unchanged.

Combining this with the translations, we see that any dilation centred on the horizon is also an isometry. The existence of these two kinds of isometries again confirms that all hemispherical planes must have the same intrinsic geometry, because any one of them can be rigidly moved to any other: translate the first so that its centre coincides with the centre of the second, then expand it till the radii are equal.

Any isometry must carry planes to planes, which means that their boundary circles in $\mathbb{C}$ must also be carried to other circles. It turns out (though we shall not be able to prove it here) that this circle-preserving property is so special that it alone suffices to completely determine[21] the complex mappings involved: *they can only be Möbius transformations!*

Conversely, *any Möbius transformation of $\mathbb{C}$ induces a unique transformation of all $\mathbb{H}^3$*, because a point p in $\mathbb{H}^3$ is uniquely determined as the intersection of three planes, which can be encoded as the three circles in which these hemispheres meet $\mathbb{C}$. The images of these three circles then determine three new hemispheres, meeting in the image of p. Furthermore, it can be shown[22] that this induced transformation is automatically a direct isometry of $\mathbb{H}^3$. To summarize this wonderful discovery of Poincaré (1883),

> *The group of direct isometries of hyperbolic 3-space, $\mathbb{H}^3$, is the Möbius group of transformations of $\mathbb{C}$.*

If you grew up in this $\mathbb{H}^3$-world, and attended $\mathbb{H}^3$-school, you would learn that angles in triangles always add up to less than two right angles, that there are infinitely many lines parallel

---

[20] For details, see VCA, Sections 3.2.6, 6.3.12.

[21] This remarkable fact was proved, in its most general form, by Carathéodory (1937).

[22] To see how Poincaré made this discovery, see Stillwell (1996, pp. 113–122). For more on the geometry underlying the result, see VCA, Section 6.3.12.

to a given line, etc. But at university you might eventually learn of a theoretical geometry that defies all your everyday experience: the angles in a triangle sum to exactly π, no matter how large the triangle! In an attempt to make intuitive sense of such bizarre phenomena, mathematicians seek to construct a model surface on which this "Euclidean" geometry actually holds true. And, remarkably, they succeed!

Returning from parable to reality, it was actually Wachter (a student of Gauss) who first discovered such a surface in 1816, called a ***horosphere***, later so named by Lobachevsky, who (along with Bolyai) independently rediscovered it. In terms of [6.6], a typical horosphere is an ordinary Euclidean sphere that rests on (touches) the horizon. Very surprisingly, if we measure distance using the hyperbolic metric (6.23), geometry within the horosphere does indeed turn out to be that of a flat Euclidean plane! See Exercises 30 and 32.

Beltrami provided a simpler way of seeing this. Using the same three-dimensional generalization of geometric inversion that allows us to swap the vertical half-planes and the hemispheres, he observed that a typical horosphere can be moved so that it becomes $Z = \text{const.} = k$. Within this horizontal surface, the hyperbolic metric (6.23) then reduces to

$$d\widehat{s} = \frac{\sqrt{dX^2 + dY^2}}{k},$$

which is clearly Euclidean.

Recall that Hilbert proved that one cannot construct a full hyperbolic plane in Euclidean space. The existence of horospheres (full Euclidean planes) in hyperbolic space is therefore our first intimation of Hyperbolic Geometry's superiority, Euclidean Geometry becoming subordinate to it. In fact Spherical Geometry is subsumed as well. For it turns out that what an inhabitant of $\mathbb{H}^3$ calls a sphere (i.e., the set of points at fixed hyperbolic distance from the centre) actually appears in our model as a Euclidean sphere, but with a different centre. Furthermore, such a hyperbolic sphere has intrinsic geometry of constant positive curvature: it's as spherical as it looks! See Exercise 31.

We have repeatedly entertained the notion of intelligent creatures living in a non-Euclidean world such as $\mathbb{H}^3$; this is less fanciful than you might suppose. In 1915 Einstein discovered that the actual spacetime that *we* inhabit is *not* flat! In fact energy and matter warp the fabric of spacetime, producing a complicated pattern of curvature, both positive and negative, varying from place to place, time to time, and direction to direction.[23]

Under normal circumstance, however, the amount of curvature is so fantastically small that angle sums of triangles differ from π by an utterly undetectable amount, creating the illusion that our world obeys the laws of Euclidean Geometry. This illusion is so perfect and so compelling that it held sway for 4000 years.

Nevertheless, the subtle pattern of spacetime curvature that Einstein discovered in 1915 is extremely important, even in everyday life, and it has a name . . . it is called *gravity!*

[23]Einstein's supremely beautiful theory is called ***General Relativity***, and it is the subject of Chapter 30.

Chapter 7

Exercises for Act II

Mapping Surfaces: The Metric

1. **Coordinate-independence of $\mathcal{K}$.** Begin with the flat Euclidean metric,

$$d\widehat{s}^2 = dx^2 + dy^2.$$

   (i) If $x = u^2 \cos v$ and $y = u^2 \sin v$, interpret the (u, v) coordinates geometrically, and deduce that they are *orthogonal*.

   (ii) Verify this orthogonality by calculating the metric in the (u, v) coordinates, and apply the curvature formula (4.10) to confirm that this new metric formula is *still* flat, as it should be: $\mathcal{K} = 0$.

   (iii) Changing coordinates again, show that the *orthogonal trajectories* of the hyperbolas $x = u^2 - v^2 = \mathrm{const.}$ are the hyperbolas $y = 2uv = \mathrm{const.}$ Repeat (ii) for these *conformal* coordinates, using (4.10) *or* (4.16) to confirm that $\mathcal{K} = 0$.

2. **Central Projection.** Consider the central projection of the sphere shown in [4.1].

   (i) Prove that an infinitesimal circle on the sphere is distorted into an ellipse whose shape is given by (4.3).

   (ii) Explain why this distortion is symmetrical: if we instead start with an infinitesimal circle in the plane, the image on the sphere is an ellipse of the same shape.

3. **Central Projection.** Rederive the central projection metric (4.2) by calculation.

4. **Central Projection.** Apply the curvature formula (4.10) to the metric (4.2) of the sphere under central projection, and confirm that $\mathcal{K} = +1/R^2$.

5. **Angular Excess of n-gon.** Show that the internal angles of a Euclidean n-gon sum to $(n-2)\pi$, thereby justifying the definition of angular excess as $\mathcal{E} \equiv [\text{Angle Sum}] - (n-2)\pi$. Now take Δ_p to be a small n-gon containing a point p on a curved surface, and divide it into triangles by joining p to the vertices. Deduce that (2.1) applies without change.

6. **Stereographic Metric.** Suppose that in [4.10] we choose $\delta\widehat{s}$ due north (instead of the illustrated choice of due west). Show that this leads to the same metric (4.22), and do so in two ways:

   (i) by calculation;

   (ii) geometrically.

7. **Antipodal Points.** The figure below shows a vertical cross section of the Riemann sphere Σ through $\widehat{p}$ and the antipodal point $\widehat{q}$. Show that the triangles $p0N$ and $N0q$ are similar. Deduce (4.27).

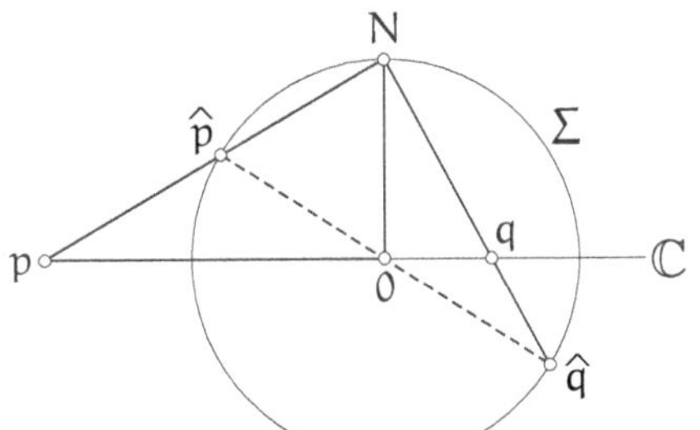

8. Conformal ("Isothermal") Coordinates. Given the metric of a general surface in general (u, v)-coordinates, (4.7),

$$d\hat{s}^2 = A^2\,du^2 + B^2\,dv^2 + 2F\,du\,dv,$$

our aim is to find *conformal* (U, V)-coordinates, such that if $Z = U + iV$, then

$$d\hat{s}^2 = \Lambda^2\,(dU^2 + dV^2) = \Lambda^2\,dZ\,d\overline{Z}.$$

Let $M = \sqrt{A^2B^2 - F^2}$ be the area magnification factor in going from the (u, v)-map to the surface, so that (4.11) becomes $d\mathcal{A} = M\,du\,dv$.

(i) Verify that the metric has the following *complex* factorization:

$$d\hat{s}^2 = \left[A\,du + \frac{(F + iM)}{A}\,dv\right]\left[A\,du + \frac{(F - iM)}{A}\,dv\right].$$

(ii) Suppose that a *complex* integrating factor Ω can be found such that

$$\Omega\left[A\,du + \frac{(F + iM)}{A}\,dv\right] = dU + i\,dV = dZ.$$

Deduce that in this case (U, V) are indeed conformal coordinates on the surface:

$$d\hat{s}^2 = \Lambda^2\,(dU^2 + dV^2) = \Lambda^2\,dZ\,d\overline{Z}, \qquad \text{where} \qquad \Lambda = \frac{1}{|\Omega|}.$$

(iii) Use $df = (\partial_u f)\,du + (\partial_v f)\,dv$ to deduce that

$$\partial_u Z = \Omega A \qquad \text{and} \qquad \partial_v Z = \Omega\left[\frac{F + iM}{A}\right].$$

(iv) Deduce that

$$[F + iM]\,\partial_u Z = A^2\,\partial_v Z.$$

(v) Multiplying both sides of the previous equation by $[F - iM]$, deduce that

$$B^2\,\partial_u Z = [F - iM]\,\partial_v Z.$$

(vi) By equating real and imaginary parts, deduce that the rates of change of U can be expressed in terms of the rates of change of V (and visa versa) as follows:

$$\partial_u U = \frac{1}{M}\left[A^2\,\partial_v V - F\,\partial_u V\right] \qquad \text{and} \qquad \partial_v U = \frac{1}{M}\left[F\,\partial_v V - B^2\,\partial_u V\right];$$

$$\partial_u V = \frac{1}{M}\left[F\,\partial_u U - A^2\,\partial_v U\right] \qquad \text{and} \qquad \partial_v V = \frac{1}{M}\left[B^2\,\partial_u U - F\,\partial_v U\right].$$

(vii) Granted that $\partial_u\partial_v\Phi - \partial_v\partial_u\Phi = 0$, deduce that $\Phi = U$ and $\Phi = V$ are both solutions of the ***Beltrami–Laplace Equation***:

$$\boxed{\partial_v\left[\frac{A^2\,\partial_v\Phi - F\,\partial_u\Phi}{M}\right] + \partial_u\left[\frac{B^2\,\partial_u\Phi - F\,\partial_v\Phi}{M}\right] = 0.}$$

This equation is ***elliptic***, and the general theory of elliptic partial differential equations now guarantees that this equation does indeed have solutions, thereby confirming the existence of conformal coordinates!

(viii) Let (x, y) be a second pair of conformal coordinates, so that

$$d\hat{s}^2 = \lambda^2(dx^2 + dy^2).$$

By setting $u = x$ and $v = y$, deduce that $A = B = \lambda$, $F = 0$, and $M = \lambda^2$. From (vi), deduce that the local mapping between the $(x + iy)$-plane and the $(U + iV)$-plane is an *amplitwist*, characterized by the celebrated ***Cauchy–Riemann equations***:

$$\partial_x U = \partial_y V \qquad \text{and} \qquad \partial_y U = -\partial_x V.$$

(For a complete discussion of these phenomena, see VCA.)

(ix) Deduce that in this case the Beltrami–Laplace Equation reduces to the ***Laplace Equation***:

$$\partial_x^2 \Phi + \partial_y^2 \Phi = 0.$$

9. Conformal Curvature Formula. Apply the conformal curvature formula (4.16) to the metric (4.22) of the sphere under stereographic projection, and confirm that $\mathcal{K} = +1/R^2$.

10. Archimedes–Lambert Projection. In the (x, y)-plane, consider the rectangle $\{0 \leqslant x \leqslant 2\pi R, -R \leqslant y \leqslant R\}$. Now imagine first taping together the left and right edges to create a cylinder, then slipping this over a sphere of radius R so that the cylinder touches the sphere along its equator, as shown in the figure below. We now draw a map of the sphere by projecting its

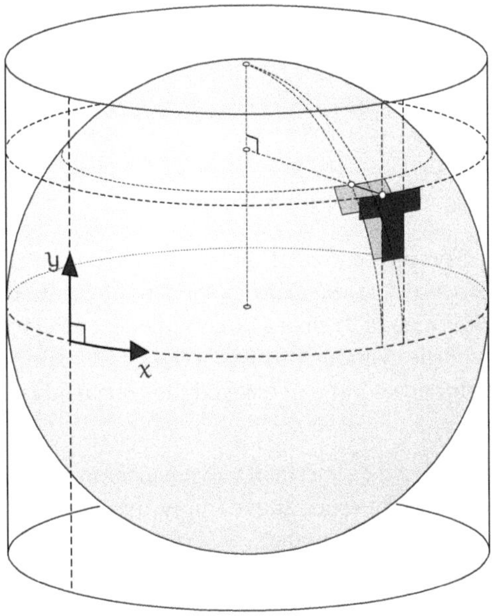

points horizontally onto the cylinder, radially outward, perpendicular to the axis of the cylinder, as shown. This is the ***Archimedes–Lambert projection***, first investigated by Archimedes

(c. 250 BCE), then rediscovered and published by Lambert about 2000 years later, in 1772. Lambert's seminal treatise was the first systematic mathematical investigation of maps, based on which properties they preserved.

(i) Show (ideally geometrically) that the metric of the sphere now takes the form

$$d\widehat{s}^2 = \frac{(R^2 - y^2)}{R^2}\, dx^2 + \frac{R^2}{R^2 - y^2}\, dy^2.$$

(ii) Apply the curvature formula (4.10) to confirm that $\mathcal{K} = +1/R^2$.

(iii) Use (4.12) to deduce that the projection *preserves area*. For example, the two illustrated T-shapes have equal area. As Archimedes realized, this implies that the area of the entire sphere must equal that or the original rectangle: $(2\pi R)(2R) = 4\pi R^2$. Archimedes was so proud of this (and related discoveries on volumes) that he asked his friends to have this particular diagram inscribed on his tomb. Almost a century-and-a-half later, in 75 BCE, Cicero sought and found the tomb, overgrown with bushes, but with the cylinder and sphere still visible.

(iv) Use the previous part to deduce (without integration) that the area $\mathcal{A}$ of the polar cap $0 \leqslant \phi \leqslant \Phi$ is

$$\mathcal{A} = 2\pi R^2 (1 - \cos\Phi).$$

11. **Central Cylindrical Projection.** Reconsider the sphere and cylinder in the diagram of the previous question, but now imagine that the cylinder extends infinitely upward and downward. We again project the sphere onto the cylinder, but this time by imagining a point source of light at the centre of the sphere, casting shadows onto the cylinder. This is called ***central cylindrical projection***.

(i) Sketch the projection and deduce that $y = R\cot\phi$.

(ii) Either using part (i), or directly geometrically, show that the metric now takes the form

$$d\widehat{s}^2 = \frac{R^2}{R^2 + y^2}\, dx^2 + \frac{R^4}{(R^2 + y^2)^2}\, dy^2.$$

(iii) Apply the curvature formula (4.10) to confirm that $\mathcal{K} = +1/R^2$.

12. **Mercator Projection.** The projections of the sphere onto the cylinder in the previous two exercises share two properties: 1) meridians ($\theta = \text{const.}$) are mapped to vertical generators of the cylinder with the same θ; 2) circles of latitude ($\phi = \text{const.}$) are mapped to horizontal circular cross sections of the cylinder. In 1569 Gerardus Mercator discovered a third projection that shared these properties but that enjoyed the crucial advantage over the other two of being *conformal*.

(i) Argue geometrically that if $\delta\widehat{s}$ is a small movement along a circle of latitude on the sphere then the two required properties above imply that $\delta\widehat{s} \asymp \sin\phi\, \delta s$, where $\delta s = \delta x$ is the horizontal movement on the cylinder.

(ii) For our map to be conformal we must insist that we have the same scale factor for infinitesimal movements in *all* directions. In particular, suppose $\delta\widehat{s}$ is instead chosen along a meridian, and let $y = f(\phi)$ be the height on the cylinder of the image of (θ, ϕ) on the sphere. Deduce that

$$f'(\phi) = -\frac{R}{\sin\phi}.$$

(iii) If we insist that points on the equator map to the x-axis, verify (or deduce, if you remember your integration techniques) that the ***Mercator projection*** is given by

$$y = f(\phi) = R \ln \left[\frac{1 + \cos \phi}{\sin \phi} \right].$$

(Note: Neither logarithms nor calculus were known in Mercator's time, so *how did he do it?!* See the excellent chapter by Osserman in Hayes et al. (2004, Ch. 18).)

(iv) If you fly a plane or sail a ship on a fixed compass heading, you travel along a ***loxodrome*** (illustrated in [6.5c]), also called a ***rhumb line***. If you unroll the cylinder onto a tabletop to obtain a standard flat Mercator navigational chart, how does your loxodromic route appear in the chart? (*Hint:* No calculation required: just use conformality!)

13. An *Impossibly* Good Map. In the last few exercises we have met a map of the sphere that preserves area, and another that preserves angles. Can we have a map that preserves both? Show that no such map can exist for *any* curved surface, not just the sphere.

The Pseudosphere and the Hyperbolic Plane

14. Pólya's Mechanical Proof of Snell's Law. Pólya (1954, pp. 149–152) provided the following ingenious *mechanical* explanation of Snell's Law, and hence yet another way for us to think about the geodesics of the hyperbolic plane. The figure below is a modified form of [5.6], in which the boundary between air and water is replaced by a frictionless rod, along which slides a ring, p. Attached to this ring are two strings of fixed length. As illustrated, each string passes over a frictionless peg at a_i and is attached to a mass m_i that hangs a distance h_i below a_i.

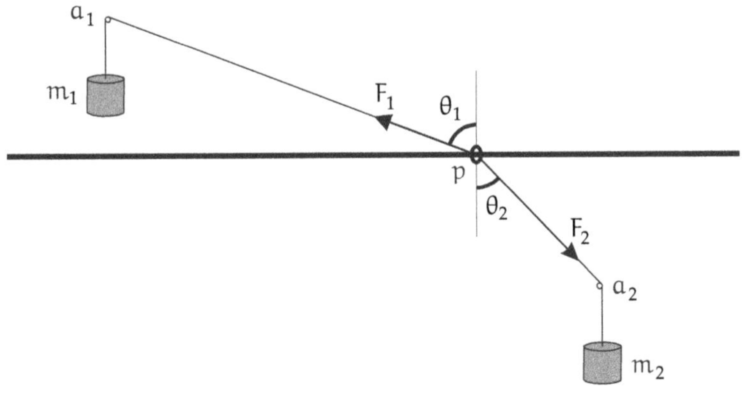

(i) By considering the potential energy of the system, deduce that equilibrium is achieved when

$$m_1 h_1 + m_2 h_2 = \text{maximum}.$$

(ii) But for equilibrium to occur, the horizontal components of the forces F_i pulling on the ring must cancel. Deduce that

$$m_1 \sin \theta_1 = m_2 \sin \theta_2.$$

(iii) Let l_i be the length of string from a_i to p, so that $l_i + h_i = \text{const.}$. Now let us choose $m_i = (1/v_i)$, where v_1 and v_2 represent the speed of light in air and in water, respectively, as in the original optical problem. Deduce that the mechanical problem in part (i) becomes mathematically identical to the original problem of finding the optical path of minimum time:

$$\frac{l_1}{v_1} + \frac{l_2}{v_2} = \text{minimum}.$$

(iv) From the solution of the mechanical problem in part (ii), deduce Snell's Law!

15. If the velocity of light in the (x, y)-plane is $v = 1/\sqrt{1-y}$, show that the geodesics are parabolas. Describe these parabolas.

16. **Tractrix Parameterization.** In (X, Y, Z)-space let us switch to ***cylindrical polar coordinates***, i.e., ordinary (r, θ) polar coordinates in the (X, Y)-plane, supplemented by Z. The Euclidean metric in this space is therefore

$$ds^2 = dr^2 + r^2 d\theta^2 + dZ^2.$$

In the plane $\theta = 0$, i.e., the (X, Z)-plane, consider the curve traced by a particle whose position at time t is

$$X = r = \operatorname{sech} t, \qquad Z = t - \tanh t.$$

(i) Show that this curve is the tractrix: the distance along the tangent to the Z-axis is fixed. What is this fixed distance? (Note: some interesting history and beautiful geometry underlies this particular representation of the tractrix; see Stillwell (2010, pp. 341–342).)

(ii) Deduce that if we rotate the curve about the Z-axis, we obtain the pseudosphere of unit radius. Show that the surface's metric is

$$d\widehat{s}^2 = \tanh^2 t\, dt^2 + \operatorname{sech}^2 t\, d\theta^2.$$

(iii) Use (4.10) to confirm that $\mathcal{K} = -1$.

(iv) Show that if we now introduce new coordinates $x = \theta$ and $y = \cosh t$ then the metric assumes the standard form (5.7).

17. **Cylindrical-Polar Area On Sphere.** If we employ the cylindrical polar coordinates of the previous exercise, the element of area on the northern hemisphere of the unit sphere is,

$$d\mathcal{A} = \frac{r\, dr\, d\theta}{\sqrt{1-r^2}}.$$

(i) Prove this by calculation.

(ii) Prove this geometrically, using Newtonian ultimate equalities.

(iii) Integrate this to obtain the area of the complete hemisphere.

18. **The Pseudosphere Has Finite Area!** Use (5.5) and (4.12) to show that the infinite pseudosphere of radius R has *finite* area $2\pi R^2$. This was discovered by Huygens in 1693.

19. **Build Your Own Pseudosphere!** Take a stack of ten sheets of paper and staple them together, placing staples along three of the edges. Use a pair of compasses (or inverted plate or bowl) to draw the largest circle that will fit comfortably inside the top sheet. Pierce through all ten sheets in the centre of the circle. With heavy scissors, cut along the circle to obtain ten identical discs, say of radius R. Repeat this whole process to double the number of discs to 20.

(i) Cut a narrow sector out of the first disc, and tape the edges together to form a shallow cone. Repeat this process with the remaining discs, steadily increasing the angle of the sector each time, so that the cones get sharper and taller. (You may find it easier to cut a radial slit and *overlap* the paper before taping it together.) Ensure that by the end of the process you are making very narrow cones, using only a quarter disc or less.

(ii) Stack these cones in the order you made them. Explain how it is that *you have created a model of a portion of a pseudosphere of radius* R.

(iii) Use the same idea to create a disc-like piece of "hyperbolic paper," such as you would get if you could simply cut out a disc from your pseudosphere. Press it against the pseudosphere and verify that you can freely move it about and rotate it on the surface. (For detailed instructions, see Henderson (2013, p. 32).)

20. Geodesics of the Pseudosphere. Using (1.7), construct a segment of a typical geodesic on the surface of the toy pseudosphere you built in the previous exercise. Extend this segment in both directions, one strip of tape at a time. Note the surprising way the geodesic only spirals a finite distance up the pseudosphere before spiralling down again.

(i) Use the Beltrami–Poincaré upper half-plane to verify mathematically that the tractrix generators are the *only* geodesics that extend indefinitely upward.

(ii) Let L be a typical geodesic, and let α be the angle between L and the tractrix generator at the point where L hits the rim $\sigma = 0$. Show that the maximum distance σ_{max} that L travels up the pseudosphere is given by $\sigma_{max} = |\ln \sin \alpha|$.

21. Conformal Curvature Formula. Show that in the case of a conformal mapping with metric (4.14) the general curvature formula (4.10) reduces to (4.16).

22. Surfaces of Revolution of Constant Curvature. Imagine a particle travelling along a curve in the (x, y)-plane at unit speed, and let its position at time t be $[x(t), y(t)]$. Now imagine rotating this plane through angle θ about the x-axis. As θ varies from 0 to 2π, the curve sweeps out a surface of revolution.

(i) Explain why $\dot{x}^2 + \dot{y}^2 = 1$, where the dot represents the time derivative.

(ii) Show geometrically that the metric of the surface is $d\hat{s}^2 = dt^2 + y^2\, d\theta^2$.

(iii) Deduce from that (4.10) that $\mathcal{K} = -\ddot{y}/y$.

(iv) Find the general solution $y(t)$ in each of the three cases of constant curvature: $\mathcal{K} = 0$, $\mathcal{K} = +1$, and $\mathcal{K} = -1$.

(v) With the assistance of part (i), calculate the velocity of the particle in each of the three cases.

(vi) Sketch some solution curves and resulting surfaces of revolution in each of the three cases; a computer may prove helpful. In particular, in the case $\mathcal{K} = +1$, confirm that one does indeed obtain surfaces like those shown in [2.5]. Likewise, verify that in the case $\mathcal{K} = -1$ one can obtain not only the pseudosphere itself, but also surfaces that look like two pseudospheres stuck together at their narrow necks.

23. Curvature of the Torus. Generate a torus T by rotating a circle C with centre o and radius r about a line l in the plane of C that is at distance $R(>r)$ from o. See the figure below.

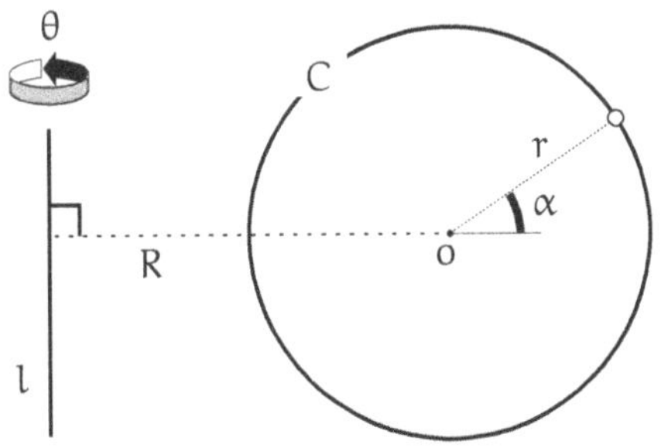

(i) If α is the illustrated angle going round the circle, and θ is the illustrated angle of rotation about the axis of symmetry l, show geometrically that the metric of T is

$$d\widehat{s}^2 = r^2 d\alpha^2 + (R + r\cos\alpha)^2 d\theta^2.$$

(ii) Use (4.10) to deduce that the curvature of the torus is

$$\mathcal{K} = \frac{1}{r(r + R\sec\alpha)}.$$

(iii) Sketch the graph of $\sec\alpha$, then of $(r + R\sec\alpha)$, and finally of $\mathcal{K}$ itself. Thereby confirm that this formula for $\mathcal{K}$ agrees with the empirical findings of Fig. [2.2].

(iv) Where on T would you expect to find $\mathcal{K} = 0$? Confirm this with the formula.

(v) We see from the formula that $\lim_{R\to\infty} \mathcal{K} = 0$. Explain this geometrically.

(vi) Use (4.12) to find the total area of T.

(vii) Check that your formula for the area agrees with the prediction of ***Pappus's Centroid Theorem***. (If you have not studied this theorem previously, first look it up in a book or on the internet.)

(viii) Show that the total curvature of the torus *vanishes*:

$$\oiint_T \mathcal{K}\, d\mathcal{A} = 0.$$

In Act III we shall see that this is no accident!

Isometries and Complex Numbers

24. Amplitwist of z^m. Generalize the geometric argument in [4.6] to deduce that the amplitwist of z^m is given by the same formula as in real calculus: $(z^m)' = m\, z^{m-1}$; but now we can *see* why it's true!

25. Metric of the Beltrami–Poincaré Disc. We know from [6.4] that Möbius transformations can map half-planes to discs, and, as we shall be able to explain later, it turns out that a specific Möbius transformation from the upper half-plane model of $\mathbb{H}^2$ [5.12] to the disc model of $\mathbb{H}^2$ [5.11] is

$$z \mapsto w = D(z) = \frac{iz+1}{z+i}.$$

To derive the metric (5.15) of the disc model, consider an infinitesimal vector dz emanating from z (in the half-plane) being amplitwisted to an infinitesimal vector $dw = D'(z)\, dz$ emanating from w (in the disc). By definition, the hyperbolic length $d\widehat{s}$ of dw is the hyperbolic length of dz. Verify (5.15) by showing that

$$\frac{2\,|dw|}{1-|w|^2} = \frac{|dz|}{\operatorname{Im} z} = d\widehat{s}.$$

26. Curvature of the Beltrami–Poincaré Disc. Use the conformal curvature formula (4.16) to verify that the Beltrami–Poincaré disc [5.11] with metric (5.15) has constant negative curvature $\mathcal{K} = -1$.

27. Möbius Rotations of the Riemann Sphere. Our aim here is to provide a constructive, semi-geometrical proof that *every rotation of the Riemann sphere is a Möbius transformation of the type (6.10)*. Our approach coincides in its essentials with the one previously published by Wilson (2008, pp. 42–44). (Note: the corresponding group of matrices is written $SU(2)$: the "S" stands for "special," meaning normalized, with unit determinant; the "U" stands for "unitary"; and the "2" stands for "2 × 2" matrices.) In the following, let R_Z^θ represent a clockwise rotation of the Riemann sphere through angle θ about the positive Z-axis of (X, Y, Z)-space, so that $R_Z^\theta(z) = e^{i\theta} z$.

(i) Find a matrix $[R_Z^\theta]$ in $SU(2)$ representing R_Z^θ.

(ii) Explain why the result (6.6) can be written

$$[R_X^\pi] = \pm \begin{bmatrix} 0 & i \\ i & 0 \end{bmatrix} \in SU(2).$$

(iii) Referring to (4.24), use a sketch to show that the complex mapping corresponding to $R_X^{(\pi/2)}$ is

$$z = \frac{X + iY}{1 - Z} \longmapsto R_X^{(\pi/2)}(z) = \frac{X - iZ}{1 - Y}.$$

(iv) To confirm that $R_X^{(\pi/2)}(z)$ is in fact a *Möbius* transformation, let us suppose for now that it is, and let us try to guess its form. Use a sketch to confirm that $-i \mapsto 0$, so the numerator must be proportional to $(z + i)$. Likewise, by finding the point that rotates to the north pole, deduce the denominator. This determines the form of $R_X^{(\pi/2)}(z)$ up to a multiplicative factor; determine this factor by noting that $0 \mapsto i$. Conclude that *if* $R_X^{(\pi/2)}(z)$ is a Möbius transformation, it can only be

$$R_X^{(\pi/2)}(z) = \frac{z + i}{iz + 1}.$$

(v) Show that this *is* in fact $R_X^{(\pi/2)}(z)$ by demonstrating that it satisfies the equation in part (iii). (*Hints:* Multiply top and bottom by $(X - iZ)$, but only multiply through in the denominator. Finally, in the resulting denominator, use the fact that $X^2 + Y^2 + Z^2 = 1$.)

(vi) Deduce that $R_X^{(\pi/2)}$ can be represented in $SU(2)$ as

$$\left[R_X^{(\pi/2)}\right] = \frac{1}{\sqrt{2}} \begin{bmatrix} 1 & i \\ i & 1 \end{bmatrix}.$$

(vii) Explain geometrically why $\left[R_X^{(\pi/2)}\right]^2 = [R_X^\pi]$, and confirm this by direct calculation.

(viii) Explain geometrically why for any angle α,

$$[R_Y^\alpha] = \left[R_X^{-(\pi/2)}\right] [R_Z^\alpha] \left[R_X^{(\pi/2)}\right],$$

and deduce that $[R_Y^\alpha] \in SU(2)$.

(ix) Finally, consider a general rotation about an arbitrary axis A. By rotating A into the (X, Y)-plane and then rotating it to the Y-axis, deduce that this too can be represented as a matrix in $SU(2)$, thereby concluding the proof.

28. Metric of $\mathbb{H}^2$ within $\mathbb{H}^3$. Let us explore different ways of expressing the metric (6.24) of a typical hyperbolic planes in $\mathbb{H}^3$, namely, a horizon-centred hemisphere.

(i) Show that if we define $u \equiv \tan\phi$, the metric (6.24) becomes

$$d\widehat{s}^2 = u^2\, d\theta^2 + \frac{du^2}{1 + u^2},$$

and use (4.10) to confirm that $\mathcal{K} = -1$.

(ii) Show that if we define a new variable ξ via $u \equiv 1/\sinh\xi$, the metric of the previous part becomes *conformal*:

$$d\hat{s}^2 = \frac{d\theta^2 + d\xi^2}{\sinh^2\xi},$$

and use (4.16) to confirm that $\mathcal{K} = -1$.

(iii) Finally, find a conformal mapping of the (θ, ξ)-plane to the (x, y)-plane such that the conformal hyperbolic metric in part (ii) takes the standard form (5.7):

$$d\hat{s}^2 = \frac{dx^2 + dy^2}{y^2}.$$

(Hint: Set $dy/y = d\xi/\sinh\xi$.)

29. **Curvature of Hemispherical $\mathbb{H}^2$ in $\mathbb{H}^3$.** Use the curvature formula (4.10) to verify that the horizon-centred hemispheres of $\mathbb{H}^3$ with metric (6.24) are indeed $\mathbb{H}^2$'s with constant negative curvature $\mathcal{K} = -1$.

30. **Horosphere: Metric and Curvature.** As discussed, a typical horosphere of $\mathbb{H}^3$ appears as a sphere resting on $\mathbb{C}$. Using the derivation of (6.24) as your inspiration, show that the metric of such a horosphere is

$$d\hat{s}^2 = \frac{\sin^2\phi\, d\theta^2 + d\phi^2}{(1 + \cos\phi)^2}.$$

Apply the curvature formula (4.10) to confirm that the horosphere is intrinsically a flat Euclidean plane: $\mathcal{K} = 0$.

31. **Spheres in $\mathbb{H}^3$.** Consider a Euclidean sphere of radius R centred at height kR above the horizon of $\mathbb{H}^3$. Using the derivation of (6.24) as your inspiration, show that the metric of such a sphere is

$$d\hat{s}^2 = \frac{\sin^2\phi\, d\theta^2 + d\phi^2}{(k + \cos\phi)^2}.$$

Apply the curvature formula (4.10) to confirm the claim in the text that if this sphere lies entirely above the horizon (i.e., $k > 1$) then intrinsically this surface is genuinely spherical, with constant positive curvature, $\mathcal{K} = k^2 - 1$. (Note: the results of the previous two exercises follow as the special cases $k = 0$ and $k = 1$, respectively.)

32. **Horosphere Metric.** Let us derive the result of Exercise 30 in another way. Define a coordinate r such that

$$dr = \frac{d\phi}{1 + \cos\phi}.$$

Show that the metric of the horosphere (see Ex. 30) can then be written as $d\hat{s}^2 = dr^2 + r^2\, d\theta^2$, which is just the metric of the Euclidean plane in polar coordinates.

33. **Penrose's Direct Labelling of Light Rays with Complex Numbers.** In the text we used the Riemann sphere to label light rays with complex numbers, via stereographic projection. Sir Roger Penrose (see Penrose and Rindler (1984, Vol. 1, p. 13)) discovered a remarkable alternative method of passing from a light ray to the associated complex number *directly*. Let the point p be one unit vertically above the origin of the (horizontal) complex plane. Now imagine that, simultaneously, p emits a flash, and $\mathbb{C}$ begins to travel straight up (in the direction $\phi = 0$) at the speed of light ($= 1$) towards p. (You may imagine that the whole of $\mathbb{C}$ flashed, creating a plane wave.) Decompose the velocity of the photon F emitted by p in the direction (θ, ϕ) into components perpendicular and parallel to $\mathbb{C}$. Hence find the time at which F hits $\mathbb{C}$. Deduce that F hits $\mathbb{C}$ at the point $z = \cot(\phi/2)\, e^{i\theta}$. Thus, *Penrose's construction is equivalent to stereographic projection!*

34. Einstein's Aberration Formula. Recall from Special Relativity that a "boost" v along the Z-axis yields the Lorentz transformation formulas,

$$\widetilde{T} = \frac{T + vZ}{\sqrt{1 - v^2}}, \qquad \widetilde{X} = X, \qquad \widetilde{Y} = Y, \qquad \widetilde{Z} = \frac{Z + vT}{\sqrt{1 - v^2}}.$$

(i) Show that this transformation can be rewritten as

$$\widetilde{T} + \widetilde{Z} = \rho(T + Z), \qquad \widetilde{X} = X, \qquad \widetilde{Y} = Y, \qquad \widetilde{T} - \widetilde{Z} = (1/\rho)(T - Z),$$

where

$$\rho = \sqrt{\frac{1 + v}{1 - v}}.$$

(ii) Deduce from (6.19) that this boost can be represented by the spin transformation,

$$\begin{bmatrix} \widetilde{\mathfrak{z}_1} \\ \widetilde{\mathfrak{z}_2} \end{bmatrix} = \begin{bmatrix} \sqrt{\rho} & 0 \\ 0 & 1/\sqrt{\rho} \end{bmatrix} \begin{bmatrix} \mathfrak{z}_1 \\ \mathfrak{z}_2 \end{bmatrix}.$$

(iii) Thereby confirm the claim made in [6.5b] and (6.21).

(iv) If (θ, ϕ) is the apparent direction of a star before you fire your spaceship engines, deduce that after you have boosted to speed v in the direction $\phi = 0$, the new direction $(\theta, \widetilde{\phi})$ of the star is given by

$$\cot \frac{\widetilde{\phi}}{2} = \rho \cot \frac{\phi}{2}.$$

This is a more elegant and memorable form of the standard ***aberration formula***, which was discovered by Einstein in 1905.

(v) Deduce that the star field appears to crowd together towards the north pole, i.e., *towards your direction of travel.*

(vi) What happens to the star field as the speed of your spacecraft approaches the speed of light?(!)

ACT III

Curvature

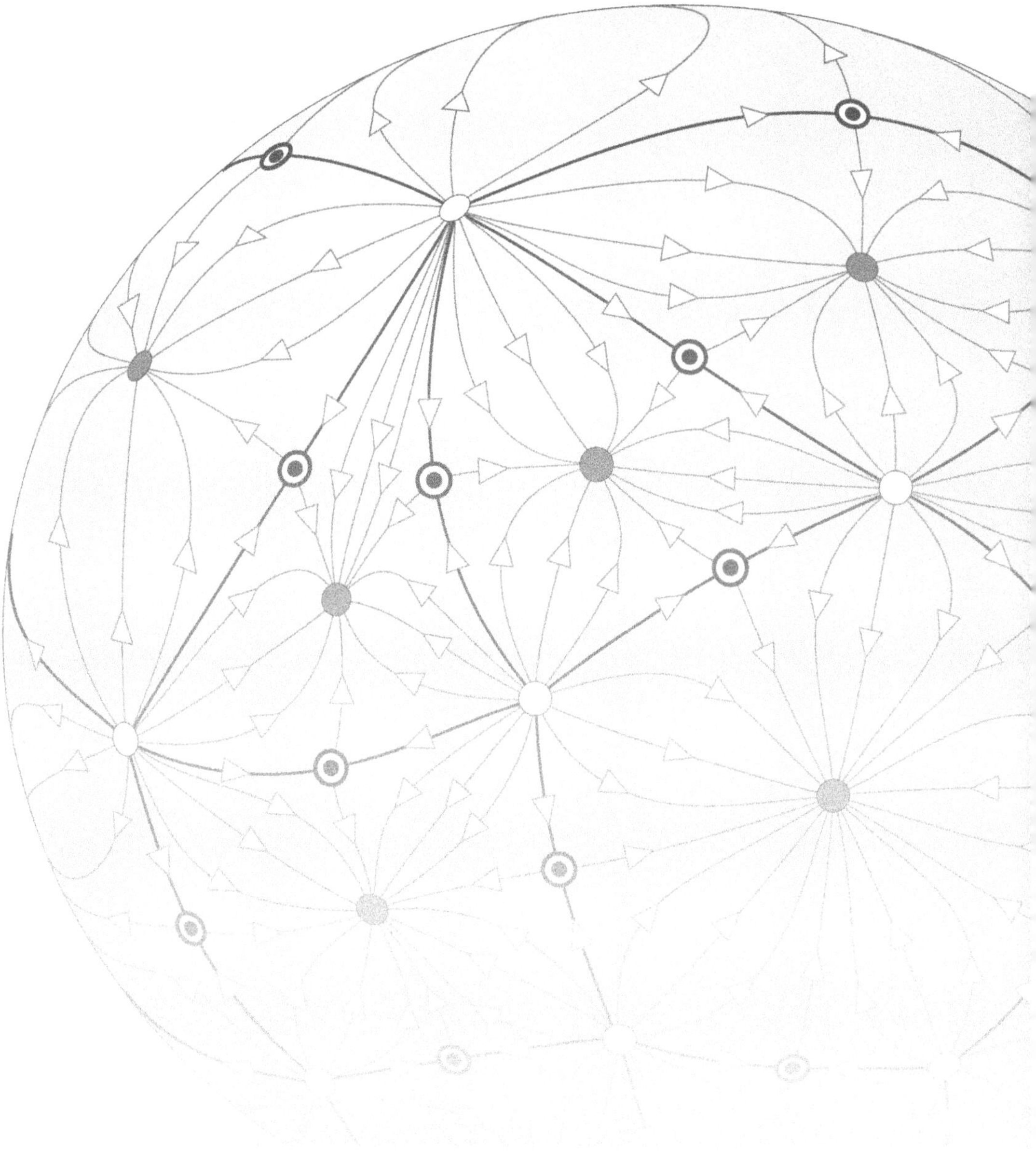

Chapter 8

Curvature of Plane Curves

8.1 Introduction

The study of curvature did not begin with surfaces. Rather, as the very name suggests, curvature was born out of the study of *curves*, specifically, curves *in the plane*.

We certainly have an intuitive idea of a plane curve being straighter in one part and more curved in another, but how can we quantify this degree of curving with mathematical precision?

Throughout the first Act we focussed on the concept of *intrinsic* (Gaussian) curvature. And at the close of Act II we provided another compelling justification for this obsession with intrinsic (versus extrinsic) geometry: it is the *only* kind of geometry of spacetime that is meaningful to us as inhabitants of spacetime, and spacetime's intrinsic curvature has a fundamental importance—it is gravity.

But whereas a patch of given 2-dimensional surface could only be deformed in very limited ways while preserving its intrinsic geometry (i.e., without stretching distances within it) the same is *not* true of a 1-dimensional curve: we can bend it (without stretching it) so as to take up the form of *any* other curve of the same length. Thus the very concept of 1-dimensional intrinsic curvature simply cannot exist! Put differently, if you were a very short 1-dimensional being living within a curve, only able to measure distance back and forth along the curve as you moved along it, the *shape* of your curve would not merely be unknowable to you, it would be *meaningless*.

Thus, from the outset, it is clear that the best we can hope for is an *extrinsic* definition of curvature: how the curve sits within the plane. The big surprise—"miracle" would be a better word—is that this extrinsic concept of 1-dimensional curvature will ultimately be seen to have a direct bearing on the *intrinsic* curvature of 2-dimensional surfaces.

As we have said, a 1-dimensional being within a curve cannot sense the curvature of his world by measurements within it. But suppose we relax our notion of what is measurable and knowable for such a being. Instead of only permitting measurements of length along the curve, suppose we admit physical concepts such as mass, velocity, and force. For now, let us continue to restrict ourselves to a *plane* curve. Picture this plane curve as a frictionless wire located in outer space, with no gravitational (or other) forces at work, and suppose we launch a frictionless bead of unit mass at unit speed along the wire.[1]

By virtue of the absence of external forces, the bead will continue to travel at unit speed throughout its journey along the wire. But Newton's First Law of Motion tells us that if *no* forces acted on the bead, it would simply travel in a straight line. What is happening here is that as the bead attempts to fly in this straight line, the wire presses on it, at right angles to the direction of motion, rotating the velocity vector so as to make it continue pointing along the wire, but without changing the length of the velocity, which is the speed.

The more tightly curved a section of wire, the quicker the velocity vector must turn as it passes through it. Newton's Second Law of Motion tells us that this rate of change of the velocity vector (acceleration) is in fact equal (for a unit mass) to the force F the curved wire exerts on the

[1] Of course while this still essentially concerns a 1-dimensional curve in a 2-dimensional plane, we are now viewing all this as embedded in our physical 3-dimensional world. Indeed, we need at least three dimensions for the bead to encircle and cling to the wire.

bead. The sharper a bend in the wire, the greater the force the wire must exert to make the bead's trajectory bend.

To make this idea more visceral, recall what it's like to travel at high (constant) speed on a ride along curved track at an amusement park or fairground. You have become effectively a 1-dimensional being, carried along the path of the ride. (We will come to curves that twist through 3-space soon, but, for now, imagine you are travelling along a *horizontal* (therefore planar) stretch of track.) Even with your eyes shut, you can still *feel* the force as you go round a bend, the sides of the car pressing against your body: the more tightly curved the bend, the greater the force.

Newton was in fact the first to introduce a purely geometric definition of curvature κ, which we shall meet in a moment, and as a consequence of his definition and his laws of motion, he was indeed able to deduce that

> *If a bead of unit mass is launched at unit speed along a wire in the form of a plane curve, the* curvature κ *of that curve is the force* F*, directed perpendicularly to the curve, exerted by the wire on the bead.* (8.1)

This connection between geometry and physics (between the geometric bending of an orbit and the force required to hold the object in that orbit) became the linchpin of Newton's *magnum opus*, arguably the single most important scientific work in history, the great *Principia* of 1687.

In it, Newton used infinitesimal geometry—the "ultimate equalities" described in the Prologue—to explain the workings of the heavens, including the elliptical orbits of the planets about the Sun as the primary focus. Only now the planets are not threaded onto wires, rather their orbits are deflected from rectilinear motion by the invisible hand of gravity, reaching out across space from the Sun, diminishing in strength in proportion to the *square* of the distance: Newton's famous ***Inverse-Square Law of Gravitation***.

8.2 The Circle of Curvature

The *Principia* lay more than two decades in the future when (shortly before his Christmas Day birthday[2] in 1664) the 21-year-old Newton began to investigate what he called the "crookednesse" of plane curves, thereby introducing the concept of curvature into mathematics for the first time.

Newton identified the ***circle of curvature*** at a point p on a curve $\mathcal{C}$ as the circle that best approximates the curve in the immediate vicinity of p, just as the tangent is the *line* that does this best (see [8.1]).

He constructed the centre c (the ***centre of curvature***) of this approximating circle as the limiting position of the intersection of the normal at p with the normal at a neighbouring point q, in the limit $q \to p$. Then pc is called the ***radius of curvature***, and $\kappa \equiv (1/pc)$ is what Newton initially dubbed the "crookednesse," but later rechristened as the ***curvature***. (IMPORTANT NOTE ON NOTATION: Here "pc" denotes the ***distance between*** the points p and c.)

Next, following Newton, we measure κ by looking at how fast the curve departs from its own tangent line (see [8.2]). By definition of the curvature κ at p, the illustrated diameter has length $ps = (2/\kappa)$. Now let q be a point on $\mathcal{C}$ near to p (where $\xi = pq$) and drop a perpendicular of length $qt = \sigma$ from q to the tangent $\mathcal{T}$ at p, and finally let $\epsilon = pt$.

Since $\mathcal{T}$ is tangent to $\mathcal{C}$, $\lim_{\epsilon \to 0}(\sigma/\epsilon) = 0$, and therefore

$$\frac{\xi^2}{\epsilon^2} = \frac{\epsilon^2 + \sigma^2}{\epsilon^2} = 1 + \left[\frac{\sigma}{\epsilon}\right]^2 \asymp 1 \quad \Longrightarrow \quad \xi \asymp \epsilon.$$

[2] In my household we (or at least *I*) refer to this event as "Newtonmas"!

IMPORTANT NOTE ON NOTATION: Here, and throughout the book, we shall employ the $\asymp$ notation, representing Newton's concept of ***ultimate equality***, as introduced and defined in the Prologue.

Also, the shaded triangle ptq is ultimately similar to the triangle sqp, so

$$\frac{\xi}{[2/\kappa]} \asymp \frac{\sigma}{\xi}.$$

This is essentially Newton's Lemma II, from Book I of the *Principia* (Newton 1687 p. 439) (see also Brackenridge and Nauenberg (2002, p. 112)). Depending on whether it is κ or σ that needs to found in terms of the other, we may combine the previous two results to deduce that

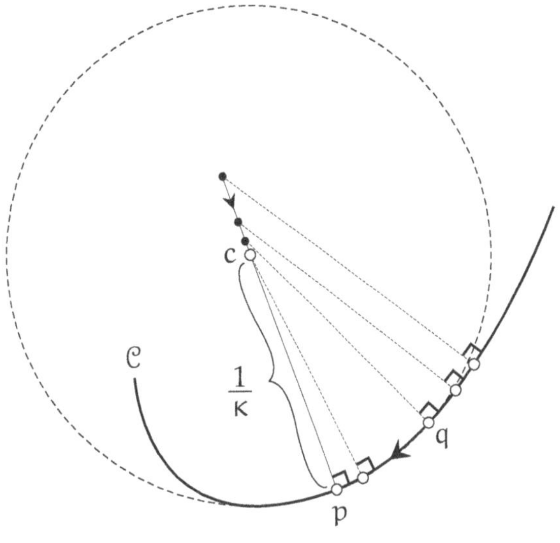

[8.1] *The* **circle of curvature** *at p is the circle that best approximates the curve there. The* **curvature** *κ is defined to be the reciprocal of its radius.*

$$\kappa \asymp \frac{2\sigma}{\epsilon^2} \quad \text{or} \quad \sigma \asymp \frac{1}{2}\kappa\epsilon^2. \tag{8.2}$$

It is convenient to attach a *sign* to the curvature: κ is *positive* if $\mathcal{C}$ is concave *up*, and negative if it is concave down. For example, it follows immediately that the parabola with Cartesian equation $y = ax^2$ has curvature $\kappa = 2a$ at the origin. Likewise, the Taylor expansion of $\cos x$ tells us [exercise] that the graph $y = \cos x$ has $\kappa = -1$ at the point $(0, 1)$, and the circle of curvature is therefore the unit circle, centred at the origin.

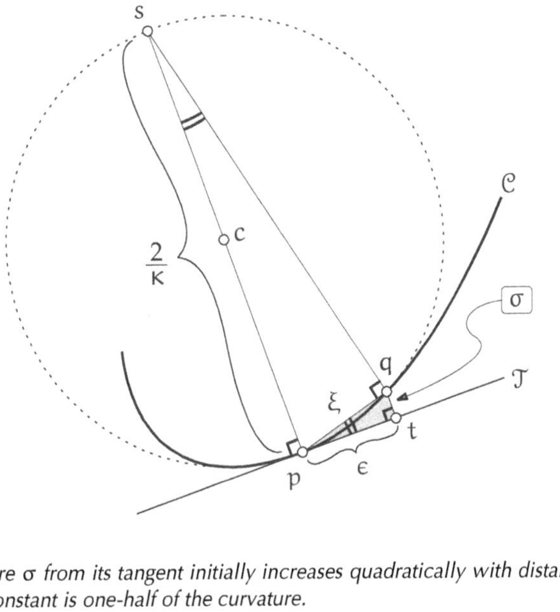

[8.2] *The curve's departure σ from its tangent initially increases quadratically with distance ϵ (like a parabola), and the proportionality constant is one-half of the curvature.*

8.3 Newton's Curvature Formula

Suppose that the curve $\mathcal{C}$ is the graph $y = f(x)$. If the x-axis is drawn parallel to the tangent line $\mathcal{T}$, then Taylor's Theorem implies [exercise] that $\sigma \asymp (1/2)f''(x_p)\epsilon^2$, where x_p denotes the x-coordinate of p. Thus (8.2) implies

$$\boxed{\kappa = f''(x_p).}$$

Note that this formula automatically complies with our convention regarding the sign of κ.

More generally, suppose that the x-axis is now drawn in an arbitrary direction, so that $\mathcal{T}$ is inclined at angle φ to it, so that $f' = \tan\varphi$. In this general case, Newton discovered that the correct generalization of the above formula is,

$$\boxed{\kappa = \frac{f''}{\left\{1 + [f']^2\right\}^{3/2}}.} \tag{8.3}$$

Here it is understood that the derivatives are evaluated at $x = x_p$.

Having now been exposed to several examples of the use of infinitesimal geometry, you should have little difficulty in following Newton's original proof of his formula, which is reproduced in Knoebel (2007, pp. 182–185). Here, however, we shall provide a different geometric argument that seeks to explain the complicated new denominator of (8.3) as a simple consequence of rotating the x-axis. See [8.3].

As before, here $\epsilon \equiv pt$, and $\sigma \equiv qt$, and we now suppose the x-axis to be drawn horizontally, instead of along the tangent $\mathcal{T}$. Taylor's Theorem now says that if $\delta x \equiv x_q - x_p$ then the height of q above $\mathcal{T}$ is given by $aq \asymp (1/2)f''(\delta x)^2$. Because the two shaded triangles are evidently similar, when aq is projected into the direction perpendicular to $\mathcal{T}$, we pick up one factor of $\cos\varphi$ and obtain $\sigma \asymp (1/2)f''(\delta x)^2 \cos\varphi$.

Next, $pa \asymp pt = \epsilon$, and therefore $\delta x \asymp \epsilon\cos\varphi$. Thus in projecting from the tangent onto the horizontal x-axis we pick up two additional factors of $\cos\varphi$, yielding three in all:

$$\sigma \asymp \tfrac{1}{2}f''\epsilon^2\cos^3\varphi.$$

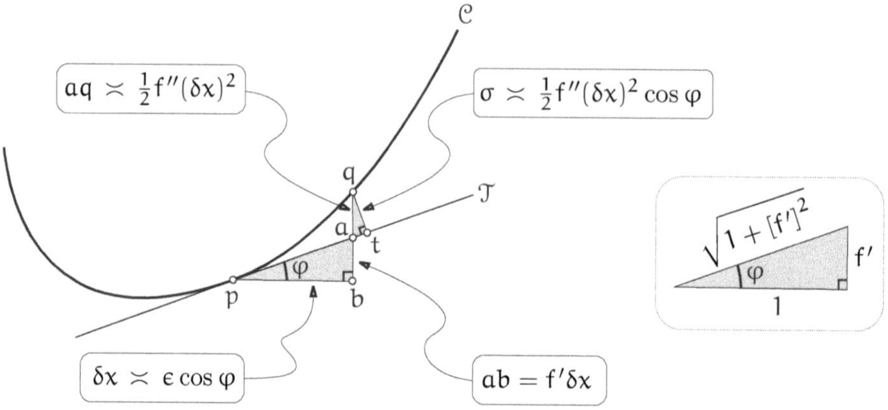

[8.3] *The simple formula $\kappa = f''$ (in the case $\varphi = 0$) must be multiplied by $\cos^3\varphi$ in the general case; as the box on the right shows, $\cos\varphi = 1/\sqrt{1+[f']^2}$.*

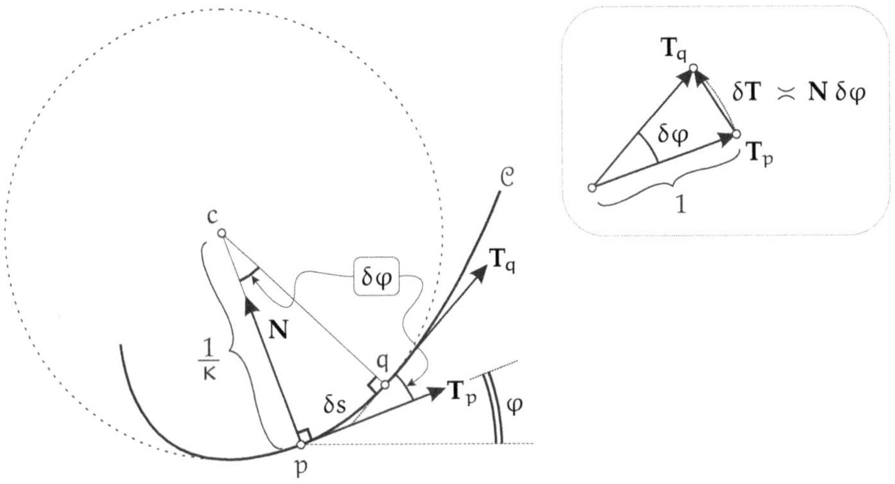

[8.4] *Curvature is the rate of turning of the tangent. If the unit tangent* $\mathbf{T}$ *turns through angle* $\delta\varphi$ *in moving distance* δs *along* $\mathcal{C}$*, then* $\kappa \asymp (\delta\varphi/\delta s)$*. Vectorially, in the box on the right, the rate of change of the unit tangent is* $(d\mathbf{T}/ds) \asymp (\delta\mathbf{T}/\delta s) \asymp (\mathbf{N}\,\delta\varphi/\delta s) \asymp \kappa\mathbf{N}$.

Finally, the boxed subfigure on the right of [8.3] demonstrates that

$$\cos\varphi = \frac{1}{\sqrt{1+[f']^2}},$$

and so Newton's formula (8.3) now follows immediately from (8.2).

8.4 Curvature as Rate of Turning

In 1761, about a century after Newton had introduced the concept of curvature, Kaestner published a simple alternative interpretation that would ultimately prove to be more amenable to generalization than Newton's.

> *Curvature is the rate of turning of the tangent with respect to arc length. In other words, if* φ *is the angle of the tangent, and* s *is arc length, then* $\kappa = (d\varphi/ds)$. (8.4)

Note that this has the immediate advantage of being determined by local measurements conducted in the immediate vicinity of a tiny piece of curve: we no longer need to draw normals and follow them off into the distance until they intersect at the centre of curvature.

To confirm that (8.4) is equivalent to Newton's definition, it suffices to verify this for a perfect circle of radius ρ and curvature $\kappa = (1/\rho)$, for this circle may then be taken to be the circle of curvature in the general case. Since the rate of turning is clearly uniform in the case of a circle, consider one complete revolution: the tangent rotates 2π while traversing the full circumference of $2\pi\rho$, and therefore the rate of turning is $(2\pi)/(2\pi\rho) = \kappa$, as was to be shown.

Fig. [8.4] reproves this equivalence, and also takes us a step further. Here $\mathbf{T}$ denotes the unit tangent vector to $\mathcal{C}$, and $\mathbf{N}$ denotes the unit normal, pointing toward the centre of curvature, c. As we move a small distance δs from p to q, $\mathbf{T}$ rotates by $\delta\varphi$. Since δs is ultimately equal to the corresponding segment of the circle of curvature of radius $(1/\kappa)$ (according to Newton's original definition), $\delta s \asymp (1/\kappa)\delta\varphi$. Therefore $\kappa = (d\varphi/ds)$, reconfirming (8.4).

But now, in addition, the subfigure on the right of [8.4] allows us to determine the change $\delta\mathbf{T} \equiv (\mathbf{T}_q - \mathbf{T}_p)$ in the vector $\mathbf{T}$ itself as we pass from p to q. By drawing both unit tangents emanating from the same point, we may visualize $\delta\mathbf{T}$ as the illustrated vector connecting their tips, which will ultimately point along $\mathbf{N}$. But the length of this vector is ultimately equal to the arc of the unit circle (shown dotted) connecting these tips, so $\delta\mathbf{T} \asymp \mathbf{N}\,\delta\varphi$. Thus,

$$\frac{\delta\mathbf{T}}{\delta s} \asymp \mathbf{N}\frac{\delta\varphi}{\delta s} \quad\Longrightarrow\quad \frac{d\mathbf{T}}{ds} = \kappa\,\mathbf{N}. \tag{8.5}$$

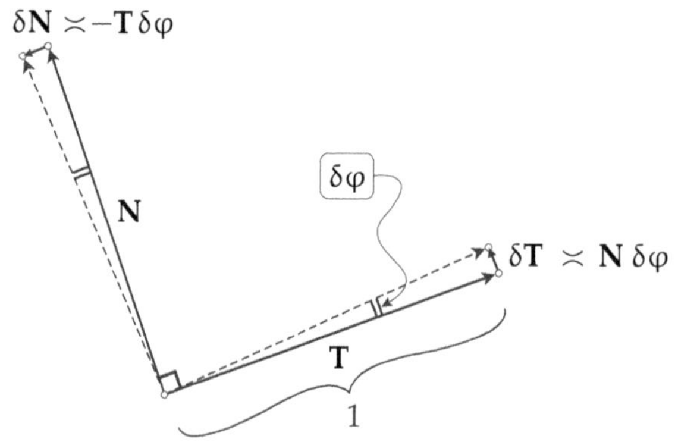

[8.5] *As* $\mathbf{T}$ *and* $\mathbf{N}$ *rotate together, their tips begin to move along* $\mathbf{N}$ *and* $-\mathbf{T}$*, respectively, and by equal amounts.*

We can now return to our opening discussion of curvature in terms of a bead of unit mass travelling at unit speed along the curve $\mathcal{C}$. Because it travels at unit speed, the velocity vector $\mathbf{v}$ of the bead is simply $\mathbf{T}$, and also $\delta s = \delta t$. Therefore in (8.5) $(d\mathbf{T}/ds) = (d\mathbf{v}/dt)$ is in fact the *acceleration* of the bead, which in turn is the force exerted by the wire. We have thus confirmed the original claim (8.1).

Of course we may just as easily think of curvature as the rate of turning of the normal $\mathbf{N}$, instead of the tangent. Indeed [8.5] shows simply and vividly that just as the tip of $\mathbf{T}$ begins to rotate in the direction of $\mathbf{N}$, so the tip of $\mathbf{N}$ begins to rotate in the direction of $-\mathbf{T}$, and by an equal amount, so

$$\frac{d\mathbf{N}}{ds} = -\kappa\,\mathbf{T}.$$

Indeed, this use of the normal in place of the tangent will prove to be of crucial importance when we redirect our attention from curves back to surfaces, for in the latter case there is no such thing as "the" tangent, but there is still a unique *normal* vector, perpendicular to the tangent plane of the surface. How this normal vector varies in the immediate vicinity of a point on the surface will indeed tell us the Gaussian curvature at that point.

(As an aside, we note once again the distinction between geometric understanding and blind calculation, which is trivially simple in the case above. Let φ denote the angle that the tangent makes with the horizontal x-axis, so that

$$\mathbf{T} = \begin{bmatrix} \cos\varphi \\ \sin\varphi \end{bmatrix} \quad \text{and} \quad \mathbf{N} = \begin{bmatrix} -\sin\varphi \\ \cos\varphi \end{bmatrix}.$$

Let a prime denote differentiation with respect to arc length, so that $\kappa = \varphi'$. Then calculation immediately yields $\mathbf{T}' = \varphi'\,\mathbf{N}$ and $\mathbf{N}' = -\varphi'\,\mathbf{T}$.)

What if the bead is instead travelling at an arbitrary (but constant) speed v and has arbitrary mass m? In that case, by definition, $\delta s \asymp v\,\delta t$, and $\mathbf{v} = v\mathbf{T}$. Thus (8.5) becomes the following generalized version of the famous result for the force needed to hold an object in a circular orbit:

$$\mathbf{F} = m\frac{d\mathbf{v}}{dt} = \kappa m v^2\,\mathbf{N}.$$

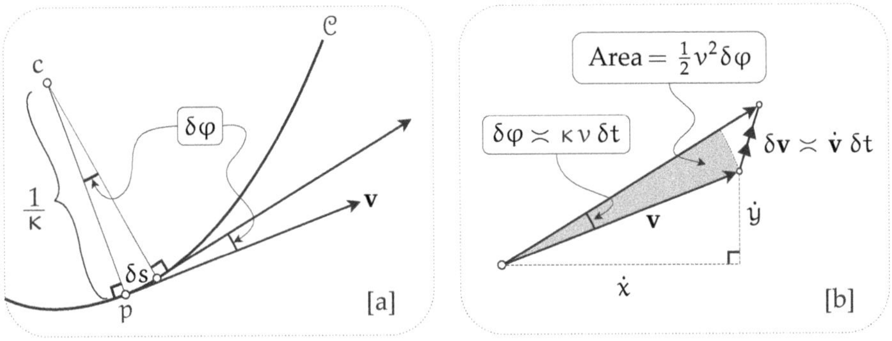

[8.6] [a] *The velocity* **v** *at* p*, and a moment* δt *later, after travelling* $\delta s \asymp v\,\delta t$ *along* $\mathcal{C}$. **[b]** *Both velocity vectors are drawn emanating from the same point, so the vector connecting their tips is* $\delta\mathbf{v} \asymp \dot{\mathbf{v}}\,\delta t$. *Looking at the area of the triangle in two different ways yields formula (8.6) for* κ.

Newton first began to glimpse this as early as 1665,[3] as he struggled to understand the force that held the moon in its orbit.

This new interpretation of curvature (as rate of turning) allows us to deal with curves that cannot be expressed as the graph of a function, and which are therefore beyond the reach of Newton's original formula (8.3). Let $\mathcal{C}$ be the orbit of a particle whose position at time t is $(x[t], y[t])$, so that the velocity is

$$\mathbf{v} = \begin{bmatrix} \dot{x} \\ \dot{y} \end{bmatrix}, \quad \text{and therefore} \quad \tan\varphi = \frac{\dot{y}}{\dot{x}},$$

where, following Newton, the dot[4] denotes differentiation with respect to time. Note that we are no longer assuming that the speed is constant: the length of **v** may vary.

By differentiating both sides of the previous equation, $\tan\varphi = (\dot{y}/\dot{x})$, and using the chain rule, it is not hard to obtain [exercise] a more general formula (also discovered by Newton):

$$\kappa = \frac{\dot{x}\,\ddot{y} - \dot{y}\,\ddot{x}}{\left[\dot{x}^2 + \dot{y}^2\right]^{3/2}}. \tag{8.6}$$

However, let us try to understand this formula in more direct, geometric terms. See [8.6a], which shows the velocity **v** at p and the velocity a moment δt later, after which the particle has travelled $\delta s \asymp v\,\delta t$ along $\mathcal{C}$. In [8.6b], both velocity vectors have been drawn emanating from the same point, so that the vector connecting their tips is the change in velocity, $\delta\mathbf{v} \asymp \dot{\mathbf{v}}\,\delta t$.

Thus the area of the shaded sector of the circle of radius $v = \sqrt{\dot{x}^2 + \dot{y}^2}$, subtending angle δφ, is given by

$$(\text{area of shaded sector}) = \tfrac{1}{2}v^2\delta\varphi \asymp \tfrac{1}{2}v^2\kappa\,\delta s \asymp \tfrac{1}{2}v^3\kappa\,\delta t.$$

But this area is ultimately equal to the area of the illustrated triangle with edge vectors

$$\mathbf{v} = \begin{bmatrix} \dot{x} \\ \dot{y} \end{bmatrix} \quad \text{and} \quad \delta\mathbf{v} \asymp \begin{bmatrix} \ddot{x} \\ \ddot{y} \end{bmatrix} \delta t,$$

[3]See Westfall (1980, pp. 148–150).

[4]Newton's *perfectly* minimalist notation is sometimes frowned upon, as the dot can be hard to see—in *this* work we have attempted to address that concern by simply *enlarging* the dot!

and by an elementary result (which itself may be proved geometrically [exercise]), this area is half the determinant of the matrix with these two columns:

$$(\text{area of triangle bounded by velocity vectors}) \asymp \tfrac{1}{2}(\dot{x}\ddot{y} - \dot{y}\ddot{x})\,\delta t.$$

Thus,

$$\tfrac{1}{2}v^3\kappa\,\delta t \asymp \tfrac{1}{2}(\dot{x}\,\ddot{y} - \dot{y}\,\ddot{x})\,\delta t,$$

thereby proving (8.6).

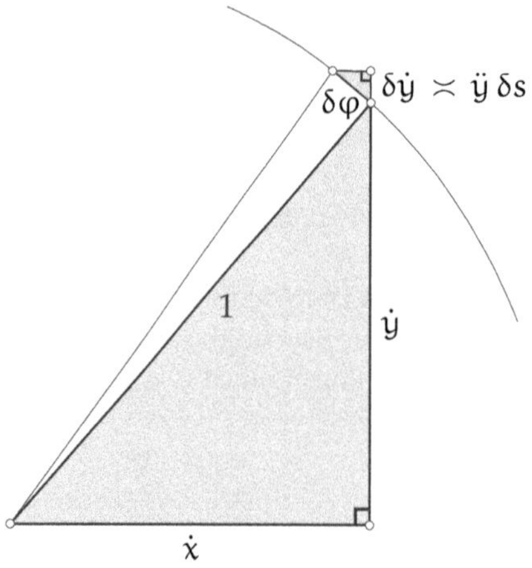

[8.7] *If the particle travels at unit speed then in a moment $\delta t = \delta s$ the tip of* **v** *traces an arc $\delta\varphi$ of the unit circle, and its height increases by $\delta\dot{y} \asymp \ddot{y}\,\delta s$. The ultimate similarity of the two shaded triangles yields (8.7).*

Note that if the particle's horizontal speed is fixed at $\dot{x} = 1$, so that its orbit is $(t, y[t])$, then we recover Newton's original formula (8.3) as a special case of (8.6). Indeed, this is precisely the manner in which Newton himself carried out the simplification; again, see Knoebel (2007, pp. 182–185).

Finally, consider another particularly important special case in which the particle travels at *unit speed*, so that

$$|\mathbf{v}| = \sqrt{\dot{x}^2 + \dot{y}^2} = 1,$$

and $\delta s = \delta t$. Thus the change from the general figure [8.6b] is that $\delta\mathbf{v}$ is now orthogonal to $\mathbf{v}$, tangent to the unit circle. As illustrated in [8.7], during the time δt the tip of $\mathbf{v}$ now traces out an arc $\delta\varphi$ of the unit circle, its height rising by $\delta\dot{y} \asymp \ddot{y}\,\delta t = \ddot{y}\,\delta s$.

From the ultimate similarity of the two shaded triangles, we deduce that

$$\frac{\delta\varphi}{\ddot{y}\,\delta s} \asymp \frac{1}{\dot{x}}.$$

Thus, recalling that $\kappa \asymp (\delta\varphi/\delta s)$, (8.6) reduces to an extremely simple formula which (rather strangely) is not readily found in standard texts: for a *unit-speed* orbit,

$$\boxed{\kappa = \ddot{y}/\dot{x}.} \tag{8.7}$$

Likewise, from the same triangles, $\kappa = -\ddot{x}/\dot{y}$. (See Ex. 1 for a less illuminating proof via calculation.)

8.5 Example: Newton's *Tractrix*

The constant negative Gaussian curvature of the pseudosphere can in fact be traced back to the curvature of Newton's tractrix, which generates it.

Given the parametric representation of the tractrix (see Ex. 16 on p. 88), its curvature can be found via a routine calculation [exercise] based on (8.6). However, a geometric analysis of the problem is more elegant *and* furnishes the answer in a form that will prove more useful in studying the pseudosphere itself. Rather than employ any of the general formulas we have developed thus far, we shall instead present an argument[5] that is tailored to this particular curve.

Let

$$\rho_1 = \textit{radius of curvature of the generating tractrix},$$
$$\rho_2 = \textit{the segment } \mathrm{pl} \textit{ of the normal from the tractrix to its axis},$$

as illustrated in [8.8]. (Later we will explain that ρ_2 is also a radius of curvature, hence the use of the same Greek letter for both distances.)

By definition, the tractrix in this figure has tangents of constant length R. At the neighbouring points p and q, [8.8] illustrates two such tangents, pa and qb, containing angle •. The corresponding normals po and qo therefore contain the same angle •. Note that ac has been drawn perpendicularly to qb.

Now let's watch what happens as q coalesces with p, which itself remains fixed. In this limit, o is the centre of the circle of curvature, pq is an arc of this circle, and ac is an arc of a circle of radius R centred at p. Thus,

$$\rho_1 \asymp \mathrm{op} \quad \text{and} \quad \frac{\mathrm{pq}}{\mathrm{op}} \asymp \bullet \asymp \frac{\mathrm{ac}}{\mathrm{R}},$$

and so

$$\frac{\mathrm{ac}}{\mathrm{pq}} \asymp \frac{\mathrm{R}}{\rho_1}.$$

Next we appeal to the defining property $\mathrm{pa} = \mathrm{R} = \mathrm{qb}$ of the tractrix to deduce that

$$\mathrm{bc} \asymp \mathrm{pq}.$$

Finally, using the fact that the triangle abc is ultimately similar to the triangle lap, we deduce that

$$\frac{\mathrm{R}}{\rho_1} \asymp \frac{\mathrm{ac}}{\mathrm{pq}} \asymp \frac{\mathrm{ac}}{\mathrm{bc}} \asymp \frac{\rho_2}{\mathrm{R}}.$$

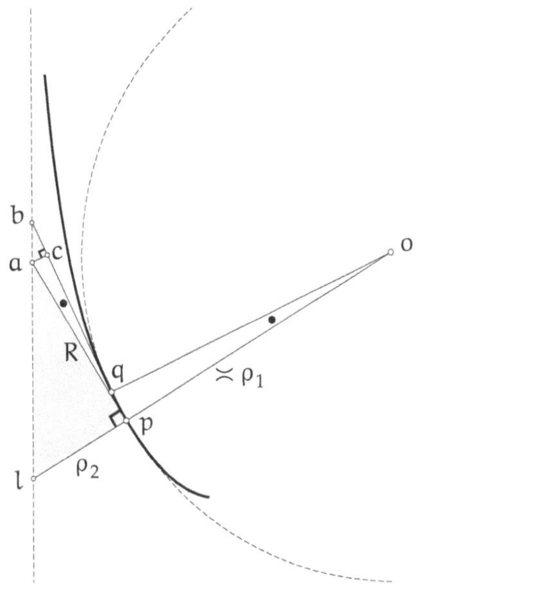

[8.8] *The tractrix has* $\kappa = (\rho_2/\mathrm{R}^2)$.

Thus,

$$\boxed{\kappa = \frac{1}{\rho_1} = \frac{\rho_2}{\mathrm{R}^2}.} \tag{8.8}$$

[5]Previously published in VCA, Section 6.3.2.

Chapter 9

Curves in 3-Space

The foregoing analysis of plane curves can be applied, with only minor modification, to curves that escape from the plane and twist their way through 3-dimensional space.

The essential insight is that even for such a twisting 3-dimensional orbit, each infinitesimal segment nevertheless lies in a plane (to which our previous analysis therefore applies), but this instantaneous plane of motion (called the ***osculating plane***) no longer remains fixed in space but instead rotates about the particle's direction of motion as the particle travels along the curve. The rate of rotation of the osculating plane is called the ***torsion***, denoted τ.

[9.1] The Frenet Frame. *Pressing the cardboard against the bent wire curve reveals the osculating plane, with* **N** *pointing toward the centre of curvature within that plane. As we move the frame along the curve, the* **torsion** τ *is the rate at which the plane's normal vector* **B** *(called the* **binormal***) rotates about* **T**.

There can be no substitute for direct physical experience of mathematical facts: do not merely try to imagine the following experiment—please *do* it!

Physically construct the contraption shown in [9.1], as follows. Cut a short section of a narrow drinking straw and tape or glue it down on an inflexible piece of stiff cardboard (or anything else flat and rigid). On the cardboard, draw a vector **T** extending the segment of straw, and draw an equally long vector **N** perpendicular to the straw, as shown. Finally, construct a third vector of equal length and attach it as the plane's normal vector, marked **B** in [9.1].

Take a sturdy piece of wire (perhaps a metal coat hanger, cut open) and bend it into any non-planar curve that suits your fancy. Now thread the wire through the section of straw, and press the cardboard against the curve so that it becomes the osculating plane. Once fitted to the curve in this way, the orthonormal set of vectors $(\mathbf{T}, \mathbf{N}, \mathbf{B})$ is called the ***Frenet frame*** of the curve.

Holding the wire fixed in space (ideally by enlisting the help of a friend), slide your Frenet frame along the curve, all the while keeping the cardboard in contact with the bent wire, as best you can, thereby ensuring that it continues to represent the osculating plane. In this manner you will experience τ as the rate at which you must rotate your piece of cardboard so as to keep it in contact with your curved wire.

Alternatively, if you are only interested in how the curvature κ varies along the curve, hold the cardboard fixed in one hand, with the thumb of that hand holding the wire down on the plane, so as to ensure it is the osculating plane. With your other hand, you may then gradually pull the wire through the straw, all the while pressing it against the cardboard with your thumb, and watch the changing curvature of the piece of curve on the fixed cardboard plane in your hand.

Returning to our previous discussion of plane curves for a moment, recall that the plane of the curve is spanned by **T** and by **N**, and further recall that the direction of **N** can be deduced from the rate of change of **T**, via (8.5): $\mathbf{T}' = \kappa\mathbf{N}$. (Recall that here the derivative is with respect to arc length, which is the same as the time derivative only if the particle travels at unit speed.)

In the present case of a twisting 3-dimensional curve, we can turn this around and *define* $\mathbf{N}$ to be the direction in which the tangent is turning:

$$\mathbf{N} \equiv \frac{\mathbf{T}'}{|\mathbf{T}'|}.$$

This normal $\mathbf{N}$ is called the ***principal normal***, to distinguish it from the infinitely many other "normals" that lie in the plane perpendicular to $\mathbf{T}$; it is distinguished by the fact that it lies in the osculating plane, and that it points directly *at* the centre of curvature (instead of away from it). Thus, a little more explicitly than before, the osculating plane, within which the particle is momentarily moving, is the plane spanned by $\mathbf{T}$ and $\mathbf{T}'$.

With these conventions in place, the acceleration of a unit-speed particle is always directed toward the centre of curvature and its magnitude is the curvature. It therefore makes sense to rechristen this acceleration as the ***curvature vector*** $\boldsymbol{\kappa}$ of this unit-speed particle:

$$\boldsymbol{\kappa} \equiv \kappa \mathbf{N}. \tag{9.1}$$

As shown in [9.1], the orientation in space of the osculating plane is conveniently encoded in its unit normal vector, denoted $\mathbf{B}$, and called the ***binormal*** of the curve:

$$\mathbf{B} \equiv \frac{\mathbf{T} \times \mathbf{T}'}{|\mathbf{T}'|} = \mathbf{T} \times \mathbf{N}.$$

As we have said, the torsion is the rate of rotation of the osculating plane about the direction of motion, $\mathbf{T}$. Equivalently, it is the rate of turning of $\mathbf{B}$ about $\mathbf{T}$.

We note the following simple but fundamental fact:

> *When a unit vector begins to rotate, its tip moves on the unit sphere, within the tangent plane to the unit sphere at that point, and therefore in a direction* perpendicular *to the vector itself.* (9.2)

As $\mathbf{B}$ rotates, its tip must therefore begin to move in a plane parallel to the osculating, cardboard $(\mathbf{T}, \mathbf{N})$-plane. Note, however, that here $\mathbf{B}$ cannot tip in the $\mathbf{T}$ direction, and its rate of change is therefore purely in the $\mathbf{N}$ direction; make sure you can see this geometrically, perhaps with the assistance of your toy Frenet frame. This can of course also be shown by calculation: see Exercise 2. Thus

$$\mathbf{B}' = -\tau \mathbf{N}.$$

NOTATIONAL NOTE: The minus sign that is included in this definition of τ does not appear to us to have anything to recommend it, but since the great majority of authors appear to include it, we shall let discretion be the better part of valour!

As for the curvature, the foregoing analysis applies with only minor change. As before, $\mathbf{T}$ spins within the osculating plane at a rate given by κ, and $\mathbf{N}$ spins with it, as in [8.5]. But now $\mathbf{N}$ does not merely rotate within the osculating plane, for it must also remain orthogonal to $\mathbf{B}$, so it rotates out of the osculating plane at the same rate as **B** rotates about **T**. In other words, applying the geometric idea [8.5] twice, and in accord with (9.2), the total rate of change of **N** is given by

$$\mathbf{N}' = -\kappa\,\mathbf{T} + \tau\,\mathbf{B}.$$

The rotation of the entire Frenet frame as we move along the curve can therefore be summarized as the following matrix equation, known as the ***Frenet–Serret Equations***:[1]

$$\begin{bmatrix} \mathbf{T} \\ \mathbf{N} \\ \mathbf{B} \end{bmatrix}' = \begin{bmatrix} 0 & \kappa & 0 \\ -\kappa & 0 & \tau \\ 0 & -\tau & 0 \end{bmatrix} \begin{bmatrix} \mathbf{T} \\ \mathbf{N} \\ \mathbf{B} \end{bmatrix} = [\Omega] \begin{bmatrix} \mathbf{T} \\ \mathbf{N} \\ \mathbf{B} \end{bmatrix}. \tag{9.3}$$

Let us retain a firm grasp on the geometry underlying the structure of $[\Omega]$: the vanishing leading diagonal is merely the algebraic manifestation of (9.2); likewise, the skew-symmetry ($[\Omega]^T = -[\Omega]$) is merely the algebraic manifestation of [8.5].

The matrix $[\Omega]$ itself tells us how the $(\mathbf{T}, \mathbf{N}, \mathbf{B})$ frame rotates from one moment to the next. If we watch the frame move along the curve for a short time δt, then

$$\text{new frame after } \delta t \asymp [I + [\Omega]\,\delta t]\,[\text{original frame}].$$

For more on the rotation of the frame, see Exercise 3.

[1]Independently discovered by Frenet in 1847 and by Serret in 1851.

Chapter 10

The Principal Curvatures of a Surface

10.1 Euler's Curvature Formula

In Act I we immediately and anachronistically launched into a discussion of Gauss's revolutionary 1827 conception of the intrinsic geometry of surfaces, and with it his associated concept of the *intrinsic* curvature $\mathcal{K}$. But, historically, the natural progression from Newton's investigation of the extrinsic curvature of plane curves was the study of the *extrinsic* curvature of surfaces: how do they bend within the surrounding space?

It was Euler in 1760 who made the first fundamental breakthrough. We will first describe his discovery, and then prove it afterwards. At a point p on the surface $\mathcal{S}$ he considered the plane curve $\mathcal{C}_\theta$ through p obtained by intersecting $\mathcal{S}$ with a plane Π_θ that rotates about the surface normal $\mathbf{n}_p$ at p. See [10.1], which illustrates two orthogonal positions of this plane Π_θ, on two different kinds of surface. Here θ denotes the angle of rotation of Π_θ, starting from an arbitrary (at least for now) initial direction. As Π_θ rotates, the shape of the intersection curve $\mathcal{C}_\theta$ changes, and therefore its curvature $\kappa(\theta)$ at p will (in general) vary too.

Before continuing, we should explain that $\kappa(\theta)$ has a *sign* attached to it, according to this convention: the vector from p to the centre of curvature c of $\mathcal{C}_\theta$ is defined to be $\frac{1}{\kappa(\theta)}\mathbf{n}$. Thus if c lies in the direction of $+\mathbf{n}$ then $\kappa(\theta)$ is positive, while if it lies in the direction of $-\mathbf{n}$ then $\kappa(\theta)$ is negative. Of course there are actually two opposite (equally valid) choices for $\mathbf{n}$. Reversing the choice of $\mathbf{n}$ reverses the sign of κ. Note that the principal normal (as defined in the previous section) of $\mathcal{C}_\theta$ is therefore $\mathbf{N} = [\text{sign of } \kappa(\theta)]\, \mathbf{n}$.

As θ varies, let κ_1 and κ_2 denote the maximum and minimum values of $\kappa(\theta)$. Euler's elegant and important discovery was that these extreme values of the curvature (the so-called ***principal curvatures***) will always occur in *perpendicular* directions, which are called the ***principal directions***. Furthermore, if we choose $\theta = 0$ to coincide with the direction that has curvature κ_1, he found

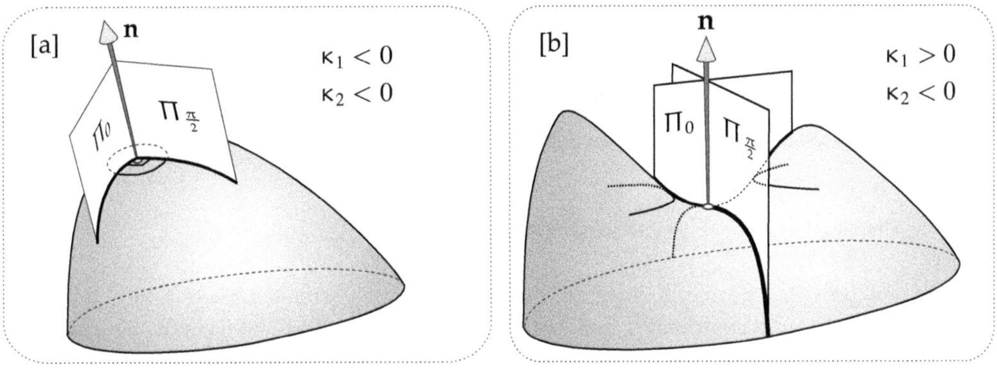

[10.1] [a] *If κ_1 and κ_2 have the same sign then the surface locally resembles a hill, and a slice parallel to (and close to) the tangent plane yields an ellipse (or misses the surface entirely).* **[b]** *If κ_1 and κ_2 have opposite sign then the surface locally resembles a saddle. A parallel slice above the tangent plane yields both branches of a hyperbola (as illustrated); slicing below the tangent plane yields both branches of an orthogonal hyperbola (not shown).*

$$\textbf{\textit{Euler's Curvature Formula:}}\quad \kappa(\theta) = \kappa_1 \cos^2\theta + \kappa_2 \sin^2\theta. \tag{10.1}$$

Although this is the standard form of the formula found in most modern texts, there is a superior way of writing it, which was in fact the form in which Euler[1] himself first published it.

Substituting $\cos^2\theta = (1+\cos 2\theta)/2$ and $\sin^2\theta = (1-\cos 2\theta)/2$, we obtain

$$\kappa(\theta) = \overline{\kappa} + \tfrac{\Delta\kappa}{2}\cos 2\theta, \tag{10.2}$$

where $\overline{\kappa} \equiv \left[\frac{\kappa_1+\kappa_2}{2}\right]$ is the ***mean curvature***, about which the value oscillates, and $(\Delta\kappa/2) \equiv (\kappa_1 - \kappa_2)/2$ is the amplitude of the oscillation. See the graph [10.2]. Note that the extremal nature of κ_1 and κ_2, together with the orthogonality of the principal directions, can be deduced directly from this formula. Furthermore, this form makes manifest the geometric fact that rotating the normal plane by π returns it to the same position, so the variation in curvature has period π.

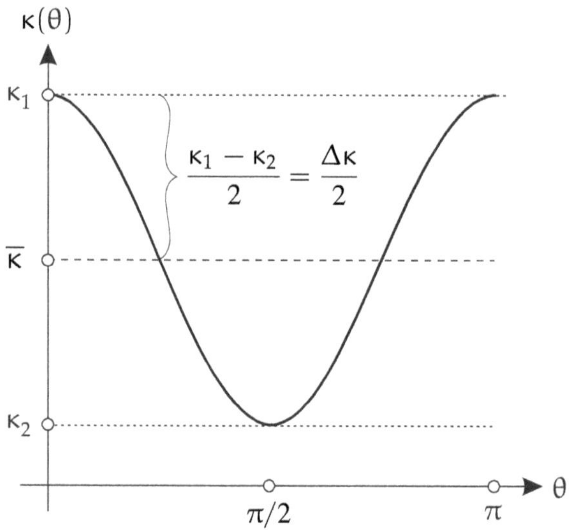

[10.2] *Euler's Curvature Formula tells us that as we vary the angle θ of the normal section $\mathcal{C}_\theta$, its curvature $\kappa(\theta)$ oscillates sinusoidally, achieving its maximum and minimum values in perpendicular directions.*

As illustrated in [10.1a], if κ_1 and κ_2 have the same sign then $\kappa(\theta)$ always shares this same sign; i.e., the surface locally resembles a hill. But if κ_1 and κ_2 have opposite signs then $\kappa(\theta)$ changes sign; i.e., $\mathcal{C}_\theta$ flips from one side of the tangent plane to the other, and, as shown in [10.1b], the surface locally resembles a saddle.

Of course it is possible that $\kappa_1 = \kappa_2$, in which case $\kappa(\theta) = \text{const.}$ at p; in this case p is called ***umbilic***, and the surface locally resembles a sphere. In general it can be shown that such umbilics can only occur in isolated places on a surface. Of course the sphere of radius R is a strong exception to this: *every* point is an umbilic, with $\kappa(\theta) = 1/R$. But it can be proved that the sphere is the *only* surface with this property.

In the exceptional case that the umbilic has $\kappa_1 = 0 = \kappa_2$, the shape of the surface surrounding the point can be much more complex. This will be discussed and illustrated in Section 12.4.

10.2 Proof of Euler's Curvature Formula

We will now provide a geometric proof[2] of Euler's Curvature Formula. Choose p to be the origin of the Cartesian (x, y, z) coordinates, and let the x and y axes be chosen to lie in the tangent plane

[1] The more familiar form (10.1) was derived from Euler's result about 50 years later, by Charles Dupin, in 1813. See Knoebel (2007, p. 188).

[2] The proof in the lovely article by Aleksandrov (1969) is similar, but it requires two calculations, which are here replaced with geometry. The now-standard idea of using the quadratic approximation relative to the tangent plane is simpler than Euler's original approach; it was discovered by Jean-Baptiste Meusnier in 1776. See Knoebel (2007, p. 194).

T_p at p. Then locally the surface can be represented by an equation of the form $z=f(x,y)$, such that $f(0,0)=0$ and $\partial_x f=0=\partial_y f$ at the origin. Expanding $f(x,y)$ into a Taylor series, we deduce that as x and y tend to zero,

$$z \asymp ax^2 + by^2 + cxy. \tag{10.3}$$

Slicing through the surface with planes $z=\text{const.}=k$ parallel to T_p, and very close to it, therefore yields intersection curves whose equations (as k goes to zero) are quadratics, $ax^2+by^2+cxy \asymp k$, and which are therefore (ultimately) conic sections.

Figure [10.1] illustrates the fact that these conics are ellipses if κ_1 and κ_2 have the same sign, and that they are hyperbolas if κ_1 and κ_2 have opposite signs. In both cases, *the conics have two perpendicular axes of symmetry that are independent of the height* k *of the slicing plane.* This follows from the homogeneous quadratic nature of the equation. For example, quadrupling the height of the slice just doubles the size of the conic, without changing its shape: $k \to 4k$ and $(x,y) \to (2x,2y)$ yield the same equation as before.

The use of this conic to quantify the amount and type of bending of $\mathcal{S}$ at p goes back to Charles Dupin in 1813, and it is called the ***Dupin indicatrix*** in his honour. The point p itself is called ***elliptic, hyperbolic***, or ***parabolic*** according to the type of the Dupin indicatrix at p.

Crucially, the symmetry of the conic sections implies that the surface itself has local mirror symmetry in two perpendicular directions.[3] We can now derive Euler's Curvature Formula and deduce that these two perpendicular planes of symmetry are in fact the *same* planes that yield the maximum and minimum curvatures; i.e., these local mirror symmetry directions are the same as the principal directions. To summarize what we shall prove,

> *An infinitesimal neighbourhood of a generic point of a surface has mirror symmetry in two perpendicular planes (both containing the surface normal), and the perpendicular directions in which these planes intersect the tangent plane are the principal directions of maximum and minimum curvature.* (10.4)

Refining our coordinate system, we now align the x and y axes with these symmetry directions. Since the equation (10.3) is now invariant under the reflections $x \mapsto -x$ and $y \mapsto -y$ it follows that $c=0$, and the local equation of the surface therefore becomes

$$z \asymp ax^2 + by^2. \tag{10.5}$$

To find the geometric meaning of the coefficients a and b we now refer back to [8.2] and view it as depicting the intersection of Π_θ with $\mathcal{S}$: the curve $\mathcal{C}$ is now $\mathcal{C}_\theta$, and the tangent $\mathcal{T}$ is now the intersection of the tangent plane T_p with Π_θ, and the deviation σ of the curve from its tangent is now simply the height z of the curve above the tangent plane.

Let $\theta=0$ correspond to the x-axis, and let $\kappa_1=\kappa(0)$ be the curvature of $\mathcal{C}_0=$ (the intersection of $\mathcal{S}$ with the xz-plane), having equation $z=ax^2$. Then the result (8.2) shows that $a=\frac{1}{2}\kappa_1$. In exactly the same way, defining $\kappa_2=\kappa(\frac{\pi}{2})$ to be the curvature of the intersection curve with the yz-plane, we find that $b=\frac{1}{2}\kappa_2$. Thus (10.5) can be expressed more geometrically as

$$z \asymp \tfrac{1}{2}\kappa_1 x^2 + \tfrac{1}{2}\kappa_2 y^2. \tag{10.6}$$

Now consider [10.3], which depicts the curve $\mathcal{C}_\theta$ for a general angle θ. (This diagram assumes (and others to follow do as well) that the Gaussian curvature is positive, but the accompanying

[3] At least in the general case, where $\kappa_1 \neq \kappa_2$; the symmetry can be much more complex in the case of an umbilic, as we shall see in Section 12.4.

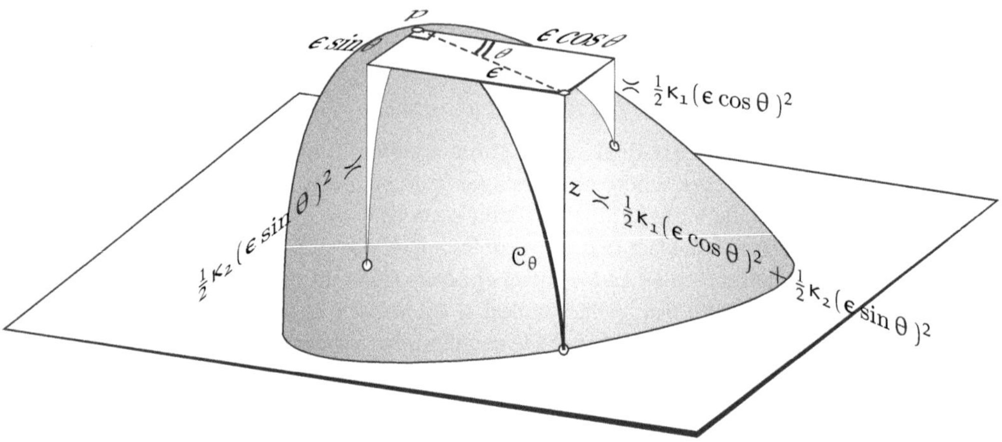

[10.3] *The principal curvatures tell us how quickly the surface falls away—downwards in this figure—from the tangent plane as we begin to travel in each of the two perpendicular principal directions. When we instead move a distance ϵ within the tangent plane in a general direction θ, the distance z that the surface falls away is simply the sum (according to (10.6)) of the falls due to each of these two components separately.*

reasoning applies equally well to negatively curved surfaces.) If we move a distance ϵ within T_p in the direction θ then we arrive at the illustrated point $x = \epsilon\cos\theta$, $y = \epsilon\sin\theta$. Thus inserting (10.6) into (8.2) yields

$$\kappa(\theta) \asymp 2\left[\frac{z}{\epsilon^2}\right] \asymp 2\left[\frac{\frac{1}{2}\kappa_1(\epsilon\cos\theta)^2 + \frac{1}{2}\kappa_2(\epsilon\sin\theta)^2}{\epsilon^2}\right] = \kappa_1\cos^2\theta + \kappa_2\sin^2\theta,$$

proving Euler's Curvature Formula, and thereby establishing the extremal nature of the curvatures κ_1 and κ_2 associated with the orthogonal directions of local mirror symmetry.

10.3 Surfaces of Revolution

If we rotate a plane curve $\mathcal{C}$ about a line L within its plane, then we obtain a surface of revolution $\mathcal{S}$ for which the principal directions are easily identified. Figure [10.4] illustrates this in the particular case in which $\mathcal{C}$ is the tractrix, L is its axis, and therefore $\mathcal{S}$ is the pseudosphere.

Clearly $\mathcal{S}$ has mirror symmetry in each plane through its axis L, so the intersection of such a plane with $\mathcal{S}$ (which is simply a copy of $\mathcal{C}$), yields one of the principal directions within $\mathcal{S}$.

At a point p on $\mathcal{S}$ let us choose this direction along the copy of $\mathcal{C}$ to correspond to $\theta = 0$, so that using the previous notation $\mathcal{C} = \mathcal{C}_0$. Thus the first principal curvature $\kappa_1 = (1/\rho_1)$ is simply (up to sign) the curvature of the original plane curve $\mathcal{C}$.

The second principal direction at p must therefore be the direction within $\mathcal{S}$ perpendicular to this plane through L. The radius of curvature ρ_2 associated with this second principal direction is in fact simply the distance pl, in the direction of the normal **n** to the surface, from p to the point l on the axis L:

$$\rho_1 = \textit{radius of curvature op of the generating curve } \mathcal{C}, \tag{10.7}$$

$$\rho_2 = \textit{the distance pl along the normal } \mathbf{n} \textit{ to the axis L}. \tag{10.8}$$

To confirm this, let us use the same notation as before and denote the intersection of $\mathcal{S}$ with this perpendicular plane as $\mathcal{C}_{(\pi/2)}$. If we rotate lp about L then it sweeps out a cone and p moves within $\mathcal{S}$ along the circular edge of this cone, which initially coincides with $\mathcal{C}_{(\pi/2)}$.

Now recall Newton's original method of locating the centre of curvature as the intersection of neighbouring normals of a curve. In the present case of $\mathcal{C}_{(\pi/2)}$, the neighbouring normals are generators of this cone, meeting at l, thereby confirming that $\kappa_2 = \pm(1/\rho_2)$, the sign depending on the choice of **n**.

Returning to [10.4], with the illustrated choice of the normal vector **n**, we find for the

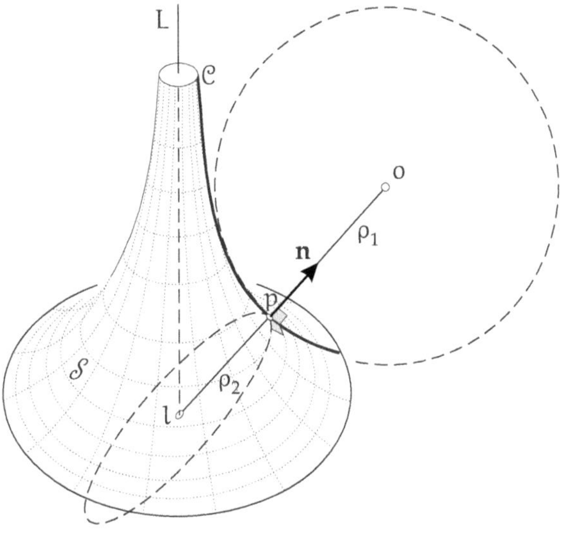

[10.4] *The principal radii of curvature ρ_1 and ρ_2 of a surface of revolution, here illustrated with the pseudosphere.*

$$\textit{pseudosphere:}\quad \kappa_1 = +\frac{1}{\rho_1} \quad \text{and}$$

$$\kappa_2 = -\frac{1}{\rho_2} = -\frac{\rho_1}{R^2}, \qquad (10.9)$$

by virtue of (8.8). (Of course, quite generally, reversing the direction of **n** reverses the sign of both principal curvatures.) Note that we have now kept our promise and have explained the notation ρ_2 that we saw in [8.8].

For our second example, consider the torus (doughnut) obtained by rotating a circle $\mathcal{C}$ of radius r and centre o about a line L at distance R from o. See [10.5]. We imagine that $\mathcal{C}$ is traced by a particle p that rotates about o, the radius op making angle α with the horizontal. From the figure, with the illustrated choice of **n**, we see that for this

$$\textit{torus:}\quad \kappa_1 = -\frac{1}{r} \quad \text{and} \quad \kappa_2 = -\frac{1}{r + R\sec\alpha}. \qquad (10.10)$$

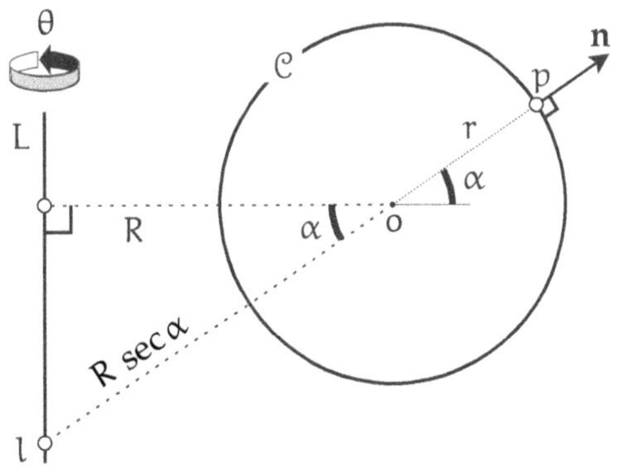

[10.5] *For the torus obtained by rotating the circle $\mathcal{C}$ about L, the principal radii of curvature are $\rho_1 = -r$ and $\rho_2 = -(r + R\sec\alpha)$.*

Leaving behind specific examples, suppose now that $\mathcal{C}$ is a general curve, but *traced at unit speed* by a particle moving in the (x, y)-plane, whose position at time t is $x = x(t)$ and $y = y(t)$. See [10.6].

Let us take L to be the horizontal x-axis, and rotate $\mathcal{C}$ about this axis to generate a surface $\mathcal{S}$. Let us choose **n** pointing to the *left* of the direction of motion, as illustrated. Using (10.7) and (8.7), we deduce the first principal curvature for a curve traced at

$$\textit{unit speed:}\quad \kappa_1 = \ddot{y}/\dot{x}. \qquad (10.11)$$

As we previously noted regarding (8.7), the result (10.11) may equally well be expressed as $\kappa_1 = -\ddot{x}/\dot{y}$. Check for yourself [exercise] that these formulas yield the correct sign for κ_1, no matter where we are on the curve.

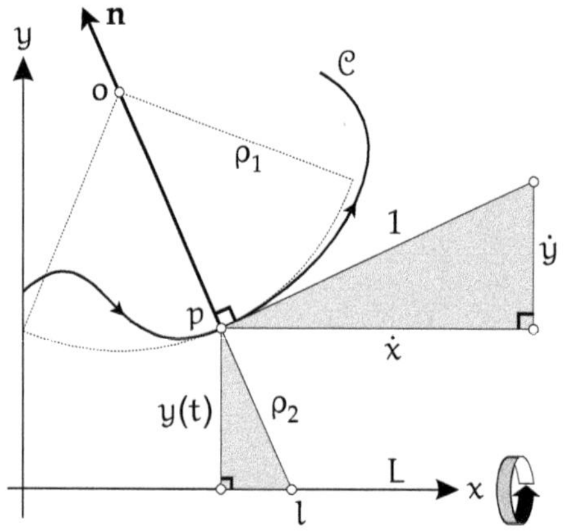

[10.6] *The principal radii of curvature ρ_1 and ρ_2 of a surface of revolution generated by rotating $\mathcal{C}$ about L (the x-axis).*

Finally, according to (10.8), $\rho_2 = pl$. By appealing to the similarity of the two shaded triangles in [10.6], we find that $(y/\rho_2) = (\dot{x}/1)$, and so we deduce the second principal curvature for a curve traced at

$$\textit{unit speed:} \quad \kappa_2 = -\dot{x}/y. \tag{10.12}$$

Check for yourself [exercise] that this formula yields the correct sign for κ_2, no matter where we are on the curve.

Finally, we note an important lesson of the above analysis:

> *Let the curve $\mathcal{C}$ be rotated about the line L to generate the surface of revolution $\mathcal{S}$. Then the parts of $\mathcal{C}$ that are concave* towards L *generate parts of $\mathcal{S}$ that have* positive *curvature, and the parts of $\mathcal{C}$ that are concave* away *from L generate parts of $\mathcal{S}$ that have* negative *curvature. Inflection points of $\mathcal{C}$ generate circles on $\mathcal{S}$ where the curvature vanishes; these circles separate the regions of opposite curvature.* (10.13)

Chapter 11

Geodesics and Geodesic Curvature

11.1 Geodesic Curvature and Normal Curvature

To us, as inhabitants of the Earth's surface, a great circle is not only analogous to a line in that it provides the shortest route between two points, but it also appears to be *straight*: it has no apparent curvature. If you walk in a "straight line" across a seemingly flat desert, you are actually walking along such a great circle, whose curvature in 3-space is 1/(radius of the Earth). How shall we reconcile these two conflicting views of one and the same curve?

The answer, in short, is that the full curvature in 3-space of a *general* curve within a *general* surface $\mathcal{S}$ can be decomposed into two components: one within the surface (visible to its inhabitants), and another perpendicular to the surface (invisible to the inhabitants). The visible component within the surface is called the ***geodesic curvature***, denoted κ_g; the invisible component perpendicular to the surface is called the ***normal curvature***, denoted κ_n.

If the osculating plane is perpendicular to the surface, i.e., contains the surface normal $\mathbf{n}$, then *all* of the curvature is "normal curvature" ($\kappa_n = \kappa$) and *none* of it is visible "geodesic curvature" within the surface ($\kappa_g = 0$). This was the case for our great circle on the surface of the Earth.

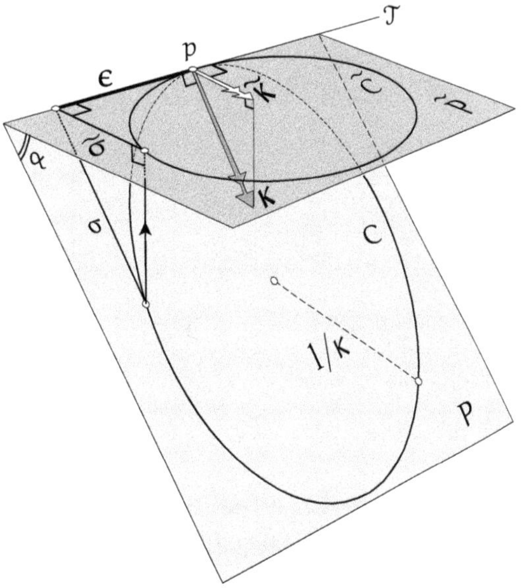

[11.1] *When the circle* C *in the plane* P *is projected orthogonally onto the plane* $\widetilde{P}$, *at angle* α *to* P, *the distance from the common tangent* $\mathcal{T}$ *contracts by* $\cos\alpha$, *and therefore the size of the curvature vector at* p *does too:* $\widetilde{\kappa} = \kappa \cos\alpha$.

Now suppose instead that you are standing in the middle of a seemingly flat desert, and you trace a very small circle of radius r in the sand at your feet. You thus obtain a seemingly planar curve with very large curvature $\kappa = 1/r$. But of course the desert is part of the curved surface of the Earth, and from this perspective all that is special about your curve is that its osculating plane almost coincides with Earth's tangent plane at your location. In this case we have almost the reverse of the former case: almost all of the curvature is now geodesic curvature, but in fact there is *still the same amount* of invisible normal curvature as before, as we shall explain shortly.

In fact the general case may be thought of as simply an appropriate mixture of the two extreme cases described just previously.

The essential point really has little to do with surfaces, *per se*, rather it has to do with how the curvature of a plane curve changes when it is projected (casts a shadow) on another plane. See [11.1], which shows a circle C of radius $(1/\kappa)$ within a plane P, the tangent at p being $\mathcal{T}$.

The figure also shows the orthogonal projection $\widetilde{C}$ of C onto a second plane $\widetilde{P}$ through $\mathcal{T}$, the angle between the planes being α. As you probably know [or exercise] $\widetilde{C}$ is in fact an ellipse, the original circle C having been compressed in a direction perpendicular to $\mathcal{T}$. Thus it is clear that the curvature $\widetilde{\kappa}$ of $\widetilde{C}$ at p is *less* than the original curvature κ.

More precisely, the figure shows that under this projection the distance from the common tangent $\mathcal{T}$ undergoes a compression by $\cos\alpha$, so that $\widetilde{\sigma} = \sigma\cos\alpha$. Thus, by virtue of (8.2),

$$\widetilde{\kappa} \asymp \frac{2\widetilde{\sigma}}{\epsilon^2} = \frac{2\sigma\cos\alpha}{\epsilon^2} \asymp \kappa\cos\alpha.$$

By thinking of C as the circle of curvature of a general curve $\mathcal{C}$, we see that this formula applies to $\mathcal{C}$ as well:

> *If a plane curve $\mathcal{C}$ (with tangent $\mathcal{T}$ at p) is projected orthogonally onto a second plane that passes through $\mathcal{T}$ and makes angle α with the first plane, then at p the projected curve $\widetilde{\mathcal{C}}$ has $\widetilde{\kappa} = \kappa\cos\alpha$.* (11.1)

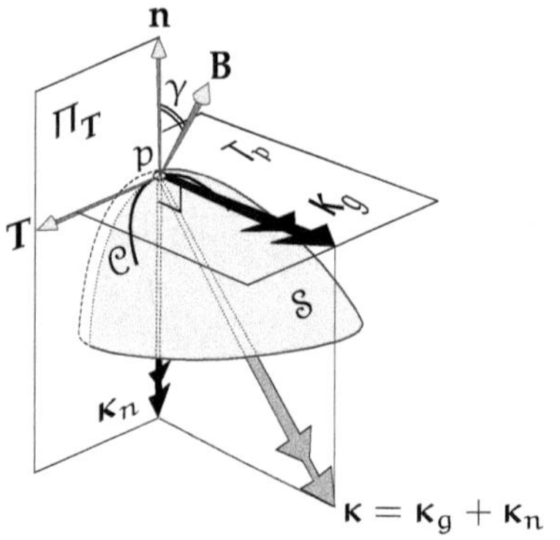

[11.2] *The net acceleration $\boldsymbol{\kappa}$ can be decomposed into the geodesic curvature vector $\boldsymbol{\kappa}_g$ tangent to the surface, and the normal curvature vector $\boldsymbol{\kappa}_n$ perpendicular to the surface: $\boldsymbol{\kappa} = \boldsymbol{\kappa}_g + \boldsymbol{\kappa}_n$.*

This result can also be understood vectorially, and perhaps even more intuitively, in terms of motion. As we have discussed, if a particle traverses $\mathcal{C}$ at unit speed then it is appropriate to rename the particle's acceleration as the ***curvature vector*** $\boldsymbol{\kappa} = \kappa\mathbf{N}$, for this acceleration vector points towards the centre of curvature and its magnitude equals the curvature.

Next, observe [exercise] that *the acceleration of the projection equals the projection of the acceleration.* Note, however, that the projected orbit $\widetilde{\mathcal{C}}$ will in general *not* be traced at constant speed, and its acceleration vector will therefore have a component tangent to $\widetilde{\mathcal{C}}$ and will therefore *not* be orthogonal to $\widetilde{\mathcal{C}}$, in contrast to $\widetilde{\boldsymbol{\kappa}}$. (Remember, $\widetilde{\boldsymbol{\kappa}}$ is the acceleration the particle *would* have had if it traversed $\widetilde{\mathcal{C}}$ at *constant unit speed*.) This breakdown of orthogonality also makes sense geometrically: in [11.1] a radius of C projects (in general) to a line in $\widetilde{P}$ that is no longer orthogonal to the image ellipse $\widetilde{C}$. That said, p is an exception to this, and as the particle and its shadow momentarily move together at unit speed through p along $\mathcal{T}$,

$$\widetilde{\boldsymbol{\kappa}} = (\text{projection of } \boldsymbol{\kappa} \text{ onto } \widetilde{P}),$$

as illustrated in [11.1]. By equating the lengths of these vectors, (11.1) follows immediately.

Now let us return to the original problem, in which $\mathcal{C}$ is a general curve on a surface $\mathcal{S}$. See [11.2]. As before, let $\mathbf{T}$ denote the unit tangent to $\mathcal{C}$ at p, let T_p be the tangent plane to $\mathcal{S}$ at p, and let us also introduce $\Pi_{\mathbf{T}}$ as the normal plane spanned by $\mathbf{n}$ and $\mathbf{T}$. Let the osculating plane be inclined at angle γ to the tangent plane T_p; equivalently, γ is the angle between the binormal $\mathbf{B}$ of $\mathcal{C}$ and surface normal $\mathbf{n}$ of $\mathcal{S}$. Thus κ_g and κ_n are the curvatures of the projections of $\mathcal{C}$ onto T_p and $\Pi_{\mathbf{T}}$, respectively. Then (11.1) implies that

$$\kappa_g = \kappa\cos\gamma \quad \text{and} \quad \kappa_n = \kappa\sin\gamma. \tag{11.2}$$

Again, this can also be understood in terms of acceleration. Imagine particles traversing the projections of $\mathcal{C}$ onto T_p and $\Pi_{\mathbf{T}}$ at unit speed. Their respective accelerations will then be the

geodesic curvature vector $\boldsymbol{\kappa}_g$ and the ***normal curvature vector*** $\boldsymbol{\kappa}_n$, pointing toward the respective centres of curvature in the tangent and normal planes, and with magnitudes equal to κ_g and κ_n, respectively. As illustrated in [11.2], the acceleration can then be decomposed into these two orthogonal components:

$$\boldsymbol{\kappa} = \boldsymbol{\kappa}_g + \boldsymbol{\kappa}_n. \tag{11.3}$$

From this (11.2) follows immediately.

11.2 Meusnier's Theorem

Consider the family of all curves (such as $\mathcal{C}$) on $\mathcal{S}$ that pass through p in the particular direction $\mathbf{T}$. The inhabitants of $\mathcal{S}$ may freely draw curves of this type that are tightly curved, slightly curved, or not curved at all: κ_g can be freely and arbitrarily specified. However, the same is *not* true of the normal curvature κ_n: the bending of the surface itself *forces* all curves within it to bend in the $\mathbf{n}$ direction (if we choose $\mathbf{n}$ in the direction of $+\boldsymbol{\kappa}_n$).

In fact, as Meusnier realized in 1779, the surface forces all such curves to bend by the *same* amount: κ_n is independent of the specific curve $\mathcal{C}$, so, in particular, κ_n must equal the curvature of the normal section in this direction. More intuitively, this says that all of these curves locally have the *same projection* onto $\Pi_{\mathbf{T}}$: near p, these projections all look like the normal section, which in turn looks (locally) like an arc of a circle within $\Pi_{\mathbf{T}}$ of radius $(1/\kappa_n)$ centred at $(1/\kappa_n)\mathbf{n}$.

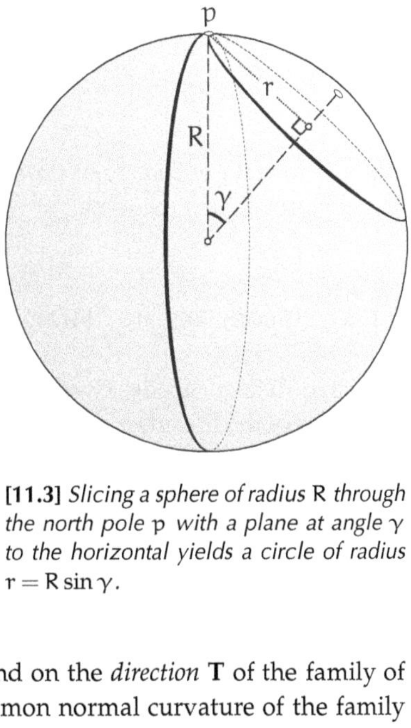

[11.3] *Slicing a sphere of radius* R *through the north pole* p *with a plane at angle* γ *to the horizontal yields a circle of radius* $r = R\sin\gamma$.

Even granted this, it's clear, however, that this curve-independent normal curvature must in general depend on the *direction* $\mathbf{T}$ of the family of curves as they pass through p. Thus we may write the common normal curvature of the family as a function $\kappa_n(\mathbf{T})$ of this direction. (For example, if $\mathbf{e}_1$ and $\mathbf{e}_2$ are the principal directions, then $\kappa_n(\mathbf{e}_{1,2}) = \kappa_{1,2}$ are the principal curvatures.) Combining this (as yet) unproven claim with (11.2), we can state

> **Meusnier's Theorem.** *All curves that pass through a point* p *of a curved surface in the same direction* $\mathbf{T}$ *have the same normal curvature* $\kappa_n(\mathbf{T})$ *as the normal section in that direction. If the osculating plane at* p *of one such curve makes angle* γ *with the tangent plane at* p*, and its curvature there is* κ_γ*, then* $\kappa_\gamma \sin\gamma = \kappa_n(\mathbf{T})$ *is independent of* γ. (11.4)

Before giving a general argument, consider the sphere of radius R, for in this special case the truth of the theorem is easy to visualize. See [11.3]. Taking p to be the north pole, the normal section is a great circle (a meridian) of radius R and hence the curvature $\kappa_n = (1/R)$. Slicing through the north pole with a plane at angle γ to the horizontal tangent plane at p, the intersection with the sphere will be a circle of radius $r = R\sin\gamma$, and hence $\kappa_\gamma = 1/r = 1/(R\sin\gamma)$.

Returning to the introductory example of a small circle drawn in the sand, imagine that γ tends to zero, in which case the radius of the circle shrinks to zero, and the length of its curvature vector tends to infinity. But, simultaneously, this curvature vector is tending towards orthogonality with the surface normal, so that less and less of it projects onto that direction. These two effects cancel each other out *exactly*, so that the projection of the curvature vector onto the surface normal has *constant* magnitude:

$$\kappa_\gamma \sin\gamma = \frac{1}{R} = \kappa_n,$$

confirming this instance of the theorem.

For the general case, imagine that $\mathcal{C}$ is traced by a particle at unit speed, and let $\mathbf{T}$ denote its velocity vector, not just at p but along the entire orbit. Then, by (11.3),

$$\dot{\mathbf{T}} = \boldsymbol{\kappa} = \boldsymbol{\kappa}_g + \boldsymbol{\kappa}_n \quad \Longrightarrow \quad \kappa_n(\mathbf{T}) = \dot{\mathbf{T}} \cdot \mathbf{n}.$$

But for the same essential reason as in [8.5], $\dot{\mathbf{T}} \cdot \mathbf{n} = -\mathbf{T} \cdot \dot{\mathbf{n}}$, so

$$\kappa_n(\mathbf{T}) = -\mathbf{T} \cdot \dot{\mathbf{n}}. \tag{11.5}$$

But $\dot{\mathbf{n}}$ is the rate of change of the surface normal in the $\mathbf{T}$ direction, which is independent of $\mathcal{C}$, thereby proving the theorem.

11.3 Geodesics are "Straight"

We have defined geodesics via their length-minimizing property within the surface. But the geodesics of the Euclidean plane (lines) may also be recognized by their *straightness*. Likewise, on the sphere we have just seen that the geodesics (great circles) not only provide the shortest routes, but they too are "straight," in the sense that none of their curvature is visible to inhabitants of the surface: $\kappa_g = 0$. In fact this connection between length minimization and intrinsic straightness is universal:

> *Geodesics appear to be straight lines to the inhabitants of the surface: they are intrinsically "straight" in the sense that their geodesic curvature vanishes: $\kappa_g = 0$ at every point of the geodesic.*

To begin to understand this, imagine a guitar string in its straight-line equilibrium position L being plucked with a pick whose position in space is p. As the pick pulls the string away from L, the string forms a triangle in the plane Π containing p and L. Once released, the net force, resulting from the string's compulsion to shorten its length, acts within Π and the resulting motion back towards L all takes place within that plane. Next, suppose that instead of pulling the string away from L into a sharp triangle in Π, we pull it away into the form of a gentle convex curve lying in Π, then the resulting forces and motion will again clearly reside within Π.

Now let us return to our curved surface and picture our geodesic as a guitar string stretched taut over the (frictionless) surface to connect two fixed points. The string is at rest: it is in equilibrium on the surface, already having slid over the surface, contracting to become as short as possible. Now focus attention on a very short segment apb of the geodesic. This segment will *almost* lie in a single plane Γ, namely, the one through a, p, and b. As a and b approach p, the limiting position of Γ is the osculating plane at p, which we shall denote Π_p. By the forgoing reasoning, the net length-shrinking force $\mathbf{F}_p$ acting on apb will ultimately lie within Π_p.

If $\mathbf{F}_p$ were to have any component tangent to the surface, the string would be free to move in this direction, thereby reducing its length. But since the string is already as short as possible, this cannot happen! Thus, having no component within the surface, the force $\mathbf{F}_p$ (lying within Π_p) must be directed perpendicularly to the surface, along the normal $\mathbf{n}_p$. Thus we have reached an important characterization[1] of the geodesic in terms of the *extrinsic* geometry of the surface:

> *At every point p of a geodesic, the osculating plane Π_p contains the surface normal $\mathbf{n}_p$, and hence the geodesic curvature vanishes: $\kappa_g = 0$.* (11.6)

Intrinsic and extrinsic geometry appear to be belong to entirely different worlds, and yet we see here that the two are strangely entangled. Later we shall witness even deeper and more mysterious connections between these two worlds.

11.4 Intrinsic Measurement of Geodesic Curvature

We have just seen that vanishing geodesic curvature characterizes geodesics, but we may also use geodesics as an intrinsic tool with which to measure the geodesic curvature of a *non*geodesic curve, for which $\kappa_g \neq 0$.

In (8.4) we saw that the curvature of a curve in the Euclidean plane may be viewed as the rate of rotation of its tangent line with respect to distance along the curve. Well, this original construction (illustrated in [8.4]) also makes perfectly good sense to the inhabitants of the surface.

From *their* perspective there is no change at all to the construction. See [11.4]. They draw tangent "lines" (dashed geodesics) to the curve at neighbouring points δs apart, find the angle $\delta\varphi$ at which these tangents intersect, then calculate the curvature (in the limit that the points merge) as $\kappa_g \asymp (\delta\varphi/\delta s)$.

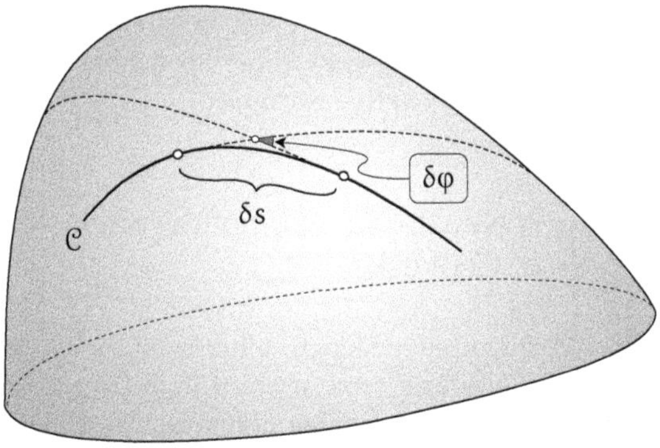

[11.4] *The surface's inhabitants can measure the geodesic curvature κ_g of the curve $\mathcal{C}$ by constructing tangent geodesics to the curve at neighbouring points δs apart, then finding the angle $\delta\varphi$ at which these tangents intersect. As the points merge, $\kappa_g \asymp (\delta\varphi/\delta s)$.*

But what they call *the* curvature of the curve, we recognize as being only one part of the curvature, its geodesic curvature κ_g; the normal curvature κ_n is invisible and unknowable to them. The only other difference between their intrinsic perspective and our extrinsic perspective (looking down on the surface) is that what they call "straight lines" we recognize as geodesics within their surface.

Of course if $\mathcal{C}$ is itself a geodesic then both tangents coincide with $\mathcal{C}$, so $\delta\varphi = 0$, and therefore $\kappa_g = 0$, as it should.

[1]This characterization was first discovered by Johann Bernoulli in 1697, who then taught it to his student, Euler.

11.5 A Simple Extrinsic Way to Measure Geodesic Curvature

In Act I we discussed the fact that if we peel away from a curved surface a narrow strip centred on a geodesic G, then when that strip is laid down on a flat plane it becomes a straight line $\widetilde{G}$. See [1.11] on page 12. Thus the intrinsic straightness ($\kappa_g = 0$) of G on the surface manifests itself as ordinary straightness in the plane: the curvature $\widetilde{\kappa}$ of the flattened strip $\widetilde{G}$ vanishes.

If we likewise remove from the surface a narrow strip centred on a *non*geodesic curve $\mathcal{C}$, for which $\kappa_g \neq 0$, then flatten it onto the plane, we obtain a plane curve $\widetilde{\mathcal{C}}$ for which its ordinary curvature $\widetilde{\kappa} \neq 0$. See [1.12] on page 13.

So how is the geodesic curvature $\kappa_g(p)$ at a particular point p on the surface related to the curvature $\widetilde{\kappa}(\widetilde{p})$ of the flattened strip at the corresponding point $\widetilde{p}$ of $\widetilde{\mathcal{C}}$ in the plane?

> *Let $\kappa_g(p)$ denote the geodesic curvature at p of a curve $\mathcal{C}$ on a curved surface, and let $\widetilde{\kappa}(\widetilde{p})$ denote the curvature at the corresponding point of the plane curve $\widetilde{\mathcal{C}}$ into which $\mathcal{C}$ is carried when a narrow strip centred on $\mathcal{C}$ is flattened onto the plane. Then*
>
> $$\kappa_g(p) = \widetilde{\kappa}(\widetilde{p}).$$

To see this, take the intrinsic construction [11.4] and imagine peeling narrow strips off the surface around the entire construction: around $\mathcal{C}$, and also around both of the small dashed segments of tangent geodesics meeting at angle $\delta\varphi$. When the small triangle of connected peeled strips is laid flat in the plane, we have returned to the original construction [8.4] for measuring curvature. The flattened geodesics have become straight lines in the plane, and since these lines are still tangent to $\widetilde{\mathcal{C}}$, they are tangent lines. But neither δs nor $\delta\varphi$ are altered in the flattening process, so $\kappa_g(p) \asymp (\delta s/\delta\varphi) \asymp \widetilde{\kappa}(\widetilde{p})$, as claimed.

11.6 A New Explanation of the Sticky-Tape Construction of Geodesics

We originally used the length-minimizing property of geodesics to explain the construction (1.7), whereby a geodesic is constructed by rolling a narrow strip of sticky tape down onto the surface, starting at an arbitrary point, and heading off in an arbitrary direction of our choosing. But, as we have been discussing, geodesics can also be characterized by their *straightness*: vanishing geodesic curvature. We now use this property to provide a new, second explanation of our geodesic construction.

Consider a narrow, straight strip of tape, $\mathcal{L}$, lying flat in the plane, with centre line L. The line L is extrinsically straight in $\mathbb{R}^3$; i.e., $\boldsymbol{\kappa} = \mathbf{0}$. It is also intrinsically straight within $\mathcal{L}$: $\boldsymbol{\kappa}_g = \mathbf{0}$. Now let us pick up $\mathcal{L}$ and isometrically bend and twist it in space into any shape we please. The centre line L remains intrinsically straight within whatever new form $\mathcal{L}$ now takes, so $\boldsymbol{\kappa}_g = \mathbf{0}$, still. Thus the curvature vector is normal to the strip all along L:

$$\boldsymbol{\kappa} = \boldsymbol{\kappa}_g + \boldsymbol{\kappa}_n = \boldsymbol{\kappa}_n.$$

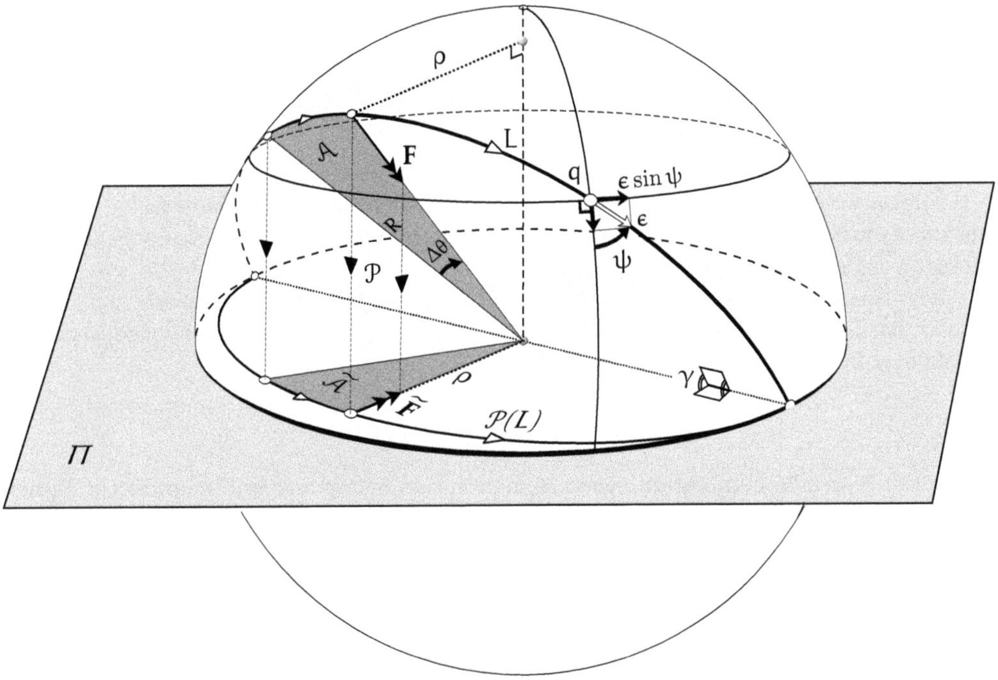

[11.5] *As the particle* q *travels along the geodesic* L *at unit speed, the radius sweeps out area at a constant rate, and therefore its projection onto* Π *does too. It follows that* $\rho \sin \psi$ *is constant along* L, *which is a special case of Clairaut's Theorem.*

Now roll the strip down onto a smooth surface $\mathcal{S}$. As we explained on page 14, only the centre line L can actually make contact with $\mathcal{S}$, but at each point of contact, p, the tangent plane of the strip coincides with the tangent plane of the surface: $\mathcal{T}_p(\mathcal{L}) = \mathcal{T}_p(\mathcal{S})$.

But we just established that $\boldsymbol{\kappa}$ is normal to the strip $\mathcal{L}$ all along L, but this means that it is also normal to the surface, $\mathcal{S}$. Thus, viewed as a curve within $\mathcal{S}$, the geodesic curvature of L vanishes, and it is indeed a geodesic of $\mathcal{S}$.

11.7 Geodesics on Surfaces of Revolution

11.7.1 Clairaut's Theorem on the Sphere

Geodesics are extremely hard to find explicitly on all but the simplest surfaces. However, there exists one general class of surfaces for which we can give an explicit geometrical recipe (called *Clairaut's Theorem*) for the paths of the geodesics: surfaces of revolution.

Of such surfaces, the sphere is one of the simplest; the Ancients already knew its geodesics—the great circles. We now look afresh at these great circles, and expose a hidden property that they possess, one that we will then be able to generalize to *all* surfaces of revolution.

Consider the sphere [11.5], thought of as the surface of revolution generated by rotating a semicircle about the vertical z-axis. As this semicircle (or indeed a general generating curve) rotates about the axis, its creates the so-called ***meridians*** of the surface. Equivalently, these are the curves of intersection of the surface with planes through its axis of symmetry, and the same is true on a general surface of revolution. On the sphere these meridians are the great circles through the poles, perhaps better known as circles of longitude. It is no accident that these meridians on the sphere are geodesics; as we already noted in the footnote on page 53, and as we shall discuss shortly, the meridians on a general surface of revolution are necessarily geodesics, too.

If we watch the particle moving along its great circle orbit L at unit speed for time Δt then the radius will rotate within the plane of L by $\Delta\theta = (\Delta t/R)$. Thus the radius sweeps out area $\mathcal{A} = \frac{1}{2}R^2\Delta\theta = \frac{1}{2}R\Delta t$, and so area is swept out at a *constant rate*: the same area $\mathcal{A}$ will be swept out in each equal period of time Δt. Historically, this situation is summarized by saying that *equal areas are swept out in equal times*.

Now let us project this orbit vertically downward onto the equatorial plane Π. This projection $\mathcal{P}$ is a linear transformation, and therefore all areas are contracted by the same factor, the determinant of the transformation. In fact it is easy to see geometrically [exercise] that if the plane of L makes an angle γ with Π, then $\det \mathcal{P} = \cos\gamma$.

It is also easy to see that the projection of the circular orbit L is an elliptical orbit $\mathcal{P}(L)$ in Π. It follows that as the particle travels round this ellipse, the radius in Π *also* sweeps out equal areas in equal times. In greater detail,

$$\frac{d\widetilde{\mathcal{A}}}{dt} = \cos\gamma \frac{d\mathcal{A}}{dt} = \frac{1}{2}R\cos\gamma.$$

Since $\mathcal{P}(L)$ is an ellipse, this steady generation of area is only possible if the projected particle moves slower when it is further away and faster when it is closer in. This variation in speed is readily confirmed and quantified by noting [exercise] the following:

If $\mathbf{q}(t)$ *is any trajectory in space, the velocity of the orthogonal projection* $\mathcal{P}(q)$ *onto a plane* Π *is the projection of the velocity:* $\frac{d}{dt}\mathcal{P}[\mathbf{q}] = \mathcal{P}\left[\frac{d\mathbf{q}}{dt}\right]$. (11.7)

The tangent plane T_q (not drawn) is spanned by the tangents to the orthogonal circles of longitude and latitude. As illustrated, the *direction* of the geodesic at q can be described by the angle ψ it makes with the meridian through q. If the particle q moves for a short time ϵ it will travel a small distance ϵ along the great circle orbit. Thus the horizontal component of the motion along the circle of latitude is $\epsilon \sin\psi$, as illustrated. It follows that the unit velocity vector to the geodesic, lying within T_p can be decomposed into a horizontal component $\sin\psi$ along the circle of latitude, and a component $\cos\psi$ along the circle of longitude (meridian).

The component along the meridian projects to radial motion in Π, generating no area. The area generation is entirely due to the horizontal component $\sin\psi$, which projects to an equal component perpendicular to the radius ρ in Π; note that ρ is the distance of the original particle from the axis of symmetry, as illustrated.

Since the rate of generation of area in Π is $\frac{1}{2}\rho\sin\psi$, we deduce that this quantity is constant along the original great circle orbit, and is given by,

$$\boxed{\tfrac{1}{2}\rho\sin\psi = \tfrac{1}{2}R\cos\gamma = \textit{const.}}$$

While this formula could have been demonstrated directly from the geometry of the sphere, the advantage of the above argument is that we will soon be able to generalize it to obtain,

Clairaut's Theorem. *Let* $\mathcal{S}$ *be a surface of revolution generated by rotating a curve* $\mathcal{C}$ *about an axis* $\mathcal{L}$. *If* ρ *is the distance from the axis* $\mathcal{L}$ *to a point* q *on a geodesic* g, *and* ψ *is the angle between the meridian* $\mathcal{C}_q$ *through* q *and the direction of* g, *then* $\rho\sin\psi$ *remains constant as* q *travels along the geodesic* g. *Conversely, if* $\rho\sin\psi$ *is constant along a curve* g *(no part of which is a parallel of* $\mathcal{S}$*), then* g *is a geodesic.* (11.8)

(Recall that a ***parallel*** is a horizontal circle on $\mathcal{S}$, the intersection of $\mathcal{S}$ with a plane perpendicular to $\mathcal{L}$.)

HISTORICAL NOTE: Alexis Claude Clairaut (1713–1765) was a French mathematician, astronomer, and geophysicist, who helped to extend Newton's results in the *Principia*. In 1752 he published an accurate, usable, approximate solution to the three-body problem of the Sun, Earth, and Moon, which Euler declared to be "... *the most important and profound discovery that has ever been made in mathematics.*" (Hankins, 1970, p. 35) The naming of the general theorem above stems from Clairaut's 1733 investigation of quadratic surfaces of revolution.

11.7.2 Kepler's Second Law

In order to understand the general version of Clairaut's Theorem, let us ask, *what is the magnitude and direction of the force that holds* $\mathcal{P}(q)$ *in its orbit* $\mathcal{P}(L)$? To answer this, we note the simple generalization of (11.7) from velocity to acceleration:

> *If* $\mathbf{q}(t)$ *is any trajectory in space, the acceleration of the orthogonal projection* $\mathcal{P}(\mathbf{q})$ *onto a plane* Π *is the projection of the acceleration:* $\frac{d^2}{dt^2}\mathcal{P}[\mathbf{q}] = \mathcal{P}\left[\frac{d^2\mathbf{q}}{dt^2}\right]$. (11.9)

Suppose for simplicity's sake that the particle orbiting on the sphere in [11.5] has unit mass, so that force is equal to acceleration. The force $\mathbf{F}$ that holds q in this orbit L is directed along the normal to the sphere, pointing directly at the centre, O, and it has constant magnitude $(v^2/R) = (1/R)$, since the particle has unit speed. Thus the force $\widetilde{\mathbf{F}}$ that holds the projection in its elliptical orbit in Π is *also* directed at O, and by virtue of the illustrated similar triangles,

$$\frac{|\widetilde{\mathbf{F}}|}{|\mathbf{F}|} = \frac{\rho}{R} \quad \Longrightarrow \quad |\widetilde{\mathbf{F}}| = (1/R^2)\rho.$$

A force field that is directed towards a single point O is called a ***central force field***. We have just established that the central force field in which the magnitude of the force directed to the centre of force is *proportional to the distance* of the particle from O results in an *elliptical orbit centred at* O *that sweeps out equal areas in equal times*. This was first proved by Newton (using a quite different argument) in the *Principia* (Proposition 10).

In practical terms, this force field and orbit can be created as follows. Imagine Π to be a frictionless sheet of ice, with a small hole at O. Take a small ice puck, and attach to it a length l of elastic string. Pass the length of string through the hole, and rest the puck on top of the hole. Take the other end of the elastic, hanging down at distance l below the hole, and attach it to a fixed point. Sitting at O, the puck experiences no pull from the string, because the string is relaxed at its natural length l. But if we move the puck a distance ρ away from O, thereby stretching the string by ρ, Hooke's Law then tells us that the stretched elastic will pull the puck back towards O with a force proportional to ρ. If we now launch the puck across the ice in an arbitrary direction with arbitrary speed, it will indeed trace an elliptical orbit centred at O, sweeping out equal areas in equal times. It is not hard to create an (ice-free) approximation of this experiment at home, and we encourage you to do so.

The fact that the orbit is an *ellipse centred at* O is specifically linked to the fact that the force varies *linearly* with distance. But, as Newton was the first to recognize and prove, the fact that equal areas are swept out in equal times is a remarkable *universal* property of *all* central force fields! We shall describe Newton's beautiful proof momentarily, but first let us pause to understand the pivotal role of this result in the *Principia*.

By analyzing the painstakingly accurate observations of the planets taken over a period of years by Tycho Brahe (1546–1601), Johannes Kepler (1571–1630) was able to discern mathematical patterns within the mass of data before him. These empirical mathematical facts are now known as Kepler's Three Laws of Planetary Motion. Kepler announced the first two laws in 1609, and the third (after tremendous struggles) in 1618:

Kepler's Laws

(I) *The orbit of a planet is an ellipse with the Sun at one of the two foci.*
(II) *A line segment joining the Sun to a planet sweeps out equal areas in equal times.*
(III) *The square of the orbital period of a planet is proportional to the cube of the semi-major axis of its orbit.*

For the next 70 years these laws would remain a mystery. Finally, in 1687, Newton succeeded in mathematically *explaining* Kepler's Laws as logical consequences of his universal, Inverse-Square Law of Gravitation—a spectacular vindication of his ideas. But in order for Newton to be able to achieve his analysis of dynamics using *geometry alone* (see Prologue) it was essential for him to be able to be able to *represent time as a geometrical quantity.*

Kepler's Second Law, or rather Newton's generalization of it to arbitrary central force fields, was therefore absolutely critical to the entire enterprise—it stands as Proposition 1 of the *Principia.* Newton's legion geometric diagrams and associated proofs in the *Principia* simply would not have been possible without this one absolutely fundamental fact: *area is the clock.*

11.7.3 Newton's Geometrical Demonstration of Kepler's Second Law

Figure [11.6] contains six copies of Newton's own diagram for Proposition 1 of the *Principia,* embellished so as to tell the story (as in a comic strip) of Newton's argument, establishing that in any central force field directed towards the fixed point *S*, the orbit ABCDEF sweeps out equal areas in equal times.

In the absence of any force, the first panel shows a particle travelling from A to B in a straight line, at uniform speed, as dictated by Newton's First Law of Motion; in this period of time, the radius from a fixed point S sweeps out the shaded area SAB. In the next equal increment of time, the particle will continue on an equal distance from B to c, still in the plane SAB, sweeping out the cross-hatched area SBc. But these two areas are *equal,* for the illustrated shear along BA brings SBc into coincidence with SAB. Thus, in the absence of force, equal areas are swept out in equal times.

Now suppose, instead, that at the moment the particle arrives at B it receives a sharp tap directed towards S. If the particle had been at rest at B initally, then this tap would have caused it to travel from B to v in the same time that it formerly travelled from B to c. The actual motion of the particle with therefore be the *sum* of these two motions, from B to C, still in the plane SAB, sweeping out the darkly shaded, cross-hatched area SBC. But this area is again *equal* to the original area SAB, for the illustrated shear parallel to SB carries SBC to SBc, and then (as before) a second shear brings SBc into coincidence with SAB.

Suppose we give the particle a second tap[2] towards S as it arrives at C, then it will travel from C to D, and, by the same reasoning as before, the area SCD will again equal the area SAB. Continuing in this manner, tapping the particle towards S at equal time intervals (at D, E, F, ...) the polygonal orbit ABCDEF sweeps out equal areas in equal times.

[2]Note that Newton's diagram appears to take the *magnitude* of each tap to be equal, but his argument is equally valid if the taps are unequal.

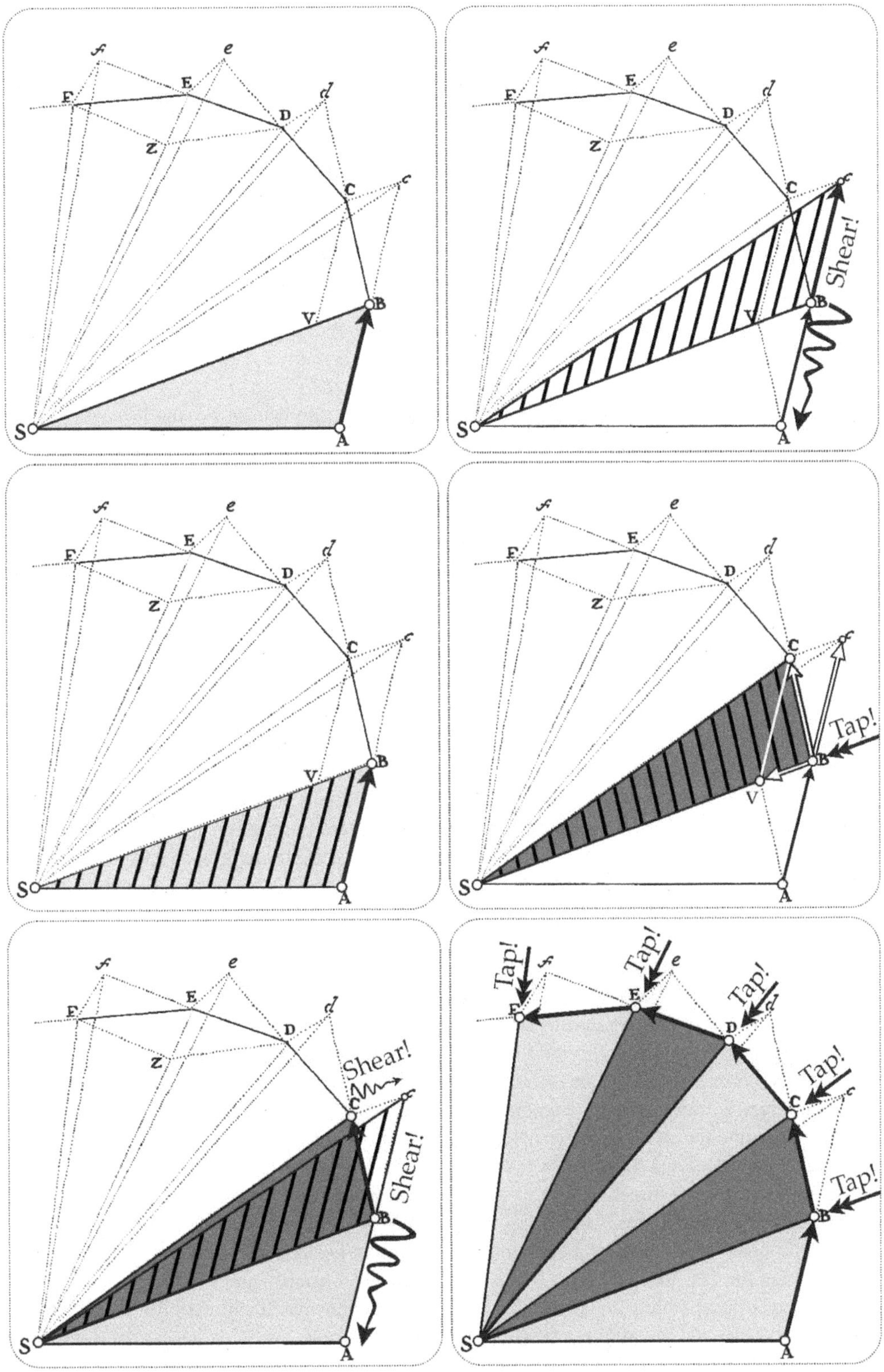

[11.6] Kepler's Second Law. *Six copies of Newton's own diagram for Proposition 1 of the* Principia *have been embellished so as to tell the story (as in a comic strip) of Newton's argument, establishing that in any central force field directed towards the fixed point S, the orbit* ABCDEF *sweeps out equal areas in equal times.*

Newton concludes,

> *Now let the number of triangles be increased and their width decreased indefinitely, and their ultimate perimeter ADF will be a curved line; and thus the centripetal force by which the body is continually drawn back from the tangent of the curve will act uninterruptedly, while any areas described will be proportional to those times Q.E.D.*

The remarkable elegance and economy of Newton's reasoning was not lost on Richard Feynman, who said[3] to his Caltech class in 1964,

> *The demonstration you have just seen is an exact copy of one in the* Principia Mathematica *by Newton, and the ingenuity and delight you may or may not have gotten from it is that already existing in the beginning of time.*

Lastly, observe [exercise] that the *converse* is also true; this is Proposition 2 of the *Principia*:

> *Every body that moves in some curved line described in a plane and, by a radius drawn to a point describes areas around that point proportional to the times, is urged by a centripetal force tending towards that same point.*

11.7.4 Dynamical Proof of Clairaut's Theorem

We are now but a step away from a satisfying explanation of Clairaut's Theorem on a general surface of revolution, as illustrated on the vase shown in [11.7].

As the point q moves along a geodesic g on this surface, the acceleration of the orbit is (by definition) always directed along the surface normal $\mathbf{n}$, and its direction therefore intersects the axis of symmetry $\mathcal{L}$. But, by virtue of (11.9), the projection of g onto Π is an orbit $\mathcal{P}(g)$ whose acceleration is therefore directed towards O, and so by Newton's generalized version of Kepler's Second Law, $\mathcal{P}(q)$ sweeps out area at a constant rate in Π. But, in the short time $\delta t = \epsilon$, the particle q travels a distance ϵ along the geodesic on the surface, and its projection sweeps out an area $\delta\mathcal{A}$ on the plane Π that is ultimately equal to the area of the white triangle with base ρ and height $\epsilon \sin\psi$. Thus $\delta\mathcal{A} \asymp \frac{1}{2}\rho\epsilon\sin\psi$, and so $\frac{d\mathcal{A}}{dt} = \frac{1}{2}\rho\sin\psi$ is constant. This completes the explanation of the first part of Clairaut's Theorem.

Note that meridians ($\psi = 0$) are exceptional geodesics, in the sense that their projections in Π move in and out radially, generating no area.

As for the converse, suppose $\rho\sin\psi$ is constant along a curve g on $\mathcal{S}$. Then $\mathcal{P}(q)$ sweeps out area in Π at a constant rate. Therefore, by Proposition 2 of the *Principia* (above), it follows that the acceleration of $\mathcal{P}(q)$ is always directed towards O. And from this it follows that the acceleration of g itself is always directed at the axis $\mathcal{L}$; equivalently, it lies within the vertical plane through q and $\mathcal{L}$, containing $\mathbf{n}$; equivalently, it is perpendicular at q to the parallel through q.

Next, assume the small segment of g containing q is *not* a parallel. Then $\mathbf{v}$ is not tangent to the parallel ($\psi \neq \frac{\pi}{2}$). Thus *the acceleration is perpendicular to two distinct directions in* $\mathcal{S}$: the direction of the parallel, and the direction $\mathbf{v}$ of g. Thus the acceleration is directed along $\mathbf{n}$, and so g is a geodesic, as was to be shown.

But suppose the segment of g containing q *is* part of a horizontal, circular parallel $\wp$, such as the top rim of the vase in [11.7]. Note that $\rho\sin\psi$ is indeed constant on $\wp$, by virtue of the fact that ρ and ψ are both separately constant, with $\psi = (\pi/2)$. But now, as illustrated, the acceleration is *horizontal* and directed towards the centre of the circular parallel, and this horizontal direction is in general *not* perpendicular to the surface, and hence $\wp$ is (in general) *not* geodesic.

---

[3]See his so-called *Lost Lecture* (Goodstein and Goodstein, 1996, p. 156), the audio recording of which is available on the internet.

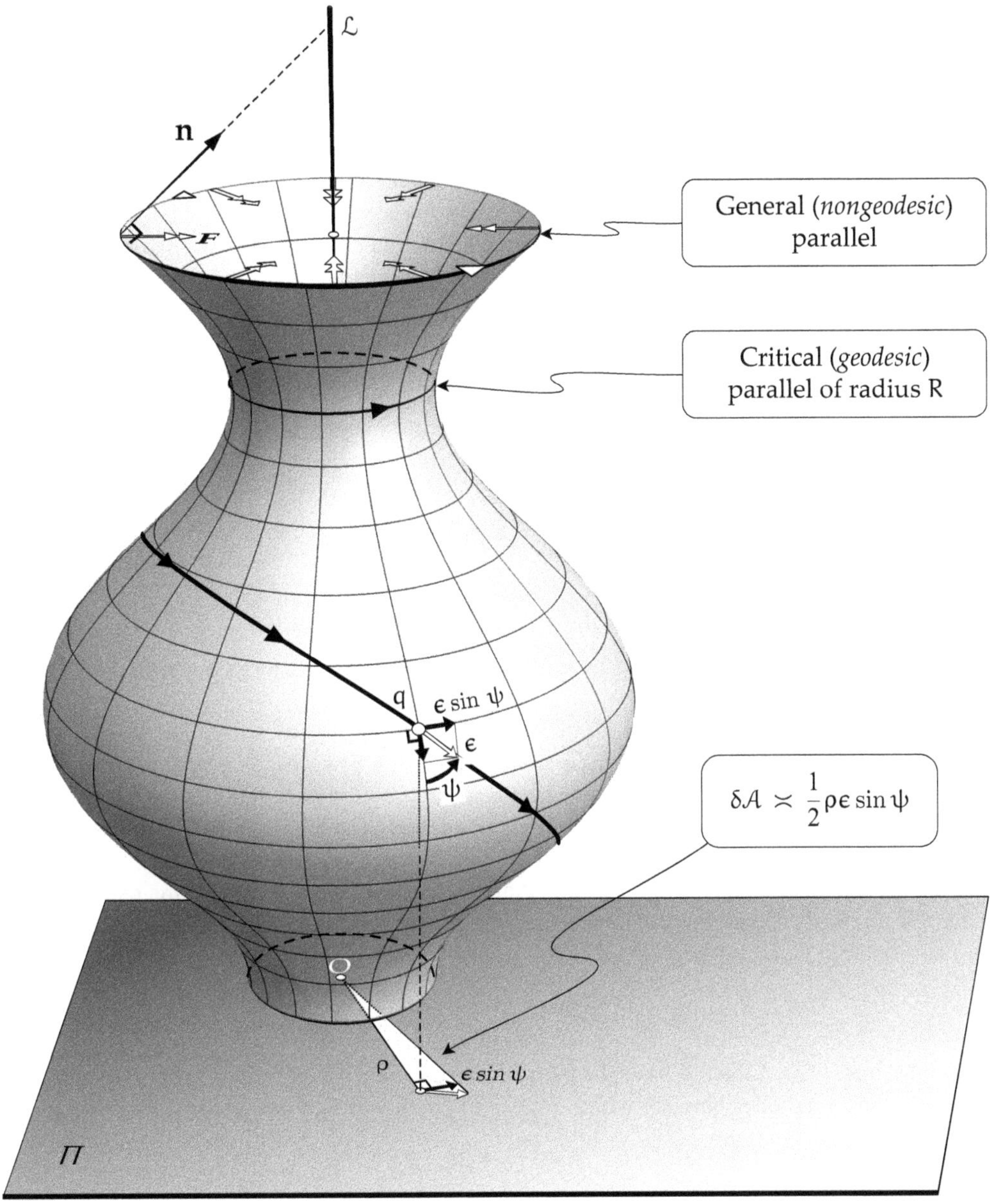

[11.7] *Clairaut's Theorem*. *In the short time* $\delta t = \epsilon$, *the particle* q *travels a distance* ϵ *along the geodesic on the surface, and its projection onto* Π *sweeps out an area* $\delta\mathcal{A}$ *that is ultimately equal to the area of the white triangle with base* ρ *and height* $\epsilon \sin\psi$. *Thus* $\frac{d\mathcal{A}}{dt} = \frac{1}{2}\rho\sin\psi$, *and therefore this quantity is constant, by virtue of Newton's generalization of Kepler's Second Law.*

However, $\wp$ *is* geodesic in the exceptional case that **n** is horizontal along $\wp$; an example is indicated in [11.7]. Such parallels $\wp$ are called *critical*, and can be characterized in various ways. For example, if the vertical cylinder with axis $\mathcal{L}$ containing $\wp$ touches $\mathcal{S}$ along $\wp$, then $\wp$ is geodesic. Most textbooks instead describe this situation by supposing that the generating curve $\mathcal{C}$ of $\mathcal{S}$ can be described by a graph $\rho = \rho(z)$, where z is vertical distance along $\mathcal{L}$. Then the parallel $\wp$ is geodesic if and only if it is "critical" in the sense that $\rho'(z) = 0$; in other words, the distance of the profile curve from $\mathcal{L}$ has a maximum, a minimum, or a point of inflection.

For readers who have studied some physics, we note that the quantity $\rho \sin \psi = \Omega$ is the ***angular momentum*** about the axis $\mathcal{L}$ of the (unit-mass) particle. The fact that Ω remains constant can be understood physically[4]: it is a consequence of the fact that the force that holds q in its geodesic orbit on $\mathcal{S}$ passes through $\mathcal{L}$ and therefore has no ***moment*** about $\mathcal{L}$. In exactly the same way that a spinning ice-skater spins faster as she pulls her arms in, so, in order to conserve its angular momentum, the particle orbiting on the surface must spin around $\mathcal{L}$ faster as it comes closer to $\mathcal{L}$. But since the particle on the surface has constant *linear* speed along the geodesic, it can only increase it *angular* speed around the axis by directing its velocity towards the horizontal.

Note that the angular momentum Ω of a geodesic g actually tells us the closest g can get to $\mathcal{L}$:

$$\rho = \frac{\Omega}{\sin \psi} \geqslant \Omega = \rho_{\mathrm{min}}.$$

For example, suppose g starts below the critical latitude (of radius R) at the throat of the vase in [11.7], and then climbs upward towards it. If the angular momentum is too large ($\Omega = \rho_{\mathrm{min}} > R$) then the g cannot ever reach the throat of the vase; instead, it bounces back down the vase. This is the case for the illustrated geodesic, which will likewise be forced to bounce back *up* again as it approaches the narrow base. To get a better feel for all this, try using a real vase, constructing geodesics using the sticky-tape construction, (1.7), page 14. See Exercise 11, page 26.

11.7.5 Application: Geodesics in the Hyperbolic Plane (Revisited)

Let us now apply Clairaut's Theorem to the pseudosphere, so as to gain a fresh view of the geodesics of the hyperbolic plane.

Since the pseudosphere becomes arbitrarily narrow the higher up we go, it follows from the discussion above that if a particle is travelling up the pseudosphere and has *any* angular momentum at all, it must eventually turn back and head back down. Thus the meridian, tractrix generators (which have zero angular momentum) are the *only* geodesics than can continue indefinitely upward.

Recall the construction of the Beltrami–Poincaré upper half-plane map of the pseudosphere, which we derived geometrically in [5.4], page 55. What we had called X in that figure is now called ρ. Thus the height y in the map (given by (5.4), page 55) is

$$y = \frac{1}{\rho},$$

where we have chosen the radius of the pseudosphere to be $R = 1$, in order to obtain the standard model of the hyperbolic plane, for which

$$\mathcal{K} = -(1/R^2) = -1.$$

Figure [11.8] shows a geodesic g (starting at a and ending at b) on the pseudosphere and its image in the map (starting at A and ending at B). Recall that we previously proved (using optics) that this image is a semicircle meeting the horizon $y = 0$ at right angles. We can now give a fresh proof of this fact using Clairaut's Theorem, and in the process give a new interpretation of the *size* of the semicircle representing g.

Recall that this map is, by construction, *conformal*. That means that the angle ψ between g and the tractrix generator (meridian) of the pseudosphere is preserved: *The image of* g *in the map makes the* same *angle* ψ *with the vertical half-line image of the tractrix.* If the unit-mass particle travelling along g has angular momentum Ω, then Clairaut's Theorem yields,

$$\rho \sin \psi = \Omega \quad \Longrightarrow \quad y = \frac{1}{\Omega} \sin \psi.$$

[4]Pressley (2010, p. 230) also provides this physical explanation.

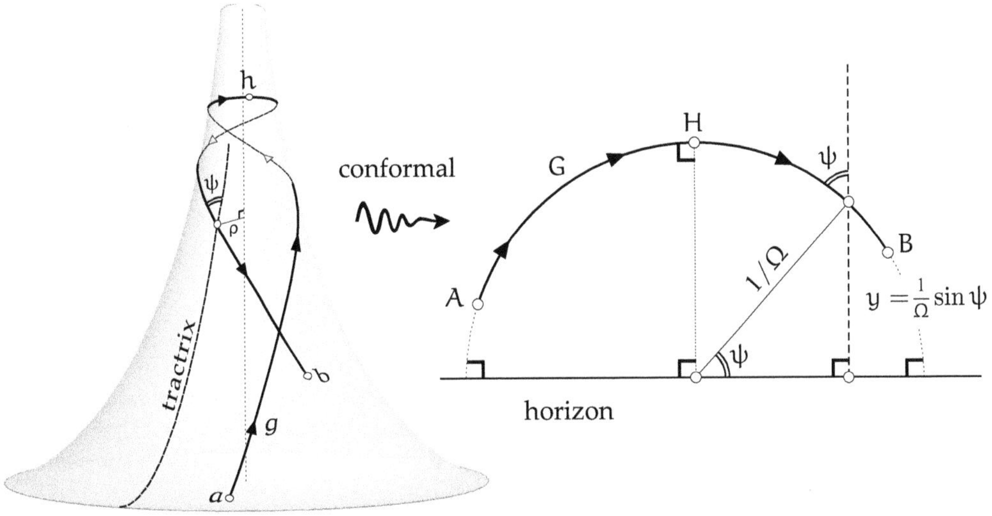

[11.8] *A unit-mass particle travels at unit speed along a geodesic* g *on the pseudosphere of radius* $R = 1$. *By Clairaut's Theorem, its angular momentum* $\Omega = \rho \sin\psi$ *is constant. But since the Beltrami–Poincaré map is* conformal, *the angle* ψ *is preserved, and it follows that the image* G *in the map has equation* $y = \frac{1}{\Omega} \sin\psi$, *which is a semicircle of radius* $(1/\Omega)$ *meeting the horizon* $y = 0$ *at right angles.*

Thus, using exactly the same reasoning as in [5.8], page 60, we conclude that,

If a unit-mass particle travels at unit speed along a geodesic g *on the pseudosphere, and its angular momentum about the axis of symmetry is* Ω, *then its image in the Beltrami–Poincaré upper half-plane travels along a semicircle meeting* $y = 0$ *at right angles. Furthermore, the radius of this semicircle is* $(1/\Omega)$. *In other words, the (Euclidean) curvature of the semicircle is the angular momentum of the particle.*

Lastly, let h be highest point that the particle can reach before its angular momentum Ω forces it to head back down the pseudosphere. Clearly, h is mapped to the highest point H on the image G of g, both h and H corresponding to $\psi = (\pi/2)$. It follows from (5.1) that the arc length σ_{max} along the segment of the tractrix generator (not shown) going from the rim straight up to h, is given by the logarithm of the radius of G:

$$\sigma_{max} = \ln \frac{1}{\Omega}.$$

Chapter 12

The Extrinsic Curvature of a Surface

12.1 Introduction

We have seen how the two principal curvatures (together with their associated principal directions) characterize the extrinsic geometry of a surface in great detail, via Euler's curvature formula, (10.1). But is there is *single number* (without any associated direction) that can characterize the overall extrinsic geometry of a surface at a point, in the same way as the Gaussian curvature $\mathcal{K}$ characterizes the intrinsic geometry?

To characterize the overall extrinsic geometry we must presumably take some kind of *average* of the principal curvatures. The two most obvious ways of doing this are to take their arithmetic mean, $\frac{\kappa_1+\kappa_2}{2}$, or else their geometric mean, $\sqrt{\kappa_1\kappa_2}$. *Both* of these averages are geometrically natural and extremely important.

The arithmetic mean is usually denoted by the letter H (or $\overline{\kappa}$) and it is simply called the ***mean curvature***:

$$\boxed{H = \overline{\kappa} \equiv \frac{\kappa_1+\kappa_2}{2}.} \tag{12.1}$$

Look again at the significance in [10.2] of H to the graph of Euler's curvature formula: it is the centre about which the curvature oscillates sinusoidally. This mean curvature H is fundamental to understanding the shape of so-called ***minimal surfaces***, which include the shapes of all possible soap films spanning a complicated curved wire frame. These minimal surfaces must, by definition, satisfy $H=0$ at every point, so that $\kappa_2=-\kappa_1$, and the surface is saddle-shaped. As noted in the Prologue, we shall not explore these fascinating surfaces in this work; instead, we refer you to the *Further Reading* section, where several excellent works are recommended.

The geometric meaning of $\sqrt{\kappa_1\kappa_2}$, or rather its *square*, $\kappa_1\kappa_2$, turns out to be even more fundamental than that of H, but we shall deliberately keep you in suspense a little longer as to what that significance might be. Forgive us, but even the great Gauss declared it to be one of the greatest punchlines in all of mathematics, so a few pages of drum-roll is in order. (If you cannot bear to wait, jump ahead to the answer: (13.3) on p. 142.)

12.2 The Spherical Map

As we know, the curvature κ of a plane curve $\mathcal{C}$ can be defined as the rate of turning of its tangent. Equivalently, κ may be viewed as the rate of turning of the normal. This latter interpretation will permit us to generalize this extrinsic definition of curvature from curves to surfaces, for the latter also admit a unique ($\pm$) normal vector.

First, however, it will be helpful if we reinterpret the rate of turning of a plane curve's normal vector as the *spread* of the directions of the normal vectors that occurs over a short piece of the curve, say of length δs.

To quantify this, imagine taking each of these unit normal vectors and transplanting them so that they all emerge from a common point $\mathcal{O}$. See [12.1]. In this way, the normal $\mathbf{N}$ may be viewed

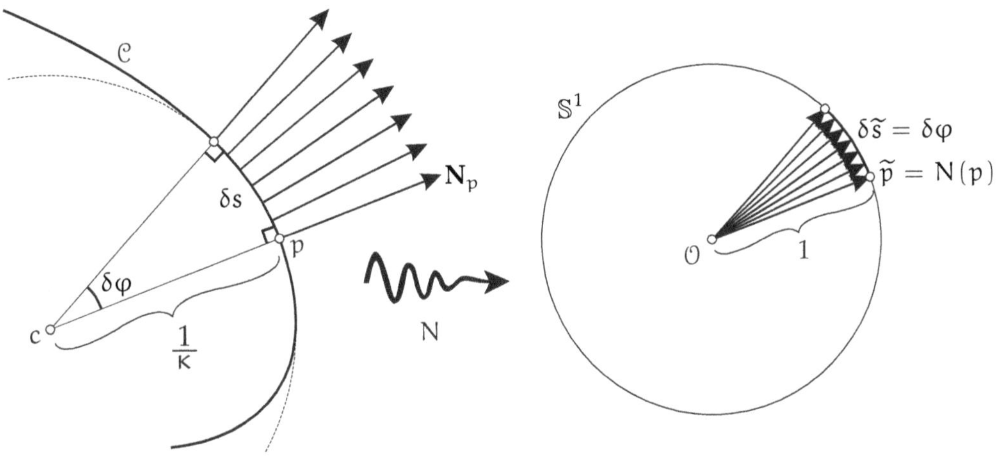

[12.1] *The map* N *is defined to send the point* p *on* $\mathcal{C}$ *to the point on the unit circle lying in the same direction as* $\mathbf{N}_p$*. Then* $\kappa \asymp (\delta\varphi/\delta s)$ *can be reinterpreted as the local length magnification factor of* N*, measuring the rate of spreading of the normals.*

as a mapping N from the point p of the plane curve $\mathcal{C}$ to the point $\tilde{p} = N(p)$ on the unit circle S^1 (centred at $\mathcal{O}$) that lies at the tip of $\mathbf{N}_p$.

Over the segment δs of $\mathcal{C}$, the directions of the normal vectors are spread over angle $\delta\varphi$, and therefore the tips of these normal vectors fill an arc of length $\delta\tilde{s} = \delta\varphi$ on the unit circle. The local spreading of the normal vectors can now be quantified by considering the local magnification of arc length under the normal map:

$$\kappa = \text{local length magnification factor of the N map} \asymp \frac{\delta\tilde{s}}{\delta s}. \tag{12.2}$$

This suggests a way to generalize the construction to a surface $\mathcal{S}$. Consider a small patch of $\mathcal{S}$ of area $\delta\mathcal{A}$ and containing the point p, the normal vector there being $\mathbf{n}_p$. See [12.2]. By analogy with [12.1], we introduce the ***spherical map***—most commonly called the ***Gauss map***, or the ***normal map***—from the surface to the unit sphere ($n : \mathcal{S} \to S^2$) sending the point p to the point $\tilde{p} = n(p)$ on the unit sphere lying in the same direction as $\mathbf{n}_p$.

HISTORICAL NOTE ON TERMINOLOGY: In essentially all other texts, the "spherical map" is instead called the "Gauss map," but this is historically inaccurate. Yes, Gauss did indeed publish it in 1827 (and privately used it years earlier), but it was Olinde Rodrigues (1795–1851) (a French banker and amateur mathematician) who first published this concept in 1815, employing it in a penetrating study of the curvature of surfaces. In recognition of this fact, Marcel Berger (one of the foremost geometric authorities of the late twentieth century) calls it the ***Rodrigues–Gauss map***, which seems to us to strike an appropriate balance. That said, we shall generally prefer the term "spherical map,"[1] by virtue of its clarity and brevity.

12.3 Extrinsic Curvature of Surfaces

We are thus led to a brand new *extrinsic* measure of surface curvature in terms of the spread of the normal vectors, which we shall temporarily denote $\mathcal{K}_{ext}$.

---

[1] A strong precedence argument can also be made for "spherical map": it was the name preferred by such legendary figures as Hilbert and S. Cohn-Vossen (1952) and Hopf (1956). It has likewise been employed by I. M. Singer, V. A. Toponogov, and other famous differential geometers of the modern era.

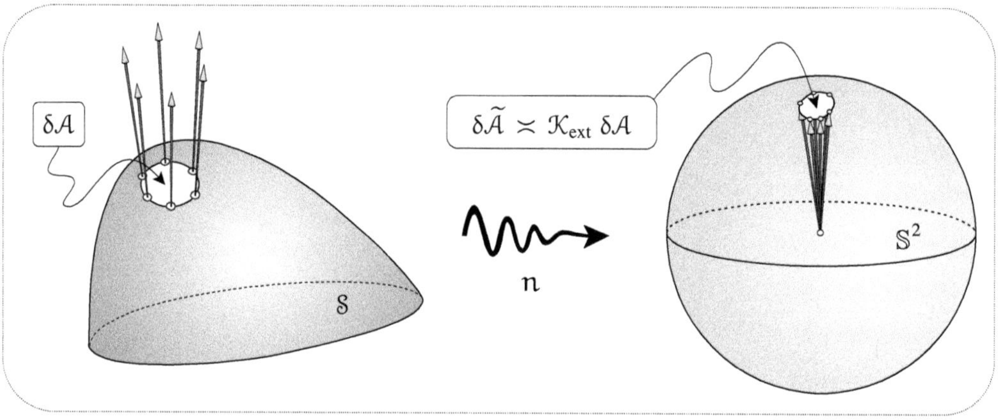

[12.2] *The extrinsic curvature $\mathcal{K}_{ext}$ is the local area magnification factor of the spherical map:* $\mathcal{K}_{ext} \asymp \frac{\delta\widetilde{\mathcal{A}}}{\delta\mathcal{A}}$.

In [12.2] we shrink the small shape down towards p and define

$$\mathcal{K}_{ext} \equiv \text{local area magnification factor of the spherical map} \asymp \frac{\delta\widetilde{\mathcal{A}}}{\delta\mathcal{A}}. \tag{12.3}$$

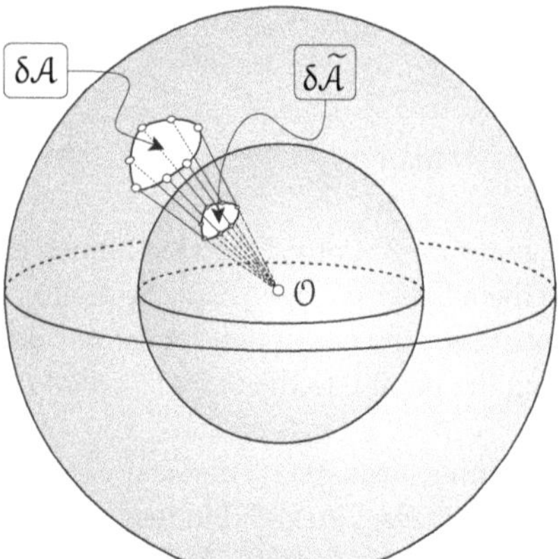

[12.3] *If S is a sphere of radius R centred at O, and the image S^2 under n is imagined to be concentric to it, then n is simply radial projection from O. Thus distances within S are compressed by* $(1/R)$ *and areas on S are compressed by* $(1/R)^2$.

For example, suppose S is a sphere of radius R centred at O, and picture the image S^2 under the spherical map as having the same centre. Then n is simply radial projection from O, as illustrated in [12.3]. Linear dimensions clearly shrink by $(1/R)$ and areas therefore shrink by $(1/R)^2$, so

The sphere of radius R has extrinsic curvature $\mathcal{K}_{ext} = (1/R^2)$. (12.4)

As a second example, let S be a cylinder of radius R with axis L, as illustrated in [12.4]. All points on the same generator of the cylinder have the same normal vector and hence the same image under the spherical map. Since all the normals of S are perpendicular to L their images under n lie on the great circle of S^2 that lies in the plane perpendicular to L. Thus any area on S is crushed down by n to an arc of this great circle, having zero area. Note that the same crushing of generators also occurs on a *cone*, thus

The cylinder and the cone both have $\mathcal{K}_{ext} = 0$. (12.5)

In contrast to (12.2), it is not immediately obvious that the magnification factor in (12.3) is uniquely defined in the general case. However, later we shall be able to show that it really is: all

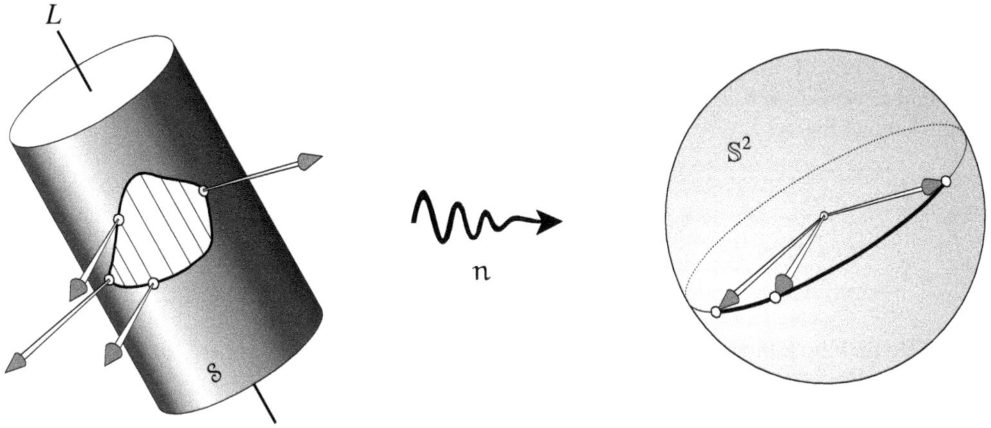

[12.4] *The spherical map* $\mathfrak{n}$ *crushes each generator of the cylinder (parallel to its axis* L*) to a single point on the great circle perpendicular to* L*, so* $\mathcal{K}_{ext} = 0$.

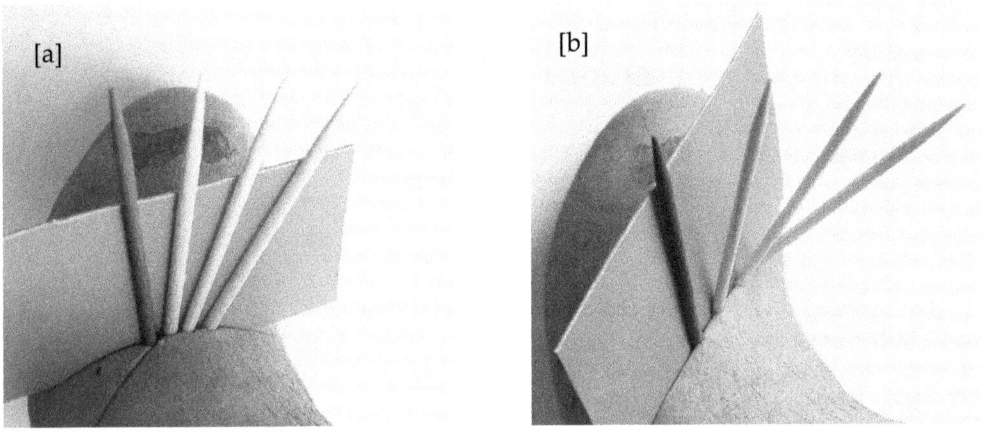

[12.5] [a] *If we move in a principal direction then the normal vector tips in that direction and initially stays within the normal plane;* **[b]** *If we move in a general direction then the normal vector immediately tips out of the normal plane.*

infinitesimal areas at a given point undergo the *same* magnification, regardless of their shape. For now, let us assume this and seek a specific shape that will reveal an explicit formula for $\mathcal{K}_{ext}$.

To help us find a special shape for which the area magnification factor is geometrically self-evident, two lemmas are needed, the first of which is this:

> *As* p *begins to move in a principal direction,* $\mathbf{n}_p$ *remains within the normal plane* Π *in that direction.* (12.6)

A bit more precisely, we mean by this that if ζ is defined to be the angle between $\mathbf{n}$ and Π, so that $\zeta(p) = 0$, then $\dot{\zeta}(p) = 0$. This follows from the local mirror symmetry of $\mathcal{S}$ in Π, for if $\mathbf{n}$ were to immediately rotate out of Π to one side or the other ($\dot{\zeta} > 0$ vs. $\dot{\zeta} < 0$) this would violate the symmetry (10.4).

Fig. [12.5a] illustrates (12.6) on a yam. On the same yam, and at the same point, [12.5b] illustrates the fact that if we instead move off in a *general* direction then $\dot{\zeta}(p) \neq 0$ and $\mathbf{n}$ immediately rotates out of Π. We strongly encourage you to conduct your own such experiments, using whatever fruits or vegetables are readily available.

If we think of the left-hand side of [12.1] as a cross-section of $\mathcal{S}$, drawn in the plane Π, and the right-hand side as a cross section of S^2 in the plane parallel to Π through $\mathcal{O}$, the second lemma follows immediately:

> *If* p *moves along a short vector* $\mathbf{v}_i$ *in the* i^{th} *principal direction of* $\mathcal{S}$*, then* $n(p)$ *ultimately moves along* $-\kappa_i \mathbf{v}_i$ *on* S^2. (12.7)

Note that the minus sign in this formula is due to our conventions: in [12.5a] $n(p)$ moves in the *same* direction as $\mathbf{v}_i$, a *positive* multiple of $\mathbf{v}_i$, but (by our convention) here $\kappa_i < 0$, because the normal section bends *away* from our chosen $\mathbf{n}$.

We are thus guided to consider the fate under the spherical map of a small rectangle whose sides are aligned with the principal directions. Let the lengths of these sides be ϵ_1 and ϵ_2, along the first and second principal directions, respectively. By virtue of (12.7), the spherical map ultimately sends this rectangle to another *rectangle* on S^2, parallel to the original, but with the sides ultimately stretched to $\kappa_1\epsilon_1$ and $\kappa_2\epsilon_2$. Thus,

$$\delta\widetilde{\mathcal{A}} \asymp (\kappa_1\epsilon_1)(\kappa_2\epsilon_2) \asymp (\kappa_1\kappa_2)\,\delta\mathcal{A},$$

and therefore (12.3) yields

$$\mathcal{K}_{\text{ext}} = \kappa_1\kappa_2. \tag{12.8}$$

For example, on a sphere of radius R, $\kappa_1 = \kappa_2 = (1/R)$, yielding $\mathcal{K}_{\text{ext}} = (1/R^2)$, in agreement with (12.4). Note that if the principal curvatures of a general surface have the same sign, then we obtain the following interpretation of their geometric mean: the extrinsic curvature of the surface is the same as that of a sphere of radius $1/\sqrt{\kappa_1\kappa_2}$.

As another example, a cylinder of radius R has $\kappa_1 = (1/R)$ and $\kappa_2 = 0$, yielding $\mathcal{K}_{\text{ext}} = 0$, in agreement with (12.5). A cone also has one vanishing principal curvature, and hence $\mathcal{K}_{\text{ext}} = 0$ in this case, too.

HISTORICAL NOTE: Both the extrinsic definition (12.3) of curvature and the explicit formula (12.8) are almost universally attributed to Gauss (1827). However, as with the spherical map (aka the Rodrigues–Gauss map) itself, both of these insights were in fact first published by Rodrigues in 1815, twelve years prior to Gauss. It appears that Gauss (along with most twentieth-century mathematicians!) was simply unaware of Rodrigues's discoveries. For more on this history, see Kolmogorov and Yushkevich (1996, p. 6) and Knoebel (2007, p. 118).

The formula (12.8) attaches a *sign* to $\mathcal{K}_{\text{ext}}$: if p is elliptic (i.e., κ_1 and κ_2 have the same sign) then $\mathcal{K}_{\text{ext}} > 0$; if it is hyperbolic (i.e., κ_1 and κ_2 have opposite signs) then $\mathcal{K}_{\text{ext}} < 0$; and if it is parabolic (i.e., one of the κ_i vanishes) then $\mathcal{K}_{\text{ext}} = 0$.

In order to make sense of this sign of $\mathcal{K}_{\text{ext}}$ in terms of its original definition (12.3), we take $\delta\mathcal{A}$ to be positive and we use a simple geometric property of the spherical map to attach a sign to $\delta\widetilde{\mathcal{A}}$, as follows.

Imagine that, with $\mathbf{n}$ pointing at your eye, you see the boundary of $\delta\mathcal{A}$ on $\mathcal{S}$ traced counterclockwise, as illustrated in [12.6]. The idea is to attach a sign to the area $\delta\widetilde{\mathcal{A}}$ on S^2 according to whether the spherical map preserves or reverses the orientation of the boundary. That is, we take $\delta\widetilde{\mathcal{A}}$ to be *positive* if its boundary is traced in the *same* counterclockwise sense (as seen from outside the sphere) that the original was, and we take it to be *negative* if it is traced in the *opposite*, clockwise sense.

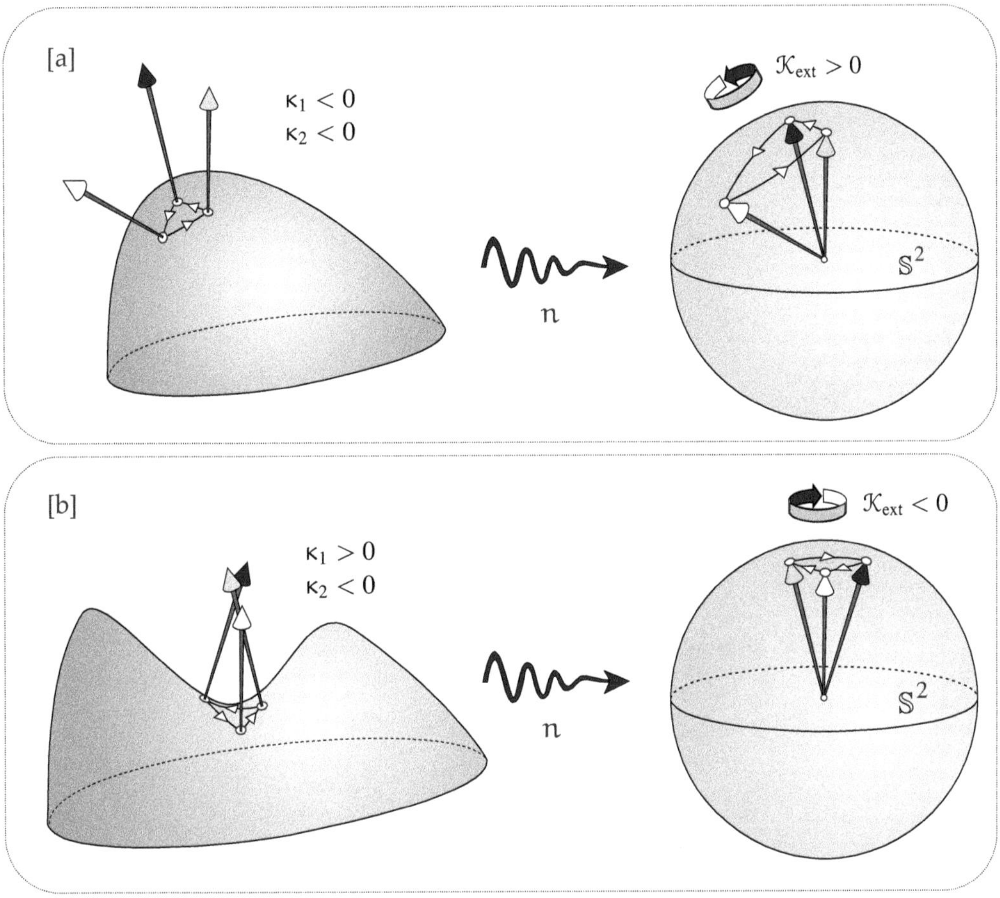

[12.6] *The sign of $\mathcal{K}_{ext}$ depends on whether the spherical map $\mathfrak{n}$ preserves orientation (**[a]**: $\mathcal{K}_{ext} > 0$) or reverses it (**[b]**: $\mathcal{K}_{ext} < 0$).*

12.4 What Shapes Are Possible?

Let us analyze systematically what shapes are possible for an arbitrary surface, at least *locally*.

In the immediate vicinity of a generic point, we have seen that the surface is described by the quadratic in (10.6) and repeated here:

$$z \asymp \tfrac{1}{2}\kappa_1 x^2 + \tfrac{1}{2}\kappa_2 y^2.$$

In general, neither principal curvature vanishes, so $\mathcal{K}_{ext} = \kappa_1\kappa_2 \neq 0$. Recall that a point is called elliptic if $\mathcal{K}_{ext} > 0$, and called hyperbolic if $\mathcal{K}_{ext} < 0$. The shape of the surface in the vicinity of such a generic point is *completely* determined by this distinction regarding the *sign* of $\mathcal{K}_{ext}$:

> *If $\mathcal{K}_{ext} > 0$ then the surface is locally a bowl, as in [12.7a].*
>
> *If $\mathcal{K}_{ext} < 0$ then the surface is locally a saddle, as in [12.7b].* (12.9)

The remaining case $\mathcal{K} = 0$ can arise in two different ways: either one principal curvature vanishes, or both do. In the former case, the point is called parabolic, and in the latter case it is called ***planar***, for reasons that will become clear shortly.

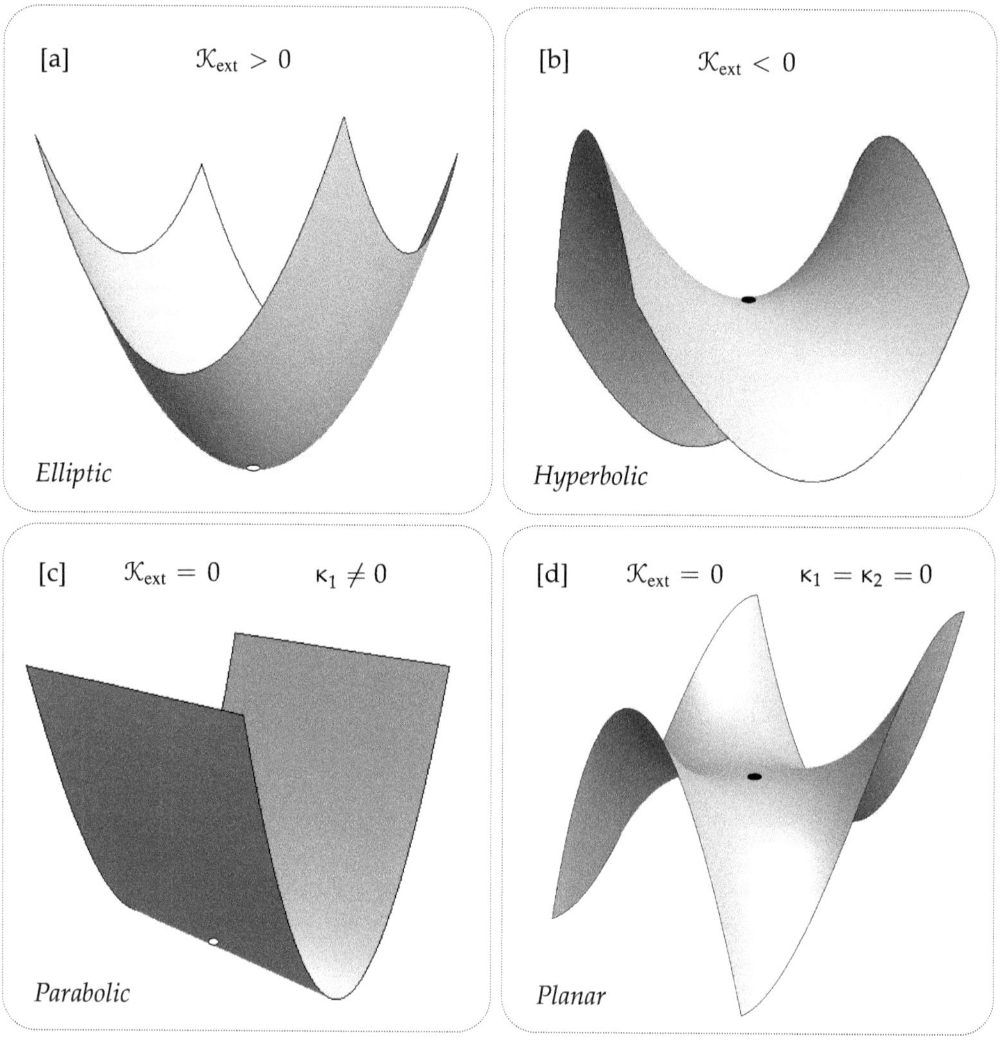

[12.7] *The local shape of the surface is* **[a]** *a bowl if $\mathcal{K}_{ext} > 0$;* **[b]** *a saddle if $\mathcal{K}_{ext} < 0$;* **[c]** *a trough if $\mathcal{K}_{ext} = 0$ (and only one of κ_1 and κ_2 vanishes). If* both *principal curvatures vanish, then the surface can be arbitrarily complex;* **[d]** *illustrates just one possibility, called the* **monkey saddle**.

In the former parabolic case, suppose that $\kappa_1 \neq 0$ and $\kappa_2 = 0$. Then the surface is locally given by $z \asymp \frac{1}{2}\kappa_1 x^2$, which is a trough (with a parabolic cross section) running in the y-direction, as illustrated in [12.7c].

The surface of a doughnut provides examples of all three cases thus far considered. Every point on the outer half of the doughnut has positive curvature, and looks like [12.7a], and every point on the inner half of the doughnut has negative curvature, and looks like [12.7b]. If we imagine the doughnut sitting on a plate, the circle of contact with the plate separates the two halves of opposite curvature, and is made up of parabolic points where $\mathcal{K}_{ext} = 0$. Sure enough, the strip of surface surrounding this circle is indeed a trough, like [12.7c]. For a diagram that focusses on this circular trough, look ahead to [13.3a], page 141.

In the final "planar" case, $\mathcal{K} = 0$ again, but now *both* principal curvatures vanish. Since $\kappa_1 = \kappa_2 = 0$, we see that z must be ultimately equal to a *cubic* (or higher-powered) homogeneous polynomial in x and y. It makes sense that such a point is called planar, because the curvature of the normal section vanishes in every direction, like a plane, and the surface departs from its tangent plane so slowly that it does indeed look very plane-like locally. But of course as we move further away, or simply look with greater precision, we can see the surface bending away from the tangent plane. As we shall now explain, the shape of the surface near such a planar point can be arbitrarily complicated.

To prepare ourselves for the more complicated behaviour that the surface can exhibit near a $\mathcal{K}_{ext}=0$ planar point, let us first take a fresh look at a regular $\mathcal{K}_{ext}<0$ saddle. For simplicity's sake, let us assume that $\kappa_1=-\kappa_2=2$. Now let us think of the tangent plane as the *complex* plane, so that each point within it can be described (with Cartesian and polar coordinates) by the complex number $x+iy=r\,e^{i\theta}$. Then,

$$\text{Height of saddle} = (x^2-y^2) = \text{Re}\,(x+iy)^2 = \text{Re}\,\left[r^2\,e^{i2\theta}\right] = r^2\cos 2\theta.$$

This new form of the height formula makes it transparent that as we circle once around the origin, the surface does *two* complete oscillations in height. If we picture an actual saddle on a horse, then the "valley" of the saddle on one side of the horse corresponds to one of these two complete oscillations of height, and it accommodates one of the rider's legs, while the second complete oscillation provides the valley on the other side, for the other leg. And directly in front and behind the rider, the saddle rises up in what we shall call "hills."

Now let us construct a saddle suitable for a *monkey*, who has two legs and a *tail*; note that the official, technical name for this surface is indeed the ***monkey saddle***! To create the three equally spaced valleys the monkey requires, we need only *cube* our complex number:

$$\text{Height of monkey saddle} = \text{Re}\,(x+iy)^3 = \text{Re}\,\left[r^3\,e^{i3\theta}\right] = r^3\cos 3\theta.$$

This monkey saddle is illustrated in [12.7d].

Note that the three valleys are now equally spaced, $(2\pi/3)$ apart, and the three hills are likewise equally spaced in between the valleys: each hill is directly opposite a valley. As you can see, this means that the normal sections necessarily have an *inflection point* at the origin, implying (without calculation) that the curvature must vanish there. Since the normal curvature vanishes in every direction, the monkey saddle does indeed have a planar point at its centre.

It is now easy to see that there exists an infinite menagerie [*sic*] of such saddles, of ever increasing complexity, of which the monkey saddle may be described as the 3-saddle. For example, if we want to create a ***cat saddle***—which we must confess is *not* standard terminology!—we need only replace the third power of the monkey saddle with the *fifth* power, to create the 5-saddle, with height given by $r^5\cos 5\theta$.

Note that it is visually clear that all these saddles have strictly *negative* curvature everywhere except at the planar point itself, where $\mathcal{K}_{ext}=0$. In fact it can be shown (see Ex. 20) that the curvature $\mathcal{K}_{ext}$ of the general n-saddle is *symmetrical* around the origin: it only depends on r and therefore has a constant negative value on each circle $r=$ *constant*.

We have not yet exhausted all possible shapes surrounding a planar point. While all the higher-order saddles have negative curvature surrounding the planar point of vanishing curvature, it is *also* possible to have the planar point be surrounded by a sea of *positive* curvature. For example, consider [exercise] the almost flat-bottomed bowl with equation $z=r^4=(x^2+y^2)^2$.

Chapter 13

Gauss's *Theorema Egregium*

13.1 Introduction

In 1827 Gauss announced the ***Theorema Egregium***—Latin for "Remarkable Theorem." Although that year witnessed the death of Beethoven, the appearance of the *Theorema Egregium* meant that it also witnessed the birth of modern Differential Geometry.

From this result sprang fundamental advances in both mathematics and physics, some of which we have touched on already, and some of which must await future chapters. It paved the road to Beltrami's crucial 1868 step in the acceptance of Hyperbolic Geometry, interpreting it as the intrinsic geometry of a saddle-shaped surface of constant negative Gaussian curvature. And Riemann's brilliant generalization of Gauss's intrinsic curvature to higher-dimensional manifolds in turn enabled Einstein in 1915 to give precise mathematical form to his supremely beautiful and supremely accurate General Theory of Relativity, in which gravitation is understood as the curvature impressed upon the intrinsic geometry of space and time by matter and energy.

But in order to properly appreciate the *Theorema Egregium* itself, we shall first describe its origins. The fascinating and insightful work of Dombrowski (1979) uses Gauss's private notebooks, letters to friends, and official publications to carefully piece together a chronology of Gauss's evolving insights into Differential Geometry in general, and into this theorem, in particular. As Dombrowski explains, Gauss's epic explorations began in 1816, when he made a *nonlocal* discovery about the curvature of surfaces that was profound and unexpected in equal measure.

13.2 Gauss's *Beautiful Theorem* (1816)

Gauss was not given to gushing. But if he was stingy in his praise of others, then at least he was not a hypocrite. A succession of his major discoveries lay hidden in his private notebooks till after his death because he deemed them insufficiently perfected to be worthy of publication. Indeed, his Latin motto was *pauca, sed matura*, "few, but ripe." One such unpublished discovery was the nonlocal result of 1816.

Agonizingly, even in his private notebook[1] Gauss left no trace of what led him to suspect the result, nor how he proved it, but with uncharacteristic exuberance he was moved to *name* it:

> "**Beautiful Theorem.** *If a curved surface on which a figure is fixed takes different shapes in space, then the surface area of the spherical image of the figure is always the same."* (13.1)

[1](Gauß, 1973, p. 372)

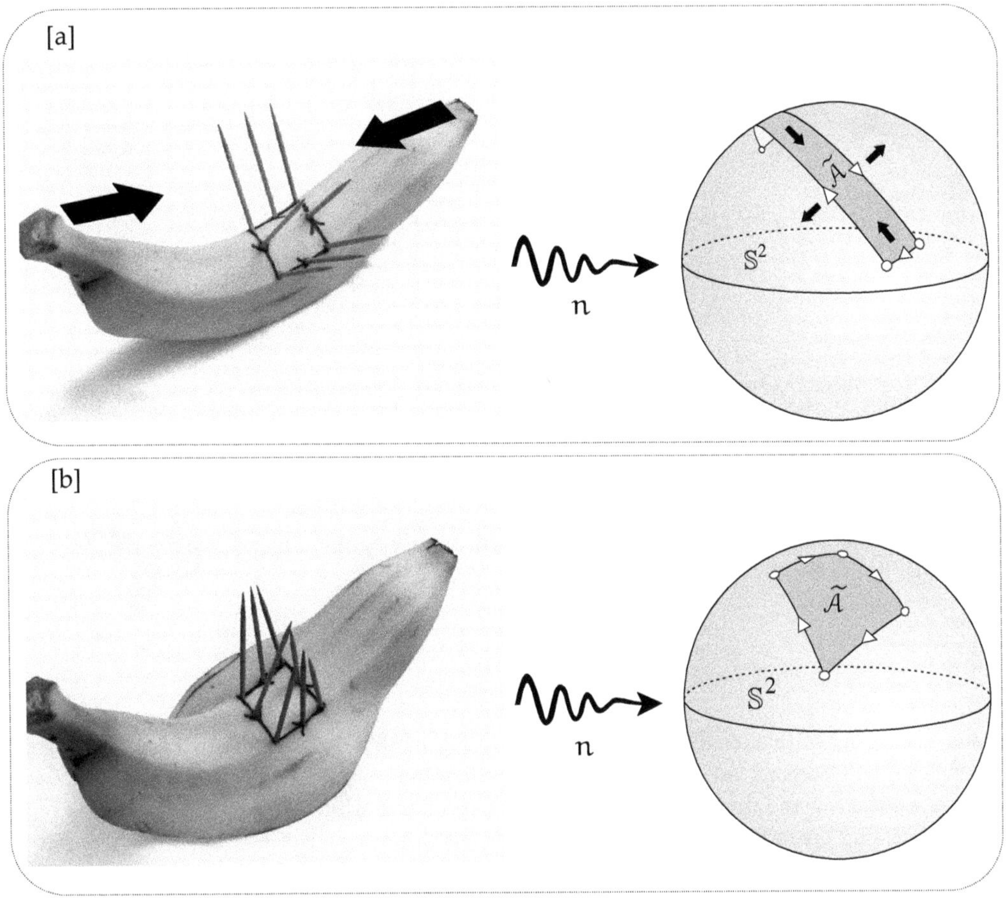

[13.1] *Gauss's* **Beautiful Theorem***:* **[a]** *Under the spherical map,* $\mathfrak{n}$*, the squarish loop on the banana is mapped to a narrow quadrilateral of area* $\widetilde{\mathcal{A}}$ *on* $\mathbb{S}^2$*, traversed in the opposite sense to the original, because of the negative curvature. Bending the banana skin by pulling the ends towards each other distorts the image on* $\mathbb{S}^2$*, yielding* **[b]***: The image on* $\mathbb{S}^2$ *is now a squarer quadrilateral, but* the area $\widetilde{\mathcal{A}}$ is exactly the same as before!

What Gauss means here by the "spherical image" is the image on $\mathbb{S}^2$ under the spherical map. Figure [13.1] illustrates the meaning of the theorem with a banana peel; we strongly encourage you to try this (or similar) experiment yourself. First, we have peeled one strip of peel from the positive-curvature side and have removed the banana itself, leaving the remainder of the peel surface intact. On the negatively curved part of peel we have drawn a squarish, counterclockwise loop, and on this loop we have erected normals (toothpicks).

Under the spherical map this loop is mapped to a narrow quadrilateral on $\mathbb{S}^2$ of area $\widetilde{\mathcal{A}}$, traversed in the *opposite* sense to the original, as illustrated in [13.1a]. Make sure to follow the loop on the banana with your eye, confirming that the tip of the normal does indeed trace such a reversed image on $\mathbb{S}^2$.

If we now pull the ends of the banana towards each other, the skin undergoes an isometry, and the normals along the boundary of our patch deform to produce a squarer image on $\mathbb{S}^2$. According to the Beautiful Theorem, this new spherical image has *exactly the same area as before!*

Such experiments provide exciting and tangible manifestations of the underlying mathematical truth, (13.1), but they *explain* nothing. In Act IV we shall introduce the concept of ***parallel transport***, and with its assistance we shall be able to provide an elegant conceptual *explanation*

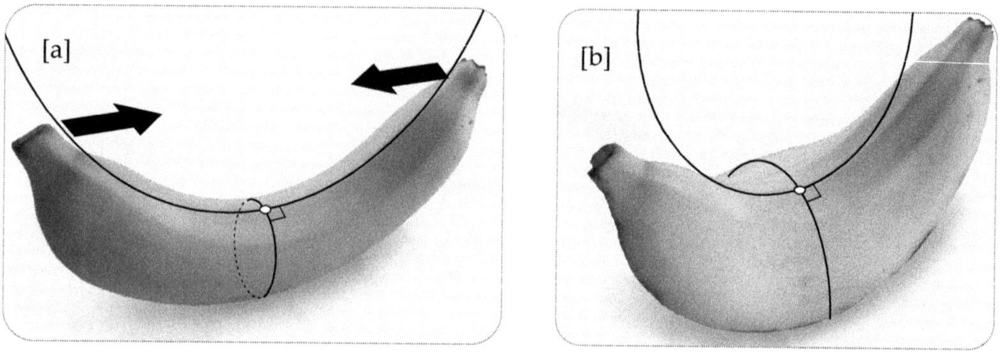

[13.2] *Gauss's* **Theorema Egregium***:* **[a]** *The two circles of curvature at the illustrated point on the banana have radii of curvature* ρ_1 *and* ρ_2*, and the extrinsic curvature there is* $\mathcal{K}_{ext} = 1/\rho_1\rho_2$*. Bending the banana skin by pulling the ends towards each other yields* **[b]***: One circle expands while the other contracts, but* the product of their radii remains constant*:* $\mathcal{K}_{ext}$ *is invariant.*

of the Beautiful Theorem; for now, though, we will simply assume it, and thereby deduce as its consequence the *Theorema Egregium*.

13.3 Gauss's *Theorema Egregium* (1827)

Gauss did not publish a single word on the subject for another *decade*, but in private he returned to Differential Geometry and to his Beautiful Theorem many times, most intensely in 1822 and 1825, even writing and then abandoning a full-length manuscript.[2] At last, in 1827 he was satisfied, publishing the result as the centrepiece of his *Disquisitiones Generales Circa Superficies Curvas* ("General Investigations of Curved Surfaces"[3]), and he allowed pent-up excitement to get the better of him. What he had privately described to himself as "beautiful" he now announced to the world as "remarkable"—the ***Theorema Egregium***.

But by this point Gauss had almost totally covered his tracks from the original nonlocal discovery of 1816, and the form of the result that he presented to the world was purely *local*. To see how he was led to this new local version of the result, take the figure occurring in the Beautiful Theorem and simply shrink it down towards a point p.

Let $\delta\mathcal{A}$ be the original area on $\mathcal{S}$ surrounding p and let $\delta\widetilde{\mathcal{A}}$ be the image area on $\mathcal{S}^2$ surrounding $n(p)$. Naturally, $\delta\mathcal{A}$ is invariant under isometries of $\mathcal{S}$, and by virtue of the Beautiful Theorem (13.1) $\delta\widetilde{\mathcal{A}}$ is *also* invariant. The "remarkable theorem" follows immediately:

> **Theorema Egregium.** *The extrinsic curvature* $\mathcal{K}_{ext} \asymp (\delta\widetilde{\mathcal{A}}/\delta\mathcal{A})$ *is invariant under isometries of* $\mathcal{S}$ *and therefore belongs to the* intrinsic *geometry of* $\mathcal{S}$*. More explicitly, while the principal curvatures individually depend on the shape of the surface in space, their product does* not*:* $\kappa_1\kappa_2$ *is invariant under isometries.* (13.2)

Figure [13.2] illustrates this with another banana peel, and again we urge you to try this experiment yourself. Pull the ends of the banana toward each other, and watch what happens at a

[2] An English translation of the aborted 1825 manuscript is appended in (Gauss 1827).

[3] English translations are available in both (Dombrowski 1979) and (Gauss 1827).

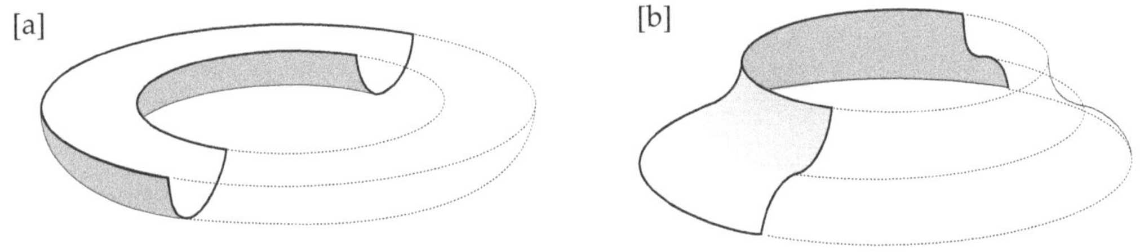

[13.3] *When the outer half of the circular trough in* **[a]** *is reflected across the plane upon which it rests, we obtain the isometric surface* **[b]**. *Yet [a]* cannot *be continuously deformed into [b] without stretching the surface. Nevertheless, the* Theorema Egregium *assures us that the two surfaces have equal curvature at corresponding points.*

particular point p: the radius of curvature ρ_1 shrinks (for the principal direction running the length of the banana), while the cross-sectional radius of curvature ρ_2 expands. But according (13.2) the product $\rho_1\rho_2$ remains perfectly constant throughout the bending.

To confirm this experimentally, take two small lengths of fairly stiff wire and bend them to fit the surface of the banana in its natural state (before removing the fruit inside) in the two principal directions at a particular point, thereby obtaining two small pieces of the principal circles of curvature there. Now lay these two small pieces of circle flat on a table and use a ruler to estimate their radii ρ_1 and ρ_2, and hence their product $\rho_1\rho_2$. Now have a friend bend the banana peel and hold it steady while you again fit the pieces of wire to the new surface at the same point as before, in the new principal directions.[4] Finally, confirm that the new value of $\rho_1\rho_2$ is the same as before, within the limits of experimental error.

The word "bending" implies continuous deformation, but this is not actually required by the theorem: there do exist isometries that cannot be carried out via gradual, continuous isometric deformation, but which nevertheless preserve the curvature by virtue of the theorem.[5]

To illustrate this we shall describe an example of Aleksandrov (1969, p. 101). Figure [13.3] is a freehand copy of Aleksandrov's original picture. Figure [13.3a] depicts a circular trough, which we imagine to rest on a plane, touching it along a circle C. To obtain the surface in [13.3b], cut the surface into two along C, then reflect the outer half in the plane. Clearly this new surface is isometric to the original surface, yet it also intuitively clear (and can be proven) that the original surface is rigid and cannot be bent into the new shape without stretching the surface in the process. Note that if we had instead performed an analogous transformation of a straight trough (a half-cylinder) then the new surface *could* have been obtained by a continuous deformation, without stretching.

A word about bending of physical "surfaces" versus mathematical ones. A physical surface, no matter how thin, cannot actually be bent without any stretching. For example, take a piece of paper and roll it into a cylinder, joining the edges together with tape. Both sides of the sheet began with the same length, but the cylinder has an outer circumference that is very slightly greater than the inner one: the outside had to be *stretched*, creating tension within the material. It is for this reason that when you remove the tape, the sheet will spontaneously spring back to its original planar form. Only in the mathematical limit of vanishing thickness can a surface be bent without any stretching.

Gauss's result (13.2) is indeed remarkable, but he took it still further. Since the result shows that $\mathcal{K}_{ext}$ is actually an *intrinsic* measure of curvature, it is natural to ask how it might be related

---

[4]We deliberately chose a point and a deformation such that the principal directions did not change. However, in general the principal directions will spin within the surface as the bending occurs.

[5]It is not clear if Gauss himself was aware of this distinction, but it is certainly true that he expressed the *Theorema Egregium* in terms of isometries, rather than bendings. It was only later authors who paraphrased Gauss as saying that $\mathcal{K}_{ext}$ was "invariant under bending."

to our original intrinsic definition (2.1) (see p. 18) of Gaussian curvature $\mathcal{K}$, as the angular excess per unit area: $\mathcal{K} \asymp \mathcal{E}(\Delta)/\mathcal{A}(\Delta)$. Gauss's answer is *very* remarkable:

> *The extrinsically defined curvature* $\mathcal{K}_{ext} = \kappa_1 \kappa_2$ *and the intrinsically defined Gaussian curvature* $\mathcal{K}$ *are numerically* equal:
>
> $$\mathcal{K}_{ext} = \mathcal{K}. \tag{13.3}$$

In light of this result, we can and will drop the distinction between these two measures of curvature, enabling us to speak simply of *the* curvature $\mathcal{K}$ of the surface.

As with the Beautiful Theorem itself, the simplest and most general proof of this wonderful result must await the introduction of the concept of ***parallel transport*** in Act IV. However, in the next chapter we shall be able to lend some plausibility to the result via a more limited argument concerning polyhedra.

In the meantime, let us confirm the validity of (13.3) for some specific surfaces for which we have already calculated both the extrinsic principal curvatures *and* the intrinsic Gaussian curvature.

- *The Cylinder and the Cone.* The intrinsically flat plane may be rolled into a cylinder or a cone, so the Gaussian curvature of these two surfaces vanishes. Combining this fact with (12.5),

$$\mathcal{K}_{ext} = 0 = \mathcal{K}.$$

- *The Sphere.* Combining the results (12.4) and (1.3) (p. 8),

$$\mathcal{K}_{ext} = (1/R^2) = \mathcal{K}.$$

- *The Pseudosphere.* Combining the results (10.9) and (5.3) (p. 53),

$$\mathcal{K}_{ext} = -(1/R^2) = \mathcal{K}.$$

- *The Torus.* Combining the results (10.10) and Exercise 23 on page 89,

$$\mathcal{K}_{ext} = \frac{1}{r(r + R\sec\alpha)} = \mathcal{K}.$$

These five surfaces are so important, both historically and mathematically, that we have afforded them the dignity of individual treatments, but in fact the truth of (13.3) for all five surfaces follows, in one fell swoop, from this:

- *The General Surface of Revolution.* Combining the results (10.11), (10.12), and Exercise 22 on page 89, in which a particle travelled along the generating curve at *unit speed*,

$$\mathcal{K}_{ext} = -\ddot{y}/y = \mathcal{K}. \tag{13.4}$$

Chapter 14

The Curvature of a Spike

14.1 Introduction

Thus far we always imagined our surface to be perfectly smooth, with a well-defined tangent plane and associated normal vector at each point. But consider [14.1], which shows the spiked surface of the durian,[1] here shown with one of its interior yellow fruits exposed.

Regardless of whether we approach this surface from the extrinsic or the intrinsic point of view, how on Earth shall we define the curvature at the tip of one of the durian's many spikes?!

Once we have answered this question we shall be able to shed some new light on the *Theorema Egregium*.

[14.1] *How can we measure the curvature of the spiked surface of the durian?*

14.2 Curvature of a Conical Spike

Suppose our spike takes the form of the tip of a cone. As we discussed in the last section, the curvature of the cylinder and the cone both vanish, at least at every point other than the tip of the cone. While the vanishing intrinsic curvature of the cylinder is easy to accept, we balk at the cone's "flatness," and with good reason.

Imagine yourself as a two-dimensional inhabitant of the cone's surface, standing at the vertex. You construct a circle centred there of radius r and measure its circumference $C(r)$. You quickly discover that $C(r)$ is *less than* the Euclidean prediction of $2\pi r$. This discrepancy clearly signals the presence of curvature, and this curvature must reside *at* the tip, and nowhere else, for the curvature at typical points has been shown to vanish.

To discuss this in more detail, suppose that the internal angle from the axis of symmetry to the cone's surface is α. See [14.2]. Intuitively, any reasonable measure of curvature should increase as α becomes smaller and the spike becomes sharper; conversely, it should diminish to zero as α approaches $\pi/2$, for in that limit the cone flattens out into a Euclidean plane.

Unlike a mathematical cone, the tip of a physical cone will not be perfectly sharp, but will instead be slightly blunt and rounded. We may gain some mathematical insight if we now imagine this rounded tip to be in the form of a polar cap of a sphere that fits smoothly onto the cone, in the

[1] This southeast Asian delicacy is justifiably known as "The King of Fruits": it tastes (and smells!) as wonderfully strange as it looks.

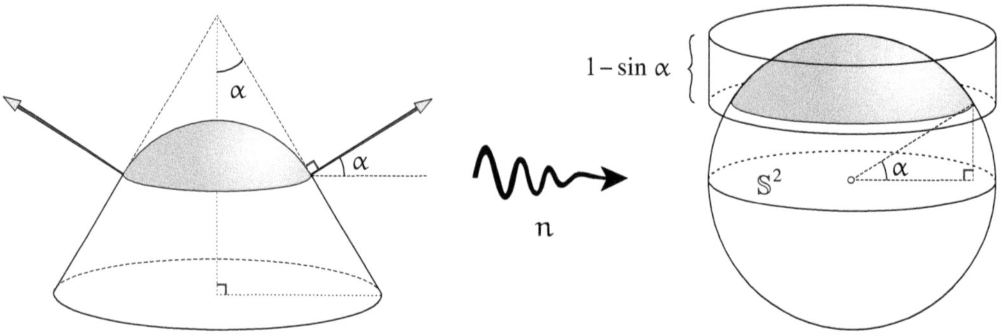

[14.2] *The spherical map sends the blunted tip of the cone to the similar polar cap on* $\mathbb{S}^2$*, the area of which equals that of the illustrated cylinder of height* $(1 - \sin\alpha)$.

sense that there is a well-defined tangent plane along the join between the cone and the polar cap. See [14.2].

As the figure illustrates, the spherical map merely expands this blunted tip into a similar polar cap on the unit sphere. It now seems reasonable to (provisionally) *define* the curvature of the spiked tip of the original mathematical cone to be the total curvature residing within the blunted tip of the physical cone. For it is clear that this definition does not depend on the *size* of the blunted tip: as we shrink its radius, making it sharper and sharper,[2] the size of the spherical image on the unit sphere does not change. Furthermore, even without calculation, it seems clear that the dependence on α of this measure of curvature conforms to our previously described intuition.

Thus we make this definition:

$$\mathcal{K}(\text{spike}) \equiv \text{total } \mathcal{K} \text{ of blunted tip} = \text{area of the (polar cap) spherical image.}$$

It is also possible to pass directly from the *original* (unblunted) cone tip to this same spherical image. Imagine a horizontal plane resting on the tip of the cone, and imagine a unit normal vector attached to the plane, pointing straight up. Taking this unit normal to be the **n** of our spherical map, this horizontal position of the plane is mapped by n to the north pole on the right of [14.2]. The plane is free to rock in any direction until it hits the side of the cone and becomes tangent to the surface, at which point n lies on the boundary of the polar cap. Thus, as the plane resting on the tip assumes all possible positions, n fills the same polar cap as before. We call this the ***generalized spherical map***.

Regardless of how we arrive at this spherical image, let us now find a formula for its area. By virtue of a result by Archimedes (see Ex. 10, page 85), the area of the polar cap equals the area of its illustrated projection onto the cylinder touching the unit sphere along its equator. Thus, since the cylinder has circumference 2π, and this segment of it has height $(1 - \sin\alpha)$, then

$$\mathcal{K}(\text{spike}) = \text{area of spherical image} = 2\pi(1 - \sin\alpha). \tag{14.1}$$

Note that this formula for the curvature does indeed achieve its maximum when $\alpha = 0$, and it also vanishes when $\alpha = (\pi/2)$, as anticipated.

There is another way way of looking at this result, a way that will shortly offer a natural generalization of curvature to *polyhedral* spikes. See [14.3]. Suppose we cut one unit along a generator

[2] You would not hesitate to press your palm down onto the blunted tip when it's 10-cm wide, but would not wish to when it's 0.01-cm wide!

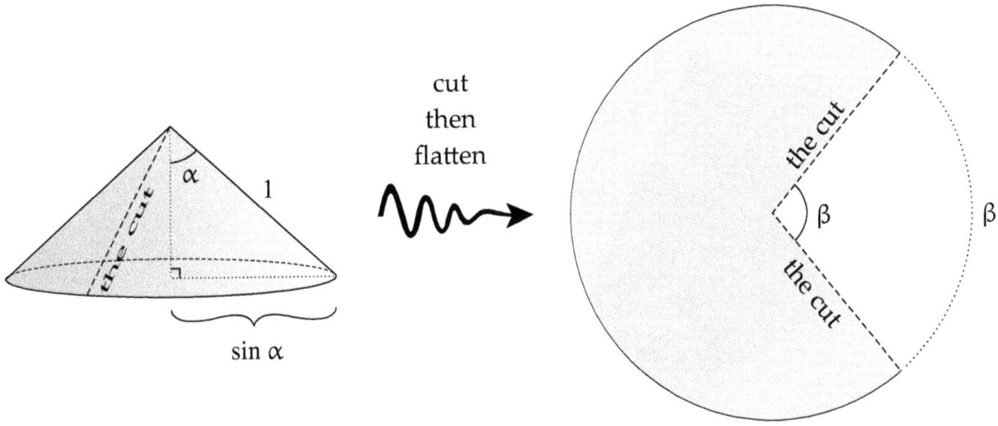

[14.3] *When the cone is cut by one unit along a generator and flattened out, the base circumference* $2\pi \sin \alpha$ *becomes the arc* $(2\pi - \beta)$. *It follows that* $\mathcal{K}(\textit{spike}) = \beta$.

of the cone, starting at the tip (the vertex), then cut horizontally around the cone. If we flatten this out we obtain a sector of a unit circle, bounded by the two sides of the cut, now split apart by angle β. Clearly β is also a measure of curvature of the original spike. In fact we will now show that it is exactly the *same* measure of curvature.

On the left of [14.3] we see that the circumference of the base is $2\pi \sin \alpha$. But this length does not change when the cone is cut and flattened to produce the sector of the unit circle on the right. Thus,

$$2\pi \sin \alpha = 2\pi - \beta.$$

Therefore (14.1) can be re-expressed as

$$\boxed{\mathcal{K}(\text{spike}) = \beta = \text{Split angle of flattened spike.}} \tag{14.2}$$

Both of these formulas (14.1) and (14.2) for $\mathcal{K}_{\text{spike}}$ employ extrinsic ideas that are unknowable to the inhabitants of the cone: the first involves the 3-dimensional concept of the normal to the surface, while the second involves the flattening of the surface within three-dimensional space. Nevertheless, it is in fact a simple matter to reinterpret the second formula *intrinsically*.

Returning to the discussion at the start of the section, the circumference $C(r)$ of a circle of radius r centred at the vertex is related to the curvature by [exercise] the following *intrinsic* formula for β:

$$\boxed{\mathcal{K}(\text{spike}) = \frac{2\pi r - C(r)}{r} = 2\pi - C(1).} \tag{14.3}$$

14.3 The Intrinsic and Extrinsic Curvature of a Polyhedral Spike

In the case of the cone we have just seen that there is a natural way of measuring the curvature of the spike both intrinsically and extrinsically, and these two measures turn out to *coincide*. As we now explain, the same is true of the vertex v of a polyhedral spike, at which we shall suppose that m flat, polygonal faces $f_1, f_2, \ldots, f_m$ come together, the outward normal of f_j being $\mathbf{n}_j$.

The generalization of the last two interpretations (14.2) and (14.3) of $\mathcal{K}(\text{spike})$ is straightforward, so we begin there. As with the cone, suppose we cut down along one of the edges, starting

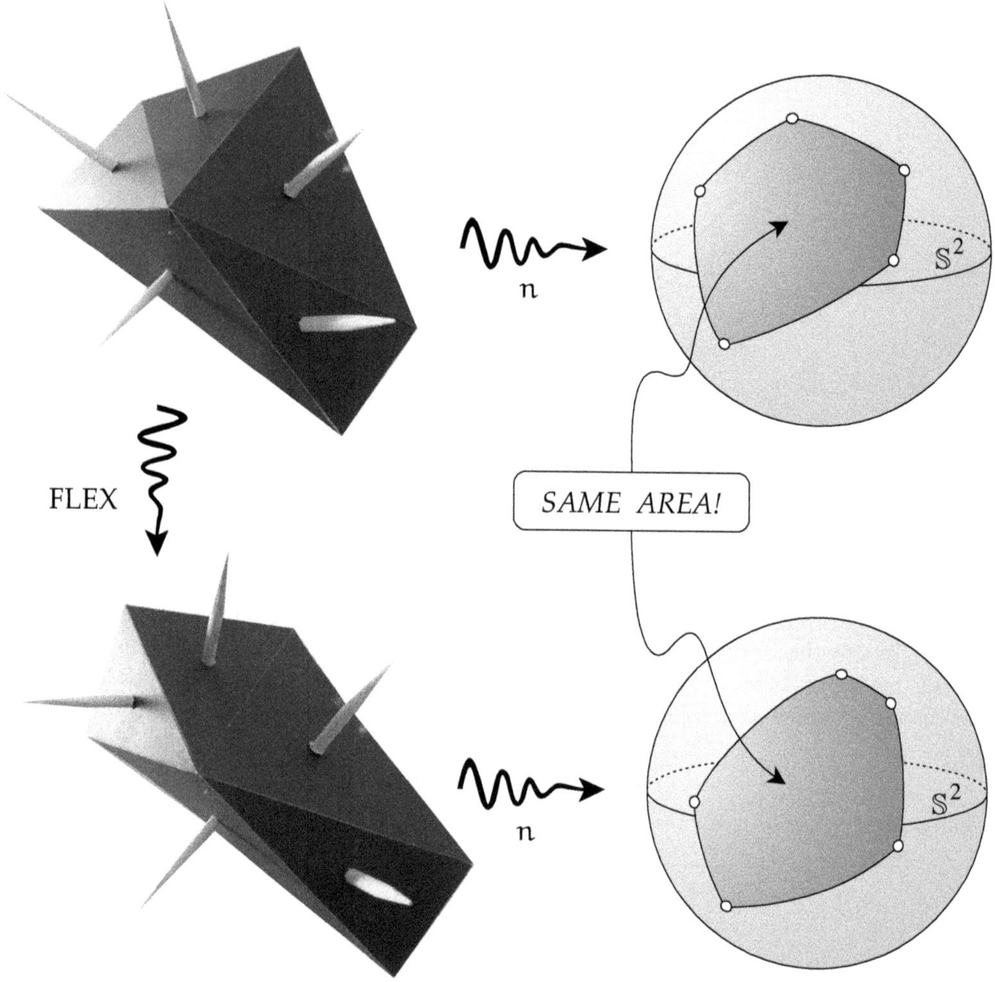

[14.4] Polyhedral Theorema Egregium. *The generalized spherical map sends the vertex of the polyhedral spike to a geodesic polygon* P_m *on* $\mathbb{S}^2$*, and as the polyhedral spike is flexed, this spherical polygon changes shape, but* its area does not change! *Compare this with the picture of the flexing banana skin in [13.1] on page 139.*

at v and ending at an arbitrary point p, then cut a sideways circuit around v across all the f_j, finally returning to p, thereby cutting off the spike from the polyhedron.

Imagine the polyhedron to be made up of stiff cardboard polygons, attached along their edges with sticky tape, so that each edge acts like a hinge. (Such a model is shown in [14.4].) Then, since it has been cut along vp, the detached spike may now be flattened out onto the plane, the two sides of the cut vp ending up split apart by angle β. If θ_j is the angle between the edges of f_j that meet at v, and $\Theta \equiv \sum \theta_j$ is the sum of the angles that meet at v, then clearly $\beta = 2\pi - \Theta$. Lastly, this can be reexpressed *intrinsically* as the circumference $C(1)$ of a unit circle centred at v drawn within the original polyhedral surface:

$$\mathcal{K}_{\text{int}}(v) = \text{split angle of flattened spike} = \beta = 2\pi - \Theta = 2\pi - C(1). \tag{14.4}$$

The extrinsic definition of curvature can also be generalized, via the generalized spherical map. As we did with the cone, imagine a plane Π (with normal $\mathbf{n}$) resting in contact with the spike. As Π rocks and assumes all possible positions, all the while remaining in contact with the spike, what image region does n fill on $\mathbb{S}^2$?

The limits of motion of Π are determined by the faces: we can rock Π until it hits and coincides with a face f_j, at which point $\mathbf{n}$ coincides with $\mathbf{n}_j$, the normal to f_j. The limits of Π's motion are thus determined by the m points n_j on S^2. Next, observe that

> *As the plane Π rocks over the vertex v, the spherical image n on S^2 fills the interior of the m-gon P_m on S^2 with vertices n_j.*

To see this, we need only imagine Π initially coincident with f_j, then rolling over the edge e in which f_j and f_{j+1} meet. As it does so, $\mathbf{n}$ must swing within the plane perpendicular to e, and therefore the image n on S^2 travels along the arc of the great circle connecting n_j to n_{j+1}. The geodesic arcs connecting the successive n_j therefore form the boundary of the region that n covers, as Π rocks over all possible positions.

We have thus arrived at a natural conception of the *extrinsic* curvature of v:

$$\mathcal{K}_{\text{ext}}(v) = \text{area on } S^2 \text{ of } m\text{-gon connecting the polyhedron's normals.} \tag{14.5}$$

14.4 *The Polyhedral Theorema Egregium*

Just as happened with the cone, the intrinsic and the extrinsic measures of curvature of the polyhedral spike turn out to *coincide*, a result which we may reasonably call the ***Polyhedral Theorema Egregium***:

$$\mathcal{K}_{\text{ext}}(v) = \mathcal{K}_{\text{int}}(v). \tag{14.6}$$

Figure [14.4] illustrates the theorem in action: as the spike is flexed, and the normals all rotate in different directions, the spherical image P_m changes, but, according to the theorem, its area is actually intrinsic to the spike, and therefore cannot change! Compare this with the picture of the flexing banana skin in [13.1] on page 139.

To understand this, consider [14.5]. The explanation, in a nutshell, is that the area of the spherical polygon only depends on its *angles*, and the figure demonstrates that these are in turn determined by the intrinsic angles of the polyhedron's faces.

On the left, two edges of the polyhedron meet at angle θ, and the planes perpendicular to these edges meet at angle $\widetilde{\theta}$. It follows from the marked right angles that $\theta + \widetilde{\theta} = \pi$. Thus, under the generalized spherical map, the corresponding edges of the spherical polygon meet at angle $\widetilde{\theta} = \pi - \theta$.

We can now verify the result in detail. Recall that in Exercise 5 on page 83 we generalized Harriot's Theorem from triangles to polygons, showing that the area $\mathcal{A}(P_m)$ of a spherical m-gon P_m is once again equal to its angular excess:

$$\mathcal{A}(P_m) = \mathcal{E}(P_m) \equiv [\text{angle sum}] - (m-2)\pi.$$

Therefore,

$$\mathcal{K}_{\text{ext}}(v) = \mathcal{A}(P_m) = 2\pi - \sum_{i=1}^{m} [\pi - \widetilde{\theta}_i] = 2\pi - \sum_{i=1}^{m} \theta_i = 2\pi - \Theta = \mathcal{K}_{\text{int}}(v),$$

and the Polyhedral *Theorema Egregium* is proved.

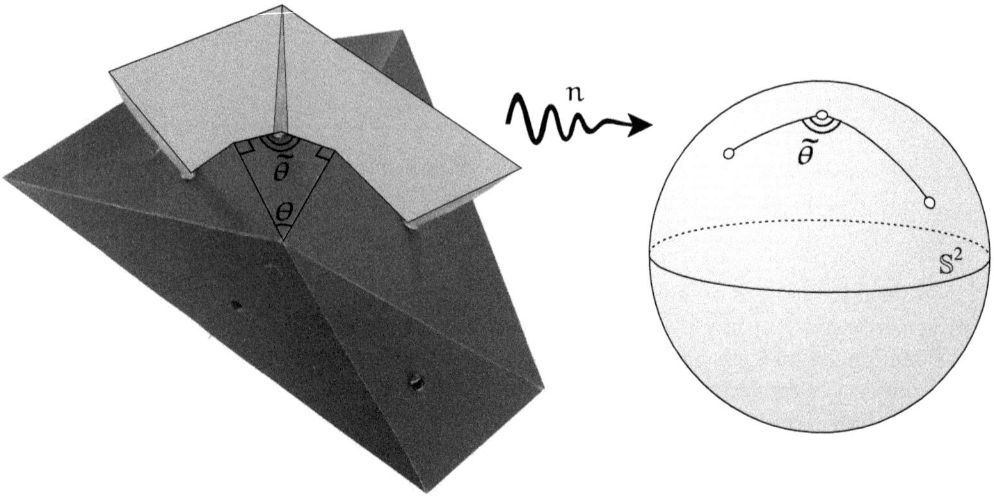

[14.5] *On the left, two edges of the polyhedron meet at angle* θ, *and the planes perpendicular to these edges meet at angle* $\widetilde{\theta}$. *It follows from the marked right angles that* $\theta + \widetilde{\theta} = \pi$. *Thus, under the generalized spherical map, the corresponding edges of the spherical polygon on the right meet at angle* $\widetilde{\theta} = \pi - \theta$.

HISTORICAL NOTES: The elegant insight above is usually attributed to Hilbert, appearing[3] in the timeless masterpiece, *Geometry and the Imagination* (Hilbert and S. Cohn-Vossen 1952, p. 195), the German original having appeared in 1932. But in fact[4] the same observation was made in 1854 by the great British physicist James Clerk Maxwell, 78 years prior to Hilbert. Maxwell's original 1854 letter to William Thomson can be found in Maxwell (2002, Vol. 1, p. 243), and his eventual paper of 1856 appears in Maxwell (2003, Vol. 1, §4).

Finally, we note that these concepts can be extended to *non*convex polyhedra and to vertices having *negative* curvature. For a beautiful and completely general investigation, see Banchoff (1970).

---

[3]However, Hilbert does not invoke our rocking plane. Instead, his justification for the transition from the n_j to P_m consists only in this: "In order to relate this to the spherical representation of surfaces, we connect the points $[n_j] \ldots$ by arcs of great circles ... [to create P_m]"

[4]I stumbled upon this little-known fact entirely by accident while leafing through my copy of Maxwell's collected works (Maxwell 2003), in search of something quite unrelated.

Chapter 15

The Shape Operator

15.1 Directional Derivatives

The Gaussian curvature $\mathcal{K}$, viewed extrinsically, measures the variation of the surface normal $\mathbf{n}$ in the vicinity of a point p, but only in a blurred, average way. Instead of looking at the overall spread of directions of $\mathbf{n}$ over a small patch of the surface $\mathcal{S}$ containing p, we now turn to a more precise method of quantifying the variation of $\mathbf{n}$, by looking at how fast it changes as we move away from p within $\mathcal{S}$ in a *specific direction.*

Let $\widehat{\mathbf{v}}$ be a unit vector emanating from p and lying within the tangent plane T_p to $\mathcal{S}$ at p. See [15.1]. We would like to define the rate of change of $\mathbf{n}$ in the direction $\widehat{\mathbf{v}}$. If we move a small distance ϵ away from p in this direction then we arrive at a point q whose position vector is therefore $\mathbf{q} = \mathbf{p} + \epsilon\widehat{\mathbf{v}}$. It is then tempting to try to define the rate of change of $\mathbf{n}$ as

$$\lim_{\epsilon \to 0} \frac{\mathbf{n}(q) - \mathbf{n}(p)}{\epsilon}.$$

But this will not do. For q actually resides within T_p, not $\mathcal{S}$, and therefore $\mathbf{n}(q)$ is not even defined.

Nevertheless, it seems clear that the concept we are groping for really is well-defined: we need only move a distance ϵ *within the surface*, instead of the tangent plane, but we must do so in the desired *direction* $\widehat{\mathbf{v}}$. One way of achieving this end is to drop a perpendicular from q to $\mathcal{S}$, meeting it at r, say. It seems clear (and can be proved) that the length pr of the geodesic segment from p to r is ultimately equal to ϵ:

$$\mathrm{pr} \asymp \epsilon.$$

Returning to our abortive definition of rate of change, we can now define

$$\boxed{\delta\mathbf{n} \equiv \mathbf{n}(r) - \mathbf{n}(p)}$$

to be the change in $\mathbf{n}$ that results from moving distance ϵ (ultimately) within the surface in the direction $\widehat{\mathbf{v}}$. Since $\mathbf{n}$ has unit length, it merely *rotates* slightly as we move from p to r. Therefore, if we picture both $\mathbf{n}(p)$ and $\mathbf{n}(r)$ as having a common origin, as illustrated in [15.2], then the movement of its tip is ultimately *orthogonal* to $\mathbf{n}$, and therefore $\delta\mathbf{n}$ is a vector that ultimately lies within the tangent plane, T_p.

Finally, the ***directional derivative*** $\nabla_{\widehat{\mathbf{v}}}$ of $\mathbf{n}$ along $\widehat{\mathbf{v}}$ can then be defined as

$$\boxed{\nabla_{\widehat{\mathbf{v}}}\mathbf{n} \equiv \lim_{\epsilon \to 0} \frac{\delta\mathbf{n}}{\epsilon}.} \tag{15.1}$$

As we have explained, this derivative lives in T_p.

This is the directional derivative per *unit* length. If instead we want the derivative along a general tangent vector $\mathbf{v} = v\,\widehat{\mathbf{v}}$ of length v, then we must simply scale up by this length:

$$\nabla_{\mathbf{v}}\mathbf{n} = \nabla_{v\widehat{\mathbf{v}}}\mathbf{n} = v\nabla_{\widehat{\mathbf{v}}}\mathbf{n}. \tag{15.2}$$

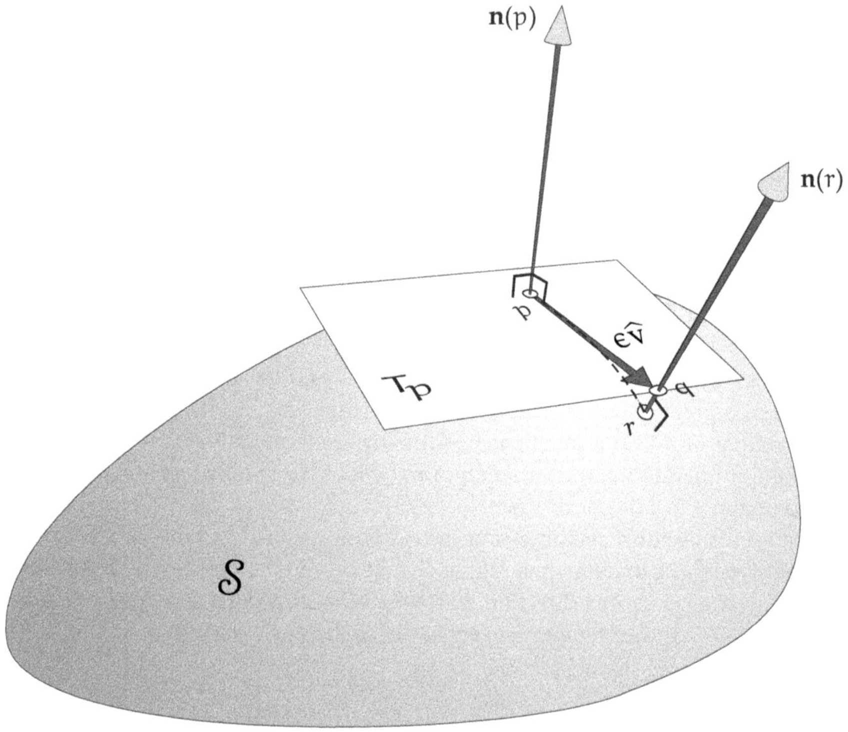

[15.1] *To find the derivative* $\nabla_{\hat{\mathbf{v}}}\mathbf{n}$ *of* $\mathbf{n}$ *in the direction* $\hat{\mathbf{v}}$*, we consider the change* $\delta\mathbf{n} \equiv \mathbf{n}(r) - \mathbf{n}(p)$ *that results from moving in that direction a distance* ϵ *within the surface. Then* $\nabla_{\hat{\mathbf{v}}}\mathbf{n} \asymp \frac{\delta\mathbf{n}}{\epsilon}$.

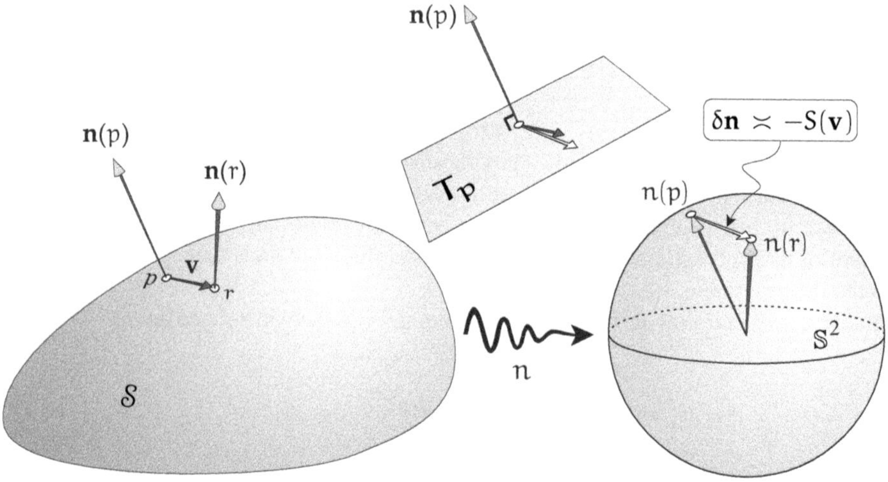

[15.2] *The* **Shape Operator** *applied to a short tangent vector* $\mathbf{v}$ *emanating from* p *in* T_p *tells us how much the normal changes from the tail to the tip of* $\mathbf{v}$*. This change* $\delta\mathbf{n} \asymp -S(\mathbf{v})$ *ultimately lies in the parallel tangent plane to* $\mathbb{S}^2$ *at* $n(p)$*. Superimposing the two tangent planes, S may be viewed as a linear transformation of* T_p *to itself, here depicted detached from* $\mathcal{S}$*, floating between the two surfaces.*

This makes greater intuitive sense if we think of $\mathbf{v}$ as the *velocity* of a particle moving over the surface, at the moment it passes through p. Then a unit vector $\widehat{\mathbf{v}}$ corresponds to a particle moving at unit speed. Thus when $\nabla_{\widehat{\mathbf{v}}}$ is applied to *any* quantity defined along the particle's trajectory (not necessarily $\mathbf{n}$) it yields the rate of change of that quantity with respect to *time*. Thus if the particle were to travel three times as fast along the same trajectory, changing $\widehat{\mathbf{v}}$ to $3\widehat{\mathbf{v}}$, then the quantity would change three times as fast along the trajectory. This is one way of looking at (15.2). Note that this concept of the directional derivative can be applied to *any* quantity (vector or scalar) that is defined on the surface, or merely along the trajectory.

There is another way of looking at the definition of the directional derivative that requires, first, that we extend the concept of "ultimate equality" to vectors. Thus, suppose that two vectors $\mathbf{a}$ and $\mathbf{b}$ depend on a small quantity ϵ that tends to zero. An obvious definition of $\mathbf{a} \asymp \mathbf{b}$ would be that each *component* of $\mathbf{a}$ is ultimately equal to the corresponding component of $\mathbf{b}$. But this definition is problematic [why?] if one or more components is identically zero. Therefore, instead, we adopt the following more satisfactory and geometrical definition, namely, that the magnitudes and directions of the two vectors are ultimately equal:

> *If two vectors* $\mathbf{a}$ *and* $\mathbf{b}$ *depend on a small quantity* ϵ *that tends to zero, we define them to be* **ultimately equal**, *written* $\mathbf{a} \asymp \mathbf{b}$, *iff* $|\mathbf{a}| \asymp |\mathbf{b}|$ *and the angle between their directions has vanishing limit.*

Granted this, we may now rewrite (15.1) as

$$\nabla_{\widehat{\mathbf{v}}}\mathbf{n} \asymp \frac{\delta\mathbf{n}}{\epsilon} \quad \Longleftrightarrow \quad \delta\mathbf{n} \asymp \epsilon\nabla_{\widehat{\mathbf{v}}}\mathbf{n} = \nabla_{\epsilon\widehat{\mathbf{v}}}\mathbf{n}.$$

Here, then, is the resulting point of view, which we shall employ *repeatedly* throughout the remainder of the book:

> *In the limit of vanishing* ϵ, *and hence vanishing* $\mathbf{v} = \epsilon\,\widehat{\mathbf{v}}$, $\nabla_{\mathbf{v}}\mathbf{n}$ *is ultimately equal to the change* $\delta\mathbf{n}$ *in* $\mathbf{n}$ *from the tail to the tip of* $\mathbf{v}$: $\nabla_{\mathbf{v}}\mathbf{n} \asymp \delta\mathbf{n}$. (15.3)

Of course, as noted earlier, this statement employs a touch of poetic license, for the tip of $\mathbf{v}$ does not actually reside within $\mathcal{S}$. But as ϵ tends to zero, so does the amount of poetic license required! For, in this limit, the distinction between q and r in [15.1] rapidly becomes insignificant.

15.2 The Shape Operator S

The ***Shape Operator***[1] S associated with the surface at the point p tells us how the normal vector changes (in which direction it tips, and how fast) as we move away from p along an arbitrary tangent vector $\mathbf{v}$, which is no longer necessarily short, but rather has arbitrary length v. More precisely, it is simply defined to be the negative of the directional derivative of $\mathbf{n}$ along $\mathbf{v}$:

$$\boxed{\mathsf{S}(\mathbf{v}) \equiv -\nabla_{\mathbf{v}}\mathbf{n}.} \tag{15.4}$$

[1]There are two other mathematical objects that encode exactly the same information as the Shape Operator: the ***Second Fundamental Form*** and the ***Weingarten map***. We shall not employ either in this book.

The insertion of the minus sign into the definition is standard; as we shall see shortly, it is motivated by our earlier definition of the sign of the curvature of normal sections.

The surface $\mathcal{S}$ and the sphere S^2 have the *same* normal vector $\mathbf{n}(p)$ at p and $n(p)$, respectively. Thus their tangent planes T_p and $T_{n(p)}$ at these two points are *parallel*, and we may imagine transporting them so as to make them coincide. Thus any tangent vector to S^2 emanating from $n(p)$ may instead be pictured as residing in T_p, emanating from p. In [15.2] we picture these two coincident planes floating between the two surfaces.

The meaning of the Shape Operator now becomes especially clear if we think of a point r very close to p being carried by the spherical map to a point $n(r)$ on S^2 close to $n(p)$. See [15.2]. If $\mathbf{v}$ is the very short vector connecting p to r and $\delta\mathbf{n}$ is the corresponding vector connecting $n(p)$ to $n(r)$ then, by virtue of (15.3),

$$S(\mathbf{v}) \asymp -\delta\mathbf{n}.$$

If we think of p as the origin of these two coincident planes, then S is in fact a *linear* transformation of T_p to itself. That is, if $\mathbf{v}$ and $\mathbf{w}$ are arbitrary tangent vectors at p, and a and b are arbitrary constants, then

$$\begin{aligned} S(a\mathbf{v}+b\mathbf{w}) &= -(\nabla_{a\mathbf{v}+b\mathbf{w}})\,\mathbf{n} \\ &= -\nabla_{a\mathbf{v}}\mathbf{n} - \nabla_{b\mathbf{w}}\,\mathbf{n} \\ &= -a\nabla_{\mathbf{v}}\mathbf{n} - b\nabla_{\mathbf{w}}\,\mathbf{n} \\ &= a\,S(\mathbf{v}) + b\,S(\mathbf{w}). \end{aligned}$$

Recall from Linear Algebra that this means that S can be represented as a rectangular array of numbers [S], called the *matrix* of S. In general, the j-th column of the matrix is the image of the j-th basis vector, or rather the numerical components of that vector. In our 2-dimensional case, this means that [S] is a 2×2 square.

Later in this chapter we shall investigate the matrix [S], but for now we illustrate this general idea of "matrixification" with an example we will need momentarily, namely, the matrix $[R_\theta]$ representing a rotation R_θ of the plane through angle θ about the origin.

Figure [15.3] illustrates the effect of R_θ on the basis vectors, from which we immediately deduce that

$$[R_\theta] = \begin{bmatrix} c & -s \\ s & c \end{bmatrix}, \tag{15.5}$$

where $c \equiv \cos\theta$ and $s \equiv \sin\theta$.

15.3 The Geometric Effect of S

Let $\mathbf{e}_1$ and $\mathbf{e}_2$ be unit vectors along the (orthogonal) first and second principal directions, respectively. What is the geometric effect of S on vectors in T_p when referred to this special orthonormal basis $\{\mathbf{e}_1, \mathbf{e}_2\}$? Our earlier result (12.7) provides the neat answer:

The principal directions are the **eigenvectors** *of the Shape Operator S, and the principal curvatures are the corresponding* **eigenvalues**:

$$S(\mathbf{e}_i) = \kappa_i \mathbf{e}_i. \tag{15.6}$$

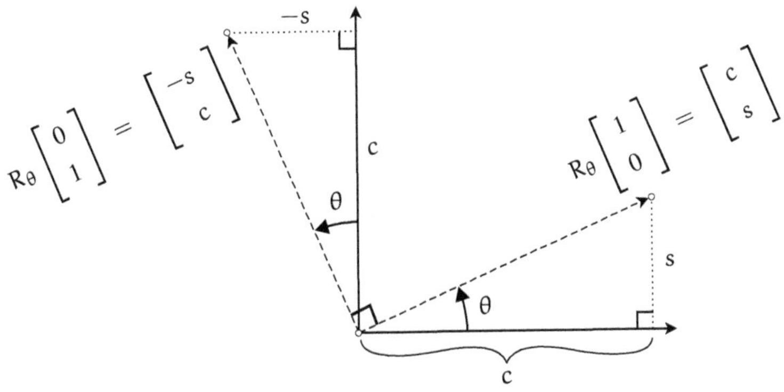

[15.3] *The first/second column of the matrix* $[R_\theta]$ *is the image of the first/second basis vector when it is rotated by* R_θ.

NOTE: It was precisely in order to cancel out the minus sign in (12.7) that a minus sign was introduced into the definition (15.4) of S.

Thus, knowing that the effect of S on the principal directions is to stretch them by their respective principal curvatures, linearity now reveals that the effect of S on a *general* tangent vector is to stretch it by these two factors in these two perpendicular directions.

If we choose our basis vectors to be $\{\mathbf{e}_1, \mathbf{e}_2\}$, then it follows that the matrix of S takes the especially simple, diagonal form,

$$[S] = \begin{bmatrix} \kappa_1 & 0 \\ 0 & \kappa_2 \end{bmatrix}. \tag{15.7}$$

Recall that a linear transformation expands the area of every shape by the *same* factor. This universal *area expansion factor* is the ***determinant*** of its matrix. This is particularly clear in the present case, where a unit square aligned with the principal axes is stretched into a rectangle of sides κ_1 and κ_2, and therefore the original square of unit area undergoes an expansion to area $|[S]| = \kappa_1 \kappa_2$, so this is the expansion factor for *all* areas. Once again we recover our familiar expression for the extrinsic version of the Gaussian curvature:

$$\mathcal{K}_{\text{ext}} = \text{area expansion factor of the Shape Operator} = |[S]| = \kappa_1 \kappa_2. \tag{15.8}$$

Recall that when we first derived this result (see (12.8), p. 134) we did so by considering the area expansion of one *particular* shape, and we merely *claimed* that the result was independent of this choice; the recognition of the Shape Operator's linearity has now proved that claim. Furthermore, given its geometric meaning, this area expansion factor (i.e., the determinant) must have the *same* value in all coordinate systems.

Notice that this matrix [S] is ***symmetric*** (also called ***self-adjoint***), meaning that it has mirror symmetry across its ***main diagonal***, which runs top-left to bottom-right. Since reflection across this diagonal is achieved by swapping rows and columns, which creates (by definition) the ***transpose*** $[S]^T$, this symmetry can be written,

$$[S]^T = [S]. \tag{15.9}$$

The symmetry (15.9) of the matrix [S] is no accident, but rather reflects (pardon the pun!) the underlying symmetry of the linear transformation S itself, and it will persist even if the basis is not aligned with the principal directions, in which case [S] will no longer be diagonal.

15.4 DETOUR: The *Geometry* of the Singular Value Decomposition and of the Transpose

In this optional detour we seek to clarify the symmetry (15.9) of the Shape Operator by first asking the following question: *If a general linear transformation* M *is represented by the matrix* [M], *what transformation* M^T *is represented by the transpose matrix* $[M]^T$? In order to answer this question, we shall begin by providing a geometric interpretation and derivation of one of the most crucial results in all of Linear Algebra, the so-called ***Singular Value Decomposition***[2]—or ***SVD*** for short.

The *geometric* interpretations and proofs we shall now present would appear to be appropriate for a first course in Linear Algebra. Yet, surprisingly, we have not been able to find these ideas described in *any* standard introductory text; indeed, we suspect that many working mathematicians may also be unfamiliar with them. Nevertheless, if you are in a hurry to proceed with Differential Geometry, please feel free to skip to the next section.

We begin by recalling the familiar fact that *if* a linear transformation of the plane has two real eigenvalues, its geometric effect is easily visualized, as follows: stretch by these factors along the two (typically nonorthogonal) eigenvector directions. But what if the transformation does *not* have real eigenvalues? A simple example of such a transformation (i.e., one that does not preserve any direction) is a *rotation*, but at least this too enjoys the virtue of being easily visualized. But how shall we make geometric sense of a general linear transformation that does not preserve any direction?

The SVD provides a wonderfully simple and vivid answer to this question, and it applies to *all* linear transformations; therefore even the transformations we thought we already understood (the ones that stretch in two nonorthogonal eigenvector directions) thereby receive new meaning.

> **Singular Value Decomposition (SVD).** *Every linear transformation of the plane is equivalent to stretching in two* orthogonal *directions (by generally different factors,* σ_1 *and* σ_2*, called the* **singular values***), followed by a rotation through angle* τ*, which we call the* **twist**. (15.10)

To understand this, consider the top half of [15.4], which shows the effect (from left to right) of a general linear transformation on an origin-centred circle C. Since the Cartesian equation of C is quadratic, the linear change of coordinates induced by the transformation will lead to another quadratic equation for the image curve. The image curve $\widetilde{C}$ is thus a conic section, and since the finite points of C are not sent to infinity, this conic must be an *ellipse*, shown top right of [15.4].

We have just used an algebraic statement of linearity; next we use the fundamental *geometric* fact that it makes no difference if we add two vectors and then map the result, or if we map the vectors first and *then* add them. Convince yourself of these two simple consequences:

- *Parallel lines map to parallel lines.*
- *The midpoint of a line segment maps to the midpoint of the image line segment.*

We now apply[3] these facts to $\widetilde{C}$.

Since all the diameters of C are bisected by the centre of C, it follows that the image chords of $\widetilde{C}$ must all pass through a common point of bisection. Thus the centre of the circle C is mapped to the centre of the ellipse $\widetilde{C}$.

---

[2]As we have noted and have sought to correct, Beltrami's name is not currently attached to his models of the hyperbolic plane. To add insult to injury, the average "mathematician in the street" is also unaware of the fact that it was the very same Eugenio Beltrami who first discovered the SVD! See Stewart (1993).

[3]The following argument previously appeared in VCA, Section 4.8.2.

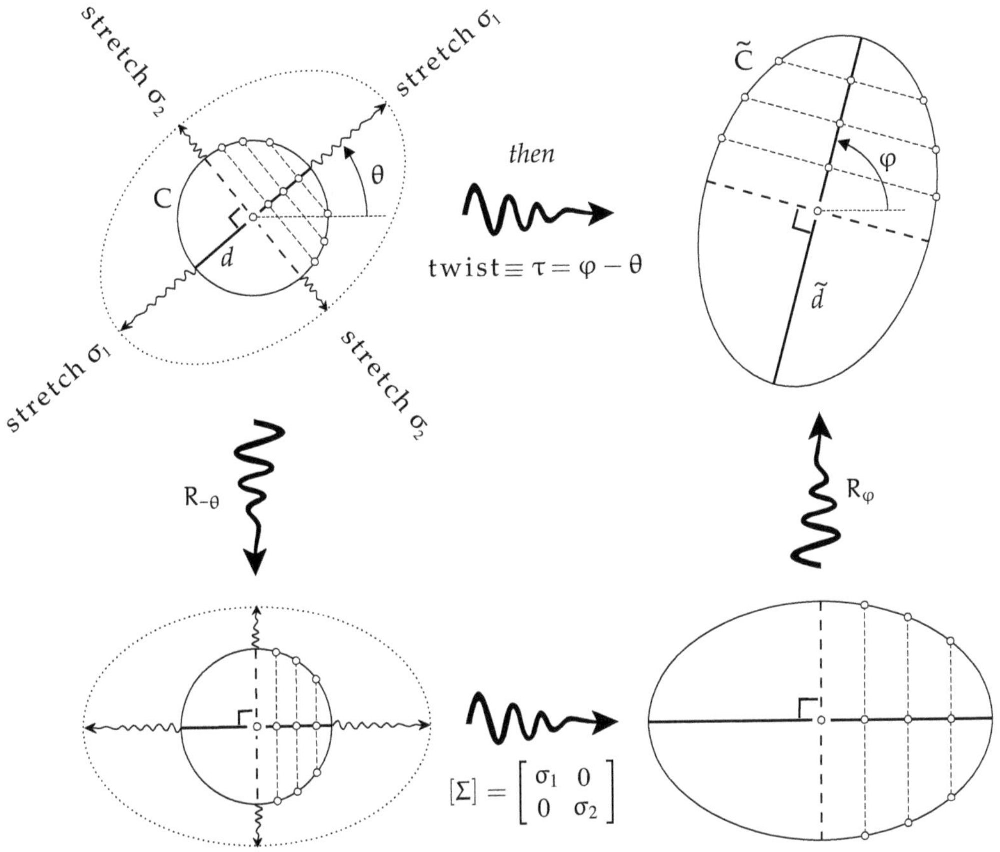

[15.4] Singular Value Decomposition (SVD): *The top half of the figure illustrates the effect, from left to right, of a general linear transformation* M *on an origin-centred circle. Preservation of midpoints enables us to prove that* M *is equivalent to two perpendicular stretches (by the "singular values"* σ_1 *and* σ_2*), followed by a rotation, which we call the* **twist** $\equiv \tau$*; here* $\tau = \varphi - \theta$*. The bottom half of the figure demonstrates that this SVD is equivalent to being able to write* $M = R_{\varphi} \circ \Sigma \circ R_{-\theta}$*, where* Σ *stretches horizontally by* σ_1 *and vertically by* σ_2*.*

Drawn in the same heavy line as its image is the particular diameter d of the circle that is mapped to the major axis $\widetilde{d}$ of the ellipse. Now consider the chords (shown dashed) of C that are perpendicular to d. Since these are all bisected by d, their images must be a family of parallel chords of $\widetilde{C}$ such that $\widetilde{d}$ is their common bisector. They must therefore be the family *perpendicular* to $\widetilde{d}$. (Convince yourself of this by drawing a family of parallel chords of $\widetilde{C}$ in a general direction.)

It is now clear that the linear transformation is a stretch in the direction of d, another stretch perpendicular to it, and finally a twist, thereby confirming the existence of the SVD. Note that we could equally well do the orthogonal stretchings *after* the twist, instead of before. But in that case the stretchings would be in the new, twisted directions, along $\widetilde{d}$, and orthogonal to it.

The bottom half of the figure demonstrates that the SVD is equivalent to being able to write

$$\boxed{M = R_{\varphi} \circ \Sigma \circ R_{-\theta},} \tag{15.11}$$

where Σ stretches horizontally by σ_1 and vertically by σ_2.

Note that this result also makes sense at the level of counting degrees of freedom. Just as the matrix has four independent entries, so the specification of our transformation also requires four bits of geometric information: the direction of d, the stretch factor in this direction, the perpendicular stretch factor, and the twist.

A vitally important special case arises if we require that the two stretch factors be set to be equal: $\sigma_1 = \sigma_2 = a$, say, in which case the image of a circle is another *circle* that is a times as big. This apparently reduces the number of degrees of freedom from four to three. However, since we are now producing an equal expansion in all directions, the direction chosen for d becomes irrelevant, and we are thus left with only *two* genuine degrees of freedom: the expansion factor a, and the twist τ.

In VCA we sought to exploit geometrically the fact that the local effect of a differentiable function $f(z)$ of a complex variable is precisely a mapping of the type just described. In this context, $a(z)$ and $\tau(z)$ both typically *vary* with position: they describe the expansion and rotation undergone by an infinitesimal neighbourhood of z.

More explicitly, recapping (4.18), page 42, every tiny complex arrow δz emanating from z is "amplitwisted" to an image arrow $\delta \widetilde{z}$ emanating from $\widetilde{z} = f(z)$, where $\delta \widetilde{z} \asymp [a e^{i\tau}]\, \delta z$. In VCA we called $a(z)$ the ***amplification*** of $f(z)$, and we called the combined effect of the local amplification and twist the ***amplitwist*** of $f(z)$, encoded as a complex number:

$$f'(z) = \text{amplitwist of } f(z) = (\text{amplification})\, e^{i(\text{twist})} = a e^{i\tau}.$$

Let us now resume our quest for the meaning of M^T. To facilitate this quest, first consider the *inverse* of the original linear transformation M (see the top of [15.4]). This inverse M^{-1} is drawn going from right to left in [15.5a]: first it undoes the twist τ of M by twisting $-\tau$, and then it undoes the stretchings of M by squashing by $1/\sigma_1$ and $1/\sigma_2$ in the same two orthogonal directions that M originally did stretching. Reversing the arrows in the bottom half of [15.4], we can also express this as

$$M^{-1} = R_\theta \circ \Sigma^{-1} \circ R_{-\varphi}.$$

At last we can reveal the geometric meaning of M^T. We shall assume that the reader is familiar with the algebraic properties of the transpose operation on matrices, and we shall use them to expose the underlying geometry. First note that taking the transpose of a rotation matrix (15.5) yields the opposite rotation: $[R_\theta]^T = [R_{-\theta}]$. Also, recall that $([P][Q])^T = [Q]^T[P]^T$. Thus (15.11) yields,

$$M^T = (R_\varphi \circ \Sigma \circ R_{-\theta})^T = R_{-\theta}^T \circ \Sigma^T \circ R_\varphi^T = R_\theta \circ \Sigma \circ R_{-\varphi},$$

and [15.5b] illustrates the meaning of this:

The linear transformation M^T *does the* opposite *twist to* M, *followed by the* same *two orthogonal expansions as* M. (15.12)

Having grasped the geometric meaning of M^T, the geometric meaning of symmetry suddenly becomes clear, too:

The linear transformation M *is symmetric (i.e.* $M^T = M$*)* iff *the twist vanishes. But if* $\tau = 0$*, then the orthogonal expansion directions become eigenvectors, so we also conclude* M *is symmetric* iff *it has orthogonal eigenvectors.* (15.13)

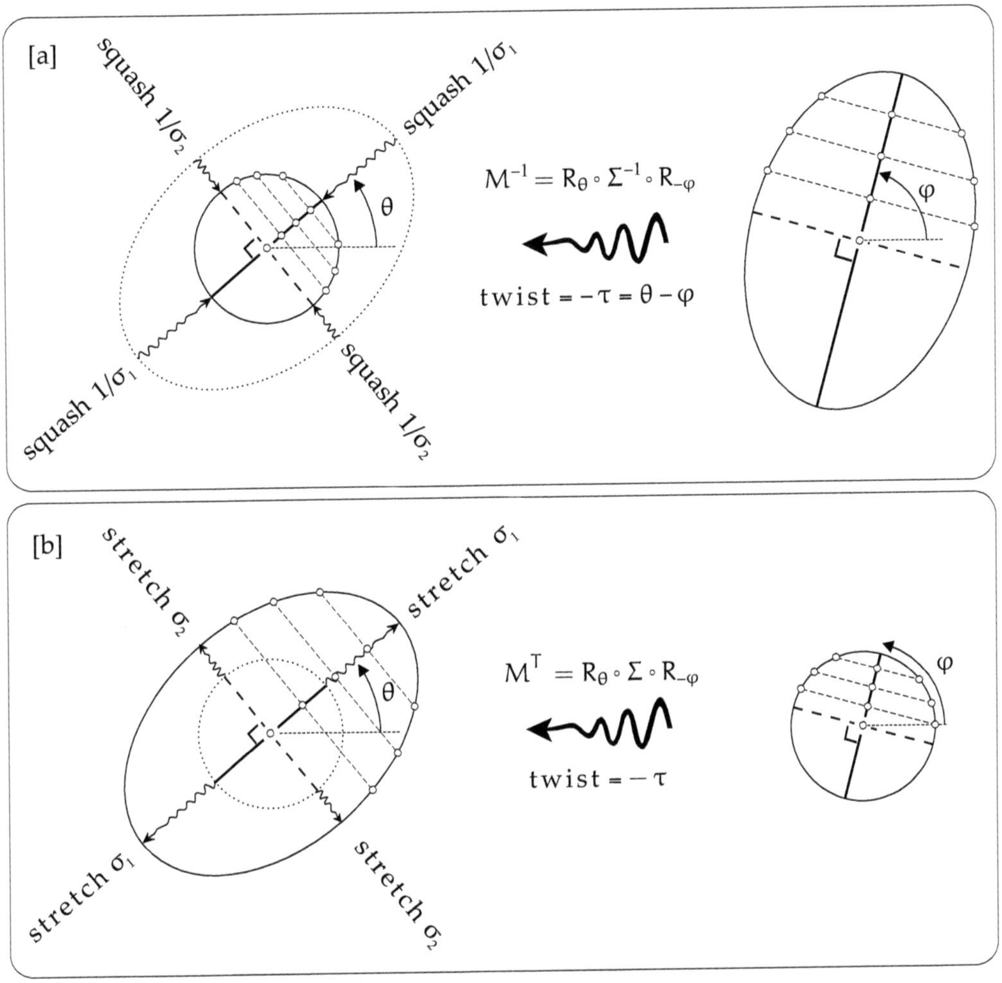

[15.5] **[a]** *The inverse* M^{-1} *(going from right to left) of the original linear transformation at the top of [15.4]: it undoes* M *by first doing the opposite twist* $-\tau$*, and then squashing by* $1/\sigma_1$ *and* $1/\sigma_2$ *in the same two orthogonal directions that* M *originally did stretching.* **[b]** *The transpose* M^T *also starts by twisting* $-\tau$*, but then it does the* same *two orthogonal stretches that* M *did.*

If $\tau = \varphi - \theta = 0$, the two rotations in the bottom half of [15.4] are equal and opposite, and so (15.11) specializes:

$$\textit{If } M^T = M \textit{ then } M = R_\theta \circ \Sigma \circ R_{-\theta}, \tag{15.14}$$

which is sometimes called the ***Spectral Theorem***.

We could, instead, have taken (15.12) as our *definition* of M^T, and then all the algebraic properties of the transpose would have become readily visualizable *geometric theorems*. For example, check that $R_\theta^T = R_{-\theta}$, which then implies $[R_\theta]^T = [R_{-\theta}]$. Likewise, check that $(P \circ Q)^T = Q^T \circ P^T$, which then implies $([P][Q])^T = [Q]^T[P]^T$.

Here is a fresh example of a result conventionally proved by calculation and now rendered transparent by geometry: M^T *expands area by the same factor as* M:

$$|M^T| = \sigma_1\sigma_2 = |M|.$$

And here is yet another result that is *proved* algebraically in every standard text, but which geometry now allows us to *understand*. If we first do M (top of [15.4]) and then do M^T ([15.5b]) then the opposite twists *cancel*, and we literally *see* that

> *The transformation* $(M^T \circ M)$ *is symmetric, having orthogonal eigenvectors (along* d *and perpendicular to it) with eigenvalues* σ_1^2 *and* σ_2^2. (15.15)

Lastly, we mention another commonly used manifestation of symmetry, leaving it to you to verify this both algebraically and geometrically:

> *A linear transformation* M *is symmetric (i.e.,* $M^T = M$*)* iff
>
> $$\mathbf{a} \cdot M(\mathbf{b}) = \mathbf{b} \cdot M(\mathbf{a}), \tag{15.16}$$
>
> *for all* **a** *and* **b**.

Armed with the geometric meaning of the transpose, Exercise 12 provides *eight more examples* of what we might call "Visual Linear Algebra." These examples underscore our overarching philosophy: direct, geometric reasoning frequently allows us to completely bypass symbolic manipulation to obtain an intuitive, *visual* grasp of mathematical reality.

15.5 The General Matrix of S

The Shape Operator S is a geometric entity, independent of any particular choice of basis vectors. But the matrix [S] that *represents* S certainly *does* depend on this choice.

Suppose that we choose an arbitrary orthonormal basis $\{\mathbf{E}_1, \mathbf{E}_2\}$, without first finding the principal directions and curvatures; how will the matrix look? While we may not yet know the principal directions, they certainly exist, so let us suppose that $\{\mathbf{e}_1, \mathbf{e}_2\}$ are obtained by rotating $\{\mathbf{E}_1, \mathbf{E}_2\}$ through some unknown angle θ. Figure [15.6] illustrates this in a case where $\kappa_1 > \kappa_2 > 0$. On the left, we see the effect of S on the unit circle, which is carried into an ellipse with semimajor axis κ_1 aligned with $\mathbf{e}_1$, and semiminor axis κ_2 aligned with $\mathbf{e}_2$.

As we see on the right, in terms of $\{\mathbf{E}_1, \mathbf{E}_2\}$ the effect of S is equivalent to the following three successive transformations:

$$S = (\text{rotate by } -\theta) \text{ then (expand by } \kappa_{1,2} \text{ along } \mathbf{E}_{1,2}) \text{ then (rotate by } \theta).$$

If you read the previous section, this will already be familiar to you as the result (15.14).

Recall from Linear Algebra that the *product* [B][A] of two matrices is (or at least *should* be!) defined to be the matrix of the composite linear transformation: $[B][A] \equiv [B \circ A]$. It follows from (15.5) that

$$\begin{aligned}
[S] &= [R_\theta] \begin{bmatrix} \kappa_1 & 0 \\ 0 & \kappa_2 \end{bmatrix} [R_{-\theta}] \\
&= \begin{bmatrix} c & -s \\ s & c \end{bmatrix} \begin{bmatrix} \kappa_1 & 0 \\ 0 & \kappa_2 \end{bmatrix} \begin{bmatrix} c & s \\ -s & c \end{bmatrix} \\
&= \begin{bmatrix} \kappa_1 c^2 + \kappa_2 s^2 & (\kappa_1 - \kappa_2) sc \\ (\kappa_1 - \kappa_2) sc & \kappa_1 s^2 + \kappa_2 c^2 \end{bmatrix}.
\end{aligned} \tag{15.17}$$

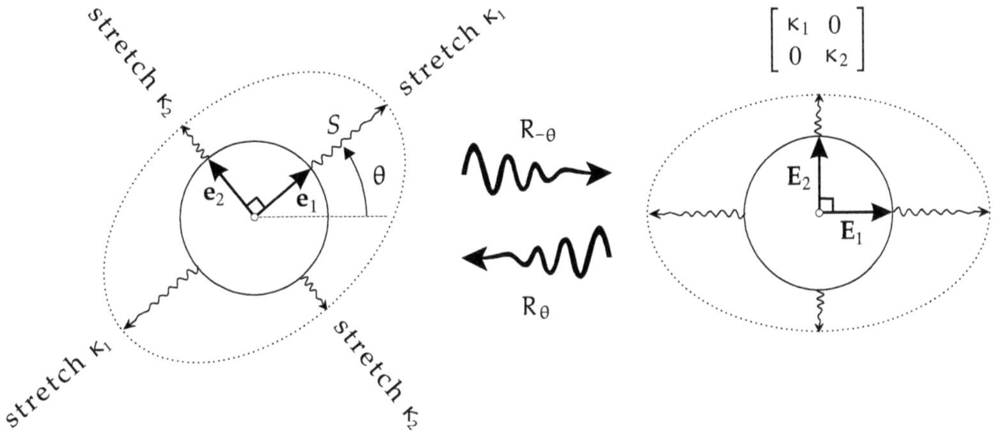

[15.6] *On the left, the Shape Operator S stretches the orthogonal principal directions by factors equal to the principal curvatures, deforming the unit circle into an ellipse aligned with the principal directions. This effect of S is equivalent to first rotating by $-\theta$, producing the figure on the right, then stretching horizontally by κ_1 and vertically by κ_2, then rotating back by θ.*

At least one of these four entries should look familiar. Let $\kappa(\mathbf{E}_1)$ denote the curvature of the normal section of $\mathcal{S}$ taken in the $\mathbf{E}_1$ direction, which makes angle $-\theta$ with the first principal direction. Then Euler's formula (10.1) informs us that the top-left entry is $\kappa_1 c^2 + \kappa_2 s^2 = \kappa(\mathbf{E}_1)$.

But *why* has this simplification occurred?

15.6 Geometric Interpretation of S and Simplification of [S]

The answer sheds new light on the meaning of the Shape Operator:

> *If $\widehat{\mathbf{v}}$ is an arbitrary unit tangent vector, then the curvature $\kappa(\widehat{\mathbf{v}})$ of the normal section in this direction is given by $\kappa(\widehat{\mathbf{v}}) = \widehat{\mathbf{v}} \bullet S(\widehat{\mathbf{v}})$. Thus,*
>
> $$\kappa(\widehat{\mathbf{v}}) = \textit{Projection of } S(\widehat{\mathbf{v}}) \textit{ onto the direction of } \widehat{\mathbf{v}}. \tag{15.18}$$

(NOTE: By Meusnier's Theorem (11.4), $\kappa(\widehat{\mathbf{v}})$ is not merely the curvature of the normal section, it is also equal to the previously introduced normal curvature $\kappa_n(\widehat{\mathbf{v}})$ of *any* curve on the surface that passes through this point travelling in the direction $\widehat{\mathbf{v}}$.)

This interpretation is illustrated in [15.7]. Note in particular that this construction correctly yields $\kappa(\mathbf{e}_1) = \kappa_1$ and $\kappa(\mathbf{e}_2) = \kappa_2$.

The essential geometric explanation for this is the same as that shown in [8.5], on page 102. In fact we previously proved the result in the form (11.5), but we repeat the proof here to illustrate our new notation. Thinking of $\widehat{\mathbf{v}}$ as the velocity of a particle travelling along the normal section at unit speed, so that $\nabla_{\widehat{\mathbf{v}}} \widehat{\mathbf{v}} = \kappa(\widehat{\mathbf{v}})\, \mathbf{n}$ is its acceleration towards the centre of curvature, we find

$$0 = \nabla_{\widehat{\mathbf{v}}}(\widehat{\mathbf{v}} \bullet \mathbf{n}) = \widehat{\mathbf{v}} \bullet \nabla_{\widehat{\mathbf{v}}} \mathbf{n} + \mathbf{n} \bullet \nabla_{\widehat{\mathbf{v}}} \widehat{\mathbf{v}} = -\widehat{\mathbf{v}} \bullet S(\widehat{\mathbf{v}}) + \kappa(\widehat{\mathbf{v}}),$$

from which the result (15.18) follows immediately.

We can give an alternative derivation if we assume Euler's formula, (10.1). Currently we are using θ to denote the angle from $\mathbf{E}_1$ to $\mathbf{e}_1$, so, to avoid confusion with our earlier discussion of

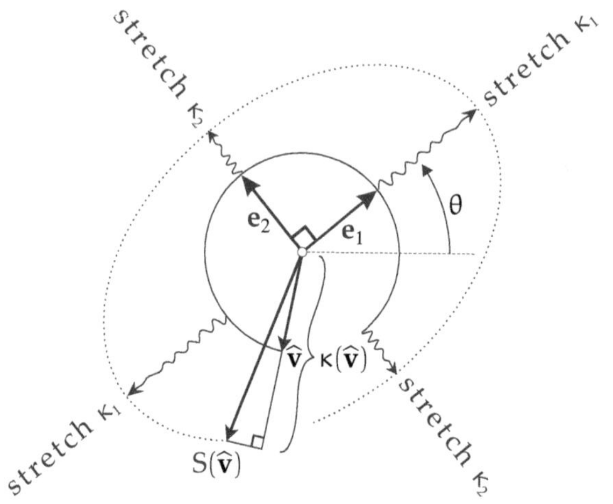

[15.7] *The curvature of a normal section of a surface taken in the direction of a general unit tangent vector* $\hat{\mathbf{v}}$ *is given by* $\kappa(\hat{\mathbf{v}}) =$ *projection of* $S(\hat{\mathbf{v}})$ *onto the direction of* $\hat{\mathbf{v}}$.

Euler's formula, let us instead use α to denote the angle that a general $\hat{\mathbf{v}}$ makes with $\mathbf{e}_1$, so that Euler's formula now reads

$$\kappa(\hat{\mathbf{v}}) = \kappa_1 \cos^2 \alpha + \kappa_2 \sin^2 \alpha.$$

Since (15.18) is a purely geometric statement, as witnessed in [15.7], it is sufficient to demonstrate its truth any particular basis. And in the principal basis $\{\mathbf{e}_1, \mathbf{e}_2\}$ a simple coordinate calculation confirms the result:

$$\hat{\mathbf{v}} \cdot S(\hat{\mathbf{v}}) = \begin{bmatrix} \cos\alpha \\ \sin\alpha \end{bmatrix} \cdot S \begin{bmatrix} \cos\alpha \\ \sin\alpha \end{bmatrix} = \begin{bmatrix} \cos\alpha \\ \sin\alpha \end{bmatrix} \cdot \begin{bmatrix} \kappa_1 \cos\alpha \\ \kappa_2 \sin\alpha \end{bmatrix} = \kappa(\hat{\mathbf{v}}),$$

by Euler's formula.

Armed with the result (15.18), we may now return to the interpretation and simplification of (15.17). Recall that even if S were an *arbitrary* linear transformation, the j-th column of its matrix would be the image of the j-th basis vector, so the first column of [S] is the vector $S(\mathbf{E}_1)$, and therefore its projections along $\{\mathbf{E}_1, \mathbf{E}_2\}$ (i.e., its components) are the dot products of this vector with these basis vectors. Likewise, this is true for the second column of [S], so

$$[S] = \begin{bmatrix} \mathbf{E}_1 \cdot S(\mathbf{E}_1) & \mathbf{E}_1 \cdot S(\mathbf{E}_2) \\ \mathbf{E}_2 \cdot S(\mathbf{E}_1) & \mathbf{E}_2 \cdot S(\mathbf{E}_2) \end{bmatrix}. \qquad (15.19)$$

Thus (15.18) explains[4] why the top-left entry of (15.17) turned out to be $\kappa(\mathbf{E}_1)$. Now look at the other diagonal entry. Whether we use Euler's formula [exercise] or refer to (15.18) and (15.19), we see that the bottom-right entry of (15.17) is $\kappa_1 s^2 + \kappa_2 c^2 = \kappa(\mathbf{E}_2)$. If you read the previous (optional) section, also note that (15.16) shows that the symmetry of [S] now follows from the symmetry of S, for $\mathbf{E}_1 \cdot S(\mathbf{E}_2) = \mathbf{E}_2 \cdot S(\mathbf{E}_1)$.

Before we do any more computation, let us pause to discuss more explicitly the geometric meaning of the entries in (15.19). As we have said, the first column is $S(\mathbf{E}_1) = -\nabla_{\mathbf{E}_1} \mathbf{n}$, which measures how the normal turns as we move off in the $\mathbf{E}_1$ direction. Let Π_1 be the normal plane in this direction, i.e., the plane spanned by $\mathbf{E}_1$ and $\mathbf{n}$. See [12.5b] on page 133 for a concrete example. Then

[4] Turning this around, we have obtained another *derivation* of Euler's formula.

the first component of $S(\mathbf{E}_1)$, namely $\mathbf{E}_1 \bullet S(\mathbf{E}_1)$, measures how fast $\mathbf{n}$ tips towards $\mathbf{E}_1$ *within* Π_1 as we begin to move along $\mathbf{E}_1$; this is determined by the curvature of the normal section, indeed it *is* the curvature $\kappa(\mathbf{E}_1)$. The second component of $S(\mathbf{E}_1)$, namely $\mathbf{E}_2 \bullet S(\mathbf{E}_1)$, measures how fast $\mathbf{n}$ rotates perpendicularly to the direction of motion, i.e., how fast it rotates *out of* Π_1.

Of course the meaning of the second column $S(\mathbf{E}_2)$ is completely analogous. The first component $\mathbf{E}_1 \bullet S(\mathbf{E}_2)$ measures how fast $\mathbf{n}$ rotates *out of* Π_2 as we begin to move along $\mathbf{E}_2$. The second component $\mathbf{E}_2 \bullet S(\mathbf{E}_2)$, measures how fast $\mathbf{n}$ rotates *within* Π_2, and this is $\kappa(\mathbf{E}_2)$.

The κ-related tipping within Π, and the rotation out of Π, are both clearly visible for the typical direction shown in [12.5b]. If, however, we move off in a *principal* direction then *all* of the tipping is in the direction of motion, within Π, and there is *no* initial rotation out of Π, and this is what we see in [12.5a].

Let us look more closely at the off-diagonal term, which measures how fast $\mathbf{n}$ swings out of $\Pi_{1,2}$. If we once again let $\Delta\kappa = (\kappa_1 - \kappa_2)$, as in (10.2), then the matrix representing the Shape Operator in a general basis $\{\mathbf{E}_1, \mathbf{E}_2\}$ is given by

$$[S] = \begin{bmatrix} \kappa(\mathbf{E}_1) & \frac{\Delta\kappa}{2}\sin 2\theta \\ \frac{\Delta\kappa}{2}\sin 2\theta & \kappa(\mathbf{E}_2) \end{bmatrix} \tag{15.20}$$

Note that this general matrix is indeed symmetric, as demanded by (15.9): $[S]^T = [S]$. Also note that if the basis coincides with the principal basis, so that $\theta = 0$, then (15.20) reduces to the diagonal form (15.7), as it should. The general matrix also reduces to this diagonal form if $\theta = (\pi/2)$, for in that case the basis is again aligned with the principal basis, only now $\{\mathbf{E}_1, \mathbf{E}_2\} = \{-\mathbf{e}_2, \mathbf{e}_1\}$. On the other hand, the off-diagonal term is *greatest* ($\mathbf{n}$ swings out of $\Pi_{1,2}$ the fastest) when $\theta = (\pi/4)$, in which case $\Pi_{1,2}$ bisect the principal directions.

15.7 [S] Is Completely Determined by Three Curvatures

Next, observe that the two equal off-diagonal entries are simply the oscillating term in Euler's formula, (10.2), phase-shifted by $\pm(\pi/4)$. Indeed, this formula tells us that the curvature of the normal section in the direction $\mathbf{E}_1 + \mathbf{E}_2$, bisecting the angle between the basis vectors, is given by

$$\kappa(\mathbf{E}_1 + \mathbf{E}_2) = \kappa(\tfrac{\pi}{4} - \theta) = \overline{\kappa} + \tfrac{\Delta\kappa}{2}\sin 2\theta. \tag{15.21}$$

Of course we could equally well work with the orthogonal direction:

$$\kappa(\mathbf{E}_1 - \mathbf{E}_2) = \kappa(-\tfrac{\pi}{4} - \theta) = \overline{\kappa} - \tfrac{\Delta\kappa}{2}\sin 2\theta. \tag{15.22}$$

But, choosing (somewhat arbitrarily) to use the first direction, (15.20) may be written

$$[S] = \begin{bmatrix} \kappa(\mathbf{E}_1) & \kappa(\mathbf{E}_1 + \mathbf{E}_2) - \overline{\kappa} \\ \kappa(\mathbf{E}_1 + \mathbf{E}_2) - \overline{\kappa} & \kappa(\mathbf{E}_2) \end{bmatrix}. \tag{15.23}$$

It might seem at first that we still need to know $\kappa_{1,2}$ (or at least their sum) in order to calculate $\overline{\kappa}$. But in fact we do not, and we shall instead see that *the Shape Operator matrix can be expressed purely in terms of the curvatures of the normal sections in three directions:* $\mathbf{E}_1$, $\mathbf{E}_2$, and $(\mathbf{E}_1 + \mathbf{E}_2)$.

Recall that the ***trace*** of a matrix is the sum of its diagonal elements, so for the original diagonal matrix (15.7) the trace is $\text{Tr}\,[S] = \kappa_1 + \kappa_2 = 2\overline{\kappa}$. But we know from Linear Algebra that if $[A]$ and $[B]$ are the matrices of *any* two linear transformation of the plane, then $\text{Tr}\,[A][B] = \text{Tr}\,[B][A]$. Thus,

$$\text{Tr}\,[R_\theta][A][R_{-\theta}] = \text{Tr}\,[A][R_{-\theta}][R_\theta] = \text{Tr}\,[A],$$

in other words,

The trace of any linear transformation is invariant under rotation of the basis vectors. (15.24)

(There is actually a *geometric* reason for this; see Arnol'd (1973, §16.3).)

Therefore, returning to the case at hand, in which $[A] = [S]$, we deduce that even in the case of a general basis, with the matrix (15.23),

$$\kappa(\mathbf{E}_1) + \kappa(\mathbf{E}_2) = \text{Tr}\,[S] = \kappa_1 + \kappa_2 = 2\,\overline{\kappa}.$$

In fact we need not even appeal to the general theorem of Linear Algebra to see this, for in our case we have already explicitly calculated $[S]$, and the truth of our assertion follows immediately from (15.17). Another specific example is obtained by adding (15.21) and (15.22).

Putting this into words, we have a result of interest in its own right:

> *The sum of the curvatures in* any *two perpendicular directions is equal to the sum of the principal curvatures.*

Thus, as claimed, the matrix (15.23) does indeed only depend on the curvatures in three directions, for $\overline{\kappa} = \frac{1}{2}[\kappa(\mathbf{E}_1) + \kappa(\mathbf{E}_2)]$, obviating the need to know the principal curvatures. Therefore, as claimed, the general matrix (15.23) of the Shape Operator can be written explicitly in terms of these three curvatures:

$$[S] = \begin{bmatrix} \kappa(\mathbf{E}_1) & \kappa(\mathbf{E}_1 + \mathbf{E}_2) - \frac{1}{2}[\kappa(\mathbf{E}_1) + \kappa(\mathbf{E}_2)] \\ \kappa(\mathbf{E}_1 + \mathbf{E}_2) - \frac{1}{2}[\kappa(\mathbf{E}_1) + \kappa(\mathbf{E}_2)] & \kappa(\mathbf{E}_2) \end{bmatrix}.$$

Finally, recall from (15.8) that the extrinsic version $\mathcal{K}_{\text{ext}}$ of the Gaussian curvature is the determinant of *any* of these forms of the matrix of $[S]$. For example, (15.23) yields

$$\mathcal{K}_{\text{ext}} = |[S]| = \kappa(\mathbf{E}_1)\kappa(\mathbf{E}_2) - [\kappa(\mathbf{E}_1 + \mathbf{E}_2) - \overline{\kappa}]^2.$$

15.8 Asymptotic Directions

Recall that *Dupin's indicatrix* $\mathcal{D}$ (p. 111) arises from the intersection of the surface with a plane $T_p(\epsilon)$ parallel to the tangent plane T_p and distance ϵ away from it, so $T_p(0) = T_p$. As ϵ increases from 0 and $T_p(\epsilon)$ just begins to move away in the normal direction, $\mathcal{D}$ is the nascent conic section of intersection with the surface.

If $\mathcal{K}(p) > 0$ then $\mathcal{D}$ is an ellipse, and p is called elliptic; if $\mathcal{K}(p) = 0$ then $\mathcal{D}$ is a parabola, and p is called parabolic; if $\mathcal{K}(p) < 0$ then $\mathcal{D}$ is a hyperbola, and p is called hyperbolic.

Here we shall focus on the hyperbolic (negative curvature) case, and shall investigate the directions of the asymptotes of the hyperbola $\mathcal{D}$, which are called the ***asymptotic directions***. As we shall now explain, the asymptotic directions have a simple geometric relationship to the Shape Operator S.

As discussed in Section 10.2, if the (x, y)-axes are aligned with the principal directions, the equation of the pair of conjugate hyperbolas $\mathcal{D}$ is

$$\kappa_1 x^2 + \kappa_2 y^2 = \pm 1,$$

the sign being determined by the sign of ϵ, i.e., whether we move the tangent plane up or down. The symmetry axes (in both the elliptic and hyperbolic cases) coincide with the principal/

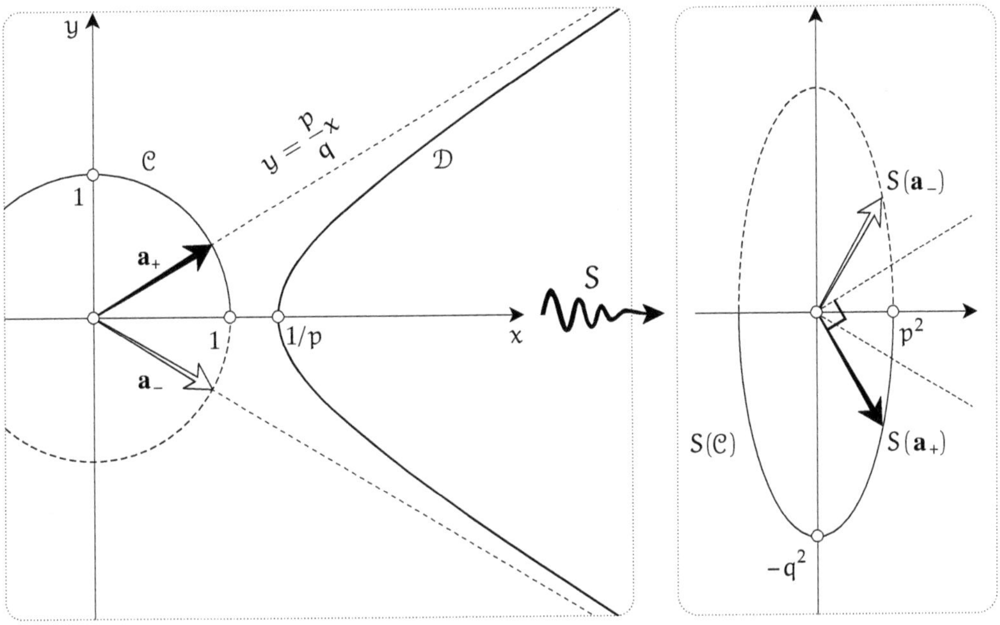

[15.8] *At a hyperbolic point, the Shape Operator maps each of the asymptotic direction vectors* $\mathbf{a}_\pm$ *of the Dupin indicatrix* $\mathcal{D}$ *into an* orthogonal *direction, and in opposite senses:* S *turns* $\mathbf{a}_\pm$ *through angle* $\mp(\pi/2)$.

coordinate axes, and in the hyperbolic case these bisect the angles between the asymptotic directions.

Let $\kappa_1 = p^2$ and $\kappa_2 = -q^2$ (with p and q positive) and let us choose the plus in the above equation for $\mathcal{D}$. Then,

$$\mathcal{D}: p^2x^2 - q^2y^2 = 1 \qquad \text{and} \qquad [S] = \begin{bmatrix} p^2 & 0 \\ 0 & -q^2 \end{bmatrix}.$$

Thus the asymptotes have equations $y = \pm\frac{p}{q}x$. See [15.8], in which we have added the unit circle $\mathcal{C}$, and its elliptical image $S(\mathcal{C})$, to help visualize the effect of the mapping S, which stretches horizontally and vertically, *and* flips across the x-axis. Thus the vectors in the asymptotic directions, and their images under S, are given by,

$$\mathbf{a}_\pm \propto \begin{bmatrix} q \\ \pm p \end{bmatrix} \qquad \Longrightarrow \qquad S(\mathbf{a}_\pm) \propto \begin{bmatrix} p \\ \mp q \end{bmatrix}.$$

As illustrated in [15.8], this means that although S maps a principal direction to the *same* direction,

The Shape Operator maps an asymptotic direction to an orthogonal *direction:*

$$\mathbf{a}_\pm \cdot S(\mathbf{a}_\pm) = 0. \tag{15.25}$$

In fact, as we see in [15.8], S turns the two asymptotic directions in opposite senses, $\mathbf{a}_+$ by $-\frac{\pi}{2}$, and $\mathbf{a}_-$ by $+\frac{\pi}{2}$.

But *why* has this happened? We have pictured $\mathcal{D}$ as the magnified intersection curve of the surface with $T_p(\epsilon)$ just as ϵ increases from 0. But, initially, when $\epsilon = 0$ this hyperbola of intersection degenerates into the asymptotes themselves. Since this intersection curve (with directions $\mathbf{a}_\pm$) lies in the tangent plane, it has vanishing normal curvature, so it follows from the geometric interpretation (15.18) of the Shape Operator that

$$\kappa(\widehat{\mathbf{a}}_{\pm}) = \widehat{\mathbf{a}}_{\pm} \cdot \mathcal{S}(\widehat{\mathbf{a}}_{\pm}) = 0,$$

thereby explaining the result.

In light of this connection with the normal curvature, the definition of ***asymptotic direction*** has been generalized to mean *any* direction for which the normal curvature vanishes. Thus a parabolic point *also* has one "asymptotic direction" [exercise: what is it?] despite the fact that the parabola $\mathcal{D}$ does not have any asymptotes.

If we choose $\mathbf{E}_1$ to be either one of the asymptotic directions, so that Π_1 is the normal plane in this direction, then as we move off within the surface along $\mathbf{E}_1$, the normal $\mathbf{n}$ must rotate about $\mathbf{E}_1$, swinging straight out of Π_1, its head moving in a direction tangent to the surface and perpendicular to $\mathbf{E}_1$, i.e., in the direction $\pm\mathbf{E}_2$. Thus $\mathcal{S}(\mathbf{E}_1) = \pm\tau\,\mathbf{E}_2$, where τ is the *rate* of spinning of $\mathbf{n}$ about $\mathbf{E}_1$, the so-called torsion[5] introduced on page 106. It follows from (15.19) that

> *At a hyperbolic point, the curvature of the surface can be expressed in terms of the torsion τ of an asymptotic curve as*
>
> $$\mathcal{K}_{ext} = |[\mathcal{S}]| = -\tau^2. \tag{15.26}$$

According to Stoker (1969, p. 101), this result is due to Beltrami and to Enneper.

15.9 Classical Terminology and Notation: The Three *Fundamental Forms*

When consulting older, classical works on Differential Geometry, you will encounter terms and notation that we do not employ in this book. In particular, you will certainly meet the three so-called ***Fundamental Forms***, denoted I, II, and III.

All three Fundamental Forms are symmetric functions of pairs of tangent vectors, $\mathbf{u}$ and $\mathbf{v}$.

First Fundamental Form: $\quad \mathrm{I}(\mathbf{u},\mathbf{v}) \equiv \mathbf{u}\cdot\mathbf{v}.$

Second Fundamental Form: $\quad \mathrm{II}(\mathbf{u},\mathbf{v}) \equiv \mathcal{S}(\mathbf{u})\cdot\mathbf{v}.$

Third Fundamental Form: $\quad \mathrm{III}(\mathbf{u},\mathbf{v}) \equiv \mathcal{S}(\mathbf{u})\cdot\mathcal{S}(\mathbf{v}).$

The Shape Operator itself does not appear in the classical literature[6]; its mathematical content is instead represented by the Second Fundamental Form. For example, the curvature of the normal section in the direction $\widehat{\mathbf{v}}$, given by (15.18), would, in yesteryears, have been written, $\kappa(\widehat{\mathbf{v}}) = \mathrm{II}(\widehat{\mathbf{v}},\widehat{\mathbf{v}})$.

WARNING: Although we shall not introduce and define ***Differential Forms*** (**Forms** for short) until Act V, we should immediately caution the reader that the three classical "Forms" are **not** *Forms* at all! While we certainly do not fault modern authors who continue to speak the classical language, *our* reliance on *genuine* Forms impelled us to turn the classical language into a dead language!

[5]We hope no confusion arises here from the letter τ serving double duty: we also recently used it to denote the (unrelated) "twist" of an SVD decomposition.

[6]The Shape Operator was first seriously championed, and brought into common usage, by Barrett O'Neill in his groundbreaking introductory text (O'Neill 2006) the first edition of which appeared in 1966.

Chapter 16

Introduction to the Global Gauss–Bonnet Theorem

16.1 Some Topology and the Statement of the Result

The ***Global Gauss–Bonnet Theorem*** is widely considered to be one of the most beautiful results in all of mathematics. Furthermore, it is also *fundamental*, having spawned the discovery of ever-more powerful generalizations, the culmination of which is perhaps (for now) the ***Atiyah–Singer Index Theorem***, discovered in 1963. *This* Theorem, in turn, caused seismic shifts in other areas of mathematics and in theoretical physics. These subsequent developments are far beyond the scope of the present work, and sadly beyond the competence of the present author, but the statement of the original form of GGB (as we shall henceforth abbreviate it) is surprisingly simple and easily understood. We need just a few preliminaries before we can state the result.

First, the ***total curvature*** $\mathcal{K}(\mathsf{P})$ of a region P of a curved surface is defined (naturally enough) to be,

$$\mathcal{K}(\mathsf{P}) \equiv \iint_{\mathsf{P}} \mathcal{K}\, d\mathcal{A}.$$

(NOTE: Even in modern times, this concept is occasionally still referred to by its old Latin name, ***Curvatura Integra***.) For example, if P is a flat piece of paper with an arbitrary simple curve as boundary, then $\mathcal{K}(\mathsf{P}) = 0$. If we now bend the paper into a different shape $\widetilde{\mathsf{P}}$ (without stretching it), say, into a portion of a cylinder or cone, then $\mathcal{K}(\widetilde{\mathsf{P}}) = 0$, by virtue of the *Theorema Egregium*.

But now suppose instead that P is made of a highly *stretchable* material, like rubber. If we stretch P over the surface of a sphere, then $\mathcal{K}(\widetilde{\mathsf{P}}) > 0$. Indeed, by choosing the sphere to be as small as we please, we can make its curvature as large as we please, and with it $\mathcal{K}(\widetilde{\mathsf{P}})$ becomes as large as we please. If we instead stretch P into the form of a portion of a pseudosphere, then $\mathcal{K}(\widetilde{\mathsf{P}}) < 0$, and again this can be made as negative as we please.

Such a continuous, one-to-one stretching, $\mathsf{P} \mapsto \widetilde{\mathsf{P}}$, which does not preserve lengths or even angles, is called a ***topological mapping*** or ***topological transformation***, or ***homeomorphism***. Just as ancient Greek geometry sought out properties of figures that are *invariant* under rigid, distance-preserving mappings, so in the nineteenth century a new area of mathematics arose, called ***topology***, wherein properties were sought that were invariant under topological transformations.

Clearly, the concept of curvature does *not* belong to topology: stretching the surface in the vicinity of a point p changes the value of $\mathcal{K}(\mathsf{p})$. Indeed, we have just seen that $\mathcal{K}(\widetilde{\mathsf{p}})$ can be made to take on any value we please, either positive or negative. And likewise the more primitive concepts of length and angle do not belong to topology. Thus, at first sight, it might appear that topology would be a rather trivial or barren area of study: how can *anything* interesting or subtle survive the arbitrarily complicated and extreme distortions of a topological mapping?!

Remarkably, nothing could be further from the truth. Out of the kindling embrace of its principal parents (Riemann and Poincaré), topology rapidly grew up to become a powerful yet beneficent Hydra, explaining and unifying phenomena in disparate and distant realms of thought.

In order to introduce our first topological invariant, we restrict our attention to *closed* surfaces that are also *orientable*. Any such surface may be pictured as the boundary of a solid object in $\mathbb{R}^3$. Such a surface is automatically ***orientable***, meaning that one may consistently decide which of the

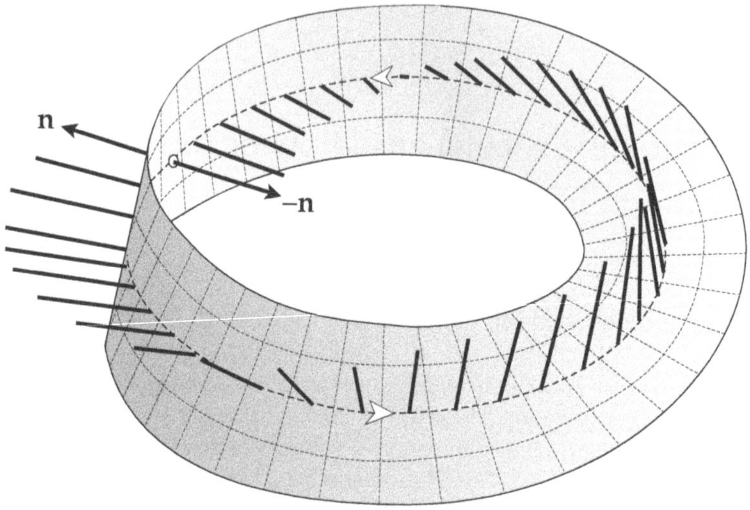

[16.1] *A* **Möbius band** *is* nonorientable*: carrying the normal* **n** *along a full circuit of the centre line, it returns to its starting point as* −**n**.

two opposite choices of **n** is *the* normal to the surface—let it point *out* of the solid object into empty space.

But isn't *every* surface orientable? No! As Möbius and Listing independently discovered in 1858, taking a paper strip and giving it a half-twist, and then gluing the ends of the strip together to form a loop, produces a surface with only *one side*! As illustrated in [16.1], starting with an arbitrary initial choice of **n**, then continuously carrying **n** along a full circuit of the centre line, we return it to its starting point as −**n**. Thus this so-called ***Möbius band***[1] is *non*orientable.

Returning to ordinary, orientable, closed surfaces, the fundamental feature that topologically distinguishes one from another is the number of *holes* it contains. This is called the ***genus*** g of the surface, and its value is illustrated in [16.2] for a few surfaces. Each pair of surfaces with the same genus is ***topologically equivalent***, or ***homeomorphic***, meaning that one may be changed into the other by a topological mapping (aka ***homeomorphism***). But two surfaces of different genus *cannot* be topologically equivalent, since g is invariant under topological mappings.

The genus was defined more precisely by Riemann, in 1851,[2] as the *maximum* number of cuts along closed, nonintersecting loops on the surface that can be performed without splitting the surface into two disconnected pieces. For example, cutting the sphere along any closed loop will split it into two, so the genus is zero. On the torus, we can cut along just *one* loop without splitting the torus into two pieces: e.g., cut along either an equator that encircles the axis of symmetry, or along a circle that goes through the hole (whose plane contains the axis of symmetry). But if we now cut the resulting surface along any loop that avoids the first, it will split it into two, so the genus of the torus is one. Try (in your mind) making loop cuts on a two-holed doughnut, as seen in [16.2c], and verify that $g = 2$.

As Möbius realized in 1863, every closed, orientable surface is topologically equivalent to a g-holed torus. We shall accept such visually plausible statements without proof, but the reader in search of precise definitions and proofs should consult the excellent topology texts we recommend in *Further Reading*, at the end of this book.

[1] Also commonly called a ***Möbius strip***.

[2] See Stillwell (1995), page 58.

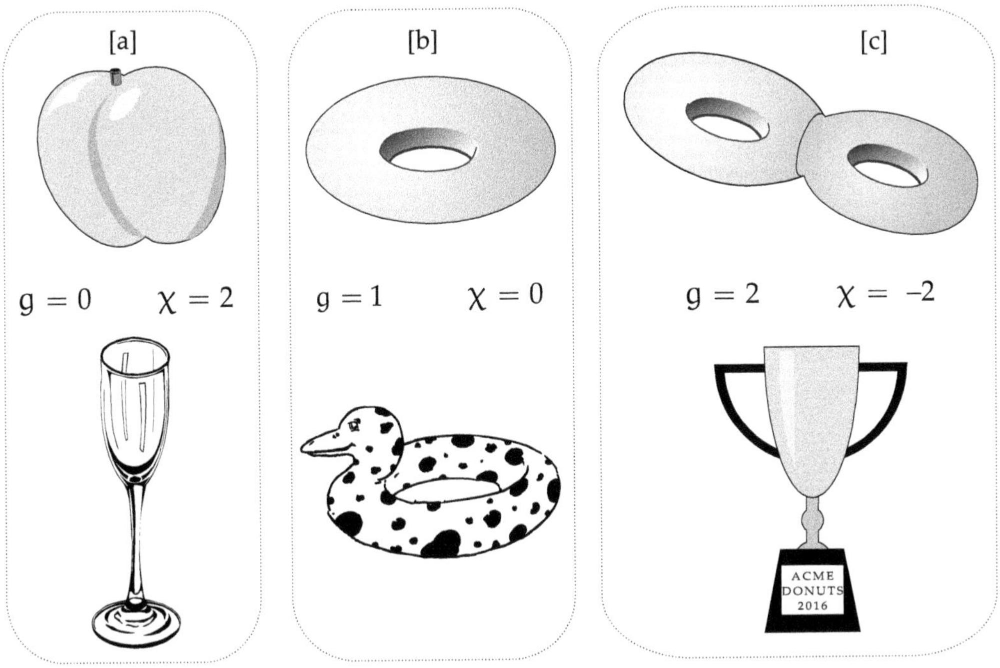

[16.2] *Each pair of surfaces is topologically the same, distinguished from the other pairs by its number of holes, the genus,* g. *Also shown are the corresponding values of the* Euler characteristic, $\chi = 2 - 2g$.

We can now state the stunning result:

Global Gauss–Bonnet Theorem (GGB). *The total curvature of a closed, orientable surface* $\mathcal{S}_g$ *depends* only *on its topological genus* g, *and is given by*

$$\mathcal{K}(\mathcal{S}_g) = 4\pi(1-g) = 2\pi\chi(\mathcal{S}_g). \tag{16.1}$$

Here the quantity

$$\chi(\mathcal{S}_g) \equiv 2 - 2g \tag{16.2}$$

serves (for now) as merely an alternative means of labelling the surfaces of different genus: see [16.2] for examples. This quantity $\chi(\mathcal{S}_g)$ is called the ***Euler characteristic*** of the surface; it arises naturally in many topological results, and actually has a meaning all its own, which will be explained in Section 18.1.

Pause for a moment to let the surprise and the beauty of the result sink in. If we take any given $\mathcal{S}_g$ (such as a simple doughnut with $g = 1$) made out of rubber or Play-Doh, then stretch it, twist it, squeeze it, and deform it in any way we please, *every resulting increase in curvature at one point of the surface must be instantaneously cancelled out by an exactly opposite decrease in curvature somewhere else on the surface.* In the case of a topological doughnut (aka, ***torus***), this total curvature remains exactly zero throughout our manipulations.

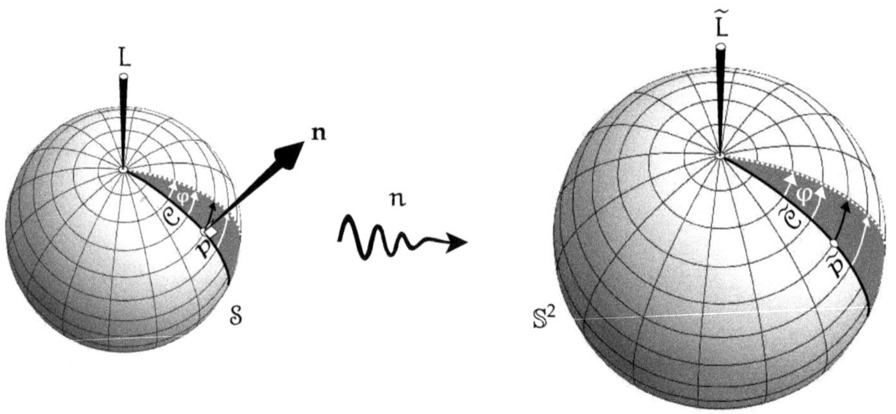

[16.3] *As the semicircle* $\mathcal{C}$ *rotates about* L *through angle* φ*, its spherical image* $\widetilde{\mathcal{C}}$ *rotates on* S^2 *at the same rate, generating a lune of equal angle* φ*. As* φ *increases to* 2π*,* $\widetilde{\mathcal{C}}$ *sweeps out the entire surface of* S^2*, so* $\mathcal{K}(\mathcal{S}) = 4\pi$.

16.2 Total Curvature of the Sphere and of the Torus

16.2.1 Total Curvature of the Sphere

If a surface $\mathcal{S}$ is topologically equivalent to a sphere (i.e., with $g = 0$), the prediction of GGB is that should have total curvature 4π. In the case of a *geometric* sphere $\mathcal{S}$, this is easily verified in multiple ways. We provide three proofs, the last of which will be shown to generalize beyond spheres.

First, each normal section of the sphere of radius R is itself a circle of radius R, i.e., of curvature $(1/R)$. So $\mathcal{K} = \kappa_1\kappa_2 = (1/R^2)$, and the total curvature

$$\mathcal{K}(\mathcal{S}) = \iint_{\mathcal{S}} \mathcal{K}\, d\mathcal{A} = \iint_{\mathcal{S}} \frac{1}{R^2}\, d\mathcal{A} = \frac{1}{R^2} 4\pi R^2 = 4\pi.$$

A second, more enlightening explanation is based on [12.2], page 132: if P is a region on $\mathcal{S}$ and $\widetilde{P} = n(P)$ is its spherical image of P, then

$$\boxed{\mathcal{K}(P) = [\text{area of spherical image of P on } S^2] = \mathcal{A}(\widetilde{P}).} \tag{16.3}$$

Now, take the image unit sphere S^2 to have the same centre as $\mathcal{S}$, then the spherical map becomes a radial projection, as illustrated in [12.3], page 132. This makes it crystal clear that $\widetilde{\mathcal{S}} = n(\mathcal{S}) = S^2$, so (16.3) implies $\mathcal{K}(\mathcal{S}) = \mathcal{A}(S^2) = 4\pi$.

Third, and finally, we picture the sphere $\mathcal{S}$ as the surface of revolution obtained by rotating a semicircle $\mathcal{C}$ of radius R about the diameter L through its ends, as depicted in [16.3].

Although we may generate the entire sphere by rotating $\mathcal{C}$ a full 2π, the figure depicts the process partway through, after $\mathcal{C}$ has rotated through angle φ, generating a so-called ***lune***.

The key observation is this, and it applies to an *arbitrary* surface of revolution, generated by rotating an *arbitrary* plane curve $\mathcal{C}$ about an *arbitrary* line within its plane:

> *If a plane curve* $\mathcal{C}$ ***and*** *its unit normal vector* **n** *are* ***together*** *rotated about an arbitrary line in the plane of* $\mathcal{C}$*, then the rotated* **n** *is the normal to the surface of revolution generated by* $\mathcal{C}$.

The explanation is readily understood from the special case in the figure. The tangent plane to the surface at a typical point p is spanned by the following two directions: (1) the direction of the rotated $\mathcal{C}$; and (2) the direction in which p moves as it rotates (i.e., perpendicularly to the plane of $\mathcal{C}$). But $\mathbf{n}$ was initially perpendicular to these two directions, so it *remains* perpendicular to (1) and (2) as $\mathcal{C}$ and $\mathbf{n}$ rotate together. Since $\mathbf{n}$ is perpendicular to two directions that span the tangent plane of $\mathcal{S}$, it *is* the normal to $\mathcal{S}$.

By virtue of (16.3), this immediately implies the following:

> *Let $\mathcal{C}$ be a plane curve, and L be a line in that plane, and let $\widetilde{L}$ be the line through the centre of S^2 that is parallel to L. Then, as $\mathcal{C}$ rotates about L, its spherical image $\widetilde{\mathcal{C}} = n(\mathcal{C})$ rotates at the same rate about $\widetilde{L}$, and the total curvature of the surface swept out by $\mathcal{C}$ is equal to the total (signed) area swept out by $\widetilde{\mathcal{C}}$ on S^2.* (16.4)

In particular, as illustrated, the spherical image of this lune is simply a lune of equal angle on S^2, and if $\mathcal{C}$ sweeps out all of $\mathcal{S}$, then its spherical image sweeps out all of S^2, once again confirming that $\mathcal{K}(\mathcal{S}) = 4\pi$.

The advantage of this last point of view is that we now gain our first real insight into GGB itself. Consider the American football depicted in [16.4], which is generated by the rotation of the curve $\mathcal{C}$ about the line L. Here we have drawn the normals at equal angular increments, which makes clear the variation in curvature: greatest around the poles, and least around the equator. Nevertheless, since the football is topologically spherical, the *total* curvature should be 4π, and we can now *see* that is. For the spherical image of $\mathcal{C}$ is the *same* semicircle $\widetilde{\mathcal{C}}$ as before (connecting the poles of S^2), and therefore as $\mathcal{C}$ rotates to generate the football, $\widetilde{\mathcal{C}}$ rotates to sweep out all of S^2, just as before, so the total curvature of the football is indeed 4π!

Clearly, the essential point is that spherical image of the football covers all of S^2. In the next chapter we will elaborate on this idea to produce our first (heuristic) proof of GGB.

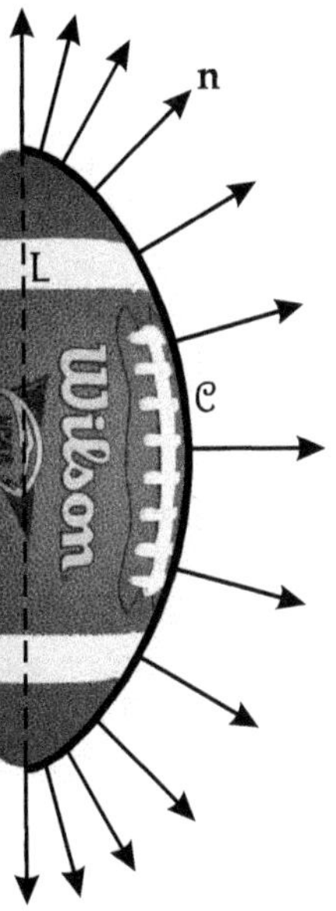

[16.4] *The spherical image of an American football is the whole of S^2, so $\mathcal{K}(\text{football}) = 4\pi$, in accordance with GGB.*

16.2.2 Total Curvature of the Torus

As we observed in (10.13), page 114, if a section of $\mathcal{C}$ is concave *away* from L, then, as it rotates about L, it generates a portion of the surface that has *negative* curvature. In this case, the area generated by $\widetilde{\mathcal{C}}$ on S^2 should be *subtracted* from the total curvature. We now illustrate this with the torus, which of course has regions of both positive and negative curvature.

According to GGB, the doughnut (torus) should have vanishing total curvature, and in Exercise 23, page 89, you actually verified this by brute force, integrating the curvature formula over the entire surface. We are now in a position to provide a real *explanation*, in fact establishing a (superficially) stronger, local (as opposed to global) result.

Suppose we wish to eat only part of the doughnut: we could cut out a wedge—the darkly shaded region in [16.5]—as one would a wedge of cake, by making two slices through the axis of symmetry. We will now apply (16.4) to [16.5] to show that this wedge of doughnut must have vanishing total curvature. The global result then follows by greed: we cut a larger and larger wedge of doughnut until we eat the whole thing!

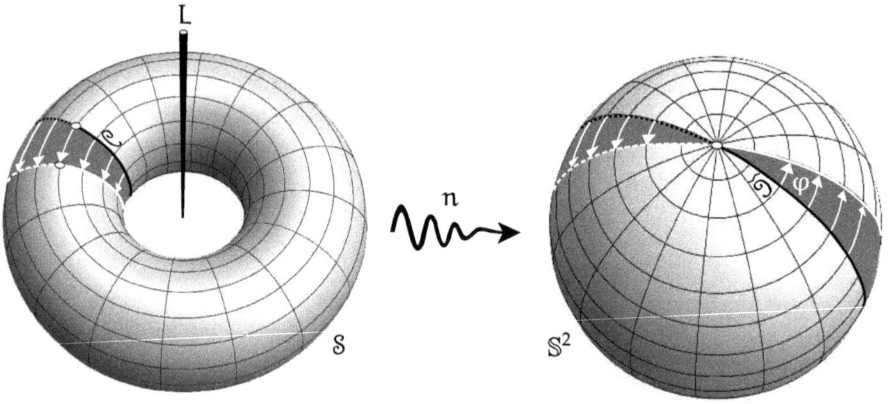

[16.5] *As the circle $\mathcal{C}$ rotates about L through angle φ, its (great circle) spherical image $\widetilde{\mathcal{C}}$ rotates on $\mathbb{S}^2$ at the same rate, generating two lunes of angle φ. As we see, the spherical map preserves orientation on the outer half of the doughnut, and reverses it on the inner half. The total curvature of the wedge of doughnut is the sum of the equal and opposite (signed) areas of the two lunes, and therefore it vanishes. Thus the total curvature of the doughnut vanishes, also.*

As the circle $\mathcal{C}$ rotates about L through angle φ, its (great circle) spherical image $\widetilde{\mathcal{C}}$ rotates on $\mathbb{S}^2$ at the same rate, generating two lunes of angle φ. As we see, the spherical map preserves orientation on the outer half of the doughnut, and reverses it on the inner half. To highlight this reversal of orientation on the negatively curved inner half of doughnut, we have placed the label $\mathcal{C}$ there, so it is mapped to a *backwards* $\widetilde{\mathcal{C}}$.

The total curvature of the wedge of doughnut is the sum of the equal and opposite (signed) areas of the two lunes, and therefore it *vanishes*. Letting φ increase to 2π, we see that the image of the outer half of the doughnut completely covers $\mathbb{S}^2$ once *positively*, while the image of the inner half completely covers $\mathbb{S}^2$ once *negatively*, so that the total is $\mathcal{K}(\textit{whole doughnut}) = 4\pi + (-4\pi) = 0$.

Observe something else. The entire circle at the top of the doughnut (which divides the outer and inner halves of opposite curvature) is mapped to a single point, namely, the north pole of $\mathbb{S}^2$. Likewise the lowest circle, upon which the doughnut would rest if it were placed on a plate, is likewise mapped to the south pole. These points of $\mathbb{S}^2$, where the two separate layers that cover $\mathbb{S}^2$ join together, are called ***branch points***.

Our next task is to find ways of visualizing the total curvature of surfaces that have *more* than one hole.

16.3 Seeing $\mathcal{K}(\mathcal{S}_g)$ via a Thick Pancake

Imagine pouring a very dense pancake batter into a frying pan to make a large, thick pancake. Before it can start to cook through and set, we quickly take a cylindrical biscuit cutter and remove g cylindrical discs from the interior of the pancake, leaving g holes. The batter starts to ooze back into the holes—thereby producing shapes like the inner halves of doughnuts—then starts to cook and harden, producing a genus g, thick pancake of the form [16.6].

The spherical map sends the entire flat bottom of the pancake to the south pole of $\mathbb{S}^2$, and it likewise sends the flat top of the pancake to the north pole. As expected, these yield no area on the sphere, so no contribution to the total curvature.

On the other hand, the outer edge of the pancake is like the outer half of a torus, and so its spherical image covers the entire sphere once, positively, contributing 4π to the total curvature. But each of the g rims to the holes in the pancake is like the inner half of a torus, so each one

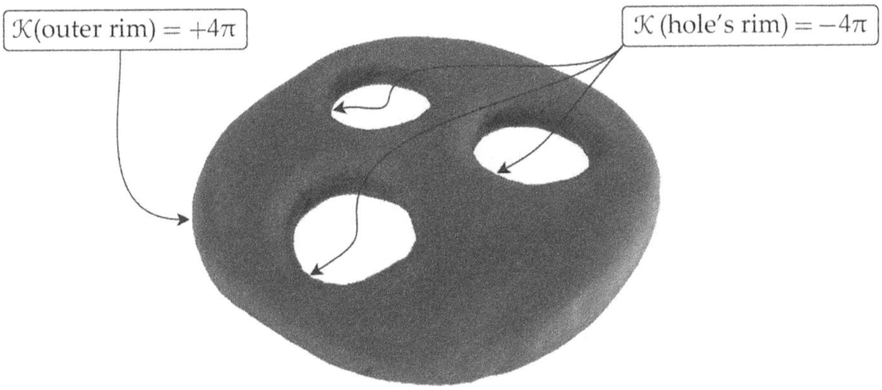

[16.6] *Here $\mathcal{S}_g$ takes the form of a thick pancake with g holes. The top and bottom are flat, and so contribute no curvature, but the outer rim has $\mathcal{K} = +4\pi$, and the rim of each of the g holes has $\mathcal{K} = -4\pi$. The total curvature is therefore $4\pi(1-g)$.*

has a spherical image that covers the entire sphere once, negatively, contributing -4π to the total curvature. Thus (16.3) allows us to *see*[3] that

$$\mathcal{K}(\mathcal{S}_g) = 4\pi + (-4\pi)g = 4\pi(1-g) = 2\pi\chi(\mathcal{S}_g).$$

16.4 Seeing $\mathcal{K}(\mathcal{S}_g)$ via Bagels and Bridges

We now descend momentarily into the first person to relate a personal but pertinent anecdote.

One day, while waiting to be served at a bagel shop near my University of San Francisco campus, I observed that only a half-dozen plain bagels remained in the display case, still joined together in a neat row in the baking tin—exactly the number I hoped to purchase! While I continued to wait, I passed the time with two thoughts: (1) A fervent (albeit atheistic) "prayer" that the three people ahead of me in the queue didn't like plain bagels, and (2) the happy certainty that, even in the absence of any detailed knowledge of the geometry of the 6-holed object of my desire, GGB absolutely *guaranteed* that its total curvature was exactly -20π.

My prayer (1) was granted, and the server gently tore each of the six bagels from the row and successively dropped them into a brown paper bag, which he then handed to me. As I left the shop it struck me that while each of the bagels in my bag *appeared* totally untampered with, their original collective total curvature of -20π had evaporated to ZERO! Seemingly without leaving a trace of his crime, the nefarious server had robbed me of *all* the negative curvature promised to me by GGB!

Clearly, the only change in the geometry had been right *at* the very small joins (or ***bridges***, as we shall call them) between the bagels, where they were torn apart, so these must have been the culprits for the dramatic -20π curvature heist. Since the six bagels were joined by *five* bridges, each one presumably added 4π when it was torn asunder. In that case, each bridge must have originally stored -4π of curvature.

We shall now confirm this theory, thereby explaining the mystery, allowing us to *see* $\mathcal{K}(\mathcal{S}_g)$ in a new way (and, incidentally, absolving the server of malfeasance!).

[3] Although I have never seen this idea written down, I know (via private conversation) that my friend Professor Tom Banchoff hit upon the same idea, long before I did, right down to thinking in terms of pancakes!

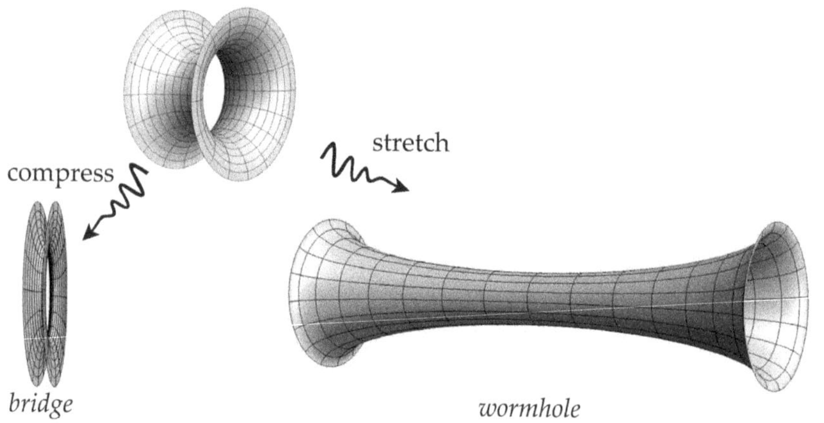

[16.7] *The surface at the top is the inner half of a torus, and has* $\mathcal{K} = -4\pi$. *This total curvature is not altered if the surface is compressed to create a bagel* **bridge** *(left), or stretched to create a* **wormhole** *(right).*

Consider [16.7]. At the top is a surface that looks like the rim of one of the holes in our thick pancake, though now turned on its side; alternatively, think of the inner half of a torus. As we have discussed, its spherical image covers S^2 once negatively, so its total curvature is $\mathcal{K} = -4\pi$.

Now suppose that we dramatically compress this surface horizontally to yield the surface on the left, resembling a bridge between two joined bagels in the baking tin. If we follow the evolution of the spherical image as the surface undergoes this compression, we see that as the region around the narrow throat of the bridge contracts, its spherical image actually expands. Nevertheless, while some parts of the spherical image expand and others contract, the compressed bridge surface, taken as a whole, has the *same* spherical image as the uncompressed original, and its total curvature is therefore still $\mathcal{K} = -4\pi$, as anticipated.

For future use, we also note that, likewise, the total curvature does not change if we instead *stretch* the surface to produce the ***wormhole*** (as we shall call it) on the right of [16.7]; it too has $\mathcal{K} = -4\pi$.

As [16.8] makes clear, the consequence of each bridge having $\mathcal{K} = -4\pi$ is that

$$\mathcal{K}(\mathcal{S}_g) = g \cdot \mathcal{K}(\text{bagel}) + (g-1) \cdot \mathcal{K}(\text{bridge}) = g \cdot 0 + (g-1) \cdot (-4\pi) = 4\pi(1-g),$$

in accord with GGB.

PUZZLE: Place three bagels at the vertices of a large triangle, then connect them together with *three* wormholes along the edges of the triangle, forming a single closed surface $\mathcal{S}$ with $\mathcal{K}(\mathcal{S}) = 3\,\mathcal{K}_{\text{wormhole}} = -12\pi$. Doesn't this violate GGB?! (Further examples along these lines can be found in Ex. 22.)

16.5 The Topological Degree of the Spherical Map

By now we hope that our examples have made it clear that we will have understood GGB *if* we can understand this:

> *Regardless of the form that the surface* $\mathcal{S}_g$ *takes,* $n(\mathcal{S}_g)$ *will always cover almost every point of the sphere* $(1-g)$ *times, provided we count the layers* algebraically, *taking into account positive and negative orientation.* (16.5)

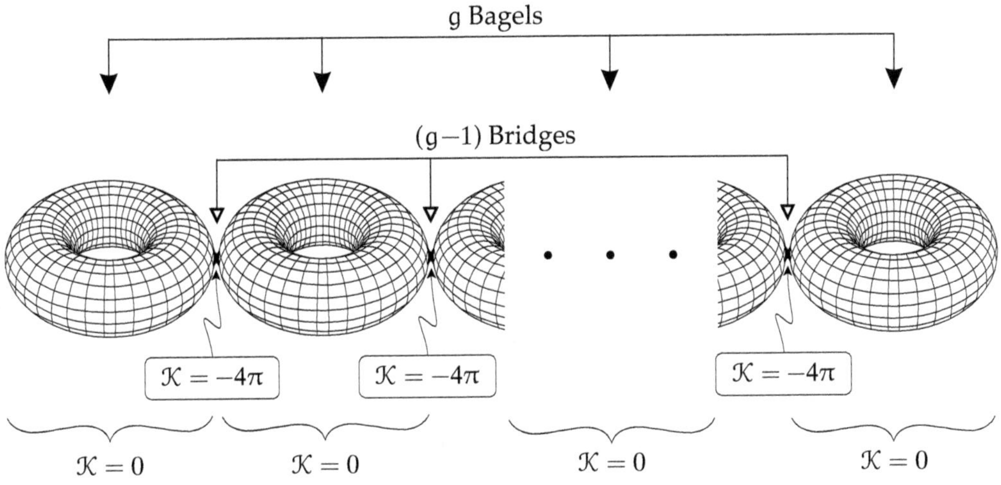

[16.8] *Here $\mathcal{S}_g$ takes the form of g bagels, still joined together in the baking tin by $(g-1)$ bridges. Each bagel has zero total curvature, while each bridge has $\mathcal{K} = -4\pi$. The total curvature is therefore $(-4\pi)(g-1) = 2\pi\chi$.*

In order to clarify this statement, this section will introduce the concept of ***topological degree***,[4] which does the precise algebraic counting of the number of times each point of S^2 is covered. The reason we say "almost every point" is to allow for branch points, as mentioned in the case of the torus.

We begin by noting that (16.5) does *not* mean that there are only $|1-g|$ sheets covering S^2, in all. Indeed, in the case of the thick pancake in [16.6], the covering $n(\mathcal{S}_g)$ of S^2 took the form of one positive covering, and g negative coverings: a total of $(g+1)$ coverings. In other words, for any given point $\widetilde{p} = n(p)$ on S^2, there are $(g+1)$ places p_i on the pancake where the normal points in the direction of $\widetilde{p}$, i.e., $n(p_i) = \widetilde{p}$, for $i = 1, 2, \ldots, (g+1)$.

Let $\mathcal{P}(\widetilde{p})$ denote the number of these points p_i on the surface $\mathcal{S}_g$ at which the curvature $\mathcal{K}(p_i)$ is *positive*, so that n is orientation-*preserving*, and a neighbourhood of the image $\widetilde{p}$ on the sphere is covered *positively*. Likewise, let $\mathcal{N}(\widetilde{p})$ denote the number points at which $\mathcal{K}(p_i)$ is *negative*, so that n is orientation-*reversing*, and $\widetilde{p}$ is covered *negatively*. For example, in the case of the thick pancake [16.6] above, $\mathcal{P}(\widetilde{p}) = +1$, independently of $\widetilde{p}$; and $\mathcal{N}(\widetilde{p}) = g$, also independently of $\widetilde{p}$. (Here, and in what follows, we exclude points for which $\mathcal{K} = 0$, as these make no contribution to covering S^2.)

We can now define the ***topological degree*** (or, more commonly, just "degree") of the spherical map, as follows:

> *Given a closed, oriented surface $\mathcal{S}_g$ of genus g, and a point $\widetilde{p}$ on S^2, the* **topological degree** *of the spherical map—written* $\deg[n(\mathcal{S}_g), \widetilde{p}]$*—is the algebraic count of the number of times $n(\mathcal{S}_g)$ covers $\widetilde{p}$, taking account of orientation:*
>
> $$\deg\,[n(\mathcal{S}_g), \widetilde{p}] \equiv \mathcal{P}(\widetilde{p}) - \mathcal{N}(\widetilde{p}), \tag{16.6}$$
>
> *where $\mathcal{P}(\widetilde{p})$ is the number of preimages of $\widetilde{p}$ for which $\mathcal{K} > 0$, and $\mathcal{N}(\widetilde{p})$ is the number of preimages of $\widetilde{p}$ for which $\mathcal{K} < 0$.*

[4] Also called the ***Brouwer degree***, after the Dutch topological pioneer, Brouwer (1881–1966), who was the first to systematically exploit the concept.

In the case that $\mathcal{S}_g$ is the thick pancake, $\mathcal{P}$ and $\mathcal{N}$ are independent of $\widetilde{p}$, and therefore so is the degree:

$$\deg[n(g\text{-holed thick pancake})] = \mathcal{P} - \mathcal{N} = 1 - g.$$

In the case that $\mathcal{S}_g$ is made up of the bridged bagels, [16.8], S^2 is covered $2g$ times by the images of the g bagels, and a further $(g-1)$ times by the images of the bridges. Thus each point of S^2 is covered by $(3g-1)$ layers. Again in this example, the number of coverings is independent of $\widetilde{p}$, and therefore so is the algebraic count of these coverings:

$$\deg[n(g\text{ bagels, bridged})] = \mathcal{P} - \mathcal{N} = g - [g + (g-1)] = 1 - g,$$

just as before.

The key to understanding GGB (at least from the current point of view) is to be able to see that this recurring result is no coincidence, but rather that the degree truly is *topological* in nature—*every* $\mathcal{S}_g$ must satisfy the same equation:

$$\boxed{\deg[n(\mathcal{S}_g)] = \mathcal{P} - \mathcal{N} = (1-g) = \tfrac{1}{2}\chi(\mathcal{S}_g).} \qquad (16.7)$$

In short, if we can prove (16.7) then we will have proved GGB, (16.1).

16.6 Historical Note

While Gauss and Bonnet certainly paved the road to GGB, neither one of them was ever even aware of this extraordinary result, let alone stated it!

But the name has stuck. Even those who grasp the name's historical inaccuracy dare not touch it now: it would seem that it matters more that we all agree what a name *means*, than that the name itself be historically accurate.

As we have discussed, in 1827 Gauss announced his discovery of the *Local* Gauss–Bonnet Theorem, (2.6), page 23, stating that if Δ is a geodesic triangle on a general surface, then its angular excess equals its total curvature:

$$\mathcal{E}(\Delta) = \mathcal{K}(\Delta).$$

In fact the ***Local Gauss–Bonnet Theorem*** refers to a generalization of Gauss's original result (by Bonnet in 1848) to the case where the sides of the triangle are no longer required to be geodesics. This adds to the right-hand side of the above equation a term representing the total geodesic curvature of the sides.[5] However, neither of these gentlemen said anything at all about closed surfaces.

It would appear that the honour of discovering GGB actually belongs, in two distinct steps, to Leopold Kronecker and to Walther Dyck.[6] First, in 1869 Kronecker introduced the concept of degree—later clarified and exploited by Brouwer—and proved that $\mathcal{K}(\mathcal{S}_g) = 4\pi \deg(n)$. Second, in 1888 Dyck proved that $\deg(n) = \frac{1}{2}\chi$, thereby completing the proof of GGB in its modern form, (16.1).

---

[5]This will proved at the end of Act IV: Exercise 6, page 336.

[6]See Hirsch (1976). Occasionally, as in Berger (2010, page 380), Werner Boy is credited with GGB. However, Boy (1903) explicitly gave credit for GGB to Kronecker and to Dyck, while Boy himself generalized GGB to nonorientable surfaces.

Chapter 17

First (Heuristic) Proof of the Global Gauss–Bonnet Theorem

17.1 Total Curvature of a Plane Loop: Hopf's *Umlaufsatz*

To begin to understand the topological nature of the degree of the spherical map (i.e., (16.7)), and with it GGB (i.e., (16.1)), we shall drop down a dimension. That is, in place of the spherical/normal map n of a 2-dimensional surface $\mathcal{S}$ in $\mathbb{R}^3$ to the 2-dimensional sphere $\mathcal{S}^2$, we shall instead consider the normal map N of a 1-dimensional curve $\mathcal{C}$ in $\mathbb{R}^2$ to the 1-dimensional circle $\mathcal{S}^1$.

Since GGB involves *closed* surfaces, we shall correspondingly restrict attention to the case where $\mathcal{C}$ is a *closed* curve. Furthermore, let us begin with the most elementary case, in which $\mathcal{C}$ is not only closed, but is also ***simple***—i.e., without any self-intersections. Let us call such a simple, closed curve a ***loop***.

Although we shall ultimately be concerned with smooth curves that everywhere have a well-defined tangent and normal, let us begin with a triangle, Δ, with internal angles θ_i and external angles φ_i, so that $\theta_i + \varphi_i = \pi$. See [17.1a].

If we imagine a particle travelling counterclockwise round Δ once, then it is clear that its velocity vector $\mathbf{v}$ executes one full revolution, and this is indeed equivalent to the fundamental fact that the angles of Δ sum to π:

$$\text{net rotation} = \varphi_1 + \varphi_2 + \varphi_3 = 2\pi \quad \Longleftrightarrow \quad \theta_1 + \theta_2 + \theta_3 = \pi.$$

Note that the net rotation statement is somehow the simpler and more fundamental of the two, in the sense that it remains the same if we generalize to n-gons, whereas the equivalent statement in terms of internal angles is now dependent upon n:

$$\text{net rotation} = \sum_{i=1}^{n} \varphi_i = 2\pi \quad \Longleftrightarrow \quad \sum_{i=1}^{n} \theta_i = (n-2)\pi.$$

The concept of *net* rotation becomes important when not all the rotation is in the same direction. This is vividly illustrated by the movement of a nut along a bolt. Suppose you observe the initial position of the nut, then shut your eyes while a friend spins the nut in a complicated combination of positive and negative rotations, moving the nut back and forth along the bolt. When you open your eyes, you have no idea exactly what rotations your friend performed, yet you *do* know what the *net* rotation has been: it is simply measured by how far the nut has moved from its starting position.

Figure [17.1b] considers this net rotation of the velocity $\mathbf{v}$, but now for a smooth loop $\mathcal{C}$, instead of a polygon. Whereas the direction of $\mathbf{v}$ executed sudden jumps of φ_i at the vertices of Δ, its direction now changes smoothly along $\mathcal{C}$, and for the illustrated curve it sometimes turns positively (counterclockwise), as at x, and sometimes turns negatively (clockwise), as at y. Nevertheless, it seems intuitively clear that as the particle travels along $\mathcal{C}$, its velocity $\mathbf{v}$ rocks back and forth along the way, but after one full positive orbit of $\mathcal{C}$, the net effect is that $\mathbf{v}$ has executed one full positive revolution. Again, in speaking of this *net* rotation of $\mathbf{v}$, we mean that negative rotation is allowed to *cancel* positive rotation.

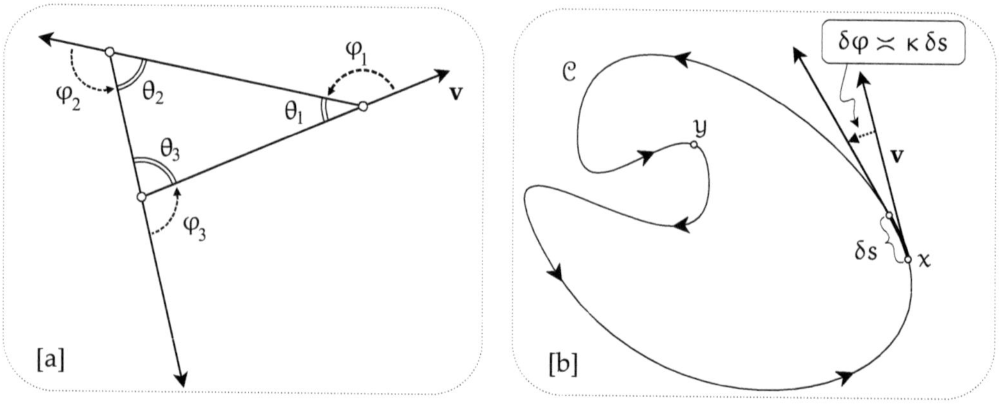

[17.1] *Hopf's* **Umlaufsatz***: As a particle traces a simple loop, its velocity executes one positive revolution. In* **[a]**, $\varphi_1+\varphi_2+\varphi_3=2\pi$*; in* **[b]**, $\oint_{\mathcal{C}} d\varphi$ *= net rotation of* $\mathbf{v}=2\pi$.

The fact that $\mathbf{v}$ executes one full revolution in one orbit of a loop is a theorem called Hopf's ***Umlaufsatz*** (from the German, "Umlauf" (circulation), and "Satz" (theorem)). And while you may not doubt its truth for an uncomplicated curve like that in [17.1b], is it really so obvious for the "simple" curve in [17.3]?! Furthermore, the result does *not* apply to curves that intersect themselves. For example, what is the net rotation [exercise] for a figure-eight curve?

While this result was in some sense known since antiquity, Watson (1917) seems to have been the first to articulate it clearly, while Hopf (1935) was the first to provide a purely topological proof. Hopf's ingenious geometric idea is described in Exercise 23. Here, however, we wish to provide a different, heuristic proof that will serve as an exact model for a corresponding proof of GGB.

[17.2] *Heinz Hopf (1894–1971). Photograph by Ernst Ammann, CC BY-SA 4.0*

We begin by making explicit connections with our previous discussion of GGB. First, recall that the curvature κ is simply the rate of rotation of $\mathbf{v}$ with arc length s (or time, if the particle travels at unit speed). Thus, the *Umlaufsatz* can be rephrased in terms of the topological invariance of the *total curvature* of the loop:

$$\oint_{\mathcal{C}} \kappa\, ds = \oint_{\mathcal{C}} d\varphi = \text{net rotation of } \mathbf{v} = 2\pi, \tag{17.1}$$

independently of the shape of $\mathcal{C}$. This bears a striking family resemblance to GGB!

To relate this to the spherical/normal map, let the unit normal to $\mathcal{C}$ be $\mathbf{N}$, and, as in [12.1], page 131, view this as a mapping N from the point p of the plane curve $\mathcal{C}$ to the point $\widetilde{p}=N(p)$ on the unit circle S^1 that lies at the tip of $\mathbf{N}_p$.

While Hopf stated his result in terms of the rotation of the tangent, we may equally well say that it is $\mathbf{N}$ that makes one net revolution after orbiting $\mathcal{C}$ once. Equivalently, we may say that, algebraically, $N(\mathcal{C})$ covers S^1 once.

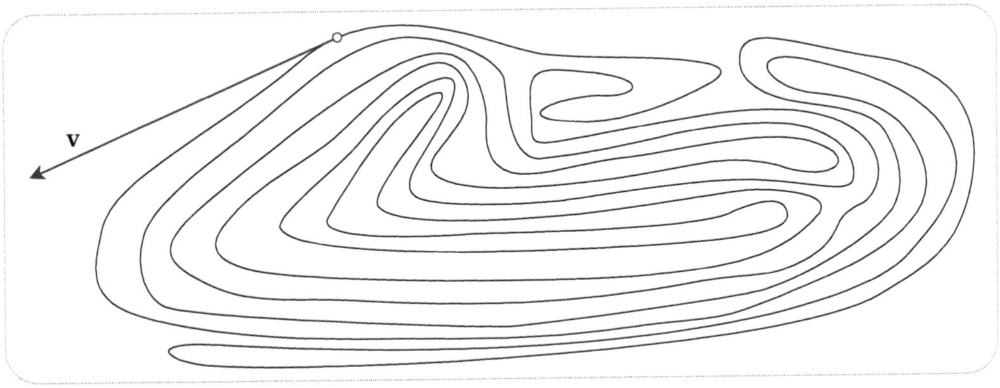

[17.3] *Is it really so obvious that* (net rotation of $\mathbf{v}$) $= 2\pi$*?*

Over the segment δs of $\mathcal{C}$, the directions of the normal vectors are spread over angle $\delta\varphi$, and therefore the tips of these normal vectors fill an arc of length $\delta\widetilde{s} = \delta\varphi$ on S^1. Thus, as we first discussed on page 131,

$$\kappa = \text{local length magnification factor of the N map} \asymp \frac{\delta\widetilde{s}}{\delta s},$$

and therefore,

$$\oint_{\mathcal{C}} \kappa \, ds = 2\pi \, [\text{Number of times } N(\mathcal{C}) \text{ covers } S^1]. \tag{17.2}$$

As with our discussion of GGB, we next introduce the ***degree*** of N, to clarify how to algebraically count the number of times $N(\mathcal{C})$ covers S^1.

Even if $\mathcal{C}$ is not a simple loop, but is allowed self-intersections, we can define the degree of the spherical/normal map N in exactly the same way as before. If $\kappa(p) > 0$ then $N(p)$ is rotating positively (counterclockwise) round S^1 as the particle travels through p on $\mathcal{C}$. Likewise, if $\kappa(p) < 0$ then $N(p)$ is instead rotating negatively (clockwise) round S^1. Then,

Given a closed curve $\mathcal{C}$, traced counterclockwise, and a point $\widetilde{p}$ on S^1, the degree *of the spherical/normal map* N *is the algebraic count of the number of times $N(\mathcal{C})$ covers $\widetilde{p}$, taking account of orientation:*

$$\deg\,[N(\mathcal{C}), \widetilde{p}] \equiv \mathcal{P}(\widetilde{p}) - \mathcal{N}(\widetilde{p}), \tag{17.3}$$

where $\mathcal{P}(\widetilde{p})$ is the number of preimages of $\widetilde{p}$ for which $\kappa > 0$, and $\mathcal{N}(\widetilde{p})$ is the number of preimages of $\widetilde{p}$ for which $\kappa < 0$.

With this more precise definition in place, the key fact is that the degree is independent of the choice of $\widetilde{p}$, so that (17.2) takes the form,

$$\oint_{\mathcal{C}} \kappa \, ds = 2\pi \deg[N(\mathcal{C})]. \tag{17.4}$$

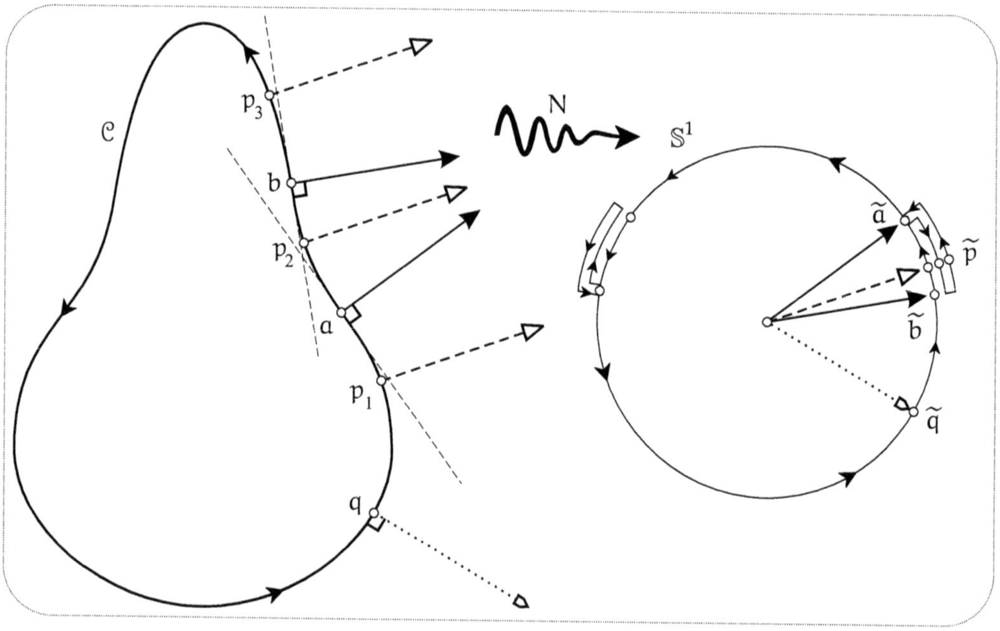

[17.4] *The* **degree** *of* N *is the algebraic count of how many times* $N(\mathcal{C})$ *covers* $\mathbb{S}^1$. *The sign of* κ *changes at the inflection points* a *and* b, *causing the image on* $\mathbb{S}^1$ *to reverse direction at* $\widetilde{a}$ *and* $\widetilde{b}$. *This in turn can be thought of a folding of the orbit back on itself. Since* $\widetilde{p}$ *is traversed three times, twice with positive motion, and once with negative motion, the net number of coverings of* $\widetilde{p}$ *is* $2-1=1$.

Hopf's *Umlaufsatz* states that $\deg[N(\text{simple loop})] = +1$, in which case (17.4) reduces to (17.1).

17.2 Total Curvature of a Deformed Circle

Let us illustrate and clarify these ideas with a concrete example. If $\mathcal{C}$ is a circle, then as x orbits $\mathcal{C}$ once, its image $\widetilde{x} = N(x)$ orbits $\mathbb{S}^1$ once, with matching angular speed. Clearly, $\deg[N(\mathcal{C})] = +1$. Now suppose that we gradually and symmetrically deform the circle, so that $\mathcal{C}$ takes the form shown on the left of [17.4], resembling the cross section of a pear.

For the illustrated point $\widetilde{q}$ on $\mathbb{S}^1$ there is precisely one preimage q on $\mathcal{C}$. But for point $\widetilde{p}$ there are instead three preimages p_1, p_2, p_3.

A helpful mental device in locating these preimages of $\widetilde{p}$ is to first observe that the tangent to $\mathbb{S}^1$ at $\widetilde{p}$ (not shown) must be parallel to the tangent to $\mathcal{C}$ at each preimage of $\widetilde{p}$. Now imagine taking this tangent line at $\widetilde{p}$ and letting it move parallel to itself towards $\mathcal{C}$, ultimately sweeping across all of $\mathcal{C}$. Note each time the moving line touches $\mathcal{C}$: these include all the preimages of $\widetilde{p}$, but they also include the preimages of the antipodal point $-\widetilde{p}$. Where [exercise] are the preimages of $-\widetilde{p}$ in [17.4]?

Restricting attention to the right hand side of $\mathcal{C}$, we observe that there are precisely two inflection points, a and b, distinguished by the fact that $\mathcal{C}$ *crosses* the tangent line at these points, and only at these points.

To see that the inflection points play a crucial role in relation to the spherical/normal map N, imagine x starting at q and travelling up the right side of $\mathcal{C}$. Then $\widetilde{x} = N(x)$ travels forward along $\mathbb{S}^1$ from $\widetilde{q}$ through $\widetilde{b}$ and $\widetilde{p}$ until it hits $\widetilde{a}$. But at $\widetilde{a}$ it bounces and travels *backwards*, passing through $\widetilde{p}$ for a *second* time before hitting $\widetilde{b}$. Now it bounces again and resumes its forward motion along $\mathbb{S}^1$, passing through $\widetilde{p}$ a *third* time. Next it arrives at $\widetilde{a}$ for a second time [exercise: where is x at this moment?] and this time it passes right through $\widetilde{a}$ and keeps going. We strongly suggest that you follow x in your mind as it completes a full orbit of $\mathcal{C}$, with $\widetilde{x}$ performing another back-and-forth motion as x traverses the left-hand side of $\mathcal{C}$.

Now comes a crucial mental leap in the visualization of this motion: *think of the motion of $\widetilde{x}$ as the orbit of a bead travelling along a continuous, unbroken thread.* Thus, as illustrated, when $\widetilde{x}$ first arrives at $\widetilde{a}$ and starts to move backwards on S^1, *it can only be because the thread is folded back on itself.* Likewise, when the bead next arrives at $\widetilde{b}$ and reverses course again, resuming its forward motion along S^1, it can only be because the thread has folded back on itself a *second* time. Thus, as illustrated, as x passes through p_1, then p_2, and finally p_3, $\widetilde{x}$ passes through $\widetilde{p}$ three times, first forward, then backward along the folded thread, then forward again along the twice-folded thread.

In reality, all three folded segments of the thread between $\widetilde{a}$ and $\widetilde{b}$ are plastered right down on top of each other on S^1, but we have lifted them away slightly in order to reveal the folds and the three *different* places on the thread that correspond to the single point $\widetilde{p}$ of S^1.

The positive curvature κ at q gave rise to positive motion through $\widetilde{q}$. Thus, since q is the *only* preimage of $\widetilde{q}$, $\mathcal{P}(\widetilde{q}) = 1$ and $\mathcal{N}(\widetilde{q}) = 0$, and therefore $\deg[N(\mathcal{C}), \widetilde{q}] \equiv \mathcal{P}(\widetilde{q}) - \mathcal{N}(\widetilde{q}) = 1$, as anticipated.

Likewise, the positive curvature κ at p_1 and p_3 gave rise to positive motion through $\widetilde{p}$, while the negative value of κ at p_2 gave rise to negative motion through $\widetilde{p}$. Thus $\mathcal{P}(\widetilde{p}) = 2$ and $\mathcal{N}(\widetilde{p}) = 1$, and therefore $\deg[N(\mathcal{C}), \widetilde{p}] \equiv \mathcal{P}(\widetilde{p}) - \mathcal{N}(\widetilde{p}) = 1$, as before.

17.3 Heuristic Proof of Hopf's *Umlaufsatz*

Let us now apply what we have learned from this example to the general case. Imagine beginning with a circle, and allowing it to continuously deform and evolve into the most general simple closed curve. Here we are thinking of the form of the curve as a function $\mathcal{C}(t)$ of time, with time t going from 0 to 1, say, and with $\mathcal{C}(0)$ being the initial circle, and $\mathcal{C}(1)$ being the final curve, such as $\mathcal{C}$ in [17.4].

But we restrict this evolution to exclude self-intersections, and to be such that κ is well-defined everywhere on the intermediate curves: this is necessary if we wish our mapping N to be continuous on each $\mathcal{C}(t)$. If a sharp corner were to develop, for example, then N would have a jump discontinuity there.

As $\mathcal{C}(t)$ evolves continuously, so too does the orbit of $\widetilde{x}$, as embodied and visualized as a stretchable, unbreakable, foldable string wrapped around S^1. Whenever an inflection point develops in $\mathcal{C}(t)$, signaling a change in the sign of κ there, the image string develops a fold.

If $\widetilde{x}$ stays away from such folds, both $\mathcal{P}$ and $\mathcal{N}$ remain separately constant, and therefore so too does their difference, $\deg[N[\mathcal{C}(t)], \widetilde{x}]$. However, if $\widetilde{x}$ crosses a point of folding, we either gain two new layers of opposite orientation, or we lose two layers of opposite orientation. But, this means that $\mathcal{P}$ and $\mathcal{N}$ both increase by 1 or they both decrease by 1. In either event, the crucial observation is that,

> *Changes in the sign of κ on $\mathcal{C}$ cause folding of $N(\mathcal{C})$ on S^1, but this folding has no effect on the number of layers (counted algebraically) that cover S^1.*

That is, $\deg[N\{\mathcal{C}(t)\}] = \mathcal{P}(\widetilde{x}) - \mathcal{N}(\widetilde{x})$ remains constant as we cross a fold. Thus $\deg[N\{\mathcal{C}(t)\}]$ is a well-defined property of the curve as a whole, independent of $\widetilde{x}$.

The value of $\deg[N\{\mathcal{C}(t)\}]$ must vary continuously with time, *and* remain an integer: thus it cannot change at all. But, initially, $\deg[N\{\mathcal{C}(0)\}] = 1$. Therefore, since the final value of the degree is the same as its initial value, $\deg[N\{\mathcal{C}(1)\}] = 1$, as Hopf asserted.

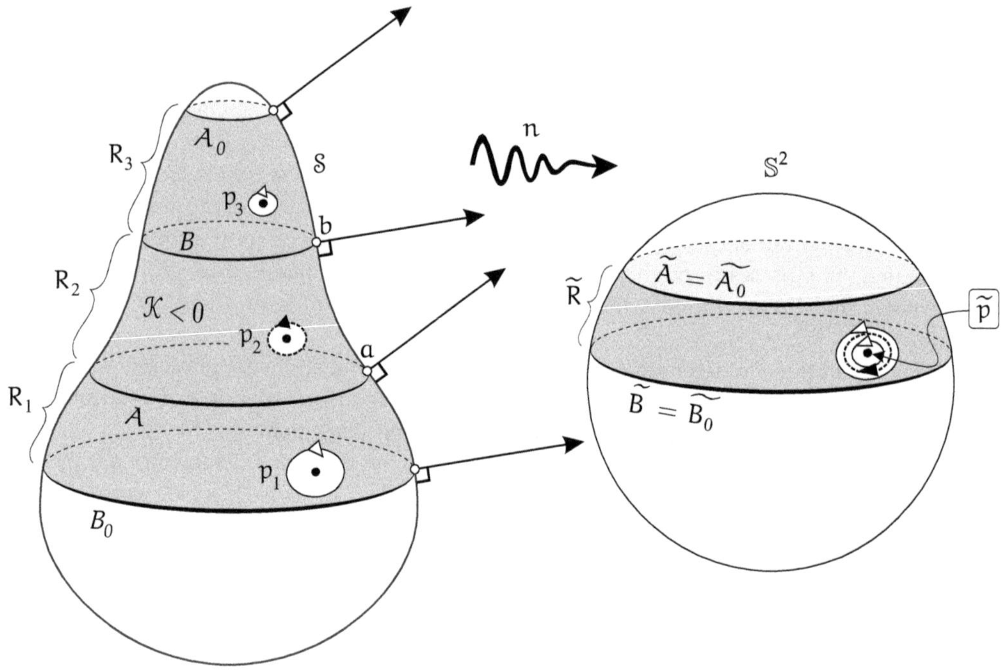

[17.5] *The* **degree** *of* $\mathfrak{n}$ *is the algebraic count of how many times* $\mathfrak{n}(\mathcal{S})$ *covers* $\mathbb{S}^2$. *The sign of* $\mathcal{K}$ *changes as we cross* A *and* B, *causing the image on* $\mathbb{S}^2$ *to reverse orientation as we cross* $\widetilde{A}$ *and* $\widetilde{B}$. *Since,* $\widetilde{p}$ *is covered three times, twice with positive orientation, and once with negative orientation, the* net *number of coverings of* $\widetilde{p}$ *is* $2-1=1$.

17.4 Total Curvature of a Deformed Sphere

In [17.4] we likened $\mathcal{C}$ to a cross section of a pear. We now turn our attention to the surface of the pear itself!

That is, let us take the right half of $\mathcal{C}$ and rotate it 2π about the vertical axis through its ends to generate the pear-like surface of revolution $\mathcal{S}$ shown in [17.5]. As is visually evident (and as was previously noted in (10.13), p. 114) the segments of $\mathcal{C}$ with $\kappa > 0$ generate elliptic regions of $\mathcal{S}$ with $\mathcal{K} > 0$, and the segments of $\mathcal{C}$ with $\kappa < 0$ generate hyperbolic regions of $\mathcal{S}$ with $\mathcal{K} < 0$. The inflection points a and b rotate to generate two circles A and B (both parabolic; i.e., with $\mathcal{K} = 0$) that separate the positively and negatively curved parts of the pear's surface.

But the figure also illustrates the fact that there are two other circles, A_0 and B_0, that have the same images under the spherical map as A and B: $\widetilde{A} = \widetilde{A_0}$ and $\widetilde{B} = \widetilde{B_0}$. These four circles divide $\mathcal{S}$ into a top region, a bottom region, and the three regions in between, which we have shaded grey and labelled R_1, R_2, R_3. All three of these regions R_i are mapped to the *same* region $\widetilde{R}$ on $\mathbb{S}^2$.

Since $\mathcal{K} > 0$ on R_1 and R_3, the spherical map is orientation-preserving in those regions, and a point in those regions is therefore mapped to a point on $\mathbb{S}^2$ that is covered *positively*. On the other hand, since $\mathcal{K} < 0$ on R_2, the spherical map is orientation-reversing in that region, and a point in that region is therefore mapped to a point on $\mathbb{S}^2$ that is covered *negatively*.

In particular, the figure illustrates the three preimages p_1, p_2, p_3 of a point $\widetilde{p}$ in $\widetilde{R}$. Thus, since $\widetilde{p}$ is covered three times, twice with positive orientation, and once with negative orientation, the *net* number of coverings of $\widetilde{p}$ is $2-1=1$.

Let us not lose sight of original formulation of GGB in terms of curvature, and our quest to understand why the total curvature of a closed surface is topologically invariant. In the case of our deformed sphere $\mathcal{S}$, the total curvature residing in the unshaded top of the pear is the area of the northern polar cap of $\mathbb{S}^2$; likewise, the total curvature in the unshaded bottom of $\mathcal{S}$ is likewise the area of the larger unshaded southern region. Each of the three shaded regions contains the *same* amount of curvature, namely, the area of $\widetilde{R}$. But since $\mathcal{K} < 0$ on R_2, n is orientation reversing there, and therefore its total curvature is the *negative* of the area of $\widetilde{R}$. Integrating the curvature over all of $\mathcal{S}$ therefore yields the whole area of $\mathbb{S}^2$, once: i.e., 4π.

17.5 Heuristic Proof of the Global Gauss–Bonnet Theorem

In the lower-dimensional case, we gained a much more intuitive understanding of the degree by thinking of the spherical image of the curve as being a stretchable, foldable thread covering $\mathbb{S}^1$. In the present case, it is likewise very helpful to *imagine* $n(\mathcal{S})$ *as being a highly stretchable, unbreakable,* foldable *membrane covering* $\mathbb{S}^2$—perhaps think of very thinly rolled out pizza dough. But understand that this special mathematical dough can not only be stretched out in whatever manner we require, but it can also *contract* just as easily, as needed.

On the left of [17.5], think of the entire surface of the pear $\mathcal{S}$ as being covered with such a thin layer of pizza dough, loosely sticking to the surface. Let us not worry for now about the *geometry* of the spherical map, and instead focus solely on the *topology* of how it maps the grey region $R_1 \cup R_2 \cup R_3$ on $\mathcal{S}$ to the grey region $\widetilde{R}$ on $\mathbb{S}^2$. See [17.6].

The arrows in this diagram indicate the sequence of topological transformations to be carried out. First, suppose we tie two pieces of string around A and B_0 to keep them in place, and then stretch B radially outward till it is the same size as B_0, and likewise stretch A_0 radially outward till it is the same size as A, stretching R_2 and R_3 into two illustrated portions of cones.

Next, leaving new A_0 in place, let us move the new B vertically downward and stick it onto B_0, *folding* R_2 on top of R_1 in the process. Note, crucially, how this folding causes the positive circulation around p_2 to be *reversed*. The fate of the large letter "T" in R_2 serves to stress this point.

Now move the new A_0 vertically downward and stick it onto A, folding R_3 on top of the new R_2 (and R_1) in the process.

At this point we have correctly embodied the topology of the spherical image, and we can imagine doing a final geometric housekeeping, pushing the top of the pear downward to make it spherical (and of unit radius), and stretching each of the three folded grey layers to get the *geometry* of the spherical map correct, so that, in particular, $\widetilde{p}_1$, $\widetilde{p}_2$, and $\widetilde{p}_3$ will all end up on top of each other, but with their orientations having been reversed between successive layers by the folding.

The details of how we arrived at this final state are irrelevant; our specific intermediate transformations were chosen merely to make this net transformation easier to imagine and draw. Also, this example involved a slight sleight of hand: if $g \neq 0$ then $n(\mathcal{S}_g)$ must have multiple layers, and we cannot imagine starting with a single layer of dough covering $\mathcal{S}_g$ and manipulating it to obtain $n(\mathcal{S})$.

At this point we can instead try to imagine directly (without intermediate steps) the effect of the spherical map on the evolving surface $\mathcal{S}$. If we start with a spherical $\mathcal{S}_0$ and gradually deform it into the pear-shaped surface in [17.5], then we can imagine the mathematical dough $n(\mathcal{S}_0)$ covering $\mathbb{S}^2$ following a corresponding evolution. As negative curvature emerges on the deformed $\mathcal{S}_0$ at parabolic curves, so the spherical image flows backwards over itself at the images of these parabolic curves, producing folds there. But as we cross such a fold in $n(\mathcal{S}_0)$ on $\mathbb{S}^2$ we always gain or lose *two* new layers of *opposite* orientation, so the algebraic sum of the number of coverings is unaltered.

This same reasoning applies if we instead start with some $\mathcal{S}_g$ (where $g \neq 0$) and allow it to evolve. For example, suppose that initially $\mathcal{S}_g$ takes the form of the g bridged bagels of [16.8]. Now suppose we press inward with our thumb on the outer (positively curved) surface of one of the bagels, creating a depression centred at a point p. As we do so, $n(\mathcal{S}_g)$ will undergo a

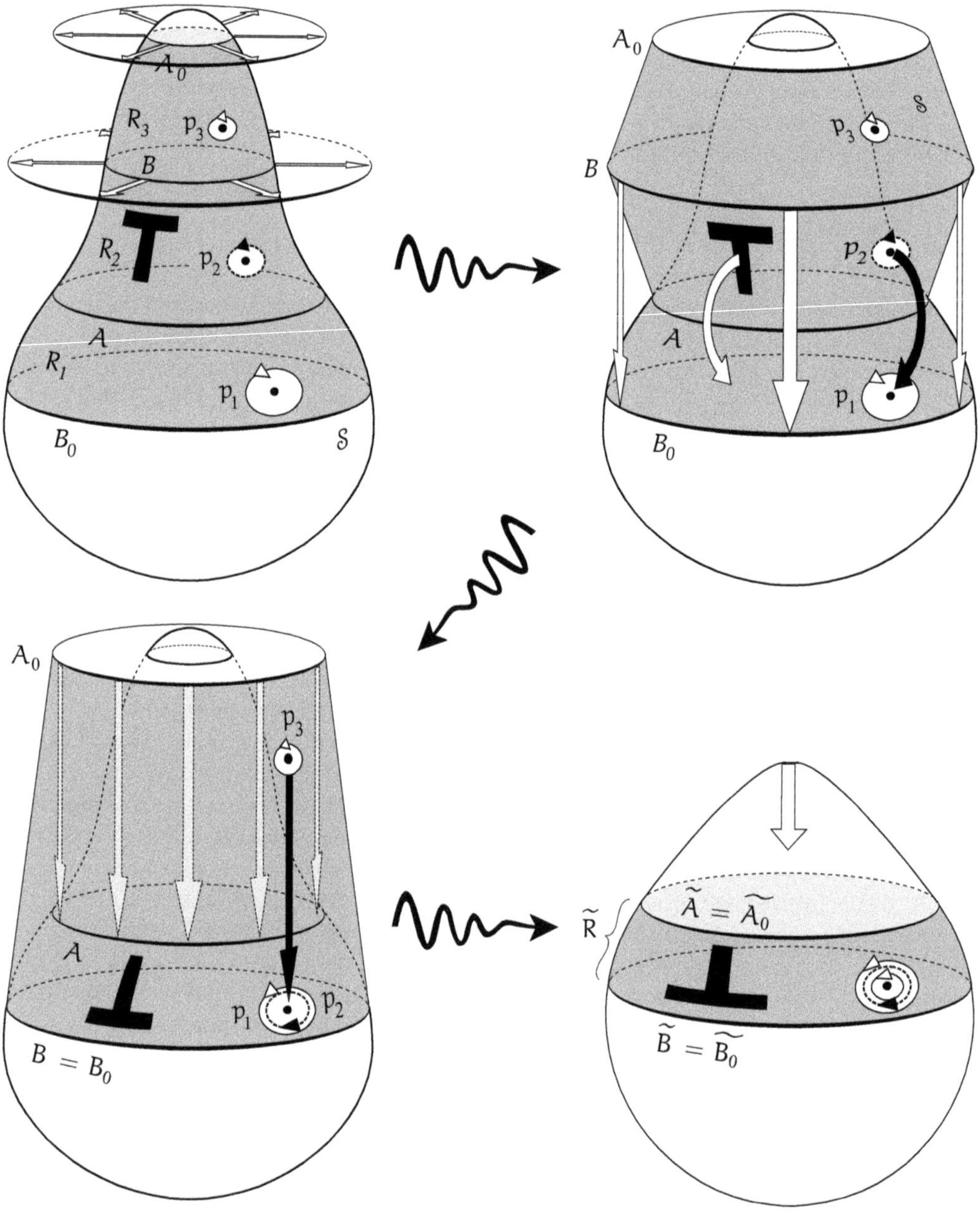

[17.6] *The topological consequence of a change of sign of $\mathcal{K}$ is folding of the spherical image. As we cross the boundary between regions of opposite curvature, the spherical image either gains or loses two layers of opposite orientation, so the algebraic count of the number of coverings of $\mathbb{S}^2$ (i.e., the degree) remains constant, and the total curvature is therefore a topological invariant.*

corresponding evolution in the vicinity of $\widetilde{p}$, lying on one of the g sheets that positively covers $\mathbb{S}^2$. That single sheet, out of the total of $(3g-1)$ sheets, will undergo an evolution like that described in the previous paragraph. This single sheet containing $\widetilde{p}$ will fold over on itself, creating two new layers of opposite orientation in a topological annulus centred at $\widetilde{p}$, but (algebraically) this folded region will continue to cover $\mathbb{S}^2$ once positively, and therefore the net number of coverings of $\mathbb{S}^2$ will remain $(1-g)$, even in the vicinity of $\widetilde{p}$.

This concludes our first (heuristic) explanation of GGB.

Chapter 18

Second (Angular Excess) Proof of the Global Gauss–Bonnet Theorem

18.1 The Euler Characteristic

Thus far we have introduced the Euler characteristic χ as merely a convenient alternative means of labelling a closed, orientable surface, $\mathcal{S}_g$, of a given genus g. However, χ actually has a definition and meaning all its own, which applies to a much wider class of objects than closed, orientable surfaces. Once we have explained this deeper meaning of χ, we may apply it to $\mathcal{S}_g$, in particular, and it is then an important *theorem* (not a definition) that $\chi(\mathcal{S}_g) = 2 - 2g$.

18.2 Euler's (Empirical) Polyhedral Formula

The story of Euler's characteristic begins on the 14th of November, 1750, not with smooth surfaces, but rather with polyhedra. On that date, Euler—pictured in [18.1]—wrote a letter to Christian Goldbach,[1] outlining a remarkable *empirical* discovery, which Euler finally succeeded in proving two years later.[2]

The discovery rested on a realization that may seem strangely obvious to modern eyes: Euler was the first to clearly recognize the very existence of a polyhedron's ***vertices***, ***edges***, and ***faces***. As he wrote in the letter,

> Therefore three kinds of bounds are to be considered in any kind of body, namely 1) points, 2) lines, and 3) surfaces, with the names specifically used for this purpose.

[18.1] *Leonhard Euler (1707–1783)*

Before Euler, mathematicians focused instead on the magnitude of the solid

[1] Goldbach is best known for a letter he wrote *to* Euler in 1742. In it, he conjectured that *every even number greater than* 2 *is the sum of two primes*; this is now known as ***Goldbach's Conjecture***. While it is believed to be true, and (as of 2017) has been verified by computer up to $4,000,000,000,000,000,000$, it remains unproven after almost 300 years.

[2] For a masterful, mathematically accurate, yet riveting account of this history and the connected mathematical ideas, see Richeson (2008). Also, see *Further Reading* at the end of this book.

angle contained between the faces that met at a vertex, rather than on the vertex itself. If we gradually change the dimensions of a polyhedron, the solid angles will vary continuously, but the discrete count V of the *number* of vertices will remain constant. This discrete count V is what Euler now seized upon.

Likewise, before Euler, nobody had given much thought to what Euler described as "the junctures where two faces come together along their sides, which, for lack of an accepted term, I call ***edges***." Euler then went on to count the *number* E of these edges, and the *number* F of faces. This shift from looking at polyhedra in terms of continuously varying lengths and angles, to looking instead at discrete topological features, was yet one more instance of Euler's genius.

Having made this subtle but profound leap, it did not take Euler long to detect a remarkable empirical pattern. Figure [18.2] illustrates the five Platonic solids, and next to each are listed the values of V, F, $(V+F)$, and E. As is now obvious, and as Euler was the first[3] to announce, the $(V+F)$ entries exceed the E entries by 2. In all cases, as Euler wrote to Goldbach, we have

Euler's original formula: $$V+F=E+2.$$

While we have merely verified this for the five Platonic solids, Euler tested many more non-Platonic examples, and in his letter to Goldbach in 1750 he stated his conviction that it was a *universal* truth about polyhedra, though he confessed he had no idea how to prove it at the time. The result is now known as ***Euler's Polyhedral Formula***. Only gradually did mathematicians discover the precise *limits* of its universality: in fact the formula only applies if the polyhedron is *topologically spherical*.

The above formula is not, however, the modern form of the result. If we move the E to the other side of the equation, we obtain a result that is trivially equivalent, algebraically speaking. However, this seemingly trivial step represents a highly *non*trivial conceptual advance (first taken by Poincaré in 1895), for the left hand side of the equation is now *a single integer that characterizes the polyhedron* $\mathcal{P}$, and *this* is our new definition of its

Euler characteristic: $$\chi(\mathcal{P}) \equiv V-E+F. \tag{18.1}$$

Armed with this concept, here is the modern form of ***Euler's Polyhedral Formula***:

$$\chi(\text{topologically spherical polyhedron})=2. \tag{18.2}$$

This is in accord with our previous definition of $\chi(\mathcal{S}_g)=2-2g$, with $g=0$, but the key difference is that now we will be able to *prove* that (18.2) follows from (18.1). In fact, we shall present two quite different proofs.

Euler himself did ultimately succeed in providing the very first proof of his own formula, and in some ways his proof is more topologically natural than the two that we shall instead present, but there are subtle obstacles that must be overcome in order to make his argument completely convincing. (See Richeson 2008, Ch. 7, for details.)

---

[3]In 1860, more than a century after Euler announced his wonderful discovery, a long-lost manuscript miraculously surfaced, *two centuries* after it had been written. It revealed that *Descartes* had made essentially the same discovery as Euler (but in a different form) as early as 1630, more than a century before Euler! The fascinating story is well told in both Stillwell (2010, p. 469) and Richeson (2008, Ch. 9).

Platonic Solid	Name	V	F	V+F	E
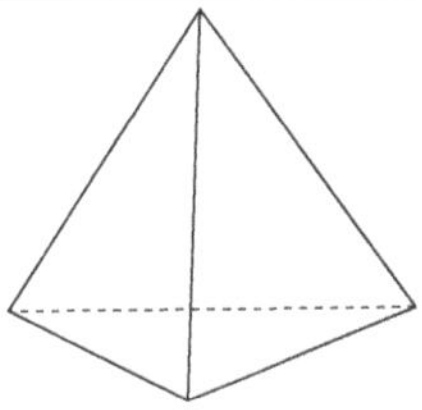	Tetrahedron	4	4	8	6
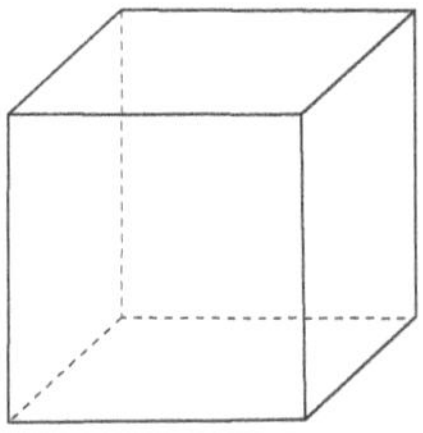	Cube	8	6	14	12
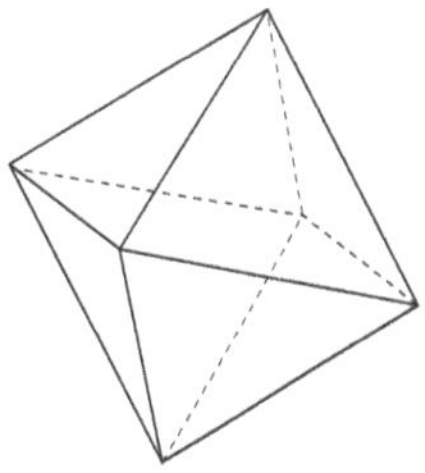	Octahedron	6	8	14	12
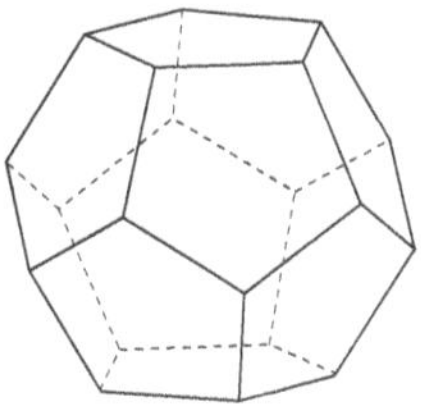	Dodecahedron	20	12	32	30
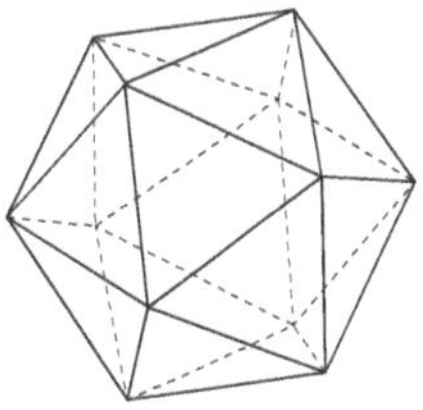	Icosahedron	12	20	32	30

[18.2] *Euler's Polyhedral Formula, empirically verified for all five Platonic solids:* $V + F = E + 2$.

Although we have not yet provided *any* proof of (18.2), we end this section by noting that it has many consequences, one of the most striking of which concerns the five Platonic solids shown in [18.2]. First, recall that Euclid's *Elements*[4] provided a *geometric* proof (outlined in Ex. 24) that these five are the *only* "regular polyhedra" that can exist. A regular polyhedron is one for which all the faces are congruent regular polygons, and the same number of these regular polygons meet at each vertex.

[4] Although Euclid is responsible for publishing the proof in his *Elements*, the proof itself is believed to be due to Theaetetus of Athens (a friend of Plato), dating from around 400 BCE.

Now suppose that we completely relax the *geometrical* constraints, and only insist that each "face"—which we now imagine to be curved, bendable, and stretchable—has the same *number* of wavy edges, and that an equal number of these irregular, bendy faces meet at each vertex. Exercise 25 uses (18.2) to demonstrate the remarkable fact that it is *still* true that there are only five *topological* possibilities, each one being topologically indistinguishable from one of the five Platonic solids of antiquity. Thus the mesmerizing *geometrical* beauty and regularity of the five Platonic solids turns out to have been an extraordinary, 2000-year-long red herring!

18.3 Cauchy's Proof of Euler's Polyhedral Formula

18.3.1 Flattening Polyhedra

From the time of the Ancient Greeks, all the way through the eighteenth century, polyhedra were viewed as solid objects; indeed, we still speak of Platonic *solids*. Cauchy, in 1813, appears to have been the first to make the conceptual leap of peeling the polyhedron off the solid it bounds, viewing it as a hollow surface in its own right.

His next crucial step towards proving Euler's Polyhedral Formula (18.2) was to *flatten* this hollow surface onto the plane. Cauchy was somewhat vague about precisely how this was to be accomplished, and Richeson (2008, Ch. 12) does an admirable job of elucidating both Cauchy's approach, and subsequent clarifications by other mathematicians.

One thing is certain, though: Cauchy's thinking was still firmly planted in *geometry*, and his polyhedra necessarily had straight edges and plane faces, and their flattened versions were also required to have straight edges. Additionally, his proof required the polyhedron to be ***convex***: this stringent geometrical requirement means that if your eye is *anywhere* inside the polyhedron, you can see the *entire* surface—there's nowhere to hide inside a convex polyhedron! But convexity is clearly *not* a topological condition, and therefore it cannot be essential to the validity of Euler's Polyhedral Formula. We therefore choose to sidestep all these historical artifacts, providing instead a more modern, purely topological version of Cauchy's argument, but one that remains faithful to his brilliant original insight.

To this end, imagine that the faces of the polyhedron are made of a bendable, stretchable *rubber sheet*, and picture the vertices and edges as nothing more than dots and connecting curves drawn in pen on this closed, topologically spherical, polyhedral balloon. Even if we start with a classical, rectilinear polyhedron, we may now continuously deform it (without cutting or joining any of its parts) and the resulting curved surface will have the same number of vertices, edges, and faces drawn upon it as did the original.

To flatten the polyhedron, imagine once again that we start with the classical rigid surface with plane faces. Now cut out and discard one of these faces, H, and place the polyhedron on a flat surface, H-side down. The left-hand side of [18.3] illustrates this for a cube. Next, return to picturing the polyhedron as a rubber sheet (now with the hole H), the vertices being dots drawn on its surface. Take the vertices that bound the hole H, and pull each of them radially outward within the plane, so that the boundary of the hole gets larger and larger, and the rest of the polyhedron is pulled down and ultimately stretched out flat within the expanding hole H. See the right-hand side of [18.3].

After flattening the polyhedron in this manner, we may further deform the edges (while keeping them in the plane) so that they take on any shape we please, and we do so with topological impunity, for such a deformation will not change V, E, or F. We shall call the resulting figure a ***polygonal net***[5].

[5]More traditional names are ***network*** or ***graph***, but the usage of these terms allows for "polygons" with only two sides, whereas we shall insist that our polygons have at least three sides.

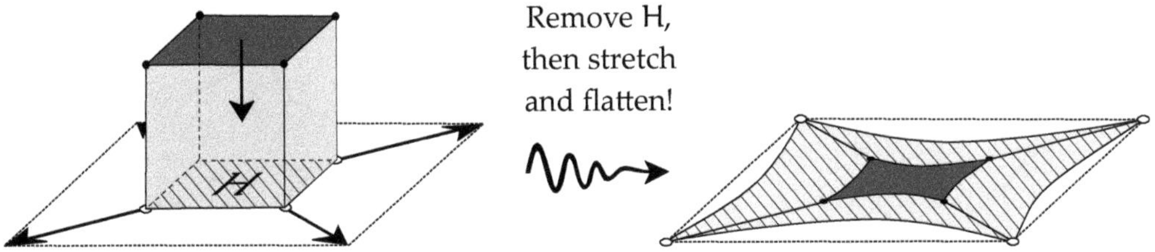

[18.3] *To flatten the cube, we remove its bottom face H, then stretch out the resulting hole (hatched) until we have pulled the remaining faces down into the plane.*

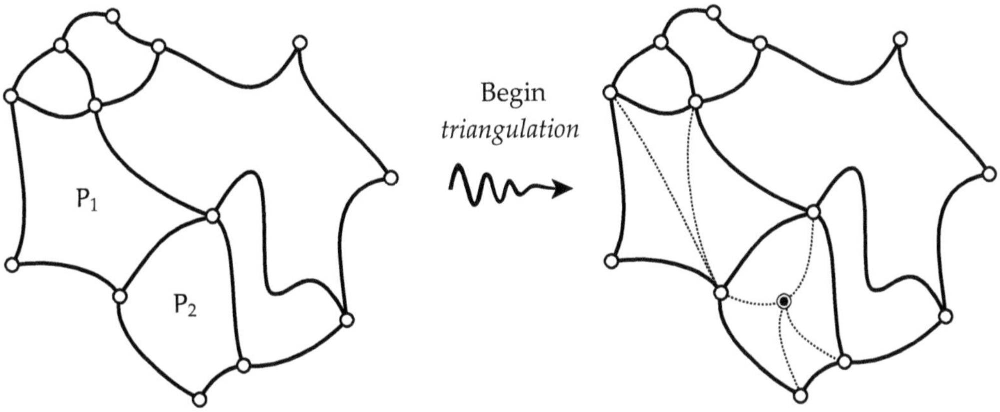

[18.4] *The Euler characteristic of a polygonal net is unaltered by triangulation.*

18.3.2 The Euler Characteristic of a Polygonal Net

Figure [18.4] illustrates a more typical example of a polygonal net, obtained by flattening a non-Platonic polyhedron.

As Cauchy reasoned, in the process of flattening the polyhedron, we have removed one face, and we have not altered the number of vertices or edges:

$$V \rightsquigarrow V, \quad E \rightsquigarrow E, \quad F \rightsquigarrow F-1 \quad \Longrightarrow \quad \chi \rightsquigarrow \chi - 1.$$

In short, flattening the polydedron has reduced its Euler characteristic by 1.

It follows that to prove Euler's Polyhedral Formula (18.2), it suffices to prove *this* elegant result:

$$\boxed{\chi(\text{polygonal net}) = 1.} \tag{18.3}$$

You may care to check that this is true for the specific example we have drawn.

To show that this is always true, we begin by observing that we may dissect each n-gon (with $n > 3$) into triangles, and that, crucially, this process of ***triangulation*** does not alter the value of χ.

The figure illustrates two approaches to the dissection. In polygon P_1 (which we suppose to be an n-gon) we draw in the $(n-3)$ curvilinear "diagonals" from one vertex to all the others, splitting this one face into $(n-2)$ triangles, for a *net* increase (pardon the pun!) of $(n-3)$ faces. Thus,

$$\text{in case } P_1\text{:} \quad V \rightsquigarrow V, \quad E \rightsquigarrow E+(n-3), \quad F \rightsquigarrow F+(n-3) \quad \Longrightarrow \quad \chi \rightsquigarrow \chi.$$

In P_2 (which we suppose to be an m-gon) we have instead added a vertex somewhere inside, and have then joined this to the m vertices, creating m new edges. This splits this one face into m

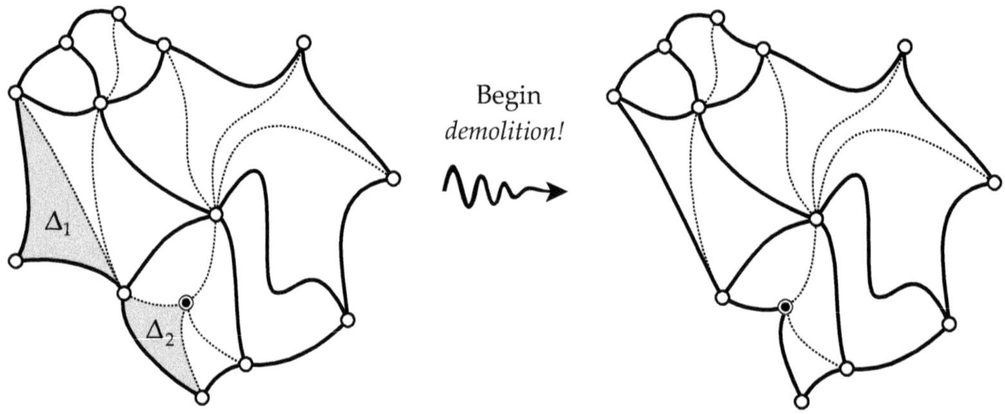

[18.5] *The Euler characteristic of a triangulation is unaltered by its demolition.*

triangles, for a net increase of $(m-1)$ faces. Thus,

$$\text{in case } P_2: \quad V \rightsquigarrow V+1, \quad E \rightsquigarrow E+m, \quad F \rightsquigarrow F+(m-1) \quad \Longrightarrow \quad \chi \rightsquigarrow \chi.$$

The left-hand side of [18.5] shows one possible way of completing the triangulation, and we have just proved that its Euler characteristic must be the same as that of the original polygonal net.

Having gone to all the trouble of constructing this triangulation, the final step in the proof of (18.3) is to *demolish* it! That is, we shall gradually erase each of the triangles, one by one, until only one triangle remains standing. In order to reduce the number of cases that need be considered, we shall agree to only nibble away triangles around the edges; we shall *not* burrow through the middle of the net, which could split it into two distinct, disconnected islands.

With this agreement in place, there are only two cases to consider: either a boundary triangle shares *one* (dotted) edge with the interior of the net (e.g., Δ_1), or else it shares *two* (dotted) edges (e.g., Δ_2).

First, suppose that we

$$\text{erase } \Delta_1: \quad V \rightsquigarrow V-1, \quad E \rightsquigarrow E-2, \quad F \rightsquigarrow F-1 \quad \Longrightarrow \quad \chi \rightsquigarrow \chi.$$

Second, suppose that we

$$\text{erase } \Delta_2: \quad V \rightsquigarrow V, \quad E \rightsquigarrow E-1, \quad F \rightsquigarrow F-1 \quad \Longrightarrow \quad \chi \rightsquigarrow \chi.$$

Ultimately, only one triangle Δ will remain standing, and its Euler characteristic will be

$$\chi(\Delta) = V - E + F = 3 - 3 + 1 = 1.$$

Since χ is invariant under the triangulation process [18.4], and under the demolition process [18.5], its initial value must equal its final value, 1, so we have proved (18.3). And with this, we have also completed Cauchy's proof of Euler's Polyhedral Formula.

18.4 Legendre's Proof of Euler's Polyhedral Formula

Recall Harriot's beautiful result (1.3), page 8, relating the angular excess $\mathcal{E}$ of a geodesic triangle on the sphere to its area $\mathcal{A}$. As you proved in Exercise 5, page 83, this can be generalized to geodesic n-gons. A Euclidean n-gon has angle sum $(n-2)\pi$, and therefore the *angular excess* $\mathcal{E}$ of a geodesic n-gon on a curved surface is

$$\mathcal{E}(n\text{-gon}) = [\text{angle sum}] - (n-2)\pi. \tag{18.4}$$

From the proven additivity of $\mathcal{E}$, Harriot's result then yields,

$$\frac{1}{R^2}\mathcal{A}(n\text{-gon}) = \mathcal{E}(n\text{-gon}) = [\text{angle sum}] - n\pi + 2\pi, \tag{18.5}$$

where R is the radius of the sphere.

Figure [18.6] provides a valuable visualization[6] of the expression for the angular excess on the right hand side of this formula. *Inside* the n-gon, (A) mark the interior angles $\theta_1, \theta_2, \ldots, \theta_n$; (B) write "$-\pi$" next to each edge; (C) write "2π" in the middle. Then $\mathcal{E}$ is the sum of everything in the picture: (A) + (B) + (C).

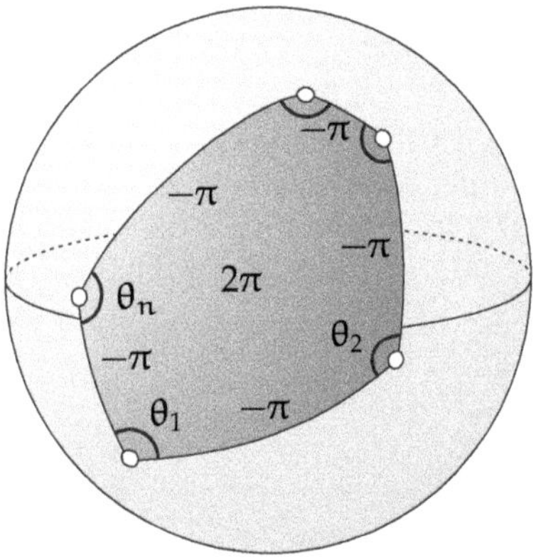

[18.6] *The value of $\mathcal{E}$ for the geodesic n-gon can be visualized as the sum of everything in the picture.*

Legendre presented an ingenious proof in 1794, the first step of which was to project the polyhedron onto a sphere. The specific way in which he carried out this projection (described in Ex. 26) required him to assume that the polyhedron was *convex*, just as Cauchy would also do, 20 years later. But, as we have noted in the context of Cauchy's proof, convexity cannot actually be relevant to a topological result. Therefore, once again, we shall sidestep this historical artifact and present a more blatantly topological version of the argument—one that does not hinge on convexity, but that stays true to Legendre's essential insight.

As before, imagine our polyhedron $\mathcal{P}$ to be a curved, topologically spherical, rubber membrane, with dots (vertices) and connecting curves (edges) drawn simply on its surface. Instead of collapsing the polyhedron, as we did in Cauchy's proof, let us this time inflate it like a balloon! We thereby arrive at a polygonal net covering the surface of an ultimately spherical balloon. Note that the edges that result from this are *not* geodesics, but, as we now explain, we can easily make them so without altering the Euler characteristic.

Imagine this sphere to be rigid, with small nails driven in at each of the vertices we have just constructed. Next, picture the current, *non*geodesic edges as being made of stretched elastic strings, temporarily held in place and attached to the nails (vertices) at their ends. Further imagine that there is no friction between these elastic strings and the sphere: they are free to slide over the surface without resistance. Now let us hammer the nails down flat onto the surface, so that any string that is not attached to this particular nail can sweep over its location without getting caught on it.

Finally, *release* the strings! They will automatically contract to create the shortest (= geodesic = great circle) routes between the nails. We have thus arrived at a *geodesic* polygonal net that completely covers the sphere, and its Euler characteristic must the same as the original value, $\chi(\mathcal{P})$.

Now we come to the second part of Legendre's ingenious idea: sum both sides of the formula (18.5) over *all* the geodesic polygons P_j that make up this geodesic, sphere-covering net, an example of which is shown in [18.7]. Let us look first at the sum of the right-hand side. The efficacy of the visualization [18.6] is now evident, for it enables us to *see* the value of this sum for the complete net, as follows.

---

[6] We hit upon this idea independently, but have since found it published by Richeson (2008, p. 95).

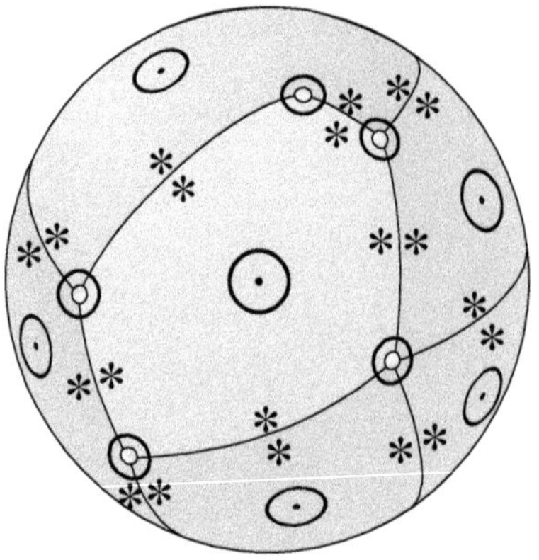

[18.7] $\sum_j \mathcal{E}(P_j)$ *is the sum of everything in this picture. Here,* $* = -\pi$, *and* $\odot = 2\pi$.

To avoid clutter, [18.7] employs two visual abbreviations: $* = -\pi$, and $\odot = 2\pi$. Focus attention on any one vertex: we are summing the interior angles θ_j of *all* the polygons, so, in particular, we are summing all the angles that surround this vertex, yielding 2π. Since each vertex contributes 2π, the total contribution of the interior angles is $2\pi V$. Next, each edge has a $*$ (i.e., $-\pi$) written *on either side* of it, yielding -2π per edge. The total contribution of the $-\pi$'s is therefore $-2\pi E$. Finally, each face carries a $\odot$ (i.e., 2π), yielding a total of $2\pi F$. Thus,

$$\sum_j \mathcal{E}(P_j) = 2\pi\,[V - E + F] = 2\pi\chi(\mathcal{P}). \tag{18.6}$$

Since the polyhedral net completely covers the surface of the sphere, summing the left-hand side of (18.5) yields

$$\frac{1}{R^2}\sum_j \mathcal{A}(P_j) = \frac{1}{R^2}[\text{area of the sphere}] = 4\pi.$$

Finally, (18.5) equates these two quantities:

$$2\pi\chi(\mathcal{P}) = 4\pi \quad \Longrightarrow \quad \chi(\mathcal{P}) = 2,$$

thereby completing Legendre's proof of Euler's Polyhedral Formula.

Despite its evident beauty, Legendre's proof feels morally wrong, for it achieves success (perversely) by means of continuously varying geometrical angles that are *meaningless* within topology. But from the point of view of Differential Geometry, it is perhaps the *right* kind of proof, for it appears to link geometry and topology in a surprising way that might help to explain GGB. As we shall see shortly, this optimism is justified.

18.5 Adding Handles to a Surface to Increase Its Genus

In order to connect these latest ideas with GGB, our next step is to show that our new definition (18.1) of the Euler characteristic does indeed imply (16.2) as a *theorem*:

$$\chi(\mathcal{S}_g) = 2 - 2g. \tag{18.7}$$

We do not believe that this result has an agreed upon name, which permits *us* to call it the ***Euler–L'Huilier formula***, in honour of Simon Antoine Jean L'Huilier, who in 1813 was the first person to state this generalization of Euler's result; see Richeson (2008, Ch. 15) for details.

We have just given two proofs of this theorem in the topologically spherical case, $g = 0$, so now we need to understand the effect of putting *holes* in our surface. Clearly, we should begin by trying to understand the simplest case, namely, the torus/doughnut with one hole, $g = 1$.

[18.8] *A coffee mug is topologically equivalent to a doughnut.* Topology Joke, *by Keenan Crane and Henry Segerman; see Segerman (2016 p. 101). Photograph provided by, and used with permission of, Professor Segerman.*

We shall tackle this problem by means of a *joke*, and it's a bad joke, at that—"probably the first time such a [thing] has ever been used for constructive purposes."[7] The joke goes like this: *A topologist is a person who cannot tell a coffee mug from a doughnut.*

The truth behind the joke is illustrated in [18.8]. Halfway through the transformation of the mug into the doughnut, the body of the mug has coalesced into a blob, still with a handle attached to it. If we now imagine this blob growing to become a sphere, we arrive at the important observation first made by Felix Klein in 1882:[8]

> *A torus is topologically equivalent to a sphere with a handle attached. More generally, $\mathcal{S}_g$ is topologically equivalent to a sphere with g handles attached.* (18.8)

(See *Further Reading* (at the end of this book) for rigorous statements and proofs of all our plausible but unproven claims regarding the topological classification of surfaces.)

To prove (18.7), it now suffices to prove that

> *Adding a handle to a closed surface $\mathcal{S}$ reduces $\chi(\mathcal{S})$ by 2.* (18.9)

Granted this, the addition of g handles clearly reduces χ by $2g$. But we have already established that a topological sphere has $\chi = 2$, so adding g handles therefore reduces this to $\chi(\mathcal{S}_g) = 2 - 2g$, thereby proving (18.7).

---

[7]Captain Kirk, addressing Mr. Spock, at the conclusion of *The Doomsday Machine*. The original quote was "such a weapon."

[8]See Stillwell (1995, p. 60).

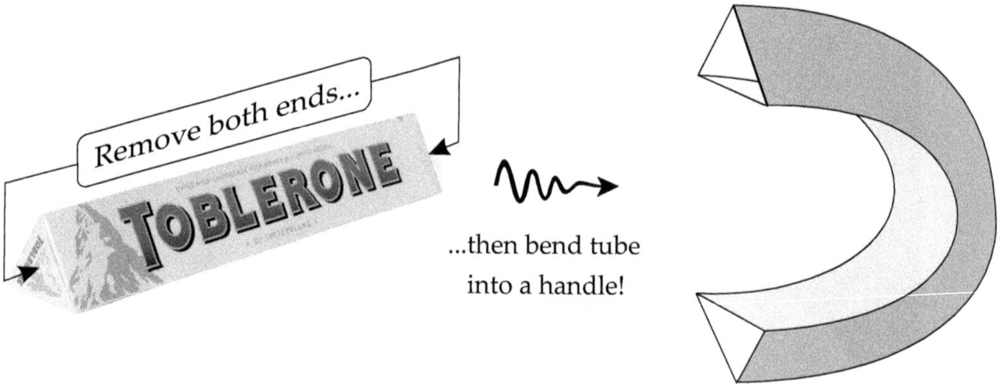

[18.9] *To create a simple handle, remove the ends from the triangular prism (represented here by a Toblerone® box), then bend the tube to create the handle (with vanishing Euler characteristic).*

To prove (18.9), we will first construct a handle, then glue it to the given surface, $\mathcal{S}$. Figure [18.9] shows the construction of an especially simple handle out of a Toblerone® box, i.e., a triangular right prism. We begin by *removing the two triangular end faces* (so $F \rightsquigarrow F-2$), thereby creating a hollow tube. By Euler's Polyhedral Formula, or [exercise] by simple counting,

$$\chi(\text{Toblerone box}) = 2 \qquad \Longrightarrow \qquad \chi(\text{hollow tube}) = 0.$$

Finally, as illustrated, we imagine the tube to be fashioned out of rubber, and we bend it into the form of a hollow handle. This does not alter its Euler characteristic, so

$$\boxed{\chi(\text{handle}) = 0.}$$

Next, as illustrated in [18.10], we create two holes in our surface $\mathcal{S}$, to which the ends of this handle will be glued. To do so, suppose that we have triangulated[9] $\mathcal{S}$. Next, remove two of these triangular faces, so $F \rightsquigarrow F-2$, and $\chi(\mathcal{S}) \rightsquigarrow \chi(\mathcal{S})-2$.

Figure [18.10] depicts the handle moving towards the now two-holed $\mathcal{S}$, moments before it gets glued to the holes in $\mathcal{S}$. The ends of the handle have six vertices and six edges, and the same is true of the holes. But *after* they are glued together, only six vertices and six edges remain: a net reduction of six vertices and six edges. But this means that *the total Euler characteristic is unaltered by the act of gluing on the handle.* (NOTE: Clearly, this is still true if the ends of the handle are n-gons (with $n > 3$), glued to matching n-gon holes.)

Thus, if $\widetilde{\mathcal{S}}$ is the new surface obtained by gluing on the handle,

$$\begin{aligned}
\chi(\widetilde{\mathcal{S}}) &= \chi(\text{two-holed } \mathcal{S} + \text{glued on handle}) \\
&= \chi(\text{two-holed } \mathcal{S}) + \chi(\text{handle prior to gluing}) \\
&= \chi(\mathcal{S}) - 2 + 0,
\end{aligned}$$

thereby proving (18.9), and therefore also (18.7).

For an elegant alternative approach, due to Hopf, see Exercise 27.

[9]The triangulation is not shown, but note that the argument does *not* require that we construct the triangulation out of *geodesic* triangles.

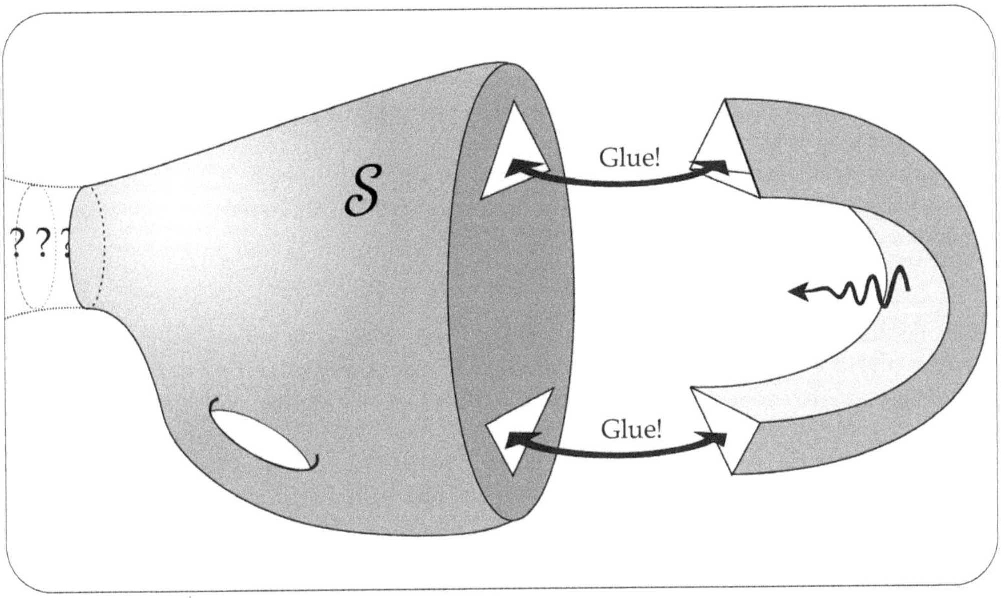

[18.10] *The general closed surface $\mathcal{S}$ extends to the left in an unknown manner, indicated by ???. We cut out two triangles from the triangulation (not shown) of $\mathcal{S}$, reducing $\chi(\mathcal{S})$ by 2, then glue on the handle, which has no effect on $\chi(\mathcal{S})$. Thus gluing on a handle reduces $\chi(\mathcal{S})$ by 2.*

18.6 Angular Excess Proof of the Global Gauss–Bonnet Theorem

Let us return to Legendre's proof of Euler's Polyhedral Formula, and let us take one step back from the result upon which it hinged: the generalization of Harriot's result, (18.5). Focus for now only on the right hand side, the "angular excess," given by (18.4):

$$\mathcal{E}(n\text{-gon}) = [\text{angle sum}] - n\pi + 2\pi.$$

Of course the whole point of this expression is to measure the excess of the interior angles of a *geodesic* n-gon over and above the prediction of Euclidean Geometry. If we drop the requirement that the sides of the n-gon are geodesics, then this interpretation ceases to apply. Furthermore, the expression is *always* devoid of *topological* meaning, for it involves angles. Nevertheless, the *expression itself* remains meaningful even if the sides of the n-gon are *not* geodesics.

Despite the untopological nature of this expression, our first crucial step on the road to a new explanation of GGB is the realization that the *sum* of this expression over the entire polygonal net *does* have topological meaning. Indeed, even if the edges are not geodesics, we see that [18.7] and its conclusion, (18.6), *remain valid*:

$$\sum_j \mathcal{E}(P_j) = 2\pi\,[V - E + F] = 2\pi\chi(\mathcal{P}).$$

For all that was needed in the proof of this result was the fact that the angles in [18.7] that surround a vertex sum to 2π.

By the same token, this part of Legendre's proof does not depend on the topologically spherical polyhedron being inflated into a perfectly round *geometric* sphere. Indeed, and this is the next crucial point, it does not even depend on the surface being *topologically* spherical: *(18.6) remains valid on a surface $\mathcal{S}_g$ of arbitrary genus* g.

As topologically cavalier as we have been (and shall continue to be!) we would be remiss if we did not point out that we cannot cast a *completely* arbitrary polygonal net over $\mathcal{S}_g$. Instead, we must exercise a modicum of caution: *we must not allow our polygons to join onto themselves.* On the torus, for example, we can certainly imagine (but must avoid) a polygon that stretches through the hole and returns to bite its own tail!

With our topological conscience now clear(er), we return to our effort to explain GGB, and to the key equation in Legendre's proof, namely, the generalized Harriot result, (18.5), for an arbitrary geodesic polygon P_j in the net:

$$\frac{1}{R^2}\mathcal{A}(P_j)=\mathcal{E}(P_j).$$

While the sum of the right-hand side has just been seen to be to be topological in nature—and always equal to $2\pi\chi(\mathcal{S}_g)$—equality with the left-hand side is *only* valid if we (i) inflate the polyhedron into a perfect sphere of radius R and (ii) ensure that all the edges of the polygonal net are *geodesic* arcs of great circles on this sphere.

Although Harriot could not have recognized it in 1603, two centuries later Gauss taught us that the sphere has constant curvature $\mathcal{K}=(1/R^2)$, and that the left-hand side of the above equation should in fact be viewed as the *total curvature* $\mathcal{K}(P_j)$ residing within P_j:

$$\mathcal{K}(P_j)=\mathcal{K}\mathcal{A}(P_j)=\frac{1}{R^2}\mathcal{A}(P_j).$$

Gauss's 1827 form of the Local Gauss–Bonnet Theorem, ((2.6), p. 23), was a tremendous generalization of this result: the angular excess of a geodesic triangle Δ on a *general* curved surface is likewise given by the total curvature within. And just as Harriot's original result generalizes to geodesic polygons, so too does Gauss's result, and in exactly the same manner:

$$\mathcal{E}(\Delta)=\iint_{\Delta}\mathcal{K}\,d\mathcal{A}\qquad\Longrightarrow\qquad\mathcal{E}(P_j)=\iint_{P_j}\mathcal{K}\,d\mathcal{A}=\mathcal{K}(P_j).$$

Finally, summing over all the geodesic polygons, and using (18.6) and (18.7), we have arrived at our second proof of GGB:

$$\boxed{\mathcal{K}(\mathcal{S}_g)=\textstyle\sum_j\mathcal{K}(P_j)=\sum_j\mathcal{E}(P_j)=2\pi\chi(\mathcal{S}_g)=2\pi(2-2g).}$$

Chapter 19

Third (Vector Field) Proof of the Global Gauss–Bonnet Theorem

19.1 Introduction

We end Act III (the traditional "climax" of our drama) with a beautiful link between geometry, topology, and *vector fields*—a climax within a climax! Here we shall sketch only those ideas that are needed to achieve a new understanding of GGB; for a much fuller account of vector fields, linked to Complex Analysis and to Physics, see chapters 10, 11, and 12 of VCA. See also *Further Reading*, at the end of this book.

19.2 Vector Fields in the Plane

Imagine a thin layer of fluid flowing over a horizontal plane. It will be to our advantage to think of this plane as the *complex* plane, $\mathbb{C}$. At each point z of $\mathbb{C}$ we therefore have a velocity vector, or rather, a complex number, $V(z)$, which we draw emanating from z. This flow $V(z)$ is called a ***vector field*** on $\mathbb{C}$.

We shall suppose that our vector field $V(z)$ is extremely well behaved, being continuous and differentiable[1] at all but a finite number of isolated points. Thus, at a normal or *regular* point, a small movement in any direction results in a correspondingly small, ultimately proportional, change in the direction and length of V.

In contrast to this, a ***singular point*** s is an exceptional place where the vector field suffers a *discontinuity*: infinitesimal movements away from s in different directions result in V pointing in completely different directions or having completely different lengths.

Figure [19.1] illustrates a typical vector field, in which the singular points a, b and c are instantly identifiable to the untrained eye, marked with black dots, for clarity.

If we follow the path of an individual particle of fluid, we obtain a ***streamline*** K (or ***integral curve***) of the flow, such that V is always tangent to K. If we trace *all* such streamlines, as we shall do in examples below, then we obtain a vivid depiction of the vector field as a whole, called the ***phase portrait***. Note that the phase portrait completely encapsulates the information about the direction $\Theta(z) = \arg[V(z)]$ of $V(z)$ at every point.

In [19.1] the vectors all have equal length, but this is certainly not true in general, as we shall illustrate in the examples that follow. This is a serious deficiency of the phase portrait, for it fails[2] to illustrate the *magnitude*, $|V(z)|$, and this magnitude represents vital information: the speed of a fluid flow, or the strength of a magnetic field, for example. Nevertheless, from the point of view of *topology*, the phase portrait is the *only* thing that matters.

---

1 In the weaker real sense, *not* the complex analytic sense of being an amplitwist.

2 In the case of the most physically important vector fields, there *is* in fact a special way to draw the streamlines so that the strength of the flow becomes visible, as the *crowding together* of the streamlines. See VCA, Section 11.3.

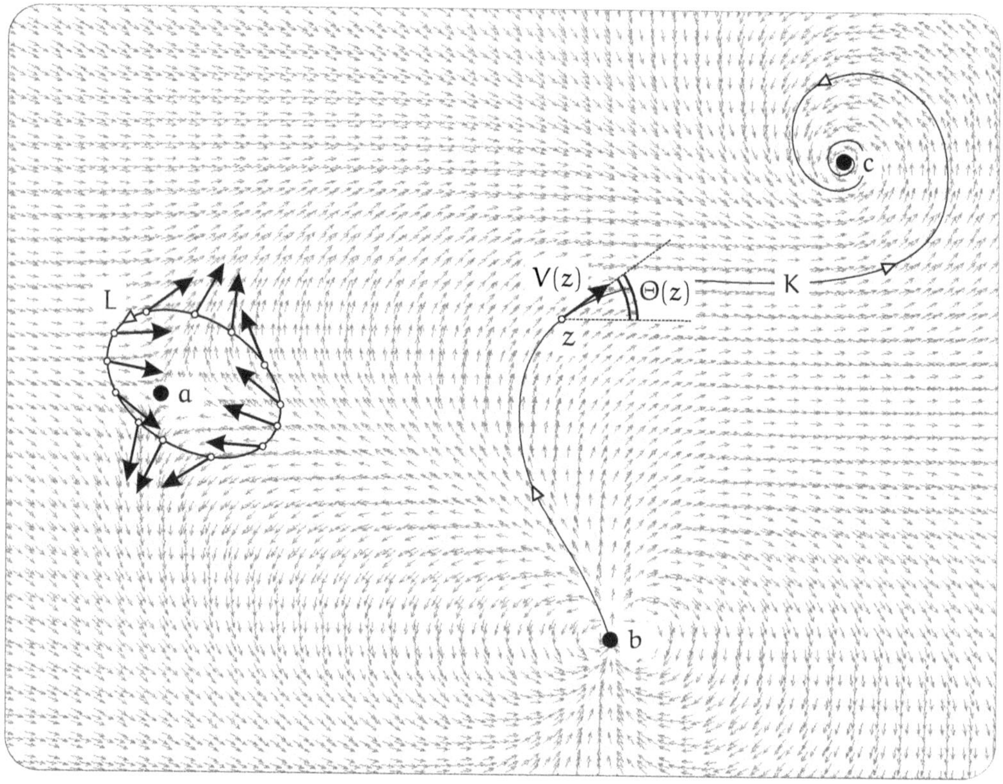

[19.1] *A typical vector field,* V(z). *Clearly visible are three singular points:* a *is called a* **saddle point** *(or* **crosspoint**); b *is called a* **dipole**; *and* c *is called a* **vortex** *(or* **focus**). *Also shown is a* **streamline**, K. *The* **index** $\mathcal{J}_V(a)$ *is the number of revolutions executed by* V(z) *as* z *travels counterclockwise once around* a, *along a loop such as* L. *Since* V *executes one clockwise (i.e., negative) revolution as we loop around* a, *we see that* $\mathcal{J}_V(a) = -1$.

19.3 The Index of a Singular Point

Just as the locations of the singular points are obvious, so too are their dramatically different *characters*. If we imagine the flow pattern [19.1] drawn on a rubber sheet, which is then stretched this way and that, we intuit that the character of a is quite different from that of b or c, and that this distinction is invariant under the stretching. In other words, it seems clear that the character of a singular point s is not so much a geometric feature as it is a *topological* feature of the flow.

Indeed, these different characters lead to different *names* for the various types of singular points. For example, in [19.1], a is called a ***saddle point*** (or ***crosspoint***), for this is the way water flows when it is poured over a horse's saddle; b is called a ***dipole***, for this is the way the magnetic field lines stream between the two poles of a short bar magnet; and c is called a ***vortex*** (or ***focus***), for this is the way water swirls around and down the drain of a kitchen sink.

We now explain how we may crystalize this vague concept of the "character" of a singular point s into a single, topologically invariant *integer* $\mathcal{J}_V(s)$, called the ***index*** of s. (NOTE: Absent any ambiguity regarding the vector field V, we may drop it and abbreviate the notation to $\mathcal{J}(s)$.)

Let us immediately state the definition, although it may not be immediately evident that it is even well defined:

> *If* s *is a singular point of a vector field* V(z)*, and* L *is any simple loop containing* s *(and no other singular points) then the* **index** $\mathcal{I}_V(s)$ *of* s *is* the net number of revolutions *that* V(z) *executes as* z *travels once round* L*, counterclockwise.* (19.1)

Let us immediately illustrate this with the singular points in [19.1]. Since the index only cares about the direction of the vectors on L, we have taken the liberty of enlarging these specific vectors for clarity. As we traverse L, it is clear that V(z) undergoes one clockwise (i.e., negative) revolution, so

$$\mathcal{I}(\text{saddle point}) = -1.$$

Try your own hand at this by examining the singular points at b and c, and verify that

$$\mathcal{I}(\text{dipole}) = +2 \qquad \text{and} \qquad \mathcal{I}(\text{vortex}) = +1.$$

Next, let us make our definition (19.1) a bit more precise. Let $\Theta(z)$ be the angle that V(z) makes with the horizontal. Of course there are infinitely many choices for this angle, differing by multiples of 2π, but suppose we choose one at the point z. Provided we insist that $\Theta(z)$ vary *continuously*, then $\Theta(z)$ is now *uniquely determined* as z moves along a directed curve J, say, that does not pass through any singular points, for then V(z) varies continuously on J.

We can now define $\delta_J\Theta$ to be the net change in the angle $\Theta(z)$ as z travels along J, from Start to Finish:

$$\delta_J\Theta \equiv \Theta(\text{Finish}) - \Theta(\text{Start})$$

Note that this is the *signed* change in the angle, so if we reverse the direction of J (denoted −J), thereby swapping Start and Finish, the sign of $\delta_J\Theta$ is also reversed:

$$\delta_{-J}\,\Theta = -\delta_J\,\Theta. \tag{19.2}$$

With this understanding and notation in place, we can restate our definition (19.1) of the index as follows:

$$\mathcal{I}_V(s) = \frac{1}{2\pi}\,\delta_L\Theta. \tag{19.3}$$

By now you probably already intuit that this definition is indeed well defined, but to *prove* that it is, consider [19.2], which illustrates the general argument with an (initially) circular loop L encircling a dipole.

Focus attention on the illustrated segment l of L, running from p = Start to q = Finish. As illustrated,

$$\delta_l\Theta = \Theta(q) - \Theta(p).$$

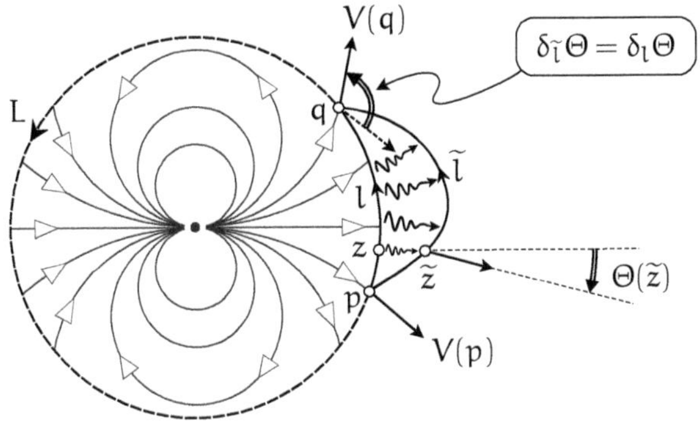

[19.2] *The circular loop* L *encircles a dipole. If the segment* l *of* L *is deformed into* $\tilde{l}$*, carrying* z *to* $\tilde{z}$*, the angle* $\Theta(z)$ *varies continuously, so* $\Theta(\tilde{z})$ *will be close to its original value,* $\Theta(z)$*. It follows that the change in* Θ *along this segment is invariant under the deformation:* $\delta_{\tilde{l}}\Theta = \delta_l\Theta$*.*

Suppose we now continuously deform $\mathfrak{l}$ a small amount into neighbouring $\widetilde{\mathfrak{l}}$, without $\mathfrak{l}$ crossing any singular points in the process.

As a point z on $\mathfrak{l}$ evolves into the new point $\widetilde{z}$ of $\widetilde{\mathfrak{l}}$, the angle $\Theta(z)$ will also vary continuously, so $\Theta(\widetilde{z})$ will be close to its original value, $\Theta(z)$. It follows that as $\widetilde{z}$ travels along $\widetilde{\mathfrak{l}}$, the change $\delta_{\widetilde{\mathfrak{l}}}\Theta$ must also be close to the change $\delta_{\mathfrak{l}}\Theta$ as z travels along $\mathfrak{l}$.

Now comes the archetypal topological argument. If $\delta_{\widetilde{\mathfrak{l}}}\Theta$ were not identically equal to $\delta_{\mathfrak{l}}\Theta$, it would have to differ from it by a multiple of 2π, but this is impossible, because $\delta_{\widetilde{\mathfrak{l}}}\Theta$ is close to $\delta_{\mathfrak{l}}\Theta$. Thus,

$$\delta_{\widetilde{\mathfrak{l}}}\Theta = \delta_{\mathfrak{l}}\Theta.$$

We may similarly deform any other segment of L, or all of it at once, so we immediately deduce that the index is indeed independent of the size and shape of L:

> *If s is a singular point of a vector field* $V(z)$, *and* L *is any simple loop containing* s *(and no other singular points) then the* **index** $\mathcal{I}_V(s)$ *does not change as* L *continuously deforms into any other such loop, so long as it does not cross any singular points in the process. In short,* $\mathcal{I}_V(s)$ *is independent of* L, *and is a property of* V *that can be attached to* s *itself.* (19.4)

If we instead view the vector field as a complex mapping $z \mapsto V(z)$ from one complex plane to another, taking the point z in the first plane to the *point* $V(z)$ in the second, then [exercise] $\mathcal{I}_V(s)$ may instead be viewed as the number of times that the image loop $V(L)$ winds around 0. This is called the ***winding number*** of $V(L)$, and it plays a fundamental role in Complex Analysis; see VCA, Chapter 7.

19.4 The Archetypal Singular Points: Complex Powers

Why did we take our plane to be the *complex* plane? There are several reasons, but here is an important one: the archetypal singular points arise naturally from the *powers* of the complex variable z:

$$P_m(z) \equiv z^m = [r\,e^{i\theta}]^m = r^m\,e^{im\theta},$$

where m is assumed to be an integer.

The only (finite)[3] singular point of $P_m(z)$ is at the origin, and it is easy to determine its index. Everywhere along the ray in the direction θ, we see that P_m points in the direction $m\theta$; only the length $|P_m(z)| = r^m$ varies as we move along the ray. Thus, if we traverse any loop that goes once round the origin counterclockwise, so that θ goes from 0 to 2π, then the angle of P_m goes from 0 to $2\pi m$. In other words, P_m executes m (positive or negative) revolutions, so $\mathcal{I}(0) = m$.

For example, reconsider the "dipole" we encountered in [19.1] at b. Figure [19.3] illustrates that this dipole field is none other than $P_2(z) = z^2$: [19.3a] shows the vector field itself on a circular loop C_r of radius r, and [19.3b] shows the phase portrait of streamlines. Make sure you can see that both these diagrams are at least qualitatively correct. (For bonus points, prove that the streamlines in [19.3b] are indeed perfect circles!)

Although the definition (19.3) of the index makes essential use of the loop L enclosing the singular point, we have already seen in (19.4) that the shape and size of L is actually a red herring: the index is actually a property of the vector field *at* the singular point itself.

To make this idea more vivid, [19.3b] illustrated the loop C_r shrinking towards the singular point. Clearly, *it is the behaviour of the vector field in an infinitesimal neighbourhood of the singular point*

[3]There is actually a second singular point at infinity; see VCA, Section 11.2.9.

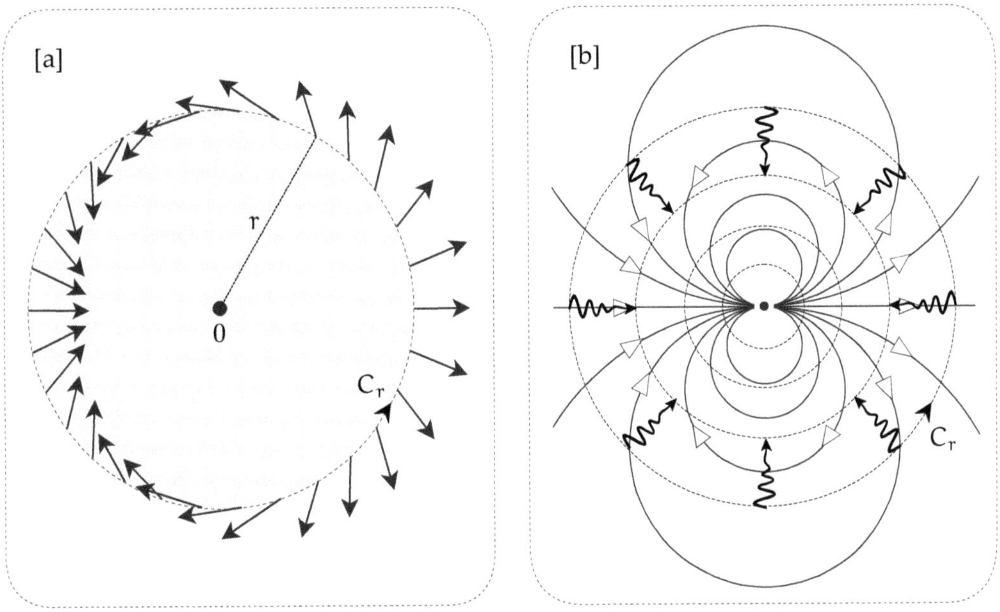

[19.3] *As we traverse the circle in* **[a]** *once (positively) round the singular point at* 0, $P_2(z) = z^2$ *executes two positive revolutions, so* $\mathcal{J}(0) = +2$. *The same result can be obtained by inspecting the phase portrait (i.e., streamlines) of the flow in* **[b]**. *As we shrink* C_r *down towards the singular point, the index does not change: the index characterizes the vector field in an infinitesimal neighbourhood of the singular point.*

that determines the index of the point:

$$\mathcal{J}_V(s) = \frac{1}{2\pi} \lim_{r \to 0} \delta_{C_r} \Theta.$$

To sum up, if m is any integer,

$$P_m(z) = z^m \quad \Longrightarrow \quad \mathcal{J}(0) = m.$$

Figure [19.4] shows these fields for $m = +3, +2, +1, 0, -1, -2$. Please take a moment to visually verify, for each field, that (i) it is indeed what it purports to be and (ii) that the index is indeed equal to the power m.

While we started our discussion of singular points with the intuitive idea of their "character," the more precise concept of index has moved us in a new direction. For example, a source, a vortex, and a sink are, from the physical point of view, all quite different in character, but they are nevertheless *indistinguishable* from the topological point of view, for all three have index $+1$.

This is explained by the following fact: multiplying a complex function $f(z)$ by a complex *constant* $k = Re^{i\varphi}$ has no effect on its indices. For $f(z) \rightsquigarrow Re^{i\varphi} f(z)$ stretches the vectors by R and *rotates them by the fixed angle* φ. Thus $kf(z)$ executes the same number of revolutions as $f(z)$ as z traverses any closed loop, and therefore the index remains the same. In particular, if we let $f(z) = z$ and let φ gradually increase from 0 to $(\pi/2)$, and then from $(\pi/2)$ to π, the field of the source in [19.4c] evolves like this: [c] $\rightsquigarrow$ [d] $\rightsquigarrow$ [e] $\rightsquigarrow$ [f].

We end this section with some general observations. First, note that *reversing the direction of any flow has no effect on the location or indices of its singular points.* As we traverse a loop around a singular point, V and $-V$ undergo exactly equal rotations, like the two ends of a compass needle, so the common singular point of the two vector fields has the same index for both.

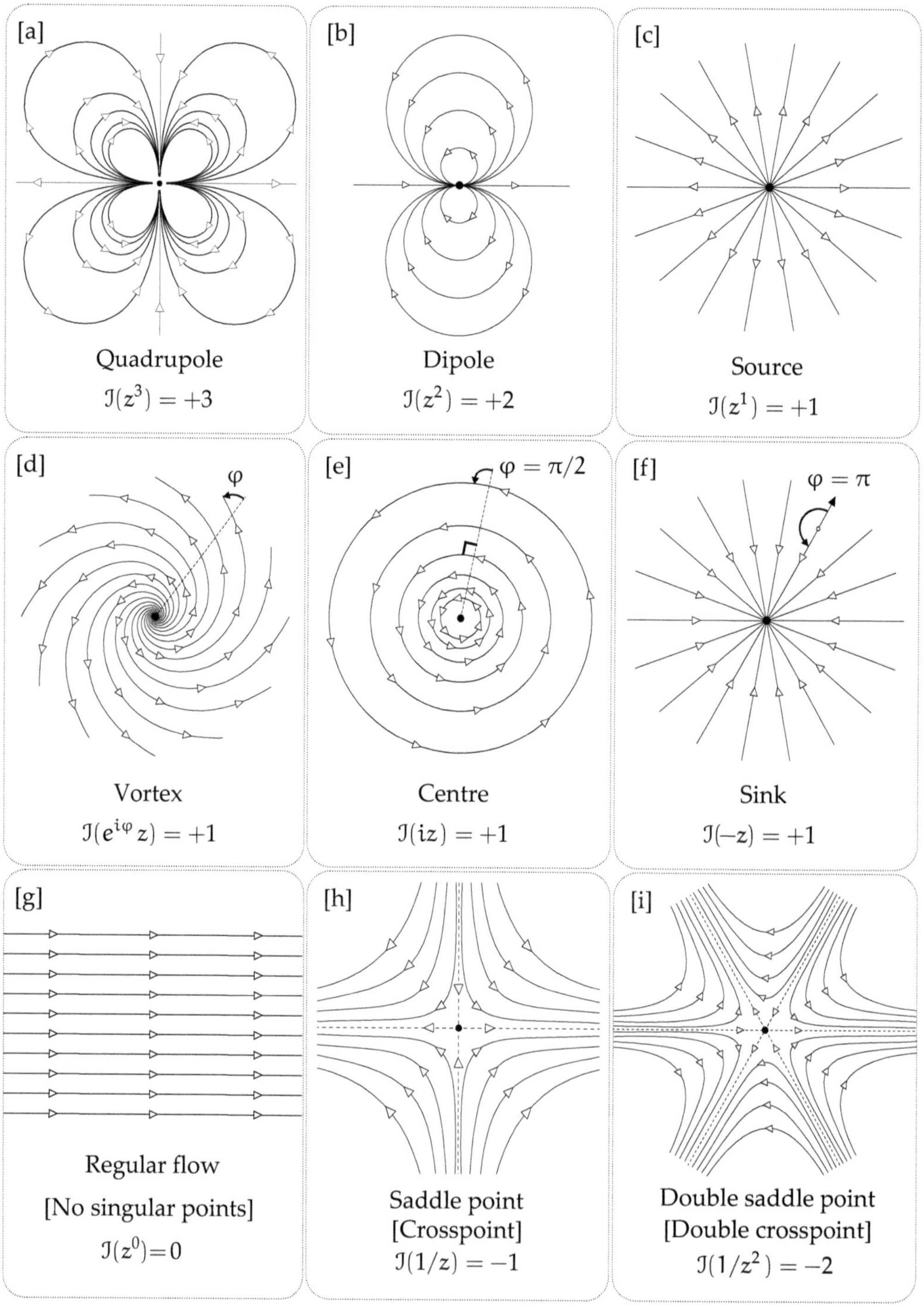

[19.4] *The vector fields in* $\mathbb{C}$ *of the integer powers of the complex variable* z, $P_m(z) = z^m$, *for* $m = +3, +2, +1, 0, -1, -2$. *The index* $\mathcal{J}$ *of the singular point at the origin is equal to* m. *The middle row shows the vector fields of* $e^{i\varphi} z$, *all with* $\mathcal{J} = +1$, *for* **[d]** *a* **Vortex** *with a general value of* φ; **[e]** *the* **centre**, $\varphi = \pi/2$; **[f]** *the* **sink**, $\varphi = \pi$.

ALTERNATIVE NAMES: *sink* = **stable node**; *source* = **unstable node**; *vortex* = **(stable or unstable) focus—[d] is unstable**; *saddle point* = **crosspoint**.

Next, note that if we define the complex conjugate vector field by

$$\overline{P}_n(z) = \overline{z^n} = \overline{z}^n,$$

then the streamlines of $P_m(z)$ are identical to those of $\overline{P}_{-m}(z)$, so their common singular point at the origin is a zero for both fields and has the same index in both fields.

More generally, for any complex function $f(z)$, the vector field $\overline{f(z)}$ has become known as the ***Pólya vector field of*** $f(z)$, in honour of George Pólya (1887–1985). Clearly the Pólya vector field has the same singular points as $f(z)$, but since $\overline{f(z)}$ has the opposite angle, these common singular points have opposite indices. The principal virtue of the Pólya vector field is that it allows for simple, intuitive, *visual and physical* interpretations of the complex contour integral of $f(z)$. For details, see Chapter 11 of VCA.

Finally, observe that if we take the reciprocal of $f(z)$ then we obtain a new vector field $1/f(z)$ that goes to infinity (the north pole of the Riemann sphere) at a singular point where $f(z) = 0$, and the index of $1/f(z)$ at such a singular point is [exercise] the *negative* of that of $f(z)$: this is particularly clear in the case of $P_m(z)$, for then $1/P_m(z)$ points in the same direction as $P_{-m}(z)$. If we now consider the Pólya vector field of this reciprocal, namely $1/\overline{f(z)}$, the indices are reversed a second time, and are therefore the *same* as those of $f(z)$.

NOTES ON THE PHYSICAL TERMINOLOGY. We have borrowed the terms, ***source***, ***vortex***, ***sink***, ***dipole***, and ***quadrupole***, from physics, to describe the streamlines of z, z^2, and z^3. While this is topologically correct, it is *not* physically correct: the physical fields corresponding to these terms are actually the Pólya vector fields $\overline{(1/z)}$, $\overline{(1/z^2)}$, and $\overline{(1/z^3)}$.

As explained in the previous paragraph, each of these fields points in the same direction as the corresponding positive power, so they have the same singular points and indices, thereby justifying our use of these terms in the current topological setting. However, the *magnitudes* of the physical fields are the *reciprocals* of those of the positive powers to which we have attached the same names.

In an attempt to clear our conscience, let us explain this in the simplest case of a *source*. Suppose water is supplied at a constant rate s through a very narrow tube to a point on the surface of a horizontal plane, which we shall take to be $\mathbb{C}$. The water will flow radially outward, symmetrically, from this *source* (now used in the *physical* sense), which we suppose to be at the origin.

While the radial velocity field $P_1(z) = z = re^{i\theta}$ depicted in [19.4c] has magnitude r, *speeding up* as we move away from the origin, it is intuitively clear that the velocity v of the radial flow from a physical source must *slow down* the further away the water travels. More precisely, the total amount of water flowing out across a circle of radius r (per unit time) must equal the amount being supplied at the origin, namely, s. Thus,

$$2\pi r\, v = s \;\Rightarrow\; v = \frac{s}{2\pi r} \;\Rightarrow\; \text{physical source} = v\, e^{i\theta} = \frac{s}{2\pi}\left[\frac{e^{i\theta}}{r}\right] = \frac{s}{2\pi}\overline{\left[\frac{1}{z}\right]}.$$

For much more on the physical interpretations of such fields, and their symbiotic relationship with Complex Analysis, see VCA, Chapters 10, 11, and 12.

19.5 Vector Fields on Surfaces

19.5.1 The Honey-Flow Vector Field

It is easy enough to imagine a vector field on a surface, instead of in the plane. Picture any smooth object left out in a rainstorm: rain water flows over the surface $\mathcal{S}$ of the object, down towards the

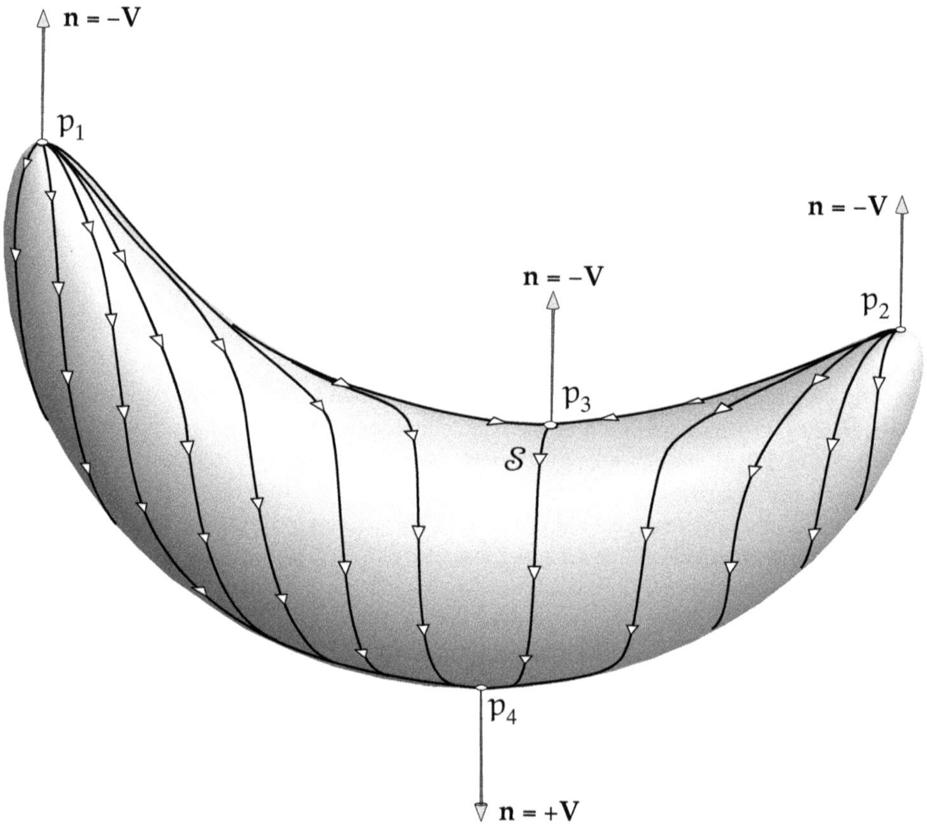

[19.5] The Honey-Flow. *Honey rains down in the gravitational direction* $\mathbf{V}$ *onto the surface* $\mathcal{S}$ *of a fried banana. The honey then flows down over its surface, creating the "honey-flow" velocity vector field* $\mathbf{v}$ *on* $\mathcal{S}$. *The singular points of* $\mathbf{v}$ *occur when the outward normal* $\mathbf{n}$ *of* $\mathcal{S}$ *is* $\mathbf{n} = -\mathbf{V}$ *(at* $p_{1,2,3}$*), or* $\mathbf{n} = +\mathbf{V}$ *(at* p_4*). Visual intuition suggests that* p_1 *and* p_2 *are sources, with* $\mathcal{I}(p_1) = \mathcal{I}(p_2) = +1$, p_3 *is a saddle point, with* $\mathcal{I}(p_3) = -1$, *and* p_4 *is a sink,* $\mathcal{I}(p_4) = +1$. *These intuitions are confirmed by the precise definition, (19.5).*

ground, and its velocity $\mathbf{v}(p)$ at each point $p \in \mathcal{S}$ is then an example of a ***vector field on the surface***, the vectors themselves being everywhere tangent to the surface.

While this rain flow $\mathbf{v}(p)$ is but one example of a vector field on $\mathcal{S}$, we shall see shortly that this *specific* one is a powerful theoretical tool, and the key to a new proof of GGB.

Figure [19.5] illustrates this flow on a particular surface, the direction of the gravitation pull being $\mathbf{V}$. In reality, the rain water flowing down over this surface might well be pulled *off* the surface, falling straight down to the ground as soon as it can, i.e., once the tangent plane becomes vertical, containing $\mathbf{V}$.

In order to allow physical intuition to continue to inform mathematical intuition, we must ensure that the fluid *adheres* to $\mathcal{S}$ and always flows over it. We therefore perform a biblical miracle: we turn water into *honey!* (By way of an ancillary miracle, let us also turn [19.5] into a fried banana!)

Accordingly, the flow $\mathbf{v}$ (resulting from this honey being dragged over the surface by a force $\mathbf{V}$ in space) will henceforth be referred to as the ***honey-flow vector field*** associated with $\mathbf{V}$. (WARNINGS: You may be shocked to learn that "honey-flow" is *not* (yet) standard mathematical terminology! Also, we shall only pay poetic homage to physics, faithfully keeping track of the *direction* of the honey-flow, but *not* its speed.)

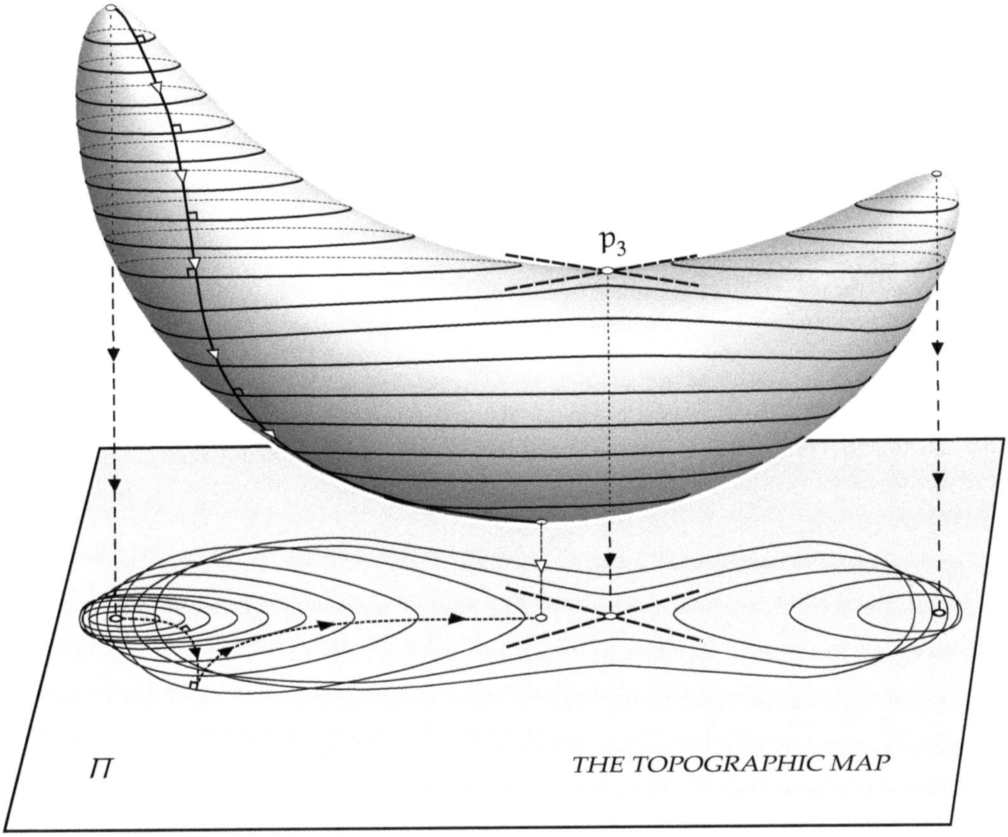

[19.6] The Topographic Map *of a surface is obtained by intersecting it with equally spaced horizontal planes, then projecting the resulting* **level curves** *vertically downward onto the map plane,* Π. *(We have not mapped the bottom half of the surface, to avoid clutter.) The level curve through the saddle point* p_3 *is a figure eight, of which only its tangents at* p_3 *are shown (dashed lines). The streamlines of the honey-flow on the surface are orthogonal to the level curves, and they* remain *orthogonal after projection down to* Π *(despite the fact that the map is* not *conformal).*

19.5.2 *Relation of the Honey-Flow to the Topographic Map*

We will give a concise, purely mathematical definition of the honey-flow **v** in Section 19.7, but, for now, let us continue to think physically, in order to gain a fresh insight. To that end, recall that we can represent the shape of the fried banana in [19.5], or any other surface, using a flat ***topographic map***, of the type used in geography.

To construct this map, we take many equally spaced horizontal planes and look at the curves on the surface where these planes intersect it, the so-called ***level curves***. Finally, we project these level curves vertically downward onto the horizontal map plane Π below, thereby creating the topographic map.

This process is illustrated in [19.6], but only for the *upper* part of the fried banana; the map of the bottom half would clutter the diagram, and is therefore best drawn in a separate figure (not shown). If the level curves are imagined to exist on a transparent surface, the topographic map is what you would see if you were to look straight down at the surface from high above. In the case of the banana in [19.6], you can experimentally verify the map by taking a long knife and slicing up the banana thinly, then examining the shape of each slice.

Note that when the intersecting plane passes through the saddle point p_3, the level curve is a figure eight (not shown); the figure does show the tangents (dashed lines) to this figure eight as it passes through p_3. Recall that these are the asymptotic directions of the surface, $\kappa = 0$, and that the pattern of hyperbolas in the vicinity of the saddle point is the Dupin Indicatrix there. See Section 15.8, page 162.

Note also that where the level curves are closest together, the surface of the banana is steepest, and where they are furthest apart, the surface is shallowest. Moving along a level curve corresponds to moving sideways on the surface, at constant height. Moving at *right angles* to the level curves therefore corresponds to travelling in the direction of steepest descent (or ascent) on the surface. Points where (infinitesimally separated) neighbouring curves in the map *intersect* correspond to different heights above a single point, i.e., to a vertical tangent plane.

Since the gravitational force pulling the honey down the surface has no horizontal component (along the level curves), we immediately deduce that the honey-flow on the surface is along the *orthogonal trajectories* of the level curves.

This has implications for the topographic map. Since $\mathbf{v}$ is orthogonal to the tangent $\mathcal{T}$ to the level curve, it must lie within the plane that is orthogonal to $\mathcal{T}$. But since $\mathcal{T}$ is horizontal (by construction) this orthogonal plane is vertical. Since the vertical planes through $\mathcal{T}$ and $\mathbf{v}$ are orthogonal, they intersect the horizontal map plane Π in orthogonal directions. So, as illustrated in [19.6],

> *On both the surface* and *in its topographic map, the streamlines of the honey-flow are the* orthogonal trajectories *of the level curves and of the topographic map.*

Note that the vertical projection from the surface down to the map is *not* conformal: in general, angles on the surface are *not* faithfully represented in the topographic map. However, despite this nonconformality, we have just proved that the right angles between the level curves and the streamlines of the honey-flow *are* preserved.

If two plane vector fields are orthogonal, then they must rotate the same amount as we traverse a curve. A shared singular point of the two fields must therefore have the same index. For example, in the topographic map, the level curves near either peak of the fried banana look like the streamlines of a centre vector field, with $\mathcal{I} = +1$. Since the streamlines of the honey-flow are orthogonal to the level curves, it follows that the (projected) honey-flow must have $\mathcal{I} = +1$, too, as it does.

Likewise, near the saddle point of the fried banana, the level curves look like the streamlines of a saddle point, with $\mathcal{I} = -1$. It follows that the honey-flow also has a saddle point with $\mathcal{I} = -1$; recall that this is *why* it is called a saddle point!

In the case of the honey-flow, we thus have a simple way of ascribing an "index" to each of its singular points, via the surface's topographic map. We now turn to a more general analysis of the concept of the index, directly on the surface itself.

19.5.3 *Defining the Index on a Surface*

Given *any* vector field on a surface, not necessarily the honey-flow, it seems intuitively clear that it must be possible to extend our definition of the index to its singular points.

For example, in [19.5] we immediately identify four singular points of the honey-flow: p_1, p_2, p_3, and p_4. These are the points where the direction of the outward surface normal either coincides with $-\mathbf{V}$, as at $p_{1,2,3}$, or else coincides with $+\mathbf{V}$, as at p_4. Clearly, the "correct" generalization of $\mathcal{I}$ should be such that p_1 and p_2 are sources, with $\mathcal{I}(p_1) = \mathcal{I}(p_2) = +1$, p_3 is a saddle point, with $\mathcal{I}(p_3) = -1$, and p_4 is a sink, with $\mathcal{I}(p_4) = +1$.

To give precise meaning to this new, generalized notion of "index," presumably we should draw a loop round the singular point on the surface, then find the net rotation of the vector field as the loop is traversed. But wait, *rotation relative to what?*

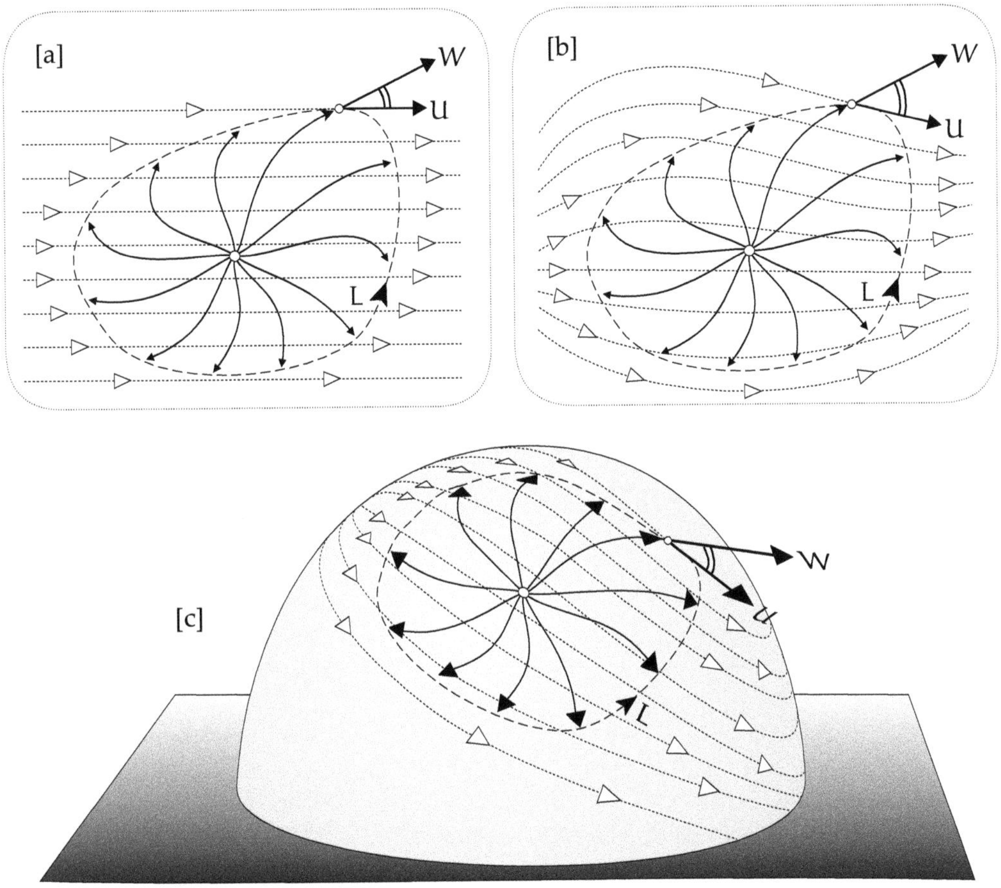

[19.7] Defining the Index on a Surface. *We usually measure the rotation of a plane vector field* W *with respect to a horizontal fiducial (or* **reference***) field* U*, as in* **[a]**. *But we may instead measure the rotation with respect to any other field without singular points, such as that in* **[b]**. *Finally, in* **[c]**, *we generalize to surfaces: draw a regular flow* U *across the region containing the singular point; then the index is the count of the revolutions of* W *relative to* U.

To answer this question, we first re-examine the familiar concept of rotation in the plane. Figure [19.7a] illustrates that the rotation of a plane vector field $W(z)$ along a loop L can be thought of as taking place relative to a ***fiducial*** (or ***reference***) vector field U having horizontal streamlines, say $U(z) = 1$. If we define $\angle UW$ to be the angle from U to W, and let $\delta_L\,(\angle UW)$ be the net change in this angle along L, then our old definition (19.3) of the index can be written as

$$\mathcal{I}_W(s) = \frac{1}{2\pi}\,\delta_L\,(\angle UW). \tag{19.5}$$

If we continuously deform the straight horizontal streamlines of U in [19.7a] to produce the curved ones in [19.7b], then, by the usual reasoning, the right-hand side of (19.5) will not change. Thus we conclude that this formula yields the correct value of the index if we replace U with *any* vector field that is nonsingular on and inside L.

Next, imagine that [19.7b] is drawn on a rubber sheet. If we continuously stretch it into the form of the curved surface in [19.7c], then not only will the right-hand side of (19.5) remain well defined, but its value will not change.

To summarize, if s is a singular point of a vector field W on a surface $\mathcal{S}$, we define its ***index*** as follows. Draw any nonsingular vector field U on a patch of $\mathcal{S}$ that covers s but no other singular points; on this patch, draw a simple loop L going round s; finally, apply (19.5)—that is, count the net revolutions of W relative to U as we traverse L.

19.6 The Poincaré–Hopf Theorem

19.6.1 Example: The Topological Sphere

Figure [19.8] shows the streamlines of two possible flows on the sphere. Notice that both possess singular points: [19.8a] has two centres, while [19.8b] has a dipole. In fact there can be *no* vector field on the sphere that is free of singular points.

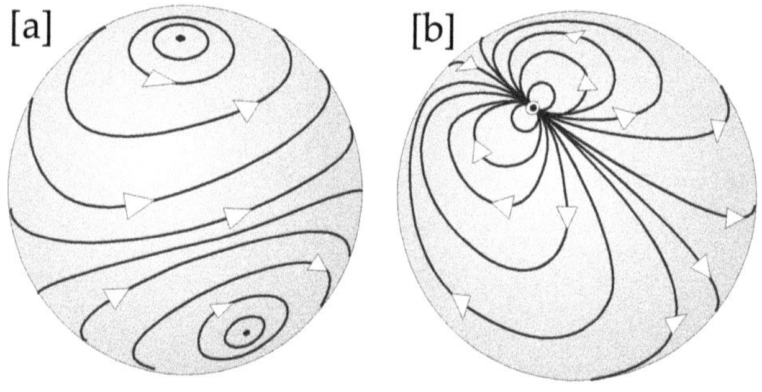

[19.8] [a] *Two centres with index sum* 2. **[b]** *A dipole with index* 2.

This is but one consequence of an extremely beautiful and important result, called the ***Poincaré–Hopf Theorem***, which we will state momentarily.

For now, to get a first whiff of the result, note that if we sum all the indices in [19.8a] we obtain

$$\mathcal{J}\,(\text{centre}) + \mathcal{J}\,(\text{centre}) = 1 + 1 = 2,$$

while if we do the same for [19.8b] we obtain

$$\mathcal{J}\,(\text{dipole}) = 2.$$

Next, reconsider the four singular points of the honey-flow depicted in [19.5]. Here the sum of the indices is $(1) + (1) + (-1) + (1) = 2$, *again!*

Try drawing your own streamlines on an orange. For example, consider flow from the north pole along the meridians to the south pole. Summing the indices of the singular points at the poles,

$$\mathcal{J}\,(\text{source}) + \mathcal{J}\,(\text{sink}) = 1 + 1 = 2,$$

again! Perhaps this is all some bizarre coincidence?

There *are* no coincidences in mathematics! In the case of the sphere, the Poincaré–Hopf Theorem states that if we sum the indices of *any* vector field on its surface, we will always get 2 for the answer. Indeed, it says that we will get this answer for any surface that is *topologically* a sphere, such as the fried banana in [19.5].

Here is the splendid general result:

> ***Poincaré–Hopf Theorem:*** *If a vector field* $\mathbf{v}$ *on a smooth surface* $\mathcal{S}_g$ *of genus* g *has only a finite number of singular points,* p_i*, then the sum of their indices equals the Euler characteristic of the surface:*

$$\boxed{\sum_i \mathcal{J}_{\mathbf{v}}(p_i) = \chi(\mathcal{S}_g) = 2 - 2g.} \tag{19.6}$$

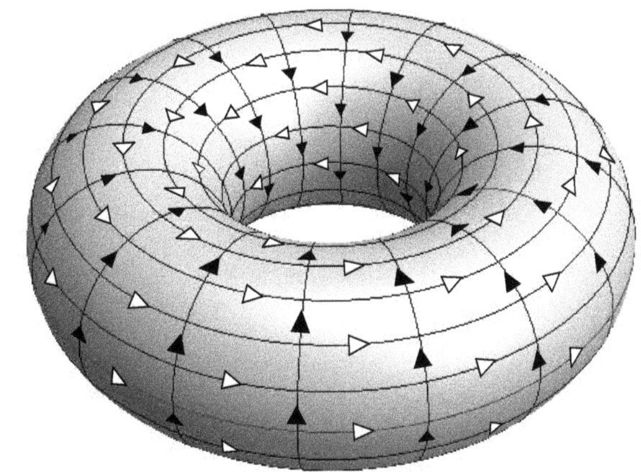

[19.9] *A flow without singular points can* only *exist on a topological doughnut. Here are two such* regular *(topologically distinct) flows on the torus: white and black.*

An immediate consequence of (19.6) is that a vector field without *any* singular points can exist *only* on a surface of vanishing Euler characteristic, i.e., a topological doughnut.

Even then, the theorem does not actually guarantee that such a vector field must exist, it merely demands that the indices of any singular points on a topological doughnut sum to zero. Nevertheless, we can easily *see* that on a doughnut there *do* exist vector fields without any singular points: [19.9] illustrates two such fields, one shown with white arrows, the other with black arrows.

All the streamlines in the white flow are topologically equivalent to each other, for any one can be continuously deformed into any other. The same is true for the black flow. However, the white and black flows are topologically *distinct* from each other: no white streamline can be deformed into a black streamline.

If we were to "add" these two flows, we could imagine a third, topologically distinct regular flow, such that each streamline went once around the doughnut and once through the hole before joining up again.

More generally, we can construct infinitely many, topologically distinct, regular flows, such that each closed streamline does m (white) circuits around the axis of symmetry, and n (black) circuits through the hole, before closing up on itself. "Adding" such an (m,n)-flow to an (m',n')-flow, by joining the end of one streamline to the start of the other, then yields the $(m+m',n+n')$-flow. This idea leads naturally to what topologists call the ***fundamental group*** of the surface, discovered by Poincaré in 1895. But we digress!

19.6.2 Proof of the Poincaré–Hopf Theorem

We can now give a very elegant derivation of the theorem (19.6), due to Heinz Hopf himself (see Hopf 1956, p. 13). The argument proceeds in two steps. First, we show that on a surface of given genus, all vector fields yield the same value for the sum of their indices. Second, we produce a concrete example[4] of a vector field for which the sum equals the Euler characteristic. This proves the result.

Consider the surface $\mathcal{S}$ in [19.10], and suppose that $\mathbf{X}$ and $\mathbf{Y}$ are two different vector fields on $\mathcal{S}$; to avoid clutter, we have only drawn these fields at a single point. If • are the singular points of $\mathbf{X}$, and ⊙ are those of $\mathbf{Y}$, we must show that

$$\sum_{\bullet} \mathcal{I}_{\mathbf{X}}[\bullet] = \sum_{\odot} \mathcal{I}_{\mathbf{Y}}[\odot].$$

We begin by partitioning $\mathcal{S}$ into curvilinear polygons (dashed curves) such that each one contains at most one • and one ⊙.

Now concentrate on just one of these polygons (darkly shaded) and its boundary K_j, taken counterclockwise as viewed from outside $\mathcal{S}$. To find the indices of the singular points of $\mathbf{X}$ and $\mathbf{Y}$

[4]We shall actually give a different example than Hopf gave in the reference (Hopf 1956) above.

within K_j, draw any nonsingular fiducial vector field **U** (again only shown at a single point) on the polygon, and then use (19.5). The difference of these indices is then

$$\begin{aligned}\mathcal{I}_{\mathbf{Y}}[K_j]-\mathcal{I}_{\mathbf{X}}[K_j] &= \frac{1}{2\pi}\left[\delta_{K_j}(\angle UY)-\delta_{K_j}(\angle UX)\right]\\ &= \frac{1}{2\pi}\delta_{K_j}(\angle XY),\end{aligned}$$

which is explicitly independent of the local fiducial vector field **U**; it depends *only* upon the illustrated (signed) angle from **X** to **Y**.

Finally, from this we deduce that

$$\begin{aligned}&\sum_{\odot}\mathcal{I}_{\mathbf{Y}}[\odot]-\sum_{\bullet}\mathcal{I}_{\mathbf{X}}[\bullet]\\ &= \sum_{j}\left(\mathcal{I}_{\mathbf{Y}}[K_j]-\mathcal{I}_{\mathbf{X}}[K_j]\right)\\ &= \frac{1}{2\pi}\sum_{\text{all polygons}}\delta_{K_j}(\angle XY)\\ &= 0,\end{aligned}$$

because (see (19.2)) *every edge of every polygon is traversed once in each direction, producing equal and opposite changes in the (signed) angle* $\angle XY$. We have thus completed the first step: the sum of the indices is independent of the vector field.

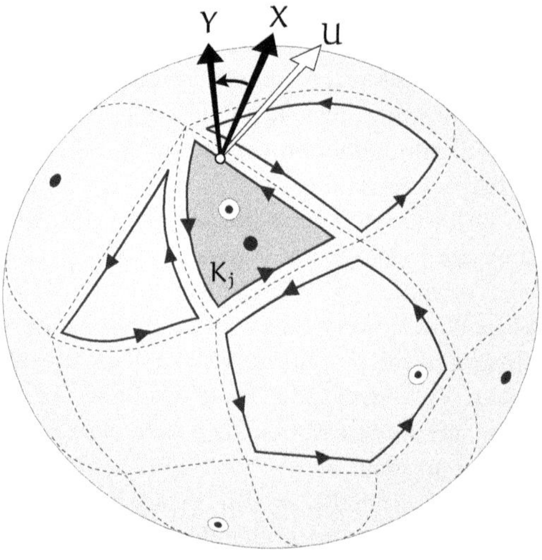

[19.10] *The difference of the indices of the singular points of* **X** *(marked* •*) and of* **Y** *(marked* ⊙*) inside* K_j *depends only on the rotation of* **Y** *relative to* **X**. *But each edge of* K_j *abuts an oppositely directed edge of a neighbouring polygon, so the sum of these rotations over all polygons must vanish.*

Since the index sum for the examples in [19.8] is 2, we now know that this is the value for *every* vector field on a topological sphere. The second step of the general argument is likewise to produce a concrete example on a surface $\mathcal{S}_g$, of arbitrary genus g, such that the sum is $\chi(\mathcal{S}_g)=(2-2g)$. Figure [19.11] is such an example (here with $g=3$)—namely, our old friend the honey-flow vector field. As the figure explains, the source at the top and the sink at the bottom each have $\mathcal{I}=+1$, and each of the g holes has two saddle points, each with $\mathcal{I}=-1$. Thus the sum of the indices of this *particular* flow is indeed $2-2g=\chi(\mathcal{S}_g)$, and so this must be the sum of the indices for *every* flow on $\mathcal{S}_g$. Done.

19.6.3 Application: Proof of the Euler–L'Huilier Formula

The Poincaré–Hopf Theorem provides a stunningly immediate proof of the Euler–L'Huilier Formula, (18.7). Recall that the latter states that if we partition a surface $\mathcal{S}_g$ of genus g into polygons, with V vertices, E edges, and F faces, then

$$\boxed{V-E+F=2-2g.}$$

As [19.12] illustrates, we may construct a consistent ***Stiefel vector field***[5] on $\mathcal{S}_g$ as follows: place a source (marked ◯) at each of the V vertices, a saddle point (marked ⊙) on each of the

[5]According to Frankel (2012, §16.2b), this is due to Hopf's student, Eduard Stiefel (1909–1978).

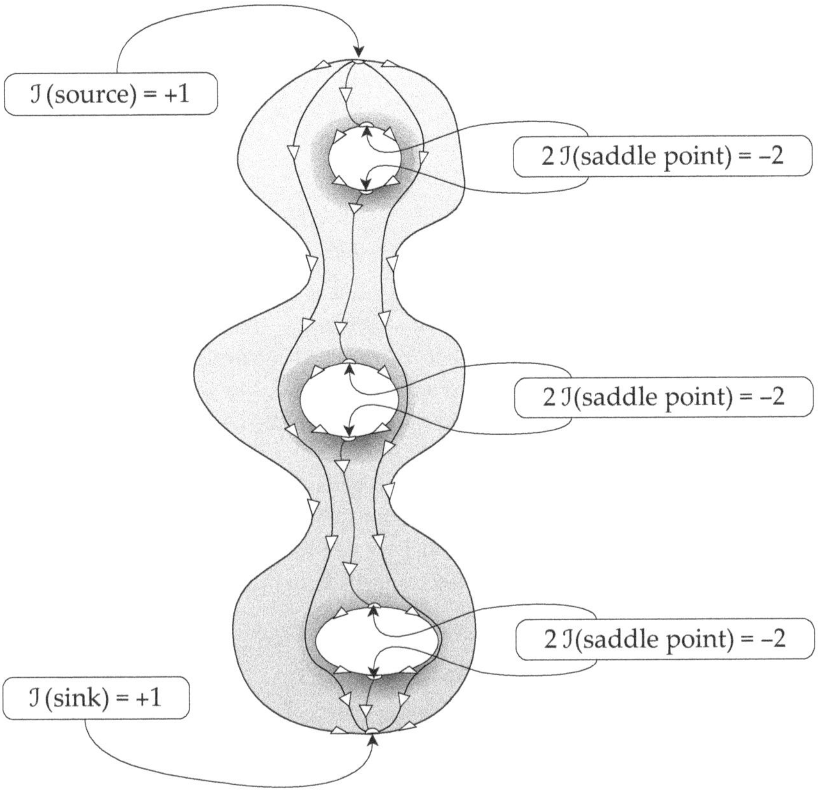

[19.11] *The honey-flow vector field on a surface of genus* g. *The source (top) and sink (bottom) each contribute* $+1$ *to the index sum, while each of the* g *holes contributes* -2. *Therefore the index sum is* $2-2g=\chi$.

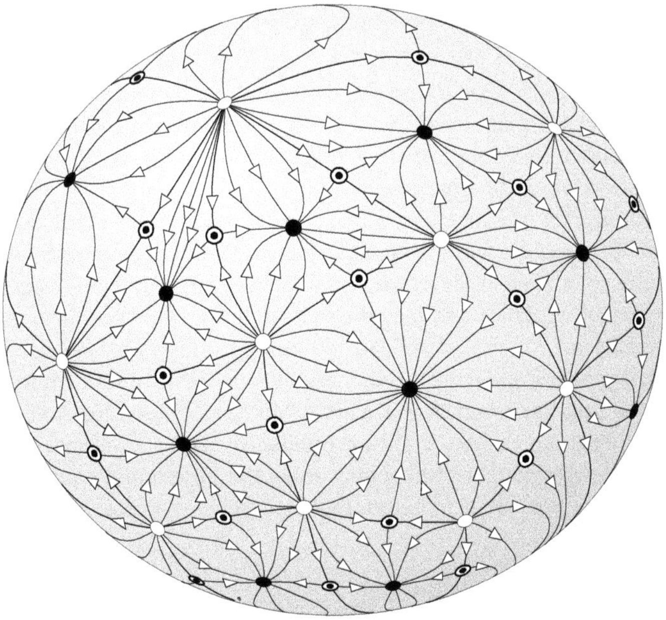

[19.12] *Given any polygonal partition of a surface* $\mathcal{S}_g$ *of genus* g, *we may construct a consistent Stiefel vector field on* $\mathcal{S}_g$ *by placing a source* (○)*—think white hole!—at each of the* V *vertices, a saddle point* (⊙) *on each of the* E *edges, and a sink* (●)*—think black hole!—inside each of the* F *faces. Thus,* $\sum \mathcal{I} = V - E + F$.

E edges, and a sink (marked ●) inside each of the F faces. Applying the Poincaré–Hopf Theorem to our constructed vector field, we instantly deduce that

$$2-2g=\sum \mathcal{I}=V\,\mathcal{I}(\bigcirc)+E\,\mathcal{I}(\odot)+F\,\mathcal{I}(\bullet)=V-E+F.$$

Done!

19.6.4 Poincaré's Differential Equations Versus Hopf's Line Fields

The 2-dimensional Poincaré–Hopf Theorem, as stated, was actually discovered by Poincaré. However, Hopf's name is correctly attached, for he enormously extended the result in two distinct directions. The first direction was to generalize the result to vector fields on n-dimensional closed manifolds; sadly, to explain this would take us too far afield, but see *Further Reading*. However, the second direction in which Hopf extended the result is quite elementary, and we shall explain it now.

Poincaré was led to investigate the topological behaviour of vector fields by his pioneering work on the *qualitative* theory of differential equations. Such equations became central to physics in 1687, when Newton announced his Second Law of Motion: if a point particle with position vector **X**, and of mass m, is subjected to a vector force **F**, then it responds by accelerating in the direction of the force, according to this law:

$$\ddot{\mathbf{X}}=\text{acceleration}=\frac{1}{m}\mathbf{F}. \tag{19.7}$$

Newton's motivation in formulating this law was the determination of the orbits of the planets—part of what is now called ***Celestial Mechanics***. As we discussed in the Prologue, in later life Newton shunned his own youthful discoveries in symbolic calculus, and in the *Principia* he instead employed elegant *geometrical* reasoning to determine the orbit of a planet, given the known (inverse square) law of the Sun's gravitational force acting upon it.

However, from the more traditional, modern point of view, (19.7) is simply a second-order differential equation, which may be solved (*in principle*) by integrating twice: once to find the velocity $\dot{\mathbf{X}}$, and a second time to find the orbit **X** itself. In the case of just two gravitating bodies (such as the Earth and Sun) Newton found the exact solution. By introducing the Moon as a *third* body, Newton was able to explain the tides on Earth, and by making *approximations*, he was able to make predictions about the variations of the tides. But Newton and his successors were unable to find an *exact* solution to this ***3-body problem***.

Finally, more than 200 years later, Poincaré—pictured in [19.13]—made decisive progress by *proving* that the 3-body problem was insoluble.[6] Of course matters only get much worse if we try to analyze the entire solar system, in which each planet is not only held in its orbit by the Sun, but also attracts every *other* planet, and they mutually determine their collective orbits. This is the so-called ***n-body problem***.

It is fitting (and perhaps somewhat ironic) that it was Poincaré's return to *geometrical* methods that secured the greatest advance in Celestial Mechanics since Newton. In three volumes (published in 1892, 1893, and 1899), Poincaré published his *New Methods of Celestial Mechanics*,[7] in which he no longer sought to explicitly solve the differential equations (which he had shown to be impossible), but rather to determine the *qualitative behaviour of the solutions* of these equations.

---

[6]To understand what we mean by "insoluble," and for a lively account of Poincaré's voyage of discovery, see Diacu and Holmes (1996).

[7]Poincaré (1899).

To see how this connects with what has gone before, let us consider, as Poincaré did, first-order differential equations in the (x, y)-plane, of the form

$$\frac{dy}{dx} = -\frac{P(x,y)}{Q(x,y)}. \tag{19.8}$$

Poincaré actually wrote these as

$$P(x,y)\,dx + Q(x,y)\,dy = 0.$$

To investigate the integral curves (the solutions), consider an infinitesimal vector $\begin{bmatrix} dx \\ dy \end{bmatrix}$ along the integral curve, and note that the previous equation then implies that

$$\begin{bmatrix} P \\ Q \end{bmatrix} \cdot \begin{bmatrix} dx \\ dy \end{bmatrix} = 0.$$

In other words,

[19.13] *Henri Poincaré (1854–1912)*

The integral curves of (19.8) are everywhere orthogonal to the vector field $\begin{bmatrix} P \\ Q \end{bmatrix}$.

For example, consider the radial vector field $\begin{bmatrix} P \\ Q \end{bmatrix} = \begin{bmatrix} x \\ y \end{bmatrix}$; this is the source in [19.4c]. Clearly, the orthogonal integral curves are origin-centred circles, so we obtain the centre shown in [19.4e]. Likewise [exercise], $\begin{bmatrix} y \\ x \end{bmatrix}$ yields the saddle point in [19.4h], and $\begin{bmatrix} y \\ -x \end{bmatrix}$ yields either a source or a sink, [19.4c,f].

Now let us turn to Hopf's extensions of Poincaré's discovery. In addition to generalizing to n-dimensional vector fields, which required fundamentally new ideas, Hopf realized that even in 2 dimensions there exist what we might call "directionless flows" that do *not* arise from differential equations of the type considered by Poincaré.

For our first example, consider [19.14a]. We have called such a pattern a "directionless flow," but Hopf actually named this concept a ***field of line elements***. However, the modern terminology, which we shall adopt going forward, is ***line field***, by analogy with *vector field*.

As you see, no arrows are attached to these "streamlines," but we may nevertheless assign a continuously varying angle Θ to the tangent line. If we travel along a simple loop L around the singular point s, the illustrated line element must return to its original position. Previously, with vector fields, the *vector* had to return home pointing in the same direction as when it left, and *therefore it had to undergo a complete number of revolutions*: the rotation was necessarily a multiple of 2π, that multiple being the index.

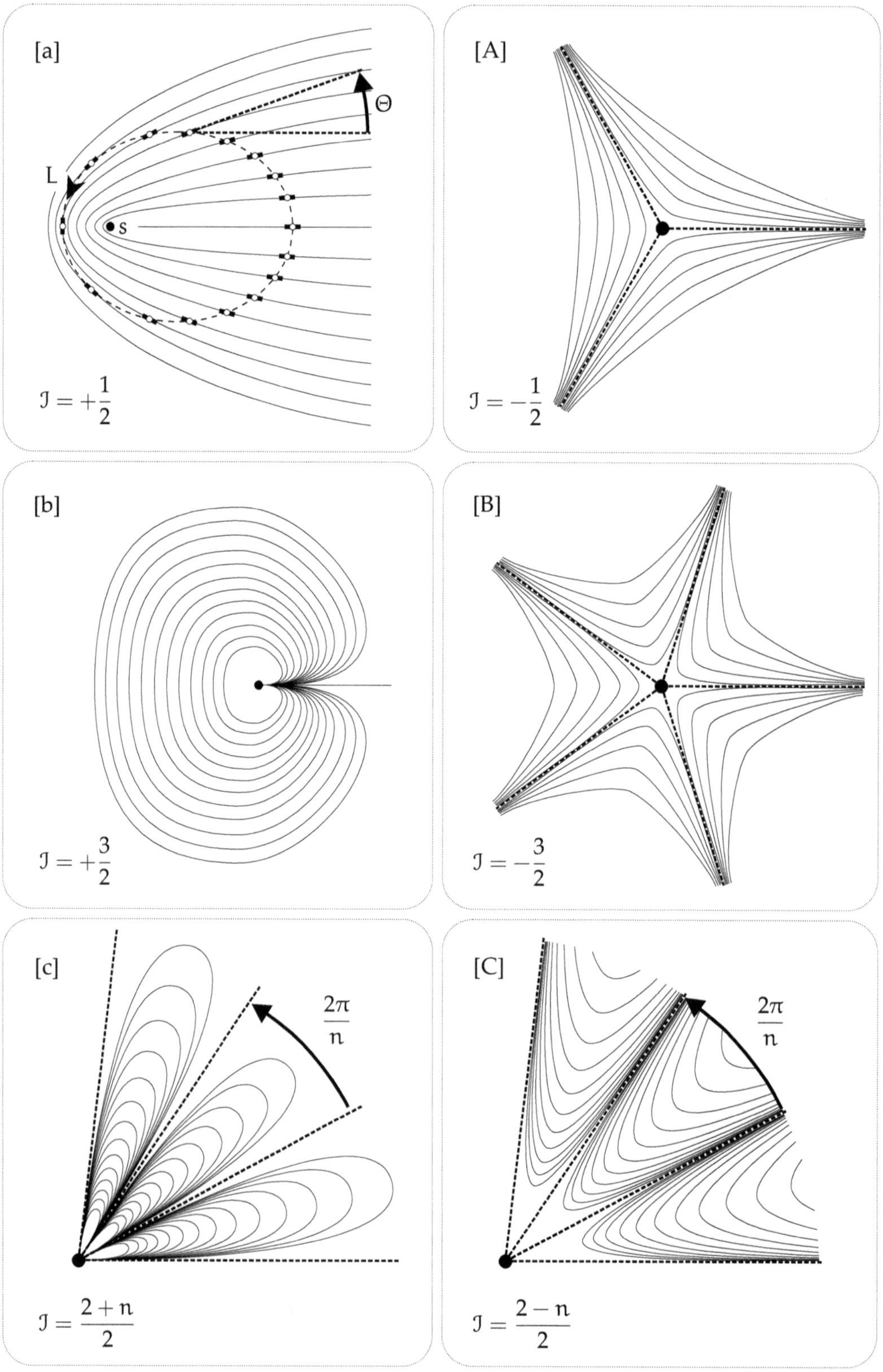

[19.14] Line Fields *do not have a direction associated with their "streamlines." As a consequence, their singular points can have* fractional *indices, which vector fields cannot have. By repeating the patterns in* **[c]** *and* **[C]**, *we may construct a singular point of arbitrary positive or negative index, respectively. If the positive integer* n *is odd, the index is a fraction, but if* n *is even then we recover the flows of the vector fields shown in [19.4], for which the indices are integers.*

But a *line element* will return to its original orientation if it undergoes a rotation through *any* multiple of π. Indeed, in [19.14a] we see that the net rotation is π. If we retain our original definition of the index, (19.3), as the rotation divided by 2π, then we see that this singular point has a *fractional index*: $\mathcal{J}=+\frac{1}{2}$. Some authors call such fractional indices, ***Hopf indices***.

Please look through all the remaining figures in [19.14] and visually confirm their indices. In particular, [19.14c] and [19.14C] show how to construct a singular point of arbitrary positive or negative index, respectively, simply by repeating the illustrated patterns until they completely surround the singular point. If n is odd, the index is a fraction, but if n is even then we recover the flows of the vector fields shown in [19.4], for which the index is an integer.

Although we cannot interpret a line field with fractional index as the streamlines of an unobstructed flow, we can instead interpret it as *a flow in the presence of 1-dimensional barriers*. For example, in [19.14a], if we think of the horizontal ray to the right of s as such a barrier, then we *can* attach arrows to the curves in a manner that is consistent with a flow around this barrier, flowing to the left above the barrier, say, and to the right below it.

From this perspective, every point of the barrier is a singular point of the vector field, for the direction of the vector flips discontinuously as we cross the barrier. However, for the illustrated, directionless line field, every such point p (other than s) is a *regular* point. For if K is a small loop around p, then you can see that Θ varies continuously on K, so $\mathcal{J}(p)=\frac{1}{2\pi}\delta_K\Theta=0$.

Now think back over the reasoning that culminated in the Poincaré–Hopf Theorem, concerning the singular points of vector fields. You will discover that it is all *equally applicable* to these more general line fields. In short,

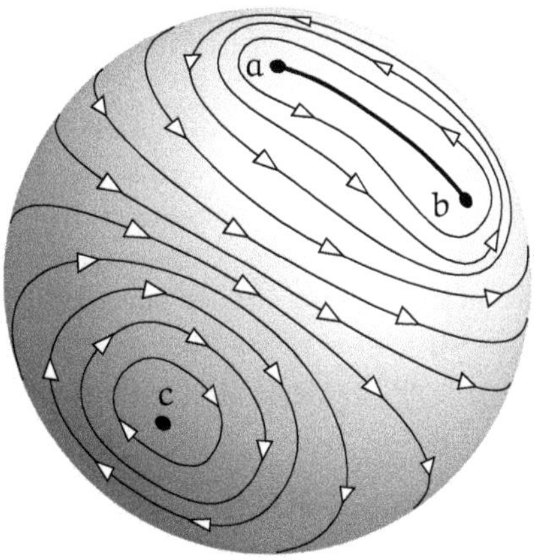

[19.15] *By installing a barrier that connects* a *and* b*, we can create a flow that encircles both points, and that has a third singular point at the centre* c*. The index sum is still equal to the Euler characteristic, despite the* fractional indices *at* a *and* b.

> *The Poincaré–Hopf Theorem (19.6) not only applies to vector fields, but also to line fields, with fractional indices.* (19.9)

Figure [19.15] is an example of this result on the sphere. Here we see a line field with three singular points, a, b, and c. We have imagined that the curve connecting a and b is a barrier, and we have *added arrows* to the curves to create a consistent flow that circulates around this barrier. Note that the flow near a and b is the one shown in [19.14a], with $\mathcal{J}=+(1/2)$.

The Poincaré–Hopf Theorem (19.6) predicts that the sum of the indices on a sphere should be $\chi=2$, and indeed it is:

$$\mathcal{J}(a)+\mathcal{J}(b)+\mathcal{J}(c)=\frac{1}{2}+\frac{1}{2}+1=2.$$

Finally, note that if we shrink the length of the barrier and allow the two $\mathcal{J}=(1/2)$ singular points a and b to approach one another, then they will ultimately coalesce into a single $\mathcal{J}=1$ centre, creating the flow shown in [19.8a].

THE HISTORY AND FUTURE OF LINE FIELDS: Despite the fact that Hopf gave a wonderfully lucid account of his ideas in Hopf (1956), I am not aware of *any* modern introductory text on Topology, Differential Topology, or Differential Geometry that even mentions this fascinating concept. Indeed, the only example we have found is now more than 50 years old: the classic text by

Stoker (1969, p. 244), who was a doctoral student of Hopf. This is despite the fact that such line fields arise naturally in mathematics. For example, the lines of curvature surrounding an umbilic point on a surface (where $\kappa_1 = \kappa_2$) typically take the form [19.14 a or A]. For a lovely illustration of case [a], see Hilbert and S. Cohn-Vossen (1952, p. 189).

While mathematicians seem to have paid scant attention to the applicability of the Poincaré–Hopf Theorem to line fields with fractional indices, the same is not true of physicists, for Nature herself has thrust such line fields upon them in multiple areas of physics, but especially in optics. The *Further Reading* section at the end of this book seeks to guide you to many of these fascinating new physical applications of Hopf's idea.

19.7 Vector Field Proof of the Global Gauss–Bonnet Theorem

Let V be an arbitrary point on S^2, with unit position vector **V**. Now imagine a plane with normal **V**, initially located far from a smooth closed surface $\mathcal{S}$, and with **V** pointing away from $\mathcal{S}$. Next, imagine the plane moving towards $\mathcal{S}$. Eventually the plane must make *contact* with $\mathcal{S}$, and at that moment it will become the tangent plane to $\mathcal{S}$ at the point(s) p of contact, and therefore **V** must coincide with the surface normal there: $\mathbf{V} = \mathbf{n}(p)$. In other words, V is the spherical image of p: $V = n(p)$. Since this is true for arbitrary **V**, it must also be true of opposite vector, $-\mathbf{V}$; the plane simply approaches from the other side of $\mathcal{S}$. To sum up,

> *If $\mathcal{S}$ is an arbitrary, closed, smooth surface, then its spherical image covers every antipodal pair of points $\pm V$ of S^2 at least once.* (19.10)

Of course, V and $-V$ may each be covered *more* than once. For example, in [19.5] we see that V is covered once, while $-V$ is covered three times.

Let us now to return to GGB, and let us briefly recap our very first "heuristic" proof in Chapter 17. Look at these coverings of V from the point of view of the total curvature, $\mathcal{K}(\mathcal{S})$. If $\mathcal{K}(p) > 0$ at p, then n preserves the orientation of a small patch of area $\delta\mathcal{A}$ surrounding p, so its spherical image of area $\delta\widetilde{\mathcal{A}} \asymp \mathcal{K}(p)\,\delta\mathcal{A}$ surrounding V on S^2 has the *same* orientation. As we discussed in Section 17.4, we count this as a *positive* covering of V. Likewise, if $\mathcal{K}(p) < 0$, then the spherical map *reverses* the orientation, and the covering is counted as *negative*. The total curvature integral automatically counts these coverings *algebraically*, taking into account orientation. As earlier, let $\mathcal{P}(V)$ and $\mathcal{N}(V)$ denote the number of positive and negative coverings of V, respectively, so that the net number of coverings of V is $[\mathcal{P}(V) - \mathcal{N}(V)]$. Thus,

Let V be a point on S^2, and let its preimages (under the spherical map n) be the points p_i of $\mathcal{S}$. Consider a small (ultimately vanishing) patch around V (with area $\delta\widetilde{\mathcal{A}}$) and let the areas of the small preimage patches around p_i be $\delta\mathcal{A}_i$. Then,

$$\textit{Total curvature within the } \delta\mathcal{A}_i \asymp \textstyle\sum_i \mathcal{K}(p_i)\,\delta\mathcal{A}_i \asymp [\mathcal{P}(V) - \mathcal{N}(V)]\,\delta\widetilde{\mathcal{A}}. \tag{19.11}$$

For example, in [19.5] we see that the curvatures at p_1 and p_2 are positive, and the coverings of $-V$ originating from these points are positive. On the other hand, the negatively curved patch containing p_3 is mapped to an orientation-reversed, negative covering of $-V$. The *net algebraic* count of the coverings of $-V$ is therefore 1. Likewise, the net number of coverings of V is 1.

Recall that this net number of coverings of $\mathbb{S}^2$ is the ***topological degree*** $\deg(n)$ of the spherical mapping, and the fundamental fact is that the degree is the same for all[8] points of $\mathbb{S}^2$. This was the essence of our heuristic explanation of GGB in Chapter 17. The final step of that explanation was the recognition that the degree depends solely on the genus g of $\mathcal{S}_g$. This is (16.7), on page 174, which we restate here:

$$\deg[n(\mathcal{S}_g)] = \mathcal{P} - \mathcal{N} = (1-g) = \tfrac{1}{2}\chi(\mathcal{S}_g). \tag{19.12}$$

As we noted in Section 16.6, this was first proved by Walther Dyck in 1888.

If we take it as *given* that the topological degree exists, so that every point of $\mathbb{S}^2$ is covered the same (algebraic) number of times, then the Poincaré–Hopf Theorem can be used to give an elegant proof of Dyck's result, and, with it, GGB. Indeed, this is the approach to GGB taken in such well-known works as Guillemin and Pollack (1974, p. 198).

We too shall reproduce that argument shortly. However, the existence of the topological degree is by no means obvious, so we now present a slightly different argument that does not depend upon it. The argument stills hinges on the Poincaré–Hopf Theorem, but it does *not* assume the existence of the topological degree. That is, if V and W are two points of $\mathbb{S}^2$, we shall *not* assume that they must be covered the same number of times (algebraically); instead, we shall entertain at least the possibility that $\deg[n(\mathcal{S}), V] \neq \deg[n(\mathcal{S}), W]$. Henceforth, we shall take it as understood that we are dealing with the spherical map of $\mathcal{S}$, so this (hypothetical and in fact impossible) nonequality can then be abbreviated to $\deg[V] \neq \deg[W]$.

To prove GGB without assuming the existence of the topological degree, reconsider the honey-flow in [19.5]. The first important observation is that the singular points of this flow occur at precisely those points where the vertical gravitational force $\mathbf{V}$ has zero component within the surface—that is, at points where the vertical force vector points directly into (or out of) the surface—that is, where the outward surface normal $\mathbf{n} = \pm\mathbf{V}$. Of course, mathematically speaking, there is nothing special about this gravitational direction $\mathbf{V}$—we may imagine gravity pulling in an arbitrary direction; alternatively, if we do not wish to tamper with the Earth's gravitational field, we may simply rotate the surface!

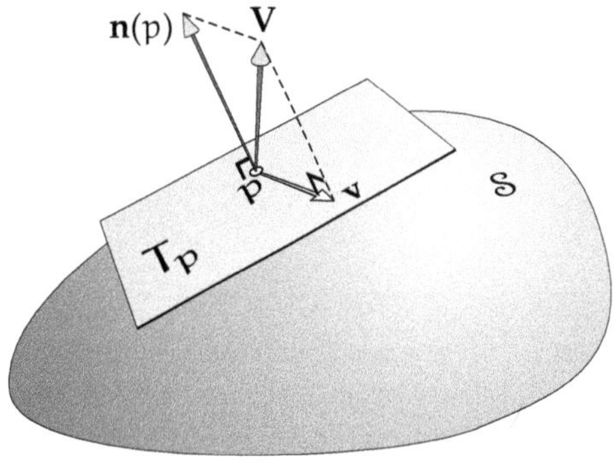

[19.16] *The* ***honey-flow*** $\mathbf{v}$ *on the surface* $\mathcal{S}$ *in the direction* $\mathbf{V}$ *is the orthogonal project of* $\mathbf{V}$ *onto the tangent plane* T_p *at* p.

To put this more mathematically, our "honey-flow" vector field is obtained by orthogonal projection of $\mathbf{V}$ onto the surface, or, more precisely, onto the tangent plane T_p of the surface. See [19.16]. Let us make this definition official:

$$\text{The } \textbf{\textit{honey-flow}} \text{ vector field in the direction } \mathbf{V} \text{ is } \quad \mathbf{v}(p) \equiv \text{proj}_{T_p} \mathbf{V}. \tag{19.13}$$

[8] As we have discussed, the exceptions are points of vanishing curvature, which contribute nothing to the total curvature.

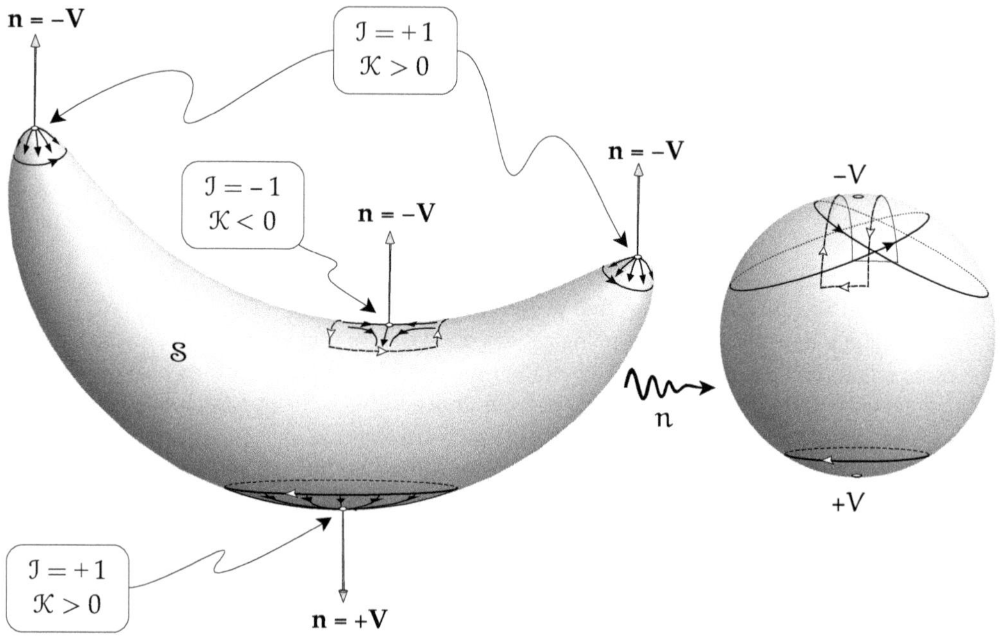

[19.17] *The singular points of the honey-flow* **v** *on* $\mathcal{S}$ *in the direction* **V** *are mapped by the spherical map* n *to either* $+V$ *or* $-V$ *on* $\mathcal{S}^2$*, and* ***the orientation of the covering on*** $\mathcal{S}^2$ ***is determined by the sign of the index*** *of* **v** *on* $\mathcal{S}$*. Here* $+V$ *is covered once, positively, by the image of the sink at the bottom of the banana—where* $\mathcal{J}=+1$ *and* $\mathcal{K}>0$*. On the other hand,* $-V$ *is covered three times: twice positively by the images of the two sources—where* $\mathcal{J}=+1$ *and* $\mathcal{K}>0$*—and once negatively by the image of the saddle point—where* $\mathcal{J}=-1$ *and* $\mathcal{K}<0$.

The singular points $\mathbf{v}(p_i)=0$ occur when $\mathbf{n}(p_i)=\pm\mathbf{V}$. In other words,

> *If honey flows over* $\mathcal{S}$*, pulled in the direction* **V** *in space, the singular points of the honey-flow* **v** *consist of the complete set of points that are sent by the spherical map to either* V *or to the antipodal point* $-V$ *on* $\mathcal{S}^2$.

Out of all possible vector fields on a surface, why have we lavished so much attention on the *honey-flow*? The answer is that there exists a crucial link between the *geometry* of the surface, and the *topology* of the honey-flow. This in turn will yield our third explanation of GGB, as a consequence of the Poincaré–Hopf Theorem.

Reconsider the singular points p_i of the honey-flow vector field **v** illustrated in [19.5]. These are shown again in [19.17], but now focusing attention on how the spherical map n sends them to $\mathcal{S}^2$.

The decisive observation is the distinction between the singular points of **v** that have positive curvature, and those that have negative curvature.

Recall (12.9) and [12.7], on page 136. If $\mathcal{K}(p_i)>0$, the surface is locally a dome facing either straight up (in which case it is a source with $\mathcal{J}(p_i)=+1$) or straight down (in which case it is a sink, again with $\mathcal{J}(p_i)=+1$). But if $\mathcal{K}(p_i)<0$, then the surface is locally a saddle, producing a corresponding saddle point in the flow, in which case $\mathcal{J}(p_i)=-1$.

This immediately implies that the spherical map preserves orientation if $\mathcal{J}(p_i)=+1$, and reverses it if $\mathcal{J}(p_i)=-1$:

> *If* p *is a singular point of the honey-flow* $\mathbf{v}$*, then its image* $(\pm V)$ *under the spherical map* n *is covered positively if* $\mathcal{I}(p) = +1$*, and negatively if* $\mathcal{I}(p) = -1$.

Now let us combine this result with the Poincaré–Hopf Theorem. If p_i is the set of singular points of $\mathbf{v}$ on $\mathcal{S}$, i.e., those whose spherical image is either $+V$ or $-V$, then

$$\chi(\mathcal{S}) = \sum_i \mathcal{I}_v(p_i) = \{\mathcal{P}(+V) - \mathcal{N}(+V)\} + \{\mathcal{P}(-V) - \mathcal{N}(-V)\}. \tag{19.14}$$

In reality, both bracketed terms on the right are equal to each other, being the topological degree of the spherical map. See (19.12). For example, in [19.17] we see that

$$\mathcal{P}(+V) - \mathcal{N}(+V) = 1 - 0 = 1$$

and

$$\mathcal{P}(-V) - \mathcal{N}(-V) = 2 - 1 = 1.$$

However, (19.14) enables us to prove GGB *without* assuming this fact. To do so, note that the singular points of the single flow $\mathbf{v}$, corresponding to honey flowing in the direction $\mathbf{V}$, actually include the singular points of *both* $\mathbf{V}$ *and* $-\mathbf{V}$. For reversing the direction of the honey flow does not alter the locations of the singular points, nor their indices.

Now if V roams over just the northern hemisphere of S^2 (the area of which is 2π) then $-V$ roams over the southern hemisphere, and therefore $\pm V$ covers the *entire* S^2. Thus, by virtue of (19.10), as V roams over just the northern hemisphere, the singular points of $\mathbf{v}$ cover the *entire* surface $\mathcal{S}$. And since we have just seen that the indices of the singular points correctly count the covers of their images under the spherical map, (19.11) and (19.14) yield our third proof of GGB, in the form

$$\boxed{\mathcal{K}(\mathcal{S}) = 2\pi\chi(\mathcal{S}).}$$

Let us return to the topic of the topological degree, and thereby obtain a bonus result. If we *assume* that the topological degree exists, then $\deg(V) = \deg(W)$, for all V and W on S^2. In particular, it follows that $\deg(+V) = \deg(-V)$. In that case, (19.14) becomes,

$$\begin{aligned}
\chi(\mathcal{S}) &= \{\mathcal{P}(+V) - \mathcal{N}(+V)\} + \{\mathcal{P}(-V) - \mathcal{N}(-V)\} \\
&= \deg(+V) + \deg(-V) \\
&= 2\deg(V) \\
&= 2\deg,
\end{aligned}$$

in which the final equality reflects the fact that the degree is independent of the choice of V. Thus, if we assume the existence of the topological degree, we recover Dyck's result, (19.12):

$$\boxed{\deg = \tfrac{1}{2}\chi(\mathcal{S}).}$$

19.8 The Road Ahead

All three of the proofs of GGB thus far presented have relied upon the interpretation of $\mathcal{K}$ as the local area expansion factor of the spherical map. This is an *extrinsic* conception of curvature, depending as it does upon the normal **n**, which is invisible and unknowable to the inhabitants of $\mathcal{S}$. While the three proofs have afforded us wonderful new insights in many different directions, something is lacking, for we know that $\mathcal{K}$ is also, in fact, an *intrinsic* property of the surface, knowable to and measurable by its inhabitants. In principle, these inhabitants could measure $\mathcal{K}$ throughout the surface, and from these purely local *geometric* measurements they could determine the *topology* of their world!

We are about to embark on Act IV, which introduces a brand new, and extremely powerful, *intrinsic* way of understanding and measuring curvature. This will allow us to finally explain some of the fundamental results we have been forced to assume up till now, such as the Local Gauss–Bonnet Theorem—and with it the *Theorema Egregium*—as well as the remarkable ("*Star Trek* phaser") formula (4.10), page 38, for the curvature in terms of the metric. Furthermore, it will allow us to remedy the noted deficiency in our three existing proofs of GGB. Indeed, using an idea of Heinz Hopf, we shall finally be able to provide a proof of GGB that is entirely *intrinsic*.

Chapter 20

Exercises for Act III

Curvature of Plane Curves

1. **Computational Proof of the Curvature Formula.** Figure [8.7] provided a geometric proof of (8.7). Prove this instead by calculation. (*Hint*: If the curve $[x(t), y(t)]$ is traced at unit speed, then $\dot{x} = \cos\varphi$, and $\dot{y} = \sin\varphi$.)

Curves in 3-Space

2. **Binormal Cannot Tip in the Direction of Motion.** Prove by calculation that as the Frenet frame $(\mathbf{T}, \mathbf{N}, \mathbf{B})$ moves along its curve, the binormal $\mathbf{B}$ only spins around $\mathbf{T}$; it cannot tip in the direction of $\mathbf{T}$. (*Hint*: Symbolically, what must be proved is that $\mathbf{B}'$ has no component in the $\mathbf{T}$ direction: $\mathbf{B}' \bullet \mathbf{T} = 0$.)

3. **The Darboux Vector.** As the Frenet frame moves along its curve, it rotates. Verify by calculation that its instantaneous *angular velocity vector* is

$$\mathbf{A} \equiv \tau\mathbf{T} + \kappa\mathbf{B}.$$

This vector $\mathbf{A}$ is sometimes referred to as the ***Darboux vector*** (see Stoker 1969, p. 62) in honour of its discoverer, Jean-Gaston Darboux (1842–1917), a pioneer of Differential Geometry, and a teacher of Cartan. *Hint*: Use (9.3) to prove that

$$\begin{bmatrix} \mathbf{T} \\ \mathbf{N} \\ \mathbf{B} \end{bmatrix}' = \mathbf{A} \times \begin{bmatrix} \mathbf{T} \\ \mathbf{N} \\ \mathbf{B} \end{bmatrix}.$$

4. **Variable Speed Frenet–Serret Equations.** The Frenet–Serret equations (9.3) assume the derivative is with respect to arc length, which is only the same as the time derivative if the curve is traced at unit speed. Suppose instead that the particle has variable speed $v = \dot{s}$.

   (i) Show that $[\Omega] \mapsto v[\Omega]$:

$$\begin{bmatrix} \mathbf{T} \\ \mathbf{N} \\ \mathbf{B} \end{bmatrix}^{\bullet} = v \begin{bmatrix} 0 & \kappa & 0 \\ -\kappa & 0 & \tau \\ 0 & -\tau & 0 \end{bmatrix} \begin{bmatrix} \mathbf{T} \\ \mathbf{N} \\ \mathbf{B} \end{bmatrix}.$$

   (ii) Prove that the acceleration is $\dot{\mathbf{v}} = [v\mathbf{T}]^{\bullet} = \dot{v}\mathbf{T} + \kappa v^2\mathbf{N}$, and interpret and explain both terms geometrically.

   (iii) Show that $\mathbf{B} = \dfrac{\mathbf{v} \times \dot{\mathbf{v}}}{|\mathbf{v} \times \dot{\mathbf{v}}|}$.

   (iv) Show that $\kappa = \left|\dfrac{\mathbf{v} \times \dot{\mathbf{v}}}{v^3}\right|$.

   (v) Show that $\tau = \dfrac{(\mathbf{v} \times \dot{\mathbf{v}}) \bullet \ddot{\mathbf{v}}}{|\mathbf{v} \times \dot{\mathbf{v}}|^2}$.

5. **The Helix.** Consider the path of a particle whose position at time t is $(R\cos\omega t, R\sin\omega t, qt)$.

   (i) Explain why this is a helix, and state the geometrical/physical interpretations of R, ω, and q.

   (ii) Prove that $v=\sqrt{(R\omega)^2+q^2}$, and explain this geometrically/physically.

   (iii) Use the previous question to calculate κ and τ.

   (iv) Explain *geometrically* why $\lim_{\omega\to\infty}\kappa=(1/R)$, and $\lim_{q\to\infty}\kappa=0$.

   (v) Use (iii) to confirm the predictions of (iv).

6. **The Frenet–Serret Approximation to a Curve.** Let $\mathbf{x}(t)$ be the position at time t of a particle that traces a curve C in space at unit speed. Let $(\mathbf{T}_0, \mathbf{N}_0, \mathbf{B}_0)$ be the Frenet–Serret frame at time $t=0$, and let κ_0 and τ_0 be the curvature and torsion at this time.

   (i) Using Taylor's Theorem and the Frenet–Serret equations (9.3), prove that the motion along C is initially given by the ***Frenet Approximation***:

$$\mathbf{x}(t) \asymp \mathbf{x}(0)+t\,\mathbf{T}_0+\kappa_0\frac{t^2}{2}\,\mathbf{N}_0+\kappa_0\tau_0\frac{t^3}{6}\,\mathbf{B}_0.$$

   (ii) Explain the first three terms geometrically.

   (iii) What does the fourth term describe, geometrically?

   (iv) In light of your answer to (iii), why *must* this term vanish when $\tau_0=0$?

   (v) Why does it make sense that this fourth term should also vanish if $\kappa_0=0$?

The Principal Curvatures of a Surface

7. **Series Expansion of Surfaces.** For each of the following equations, use a computer to draw the surface. Use series expansions to calculate the quadratic approximation near the origin, then use (10.6), page 111, to deduce the principal curvatures and $\mathcal{K}$ there. Visually confirm that your calculations are at least qualitatively correct.

   (i) $z=\exp(x^2+4y^2)-1$.

   (ii) $z=\ln\cos y-\ln\cos 2x$.

8. **Variable-Speed Formulas for the Curvature of a Surface of Revolution.** In the text, we pictured a particle moving at constant unit speed along the generating curve of a surface of revolution, in which case the principal curvatures are given by (10.11) and (10.12). If the speed $v(t)$ is *not* held constant, recall that (10.11) becomes (8.6):

$$\kappa_1=\frac{\dot{x}\,\ddot{y}-\dot{y}\,\ddot{x}}{v^3}.$$

   (i) By modifying [10.6], page 114, deduce that the second principal curvature is given by

$$\kappa_2=-\frac{\dot{x}}{yv}.$$

   (ii) Deduce that

$$\mathcal{K}=-\frac{\dot{x}\left[\dot{x}\,\ddot{y}-\dot{y}\,\ddot{x}\right]}{y\left[\dot{x}^2+\dot{y}^2\right]^2}.$$

(iii) If the particle only moves to the right, never retracing any x-value, then we are free to adjust the speed along the curve so that x represents *time*: $x(t) = t$, in which case $y(t) = y(x)$, and the time derivative becomes the x-derivative: $\dot{y} = y'$. Show that the curvature formula in (ii) then reduces to

$$\mathcal{K} = -\frac{y''}{y\,[1+(y')^2]^2}. \tag{20.1}$$

9. Polar Formula for the Curvature of a Surface of Revolution. In the previous exercise, the surface of revolution was defined using the distance of the generating curve from the rotation axis, which was taken to be the x-axis. Let us now, instead, describe such a surface by giving its height above a plane perpendicular to its axis of symmetry, which we now take to be the z-axis. If r is the usual polar coordinate in the (x, y)-plane, consider the surface of revolution $z = f(r)$.

(i) By simply changing names appropriately, and using the chain rule, show that (20.1) becomes

$$\mathcal{K}(r) = \frac{f'(r)\,f''(r)}{r\left\{1+[f'(r)]^2\right\}^2}.$$

(ii) Find $f(r)$ for a sphere or radius R centred at the origin. Check that the formula in (i) gives the correct value of $\mathcal{K}$.

(iii) Manually sketch the surface $z = \exp[-r^2/2]$, calculate $\mathcal{K}$, and find the regions of positive, negative, and zero curvature. (A computer-generated graph of the surface should visually confirm your answers.)

10. In the text we noted that it is visually evident that the surface $z = r^4$ has a planar point (with $\mathcal{K} = 0$) at the origin, and that the surface has positive curvature everywhere else. Prove this is correct by using the formula in the previous question to calculate $\mathcal{K}(r)$.

Gauss's *Theorema Egregium*

11. Why Paper Folds into a Straight Line. We take it for granted that when we fold a piece of paper, the fold automatically forms itself into a *straight line*, but *why* does this happen? Show (without any calculation) that this is a direct consequence of the *Theorema Egregium*! (I owe this delightful insight to my colleague Dr. Robert Wolf, a former student of Chern.)

The Shape Operator

12. Visual Linear Algebra. Do all of the following by reasoning directly and *geometrically* about the linear transformations themselves, *not* by applying the "Devil's machine" (see Prologue) to the matrices that represent them! We hope that these examples (together with those in the text) will inspire you to bring this geometrical perspective to bear whenever you next encounter Linear Algebra.

(i) Verify that the geometric interpretation [15.4] of the SVD generalizes to $\mathbb{R}^3$, and therefore the interpretation [15.5] of M^T does too.
NOTE: Although visualization becomes harder, the same is true in $\mathbb{R}^n$.

(ii) Recall that an ***orthogonal*** linear transformation is one that preserves lengths, and therefore angles (in general), and orthogonality (in particular)—examples include rotations and reflections. In $\mathbb{R}^3$, use (i) to explain why any orthogonal transformation R has the property that $R^{-1} = R^T$.

(iii) Recall that a matrix [M] is called ***skew symmetric*** if $[M]^T = -[M]$. In $\mathbb{R}^2$, use (15.12) to give a *geometric* characterization (in terms of the twist, τ) of the underlying skew-symmetric linear transformation M.

(iv) A symmetric linear transformation P is called ***positive definite*** if $\mathbf{x} \bullet (P\mathbf{x}) > 0$, for all $\mathbf{x}$. Use (15.13) to show that P is positive definite if (and only if) all of its eigenvalues are positive.

(v) Arguing directly, or, alternatively, building on (iv), show that a symmetric, positive-definite linear transformation must have positive determinant (and hence be invertible). Show that the converse is *false*, by finding a simple counterexample in $\mathbb{R}^3$. (*Hint*: Geometrically, the determinant is the *(signed) volume-expansion factor* of the linear transformation, the sign being postive (+) or negative (−) according to whether orientation is preserved or reversed, respectively.)

(vi) Given (iv) above, it follows from (15.15) that $M^T M$ and MM^T are always symmetric and positive definite. Conversely, show that *any* symmetric, positive-definite linear transformation P can be factorized as $P = M^T M$, in infinitely many ways.

(vii) Building on (vi), show that just as any positive number has a real square root, so any positive-definite linear transformation P has a square root Q, such that $P = Q^2$. In greater detail, show that $Q = R^{-1} D R$, where R is orthogonal, and [D] is the diagonal matrix whose entries are the square roots of the eigenvalues of P.

(viii) A symmetric linear transformation S is called ***positive semidefinite*** if $\mathbf{x} \bullet (S\mathbf{x}) \geqslant 0$, for all $\mathbf{x}$. If M is any linear transformation of $\mathbb{R}^3$, show that $M = RS$, where R is orthogonal, and S is symmetric and positive semidefinite. This is called the ***Polar Decomposition*** of M.

13. Vanishing Shape Operator $\Longleftrightarrow$ Flat. Show, both geometrically and by calculation, that if the Shape Operator vanishes identically, then the surface is a portion of a plane.

14. Gaussian Curvature in terms of the Shape Operator. If $\mathbf{u}$ and $\mathbf{v}$ are short tangent vectors at a point p on a surface with Shape Operator S, show geometrically that $S(\mathbf{u}) \times S(\mathbf{v}) = \mathcal{K}(p)\ [\mathbf{u} \times \mathbf{v}]$. (NOTE: The equation itself is valid even if the tangent vectors are *not* short, but the geometric *explanation* relies on thinking in terms of ultimate equalities.)

15. Cartesian Formulas for the Shape Operator, Curvature, and Mean Curvature.
Let T_p be the tangent plane of a general surface $\mathcal{S}$ at a point p. Introduce Cartesian coordinates (x, y) into T_p, and take $p = (0, 0)$, so that the normal $\mathbf{n}$ there points along the z-axis. Then, locally, $\mathcal{S}$ is given by

$$z = f(x, y) \quad \text{where} \qquad f(0,0) = 0 \quad \text{and} \quad \partial_x f = 0 = \partial_y f \ \text{ at } (0,0).$$

(i) Note that if we define $-F(x, y, z) \equiv f(x, y) - z$, then F is constant on $\mathcal{S}$. Deduce that ∇F is normal to $\mathcal{S}$, and therefore,

$$-\mathbf{n} = \frac{1}{\sqrt{1 + (\partial_x f)^2 + (\partial_y f)^2}} \begin{pmatrix} \partial_x f \\ \partial_y f \\ -1 \end{pmatrix}.$$

(ii) Deduce that the matrix of the Shape Operator at $(0,0)$ is

$$[S] = \begin{bmatrix} \partial_x^2 f & \partial_y \partial_x f \\ \partial_x \partial_y f & \partial_y^2 f \end{bmatrix}, \tag{20.2}$$

from which it follows immediately that

$$\mathcal{K} = \kappa_1 \kappa_2 = \det[S] = (\partial_x^2 f)(\partial_y^2 f) - (\partial_x \partial_y f)^2. \tag{20.3}$$

Likewise, $\overline{\kappa} \equiv$ ***mean curvature*** $= \left[\frac{\kappa_1 + \kappa_2}{2}\right] = \frac{1}{2}\mathrm{Tr}\,[S] = \frac{1}{2}\left[\partial_x^2 f + \partial_y^2 f\right]$, which may be written

$$\overline{\kappa} = \tfrac{1}{2}\nabla^2 f,$$

where $\nabla^2 = \partial_x^2 + \partial_y^2$ is the ***Laplacian operator***, first encountered on page 41.

16. Nonprincipal Coordinates. Let us reanalyze the saddle surface, this time *without* first aligning the coordinate axes with the principal directions, as we did in the text.

(i) Sketch the surface $z = xy$ near the origin. (*Hint*: Recalling that $\sin 2\theta = 2\sin\theta\cos\theta$, find the height of the surface above and below the points of the origin-centred circle of radius r, for which $x = r\cos\theta$ and $y = r\sin\theta$.)

(ii) Use (10.4), p. 111, to identity the principal directions.

(iii) Use (20.2) to deduce that the Shape Operator has the matrix

$$[S] = \begin{bmatrix} 0 & 1 \\ 1 & 0 \end{bmatrix},$$

and deduce that $\mathcal{K} = -1$ and $\overline{\kappa} = 0$.

(iv) What is the geometric transformation S represented by $[S]$?

(v) Use (iv) to determine the eigenvectors and eigenvalues of S, geometrically (*not* by calculation). (NOTE: The eigenvectors should point in the same directions you found in (ii).)

(vi) Guided by either (ii) or (v), rotate the (x,y)-axes to obtain new (X,Y)-axes aligned with the principal directions, and show that the original equation now becomes, $z = \frac{1}{2}(X^2 - Y^2)$.

(vii) By comparing the new equation in (vi) with (10.6), page 111, confirm the values of $\mathcal{K}$ and $\overline{\kappa}$ you found in (iii).

(viii) Use (20.2) to write down $[S]$ in the new (X,Y)-coordinate system of (vi), and confirm that $\mathcal{K}$ and $\overline{\kappa}$ have not changed.

(ix) Verify that the two different matrices $[S]$ in (iii) and (viii) (representing the same linear transformation S) conform to the general matrix formula (15.20), page 161.

17. Use (20.2) and (20.3) to verify your conclusions in Exercise 7.

18. Curvature Formula in Polar Coordinates. If $z=f(r,\theta)$, show that the curvature is given by this disappointingly complicated formula:

$$\mathcal{K}=\frac{r^2\,\partial_r^2 f\,(\partial_\theta^2 f+r\,\partial_r f)-[\partial_\theta f-r\,\partial_r\partial_\theta f]^2}{\left\{r^2\left[1+(\partial_r f)^2\right]+(\partial_\theta f)^2\right\}^2} \tag{20.4}$$

19. Nonisometric Surfaces with Equal Curvatures. As we discussed in Act I, Minding proved that if two surfaces have the same *constant* $\mathcal{K}$, then they are locally isometric to each other. Thus, according as this constant curvature $\mathcal{K}>0$, $\mathcal{K}=0$, $\mathcal{K}<0$, the surface is locally isometric to a sphere, plane, or pseudosphere, respectively. We now provide an example to show that surfaces of equal but *variable* curvature need not be isometric. Thus the converse of the *Theorema Egregium* is false.

(i) Using the same notation as in the previous exercise, find the shapes of these two surfaces: $\mathcal{S}_1$ with $f_1(r,\theta)=\ln r$, and $\mathcal{S}_2$ with $f_2(r,\theta)=\theta$. (*Hint*: $\mathcal{S}_2$ is called the ***helicoid***.) Check your answers by using a computer to draw them.

(ii) Let us say that a point in $\mathcal{S}_1$ and a point in $\mathcal{S}_2$ "correspond" if they have the same (r,θ)-coordinates. By finding the formulas for the metrics of both surfaces, show that this correspondence is *not* an isometry.

(iii) Nevertheless, use (20.4) to deduce that corresponding points of the two surfaces have the *same* curvature!

$$\mathcal{K}_1(r,\theta)=\frac{-1}{(r^2+1)^2}=\mathcal{K}_2(r,\theta).$$

20. Curvature of the *n*-Saddle. Given that the height of the n-saddle is $z=f(r,\theta)=r^n\cos n\theta$, use (20.4) to prove that its (negative) curvature is constant on each circle $r=$*const.*, and is given by the following formulas.

(i) The common 2-saddle (surrounding a generic hyperbolic point) has

$$\mathcal{K}=-\frac{4}{(1+4r^2)^2}.$$

(ii) The monkey 3-saddle has

$$\mathcal{K}=-\frac{36r^2}{(1+9r^4)^2}.$$

(iii) The n-saddle has

$$\mathcal{K}=-\frac{(n-1)^2n^2r^{2n}}{(r^2+n^2r^{2n})^2}.$$

Introduction to the Global Gauss–Bonnet Theorem

21. Genus-Dependent Predictions of GGB.

(i) Use GGB to deduce that any topological sphere *must* have regions of positive curvature.

(ii) Use GGB to deduce that any topological doughnut *must* have regions of positive curvature.

(iii) In fact *every* smooth closed surface $\mathcal{S}_g$ with $g\geqslant 2$ must also have a point of positive curvature, though GGB alone no longer guarantees this. Let $S(r)$ be a sphere of radius r centred at an arbitrary point inside $\mathcal{S}_g$. Imagine we start with a sufficiently large value of

r that $S(r)$ contains $\mathcal{S}_g$. Now shrink the sphere until it makes first contact with $\mathcal{S}_g$, when $r = R$, say. Argue that at the point of contact with $S(R)$, $\mathcal{K} > (1/R^2) > 0$.
Hint: Consider the principal directions at the point of contact.

22. Holy Broken Gauss–Bonnet, Batman! Recall our definition of a ***wormhole*** in [16.7], and also recall that it has total curvature, $\mathcal{K} = -4\pi$.

(i) If n small bagels are placed at the vertices of a large regular n-gon, and are joined together with wormhole connectors along the edges of the n-gon into a surface $\mathcal{S}$, explain why $\mathcal{K}(\mathcal{S}) = -4\pi n$. Why is this *not* a violation of GGB, as it appears to be?

(ii) Suppose $n = 4$, so that the bagels are at the corners of a square, joined along the four edges. Now add a *fifth* wormhole connector along one of the diagonals, connecting two previously disconnected bagels. What is the total curvature now, and why is this *not* a violation of GGB, as it appears to be?

First (Heuristic) Proof of GGB

23. Hopf's Umlaufsatz. Let us attempt to reduce Hopf's (1935) ingenious argument to its essence. Imagine that the (smooth and "simple") closed curve C of the theorem is a convoluted hiking/running path snaking through sand dunes. See map below. The entrance is at the southern-most point (downward on the page); the rest of C lies to the north (upwards). As illustrated, let $\Theta(t)$ be the angle of the line $L(t)$ connecting you (◎) and your friend, Mary (⊛), at time t. Initially, Mary is standing next to you, so $L(0)$ (dashed line) is horizontal, with $\Theta(0) = 0$.

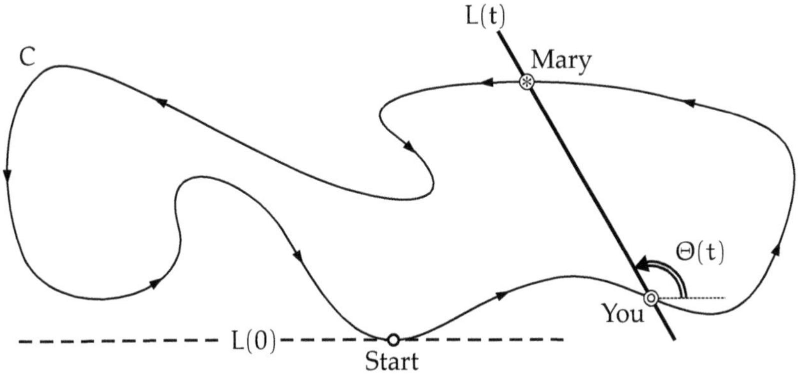

(i) Mary decides to do a warm-up lap, but you stay at the start and just watch her run. As she runs the length of C, counterclockwise, you turn your head to follow. Explain why your line of sight $L(t)$ must execute a *net* rotation of π by the time Mary returns to you: $\delta\Theta = \pi$. Next, Mary stays put and rests while *you* run a lap, so L rotates another π. Thus, after you and Mary have each run one lap, L has executed one full revolution: $\delta\Theta = 2\pi$.

(ii) Next, Mary runs another lap, but this time you don't wait till she finishes before you start to move. You start walking *very slowly* along the track behind her. You only cover a small distance before Mary has returned to the start line. Once she's there, you jog along the remainder of the path back to her. Explain why the net rotation of L must *still* be $\delta\Theta = 2\pi$.

(iii) Next, you decide to race Mary, despite the fact that she *always* wins. Sure enough, you never even get close. Explain why the net rotation of L after you and Mary have each completed your lap is *still* $\delta\Theta = 2\pi$. (*Hint*: Imagine *gradually* increasing your speed in (ii), and picture the *continuous* evolution of the graph of $\Theta(t)$: (i) $\rightsquigarrow$ (ii) $\rightsquigarrow$ (iii).)

(iv) You decide to have a final race, and this time you give it all you've got, and you succeed in staying right on Mary's heels the whole time. As you stare at her back, eyes front along the path, your line of sight L is now the *tangent line* to C. And since the net rotation of L must *still* be $\delta\Theta = 2\pi$, Hopf's proof is complete!

Second (Angular Excess) Proof of GGB

24. There Are Only Five Regular Polyhedra. If each face of a regular polyhedron is a regular n-gon, and m such faces meet at each vertex, use the following steps to show that the *only* possibilities are the five Platonic solids illustrated in [18.2], namely, $(n, m) = (3,3)$ or $(4,3)$ or $(3,4)$ or $(5,3)$ or $(3,5)$.

(i) Explain why $m \geqslant 3$.

(ii) If θ_n is the internal angle of the regular n-gon face, explain why $m\theta_n < 2\pi$. (*Hint*: Imagine cutting out the m faces that meet at a vertex v, now cut along one of the edges containing v, and finally flatten out the surface onto the plane.)

(iii) Write down θ_n for $n = 3, 4, 5, 6$ and exhaust the possibilities.

25. There Are Only Five *Topologically* Regular Polyhedra. Let us use the same (n, m) notation as in the previous exercise, only now the faces of our topological polyhedron can be curved, irregular, and/or with wavy edges. As usual, let V, E, and F denote the number of vertices, edges, and faces.

(i) Explain why $E = Fn/2$.

(ii) Explain why $V = Fn/m$.

(iii) By substituting the previous two equations into Euler's Polyhedral Formula, deduce that

$$F = \frac{4m}{2(m+n) - mn}.$$

(iv) Deduce that $\Omega(n, m) \equiv 2(m+n) - mn > 0$.

(v) Think of (n, m) as grid points in $\mathbb{R}^2$. In the relevant region ($n \geqslant 3$ and $m \geqslant 3$), mark each grid point with the value $\Omega(m, n)$, noting and using the symmetry about the line $n = m$.

(vi) Deduce that the only solutions (n, m) are the same ones we found in the previous exercise: the five Platonic solids, but now topologically deformed.

This result is due to Simon Antoine Jean L'Huilier (1811). See Robin Wilson in James 1999, page 516.

26. Legendre's Projection of Convex Polyhedra. As noted in the text, Legendre's proof of Euler's Polyhedral Formula assumed that the polyhedron was *convex*. The proof we presented did *not* require this assumption, but Legendre's assumption of convexity gave him an elegant advantage: it allowed him to instantly jump from the original polyhedron $\mathcal{P}$ to a topologically equivalent *geodesic* polygonal partition of the sphere (see [18.7]), as follows. Imagine that the convex polyhedron $\mathcal{P}$ is a wire frame, and a light bulb B is placed somewhere inside it. If $\mathcal{S}$ is a sphere centred at B and enclosing $\mathcal{P}$, prove that the *shadow* of the frame on the sphere is the desired *geodesic* partition.

27. **Hopf's Proof That $\chi(\mathcal{S}_g)=2-2g$.** In the text, we proved $\chi(\mathcal{S}_g)=2-2g$ by first showing that $\chi(\mathbb{S}^2)=2$, and then showing that gluing on a "Toblerone® handle" (see [18.9]) reduces χ by 2 (see [18.10]). Here is an elegant alternative argument, due to Hopf (1956, pp. 8–10).

    (i) Glue together two Toblerone® handles to create a topological torus, $\mathcal{S}_1$, and deduce that $\chi(\mathcal{S}_1)=0$.

    (ii) Remove one triangle from this torus, and remove one triangle from a general surface $\mathcal{S}_g$ of genus g. Now glue the edges of the two triangular holes to each other, thereby creating a new closed surface $\mathcal{S}_{g+1}$ of one higher genus. Deduce that $\chi(\mathcal{S}_{g+1})=\chi(\mathcal{S}_g)-2$.

    (iii) Use (i) and (ii) to deduce that $\chi(\mathbb{S}^2)=2$, and that $\chi(\mathcal{S}_g)=2-2g$.

28. **Dual Polyhedra.** Place a vertex in the centre of each face of a cube, and join them together to create an octahedron; this is the ***dual*** of the cube, each face having become a vertex, and each vertex having become a face. What if we keep going, by taking the dual of the octahedron, i.e., the dual of the dual of the cube? With the assistance of [18.2], repeat this exercise for the three remaining Platonic solids.

Third (Vector Field) Proof of GGB

29. **Dipole Streamlines Are Circular.** Prove that the dipole streamlines in [19.3] are *circles*,

    (i) by calculation

    (ii) geometrically.

30. **Honey-Flow on $\mathcal{S}_g$.** Take $\mathcal{S}_g$ in [19.11] and rotate it by a right angle about an axis perpendicular to the page, so that the holes are aligned horizontally, instead of vertically. Sketch the new honey-flow, and deduce that there are a total of $6g-2$ singular points. Confirm that the sum of their indices is indeed $\chi=2-2g$, in accord with the Poincaré–Hopf Theorem.

31. **Existence of the Stiefel Vector Field.** Explain, as explicitly as possible, *why* the construction in [19.12] is guaranteed to create a consistent vector field on $\mathcal{S}_g$, regardless of the specific partition into polygons.

32. **A Vector Field on $\mathcal{S}_g$ with *One* Singular Point?** The dipole field on the sphere [19.8b] has only one singular point, and its index is 2 ($=\chi$), in accord with the Poincaré–Hopf Theorem. On a torus ($g=1$) we have seen we need not have any singular points in the flow. But if $g\geqslant 2$, there must be singular points, for the sum of their indices is $\chi\neq 0$. In this general case, is it always possible to generalize the dipole field on the sphere, to find a flow with only *one* singular point of index χ? (*Hint*: A new field can be created by coalescing two or more of its singular points into one. For example, we can imagine creating the dipole field in [19.8b] by coalescing the two vortices in [19.8a].)

33. **Vector Fields on $\mathcal{S}_g$ with $-\chi$ Saddle Points.** Show (by drawing it!) that it is possible to construct a vector field on $\mathcal{S}_g$ (with $g\geqslant 2$) such that there are precisely $(2g-2)=-\chi$ singular points, each with $\mathcal{I}=-1$. (*Hints*: (A) First tackle $g=2$; the generalization to higher genus turns out to be obvious. (B) Picture $\mathcal{S}_2$ as two bagels still joined together. Next, imagine this surface crushed almost flat under a great weight, so that it resembles two joined annuluses. Finally, confirm (by drawing it!) that it is possible to place one saddle point on the "top" surface, and one on the "bottom," *and* to have the top and bottom fields agree where they meet, at the edges.)

ACT IV

Parallel Transport

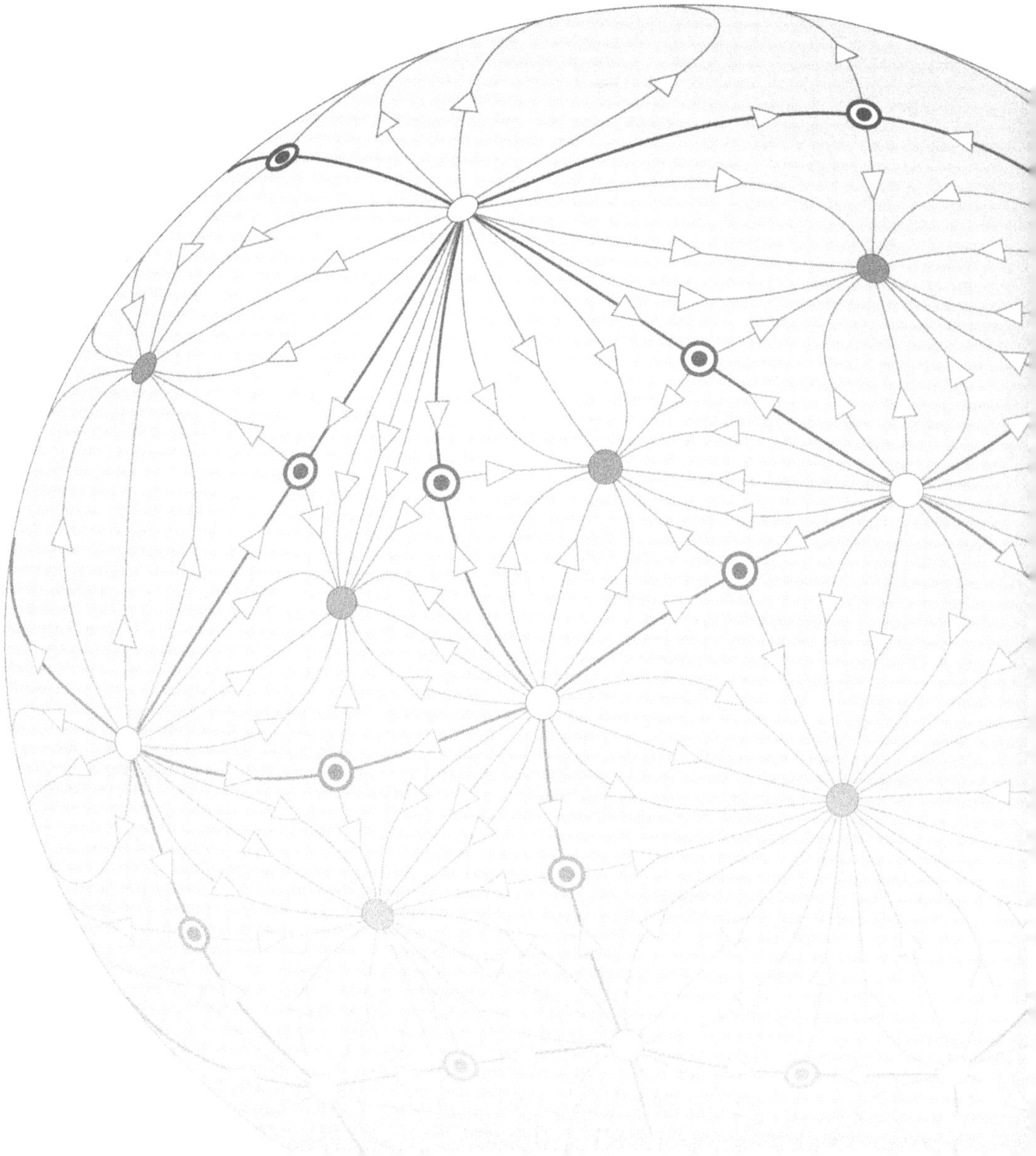

Chapter 21

An Historical Puzzle

Parallel transport[1] is now understood to be one of the most fundamental and powerful concepts of Differential Geometry, but it was remarkably late to the game.

By 1915 most of the fundamental concepts and computational tools of modern Differential Geometry in n-dimensional space were already in place, and not a moment too soon.

Einstein had spent the previous *decade* struggling to reconcile his 1905 Special Theory of Relativity with gravity. He had known since 1905 that no physical effect could travel faster than light, but Newton's Inverse-Square Law—unquestioned since 1687—insisted that if a giant solar flare erupted from the surface of the Sun, its tiny gravitational tug on the Earth would be felt *instantly*, despite the fact that the light from the flare would take eight minutes to reach Earth—a naked contradiction!

Only very gradually did Einstein's physical intuition drive him towards Differential Geometry, where he was miraculously blessed to find that in 1901 Gregorio Ricci (1853–1925) and Tullio Levi-Civita (1873–1941) had jointly published *"the marvelously prepared and almost predestined instrument for the exposition of his theory"*[2]—that instrument was ***Tensor Calculus***.[3]

On the 25th of November, 1915, Einstein finally ended his terrible, decade-long struggle, harnessing Tensor Calculus to write down his famous ***Field Equation*** that reconciled gravity with Special Relativity; he christened this union the ***General Theory of Relativity***. Einstein had succeeded in understanding the true nature of gravity: it is the ***Riemannian curvature***[4] impressed upon 4-dimensional spacetime by matter and energy; free-falling particles then respond to the gravitational field by travelling through curved spacetime along *geodesics*.[5]

As of the time of this writing, in 2020, every single testable prediction of the theory has been confirmed, including the astonishing, Nobel Prize–winning[6] confirmation—on the 14th of September, 2015—that *gravitational waves exist*, which Einstein had predicted almost exactly a century earlier, in 1916!

Little appreciated is the fact that in several instances these experimental confirmations have attained a breathtaking degree of accuracy, rivalling or exceeding those of the former gold standard: Quantum Electrodynamics. Indeed, our daily use of Global Positioning System (GPS) technology[7] would simply be impossible without taking into account the *precise* time-warping effect of gravity predicted by General Relativity! Einstein's 1915 discovery therefore stands as

[1]Sometimes instead referred to as "parallel displacement" or "parallel propagation."

[2]Levi-Civita (1931). For the history of the slow acceptance of Tensor Calculus before 1915, and the dramatic change post-1915, see Bottazzini (1999).

[3]This purely mathematical discovery was originally called the ***Absolute Differential Calculus***, but later became known as ***Ricci Calculus***, and finally as ***Tensor Calculus***. Élie Cartan's more powerful and elegant version of Tensor Calculus constitutes the dénouement of our drama (Act V) and is called the ***Exterior Calculus of Differential Forms***.

[4]This intrinsic measure of curvature will be described at the conclusion of Act IV: it generalizes Gaussian curvature to n-dimensional spaces.

[5]Though geodesics now *maximize* the "distance" measured using the metric of curved spacetime, which generalizes the Minkowski spacetime interval $\beth$; see (6.15), page 75.

[6]The 2017 Nobel Prize in Physics was awarded to Rainer Weiss, Barry C. Barish, and Kip S. Thorne for their joint construction of the Laser Interferometer Gravitational-Wave Observatory (LIGO) detector, which made the discovery possible.

[7]See Taylor and Wheeler (2000, A-1).

not only a supremely beautiful scientific triumph of the human intellect, but also as one of the *best-tested* physical theories we have.

But there is a serious puzzle[8] here, one that is not widely known or recognized. Einstein's success was all the more remarkable, and remains all the more *puzzling*, because he achieved it *before* Levi-Civita—pictured in [21.1]—discovered[9] the concept of ***parallel transport***, which did not occur until *1917!* Without this concept, I personally do not know of any way to make complete geometrical sense of Einstein's 1915 discovery.

[21.1] *Tullio Levi-Civita (1873–1941)*

So what *is* parallel transport? Like the concept of the geodesic, it is a queer amphibian, equally at home in the world of extrinsic geometry and in the world of intrinsic geometry, and we shall now describe what it looks like in each of these worlds, and attempt to understand how it is able to straddle the two.

[8]For readers familiar with General Relativity, there is in fact a second, even more spectacular puzzle: in November 1915, *Einstein did not know the Differential Bianchi Identity*—stated later as (29.17)—and therefore he did not realize that energy–momentum conservation followed automatically from his law! See Pais (1982, p. 256).

[9]Levi-Civita (1917).

Chapter 22

Extrinsic Constructions

22.1 Project into the Surface as You Go!

Here is the heuristic idea of what Levi-Civita sought to achieve in his fundamental paper of 1917. To ***parallel transport*** a tangent vector $\mathbf{w}$ to a surface $\mathcal{S}$ along a curve K connecting two points a to b within $\mathcal{S}$, we want to always keep $\mathbf{w}$ pointing in the same direction (and having the same length) *while remaining tangent to* $\mathcal{S}$ as it moves along K, the direction of $\mathbf{w}$ at each moment always being parallel to its direction a moment before.

But this is seemingly *impossible!* If we move even a small distance ϵ from p to q along K, then the vector $\mathbf{w}$ in the tangent plane T_p, when rigidly moved parallel to itself (in $\mathbb{R}^3$) to q, will generally *not* lie in T_q, but will instead stick out of $\mathcal{S}$.

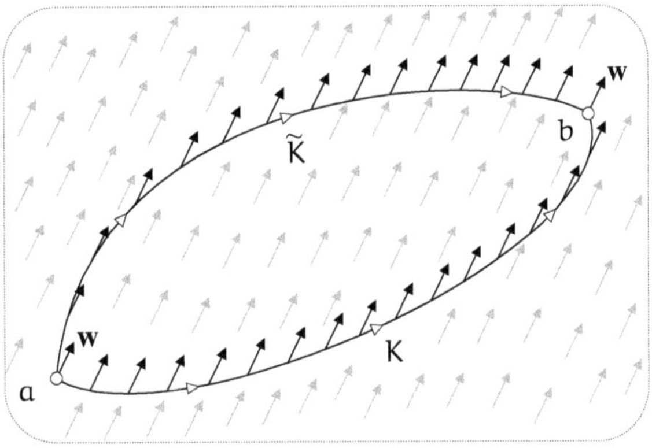

[22.1] *In the Euclidean plane there is a* global *concept of parallelism, manifested by the flow with constant velocity vector* $\mathbf{w}$. *To parallel transport* $\mathbf{w}$ *along* K, *we need only restrict this vector field to* K. *Thus we always obtain the* same *vector* $\mathbf{w}$ *at* b, *whether we travel along* K *or* $\tilde{K}$. *This path-independence holds* only *in this case of vanishing curvature.*

In the Euclidean plane no such problem arises, and there is a simple *global* sense of parallelism, resulting from the fact that through any point not lying on a line L (in the direction $\mathbf{w}$), there is a unique line that is parallel to L, in the same direction $\mathbf{w}$. This global parallelism can be pictured as a uniform flow across the plane, with velocity $\mathbf{w}$; see [22.1]. The parallel-transported vector along K is simply the restriction to K of this constant velocity field. It is therefore trivially clear that we always obtain the *same* parallel-transported vector $\mathbf{w}$ at b, regardless of whether we get there by travelling along K, or along some other route $\tilde{K}$.

Levi-Civita's key observation was that this path-independence of the parallel-transported vector *only* occurs if the space is flat; if the space is curved then parallel transport (as it will be defined momentarily) along two different paths will yield *different* final vectors. This crucial phenomenon is called ***holonomy***, and is the subject of an upcoming chapter by that name. As we shall see, this holonomy can be used to *measure* curvature, and not only the Gaussian curvature of surfaces, but also the Riemannian curvature of higher-dimensional curved spaces, like Einstein's curved, 4-dimensional spacetime.

If we try to repeat the Euclidean construction in the hyperbolic plane, $\mathbb{H}^2$, then we immediately run into trouble, for if we are given a point and a line L though it, we know that there are now *infinitely many* "parallel" lines running through a neighbouring point; which one should we choose?!

Later we shall return to this topic of how to carry out parallel transport in $\mathbb{H}^2$, or equivalently on the pseudosphere, but let us now immediately jump in at the deep end and see how to do parallel transport on a *general* curved surface.

Our first extrinsic method of parallel transport is illustrated in [22.2], using the notation that was introduced in the opening paragraph, above. If we move the tangent vector $\mathbf{w}$ at p parallel to itself (as a vector in $\mathbb{R}^3$) a small (ultimately vanishing) distance ϵ along the curve K, then $\mathbf{w}$ is no longer tangent to $\mathcal{S}$ at q: it sticks out of the tangent plane T_q, but only slightly, at angle $\Theta(\epsilon)$. The best approximation to $\mathbf{w}$ that *is* tangent to $\mathcal{S}$ at q is $\mathbf{w}_{||}$, obtained by projecting $\mathbf{w}$ orthogonally down onto T_q. If $\mathcal{P}$ denotes this projection, and $\mathbf{n}(q)$ is the unit normal at q, then, as explained by [22.2],

$$\mathbf{w}_{||} = \mathcal{P}[\mathbf{w}] = \mathbf{w} - (\mathbf{w} \bullet \mathbf{n})\mathbf{n}. \tag{22.1}$$

To parallel transport $\mathbf{w}$ along K, we must imagine breaking K into an enormous number of tiny steps of length ϵ, repeating this process of translating in $\mathbb{R}^3$ and projecting back into the surface, over and over, one ϵ-step at a time, until we finally arrive at the end of K. Even then, this is merely an *approximation* of parallel transport—perfect parallel transport is only achieved when we take the *limit* that ϵ vanishes, in which case we *continually project into the surface as we go.*

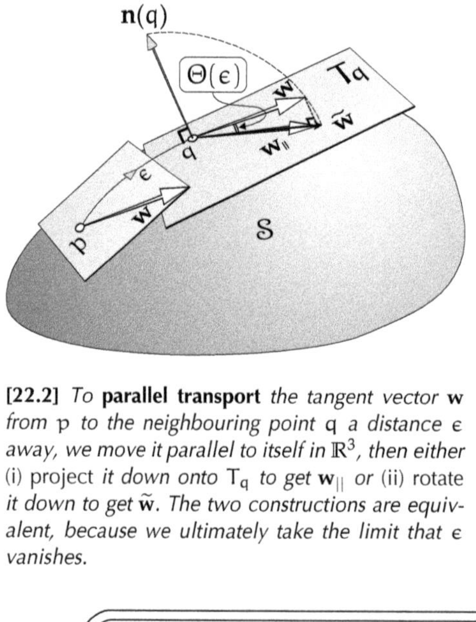

[22.2] *To* **parallel transport** *the tangent vector* $\mathbf{w}$ *from* p *to the neighbouring point* q *a distance* ϵ *away, we move it parallel to itself in* $\mathbb{R}^3$*, then either* (i) project *it down onto* T_q *to get* $\mathbf{w}_{||}$ *or* (ii) rotate *it down to get* $\widetilde{\mathbf{w}}$*. The two constructions are equivalent, because we ultimately take the limit that* ϵ *vanishes.*

Note that the projection $\mathbf{w}_{||}$ in [22.2] is slightly shorter than the original vector $\mathbf{w}$. If we instead simply *rotate* $\mathbf{w}$ straight down onto T_q then we obtain $\widetilde{\mathbf{w}}$, pointing in the same direction as $\mathbf{w}_{||}$ but with $|\widetilde{\mathbf{w}}| = |\mathbf{w}|$. (The length $|\mathbf{w}|$ is irrelevant to the construction, but here we have chosen it to have the same unit length as $\mathbf{n}$, for ease of visualization.) But, since $\lim_{\epsilon \to 0} \Theta(\epsilon) = 0$, we see that

$$\boxed{\mathbf{w}_{||} \asymp \widetilde{\mathbf{w}}.}$$

Finally, because perfect parallel transport takes this limit of vanishing ϵ, we deduce that *parallel transport preserves length.*

We have thus arrived at two *equivalent* extrinsic methods of parallel transport:

> *To* **parallel transport** *the tangent vector* $\mathbf{w}$ *to* $\mathcal{S}$ *along a curve* K*, move* $\mathbf{w}$ *parallel to itself in* $\mathbb{R}^3$ *while continually either* (i) projecting *into* $\mathcal{S}$ *as you go or* (ii) rotating *down to* $\mathcal{S}$ *as you go. The length of the vector remains constant as it is parallel transported.* (22.2)

Method (ii) is much easier in practice, and we strongly encourage you to try it out yourself on any curved object you have at hand—fruits and vegetables are convenient, and they have a crucial advantage that will soon become apparent: they can be peeled!

Draw any curve K you please on your surface, connecting point a to point b, press one end of a toothpick down against the surface at a, in any tangent direction you like. Now move it parallel to itself (in space) a small distance along K, then press that same end straight down against the surface (so that it becomes tangent again), ... repeat, ... repeat, ... repeat, ... until you arrive at b!

Figure [22.3] illustrates this on a pomelo. Although we may choose any direction at a that is tangent to $\mathcal{S}$, here we have chosen $\mathbf{w}$ to be the initial velocity along K, in order to make this point:

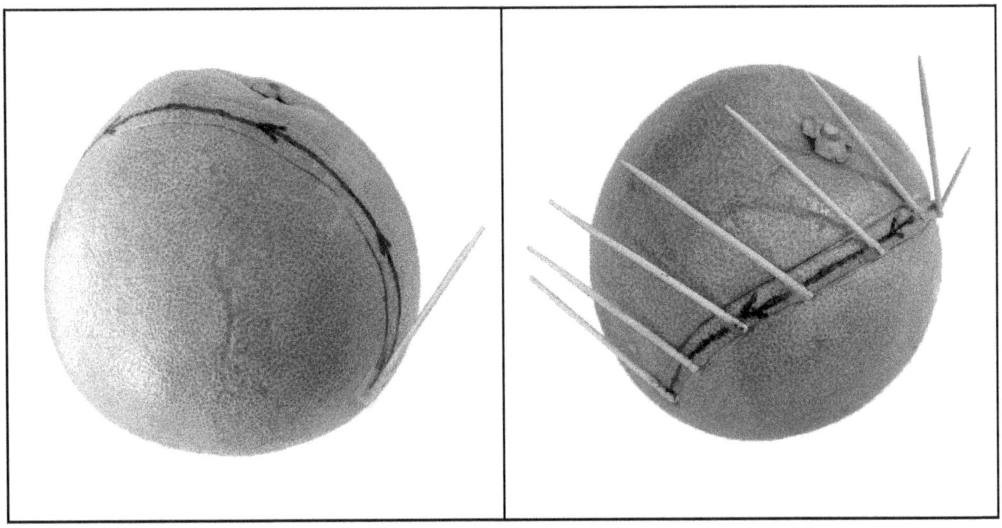

[22.3] *Parallel transport on a pomelo: continually press the toothpick straight down onto the surface while moving the toothpick parallel to itself along the curve.*

in general, $\mathbf{w}_{||}$ *does not remain tangent* to K, though it certainly *does* remain tangent to the surface, by construction.

Here we have chosen the path K so as to be close to a vertical section of the surface of the pomelo, and in the photograph on the right we have deliberately taken the picture from almost directly overhead, so that it *looks* remarkably straight. It is therefore perhaps very surprising and confusing how fast the toothpick rotates from initially being tangent to K, to being almost perpendicular to K by the end. This because our photograph does not reveal just how *curved* K is with respect to the *intrinsic* geometry. The intrinsic, geodesic curvature of this particular path K will be revealed in Section 22.3, where a different method of parallel transport will be revealed.

22.2 Geodesics and Parallel Transport

Under what circumstances *would* parallel transport of the initial tangent vector along a curve G *maintain* tangency to G? The answer is that this happens precisely when G is a *geodesic*!

To see why, recall that the local definition of a geodesic G is as a unit-speed curve for which the acceleration is always directed along the surface normal, $\mathbf{n}$. That means that if we look at the velocity at neighbouring points p and q, distance/time ϵ apart, then the difference of the two velocities is ultimately $\dot{\mathbf{v}}\epsilon$ and is ultimately directed along $\mathbf{n}$. Figure [22.4] depicts this new, special case of [22.2], now with $\mathbf{v}$ as the velocity, pointing along the segment ϵ. We see that this means that the new velocity at q is ultimately the original velocity $\mathbf{v}$ at p, *parallel-transported along itself.*

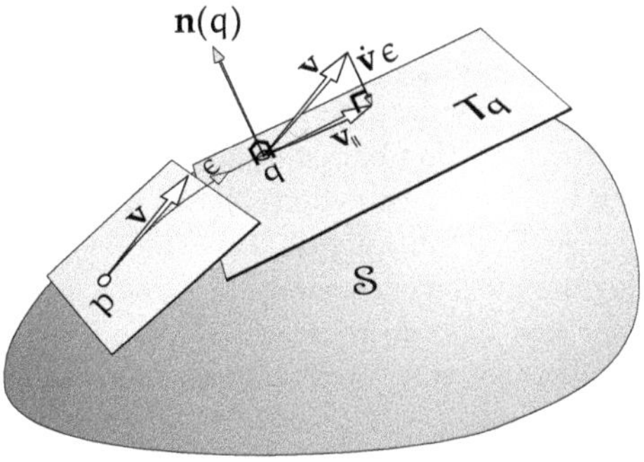

[22.4] Constructing a Geodesic via Parallel Transport. *Launch a particle from* p *with velocity* $\mathbf{v}$, *and define* G *to be the curve obtained by parallel-transporting* $\mathbf{v}$ along itself, *so that the new velocity at the neighbouring point* q *(distance/time* ϵ *away) is* $\mathbf{v}_{||}$. *Then the acceleration* $\dot{\mathbf{v}}$ *is given by* $\dot{\mathbf{v}}\epsilon \asymp \mathbf{v}_{||} - \mathbf{v}$. *But then* $\dot{\mathbf{v}} \propto \mathbf{n}$, *and therefore* G *is geodesic.*

Conversely, suppose we are given a point p and that we launch a particle from there, out

[22.5] **[a]** *A geodesic segment G on a yellow squash is first constructed by stretching a string across its surface.* **[b]** *The string is removed, but a toothpick is left at the initial point, tangent to G. As predicted, parallel transporting the initial toothpick along itself yields toothpicks that are everywhere tangent to the* same *geodesic curve G.*

across the surface, in an arbitrary direction **v** within $\mathcal{S}$. The same geometry as before tells us that we may *construct* the geodesic motion that ensues by performing parallel transport: carry **v** a short distance ϵ within $\mathcal{S}$ along itself, then press it down onto the surface to obtain the new velocity there . . . repeat!

Figure [22.5] illustrates this new method of constructing geodesics. In order to check the correctness of our ultimate solution, we first construct a geodesic segment G on our surface (a squash) using our original trick of stretching a string across it, here tied to two (normal) toothpicks at the ends. If we remove the string, but leave behind an initial toothpick tangent to the start of the string, we can then take this initial-velocity toothpick and parallel transport it along itself. As you see, and as we urge you to try for yourself, this construction yields the *same* geodesic segment G as the string did.

22.3 Potato-Peeler Transport

We will now describe a third extrinsic method of parallel transport that will quickly prove its mettle,[1] yielding important *properties* of parallel transport that are not immediately evident from our first two constructions seen in (22.2).

In [22.2] imagine that we use a potato peeler to peel off a narrow strip of $\mathcal{S}$ surrounding the movement ϵ along K. Now, instead of rotating **v** down to the surface, imagine that we take this short strip of peel and bend it up, pressing it flat onto T_p, bringing $\mathbf{v}_{||}$ to **v**, instead of the other way around. If we keep peeling a narrow strip all along K, we arrive at our new construction:

[1]This new construction is not readily found in modern textbooks, and I recall being elated when I first hit upon it, more than 30 years ago. Other authors likewise believed they had found something new: e.g., Koenderink (1990), Casey (1996), and Henderson (2013). But when the time came to write this book, I turned to original sources, and there I found that what we had *re*discovered was *first* discovered, more than a century ago, by Levi-Civita himself! To read Levi-Civita's own account of his idea, translated into English, see Levi-Civita (1926, p. 102). For more on the history of the discovery, see Goodstein (2018, Ch. 12).

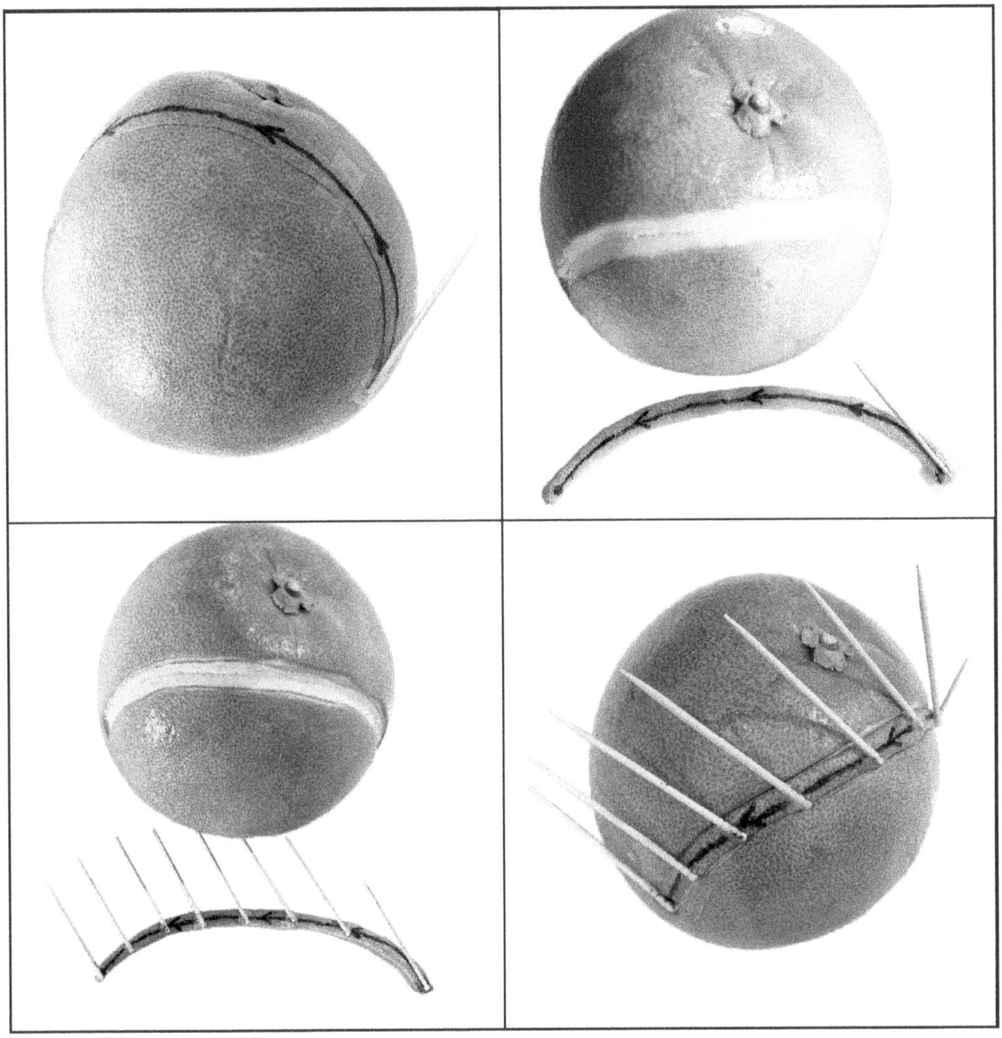

[22.6] "Potato-Peeler" Parallel Transport *on a pomelo: remove a narrow strip of peel along the curve, flatten it out on the table, do Euclidean parallel transport within the flat tabletop, then finally reattach the peel and vectors to the surface.*

> **Potato-Peeler Parallel Transport.** *To* **parallel transport** *the tangent vector* **v** *to* $\mathcal{S}$ *along a curve* K*, peel a narrow (ultimately vanishingly thin) strip of* $\mathcal{S}$ *containing* K *as its centre line. Lay this strip flat in the plane, then do ordinary Euclidean parallel transport of* **v** *along the flattened* K*. Finally, reattach* K *(and the constructed vectors) to* $\mathcal{S}$ *in its original location.* (22.3)

Figure [22.6] reveals how we cheated and actually used *this* method (rather than those in (22.2)) to carry out the parallel transport previously shown in [22.3].

While a potato peeler conjures up the right mental image, it is *not* in fact the best tool for the job. To obtain a tidy, *narrow* strip, take a small sharp knife and make a very shallow incision along a curve just to one side of K, cutting beneath K; now do the same from the other side, thereby cutting out the desired narrow strip of peel, with a shallow V-shaped cross-section.

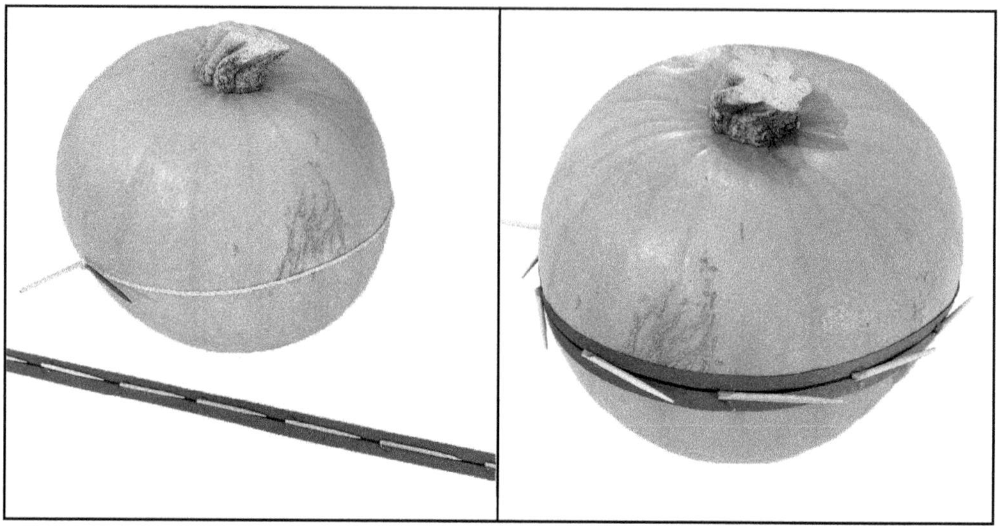

[22.7] [a] *A geodesic segment G on a pumpkin is first constructed by stretching a string across its surface. Meanwhile, on the tabletop, a toothpick pointing along a straight strip of tape is parallel transported along that strip, automatically remaining tangent to the centre line.* **[b]** *The string is removed, and the tape (with attached toothpicks) is unrolled onto the surface, starting at the same point and heading off in the same direction. As predicted, the tape automagically rolls down onto the surface along the* same geodesic *curve G, and the toothpicks are parallel-transported along it.*

Two important properties of parallel transport become immediately clear from the new construction.

First, if two vectors are parallel transported along a curve in the Euclidean plane, then clearly the angle between them remains constant. The new construction immediately implies that the same must therefore be true on a curved surface:

> *If two vectors are parallel transported along a curve on a curved surface, then the angle between them remains constant.* (22.4)

Conversely, if only one of the vectors is known to be parallel-transported, and the angle with the second is kept fixed, then that second vector is necessarily parallel-transported, also.

Second, recall from (1.6), page 13 and from (1.7), page 14, that if a *geodesic* is peeled from the surface and laid flat on the table, it becomes a *straight line*; conversely, if a straight and narrow strip of sticky tape is gradually rolled onto the surface, it will automatically generate a geodesic. In the Euclidean plane, the direction vector of a line continues to point along the line as it is parallel-transported along the line. We thus have a much more intuitive, immediate, *visual* proof the previously established fact that,

> *If G is a geodesic traced at unit speed, launched with initial velocity* **v**, *then the velocity at any future time is obtained by parallel-transporting the initial velocity along G. Conversely, given an arbitrary initial point, and an arbitrary initial velocity* **v** *tangent to $\mathcal{S}$, parallel-transport of* **v** *along* **v** *yields the unique geodesic arising from these initial conditions.* (22.5)

This is illustrated in [22.7]. Again, we urge you to try this for yourself.[2]

Figure [22.8] simply illustrates the same idea again. The strip form [22.7] has been removed from the pumpkin and rolled down onto the surface of the yellow squash shown in [22.5], along the same geodesic as before. But whereas in that figure we were forced to manually carry out the parallel transport and associated generation of the illustrated geodesic, *this* time both things happen automatically.

[22.8] *The same strip shown in [22.7] is here rolled down onto the surface shown in [22.5]. Unlike the manual construction in [22.5], here both the generation of the geodesic and the parallel transport of its velocity vector along it are* automatic.

---

[2]We recommend using masking tape (aka painter's tape) because it comes in bright colours, and once a strip has been created, it can be detached and reattached repeatedly, with ease. A simple way to create narrow strips (from the usually wide roll of tape) is to stick a length of tape down onto a kitchen cutting board, then use a sharp knife to cut down its length, creating strips as narrow as you please. As for the toothpicks, we used a hot glue gun to attach them, but just at their bases, so that they were free to become tangent to the surface once the strip was attached

Chapter 23

Intrinsic Constructions

23.1 Parallel Transport via Geodesics

In the Euclidean plane, here is an easy, obvious way to parallel-transport a vector $\mathbf{w}$ along a straight line L (with direction $\mathbf{v}$): *keep the angle between* $\mathbf{w}_{||}$ *and* $\mathbf{v}$ *constant as it moves along* L. See [23.1a].

Well, on a curved surface the analogue of a line L is a geodesic G, so it is natural to guess that the Euclidean construction generalizes as follows:

> *To parallel transport a tangent vector* $\mathbf{w}$ *to a surface* $\mathcal{S}$ *along a geodesic* G *with velocity* $\mathbf{v}$*, simply keep the angle between* $\mathbf{w}_{||}$ *and* $\mathbf{v}$ constant *as it moves along* G. (23.1)

This *intrinsically* defined method does indeed produce the *same* parallel-transported vector as the three extrinsic methods above, as follows immediately [exercise] by combining (22.4) and (22.5). This is illustrated in [23.2].

Let us repeat our mantra: please try this construction for yourself! Here we have used a stretched string to emphasize the intrinsic nature of this construction, but in practice it is often easier to construct the geodesic with narrow-sticky-tape construction, for the tape will continue to roll down as a geodesic even in parts of the surface that bend towards you, across which a string cannot be stretched, at least not from the outside of the surface.

To (intrinsically) parallel transport $\mathbf{w}$ along an arbitrary curve K in the Euclidean plane, the first step is to approximate K by a sequence $K^{\star}$ of short, straight line segments, $\{L_i\}$, all of which have length less than ϵ, say. See [23.1b]. Keeping the angle constant along each successive L_i, we have parallel transported $\mathbf{w}$ along $K^{\star}$. Finally, taking the limit that ϵ vanishes, $K^{\star}$ becomes K, and we have achieved intrinsic parallel transport along K.

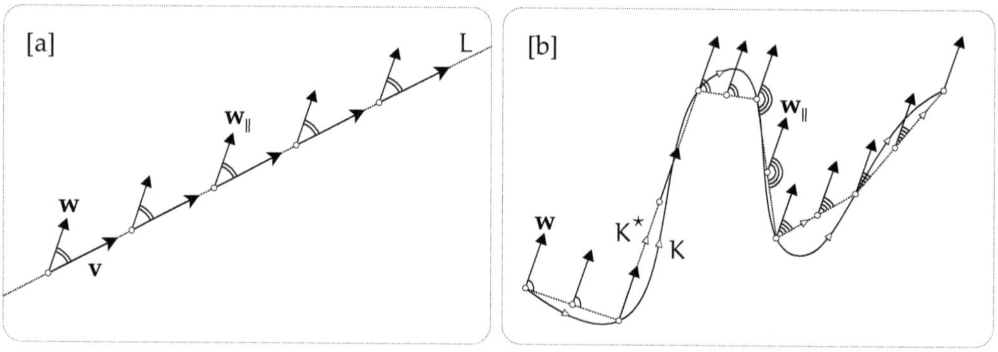

[23.1] **[a]** *In the Euclidean plane, to parallel transport* $\mathbf{w}$ *along the line* L *in the direction* $\mathbf{v}$*, keep the angle between* $\mathbf{w}_{||}$ *and* $\mathbf{v}$ constant. **[b]** *To parallel transport along a general curve* K*, approximate it with a sequence of line segments,* $K^{\star}$*, and keep the angle constant with each one, in succession. Finally, let the length of each segment go to zero, so that* $K^{\star}$ *becomes* K.

Of course this is all somewhat theatrical in the Euclidean plane, where we have a global concept of parellelism, but the point is that it is now clear what we must do on a curved surface:

> *To parallel transport a tangent vector* $\mathbf{w}$ *to a surface* $\mathcal{S}$ *along a general curve* K*, approximate* K *by a sequence* K^* *of geodesic segments* $\{G_i\}$*, each of length less than* ϵ*, then carry* $\mathbf{w}$ *along each* G_i*, maintaining a* constant angle *between the direction of* G_i *and* $\mathbf{w}_{||}$*. Finally, take the limit that* ϵ *vanishes, so that* K^* *becomes* K.

23.2 The Intrinsic (aka, "Covariant") Derivative

When we parallel-transport a vector along a curve K in a surface $\mathcal{S}$, it appears to be *unchanging* to an inhabitant of $\mathcal{S}$ who walks along K watching it. This idea of constancy opens the door to an *intrinsic* measure of how fast a vector *changes* as it moves along a curve.

Suppose a particle moves at unit speed along a curve K on a surface $\mathcal{S}$, its position at time t being $p(t)$, and its unit velocity being $\mathbf{v}(t)$. Furthermore, suppose that emanating from $p(t)$ we have a vector $\mathbf{w}(t)$ that is tangent to $\mathcal{S}$, so that $\mathbf{w}$ is defined everywhere along K (though not necessarily anywhere else). What is the *intrinsic* rate of change of $\mathbf{w}(t)$?

Well, if p moves for a short time ϵ to $q = p(t+\epsilon)$, then the *extrinsically* defined rate of change $\nabla_{\mathbf{v}}\mathbf{w}$ is given by

$$\epsilon\nabla_{\mathbf{v}}\mathbf{w} = \epsilon\mathbf{w}'(t) \asymp \mathbf{w}(q) - \mathbf{w}(p),$$

where the difference of the two vectors on the right makes perfectly good sense in $\mathbb{R}^3$. But $\mathbf{w}(q)$ lives in T_q, while $\mathbf{w}(p)$ lives in T_p, so this difference lives in neither, and so is certainly *not* intrinsic to $\mathcal{S}$.

[23.2] *A geodesic segment* G *on a pumpkin is first constructed by stretching a string across its surface. Any initial tangent vector to the surface can then be parallel transported along* G *simply by keeping the angle between it and* G *constant.*

It is precisely the parallel-transport depicted in [22.2] that allows us to bring $\mathbf{w}(p)$ to q as $\mathbf{w}_{||}(p \rightsquigarrow q)$, keeping it "constant." We can now compare this old value $\mathbf{w}_{||}(p \rightsquigarrow q)$ to the new value $\mathbf{w}(q)$, to see how much it has changed. Let $D_{\mathbf{v}}$ denote this *intrinsic* rate of change (operator) along K, which we shall call the ***intrinsic derivative***, then

$$\epsilon D_{\mathbf{v}}\mathbf{w} \asymp \mathbf{w}(q) - \mathbf{w}_{||}(p \rightsquigarrow q).$$

Since both these vectors lie in T_q, so does their difference. Since $D_{\mathbf{v}}\mathbf{w}$ is tangent to $\mathcal{S}$, it may be viewed as intrinsic to $\mathcal{S}$.

WARNING: The standard name for $D_{\mathbf{v}}$ is the ***covariant derivative***. Historically, the word "covariant" had to do with how the coordinate expression of $D_{\mathbf{v}}$ transforms under a change of coordinates, but since "covariant" means little or nothing to the ears of modern students, we would argue (by example!) that the time has come for a new, *self-explanatory* name. That said, we

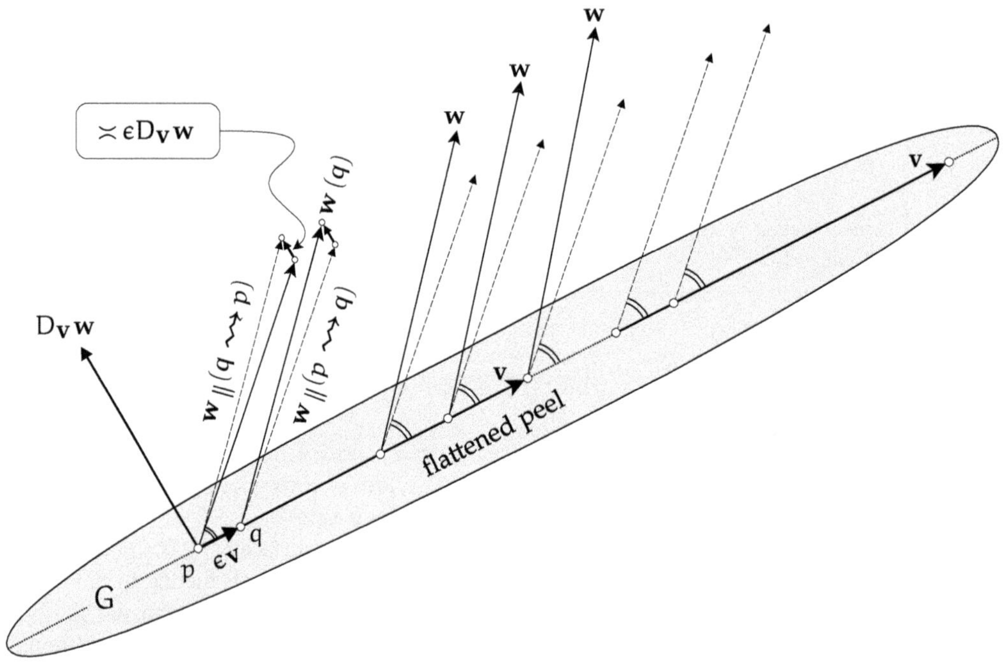

[23.3] *The* **Intrinsic (aka "Covariant") Derivative**, $D_{\mathbf{v}}\mathbf{w}$, *which measures the intrinsic rate of change of* $\mathbf{w}$ *along* $\mathbf{v}$*, is especially easy to visualize if* $\mathbf{v}$ *is the velocity of a* geodesic G*, for then the strip of peel surrounding* G *flattens into a straight line, and parallel transport of* $\mathbf{w}_{||}$ *within the surface becomes ordinary Euclidean parallel transport within the plane.*

would be remiss if we did not warn the reader that (as of this writing in 2019) *essentially every other Differential Geometry (and physics) book instead refers to* $D_{\mathbf{v}}$ *as the* **covariant derivative**. *Finally, we note that this is also called the* **Levi-Civita Connection**.

In order to give a vivid picture of this intrinsic derivative, suppose for simplicity's sake that the curve is a *geodesic*, G. If we peel the strip surrounding G and press it onto the table, it becomes the straight line illustrated in [23.3], and both the tangent vectors $\mathbf{v}(t)$ to G, and the tangent vector field $\mathbf{w}(t)$ to $\mathcal{S}$ (defined along G) all get pressed down onto this table top, $\mathbb{R}^2$, just as in [22.6] and [22.7]. Parallel transport of $\mathbf{w}$ along G (as in [23.2]) now amounts to ordinary Euclidean parallel transport along this line in $\mathbb{R}^2$, keeping the angle between $\mathbf{w}_{||}$ and $\mathbf{v}$ constant, as in [23.1a].

As is evident in the figure, here $\mathbf{w}$ is growing in length and rotating counterclockwise as it moves along G. The job of $D_{\mathbf{v}}\mathbf{w}$ is to *quantify* the rate at which this is happening, i.e., the rate at which $\mathbf{w}(t)$ departs from its initial value (as represented by $\mathbf{w}_{||}$) just as we *begin* to move away from p along G.

Alternatively, and this is in fact the more standard definition, we may parallel transport $\mathbf{w}(q)$ *back* to p to obtain $\mathbf{w}_{||}(q \rightsquigarrow p)$, and then compare it to the original vector $\mathbf{w}(p)$, as illustrated in [23.3]. Thus the ***intrinsic derivative*** $D_{\mathbf{v}}\mathbf{w}$ at p may be pictured as emanating from p, lying with T_p, given by

$$D_{\mathbf{v}}\mathbf{w} \asymp \frac{\mathbf{w}_{||}(q \rightsquigarrow p) - \mathbf{w}(p)}{\epsilon}. \tag{23.2}$$

Figure [23.3] illustrates this geometric construction with $\epsilon = 0.1$: we draw the change in $\mathbf{w}$ resulting from moving one tenth of $\mathbf{v}$, then stretch this change vector by a factor of ten. Of course the *exact* value of $D_{\mathbf{v}}\mathbf{w}$ is obtained only when we take the limit of this figure as ϵ vanishes.

In the general case, where K is not geodesic, everything we have just visualized remains the same, provided that we only peel off the part of the surface surrounding the very short segment of K connecting p and q.

Note that the intrinsic constancy of $\mathbf{w}_{||}$ along K can now be neatly expressed by saying that its intrinsic derivative vanishes:

$$D_{\mathbf{v}}\mathbf{w}_{||} = 0 \quad \Longleftrightarrow \quad \mathbf{w}_{||} \text{ is parallel-transported along } \mathbf{v}.$$

Here is an *extrinsic* way of looking at the intrinsic derivative. Instead of thinking of $D_{\mathbf{v}}$ as measuring the rate of change of the projection onto the tangent plane, we may instead take the projection of the rate of change $\nabla_{\mathbf{v}}$ itself. Once again letting $\mathcal{P}$ denote orthogonal projection onto the tangent plane, this generalization of (22.1) takes the form

$$D_{\mathbf{v}}\mathbf{w} = \mathcal{P}\left[\nabla_{\mathbf{v}}\mathbf{w}\right] = \nabla_{\mathbf{v}}\mathbf{w} - (\mathbf{n} \bullet \nabla_{\mathbf{v}}\mathbf{w})\,\mathbf{n}. \tag{23.3}$$

Note that the extrinsic term may also be expressed in terms of the Shape Operator:

$$D_{\mathbf{v}}\mathbf{w} = \nabla_{\mathbf{v}}\mathbf{w} - [\mathbf{w} \bullet \mathcal{S}(\mathbf{v})]\,\mathbf{n}.$$

In other words, to obtain $D_{\mathbf{v}}\mathbf{w}$ we take the full rate of change $\nabla_{\mathbf{v}}\mathbf{w}$ in $\mathbb{R}^3$, then subtract out the part that is not tangent to the surface, thereby leaving behind the part that *is* intrinsic to the surface.

The formula (23.3) may be used to prove [exercise] that the intrinsic derivative $D_{\mathbf{v}}$ obeys all the same linearity and Leibniz (product) rules as the full vector derivative $\nabla_{\mathbf{v}}$, so that if a and b are constants, and f and g are scalar functions defined along K, and $\mathbf{x}$, $\mathbf{y}$, $\mathbf{z}$ are tangent to $\mathcal{S}$ along K, then

$$\begin{aligned}
D_{\mathbf{v}}[a\mathbf{x} + b\mathbf{y}] &= aD_{\mathbf{v}}\mathbf{x} + bD_{\mathbf{v}}\mathbf{y}, \\
D_{[f\mathbf{x}+g\mathbf{y}]}\,\mathbf{z} &= fD_{\mathbf{x}}\mathbf{z} + gD_{\mathbf{y}}\mathbf{z}, \\
D_{\mathbf{v}}[f\mathbf{x}] &= f\,D_{\mathbf{v}}\mathbf{x} + [D_{\mathbf{v}}f]\,\mathbf{x} = f\,D_{\mathbf{v}}\mathbf{x} + f'\mathbf{x}, \\
D_{\mathbf{v}}[\mathbf{x} \bullet \mathbf{y}] &= [D_{\mathbf{v}}\mathbf{x}] \bullet \mathbf{y} + \mathbf{x} \bullet [D_{\mathbf{v}}\mathbf{y}].
\end{aligned}$$

(Here the prime denotes differentiation with respect to either time or distance along K.) However, instead of establishing these facts by calculation, it is much simpler to think of flattening onto the tabletop the strip surrounding K, together with the vector fields $\mathbf{x}$, $\mathbf{y}$, $\mathbf{z}$, for then $D_{\mathbf{v}}$ simply *is* $\nabla_{\mathbf{v}}$.

The intrinisic derivative sheds new light on our earlier discussion of *geodesic curvature*. Recall from [11.2], page 116, that the full acceleration $\boldsymbol{\kappa} \equiv \nabla_{\mathbf{v}}\mathbf{v}$ of a particle travelling at unit speed over the surface can be broken down into two components:

$$\nabla_{\mathbf{v}}\mathbf{v} = \boldsymbol{\kappa} = \boldsymbol{\kappa}_g + \boldsymbol{\kappa}_n.$$

The first component $\boldsymbol{\kappa}_g$ is the ***geodesic curvature vector***: it is the component of the acceleration tangent to $\mathcal{S}$. It is automatically perpendicular to the trajectory, and it points towards the centre of curvature *as perceived by inhabitants of* $\mathcal{S}$, the magnitude $|\boldsymbol{\kappa}_g|$ being the curvature of this circle; in other words, it measures the part of the curvature of the trajectory that is intrinsic to $\mathcal{S}$. In contrast to this, the ***normal curvature vector*** $\boldsymbol{\kappa}_n$ is directed along $\mathbf{n}$ and is invisible to these inhabitants.

Intuitively, κ_g should be the intrinsic rate of rotation of $\mathbf{v}$ within the surface:

$$\boxed{\kappa_g = D_{\mathbf{v}}\mathbf{v}.}$$

That this is indeed true follows immediately from (23.3), since $\kappa_n = (\mathbf{n} \cdot \nabla_{\mathbf{v}}\mathbf{v})\,\mathbf{n}$.

If $\theta_{||}$ denotes the angle between $\mathbf{v}$ and any vector $\mathbf{w}_{||}$ that is parallel transported along the trajectory, then (see Ex. 4) *the geodesic curvature is simply the rate of turning of the velocity vector within* $\mathcal{S}$, *relative to the "constant" vector* $\mathbf{w}_{||}$:

$$\boxed{|\kappa_g| = |D_{\mathbf{v}}\theta_{||}| = |\theta'_{||}|.} \tag{23.4}$$

A geodesic is a curve that seems *straight* to the inhabitants of $\mathcal{S}$: in other words, $\kappa_g = 0$. Thus the ***geodesic equation*** takes the form,

$$\boxed{\kappa_g = 0 \quad \Longleftrightarrow \quad D_{\mathbf{v}}\mathbf{v} = \mathbf{0},} \tag{23.5}$$

which is simply another way of looking at the fact that $\mathbf{v}$ is parallel transported along itself, and appears *constant* to the inhabitants of $\mathcal{S}$.

On the other hand, if a curve within $\mathcal{S}$ is *not* geodesic, then (see Ex. 4) when a strip surrounding it is peeled off and pressed flat onto the plane, it will appear *curved*, and its curvature within the plane is none other than κ_g!

Chapter 24

Holonomy

24.1 Example: The Sphere

Consider [24.1], which depicts a geodesic triangle Δ on the sphere of radius R, the angle between the meridians being Θ, and the third side being a segment of the equator.

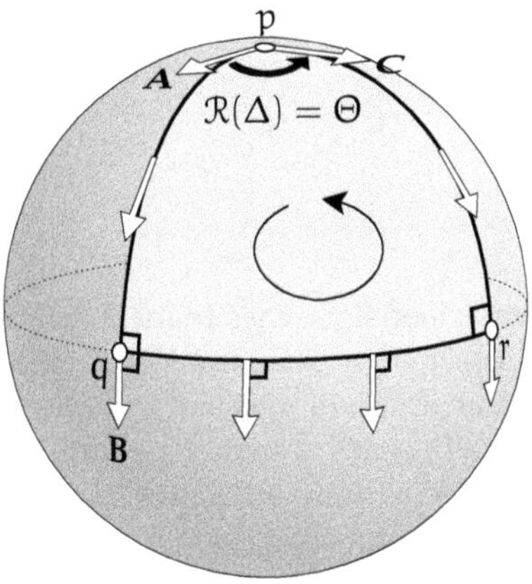

[24.1] *When* **A** *is parallel transported round the geodesic triangle* Δ*, it returns to* p *rotated by the* **holonomy** *of* Δ*, namely,* $\mathcal{R}(\Delta) = \Theta$.

Suppose we start with the south-pointing vector **B** at q on the equator, and then use (23.1) to parallel transport it to p, along two different geodesic routes. If we carry it due north along the meridian geodesic qp, we obtain **A**. But if we instead parallel transport it due east along the geodesic equator segment qr (maintaining the right angle with the geodesic), and then carry it due north along the meridian geodesic rp, we obtain the quite different vector **C**. This discrepancy between the result of parallel transport along different routes is the ***holonomy*** we alluded to earlier, which Levi-Civita discovered in 1917.

It turns out to be fruitful to look at this holonomy in a slightly different way. Instead of carrying **B** to p along two different routes, suppose we start with **A** at p and then parallel transport it counterclockwise around the closed loop, $\mathrm{p} \rightsquigarrow \mathrm{q} \rightsquigarrow \mathrm{r} \rightsquigarrow \mathrm{p}$. The figure shows that it returns to p having undergone a counterclockwise rotation of $\mathcal{R}(\Delta) = \Theta$. This is the holonomy of Δ, and we can now introduce the general definition:

> *The* **holonomy** $\mathcal{R}(\mathrm{L})$ *of a simple closed loop* L *on a surface* $\mathcal{S}$ *is the net rotation of a tangent vector to* $\mathcal{S}$ *that is parallel transported around* L. (24.1)

Note that this definition does not specify *which* tangent vector is to be parallel-transported. The reason is that (22.4) implies that *all* tangents vectors must rotate rigidly together, through the *same* angle $\mathcal{R}(\mathrm{L})$. Thus,

> *We may think of the holonomy as the rotation of the entire tangent plane as it is parallel transported around the loop.*

The definition of holonomy also does not specify *where* on L we should begin. To see why this, too, does not matter, suppose that instead of starting with **A** at p, we start with **B** at q, then parallel transport it $\mathrm{q} \rightsquigarrow \mathrm{r} \rightsquigarrow \mathrm{p} \rightsquigarrow \mathrm{q}$. Using (23.1), you may visually confirm that upon return to q, the vector has undergone the *same* rotation $\mathcal{R}(\Delta) = \Theta$ as before. More generally [exercise], convince yourself that *the holonomy is independent of the starting point of the loop* (as well as the initial tangent vector).

Next, note that in [24.1] the counterclockwise *sense* of the holonomy on the sphere *matches* the sense in which we traverse Δ. This will turn out to be because the sphere has positive curvature.

If we had instead transported the vector on a surface of *negative* curvature, then the rotation would have been *opposite* to the direction of transport. At this point we strongly encourage you to verify this empirically, using the sticky-strip construction of geodesics to create a geodesic triangle on a negatively curved patch of a suitable fruit or vegetable. You may then easily parallel-transport a toothpick around the triangle by maintaining a constant angle with each successive edge.

Not only is the sign of $\mathcal{R}(L)$ determined by the curvature within Δ, its *magnitude* is too! The constant curvature of the sphere is $\mathcal{K} = (1/R^2)$, so the total curvature residing within Δ is

$$\mathcal{K}(\Delta) = \iint_\Delta \mathcal{K}\, d\mathcal{A} = \frac{1}{R^2} \iint_\Delta d\mathcal{A} = \frac{1}{R^2}[R^2\,\Theta] = \Theta,$$

and therefore, for $L = \Delta$,

$$\boxed{\mathcal{R}(L) = \mathcal{K}(L).} \tag{24.2}$$

As we shall see, this is no accident—it is true for *any* simple loop L on *any* surface $\mathcal{S}$! Establishing this result (in the next chapter) will furnish us with a seemingly universal key, capable of unlocking some of the deepest mysteries we have encountered. It will unlock the *Theorema Egregium*. It will unlock the ***intrinsic*** nature of the Global Gauss–Bonnet Theorem. It will unlock the metric curvature formula, (4.10), the "*Star Trek* phaser" delivered from our future. And its generalization to higher dimensions will unlock the Riemannian curvature that lies at the heart of Einstein's curved-spacetime theory of gravity.

In fact the list goes on, extending far beyond the confines of this book. It includes Sir Michael Berry's remarkable 1983 discovery (see Shapere and Wilczek (1989) and Berry (1990)) of what is now called the ***Berry phase*** in quantum mechanics, as well as other "geometric phases" in physics. For a lovely selection of applications of holonomy to physics, see Berry (1991) (but be warned that what we call ***holonomy***, physicists sometimes call ***anholonomy***).

24.2 Holonomy of a General Geodesic Triangle

Figure [24.2] depicts a general geodesic triangle Δ on a general surface, the interior angles being θ_i and the exterior angles being φ_i, so

$$\theta_i + \varphi_i = \pi. \tag{24.3}$$

We know that the holonomy $\mathcal{R}(\Delta)$ is independent of the vector that is parallel transported around it, and we now use this freedom to make a choice that will make the answer vividly clear: we choose the tangent vector **v** to the first edge of Δ.

Parallel transport keeps **v** tangent to the first edge, so when it reaches the end, it makes angle φ_2 with the second edge. Since this second edge is also geodesic, the angle φ_2 is maintained as **v** is parallel transported along it. Thus, when it arrives at the end of the second edge, it makes angle $(\varphi_2 + \varphi_3)$ with the third and final edge, and this angle is maintained as it moves along that edge, finally returning to its starting point as $\mathbf{v}_{||}$, making angle $(\varphi_1+\varphi_2+\varphi_3)$ with the initial edge. Thus we can *see* that the holonomy is

$$\mathcal{R}(\Delta) = 2\pi - (\varphi_1+\varphi_2+\varphi_3). \tag{24.4}$$

In our discussion of Hopf's *Umlaufstatz*, we began by noting (see [17.1], p. 176) that if a particle travels round a Euclidean triangle Δ then the rotation of the velocity vector is $(\varphi_1+\varphi_2+\varphi_3)=2\pi$. Thus the holonomy formula (24.4) measures how much this total rotation $(\varphi_1+\varphi_2+\varphi_3)$ of the velocity *differs* from the Euclidean prediction of 2π.

Throughout this book we have used a different way of measuring the degree to which a geodesic triangle on a curved surface departs from a Euclidean one, namely, its ***angular excess***, $\mathcal{E}(\Delta)$. But in fact these two conceptually different measures of the curvature within Δ are *equal*! To see this, we combine (24.3) with (24.4):

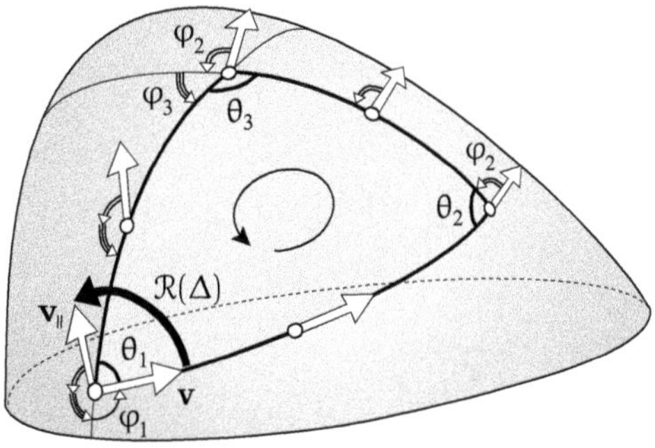

[24.2] *The tangent vector* $\mathbf{v}$ *to the first edge of a general geodesic triangle* Δ *on a general surface is parallel transported around* Δ*, returning to its starting point as* $\mathbf{v}_{||}$*, rotated by the holonomy* $\mathcal{R}(\Delta)$*.*

$$\mathcal{R}(\Delta) = 2\pi - [(\pi-\theta_1)+(\pi-\theta_2)+(\pi-\theta_3)] = \theta_1+\theta_2+\theta_3-\pi,$$

so,

$$\mathcal{R}(\Delta) = \mathcal{E}(\Delta). \tag{24.5}$$

All of the above generalizes easily from a geodesic 3-gon to a geodesic m-gon, P_m. First, [24.2] clearly generalizes to yield

$$\mathcal{R}(P_m) = 2\pi - \sum_{i=1}^{m} \varphi_i. \tag{24.6}$$

On the other hand, the angular excess of P_m is given by (18.4), page 189:

$$\mathcal{E}(P_m) = \sum_{i=1}^{m} \theta_i - (m-2)\pi.$$

Once again using (24.3), we find that the two seemingly different measures of the curvature within P_m are actually equal:

$$\mathcal{R}(P_m) = \mathcal{E}(P_m). \tag{24.7}$$

24.3 Holonomy Is Additive

Recall from [2.8], page 23, that if we split Δ into two geodesic triangles Δ_1 and Δ_2, then the angular excess $\mathcal{E}$ is *additive*:

$$\mathcal{E}(\Delta) = \mathcal{E}(\Delta_1) + \mathcal{E}(\Delta_2).$$

It follows from (24.5) that the holonomy $\mathcal{R}$ is additive, too. However, we consider $\mathcal{R}$ to be the more fundamental of the two concepts, so rather than view its additivity as being *inherited* from $\mathcal{E}$, we should try to understand this *directly*.

Figure [24.3] is such a direct, geometric proof that $\mathcal{R}$ is additive:

$$\boxed{\mathcal{R}(\Delta) = \mathcal{R}(\Delta_1) + \mathcal{R}(\Delta_2).} \tag{24.8}$$

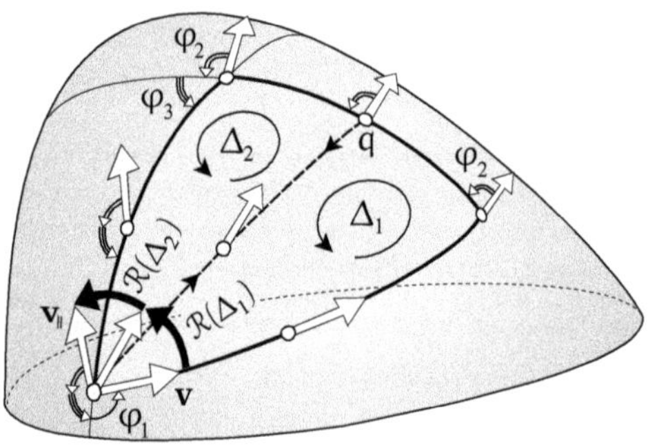

[24.3] Holonomy Is Additive. *The geodesic triangle Δ is split into Δ_1 and Δ_2 by the insertion of the dashed geodesic. The tangent vector* $\mathbf{v}$ *to the first edge of Δ_1 is parallel transported around Δ_1 and then around Δ_2. We see that the parallel transport back and forth along the dashed geodesic "cancels," and therefore $\mathcal{R}(\Delta) = \mathcal{R}(\Delta_1) + \mathcal{R}(\Delta_2)$.*

Here we have inserted the dashed geodesic into our original geodesic triangle Δ, thereby splitting it into the two geodesic triangles Δ_1 and Δ_2. The tangent vector $\mathbf{v}$ to the first edge of Δ_1 is parallel-transported around Δ_1, returning home rotated by $\mathcal{R}(\Delta_1)$. It is then parallel-transported around Δ_2, returning home rotated by $\mathcal{R}(\Delta_2)$. So the total rotation after parallel translation around both is $[\mathcal{R}(\Delta_1) + \mathcal{R}(\Delta_2)]$, as illustrated.

But because the final edge of Δ_1 (starting at q) is also the first edge of Δ_2 (ending at q), we parallel-transport the vector along the dashed geodesic twice, in succession, in opposite directions, so the vector returns to q *unchanged*. Successively parallel-transporting a vector around Δ_1 and then Δ_2 is therefore equivalent to parallel-transporting it around Δ, as was to be shown.

In this sense, parallel-transportation back and forth along the same curve "cancels," even if that curve is not a geodesic.

24.4 Example: The Hyperbolic Plane

We end this chapter by applying the concept of parallel transport to the pseudosphere (via the Beltrami–Poincaré half-plane model), to obtain a new, simple *intrinsic* geometric demonstration[1] of the constant negative curvature of the hyperbolic plane. In order to do this we shall *assume*, for now, the fundamental fact (24.2)—which will be proved in the next chapter—that the holonomy of a loop measures the total curvature within it.

Before we begin the demonstration, we make the important observation: just as we originally used the angular excess to give an intrinsic definition of the curvature $\mathcal{K}(p)$ *at* a point [see (2.1), p. 18], so (24.2) can now be used in the same way to find the curvature *at* a point p.

[1] The following argument previously appeared in Needham (2014).

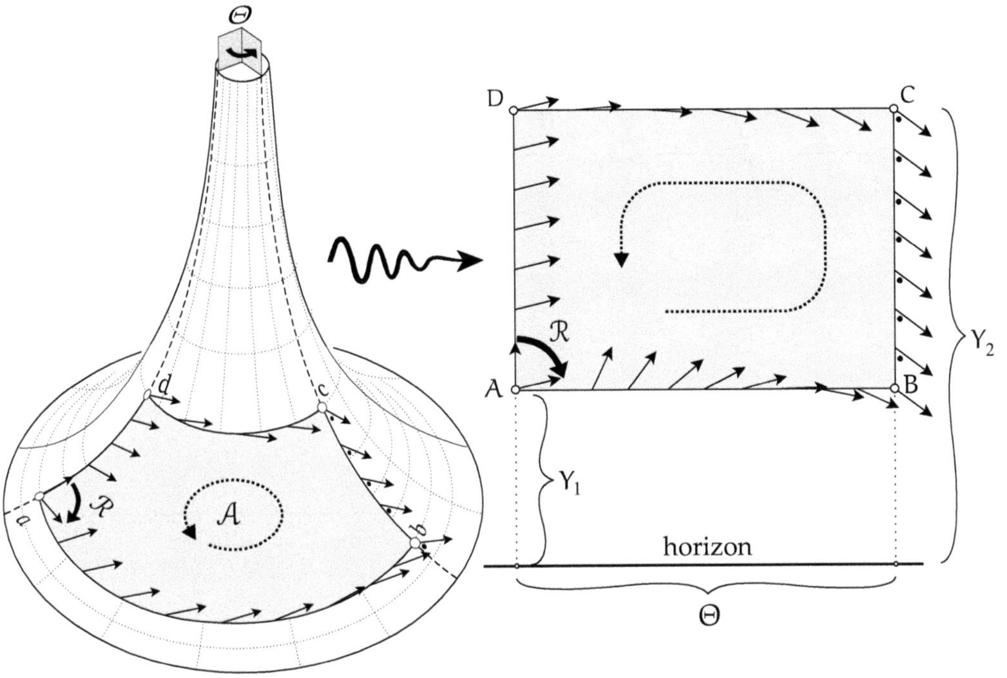

[24.4] *The "rectangle"* abcd *(with area* $\mathcal{A}$*) on the pseudosphere (left) is conformally mapped to* ABCD *in the Beltrami–Poincaré upper-half-plane (right). When the illustrated vector at* a *is parallel-transported counterclockwise around* abcd*, it returns rotated clockwise by* $\mathcal{R}$*. The conformality of the map ensures that the parallel-transported vector in the map undergoes the* same *rotation* $\mathcal{R}$*.*

If L_p is a small loop around p, then we can apply (24.2) to L_p as it shrinks down to p to find the curvature *at* p:

$$\mathcal{K}(p) = \lim_{L_p \to p} \frac{\mathcal{R}(L_p)}{\mathcal{A}(L_p)} = \textit{holonomy per unit area at } p. \tag{24.9}$$

We can now return to the problem at hand. On the pseudosphere of radius R, consider the "rectangle" abcd (traced counterclockwise) bounded by the vertical segments ad and bc of geodesic tractrix generators (Θ being the angle from the first to the second) together with the *non*geodesic horizontal circular arcs ab, cd. See the left side of [24.4]. As illustrated, let us parallel-transport a vector round abcd to discover the total curvature within it.

The right side of [24.4] depicts the conformal image in the Beltrami–Poincaré model: abcd is mapped to the rectangle with vertices $A=(x,Y_1)$, $B=(x+\Theta,Y_1)$, $C=(x+\Theta,Y_2)$, $D=(x,Y_2)$. Thus, using (5.6), page 55, the area $\mathcal{A}$ of the rectangle abcd on the pseudosphere is

$$\mathcal{A} = \int_{x=0}^{x=\Theta} \int_{Y_1}^{Y_2} \frac{R^2\,dx\,dy}{y^2} = R^2\Theta\left[\frac{1}{Y_1} - \frac{1}{Y_2}\right]. \tag{24.10}$$

At a we have chosen an initial vector pointing up the pseudosphere, along ad. As we parallel-transport it along ab, it rotates clockwise relative to the direction of motion; along bc it maintains the constant illustrated angle • with the direction of motion (because it is a geodesic); along cd it rotates counterclockwise relative to the direction of motion, *but not as much as it did on* ab; finally,

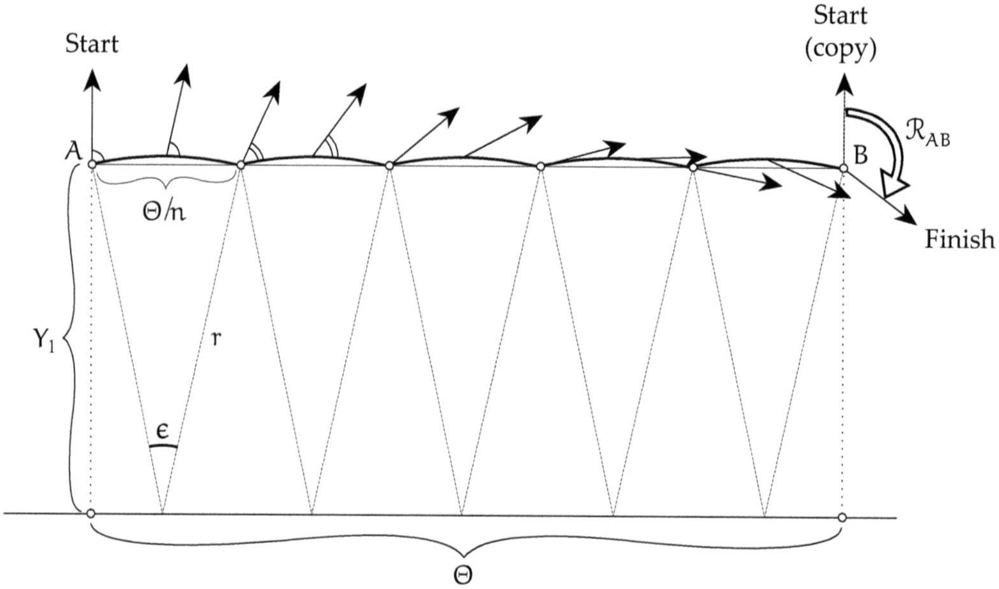

[24.5] *The initially vertical vector at* A *is parallel transported along the horizontal Euclidean line segment* AB *in the Beltrami–Poincaré half-plane. To do so, we approximate* AB *with* n *geodesic segments* (arcs of circles centred on the horizon, $y=0$), *then maintain constant angle with each geodesic segment, in succession. Finally, we let* n *go to infinity.*

it maintains constant angle with the geodesic da, returning to a having undergone a negative net rotation of $\mathcal{R}$.

Because the Beltrami–Poincaré map is conformal, when the vector is transported around ABCD it undergoes the *same* net rotation $\mathcal{R}$. But, as we now explain, the crucial advantage of the map is that it enables us to *see*[2] what this rotation actually is.

Divide the nongeodesic horizontal segment AB of Euclidean length Θ into n small segments of length (Θ/n). Next, as illustrated in [24.5], approximate these segments with geodesic segments: recall that these are arcs of circles centred on the horizon. Let ϵ be the angle that each such arc subtends on the horizon, as illustrated.

As the initially-vertical *Start* vector is parallel-transported along the first geodesic segment, its angle with that segment remains constant, and it therefore rotates through angle $-\epsilon$. Likewise for each successive segment, so that after all n segments have been traversed the total rotation from *Start* to *Finish* is $-n\epsilon$. But since

$$r\epsilon \asymp \frac{\Theta}{n} \qquad \text{and} \qquad r \asymp Y_1,$$

we deduce that the total angle through which the vector is rotated in the map is

$$\mathcal{R}_{AB} \asymp -n\epsilon \asymp -\frac{\Theta}{r} \asymp -\frac{\Theta}{Y_1}.$$

The same reasoning yields $\mathcal{R}_{CD} = (\Theta/Y_2)$. And since the vector does not rotate along either of the geodesics BC or DA, we deduce that the net rotation upon returning to A is

$$\mathcal{R} = \mathcal{R}_{AB} + \mathcal{R}_{CD} = -\frac{\Theta}{Y_1} + \frac{\Theta}{Y_2} = \left[-\frac{1}{R^2}\right]\mathcal{A},$$

by virtue of (24.10).

[2] A small amount of calculation is called for, but this is greatly simplified by our ability to manipulate "ultimate equalities" (involving $\asymp$) exactly as though they were ordinary equalities (involving $=$).

Thus,

$$\textit{rotation per unit area} = -\frac{1}{R^2}\,.$$

The fact that this answer is independent of the size, shape, and location of the rectangle proves, via (24.9), that the hyperbolic plane does indeed have *constant negative intrinsic curvature* $-1/R^2$, as was to be shown.

If you would like to try your own hand at performing parallel transport on specific surfaces (the cone and the sphere), try Exercise 5.

Chapter 25

An Intuitive Geometric Proof of the *Theorema Egregium*

25.1 Introduction

Several proofs of the *Theorema Egregium* have been found since Gauss first shocked the world with it in 1827. By far the most common proof, to be found in essentially all textbooks, is a variant of Gauss's original proof: perform a lengthy computation, sleep through the journey, then awake to find yourself staring at a formula for the *extrinsic* curvature $\mathcal{K}_{ext}$ that *simply turns out* to depend only on the metric, thereby establishing that $\mathcal{K}_{ext} = \kappa_1 \kappa_2$ is in fact *intrinsic* to the surface.

There is no question that this is a watertight *proof*, just as it was when Gauss first wrote it down in 1827, but we are left clueless as to *why* the result is true!

In his authoritative (and excellent) book, *A Panoramic View of Riemannian Geometry* (Berger 2003, p. 106), Professor Marcel Berger (1927–2016)—one of the foremost geometric authorities of the late twentieth century—provides summaries of the various known proofs, and concludes, "*To our knowledge there is no simple geometric proof of the Theorema Egregium today.*"[1] The purpose of this chapter is to rectify this situation by providing such a simple geometric proof.[2]

Rather than trying to understand the *Theorema Egregium* directly, we shall instead seek to understand Gauss's initial discovery, his so-called "Beautiful Theorem," (13.1). For convenience's sake, we restate the result here, again quoting Gauss's own words, exactly as he recorded the discovery in his private notebook of 1816:

> "**Beautiful Theorem**. *If a curved surface on which a figure is fixed takes different shapes in space, then the surface area of the spherical image of the figure is always the same.*"

In Section 13.3 we explained that the local *Theorema Egregium* follows easily and intuitively from the Beautiful Theorem, simply by letting the figure shrink down towards a point. So, to understand the Beautiful Theorem is to understand the *Theorema Egregium*.

But, as we noted in that same section, Dombrowski (1979) found that even in his private notebooks Gauss did not leave any trace of how he discovered the result, nor did he give any indication of a proof (assuming that he had one).

As should be clear from the present context, *we* shall prove the Beautiful Theorem using parallel transport, a concept whose discovery by Levi-Civita lay more than a century in the future in

[1] As soon as I discovered the simple geometric proof that I am about to present, I sent it to Banchoff, Berger, and Penrose, none of whom had ever seen it before. In particular, Berger, whose first language was French, emailed back, "I am glad to be able to congratulate you for this amazing proof, bypassed for above more than a century, for not only [is it] one of the most beautiful result[s] in geometry, but also this result started the whole Chern, etc. business."

[2] As this book neared completion, I learned that the same insight was previously published by Professor David W. Henderson (2013, Problem 6.3), of Cornell University. I was about to write to him when I learned that he had been struck by a car and killed, just days earlier, on December 20th, 2018. So, in the interest of keeping the record straight, the credit for this discovery should go to the late Professor Henderson (unless there is a precursor that I am not aware of). Regardless of its provenance, I hope that my *visualization* here (which Henderson does not provide) will help to make this simple and intuitive proof much more widely known. (And, on an extremely personal note, priority isn't everything: the startling, unanticipated flash of clarity—in the Sierra mountains, surrounded by pristine snow—was one of the happiest moments of my life.)

1816. Thus, as intuitive and simple as we hope you will find our parallel-transport proof, it could *not* have been Gauss's original approach.

We stress that Gauss discovered the Beautiful Theorem fully 11 years prior to developing his 1827 calculational apparatus for analyzing general surfaces. Thus the enigma remains, and, with it, the tantalizing possibility that a quite different, even simpler approach may exist—but this may have to await the second coming of Gauss.

25.2 Some Notation and Reminders of Definitions

Let $\mathcal{S}$ be the surface with unit normal vector field $\mathbf{n}$, S^2 be the unit sphere, and $n: \mathcal{S} \mapsto S^2$ be the spherical map.

Let $\mathcal{K}_{ext}$ be the *extrinsic* curvature, defined as the local (signed) area expansion factor of the spherical map, as introduced in (12.3), p. 132.

Let a tilde $(\widetilde{\ })$ denote a quantity defined on S^2. So, for example, $\mathcal{A}$ denotes area on $\mathcal{S}$, while $\widetilde{\mathcal{A}}$ denotes area on S^2.

Let $\mathcal{E}(\Delta)$ be the angular excess of a geodesic triangle Δ on $\mathcal{S}$, or more generally let $\mathcal{E}(P_m)$ denote the angular excess of a geodesic m-gon, P_m, given by (18.4), p. 189.

Let L be a simple loop bounding a region Ω of $\mathcal{S}$, and let $\mathcal{R}(L)$ denote the net rotation (holonomy angle) of a tangent vector $\mathbf{w}$ to $\mathcal{S}$ that is parallel transported counterclockwise round L, the sense of rotation being determined by having $\mathbf{n}$ pointing at our eye. For ease of visualization, let us assume that $\mathcal{K}_{ext}$ has a *single sign* (either always positive, or else always negative) throughout Ω, so that $n(\Omega)$ does not contain any *folds*; see the discussion in Section 17.5.

Likewise, let $\widetilde{\mathcal{R}}(\widetilde{L})$ denote the holonomy of a tangent vector to S^2 that is parallel transported around the image $\widetilde{L} \equiv n(L)$ on S^2 of L.

Finally, let $\mathcal{K}_{ext}(\Omega)$ denote the total amount of *extrinsic* curvature within Ω,

$$\mathcal{K}_{ext}(\Omega) = \iint_{\Omega} \mathcal{K}_{ext}\, d\mathcal{A},$$

and let $\mathcal{K}(\Omega)$ denote the total amount of *intrinsic* curvature within Ω,

$$\mathcal{K}(\Omega) = \iint_{\Omega} \mathcal{K}\, d\mathcal{A} = \mathcal{R}(L).$$

25.3 The Story So Far

Let us briefly recap what we actually *know* so far. This is important, because throughout this book we have felt at liberty to quote (and use) future results long before we were in a position to prove or explain them. That said, we have at all times scrupulously avoided so much as a whiff of circular reasoning.

Nevertheless, it is quite possible that in the course of our time-travelling exposition the line between what is already established and what has yet to be proved may have become blurred. Here, then, are the few facts that we shall need in the following explanation of the Beautiful Theorem, all of which *have* been properly established:

- Since $\mathcal{K}_{ext}$ is defined to be the local (signed) area expansion factor of the spherical map, *the total amount of extrinsic curvature within Ω on $\mathcal{S}$ is the area of its image $\widetilde{\Omega}$ on S^2*:

$$\mathcal{K}_{ext}(\Omega) = \iint_{\Omega} \mathcal{K}_{ext}\, d\mathcal{A} = \iint_{\widetilde{\Omega}} d\widetilde{\mathcal{A}} = \widetilde{\mathcal{A}}(\widetilde{\Omega}).$$

- As we proved in (12.8), page 134, the extrinsic curvature can be expressed as the product of the principal curvatures:

$$\mathcal{K}_{\text{ext}} = \kappa_1 \kappa_2.$$

Recall that we initially proved this by examining the effect of n on a small rectangle aligned with the principal directions. However, later ((15.8), p. 153) we established that *all* shrinking shapes ultimately undergo the *same* expansion, $\kappa_1 \kappa_2$.

Thus the area on S^2 of the image $\widetilde{\Omega} = n(\Omega)$ can also be expressed as,

$$\widetilde{\mathcal{A}}(\widetilde{\Omega}) = \mathcal{K}_{\text{ext}}(\Omega) = \iint_{\Omega} \kappa_1 \kappa_2 \, d\mathcal{A}.$$

- Harriot's 1603 result ((1.3), p. 8) on a sphere of radius R implies (with $R=1$) that on S^2 we have $\widetilde{\mathcal{E}}(\widetilde{\Delta}) = \widetilde{\mathcal{A}}(\widetilde{\Delta})$, and this easily generalizes to a geodesic m-gon, $\widetilde{P}_m$:

$$\widetilde{\mathcal{E}}(\widetilde{P}_m) = \widetilde{\mathcal{A}}(\text{interior of } \widetilde{P}_m).$$

- But on S^2 we also have proved in (24.5) that $\widetilde{\mathcal{R}}(\widetilde{\Delta}) = \widetilde{\mathcal{E}}(\widetilde{\Delta})$, and in (24.7) we generalized this to geodesic m-gons. Thus if $\widetilde{P}_m$ is a geodesic m-gon on S^2 then the *net rotation [holonomy] of a vector that is parallel-transported around* $\widetilde{P}_m$ *is the same as the area it encloses*:

$$\widetilde{\mathcal{R}}(\widetilde{P}_m) = \widetilde{\mathcal{A}}(\text{interior of } \widetilde{P}_m).$$

- If $\widetilde{L}$ is the (generally nongeodesic) image on S^2 of the simple loop L on $\mathcal{S}$, we may approximate $\widetilde{L}$ with a geodesic m-gon and then let $m \to \infty$.

Thus, combining the above results, we may summarize the relevant known facts as follows:

> ***The Story So Far.*** *If the simple loop* L *on* $\mathcal{S}$ *(enclosing the region* Ω *of single-signed extrinsic curvature) is mapped by* n *to* $\widetilde{L}$*, then the total amount of extrinsic curvature within* Ω *on* $\mathcal{S}$ *is the (signed) area of* $\widetilde{\Omega}$ *on* S^2*, bounded by* $\widetilde{L}$*, and this is in turn is equal to the* holonomy *of a tangent vector to* S^2 *that it parallel-transported around* $\widetilde{L}$*:*
>
> $$\iint_{\Omega} \kappa_1 \kappa_2 \, d\mathcal{A} = \mathcal{K}_{ext}(\Omega) = \widetilde{\mathcal{A}}(\widetilde{\Omega}) = \widetilde{\mathcal{R}}(\widetilde{L}). \qquad (25.1)$$

25.4 The Spherical Map Preserves Parallel Transport

Consider [25.1], which shows the simple loop L on $\mathcal{S}$ (a yellow squash, on the left) and its image on S^2 (a pomelo, on the right) under the spherical map, $\widetilde{L} = n(L)$. The tangent planes at p and $n(p)$ are parallel, so any tangent vector field $\mathbf{w}$ defined along L on $\mathcal{S}$ is—after being transplanted from p to $n(p)$—automatically also a tangent vector field to S^2 along $\widetilde{L}$.

While the shape of L is immaterial to the following argument, here we have chosen to construct an intrinsic circle, using a stretched piece of string (not shown) to trace the locus of points at constant distance from a fixed point. To help visualize the effect of the spherical map on L, we have erected normals at four equally spaced points along L; their image points on S^2 can be identified

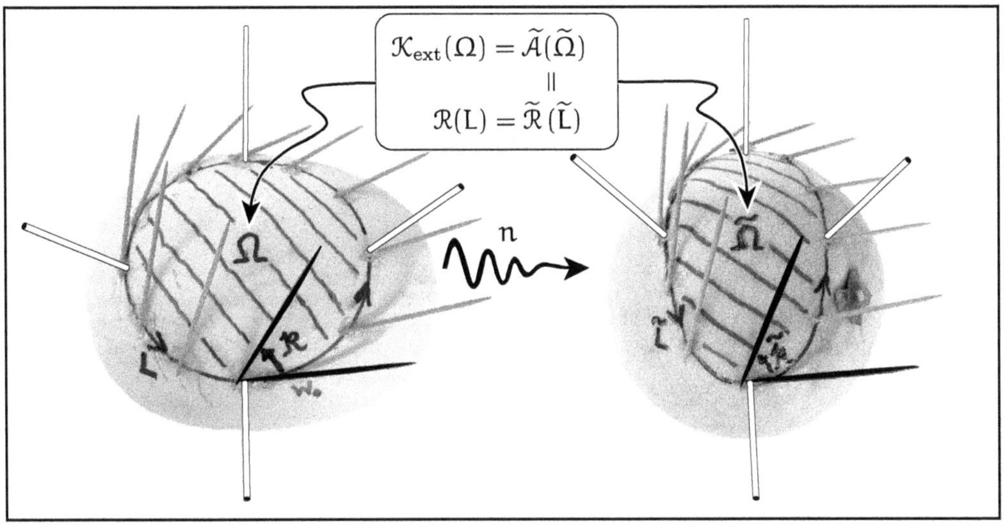

[25.1] Geometric Proof of the Beautiful Theorem. *The total extrinsic curvature* $\mathcal{K}_{ext}(\Omega)$ *within* Ω *on* $\mathcal{S}$ (left) *equals the area* $\widetilde{\mathcal{A}}(\widetilde{\Omega})$ *of its spherical image* $\widetilde{\Omega}$ *on* $\mathbb{S}^2$ (right). *Furthermore,* $\widetilde{\mathcal{A}}(\widetilde{\Omega}) = \widetilde{\mathcal{R}}(\widetilde{L})$. *But* **the spherical map preserves parallel transport**, *so* $\widetilde{\mathcal{R}}(\widetilde{L}) = \mathcal{R}(L)$. *Thus the total* **extrinsic** *curvature* $\mathcal{K}_{ext}(\Omega)$ *is equal to the* **intrinsically** *defined holonomy* $\mathcal{R}(L)$ *on* $\mathcal{S}$.

by the fact that they have the *same* normals, as shown. Due to the disparity in the principal curvatures, note that the spherical image $\widetilde{L}$ of this circle is stretched out rather dramatically into an oval on $\mathbb{S}^2$.

Now suppose that $\mathbf{w}$ is not an arbitrary tangent field to $\mathcal{S}$ along L, but, as illustrated, is instead obtained by *parallel-transporting* an initial tangent vector $\mathbf{w}_0$ along L to generate $\mathbf{w}_{||}$. The choice of $\mathbf{w}_0$ is immaterial to the argument, but in [25.1] we have chosen it to be tangent to L.

We now come to the surprisingly simple and elegant crux of the matter. By the definition of parallel transport, the rate of change of $\mathbf{w}_{||}$ is always perpendicular to $\mathcal{S}$ as we travel along L, i.e., the rate of change is along $\mathbf{n}$. But this means that the rate of change of $\mathbf{w}_{||}$ is *also* perpendicular to $\mathbb{S}^2$ at $n(p)$, for the normal $\mathbf{n}$ to $\mathcal{S}$ at p is the *same* as the normal to $\mathbb{S}^2$ at $n(p)$. In other words, the *same* vector $\mathbf{w}_{||}$—transplanted from p to $n(p)$—is automatically parallel-transported along $\widetilde{L}$ on the sphere!

Let us make the same point differently, in the hope that (one way or the other) this critically important result will become crystal clear in the process. Our recipe [22.2] for parallel transport calls for us to move $\mathbf{w}_{||}$ along L parallel to itself in $\mathbb{R}^3$, while continually projecting back into the surface (or, technically, into the tangent plane T_p) as we go, thereby obtaining $\mathbf{w}_{||}$. Now transplant (by simple translation in $\mathbb{R}^3$) both $\mathbf{w}_{||}$ and T_p from p on $\mathcal{S}$ to $n(p)$ on $\mathbb{S}^2$. The transplanted tangent plane simply *is* the tangent plane to $\mathbb{S}^2$ at $n(p)$, so projecting into it yields parallel transport on $\mathbb{S}^2$, too.

In summary, and as illustrated,

The Spherical Map Preserves Parallel Transport. *As* $\mathbf{w}_{||}$ *at* p *is parallel-transported with respect to the surface* $\mathcal{S}$ *along* L, *the* exact same *vector* $\mathbf{w}_{||}$ *at* $n(p)$ *is automatically* also *parallel-transported with respect to* $\mathbb{S}^2$ *along* $\widetilde{L} = n(L)$. (25.2)

25.5 The Beautiful Theorem and *Theorema Egregium* Explained

Upon executing a full circuit of L, *we can now literally* see *that the net rotation (holonomy) on the surface is equal to the net rotation on the sphere:*

$$\mathcal{R}(L) = \widetilde{\mathcal{R}}(\widetilde{L}).$$

But, according to "The Story So Far," (25.1), the net rotation on the sphere is simply the area enclosed on the sphere, which in turn is the total amount of *extrinsic* curvature enclosed by L back on the surface $\mathcal{S}$:

$$\mathcal{R}(L) = \mathcal{R}(\widetilde{L}) = \widetilde{\mathcal{A}}(\widetilde{\Omega}) = \mathcal{K}_{ext}(\Omega) = \iint_{\Omega} \kappa_1 \kappa_2 \, d\mathcal{A}.$$

Since $\mathcal{R}(L)$ is defined intrinsically on $\mathcal{S}$, it is invariant under isometries of $\mathcal{S}$, and we have therefore proved Gauss's original Beautiful Theorem of 1816.

If we take L_p to be a small loop on $\mathcal{S}$ surrounding p, and let Ω_p be the region it encloses, and then shrink L_p down to p, we recover a more standard local statement of the *Theorema Egregium*:

$$\kappa_1 \kappa_2 = \lim_{L_p \to p} \frac{\mathcal{R}(L_p)}{\mathcal{A}(\Omega_p)}.$$

Alternatively, suppose we take L_p to be a small geodesic triangle Δ_p containing p, and let $\mathcal{A}(\Delta_p)$ denote the area of the interior of Δ_p. Then, by virtue of (24.5), we have also recovered our original form of the *Theorema Egregium*, in terms of our original intrinsic definition ((2.1), p. 18) of curvature:

$$\kappa_1 \kappa_2 = \lim_{\Delta_p \to p} \frac{\mathcal{E}(\Delta_p)}{\mathcal{A}(\Delta_p)} = \mathcal{K}(p).$$

Since the quantity on the right of either of the previous two equations is intrinsic to $\mathcal{S}$, it follows that the *extrinsically* defined curvature $\kappa_1 \kappa_2$ on the left must *also* be—with thrilling unexpectedness!—*invariant under isometries.*

Thus we have arrived (at last!) at a satisfying geometric *explanation* of the empirical phenomena that we first observed in [13.1], page 139, and in [13.2], page 140.

Chapter 26

Fourth (Holonomy) Proof of the Global Gauss–Bonnet Theorem

26.1 Introduction

Recall that as the curtain fell on Act III, no fewer than three distinct explanations of GGB had played out, but all of them had been forced to rely on *extrinsic* geometry.

Only now, armed with parallel transport, are we finally in a position to fulfill the promise of an *intrinsic* proof of GGB. The elegant argument that follows is entirely due to Hopf (1956, pp. 112–113), but we shall explain some important details that Hopf's original presentation took for granted, and, more significantly, we shall explicitly *visualize* the proof in a way that Hopf's exposition did not.

A significant bonus of Hopf's intrinsic proof of GGB is that it will simultaneously provide us with a brand new proof of the Poincaré–Hopf Theorem, (19.6), page 206.

26.2 Holonomy Along an *Open* Curve?

Holonomy, as we have defined it in (24.1), is a concept that *only* makes sense for a *closed loop*, L. We parallel transport an initial vector $\mathbf{w}_0$ around L, generating $\mathbf{w}_{||}$ along the way, and *when* $\mathbf{w}_{||}$ *returns to its starting point*, we can compare it to the initial vector $\mathbf{w}_0$ to see the angle $\mathcal{R}(L)$ through which it has been rotated.

The first step in Hopf's argument is to generalize the holonomy concept to an *open* curve K. Clearly we should define this to be the rotation within the surface of $\mathbf{w}_{||}$ as it is parallel-transported along K. But rotation relative to *what*?

Recall that we previously faced a similar problem in trying to define the index $\mathcal{J}_F(s)$ of a singular point s of a vector field **F** on a surface $\mathcal{S}$. As we illustrated in [19.7], page 205, our answer was to introduce a *fiducial vector field* **U**, the only requirement being that **U** not have any singular points on or inside the loop L surrounding s. This allowed us to define the index as the *number of revolutions of* **F** *relative to* **U** as we travel round s along the loop L, namely, (19.5), page 205:

$$2\pi \mathcal{J}_F(s) = \delta_L (\angle UF).$$

As we proved, this definition of the index is indeed well-defined, which is to say that *the choice of the vector field* **U** *does not matter*, despite the fact that the precise variation in the angle $\angle UF$ along L certainly *does* depend on the specific choice of the **U** field.

Let us now apply this same idea to the concept of holonomy. Reconsider [24.2], but now thinking of the geodesic triangle Δ as made up of its successively traversed (geodesic) edges, K_j:

$$\Delta = K_1 + K_2 + K_3,$$

as illustrated in [26.1]. (NOTE: We are reusing this example because having geodesic edges makes parallel transport easy to visualize, but the following reasoning applies equally well to *any* (nongeodesic) curve.)

Figure [26.1] reproduces [24.2], but it introduces a fiducial vector field **U**, enabling us to generalize the concept of holonomy. We define $\mathcal{R}(\mathsf{K})$ for the *open* curve K to be the net change in the angle $\angle \mathbf{U}\mathbf{w}_{||}$—from **U** to the parallel-transported vector $\mathbf{w}_{||}$—as we travel along K:

$$\boxed{\mathcal{R}_{\mathsf{U}}(\mathsf{K}) \equiv \delta_{\mathsf{K}}\,(\angle \mathsf{U}\mathsf{w}_{||}).} \tag{26.1}$$

NOTES: The subscript U is essential here, for $\mathcal{R}_{\mathsf{U}}(\mathsf{K})$ *does indeed depend on our arbitrary choice of* **U**; thus $\mathcal{R}_{\mathsf{U}}(\mathsf{K})$ has no true mathematical meaning; it is merely a stepping stone in the argument that is to follow. That said, if two different vectors are parallel transported along K, the angle between them remains fixed (by virtue of (22.4)) and therefore they both rotate the same amount relative to **U**. Thus, as the notation indicates, $\mathcal{R}_{\mathsf{U}}(\mathsf{K})$ *is independent of the choice of* $\mathbf{w}_{||}$.

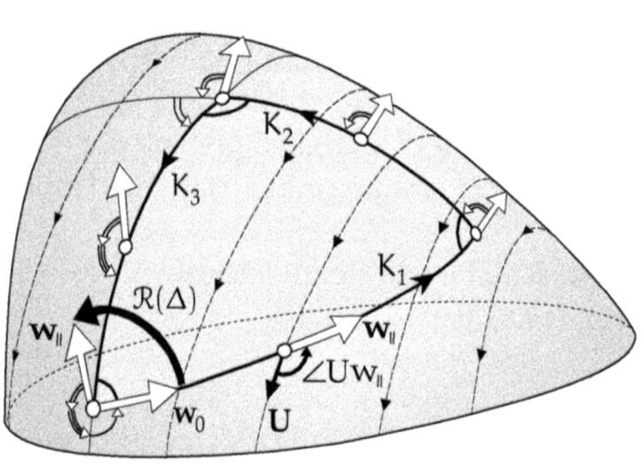

[26.1] *The holonomy $\mathcal{R}(\Delta)$ can be found by covering Δ with a nonsingular fiducial vector field* **U**, *and then summing over the edges K_j the change $\mathcal{R}_U(\mathsf{K}_j)$ in the illustrated angle $\angle \mathsf{U}\mathsf{w}_{||}$.*

The holonomy $\mathcal{R}(\Delta)$ of the closed triangular loop Δ can then be expressed as the sum of the "holonomies" of its edges:

$$\mathcal{R}(\Delta) = \mathcal{R}_{\mathsf{U}}(\mathsf{K}_1) + \mathcal{R}_{\mathsf{U}}(\mathsf{K}_2) + \mathcal{R}_{\mathsf{U}}(\mathsf{K}_3).$$

While each of the three individual terms on the right *does* depend on the arbitrary choice of **U**, their sum $\mathcal{R}(\Delta) = \mathcal{K}(\Delta)$ is *independent of* **U**.

We note that this approach provides a better definition of holonomy than we had before, solving a potential problem that we glossed over in our initial discussion: what if $\mathcal{R}(\Delta) > 2\pi$? Then if we naively compare $\mathbf{w}_{||}$ to $\mathbf{w}_0$ we will only (and incorrectly) perceive the rotation $\mathcal{R}(\Delta)$ as being the *excess* over 2π. But the use of a fiducial vector field **U** allows us to keep continuous track of the rotation of $\mathbf{w}_{||}$, and the above formula will yield the true value of $\mathcal{R}(\Delta)$.

26.3 Hopf's Intrinsic Proof of the Global Gauss–Bonnet Theorem

Now suppose that $\mathcal{S}_g$ is a closed surface of genus g, and suppose that **F** is a vector field on $\mathcal{S}$ with a finite number of singular points, s_i. Our ultimate aim will be to prove that

$$\boxed{\mathcal{K}(\mathcal{S}_g) \equiv \iint_{\mathcal{S}_g} \mathcal{K}\, d\mathcal{A} = 2\pi \sum_i \mathcal{I}_{\mathsf{F}}(s_i).} \tag{26.2}$$

But, before we do so, let us explain how the Poincaré–Hopf Theorem and GGB both follow immediately from this.

First, since the left-hand side of (26.2) is independent of **F**, it follows that the sum of the indices on the right-hand side must have the same value for *all* vector fields. But, by examining any of our previous examples of vector fields on $\mathcal{S}_g$, such as the honey-flow shown in [19.11], page 209, this universal index sum must equal $\chi(\mathcal{S}_g) = 2 - 2g$, thereby proving the Poincaré–Hopf Theorem. Thus, if we can prove (26.2), we will *also* have our proof of GGB:

$$\mathcal{K}(\mathcal{S}_g) = 2\pi\chi(\mathcal{S}_g).$$

To begin to understand (26.2), consider [26.2] which shows the same geodesic triangle Δ as before, and the same vector $\mathbf{w}_{||}$ parallel-transported around its boundary. But now the figure also imagines a vector field $\mathbf{F}$ on the surface, with a singular point s within Δ. (Here we have specifically pictured a source, but the argument applies to any singular point.)

From this figure we deduce that,

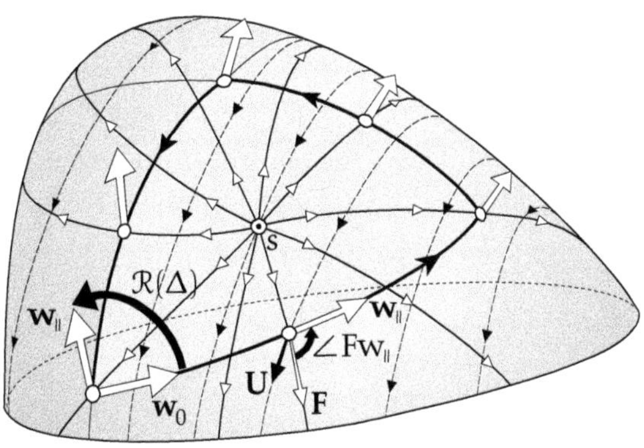

[26.2] *The difference $\mathcal{R}(\Delta) - 2\pi\mathcal{I}_F(s)$ can be found by summing over the edges K_j the change $\Phi(K_j)$ in the illustrated angle $\angle \mathbf{F}\mathbf{w}_{||}$, i.e., the rotation of $\mathbf{w}_{||}$ relative to $\mathbf{F}$.*

$$\begin{aligned}
\mathcal{K}(\Delta) - 2\pi\mathcal{I}_F(s) &= \mathcal{R}(\Delta) - \delta_\Delta(\angle \mathbf{UF}) \\
&= \sum_j \left[\mathcal{R}_U(K_j) - \delta_{K_j}(\angle \mathbf{UF})\right] \\
&= \sum_j \left[\delta_{K_j}(\angle \mathbf{U}\mathbf{w}_{||}) - \delta_{K_j}(\angle \mathbf{UF})\right] \\
&= \sum_j \delta_{K_j}(\angle \mathbf{F}\mathbf{w}_{||}).
\end{aligned}$$

The final expression *measures the rotation of* $\mathbf{w}_{||}$ *relative to* $\mathbf{F}$; it is manifestly *independent* of the arbitrary fiducial vector field $\mathbf{U}$.

To simplify matters, let us adopt Hopf's notation, and *define* $\Phi(K_j)$ *to be the net rotation along* K_j *of* $\mathbf{w}_{||}$ *relative to* $\mathbf{F}$*, i.e., the net change in the angle between the vector field* $\mathbf{F}$ *and the parallel-transported vector* $\mathbf{w}_{||}$ *as we traverse* K_j:

$$\Phi(K_j) \equiv \delta_{K_j}(\angle \mathbf{F}\mathbf{w}_{||}).$$

NOTE: $\Phi(K_j)$ *is independent of the choice of* $\mathbf{w}_{||}$, for the same reason that $\mathcal{R}_U(K_j)$ was.

Then the previous result may be written

$$\mathcal{K}(\Delta) - 2\pi\mathcal{I}_F(s) = \sum_j \Phi(K_j). \tag{26.3}$$

Clearly, this result holds equally well if Δ is replaced with a polygon, and we also remind the reader that this conclusion does *not* require that the edges be geodesics.

As usual, let $(-K_j)$ denote K_j traversed in the opposite direction. Since parallel transport does not depend on the direction in which a curve is traversed, we deduce that

$$\boxed{\Phi(-K_j) = -\Phi(K_j).}$$

The remainder of the proof of (26.2) now follows the exact same path as our original proof of the Poincaré–Hopf Theorem. Consider [26.3], which is simply a relabelled copy of [19.10], page 208. We partition $\mathcal{S}_g$ into polygons P_l with edges K_j (which, again, need *not* be geodesic), such that each polygon contains at most one singular point s_i of **F**.

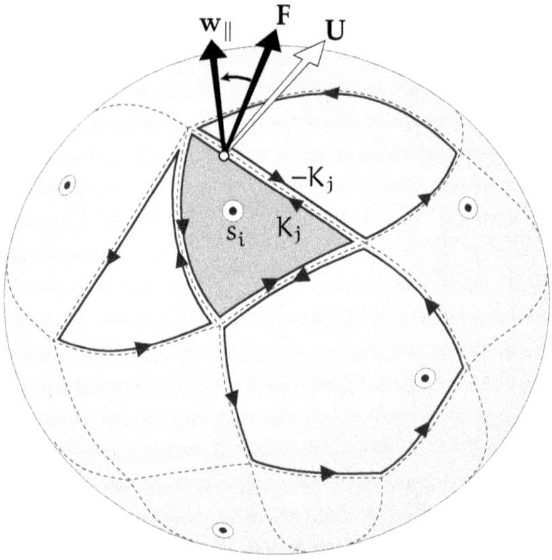

[26.3] *The surface* $\mathcal{S}_g$ *is partitioned into polygons* P_l *such that at most one singular point* s_i *of* **F** *lies in each one. The difference* $\mathcal{K}(\mathcal{S}_g) - 2\pi \sum_i \mathcal{I}_F(s_i)$ *can be found by summing over all the edges* K_j *(of all the polygons* P_l*) the change* $\Phi(K_j)$ *in the illustrated angle* $\angle F w_{\|}$*. But each edge of* K_j *abuts an oppositely directed edge of a neighbouring polygon, so the sum of these rotations over all polygons* vanishes. *Therefore,* $\mathcal{K}(\mathcal{S}_g) = 2\pi \sum_i \mathcal{I}_F(s_i)$.

If we now sum (26.3) over all polygons, we find

$$\begin{aligned} \mathcal{K}(\mathcal{S}_g) - 2\pi \sum_i \mathcal{I}_F(s_i) &= \sum_l \mathcal{K}(P_l) - 2\pi \sum_i \mathcal{I}_F(s_i) \\ &= \sum_{\text{all polygons}} \Phi(K_j) \\ &= 0, \end{aligned}$$

because each edge K_j belongs to two adjacent polygons, and is therefore traversed twice in opposite directions, contributing $\Phi(K_j) + \Phi(-K_j) = 0$.

This completes Hopf's intrinsic proof of (26.2) and, with it, both the Poincaré–Hopf Theorem and the Global Gauss–Bonnet Theorem.

Chapter 27

Geometric Proof of the Metric Curvature Formula

27.1 Introduction

The sole purpose of this chapter is to use parallel transport to provide (at long last!) a geometric proof of the "*Star Trek* phaser" formula, (4.10), page 38, for the curvature in terms of the metric, which we repeat here:

$$\mathcal{K} = -\frac{1}{AB}\left(\partial_v\left[\frac{\partial_v A}{B}\right] + \partial_u\left[\frac{\partial_u B}{A}\right]\right). \tag{27.1}$$

We are now living more than 20 chapters in the future of the point at which this formula was first announced, so we begin by reminding the reader of the meaning of the (u, v) coordinates, and of the metric components A and B.

The construction of a general coordinate system (as illustrated in [4.3], page 35) can always be specialized to an *orthogonal* coordinate system (u, v) on a surface $\mathcal{S}$, as illustrated in [27.1]. We first draw a family of nonintersecting curves covering the patch of surface we are analyzing, so that one (and only one) curve from our family passes though each point $\widehat{p}$ on the patch. We now number these curves arbitrarily (but differentiably) to create the u-coordinate, and call these curves, with their assigned u-values, the u-*curves*.

To complete the orthogonal coordinate system, we now draw the ***orthogonal trajectories*** of the u-curves, and label them differentiably with the v-coordinate, and call these v-*curves*.

Thus, as illustrated, the point $\widehat{p}$ on $\mathcal{S}$ can be labelled by the unique u-curve (say $u = U$) and v-curve (say $v = V$) that intersect there. So, in the map, $\widehat{p}$ can now be represented as $p = (U, V)$. Thus, as illustrated, u-curves are represented in the map by vertical lines, while v-curves are represented by horizontal lines.

NOTE: Here we are reverting to our earlier notation, distinguishing objects in the map from corresponding objects in the surface by attaching a "hat" $(\widehat{\ })$ to the latter. So, for example, an element of area in the map will be denoted $\delta\mathcal{A}$, while the corresponding area on the surface will be denoted $\delta\widehat{\mathcal{A}}$.

Let X measure distance along v-curves on $\mathcal{S}$, which correspond to horizontal lines in the map. Likewise, let Y measure distance along u-curves, which correspond to vertical lines in the map. We can now use X and Y to explain the meanings of A and B. If we move a small (ultimately vanishing) distance δu along a horizontal line in the map, then the corresponding point on the surface moves an ultimately *proportional* distance δX along the corresponding v-curve:

$$\delta X \asymp A\,\delta u \quad \Longleftrightarrow \quad \text{A is the local horizontal expansion factor.}$$

Likewise, if we move a distance δv along a vertical line in the map, then the corresponding point on the surface moves a distance δY along the corresponding u-curve, and

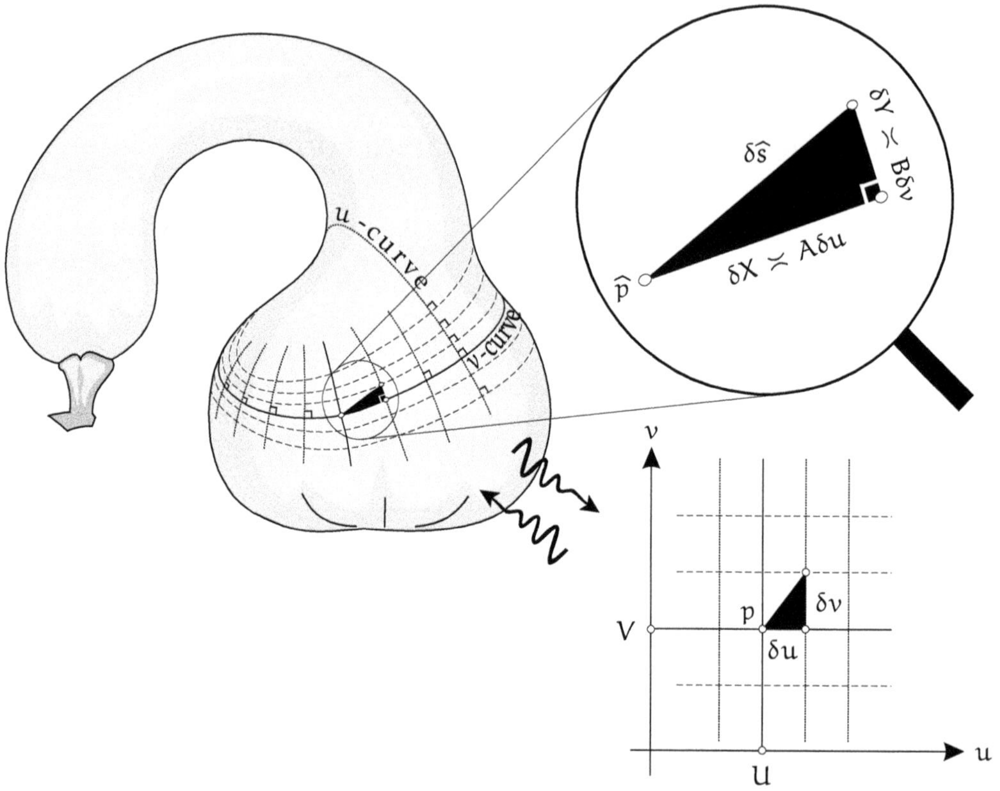

[27.1] The Metric in Orthogonal Coordinates. *Having drawn a family of (nonintersecting) "u-curves" (u = const.), we construct the v-curves as their* orthogonal trajectories. *If $\widehat{p}$ is the intersection of the curves u = U and v = V, then it represented by the point p = (U, V) in the map. A small horizontal movement δu in the map produces an ultimately proportional movement δX ≍ A δu on the surface (along a v-curve), and a vertical movement δv in the map produces an ultimately proportional* orthogonal *movement δY ≍ B δv on the surface (along a u-curve).*

$$\delta Y \asymp B\,\delta v \quad \Longleftrightarrow \quad \text{B is the local vertical expansion factor.}$$

The *metric* then tells us the true distance $d\widehat{s}$ within the surface in terms of apparent distances within the map, and Pythagoras's Theorem implies that in our *orthogonal* (u, v)-coordinate system it takes the form (4.9):

$$d\widehat{s}^2 = A^2\,du^2 + B^2\,dv^2.$$

27.2 The Circulation of a Vector Field Around a Loop

To avoid disrupting the flow of the proof to follow, we first address a lemma, namely, how to calculate the circulation of a vector field **V** around a small (ultimately vanishing) loop.

The required concept—called the ***curl*** of **V**—will be familiar to those who have studied undergraduate Vector Calculus, but a reminder can do no harm; furthermore, our exposition will be more geometrical than is customary.

Let $\mathbf{V} = \begin{bmatrix} P(u,v) \\ Q(u,v) \end{bmatrix}$ be a vector field in the (u, v)-plane, and let $\mathbf{r} = \begin{bmatrix} u \\ v \end{bmatrix}$ be the position vector of a point that traces (counterclockwise) a closed simple loop L. Then, as clarified in [27.2a],

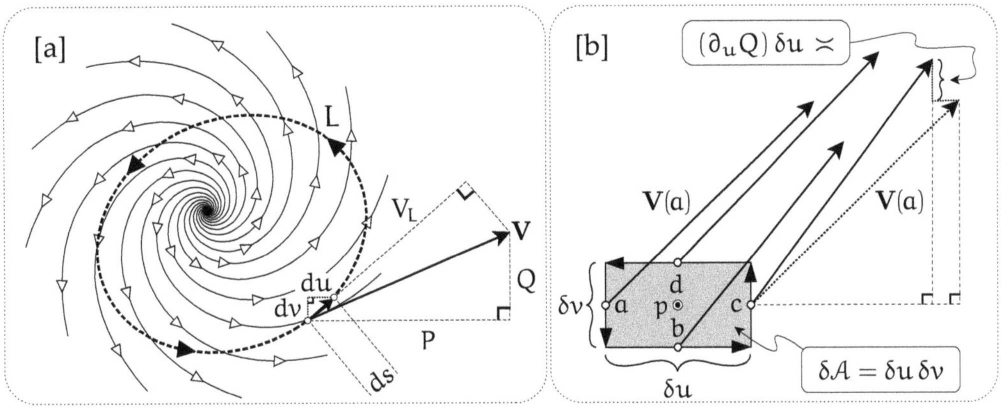

[27.2] [a] *The* **circulation** $\mathcal{C}_L(\mathbf{V}) = \oint_L V_L\, ds = \oint_L [P\, du + Q\, dv]$. **[b]** *As the shaded rectangle shrinks, the circulation around its boundary is ultimately equal to* $\{\partial_u Q - \partial_v P\}\,\delta\mathcal{A}$.

we define the ***circulation*** $\mathcal{C}_L(\mathbf{V})$ ***of*** $\mathbf{V}$ ***around*** L to be the integral of the component of $\mathbf{V}$ in the direction of L:

$$\mathcal{C}_L(\mathbf{V}) \equiv \oint_L \mathbf{V} \cdot d\mathbf{r} = \oint_L V_L\, ds = \oint_L [P\, du + Q\, dv], \tag{27.2}$$

where V_L is the (signed) projection of $\mathbf{V}$ onto the direction $d\mathbf{r}$ of L, and where $ds = |d\mathbf{r}|$. For the illustrated vortex, it is clear that V_L is always positive on L, so the circulation $\mathcal{C}_L(\mathbf{V})$ is positive.[1]

Now consider [27.2b], which illustrates a small (ultimately vanishing) coordinate rectangle R centred at p, with sides δu and δv, and hence with area $\delta\mathcal{A} = \delta u\,\delta v$. To calculate $\mathcal{C}_R(\mathbf{V})$, we perform a Riemann sum, and, as illustrated, we choose to use the *midpoints* of the sides. Of course in the limit that R vanishes, this specific choice does not matter, but we note that this choice of midpoints makes the approximation *much* more accurate than a random choice, the error for each side dying away as the *cube*[2] of the side.

The choice of midpoints also makes it especially easy to visualize and calculate the circulation. The contribution to the circulation from the upward vertical edge through c is ultimately equal to $Q(c)\,\delta v$. And the contribution from the downward edge through a is likewise $[-Q(a)]\,\delta v$. Thus the net contribution from the two vertical edges is governed by the illustrated *difference* in the vertical components of $\mathbf{V}$, namely, $\{Q(c) - Q(a)\}\,\delta v$. But, as illustrated, this change in Q is in turn governed by its partial derivative, which we may take to be evaluated at the centre p of R, as R shrinks down towards it. Thus the net contribution of the two vertical edges to the circulation is ultimately equal to $(\partial_u Q\,\delta u)\,\delta v$.

Exactly the same reasoning applies to the two horizontal edges, and therefore the total circulation around R is given by

$$\begin{aligned} \mathcal{C}_R(\mathbf{V}) &= \oint_R [P\, du + Q\, dv] \\ &\asymp Q(a)\,(-\delta v) + P(b)\,(\delta u) + Q(c)\,(\delta v) + P(d)\,(-\delta u) \\ &= \{Q(c) - Q(a)\}\,\delta v - \{P(d) - P(b)\}\,\delta u \end{aligned}$$

[1]In many important physical circumstances, the value of $\mathcal{C}_L(\mathbf{V})$ is *independent* of the precise shape of L; if [27.2a] were such an example, $\mathcal{C}_L(\mathbf{V})$ would simply measure the total strength of the vortex. See VCA, Chapter 11.

[2]See VCA, Section 8.2.3.

$$\begin{aligned} &\asymp (\partial_u Q\, \delta u)\, \delta v - (\partial_v P\, \delta v)\, \delta u \\ &= \{\partial_u Q - \partial_v P\}\, \delta \mathcal{A}. \end{aligned}$$

As this formula shows, we can now *define the* **curl** *of* **V** *to be the local circulation per unit area*:

$$\boxed{\operatorname{curl}\begin{bmatrix} P \\ Q \end{bmatrix} \asymp \frac{\mathcal{C}_R(\mathbf{V})}{\delta \mathcal{A}} \asymp \partial_u Q - \partial_v P.} \tag{27.3}$$

27.3 Dry Run: Holonomy in the Flat Plane

In order to get our feet wet, let us see how to determine the holonomy per unit area ($\asymp \mathcal{K}$) for a small loop in the flat plane. Of course the answer had better turn out to be $\mathcal{K}=0$!

In the previous chapter we generalized ***holonomy*** to an *open curve* $\widehat{K}$, defining it in (26.1) to be the rotation of any parallel-transported vector $\mathbf{w}_{||}$ along $\widehat{K}$, relative to an arbitrary (but singularity-free) fiducial vector field $\mathbf{U}$:

$$\mathcal{R}_U(\widehat{K}) \equiv \delta_{\widehat{K}}\,(\angle U w_{||}).$$

Recall that while this is independent of $\mathbf{w}_{||}$, it does not directly tell us *anything* about the geometry of the surface—it merely measures the rotation of $\mathbf{w}_{||}$ relative to the *arbitrarily* chosen $\mathbf{U}$ field. However, if $\widehat{K}$ becomes a *closed loop* $\widehat{L}$, then $\mathcal{R}_U(\widehat{L}) = \mathcal{R}(\widehat{L})$ becomes *independent* of $\mathbf{U}$, and it equals the total amount of curvature within $\widehat{L}$.

Let us take our surface $\mathcal{S}$ to be the flat plane, and let us apply the idea above, using ordinary polar coordinates: $u=r$, $v=\theta$, $A=1$, and $B=r$:

$$d\widehat{s}^2 = A^2\,du^2 + B^2\,dv^2 = dr^2 + r^2\,d\theta^2.$$

On the left of [27.3] we see a small (ultimately vanishing) rectangular loop $L = efghe$ in the polar-coordinate map plane, with sides δr and $\delta\theta$. On the right is its image $\widehat{L} = \widehat{e}\widehat{f}\widehat{g}\widehat{h}\widehat{e}$ on the (flat!) surface $\mathcal{S}$. Let us choose the fiducial vector field $\mathbf{U}$ to be radial, pointing outward along the rays $\theta = const.$, and let us find the holonomy along each of the four legs of the closed loop $\widehat{L}$. Starting at $\widehat{e}$, let us choose our initial vector $\mathbf{w}$ to be $\mathbf{U}$ at $\widehat{e}$, and now let us parallel transport it as $\mathbf{w}_{||}$ around the loop.

On the first leg of our journey, $\mathbf{w}_{||}$ does not rotate relative to $\mathbf{U}$, so $\delta\mathcal{R}_U(\widehat{e}\widehat{f}) = 0$. On the second leg, as we see clearly in [27.3], the rotation of $\mathbf{w}_{||}$ is $\delta\mathcal{R}_U(\widehat{f}\widehat{g}) = -\delta\theta$. On the third, radial leg $\widehat{g}\widehat{h}$ there is no rotation. Finally, as we return home along $\widehat{h}\widehat{e}$, the rotation of $\mathbf{w}_{||}$ relative to $\mathbf{U}$ is $\delta\theta$.

Thus, the rotations of $\mathbf{w}_{||}$ relative to $\mathbf{U}$ cancel, and therefore the holonomy (which is independent of $\mathbf{U}$)—and with it the curvature—does indeed vanish, as it should:

$$\begin{aligned} \mathcal{R}(\widehat{L}) &= \delta\mathcal{R}_U(\widehat{e}\widehat{f}) + \delta\mathcal{R}_U(\widehat{f}\widehat{g}) + \delta\mathcal{R}_U(\widehat{g}\widehat{h}) + \delta\mathcal{R}_U(\widehat{h}\widehat{e}) \\ &= 0 + (-\delta\theta) + 0 + \delta\theta \\ &= 0. \end{aligned}$$

But there is another way to determine the rotation of $\mathbf{w}_{||}$ relative to $\mathbf{U}$, and it is this method that we shall generalize to derive the general formula for $\mathcal{K}$: we shall use local measurements of *distances* within the coordinate grid—in other words, the *metric*.

The apparent rotation along $\widehat{f}\widehat{g}$ arises because the far side of the quadrilateral at radius $r + \delta r$ is longer than the near side, at radius r. How much longer? Well, the far side has length $(r + \delta r)\,\delta\theta$, while the near side has length $r\,\delta\theta$, so, as illustrated, the increase in length is $\delta r\,\delta\theta$.

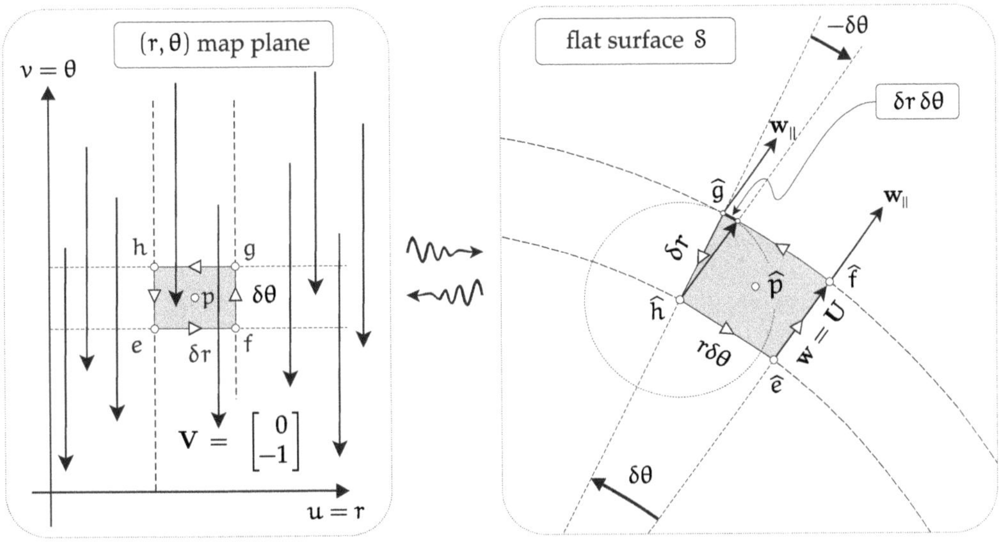

[27.3] *The fact that* $\delta\mathcal{R}_U(\widehat{h}\widehat{e}) = \delta\theta$ *can be deduced from the* metric *fact that this side is shorter than the opposite side by* $\delta r\,\delta\theta$. *The circulation of* $\mathbf{V}$ *around the loop in the map plane yields the (vanishing) holonomy on the (flat) surface on the right.*

If we think of this as ultimately equal to a small arc of the illustrated circle (of radius δr) centred at $\widehat{h}$ passing through $\widehat{g}$, then this increase in length subtends an angle at $\widehat{h}$ that is the rotation that we seek:

$$-\delta\mathcal{R}_U(\widehat{f}\widehat{g}) \asymp \frac{\text{arc}}{\text{radius}} \asymp \frac{\text{increase in side length}}{\text{orthogonal side}} = \frac{\delta r\,\delta\theta}{\delta r} = \delta\theta.$$

More generally, if we call the side length $\delta Y \asymp B\,\delta v$, then the increase $\delta^2 Y$ resulting from increasing u by δu will be

$$\delta^2 Y \asymp [\partial_u B\,\delta u]\,\delta v.$$

In the present case, $B = r$, $u = r$, and $v = \theta$, and therefore

$$\delta^2 Y \asymp [\partial_u B\,\delta u]\,\delta v = [\partial_r r\,\delta r]\,\delta\theta = \delta r\,\delta\theta,$$

as it should.

Before we turn to the general case of a curved surface, let us reinterpret what we have done, but now in the (r, θ) map plane on the left of [27.3]. On each leg of the loop L, the rotation on $\mathcal{S}$ is given by

$$\delta\mathcal{R}_U(\text{edge}) = 0\,\delta r + (-1)\,\delta\theta = \begin{bmatrix} 0 \\ -1 \end{bmatrix} \bullet \begin{bmatrix} \delta r \\ \delta\theta \end{bmatrix}.$$

The vector field $\mathbf{V} = \begin{bmatrix} 0 \\ -1 \end{bmatrix}$ is illustrated in the map plane of [27.3]. We now recognize the fact that the holonomy of the loop on $\mathcal{S}$ is equal to the circulation of $\mathbf{V}$ around the loop in the map plane. The flow of $\mathbf{V}$ is perpendicular to the top and bottom edges, so they contribute nothing to the circulation. But $\mathbf{V}$ does flow directly along the edge he, and it flows directly against the edge fg, and these two contributions are exactly equal and opposite to each other.

In summary,

$$\mathcal{R}(\widehat{L}) = \mathcal{C}_L(\mathbf{V}).$$

As we shall now see, in the general case, too, it is possible to visualize the holonomy of a loop on a curved surface $\mathcal{S}$ as the circulation of a vector field $\mathbf{V}$ around the corresponding loop in the (u, v) map plane, but in this case the circulation will generally *not* vanish, corresponding to the presence of curvature within the loop on $\mathcal{S}$.

27.4 Holonomy as the Circulation of a Metric-Induced Vector Field in the Map

In this section we will generalize the above argument and derive a formula for the holonomy $\mathcal{R}(\widehat{L})$ of a simple loop $\widehat{L}$ on $\mathcal{S}$ in terms of the circulation $\mathcal{C}_L(\mathbf{V})$ of a special vector field $\mathbf{V}$—determined by the metric—around the corresponding loop L in the (u, v)-map.

Let R denote the small rectangle (with boundary $L = efgh$) shown in the map plane of [27.3] (with centre p) only now with general (u, v)-coordinates, and so with sides δu and δv. This maps to a curvilinear quadrilateral $\widehat{R}$ on the curved surface $\mathcal{S}$ with corresponding "centre" $\widehat{p}$.

Since the sides of $\widehat{R}$ are ultimately $\delta X \asymp A\,\delta u$ and $\delta Y \asymp B\,\delta v$, the area $\delta\widehat{\mathcal{A}}$ of $\widehat{R}$ is ultimately related to the area $\delta\mathcal{A} = \delta u\,\delta v$ of R by,

$$\delta\widehat{\mathcal{A}} \asymp \delta X\,\delta Y \asymp (A\delta u)(B\delta v) = (AB)\,\delta\mathcal{A}. \tag{27.4}$$

In other words, as we pass from the map to the surface, *the local area expansion factor is* AB.

As R shrinks down to the centre point p, $\widehat{R}$ shrinks down to the point $\widehat{p}$, so that to the naked eye it will appear to be a true, plane rectangle. However, in order to investigate curvature we must place $\widehat{R}$ under a powerful microscope of magnification $(1/\delta u)$—or $(1/\delta v)$—in which case it becomes clear that the lengths of opposite edges of this "rectangle" differ from each other, if only very slightly.

At this point, to follow along more vividly, we *strongly* encourage you to draw a small "rectangle" $\widehat{R}$ on an orange, apple, or grapefruit, perhaps using ordinary spherical polar coordinates to create your grid, by drawing circles of longitude and latitude. You may then parallel transport a toothpick along one of the edges of $\widehat{R}$, and watch how it appears to rotate relative to the coordinate grid you have chosen to draw on your fruit's surface.

Figure [27.4] shows the vertices of $\widehat{R}$ orthogonally projected onto the tangent plane of $\mathcal{S}$ at any interior point of $\widehat{R}$, say $\widehat{p}$. We have connected these with straight lines to form a quadrilateral. Note that while the (u, v)-coordinate curves on the surface are always orthogonal, the sides of this projected quadrilateral are not orthogonal. But as R shrinks, so too does $\widehat{R}$, and its form is ultimately a rectangle. Here we have deliberately drawn a rather extreme deviation from this limiting case, in order to make the geometrical reasoning easier to follow.

Using the same notation as in the previous subsection, let $\delta^2 X = \delta[\delta X]$ denote the increase in δX, as illustrated. This δ^2-notation serves to remind us that we are now dealing with a *second-order* infinitesimal, as is made manifest by this:

$$\delta^2 X \asymp [\partial_v A\,\delta v]\,\delta u.$$

As [27.4] demonstrates, $\delta^2 X$ is ultimately equal to the arc of the circle of radius $\delta Y \asymp B\,\delta v$, subtending the angle $\delta\mathcal{R}_1$ at $\widehat{e}$. Therefore the small holonomy resulting from the change δu in the map is ultimately given by

$$\delta\mathcal{R}_1 \asymp \frac{\text{arc}}{\text{radius}} \asymp \frac{\delta^2 X}{\delta Y} \asymp \frac{(\partial_v A\,\delta v)\,\delta u}{B\,\delta v} = \left(\frac{\partial_v A}{B}\right)\delta u,$$

in which both $\partial_v A$ and B are understood to be evaluated at the centre p of R, as R shrinks down to p.

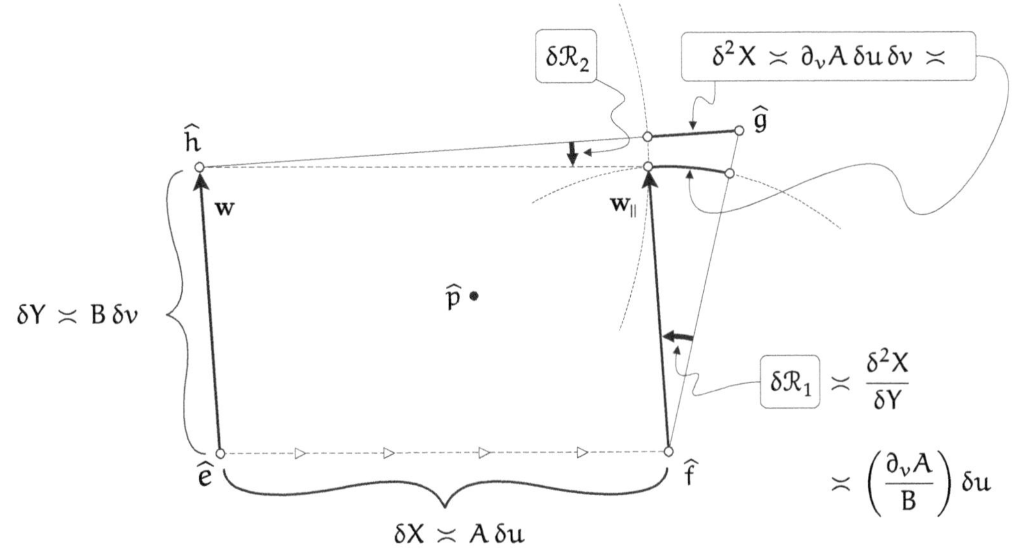

[27.4] Geometric Proof of the Metric Curvature Formula. *Parallel transport along* $\widehat{e}\widehat{f}$ *induces a rotation* $\delta\mathcal{R}_1$*—relative to the* (u,v)*-curves—given by* $\delta\mathcal{R}_1 \asymp \left(\frac{\partial_v A}{B}\right)\,\delta u$.

Exactly the same reasoning applies to the holonomy resulting from the change δv in the map, simply by performing the simultaneous exchanges $u \leftrightarrow v$ and $A \leftrightarrow B$.

However, if $\partial_u B$ is *positive*—as we have drawn it to be in [27.4]—then the resulting rotation $\delta\mathcal{R}_2$ of the parallel transported vector relative to the coordinate grid is now *clockwise*, as illustrated, and therefore we must introduce a minus sign into the formula:

$$\delta\mathcal{R}_2 \asymp -\left(\frac{\partial_u B}{A}\right)\,\delta v.$$

The linearity of the intrinsic derivative implies that the net rotation resulting from the changes δu and δv is the sum of the separate rotations, and is therefore given by

$$\delta\mathcal{R} \asymp \left[\frac{\partial_v A}{B}\right]\delta u - \left[\frac{\partial_u B}{A}\right]\delta v = \begin{bmatrix} (\partial_v A)/B \\ -(\partial_u B)/A \end{bmatrix} \cdot \begin{bmatrix} \delta u \\ \delta v \end{bmatrix}. \quad (27.5)$$

To obtain the holonomy of a simple closed loop $\widehat{L}$ on $\mathcal{S}$, we must integrate this formula along its image L in the (u,v)-map. We have thus arrived at this important result:

> *Let* $\mathcal{S}$ *be a surface with metric* $d\widehat{s}^2 = A^2\,du^2 + B^2\,dv^2$, *and let* $\widehat{L}$ *be a simple loop on* $\mathcal{S}$, *represented by the simple loop* L *in the* (u,v)*-map plane. If in the map plane we define the vector field*
>
> $$\mathbf{V} \equiv \begin{bmatrix} (\partial_v A)/B \\ -(\partial_u B)/A \end{bmatrix}, \quad (27.6)$$
>
> *then* the holonomy of $\widehat{L}$ on $\mathcal{S}$ is equal to the circulation of $\mathbf{V}$ around L in the map*:*
>
> $$\mathcal{R}(\widehat{L}) = \mathcal{C}_L(\mathbf{V}).$$

You may easily verify [exercise] that this general result is in accord with the special case illustrated in [27.3].

27.5 Geometric Proof of the Metric Curvature Formula

Now let us take the limit that the loop $\widehat{\mathrm{L}}$—with area $\delta\widehat{\mathcal{A}}$—shrinks down towards $\widehat{\mathrm{p}}$. The curvature $\mathcal{K}(\widehat{\mathrm{p}})$ is ultimately equal to the holonomy per unit area, so by combining (27.4) and (27.6) we obtain,

$$\mathcal{K}(\widehat{\mathrm{p}}) = \lim_{\widehat{\mathrm{L}}\to\widehat{\mathrm{p}}} \frac{\mathcal{R}(\widehat{\mathrm{L}})}{\delta\widehat{\mathcal{A}}} \asymp \frac{1}{\mathrm{AB}}\left[\frac{\mathcal{C}_{\mathrm{L}}(\mathbf{V})}{\delta\mathcal{A}}\right].$$

But (27.3) now tells us that the bracketed term on the right—namely, the local circulation per unit area—is none other than the *curl* of **V**.

This allows us to complete our geometric proof of the metric curvature formula, (27.1), one of the most elegant and *explicit* manifestations of Gauss's *Theorema Egregium* of 1827:

$$\begin{aligned}
\mathcal{K}(\widehat{\mathrm{p}}) &= \frac{1}{\mathrm{AB}}\left\{\text{curl of}\begin{bmatrix} (\partial_{\mathrm{v}}\mathrm{A})/\mathrm{B} \\ -(\partial_{\mathrm{u}}\mathrm{B})/\mathrm{A}\end{bmatrix}\right\} \\
&= \frac{1}{\mathrm{AB}}\left\{\partial_{\mathrm{u}}\left[-\frac{\partial_{\mathrm{u}}\mathrm{B}}{\mathrm{A}}\right] - \partial_{\mathrm{v}}\left[\frac{\partial_{\mathrm{v}}\mathrm{A}}{\mathrm{B}}\right]\right\} \\
&= -\frac{1}{\mathrm{AB}}\left(\partial_{\mathrm{v}}\left[\frac{\partial_{\mathrm{v}}\mathrm{A}}{\mathrm{B}}\right] + \partial_{\mathrm{u}}\left[\frac{\partial_{\mathrm{u}}\mathrm{B}}{\mathrm{A}}\right]\right).
\end{aligned}$$

Of course in proving this formula, we have also proved the important special case in which the map is *conformal*, so that $\mathrm{A} = \mathrm{B} = \Lambda$. In this case the formula reduces to the even more elegant form, (4.16):

$$\boxed{\mathcal{K} = -\frac{\nabla^2 \ln \Lambda}{\Lambda^2}.}$$

Chapter 28

Curvature as a Force between Neighbouring Geodesics

28.1 Introduction to the Jacobi Equation

This chapter introduces a brand new interpretation of curvature, one that is critically important in Einstein's curved-spacetime theory of gravity. The idea is to look at the separation of neighbouring unit-speed geodesics, and to examine their *relative acceleration*, which may be towards each other (attraction), or away from each other (repulsion).

We first examine this phenomenon in the three principal cases of *constant* curvature, and we then go on to analyse a general surface of variable curvature. In this general case, we will provide two different geometric proofs of the fundamental equation that governs this relative acceleration, called the ***Equation of Geodesic Deviation***, or the ***Jacobi Equation***, named after Carl Gustav Jacob Jacobi—pictured in [28.1]—who discovered it in 1837.

[28.1] *Carl Gustav Jacob Jacobi (1804–1851).*

As we shall see, the essential insight is that if two neighbouring geodesics pass through a region of *positive* curvature, they are *attracted* to each other. If these two geodesics are launched from a common point of origin o, in slightly different directions, this attractive force arising from the positive curvature may cause them to be focused back to a *second* intersection point, called a ***conjugate point*** of o.

On the other hand, neighbouring geodesics travelling through a region of *negative* curvature are *repelled* by each other, accelerating apart.

In both cases, the force of attraction or repulsion is directly *proportional to the separation* of the geodesics, and (locally) the proportionality "constant" *is equal to the curvature* of the surface at the location of the particle!

This is the essence of Jacobi's discovery.

28.1.1 Zero Curvature: The Plane

Consider [28.2a], which shows a particle with position $\mathrm{p}(t)$ travelling at unit speed along a geodesic straight line $\mathcal{L}$ in the plane, with unit velocity $\mathbf{v} = \dot{\mathbf{p}}$. It also shows a neighbouring *parallel* geodesic $\widetilde{\mathcal{L}}$, traced by $\widetilde{\mathrm{p}}$, with velocity $\widetilde{\mathbf{v}}$.

As illustrated,

$$\text{Perpendicular connecting vector } \boldsymbol{\xi} \equiv \overrightarrow{\mathrm{p}\widetilde{\mathrm{p}}} \qquad \Longrightarrow \qquad \delta\mathbf{v} \equiv \widetilde{\mathbf{v}} - \mathbf{v} = \dot{\boldsymbol{\xi}}.$$

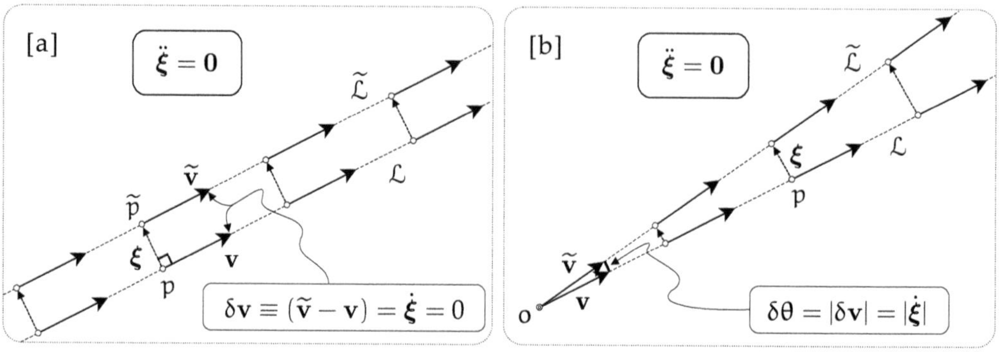

[28.2] **[a]** *In the flat plane, the separation* $\boldsymbol{\xi}$ *of neighbouring parallel lines is constant, so* $\dot{\boldsymbol{\xi}} = \mathbf{0}$, *and so (trivially)* $\ddot{\boldsymbol{\xi}} = \mathbf{0}$. **[b]** *If the lines instead diverge, separated by angle* $\delta\theta$, *then* $|\dot{\boldsymbol{\xi}}| = \delta\theta$, *and so* $\ddot{\boldsymbol{\xi}} = \mathbf{0}$, *again.*

Obviously in this case $\widetilde{\mathbf{v}} = \mathbf{v}$, so $\delta\mathbf{v} = \dot{\boldsymbol{\xi}} = \mathbf{0}$, but in the general case the equation simply says that the difference of the two velocities is the velocity with which $\widetilde{\mathcal{L}}$ departs from $\mathcal{L}$.

The focus of our attention in this chapter will not be on the relative velocity, but rather on the relative *acceleration*:

$$\ddot{\boldsymbol{\xi}} = \frac{d(\delta\mathbf{v})}{dt} = \textit{relative acceleration}. \tag{28.1}$$

In [28.2a] we find (trivially) that $\ddot{\boldsymbol{\xi}} = \mathbf{0}$.

Next, consider [28.2b], in which $\mathcal{L}$ and $\widetilde{\mathcal{L}}$ are now rays emanating from their common point of origin o, separated by a small angle $\delta\theta$. These geodesics are spreading out, diverging from each other, but they are doing so at a *steady rate* $\delta\theta$. The diagram explains this by geometrically constructing $\delta\mathbf{v}$ as the connecting vector from the tip of $\mathbf{v}$ to the tip of $\widetilde{\mathbf{v}}$, which is ultimately equal to the arc of the circle of radius $|\mathbf{v}| = 1$, subtending the angle $\delta\theta$ at o.

We can also deduce this symbolically. If $r = \text{op}$, then the unit speed of p along $\mathcal{L}$ can be written $|\mathbf{v}| = \dot{r} = 1$. Since $|\boldsymbol{\xi}| \asymp r\,\delta\theta$, the speed of separation $|\dot{\boldsymbol{\xi}}| \asymp |(\dot{r}\,\delta\theta)| = \delta\theta$.

Thus, even though these geodesics diverge from each other, they do so without any force *pushing* them apart—*their relative acceleration vanishes*:

$$\ddot{\boldsymbol{\xi}} = \mathbf{0}.$$

As we shall come to see, this vanishing of the relative acceleration of neighbouring geodesics is a new manifestation of the plane's vanishing curvature.

28.1.2 *Positive Curvature: The Sphere*

Consider [28.3]. We launch two particles from the north pole N of a sphere of radius R, the angle between them being $\delta\theta$. Recall that if the particles are stuck to the surface but are not subjected to any sideways forces *within* the surface, they will automatically follow geodesics of the surface, in this case, great circles.

After time t the particle with position $p(t)$ on one of these geodesics will have travelled distance $r = t$ over the surface, subtending angle ϕ at the centre of the sphere. Thus $p(t)$ lies on the illustrated circle of latitude of intrinsic radius $r = t = R\phi$, but with the illustrated *extrinsic* radius,

$$\rho = R \sin\phi = R \sin\left(\frac{t}{R}\right).$$

Thus the length ξ of the illustrated connecting vector $\boldsymbol{\xi}$ is

$$|\boldsymbol{\xi}| = \xi = \rho\,\delta\theta = R\,\delta\theta \sin\left(\frac{t}{R}\right).$$

Differentiating twice yields (first) the relative velocity, and (second) the relative acceleration as follows:

$$\frac{d\xi}{dt} = \left[\frac{1}{R}\right] R\,\delta\theta \cos\left(\frac{t}{R}\right)$$

$$\Longrightarrow \quad \frac{d^2\xi}{dt^2} = -\left[\frac{1}{R^2}\right] R\,\delta\theta \sin\left(\frac{t}{R}\right).$$

But since the curvature of the sphere is $\mathcal{K} = +(1/R^2)$, this result can be written in the following remarkably compact and elegant form, called the ***Equation of Geodesic Deviation***, or the

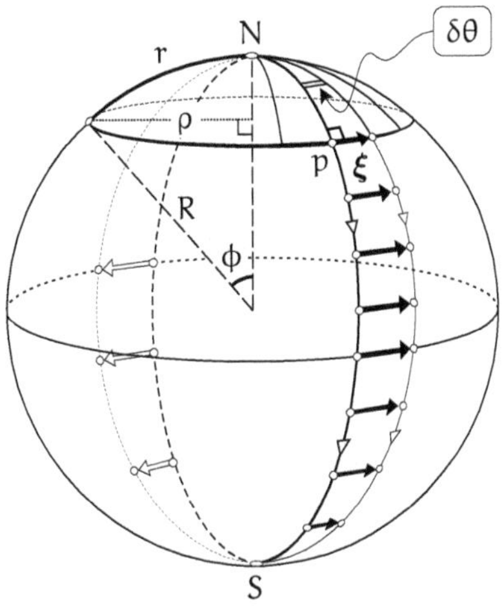

[28.3] The Jacobi Equation. *Two particles are launched with unit speed from the north pole* N *of the sphere of radius* R*, a small angle* δθ *apart. After time* t*, their separation is* $\xi = R\,\delta\theta \sin(t/R)$. *Their relative acceleration is therefore given by the Jacobi Equation:* $\ddot{\boldsymbol{\xi}} = -\mathcal{K}\boldsymbol{\xi}$.

$$\textit{Jacobi Equation:} \qquad \boxed{\ddot{\boldsymbol{\xi}} = -\mathcal{K}\,\boldsymbol{\xi}.} \tag{28.2}$$

We have merely proved the special case of this equation in the case where $\mathcal{K}$ is constant (and positive), but soon we will be able to prove that it is true on a general surface of *variable* curvature. In the general case we must take $\mathcal{K}$ in the equation to be $\mathcal{K}(p)$, i.e., the curvature *at* the location of the particle as it travels across the surface along the geodesic.

There is much more to be said about this example before we move on. First, for some readers, (28.2) may already be ringing a loud, harmonic bell! For this Equation of Geodesic Deviation also goes by a quite different name—it is the equation of the ***harmonic oscillator***, which is ubiquitous in physics, in both the classical and quantum realms.

It is easy enough to create your own personal harmonic oscillator at home. Take a rubber band (or spring) and attach one end to the underside of a table. Now attach a small object (which we shall take to have unit mass) to the other end, and gently lower it until it hangs in equilibrium, the downward pull of gravity being balanced by the upward pull of the stretched rubber band or spring.

Now launch the weight straight down from this equilibrium position, thereby stretching the rubber band or spring. In 1676[1] Robert Hooke (a contemporary and a rival of Newton) discovered that the force pulling the weight back up is *proportional* to the displacement $\boldsymbol{\xi}$ from the equilibrium position—this is called ***Hooke's Law***, and the proportionality constant k is called the ***stiffness*** of the rubber band or spring.

[1] In fact Hooke merely *laid claim* to his discovery in 1676, without revealing *what* he had discovered, publishing it as an incomprehensible Latin anagram: *"ceiiinosssttuv"*! Only in 1678 did he reveal the *solution* to his anagram: *ut tensio, sic vis* ("as the extension, so the force").

Newton's Second Law of Motion now yields *exactly* the same equation of motion as Jacobi's Equation, (28.2), but with *the surface's curvature replaced by the spring's stiffness:*

$$\ddot{\xi} = -k\,\xi, \qquad \text{with solution} \qquad \xi = \xi_0 \sin(\sqrt{k}\,t),$$

where ξ_0 denotes the maximum extension, occurring at the moment that the spring halts the downward motion and starts pulling the weight back up.

The appropriateness of the name *harmonic oscillator* is clear from this experiment and from the mathematical solution above: the weight *oscillates* up and down, sinusoidally, achieving its maximum speed (equal to its launch speed) each time it passes through the original equilibrium position, $\xi = 0$.

Now, with this *mathematically perfect* physical analogy in mind, let us return to [28.3]. First, notice that immediately after we launch the two particles, they appear to diverge at the same (acceleration-free) rate they would in the plane, as illustrated in [28.2b]. This is clear geometrically, but can also be confirmed by noting that $\sin\epsilon \asymp \epsilon$, and therefore, as r goes to zero,

$$\xi = R\,\delta\theta \sin\left(\frac{r}{R}\right) \asymp R\,\delta\theta\left(\frac{r}{R}\right) = r\,\delta\theta.$$

Despite the fact that the particles are moving freely over the surface, as their separation grows the curvature effectively exerts an *attractive force* that is *exactly* like a spring, and this mysterious curvature "force" starts to slow their separation.

As the separation grows, so does the force of attraction, in direct proportion, and hence the speed of separation is reduced more rapidly, until, finally, the relative speed drops to zero, just as the particles reach maximum separation at the equator. (This corresponds to the weight being at its lowest point, with the spring at maximum extension.) The attraction is now at maximum strength, and it starts pulling the geodesics back together again, until they are finally focussed to a *second* intersection point at the south pole, S—this is called the ***conjugate point*** of N.

As the particles converge and intersect at S (where $\xi = 0$) they achieve maximum relative speed, just as our oscillating weight achieves maximum speed as it passes through the equilibrium point, $\xi = 0$.

This journey from N to S is only one-*half* of a complete ξ-oscillation, the remaining half being completed on the back side of the sphere, with the direction of ξ now reversed, as illustrated, switching over from pointing to the left of p's trajectory to its right.

Once the particles have returned to N, the ξ-oscillation begins anew, and repeats forever.

28.1.3 Negative Curvature: The Pseudosphere

Just as geodesics are attracted to each other on the positively curved sphere, so they are repelled by each other on the negatively curved pseudosphere. To make this especially clear, [28.4a] shows two neighbouring geodesics being launched from the rim in the "*same*" direction (in the sense that $\dot{\xi}(0) = 0$). Soon, however, the repulsive negative curvature makes itself felt, and we see the geodesics flying apart.

But in order to *calculate* the relative acceleration with ease, we shall instead consider [28.4b], which shows two neighbouring particles on the rim being launched straight up the pseudosphere at unit speed, so that they are *approaching* each other as they travel along their respective tractrix generators, the arc length being equal to the time: $\sigma = t$.

The initial velocities of the particles are horizontal, both pointing at the centre o of the rim, distance R away. Since they are travelling at unit speed, in the absence of force they should therefore collide at time $t = R$. But clearly they don't collide, … ever! Yes, the two tractrix generators

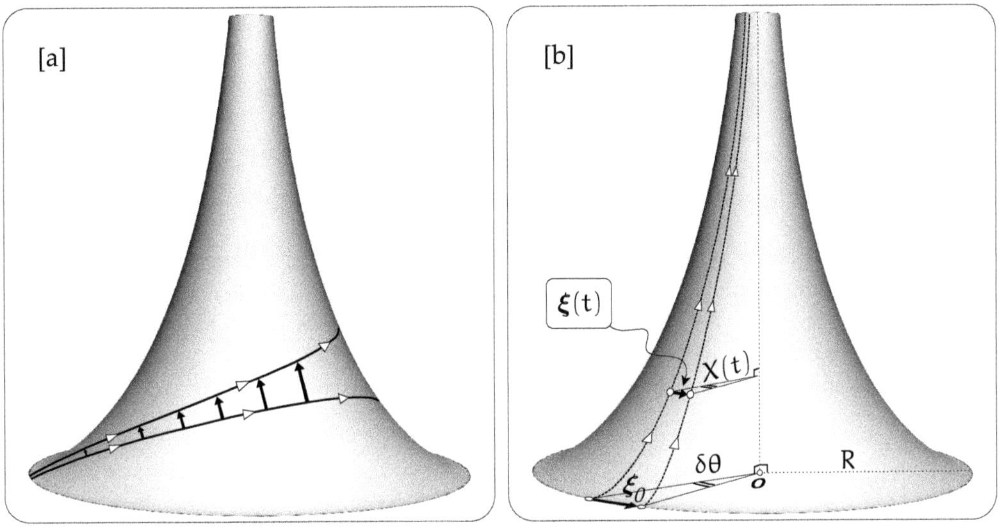

[28.4] [a] *Two neighbouring geodesics start off in the same direction, but the negative curvature forces them apart.* **[b]** *The connecting vector* $\boldsymbol{\xi}$ *between neighbouring tractrix generators is subject to* $\dot{\boldsymbol{\xi}} = -\left(\frac{1}{R}\right)\boldsymbol{\xi}$, *and so* $\ddot{\boldsymbol{\xi}} = +\left(\frac{1}{R^2}\right)\boldsymbol{\xi} = -\mathcal{K}\,\boldsymbol{\xi}$.

approach each other, but their approach speed *reduces* as they travel up the pseudosphere, thanks to the repulsive effective force arising from the negative curvature.

If $\delta\theta$ is the illustrated angle through which one of these generators must be rotated about the symmetry axis to obtain the other, then their initial small separation will be $|\boldsymbol{\xi}_0| \asymp R\,\delta\theta$. If $X(t)$ denotes the distance of the particles from the symmetry axis at time t, then

$$|\boldsymbol{\xi}(t)| \asymp X(t)\,\delta\theta \asymp \left[\frac{|\boldsymbol{\xi}_0|}{R}\right] X(t). \qquad (28.3)$$

Next, we remind the reader of the defining geometrical property of the tractrix, illustrated in [28.5] (which merely reproduces our original figure, [5.2], page 52). This figure implies that

$$\frac{-dX}{dt} = \frac{X}{R}.$$

This equation leads *directly* to the Jacobi Equation, without even needing to solve it first!

To see how, we rephrase the equation in terms of the relative velocity of our particles: (28.3) implies

$$\dot{\boldsymbol{\xi}} = -\left(\frac{1}{R}\right)\boldsymbol{\xi}.$$

Thus the initial approach speed is $|\boldsymbol{\xi}_0|/R$, and therefore (as we observed earlier) in the absence of the force the initial gap $|\boldsymbol{\xi}_0|$ would shrink to zero in time $t = R$.

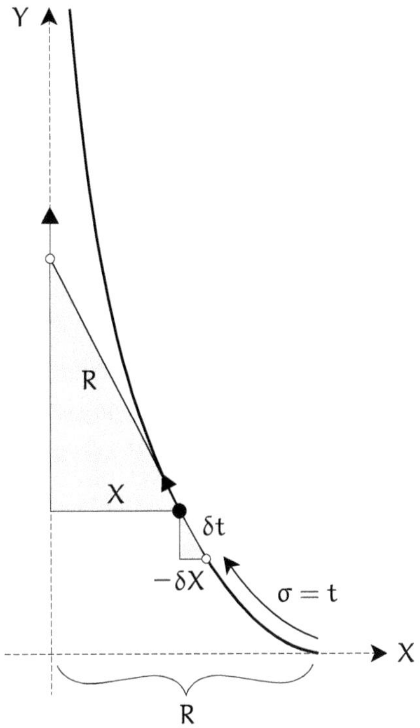

[28.5] The Tractrix *is defined by the property that the distance along its tangent to a fixed line has constant length,* R.

The fact that the particles do *not* in fact collide is *due to the repulsive force arising from the negative curvature*. This is confirmed by differentiating a second time, once again arriving at Jacobi's Equation Of Geodesic Deviation:

$$\ddot{\xi} = -\left(\frac{1}{R}\right)\dot{\xi} = +\left(\frac{1}{R^2}\right)\xi = -\mathcal{K}\,\xi.$$

The last equality follows from the fact that the pseudosphere has constant negative curvature, $\mathcal{K} = -(1/R^2)$.

The above argument can easily be generalized to prove that the Jacobi Equation holds on *all* surfaces of revolution, but we leave this for you to confirm, in Exercise 8. Instead, we now move on to establish the result in the *general* case, for a surface without any symmetries.

28.2 Two Proofs of the Jacobi Equation

28.2.1 Geodesic Polar Coordinates

Figure [28.3] may be viewed as providing a novel interpretation of ordinary spherical polar coordinates on $\mathbb{S}^2$.

We simultaneously launch particles in all possible directions from the north pole, so that every point of $\mathbb{S}^2$ (other than the poles) is uniquely specified by *which* particle hits it, and *when*. In greater detail, *which* particle hits the point can be specified by the longitude angle θ that the launch direction makes with some particular[2] (but arbitrary) direction, defined to be $\theta = 0$. *When* the particle hits the point is specified by the latitude angle $\phi = \sigma = t$ (assuming now that $R = 1$).

Geodesic polar coordinates are the natural generalization of this idea to an arbitrary surface, $\mathcal{S}$. See [28.6], which illustrates the idea in the case that $\mathcal{S}$ is a torus. From an arbitrary point[3] o on $\mathcal{S}$, launch particles at unit speed in all directions, so that they travel outward along geodesics, and choose one of these to be $\theta = 0$. Provided the region surrounding o is not so large that it contains any conjugate points, then every point within it is hit by a unique particle at a unique time. If that particle is launched in direction θ and hits the point at time t, then we can assign to that point the unique ***geodesic polar coordinates*** (t, θ).

Both in the plane and on the sphere, we know that the intrinsic circle $K(\sigma)$ with intrinsic radius $\sigma = t = const.$ is *orthogonal to its "radii,"* i.e., the geodesic segments of length σ emanating from its centre o. It turns out that this remains true on a *general* surface, as illustrated in [28.6]. Gauss (1827, §15) proved this on his way to proving other results, and so it has become known as

> **Gauss's Lemma.** *If particles are launched in all directions from a point on a general surface, travelling out a distance σ along geodesics, they will form (by definition) a* geodesic circle $K(\sigma)$ *with intrinsic radius σ.* This geodesic circle $K(\sigma)$ cuts its geodesic radii at right angles. (28.4)

Vector Calculus would not be developed by Gibbs and Heaviside[4] till the 1880s, so in 1827 Gauss was forced to prove this by means of a full page of computation. (A much shorter and

[2]On planet Earth we call $\theta = 0$ the ***prime meridian***, and define it to be the one passing through the Royal Observatory at Greenwich, London, England. Since 1999, this direction has been marked by particles of *light* (from a powerful green laser) launched northward across the London night sky!

[3]We now drop the hat for points on $\mathcal{S}$, because we no longer need to distinguish them from points in the map.

[4]See the *Chronology* in Crowe (1985, pp. 256–259)

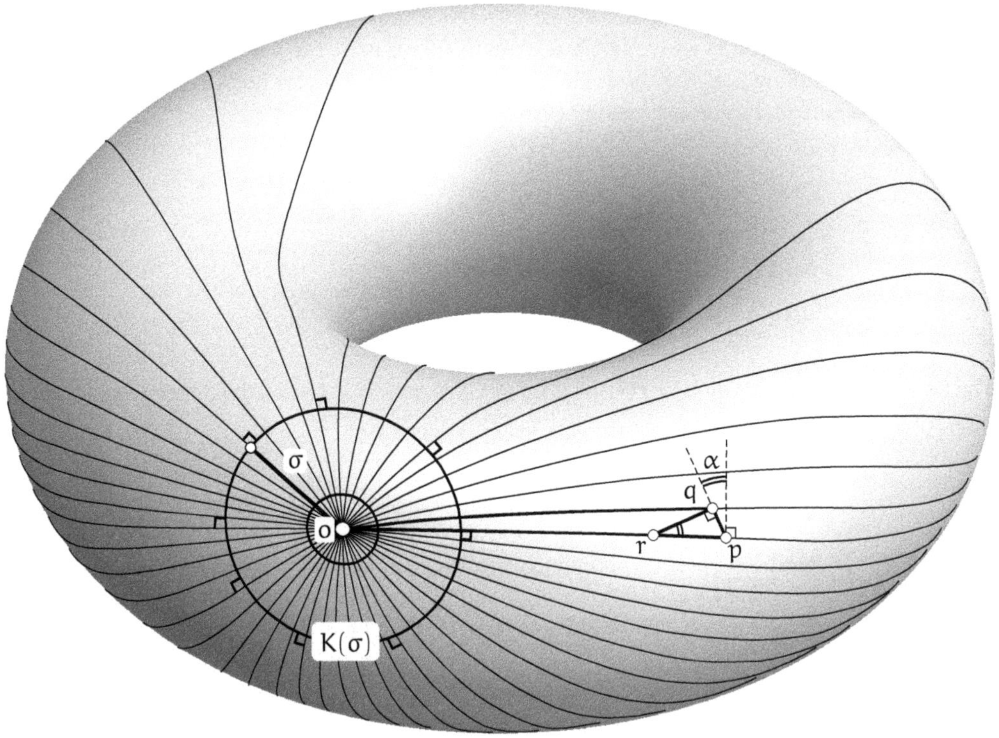

[28.6] Gauss's Lemma: A geodesic circle $K(\sigma)$ of intrinsic radius σ cuts its geodesic radii at right angles. *Gauss's second, geometrical proof of this result assumes that the two neighbouring points* p *and* q *lie on* $K(R)$*, and yet* $\alpha \neq 0$*. As explained in the text, this leads to a contradiction.*

more transparent version of this now-standard computation—based upon the *extrinsic* fact that the acceleration of the geodesics is normal to the surface—can be found in Ex. 9.) But immediately after completing his computational proof, Gauss did something completely uncharacteristic: he offered a *second* proof that was instead *intuitive and geometrical*. Furthermore, this second proof was *intrinsic*.

We can only speculate that Gauss did this because he thought this second proof was especially important, or because he was especially proud of his geometrical ingenuity. In any event, here is the evidence, in Gauss's own words: "We have thought it worthwhile to deduce this theorem from the fundamental property of shortest lines; but the truth of the theorem can be made apparent without any calculation by means of the following reasoning."

We shall now present this reasoning, which appears to have been completely lost in the mists of time—indeed, we have yet to discover a single instance of its reincarnation in any modern text. This is perhaps due to the fact that Gauss expressed his reasoning in terms of *infinitesimals*, thereby wrongfooting modern mathematicians. In place of Gauss's infinitesimals, we shall instead employ Newtonian *ultimate equalities* between small, ultimately vanishing quantities. Also, in contrast to the ever-unsolicitous Gauss (1827, §15), we have supplied a *diagram* in order to *see* what Gauss meant!

Suppose we launch two neighbouring geodesics from o, the small (ultimately vanishing) angle between them being $\delta\theta$. These are the two darker geodesic segments shown in [28.6] that terminate at p and q. Suppose that p and q both lie on $K(R)$—i.e, the geodesic segments op and oq have *the same length* $\sigma = R$. But now imagine (as illustrated) that pq is *not* ultimately orthogonal to op and oq (as $\delta\theta \to 0$), contrary to Gauss's Lemma. Following Gauss, we now show that this is impossible, thereby proving (28.4).

Suppose that $\angle \text{opq} \asymp (\pi/2) - \alpha$, as illustrated. Now draw the geodesic through q orthogonal to pq, and let its intersection with op be r, as shown. Then $\angle \text{prq} \asymp \alpha$, and therefore $\text{rq} \asymp \text{rp} \cos \alpha$. (Here, and in the following, "rq" is serving double duty, standing for both the segment, and the *length* of the segment, the context making clear which is meant.)

Therefore,

$$\begin{aligned} \text{length of o} \rightsquigarrow \text{r} \rightsquigarrow \text{q} &= \text{or} + \text{rq} \\ &\asymp (R - \text{rp}) + \text{rp} \cos \alpha \\ &= R - \text{rp}(1 - \cos \alpha) \\ &< R. \end{aligned}$$

Thus this indirect route from o to q (via r) is *shorter than the shortest route*, the direct geodesic route, oq. Gauss has his contradiction!

Having established the orthogonality of the (t, θ) coordinates, the metric takes the form

$$d\hat{s}^2 = dt^2 + \rho^2(t, \theta)\, d\theta^2.$$

Thus if $\delta\theta$ is the angular separation of two neighbouring geodesics launched from o, then

$$|\xi| \asymp \rho\, \delta\theta \quad \Longrightarrow \quad |\ddot{\xi}| \asymp \ddot{\rho}\, \delta\theta. \tag{28.5}$$

But if we substitute $u = t$, $v = \theta$, $A = 1$, and $B = \rho$ into (27.1), we obtain

$$\begin{aligned} \mathcal{K} &= -\frac{1}{AB}\left(\partial_v \left[\frac{\partial_v A}{B}\right] + \partial_u \left[\frac{\partial_u B}{A}\right]\right) \\ &= -\frac{1}{\rho}\left(\partial_\theta \left[\frac{\partial_\theta 1}{\rho}\right] + \partial_t \left[\frac{\partial_t \rho}{1}\right]\right) \\ &= -\frac{\ddot{\rho}}{\rho}. \end{aligned} \tag{28.6}$$

Combining this result with (28.5), we have proved the Jacobi Equation, (28.2):

$$\ddot{\xi} = -\mathcal{K}\, \xi.$$

Finally, we remark that (28.6) also allows us to finally prove ***Minding's Theorem*** (alluded to on p. 21), which states that if two surfaces have the same *constant* curvature, then they are locally isometric. (See Ex. 7 for the details.)

28.2.2 *Relative Acceleration = Holonomy of Velocity*

While our first proof of the Jacobi Equation was extremely brief (once Gauss's Lemma was established), it did require the full force of the ("*Star Trek* phaser") formula, (27.1). As we shall now see, there was actually no need to "set phasers to kill"! In its place, we now present a second proof that is *direct, geometrical, and intuitive*.

Indeed, the title of this subsection *is* that proof, albeit in aphoristic form. We now spell out the details, using [28.7], which uses the same notation as in [28.2]. This figure reuses the surface from our proof of Clairaut's Theorem, [11.7], but seen now with new eyes. Specifically, we see that the meridian geodesic generators $\mathcal{L}$ and $\widetilde{\mathcal{L}}$ (through a and the neighbouring point $\widetilde{a}$) are attracted to each other initially in the region of positive curvature, but are then repelled from each other as they enter the negative curvature region in the neck of the vase. (NOTE: The following argument is completely general, and does not rely on or make any use of the symmetry of this particular surface.)

Let $\boldsymbol{\xi}$ connect two neighbouring geodesics, $\mathcal{L}$ and $\widetilde{\mathcal{L}}$, launched with unit speed, perpendicular to a short geodesic segment $a\widetilde{a}$, as illustrated. This ensures that $\mathcal{L}$ and $\widetilde{\mathcal{L}}$ start out parallel, in the sense that initially they remain the same distance apart, with zero rate of separation: $\dot{\boldsymbol{\xi}} = \mathbf{0}$. (NOTE: We have drawn the separation fairly large in order to make the subsequent geometric reasoning easier to follow, but ultimately we will let the gap $a\widetilde{a}$ go to zero, so $\boldsymbol{\xi}$ will ultimately be a genuine tangent vector that can be pictured as lying *within* the surface.)

After time δt both particles will have travelled distance δt across the surface, arriving at b and $\widetilde{b}$, respectively. By then the relative velocity $\dot{\boldsymbol{\xi}}$ between $\mathcal{L}$ and $\widetilde{\mathcal{L}}$ has increased from zero to $\delta\mathbf{v} \asymp \ddot{\boldsymbol{\xi}}\,\delta t$, by virtue of the definition, (28.1).

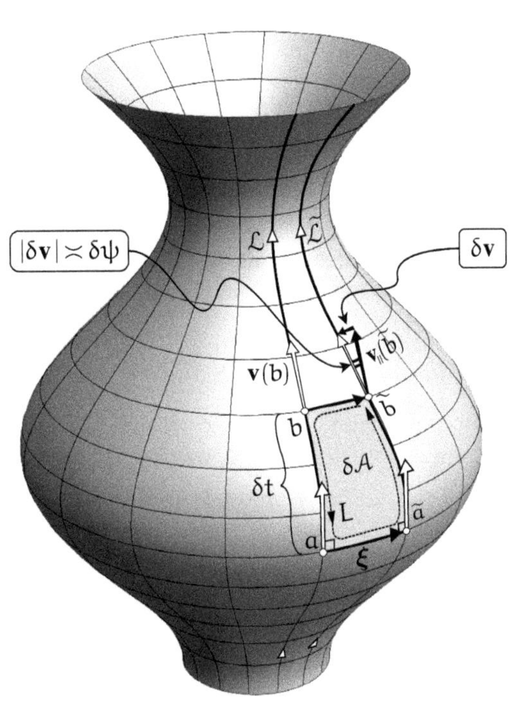

[28.7] Geometric Proof of the Jacobi Equation. *Parallel transporting* $\mathbf{v}_{||}(\widetilde{b})$ *around* L *returns it to* $\widetilde{b}$ *rotated by the holonomy* $\delta\psi = \mathcal{R}(L) \asymp \mathcal{K}\,\delta\mathcal{A} \asymp \mathcal{K}\,|\boldsymbol{\xi}|\,\delta t$. *But* $\delta\psi \asymp |\delta\mathbf{v}| \asymp |\ddot{\boldsymbol{\xi}}|\,\delta t$, *and the Jacobi Equation follows.*

Presumably $\delta\mathbf{v}$ should here be interpreted as the difference $[\widetilde{\mathbf{v}}(\widetilde{b}) - \mathbf{v}(b)]$ between the new velocities. However, this does not make *intrinsic* sense, since we cannot subtract tangent vectors to $\mathcal{S}$ that emanate from different points and lie within different tangent planes. In order to compare them, let us join b to $\widetilde{b}$ with a short geodesic[5] segment $b\widetilde{b}$ and then parallel transport $\mathbf{v}(b)$ along $b\widetilde{b}$ to obtain $\mathbf{v}_{||}(\widetilde{b})$ at $\widetilde{b}$, as illustrated. We thereby obtain an *intrinsic* measure of the relative acceleration:

$$\ddot{\boldsymbol{\xi}}\,\delta t \asymp \delta\mathbf{v} = \widetilde{\mathbf{v}}(\widetilde{b}) - \mathbf{v}_{||}(\widetilde{b}).$$

As illustrated, let $\delta\psi$ be the angle between these two vectors at $\widetilde{b}$. Since the arc (not shown) of the unit circle centred at $\widetilde{b}$ through the tips of these vectors is ultimately equal to the chord connecting those tips, we see that

$$|\delta\mathbf{v}| \asymp \delta\psi.$$

Now we come to the crux of the matter. As illustrated, let us define L to be the counterclockwise loop around the boundary of the shaded quadrilateral: $L \equiv \widetilde{b}ba\widetilde{a}\widetilde{b}$. Starting at $\widetilde{b}$, let us parallel transport $\mathbf{v}_{||}(\widetilde{b})$ around L.

First, since $\mathbf{v}_{||}(\widetilde{b})$ was originally obtained by parallel transporting $\mathbf{v}(b)$ along $b\widetilde{b}$, it follows that parallel transporting it back again along $\widetilde{b}b$ will return it to its original state, $\mathbf{v}(b)$. Next, since the velocity of a geodesic is parallel transported along itself, parallel transport along ba yields $\mathbf{v}(a)$ at a. Next, parallel transport along the geodesic $a\widetilde{a}$ maintains orthogonality with it, so it arrives at $\widetilde{a}$ as $\widetilde{\mathbf{v}}(\widetilde{a})$. Finally, parallel transport along the geodesic $\widetilde{a}\widetilde{b}$ returns it to its starting point as $\widetilde{\mathbf{v}}(\widetilde{b})$, so parallel transport around L has rotated $\mathbf{v}$ by $\delta\psi$.

[5] This choice makes the argument easier to follow, but we will ultimately take the limit that the segment shrinks to zero, so the precise nature of the segment does not matter.

In other words,

> *The relative acceleration of the geodesics is measured by the rotation $\delta\psi$ of the velocity vector, but this is none other than the* holonomy *of the velocity when it is parallel transported around* L. *In brief,* $\delta\psi = \mathcal{R}(L)$!

Since the area $\delta\mathcal{A}$ inside L is given by $\delta\mathcal{A} \asymp |\boldsymbol{\xi}|\,\delta t$, the proof of the Jacobi Equation comes down to the following sequence of ultimate equalities, each of which should now be visually apparent:

$$|\ddot{\boldsymbol{\xi}}|\,\delta t \asymp |\delta\mathbf{v}| \asymp \delta\psi = \mathcal{R}(L) \asymp \mathcal{K}\,\delta\mathcal{A} \asymp \mathcal{K}|\boldsymbol{\xi}|\,\delta t.$$

Cancelling δt, and taking into account the fact that $\delta\mathbf{v}$ points in the *opposite* direction to $\boldsymbol{\xi}$ for the illustrated case of *positive* $\mathcal{K}$, we have arrived again at the Jacobi Equation:

$$\ddot{\boldsymbol{\xi}} = -\mathcal{K}\,\boldsymbol{\xi}.$$

28.3 The Circumference and Area of a Small Geodesic Circle

In Section 2.2 we showed that, on the sphere, the inhabitants of the surface could determine the curvature of their world by examining carefully either the circumference $C(r)$ or the area $\mathcal{A}(r)$ of a small circle $K(r)$ of intrinsic radius r. In both cases, the key to finding the curvature was to detect how these quantities *departed from* their Euclidean values of $C(r) = 2\pi r$ and $\mathcal{A}(r) = \pi r^2$.

Specifically, we proved that in the limit that r vanishes, the curvature is given by (2.4) and (2.5), which we repeat here, for the convenience of the reader:

$$\mathcal{K} \asymp \frac{3}{\pi}\left[\frac{2\pi r - C(r)}{r^3}\right], \tag{28.7}$$

and

$$\mathcal{K} \asymp \frac{12}{\pi}\left[\frac{\pi r^2 - \mathcal{A}(r)}{r^4}\right]. \tag{28.8}$$

We claimed at the time that these formulas apply to *all* surfaces, not merely the sphere, and we are now finally in a position to prove this.

With the same geodesic polar coordinates as before, let $\xi(r)$ denote the separation of the geodesics launched at angle θ_0 and $\theta_0 + \delta\theta$, so that if we define

$$g(r) \equiv \rho(\theta_0, r) \quad \text{then} \quad \xi(r) \asymp g(r)\,\delta\theta.$$

As $r \to 0$, we know this reduces to the Euclidean formula, $\xi(r) \asymp r\,\delta\theta$, but it is precisely the small *departure* from this Euclidean result that we must detect in order to measure the non-Euclidean-ness, aka the curvature.

To that end, we expand $g(r)$ into a Maclaurin series:

$$g(r) = r + \frac{1}{2} g''(0)\, r^2 + \frac{1}{6} g'''(0)\, r^3 + \cdots .$$

But (28.6) tells us that $g''(r) = -\mathcal{K} g(r)$, so $g''(0) = 0$, and also

$$g'''(0) = [-\mathcal{K} g]'(0) = -\mathcal{K}'(0) g(0) - \mathcal{K}(0) g'(0) = -\mathcal{K}(0).$$

Thus,

$$\boxed{C(r) = 2\pi\, g(r) = 2\pi r - \tfrac{\pi}{3}\, \mathcal{K}(0)\, r^3 + \cdots ,}$$

and (28.7) follows immediately.

Finally, the area $\delta\mathcal{A}$ of an annulus of geodesic width δr is given by

$$\delta\mathcal{A} \asymp C(r)\, \delta r,$$

so integration of the formula for $C(r)$ readily yields [exercise] the area formula, (28.8).

Chapter 29

Riemann's Curvature

29.1 Introduction and Summary

In this chapter we shall see how the forgoing insights concerning 2-dimensional surfaces can be extended naturally to n-dimensional spaces, called ***manifolds***.

Just as our 2-dimensional surfaces are locally described by their tangent planes, each of which has the structure of $\mathbb{R}^2$, so the immediate vicinity of a point in an n-dimensional manifold locally resembles $\mathbb{R}^n$, but distances between neighbouring points are measured with a metric[1] that is not Euclidean.

• **First**, we shall extend to an ***n-manifold*** (shorthand for an n-dimensional manifold) our very first measure of the intrinsic curvature $\mathcal{K}(p)$ of a 2-surface, namely, the ***local angular excess per unit area*** of a small geodesic triangle Δ as it shrinks to a point p:

$$\mathcal{K}(p) = \lim_{\Delta \to p} \frac{\mathcal{E}(\Delta)}{\mathcal{A}(\Delta)}.$$

• **Second**, it turns out that the angular excess is too blunt an instrument to give direct insight into the more subtle curvature(s!) of an n-manifold. However, the *holonomy* resulting from parallel-transporting a vector around a small shrinking loop *is* able to fully reveal this more intricate curvature structure, in a very direct way. Clearly, then, the next step towards defining the curvature of an n-manifold must be to understand ***parallel transport*** in such a space.

Generalizing parallel transport from 2-surfaces to n-manifolds is *nontrivial*, and yet it is given scant attention in almost every standard text we have examined. To redress the balance, we shall provide *three* different geometrical constructions (all leading to the same result) that can be used to generalize Levi-Civita's parallel transport to such an n-manifold.

• **Third**, we use parallel transport to define the ***intrinsic*** (aka ***"covariant"***) ***derivative*** within an n-manifold. The good news is that—when expressed in *intrinsic* terms—the passage from a 2-surface to an n-manifold requires absolutely no change to our original definition (23.2); only the *notation* changes, writing $\nabla_{\mathbf{v}}$ in place of $D_{\mathbf{v}}$.

• **Fourth**—and this is the beating heart of the chapter—we use parallel transport to generalize holonomy from 2-dimensional surfaces to n-manifolds.

Riemann, pictured in [29.1], discovered that in place of Gauss's single number $\mathcal{K}$ characterizing a 2-surface, his generalized *intrinsic* curvature of an n-manifold is specified by an array of[2]

---

[1]The fully general conception of a manifold does not actually require that distance be defined within it. For our purposes, however, the metric is *the* central structure to be investigated, so we must insist that *our* manifolds have metrics! The technical name for a manifold with a positive-definite metric is a ***Riemannian manifold***, and a manifold with metric that can yield *negative* ("distance")^2, as is the case for spacetime, is called ***pseudo-Riemannian***. Rather than constantly stating explicitly that our manifolds are either Riemannian or pseudo-Riemannian, we shall merely let it be *understood* that our manifolds have metrics: they will *always* be either Riemannian or pseudo-Riemannian.

[2]This is proved in Exercise 11.

$$\boxed{\frac{1}{12}n^2(n^2-1)} \qquad (29.1)$$

distinct curvature components: these form the numerical description of a geometrical object called the ***Riemann tensor***.

In a 2-dimensional surface the Riemann tensor therefore reduces (with $n=2$) to a single component, and this is simply $\mathcal{K}$. The room in which you now sit *appears* to be 3-dimensional, and therefore the space around you (with $n=3$) is described by six curvature components. However, as you sit still in your chair, you are actually hurtling into the future, along the fourth dimension, *time!* In Einstein's 4-dimensional, curved spacetime, there are 20 curvature components, and, as we shall see in the next chapter, these describe the gravitational field!

[29.1] *Bernhard Riemann (1826–1866).*

- **Fifth**, we shall see how to generalize the ***Jacobi Equation*** to an n-manifold.
- **Sixth**, we shall describe a particularly important, geometrically meaningful *average* of Riemann's curvatures, called the ***Ricci curvature tensor***. In Einstein's curved, 4-dimensional spacetime, this captures exactly *half* of the complete curvature information (10 components out of 20 total).

29.2 Angular Excess in an n-Manifold

Our study of 2-surfaces has provided the essential springboard to understand n-manifolds. But a 2-surface is simply too special to manifest the full variety of phenomena that can and do occur in n-manifolds.

However, very happily, it turns out that essentially *all*[3] *the fundamentally new features and concepts of n-manifolds are revealed if we merely go up* one *dimension, to a readily-visualizable* **3***-manifold,* such as the space you currently sit in. We shall therefore focus our attention initially on this very concrete case, but in the next chapter we shall increase the dimension yet again, from three to four, in order to understand Einstein's curved spacetime.

We have gained many important insights into our 2-surfaces by viewing them as embedded in $\mathbb{R}^3$. And while it is certainly possible to conceive of our 3-manifold as likewise embedded in a higher dimensional space, there is no longer any visual advantage to doing so. Furthermore, with our eye on the prize of understanding Einstein's curved spacetime, our focus will be solely on *intrinsic* properties of an n-manifold, as determined by its intrinsic metric. Therefore, *henceforth, we shall always assume that we are creatures living within the n-manifold, and* only *intrinsic measurements within it are meaningful.*

[3]There are a few phenomena that only appear once we reach dimension $n=4$. For example, the vitally important Weyl curvature tensor, described in Exercise 15, only exists if $n \geqslant 4$.

Our very first measure of intrinsic curvature was via the angular excess $\mathcal{E}(\Delta)$ of a geodesic triangle Δ. But to construct a geodesic triangle Δ, the first thing we need are the geodesics themselves! On a 2-surface, we constructed a geodesic by stretching a string over[4] the surface. In a 3-manifold, we simply stretch the string between two points, thereby obtaining a seemingly straight line between them. Alternatively, we may shoot a laser beam from one point to the other. But we no longer assume that our 3-manifold is the Euclidean $\mathbb{R}^3$; *instead, it is equipped with a general, non-Euclidean metric*, so stretched strings and laser beams can no longer be relied upon to behave as expected. Indeed, as we shall discuss in the next chapter, the bending of "straight" light rays was one of the first experimental confirmations (in 1919) of Einstein's prediction that the geometry of physical space is curved.

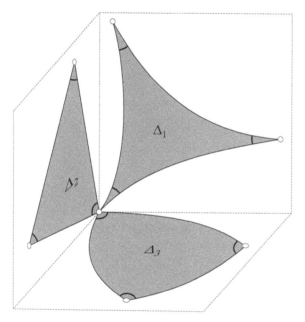

[29.2] *Geodesic triangles constructed in three perpendicular coordinate planes of a 3-manifold reveal only* part *of the curvature structure of the space. Here,* $\mathcal{E}(\Delta_1) < 0$, $\mathcal{E}(\Delta_2) = 0$, *and* $\mathcal{E}(\Delta_3) > 0$.

In a 2-surface, there was only one possible choice for the plane of our Δ. But in our 3-manifold, there are clearly infinitely many planes to choose from, one for each choice of the normal vector. The curvature within such a slice of space is called a ***sectional curvature***, and is the subject of Section 29.5.8. Figure [29.2] illustrates three such sectional curvatures in three orthogonal coordinate planes in a 3-manifold: they are all independent, both as regards their magnitudes and their signs. Here we have illustrated three geodesic triangles Δ_i with a common vertex, but each lying in a different coordinate plane. As illustrated, here $\mathcal{K}(\Delta_1) < 0$, $\mathcal{K}(\Delta_2) = 0$, and $\mathcal{K}(\Delta_3) > 0$.

But we have already stated that the Riemann tensor of a 3-manifold has six curvature components, so the angular excess in these three coordinate planes has only been able to detect *half* of the curvature structure. Things only get worse as the dimension increases: in a 4-manifold, the angular excess in the six different coordinate planes only yields six curvatures, whereas the Riemann tensor has 20 curvature components.

To explore curvature, in all its glory, it is essential that we instead turn to holonomy. To that end, we must now grapple with the *nontrivial* task of generalizing parallel transport from a 2-surface to an n-manifold. By virtue of its fundamental importance, we now offer not one, but *three* different geometrical constructions!

29.3 Parallel Transport: Three Constructions

29.3.1 Closest Vector on Constant-Angle Cone

Recall that in order to parallel transport a vector $\mathbf{w}(p)$ along a curve in a 2-surface, the fundamental *intrinsic* construction takes this curve to be a *geodesic* G. (To parallel transport along a *general*

[4] To cope with negative curvature, recall that we imagined the string trapped between two parallel layers of surface.

curve, we break it down into geodesic segments.) To achieve parallel transport along G, it is only necessary to *keep the angle α between* $\mathbf{w}_{||}$ *and* G *constant* (and keep the length of $\mathbf{w}_{||}$ constant).

In a 3-manifold, this construction immediately runs into trouble. See [29.3]. Suppose we try to carry $\mathbf{w}(p)$ a short distance ϵ along G from p to q to create the parallel vector $\mathbf{w}_{||}(p \rightsquigarrow q)$, and all we know is that α remains constant, and the length $\mathbf{w}_{||}$ remains constant. Then $\mathbf{w}_{||}(p \rightsquigarrow q)$ could lie anywhere on the illustrated *cone* $\mathcal{C}$ of directions; which generator of $\mathcal{C}$ should we pick?

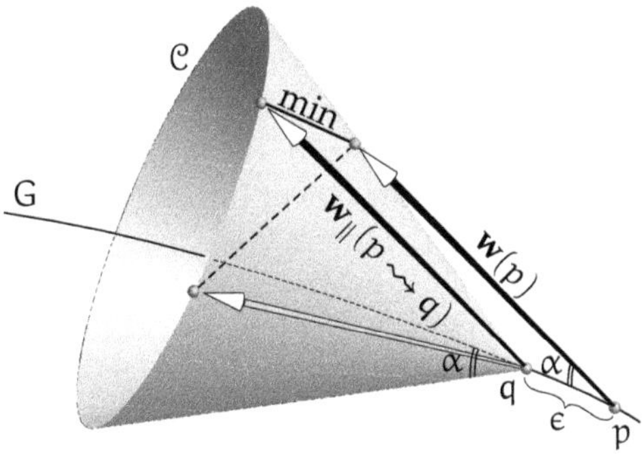

[29.3] *Suppose that* $\mathbf{w}(p)$ *makes angle* α *with the geodesic* G*. To parallel-transport* $\mathbf{w}(p)$ *distance* ϵ *along the geodesic* G *to* q*, construct the cone* $\mathcal{C}$ *of vectors at* q *making angle* α *with* G*, then pick the one whose tip is closest to the tip of* $\mathbf{w}(p)$*.*

For this construction, and for each of the other constructions that follow, it suffices to ask ourselves how we would do the analogous construction in $\mathbb{R}^3$, for as $\epsilon \to 0$ the two constructions must agree. Well, in $\mathbb{R}^3$, the generator of $\mathcal{C}$ that is parallel to $\mathbf{w}(p)$ is the one that is *closest*. That is, as illustrated,

> $\mathbf{w}_{||}(p \rightsquigarrow q)$ is *the* generator of $\mathcal{C}$ whose tip is *closest* to the tip of $\mathbf{w}(p)$.

By repeating this process over and over, we may parallel-transport $\mathbf{w}(p)$ as far as we wish along G.

29.3.2 Constant Angle within a Parallel-Transported Plane

The next construction, which we have only ever seen described by Arnol'd (1989, pp. 305–306), again draws its inspiration from $\mathbb{R}^3$.

To parallel-transport $\mathbf{w}(p)$ along a Euclidean straight line G in $\mathbb{R}^3$, with tangent $\mathbf{v}$, let $\Pi(p)$ be the plane spanned by $\mathbf{w}(p)$ and $\mathbf{v}(p)$, the angle between the two vectors being α, as before. Now parallel-transport $\Pi(p)$ along G to obtain $\Pi_{||}$. Finally, parallel-transport $\mathbf{w}(p)$ along G as the unique vector in $\Pi_{||}$ making angle α with $\mathbf{v}$.

Of course in $\mathbb{R}^3$ this is a perverse waste of time, for we have an absolute parallelism that allows us to move $\mathbf{w}(p)$ along G without the assistance of $\Pi_{||}$. *However*, this now suggests a new construction in a curved 3-manifold (or n-manifold). See [29.4].

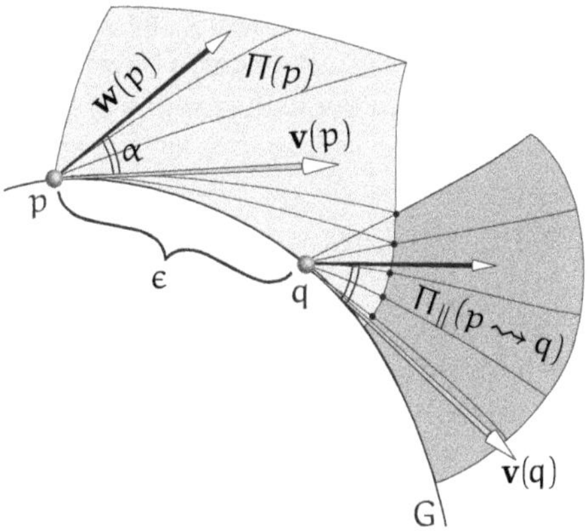

[29.4] *We construct the "plane"* $\Pi(p)$ *by launching geodesics from* p *in all directions spanned by* $\mathbf{w}(p)$ *and* $\mathbf{v}(p)$*. We then (approximately) parallel-transport* $\Pi(p)$ *to* q *to form* $\Pi_{||}(p \rightsquigarrow q)$*, generated by geodesics emanating from* q *and passing through neighbouring points of* $\Pi(p)$*.*

In $\mathbb{R}^3$, $\Pi(p)$ was made up of all the straight line emanating from p whose directions were linear combinations of $\mathbf{v}(p)$ and $\mathbf{w}(p)$. By analogy, in our curved 3-manifold, let $\Pi(p)$ now denote the "plane" made up of all the *geodesics* emanating from p whose directions are linear combinations of $\mathbf{v}(p)$ and $\mathbf{w}(p)$; of course this "plane" is actually a *curved 2-surface*. Note that this surface $\Pi(p)$ necessarily contains G, the geodesic in the direction $\mathbf{v}(p)$.

As in the previous construction, let q lie ϵ along G. By construction, $\Pi(p)$ necessarily contains q *and* the new direction of G at q, namely $\mathbf{v}(q)$. If we zoom in on the small region of $\Pi(p)$ surrounding q, it looks like a Euclidean plane. We can now construct a new surface $\Pi_{||}(p \rightsquigarrow q)$—the parallel translation of $\Pi(p)$ to q—by launching geodesics out from q in all directions within this small planar region of $\Pi(p)$ surrounding q. Figure [29.4] illustrates a few of these geodesics (dashed), launched from q and passing through nearby (black dot) points of $\Pi(p)$. To be more precise, the geodesic generators of $\Pi_{||}(p \rightsquigarrow q)$ are obtained by taking the *limit* that the black dots are pulled into q.

Repeating this construction over and over, we may approximate parallel-transport of $\Pi(p)$ down the length of G. Finally, letting $\epsilon \to 0$, we obtain a continuously varying, parallel-transported 2-surface $\Pi_{||}$ along the length of G, and then,

> *To parallel transport $\mathbf{w}(p)$ along the geodesic G, we keep its length constant, and maintain its angle α with G, while always keeping it tangent to the parallel-transported 2-surface $\Pi_{||}$.*

29.3.3 Schild's Ladder

Our final construction is also the simplest. It was first stated by Alfred Schild (1921–1977) in an (unpublished) Princeton lecture in 1970, and was later named ***Schild's Ladder*** in Misner, Thorne, and Wheeler (1973).

On a 2-surface, it has often been helpful for us to imagine that a very short tangent vector (that actually lives in the $\mathbb{R}^2$ tangent plane to the surface) instead lies *in* the surface. Likewise, in an n-manifold, a very short vector in the $\mathbb{R}^n$ tangent space can be imagined to be *in* the manifold. Therefore, since parallel-transport (by definition) preserves length, if we wish to parallel-transport a long tangent vector at p, we may first *shrink it by some large factor* N in the $\mathbb{R}^n$ tangent space at p, then picture it as living *in* the manifold, then parallel-transport it, then finally *expand it by the same large factor* N in the $\mathbb{R}^n$ tangent space at the destination, thereby restoring its original length.

Schild's Ladder construction uses this freedom to imagine that the vector $\mathbf{w}(p)$ that is to be parallel-transported along the geodesic G is very short, and can therefore be pictured as a very short geodesic segment *in* the manifold. With this understanding, the construction of *Schild's Ladder* is spelled out in the caption of [29.5], which illustrates it.

29.4 The Intrinsic (aka "Covariant") Derivative $\nabla_{\mathbf{v}}$

Previously, we used $\nabla_{\mathbf{v}}$ to denote the directional derivative in $\mathbb{R}^3$, and we used $D_{\mathbf{v}}$ to denote the intrinsic derivative within the 2-surface. In *extrinsic* terms, we saw in (23.3) that $D_{\mathbf{v}}$ could be viewed as the projection $\mathcal{P}$ of $\nabla_{\mathbf{v}}$ into the surface, along its normal $\mathbf{n}$:

$$D_{\mathbf{v}}\mathbf{w} = \mathcal{P}[\nabla_{\mathbf{v}}\mathbf{w}] = \nabla_{\mathbf{v}}\mathbf{w} - (\mathbf{n} \cdot \nabla_{\mathbf{v}}\mathbf{w})\,\mathbf{n}.$$

However, as we have noted, we shall *not* consider our n-manifold to be embedded in a higher-dimensional space, so no analogue of the above formula will be sought or found. Instead, we will think entirely *intrinsically*.

Fortunately, we already have an intrinsic definition of this intrinsic derivative in (23.2), page 242, and we even drew a picture of it in [23.3]. Let us briefly review that construction.

Our task is to find the rate of change of a vector field $\mathbf{w}$ as we move away from a point p in the direction of a unit vector $\mathbf{v}$. We move a small (ultimately vanishing) distance ϵ away from p along $\epsilon\mathbf{v}$, arriving at q. To see how much the new vector $\mathbf{w}(q)$ has changed from its original value $\mathbf{w}(p)$,

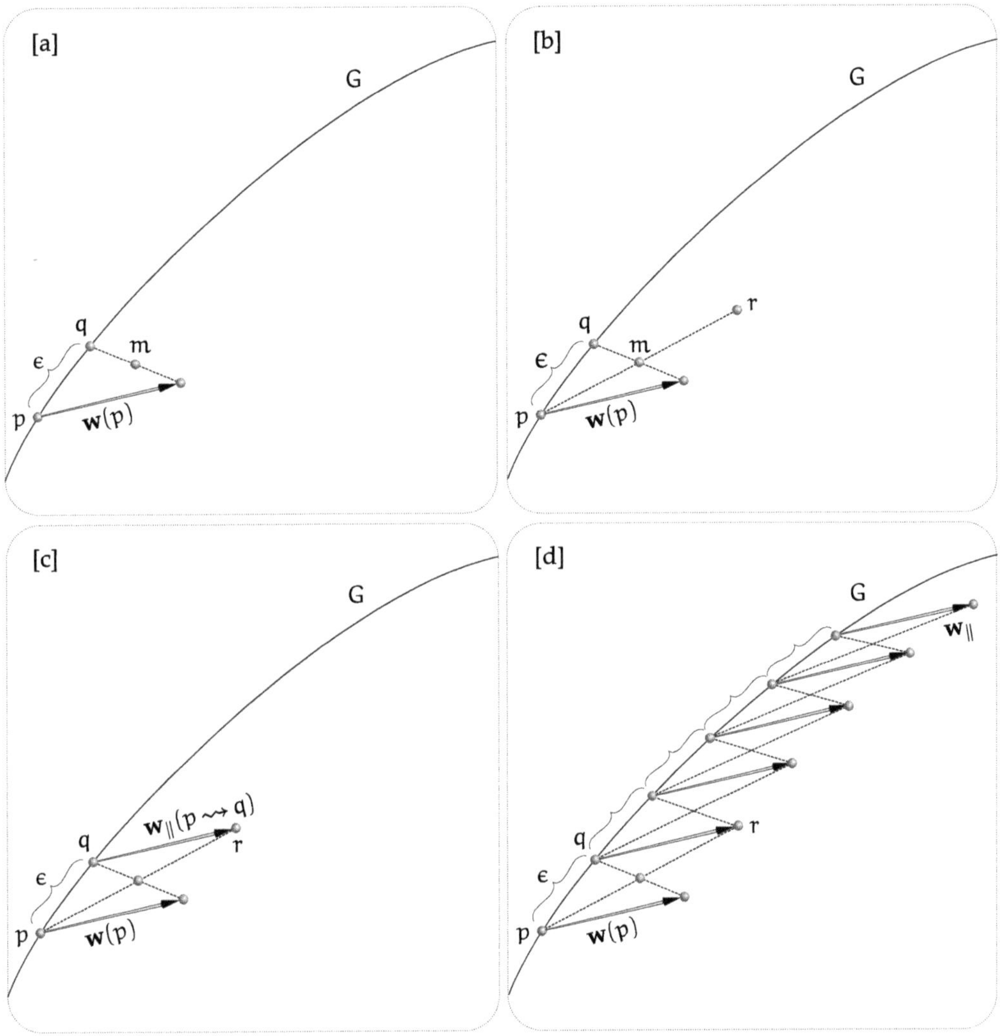

[29.5] Schild's Ladder. **[a]** *Travel distance* ϵ *along the geodesic* G *to construct the point* q. *Connect the tip of* $\mathbf{w}(\mathrm{p})$ *to* q *with a geodesic segment, and mark its midpoint* m. **[b]** *Connect* p *to* m *with another geodesic segment, and then extend it an equal distance beyond* m *to construct the segment terminating in* r. **[c]** *Join* q *to* r *to create the first rung of Schild's Ladder* $\asymp \mathbf{w}_{||}(\mathrm{p} \rightsquigarrow \mathrm{q})$. **[d]** *Repeat the construction to add more rungs to Schild's Ladder. Finally, let* $\epsilon \to 0$, *thereby parallel transporting* $\mathbf{w}_{||}$ *along* G.

we parallel-transport it back from q *to* p *as* $\mathbf{w}_{||}(\mathrm{q} \rightsquigarrow \mathrm{p})$. We then find the change, $[\mathbf{w}_{||}(\mathrm{q} \rightsquigarrow \mathrm{p}) - \mathbf{w}(\mathrm{p})]$, and finally we divide by ϵ to find the *rate* of change.

Now that we know how to do parallel transport [three ways!] in an n-manifold, we can define the intrinsic (aka "covariant") derivative in an n-manifold *exactly* as just described. Only the notation changes: henceforth, *we adopt the standard choice of a* ***bold*** *nabla symbol—*$\boldsymbol{\nabla}$*—to represent the* ***intrinsic derivative***,

$$\boldsymbol{\nabla}_{\mathbf{v}}\mathbf{w} \asymp \frac{\mathbf{w}_{||}(\mathrm{q} \rightsquigarrow \mathrm{p}) - \mathbf{w}(\mathrm{p})}{\epsilon}. \tag{29.2}$$

For our upcoming purposes, it will prove much more useful to think of the intrinsic derivative not in terms of a *rate* of change, but rather more directly in terms of the actual change itself:

$$\mathbf{w}_{||}(q \rightsquigarrow p) - \mathbf{w}(p) \asymp \epsilon \nabla_{\mathbf{v}} \mathbf{w} = \nabla_{\epsilon \mathbf{v}} \mathbf{w}.$$

Let $\delta_{pq}\mathbf{w}$ denote the small intrinsic change in $\mathbf{w}$ along the vector $\boldsymbol{\epsilon} \equiv \epsilon \mathbf{v}$ from p to q, then

$$\delta_{pq}\mathbf{w} = \text{intrinsic change in } \mathbf{w} \text{ from tail to tip of } \boldsymbol{\epsilon} \asymp \nabla_{\boldsymbol{\epsilon}} \mathbf{w}. \qquad (29.3)$$

It follows directly from this definition, that

$$\nabla_{\mathbf{v}} \mathbf{w}_{||} = 0 \quad \Longleftrightarrow \quad \mathbf{w}_{||} \text{ is parallel-transported along } \mathbf{v}.$$

If $\mathbf{v}$ is the velocity of a geodesic, then it is parallel-transported along itself, so the ***Geodesic Equation*** (23.5) now takes the form,

$$\nabla_{\mathbf{v}} \mathbf{v} = 0. \qquad (29.4)$$

We should note that if we allow the particle to speed up or slow down as it travels along the geodesic, then we obtain a more general form of the geodesic equation, namely, $\nabla_{\mathbf{v}} \mathbf{v} \propto \mathbf{v}$, which says that the *direction* of $\mathbf{v}$ is intrinsically constant, but its magnitude is permitted to change.

29.5 The Riemann Curvature Tensor

29.5.1 Parallel Transport Around a Small "Parallelogram"

Just as we did in the case of 2-manifolds (surfaces) we can now study curvature in an n-manifold by parallel-transporting a unit vector $\mathbf{w}_{||}$ around a small loop L. More specifically, we shall *attempt* to construct L as a *parallelogram* whose edges are built out of two (unit) vector fields $\mathbf{u}$ and $\mathbf{v}$.

Therefore, as illustrated in [29.6], starting at o, we lay out short vectors in these two directions, $\mathbf{u}(o)\,\delta u$ (connecting o to a) and $\mathbf{v}(o)\,\delta v$ (connecting o to p). In order to attempt to create a parallelogram, we now lay out $\mathbf{v}(a)\,\delta v$ (connecting a to b) and $\mathbf{u}(p)\,\delta u$ (connecting p to q). But the problem is that, in general, $q \neq b$: *the "parallelogram" fails to close!*

In the next subsection we shall find the formula for the extremely small "gap-closing" vector c connecting b to q. For now, though, simply suppose that we know how to close this gap, and are therefore able to parallel-transport $\mathbf{w}$ around the *closed* loop, L = oabqpo.

Within a 2-surface, the initial vector $\mathbf{w}_o$ that is parallel-transported can only lie in the same plane as $\mathbf{u}$ and $\mathbf{v}$. The *fundamentally new feature* in a 3-manifold (or more generally an n-manifold) is that $\mathbf{w}$ *can now stick out of the plane of the small loop around which it is parallel-transported.*

Having set the scene, let us parallel-transport $\mathbf{w}_o$ around L as $\mathbf{w}_{||}$, returning it to o as $\mathbf{w}_{||}(o)$. We can then define the ***vector holonomy***[5] to be the net change in $\mathbf{w}_{||}$ induced by the curvature:

$$\delta\mathbf{w}_{||} \equiv \mathbf{w}_{||}(\text{upon returning to o}) - \mathbf{w}(\text{as it departs from o})$$

Within a 2-surface, we introduced the holonomy operator $\mathcal{R}(L)$, which (when applied to $\mathbf{w}$) gave the net rotation (holonomy) of $\mathbf{w}$ after it was parallel-transported around L. Now, in a 3-manifold, or indeed an n-manifold, there are infinitely many different planes within which L may lie. Furthermore, in a 2-surface the entire tangent plane rotated rigidly as the vectors within

[5] We have invented this term because we are not aware of a standard name for this concept in the literature.

it were parallel-transported, so we did not need to pay attention to *which* vector $\mathbf{w}$ was being transported: they all rotated the same amount, $\mathcal{R}(L)$. But in a 3-manifold, $\mathbf{w}$ can stick out of the plane of L, and in an n-manifold, there can be many independent ways it can point. Crucially, *the vector holonomy now* does *depend on which vector is parallel-transported around the loop.*

For both these reasons, we must refine and generalize our previous notation, and introduce the ***Riemann curvature operator*** $\mathfrak{R}$ associated with the edges of the parallelogram L, which then acts on the vector that is parallel-transported, yielding the vector holonomy:

$$-\delta\mathbf{w}_{||} \equiv \mathfrak{R}(\mathbf{u}\,\delta u, \mathbf{v}\,\delta v)\,\mathbf{w}.$$

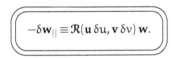

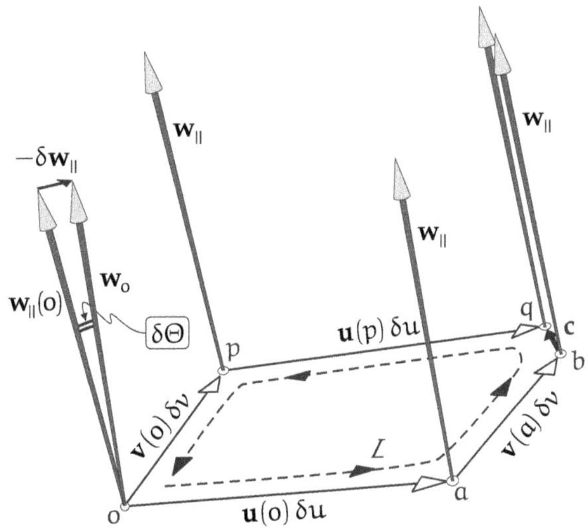

[29.6] Geometric Meaning of Riemann's Curvature. *Starting at o, we create a small parallelogram out of two vector fields* $\mathbf{u}$ *and* $\mathbf{v}$*, then parallel-transport an initial vector* $\mathbf{w}_o$ *around the loop L as* $\mathbf{w}_{||}$*. When it returns to o as* $\mathbf{w}_{||}(o)$*, it has changed by the illustrated* **vector holonomy** $\delta\mathbf{w}_{||} = -\mathfrak{R}(\mathbf{u}\,\delta u, \mathbf{v}\,\delta v)\,\mathbf{w}$*, where* $\mathfrak{R}$ *is the* **Riemann curvature operator**.

As we know, the curvature $\mathcal{K}$ of a 2-surface is completely captured by the holonomy per unit area. In an n-manifold we can likewise look at the angle $\delta\Theta$ between the initial vector $\mathbf{w}_o$ and the parallel-transported vector $\mathbf{w}_{||}(o)$ when it returns to o, as illustrated in [29.6]. Since $\mathbf{w}$ is a unit vector, the distance $|\delta\mathbf{w}|$ that its tip rotates is ultimately equal to $\delta\Theta$, the angle of rotation.

For simplicity's sake, suppose that $\mathbf{u}$ and $\mathbf{v}$ are orthogonal, so that our parallelogram is a rectangle of area $\delta\mathcal{A} = \delta u\,\delta v$. We can then generalize $\mathcal{K}$ and extract a scalar curvature $\mathcal{K}(\mathbf{u},\mathbf{v};\mathbf{w})$, once again defined as the rotation per unit area:

$$\mathcal{K}(\mathbf{u},\mathbf{v};\mathbf{w}) \asymp \frac{\delta\Theta}{\delta\mathcal{A}} \asymp \frac{|\delta\mathbf{w}_{||}|}{\delta\mathcal{A}} \asymp |\mathfrak{R}(\mathbf{u},\mathbf{v})\,\mathbf{w}|\,.$$

However, it is clear that this is no longer a satisfactory measure of the curvature, for we have completely lost the crucial information about the *direction* of the vector holonomy $\delta\mathbf{w}_{||}$. As we shall see, the Riemann tensor is the geometric object that encodes the *full* curvature information: both the angle $\delta\Theta$ by which $\mathbf{w}_{||}$ tips away from $\mathbf{w}_o$, *and* the direction in which it does so.

29.5.2 Closing the "Parallelogram" with the Vector Commutator

The vector $\mathbf{c}$ that closes up our near-miss parallelogram is extremely small, being (at worst[6]) of order $\delta u\,\delta v$, so ignoring it would still yield an excellent approximation to the curvature. But to obtain a *mathematically perfect* description of the Riemann tensor, we must parallel transport our vector around a *closed* loop, and therefore we *must* close the "parallelogram" with $\mathbf{c}$.

As we see geometrically, directly from [29.7], this very short vector that closes up the gap in the parallelogram can be expressed as the ***commutator***[7] of its edges:

---

[6]For vector fields tied to a coordinate grid, the gap disappears completely.

[7]Also called the ***Lie Bracket***, after the great Norwegian mathematician, Sophus Lie (pronounced "lee") (1842–1899).

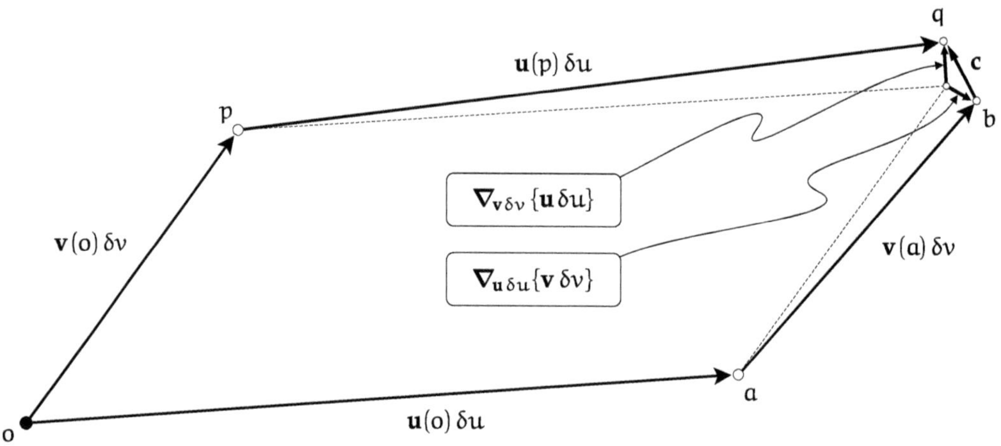

[29.7] *Geometric proof that the vector* **c** *that closes up the gap in the "parallelogram" is the* **commutator** *of its edges:* $\mathbf{c} \asymp \nabla_{\mathbf{v}\delta v}\{\mathbf{u}\,\delta u\} - \nabla_{\mathbf{u}\delta u}\{\mathbf{v}\,\delta v\} = [\mathbf{v},\mathbf{u}]\,\delta u\,\delta v$.

$$\mathbf{c} \asymp [\mathbf{v}\,\delta v, \mathbf{u}\delta u] = [\mathbf{v},\mathbf{u}]\,\delta u\,\delta v,$$

where

$$[\mathbf{v},\mathbf{u}] \equiv \nabla_{\mathbf{v}}\mathbf{u} - \nabla_{\mathbf{u}}\mathbf{v}.$$

We note a simple fact which we will need very shortly, namely, that the commutator is *antisymmetric*:

$$[\mathbf{v},\mathbf{u}] = -[\mathbf{u},\mathbf{v}].$$

29.5.3 The General Riemann Curvature Formula

We can now return to the problem of finding the change $\delta\mathbf{w}_{||}$ in $\mathbf{w}_{||}$ when it is parallel-transported around the closed loop, $L = oabqpo$.

To motivate the following argument, let us briefly review the first crucial step in Hopf's intrinsic proof of GGB—the introduction of the "holonomy" of an *open* curve. As illustrated in [26.1], page 258, we introduced a fiducial vector field **U**, enabling us to define $\mathcal{R}_U(K)$ for the *open* curve K to be the net change in the angle $\angle U w_{||}$—from **U** to the parallel-transported vector $\mathbf{w}_{||}$—as we travel along K:

$$\mathcal{R}_U(K) \equiv \delta_K\,(\angle U w_{||}).$$

As we noted at the time, the subscript U is essential, for $\mathcal{R}_U(K)$ *does indeed depend on our arbitrary choice of* **U**. Thus $\mathcal{R}_U(K)$ has no true mathematical meaning; it was merely a stepping stone in finding the geometrically meaningful holonomy of a *closed* loop.

Indeed, the holonomy $\mathcal{R}(L)$ of a closed polygonal loop L could then be expressed as the sum of the "holonomies" of its edges, and while each of the individual holonomies *did* depend on the arbitrary choice of **U**, their sum $\mathcal{R}(L) = \mathcal{K}(L)$ measured the curvature within, and was *independent of* **U**.

In a 2-surface we did *not* need to specify the vector $\mathbf{w}$ that was parallel-transported, because the entire tangent plane rotated rigidly as we moved across the surface. (This is *no longer true* in a 3-manifold: the initial choice of the direction of $\mathbf{w}$ *does* affect its vector holonomy around a closed loop.)

Still in a 2-surface, for the moment, we may choose $\mathbf{w}$ freely, so let us choose $\mathbf{w} = \mathbf{U}$. Now consider the *negative* of this holonomy—the change in $\mathbf{U}$ relative to parallel-transported vector $\mathbf{U}_{||}$—along a very a short vector $\boldsymbol{\epsilon}$. Taking $\mathbf{U}$ to have unit length, the holonomy becomes none other than the *intrinsic derivative* of $\mathbf{U}$:

$$-\mathcal{R}_{\mathbf{U}}(\boldsymbol{\epsilon}) = \delta_{\boldsymbol{\epsilon}}\,(\angle \mathbf{U}_{||}\mathbf{U}) \asymp \nabla_{\boldsymbol{\epsilon}}\,(\angle \mathbf{U}_{||}\mathbf{U}) \asymp |\nabla_{\boldsymbol{\epsilon}}\,\mathbf{U}|.$$

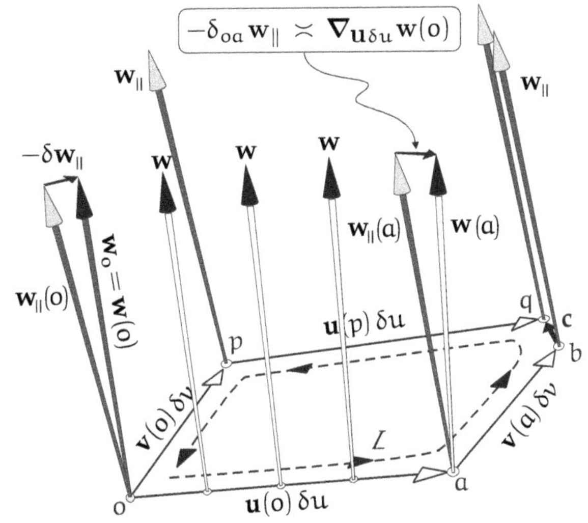

[29.8] The Riemann Curvature Tensor. *To calculate the vector holonomy, we introduce a fiducial vector field* $\mathbf{w}$ *that is completely arbitrary,* except *that* $\mathbf{w}(o) = \mathbf{w}_o$*; these are the black and white vectors (only those along* oa *are shown). When summed over all five edges, the changes in* $\mathbf{w}$ *relative to* $\mathbf{w}_{||}$ *yield the* negative *of the vector holonomy of the loop.*

We can now return to the problem at hand and apply this insight to the evaluation of $-\delta\mathbf{w}_{||}$ within the 3-manifold [29.6]. As illustrated in [29.8], we now introduce a fiducial vector field $\mathbf{w}$ throughout the region containing the parallelogram—these are the black-and-white vectors, which we have only drawn along oa, to avoid clutter. This vector field can be completely arbitrary, *except* that now it *does* matter which vector $\mathbf{w}_o$ is parallel-transported, so we insist that $\mathbf{w}(o) = \mathbf{w}_o$. Then, using (29.3), the negative of the vector holonomy along the first edge of the parallelogram is[8]

$$-\delta_{oa}\mathbf{w}_{||} = \mathbf{w}(a) - \mathbf{w}_{||}(a) \asymp \nabla_{\mathbf{u}\delta u}\mathbf{w}(o) = \delta u\,\nabla_{\mathbf{u}}\mathbf{w}(o).$$

Let us now repeatedly apply (29.3) to sum these changes in the arbitrary vector field $\mathbf{w}$, relative the parallel-transported vector $\mathbf{w}_{||}$, along the five sides of the closed-up parallelogram. When we return to o, of course $\mathbf{w}(o)$ has the same value as when we left o, so the net change in $\mathbf{w}$ relative to $\mathbf{w}_{||}$ is *entirely* due to the absolute change in $\mathbf{w}_{||}$, and is *independent* of the arbitrary choice of the vector field $\mathbf{w}$—this net change is the vector holonomy we seek.

We spell out every step of this summation, and encourage you to refer back to [29.8] to make geometrical sense of each line. Having absorbed each separate step of this crucial argument, it is worth your time to step back and grasp it as a single, complete idea:

$$\begin{aligned}
-\delta\mathbf{w}_{||} &= -\left[\delta_{oa}\,\mathbf{w}_{||} + \delta_{ab}\,\mathbf{w}_{||} + \delta_{bq}\,\mathbf{w}_{||} + \delta_{qp}\,\mathbf{w}_{||} + \delta_{po}\,\mathbf{w}_{||}\right] \\
&\asymp \delta u\,\nabla_{\mathbf{u}}\mathbf{w}(o) + \delta v\,\nabla_{\mathbf{v}}\mathbf{w}(a) + \nabla_{\mathbf{c}}\mathbf{w}(b) - \delta u\,\nabla_{\mathbf{u}}\mathbf{w}(p) - \delta v\,\nabla_{\mathbf{v}}\mathbf{w}(o) \\
&= \delta v\,\{\nabla_{\mathbf{v}}\mathbf{w}(a) - \nabla_{\mathbf{v}}\mathbf{w}(o)\} - \delta u\,\{\nabla_{\mathbf{u}}\mathbf{w}(p) - \nabla_{\mathbf{u}}\mathbf{w}(o)\} + \delta u\,\delta v\,\nabla_{[\mathbf{v},\mathbf{u}]}\mathbf{w}(o) \\
&\asymp \delta v\,\{\delta u\,\nabla_{\mathbf{u}}\nabla_{\mathbf{v}}\mathbf{w}\} - \delta u\,\{\delta v\,\nabla_{\mathbf{v}}\nabla_{\mathbf{u}}\mathbf{w}\} - \delta u\,\delta v\,\nabla_{[\mathbf{u},\mathbf{v}]}\mathbf{w}
\end{aligned}$$

[8]Here, and in the following, a more accurate approximation would be obtained by evaluating the derivative at the *midpoint* of the segment. However, since we shall shrink the entire figure down to the point o, and are dealing with ultimate equality in this limit, we think the argument may be easier to follow without taking this extra step.

$$= \quad \delta u\, \delta v \left\{ \nabla_{\mathbf{u}}\nabla_{\mathbf{v}} - \nabla_{\mathbf{v}}\nabla_{\mathbf{u}} - \nabla_{[\mathbf{u},\mathbf{v}]} \right\} \mathbf{w}(o)$$

$$= \quad \delta u\, \delta v\, \mathcal{R}(\mathbf{u},\mathbf{v})\, \mathbf{w}(o).$$

We have arrived at a major event in our drama—**vector holonomy in terms of Riemann's curvature**:

$$-\frac{\delta \mathbf{w}_{||}}{\delta u\, \delta v} \asymp \mathcal{R}(\mathbf{u},\mathbf{v})\, \mathbf{w} = \left\{ [\nabla_{\mathbf{u}}, \nabla_{\mathbf{v}}] - \nabla_{[\mathbf{u},\mathbf{v}]} \right\} \mathbf{w}. \tag{29.5}$$

We now see that the ***Riemann curvature operator*** introduced in [29.6] is, in fact,

$$\mathcal{R}(\mathbf{u},\mathbf{v}) = [\nabla_{\mathbf{u}}, \nabla_{\mathbf{v}}] - \nabla_{[\mathbf{u},\mathbf{v}]}. \tag{29.6}$$

The minus sign on the left-hand side of (29.5) is important, so let us make a point of reminding the reader of its origin and geometrical meaning. The operator on the right-hand side tells us the net change in the (arbitrary) fiducial vector field $\mathbf{w}$ (*relative to* $\mathbf{w}_{||}$) as we traverse the loop. But the geometrically meaningful quantity we seek is the vector holonomy $\delta\mathbf{w}_{||}$, which is the *opposite*: the change in $\mathbf{w}_{||}$ *relative to* $\mathbf{w}$.

Note that $\mathcal{R}$ is *antisymmetric*:

$$\mathcal{R}(\mathbf{u},\mathbf{v}) = -\mathcal{R}(\mathbf{v},\mathbf{u}) \quad \Longrightarrow \quad \mathcal{R}(\mathbf{u},\mathbf{u}) = 0. \tag{29.7}$$

The truth of the first equation is geometrically clear, for it simply asserts that if we reverse the direction in which we traverse the parallelogram, the vector holonomy is reversed. Likewise, the second equation may be viewed as saying that if *any* vector is parallel-transported back and forth a short distance along an arbitrary vector $\mathbf{u}$, then it returns unchanged.

Only a minor change of notation is now needed to finally arrive at the standard definition of the famous ***Riemann curvature tensor*** $\mathbf{R}$—it is a vector-valued function $\mathbf{R}(\mathbf{u},\mathbf{v};\, \mathbf{w})$ of three input vectors:[9]

$$\mathbf{R}(\mathbf{u},\mathbf{v};\, \mathbf{w}) \equiv \mathcal{R}(\mathbf{u},\mathbf{v})\, \mathbf{w} = \left\{ [\nabla_{\mathbf{u}}, \nabla_{\mathbf{v}}] - \nabla_{[\mathbf{u},\mathbf{v}]} \right\} \mathbf{w}. \tag{29.8}$$

CONVENTION WARNING: The arrangement of vectors in the slots of $\mathbf{R}$ differs amongst authors. For example, Misner, Thorne, and Wheeler (1973) *agree* with our definition of $\mathcal{R}(\mathbf{u},\mathbf{v})\, \mathbf{w}$ above, but, on the left-hand side, their $\mathbf{R}(\mathbf{u},\mathbf{v};\, \mathbf{w})$ rearranges the slots in such a way that their $\mathbf{R}$ turns out to be the *negative* of ours.

---

[9]It is *not* entirely standard to use a semicolon to separate the parallelogram vectors from the vector that is parallel-transported around it, but we think it helpful (as does Penrose!).

29.5.4 Riemann's Curvature Is a Tensor

As we shall discuss in greater detail in Act V, a ***tensor*** is (by definition) a *linear* function of its multiple vector[10] inputs. As we now explain, $\mathbf{R}$ is indeed ***multi-linear***:

> $\mathbf{R}(\mathbf{u}, \mathbf{v};\ \mathbf{w})$ *is a linear function of each of its three inputs, when the other two are held fixed.*

Furthermore, *despite* the fact that formula (29.8) is built out of derivatives, the Riemann curvature *only* depends on the values of the three vectors *at* the point at which $\mathbf{R}$ is evaluated—it is *independent* of how the vectors vary in the vicinity of that point. At the computational level, this appears downright *paradoxical*, and yet it does make geometrical sense, for $\mathbf{R}$ is telling us about the curvature of the *space itself*, and the vectors merely serve to single out particular parts (or ***components***) of this curvature.

To summarize, here are

> **The two defining properties of a *tensor*:**
>
> - *The output depends* linearly *on each input vector.*
> - *The output* only *depends on the input vectors* at *the point of evaluation.*

First, let us establish linearity in the $\mathbf{w}$-slot. Suppose that $\mathbf{w} = k_1\mathbf{w}_1 + k_2\mathbf{w}_2$, where k_1 and k_2 are constants. If $\mathbf{w}$, $\mathbf{w}_1$, and $\mathbf{w}_2$ are each parallel-transported (starting at o) around a small (ultimately vanishing) parallelogram L with edges along $\mathbf{u}$ and $\mathbf{v}$, then

$$\delta\mathbf{w}_{||} = k_1\delta[\mathbf{w}_1]_{||} + k_2\delta[\mathbf{w}_2]_{||},$$

so the geometrical interpretation (29.5) implies that

$$\mathbf{R}(\mathbf{u}, \mathbf{v};\ k_1\mathbf{w}_1 + k_2\mathbf{w}_2) = k_1\,\mathbf{R}(\mathbf{u}, \mathbf{v};\ \mathbf{w}_1) + k_2\,\mathbf{R}(\mathbf{u}, \mathbf{v};\ \mathbf{w}_2).$$

Of course this can also be instantly confirmed by computation: just use the formula (29.8) and the simple fact that the intrinsic derivatives act linearly, e.g.,

$$\nabla_{\mathbf{v}}[k_1\mathbf{w}_1 + k_2\mathbf{w}_2] = k_1\nabla_{\mathbf{v}}\mathbf{w}_1 + k_2\nabla_{\mathbf{v}}\mathbf{w}_2.$$

To start to test the *second* requirement of tensorhood, suppose that we change $\mathbf{w}^{\text{OLD}}$ in the vicinity of o into $\mathbf{w}^{\text{NEW}} = f\,\mathbf{w}^{\text{OLD}}$, where f is an arbitrary (differentiable) function of position, *except* that $f(o) = 1$, so that $\mathbf{w}^{\text{NEW}}(o) = \mathbf{w}^{\text{OLD}}(o)$.

At this point, we strongly encourage you to pause in order to actually *carry out* the somewhat lengthy (but straightforward) *computation* that proves that, when evaluated at o,

$$\mathbf{R}(\mathbf{u}, \mathbf{v};\ \mathbf{w}^{\text{NEW}}) = \mathbf{R}(\mathbf{u}, \mathbf{v};\ \mathbf{w}^{\text{OLD}}).$$

If you just did this calculation, you saw that four terms arose that involved derivatives of f, but then, "miraculously," they all cancelled each other out! (NOTE: This is *not* cause for celebration—see the definition of a "false miracle" in the Prologue!)

---

[10] In Act V we will generalize this to the *standard* definition, whereby we allow the input of both vectors *and* 1-forms (introduced at the start of Act V).

Fortunately, this can instead be understood *directly* from the geometrical perspective, because varying the fiducial vector field $\mathbf{w}$ along L in *any* manner (not merely scaling it by f) has absolutely no effect on the geometrically meaningful parallel transport of the initial vector around L, and hence absolutely no effect on the vector holonomy! Thus $\mathbf{R}(\mathbf{u}, \mathbf{v};\, \mathbf{w})$ does indeed *only* depend on the value of $\mathbf{w}$ *at* o.

Now let us turn our attention to the first two slots, and let us start with the second defining property of a tensor. In the limit that loop shrinks to o, only the values $\mathbf{u}(o)$ and $\mathbf{v}(o)$ are needed to construct the loop around which $\mathbf{w}$ is parallel-transported, so the vector holonomy, and therefore $\mathbf{R}$, *only* depends on the values of these input vectors *at* o.

To witness the decisive advantage of the geometrical perspective, but also as a character-building exercise in the computational use of formula (29.8), let us play the same game with $\mathbf{u}$ that we previously did with $\mathbf{w}$. That is, let us change $\mathbf{u}^{\text{OLD}}$ in the vicinity of o into $\mathbf{u}^{\text{NEW}} = f\,\mathbf{u}^{\text{OLD}}$, where f is an arbitrary function. Now try your hand at proving (directly from the formula) that

$$\mathbf{R}(\mathbf{u}^{\text{NEW}}, \mathbf{v};\, \mathbf{w}) = \mathbf{R}(f\,\mathbf{u}^{\text{OLD}}, \mathbf{v};\, \mathbf{w}) = f\,\mathbf{R}(\mathbf{u}^{\text{OLD}}, \mathbf{v};\, \mathbf{w}).$$

Thus, if we again insist that $f(o) = 1$, to ensure that $\mathbf{u}^{\text{NEW}}(o) = \mathbf{u}^{\text{OLD}}(o)$, then

$$\mathbf{R}(\mathbf{u}^{\text{NEW}}, \mathbf{v};\, \mathbf{w})(o) = \mathbf{R}(\mathbf{u}^{\text{OLD}}, \mathbf{v};\, \mathbf{w})(o),$$

as required of a tensor.

Finally, let us turn to the linearity in the first two slots. Because of the antisymmetry (29.7), it suffices to prove linearity in either slot. Here there is essentially no difference between the geometrical and computational perspectives, and we leave the very short calculation to you, which establishes that,

$$\mathbf{R}(k_1\mathbf{u}_1 + k_2\mathbf{u}_2, \mathbf{v};\, \mathbf{w}) = k_1\,\mathbf{R}(\mathbf{u}_1, \mathbf{v};\, \mathbf{w}) + k_2\,\mathbf{R}(\mathbf{u}_2, \mathbf{v};\, \mathbf{w}).$$

GRAND CONCLUSION: Riemann's curvature is indeed a *tensor*!

29.5.5 Components of the Riemann Tensor

In order to give a numerical description of geometrical objects, we introduce a set of *orthonormal basis vectors*, $\{\mathbf{e}_i\}$, for the $\mathbb{R}^n$ tangent space at each point of the n-manifold. Then, for example, a geometrical vector $\mathbf{u}$ is represented by its numerical *components*, $\{u^i\}$, where $\mathbf{u} = \sum_i u^i \mathbf{e}_i$. WARNING: When dealing with components, we must remember that the superscripts are labels, *not* powers!

More complex geometrical objects, such as the Riemann tensor, require *multiple* distinct indices to describe their components. This leads to a profusion of summations. But it turns out that we can always arrange for the index that is being summed over to appear once as a superscript and once as a subscript.

Einstein therefore introduced a simple, clutter-clearing convention, called the ***Einstein summation convention***, whereby the sum over such a pair of indices is simply *understood*, and the summation sign is omitted. For example,

$$\mathbf{u} = \sum_{i=1}^{n} u^i \mathbf{e}_i \quad \Longleftrightarrow \quad \textit{Einstein summation convention:} \quad \mathbf{u} = u^i \mathbf{e}_i.$$

To find the components of the Riemann tensor, we decompose each of its three vector inputs into components:

$$\mathbf{u} = u^i \mathbf{e}_i, \qquad \mathbf{v} = v^j \mathbf{e}_j, \qquad \text{and} \qquad \mathbf{w} = w^k \mathbf{e}_k.$$

Thus,

$$\begin{aligned}\mathbf{R}(\mathbf{u},\mathbf{v};\,\mathbf{w}) &= \mathbf{R}(u^i\mathbf{e}_i,\,v^j\mathbf{e}_j;\,w^k\mathbf{e}_k)\\ &= \mathbf{R}(\mathbf{e}_i,\,\mathbf{e}_j;\,\mathbf{e}_k)\,u^i\,v^j\,w^k\end{aligned}$$

We can now define $R_{ijk}{}^{l}$, the ***components of the Riemann tensor***, as the components of the vector that results when the Riemann tensor acts on three basis vectors:

$$\mathbf{R}(\mathbf{e}_i,\,\mathbf{e}_j;\,\mathbf{e}_k) \equiv R_{ijk}{}^{l}\,\mathbf{e}_l.$$

Thus the effect of **R** on three general vectors can easily be expressed in terms of these components:

$$\mathbf{R}(\mathbf{u},\mathbf{v};\,\mathbf{w}) = \left[R_{ijk}{}^{l}\,u^i\,v^j\,w^k\right]\,\mathbf{e}_l.$$

For later use, we also define,

$$R_{ijkm} \equiv \mathbf{R}(\mathbf{e}_i,\,\mathbf{e}_j;\,\mathbf{e}_k)\bullet\mathbf{e}_m. \tag{29.9}$$

In our chosen *orthonormal basis*, $\mathbf{e}_l \bullet \mathbf{e}_m = 0$ if $l \neq m$, and it equals 1 when $l = m$, so

$$R_{ijkm} = R_{ijk}{}^{m}. \tag{29.10}$$

29.5.6 *For a Given* $\mathbf{w}_o$, *the Vector Holonomy* Only *Depends on the* Plane *of the Loop and Its* Area

Within a 2-surface, the holonomy $\mathcal{R}(L) \asymp \mathcal{K}\,\delta\mathcal{A}$ of a small loop L *only* depended on its area, $\delta\mathcal{A}$, and was independent of it shape. In an n-manifold, there are now many, independent choices for the plane of L. Let us therefore define,

$$\Pi(\mathbf{u},\mathbf{v}) \equiv \text{The plane spanned by } \mathbf{u} \text{ and } \mathbf{v}.$$

This choice of $\Pi(\mathbf{u},\mathbf{v})$ certainly dictates the vector holonomy. However, for a given $\mathbf{w}_o$, the vector holonomy *only* depends on Π itself, and on the area of the loop:

If we parallel transport $\mathbf{w}_o$ *around a small (ultimately vanishing) parallelogram in a plane* Π, *then the vector holonomy is proportional to the* area $\delta\mathcal{A}$ *of the parallelogram, and is* independent *of its shape.* (29.11)

We will prove this by calculation, in order to illustrate the power of our brand-new tool of tensor components. Let $\mathbf{e}_1$ and $\mathbf{e}_2$ be chosen within the plane of L. Then, using the Einstein summation convention,

$$\mathbf{u}\,\delta u = \delta u^1\,\mathbf{e}_1 + \delta u^2\,\mathbf{e}_2 = \delta u^i\,\mathbf{e}_i \qquad \text{and} \qquad \mathbf{v}\,\delta v = \delta v^1\,\mathbf{e}_1 + \delta v^2\,\mathbf{e}_2 = \delta v^j\,\mathbf{e}_j.$$

Thus the area of the loop is

$$\delta\mathcal{A} = \begin{vmatrix} \delta u^1 & \delta v^1 \\ \delta u^2 & \delta v^2 \end{vmatrix} = \delta u^1\,\delta v^2 - \delta u^2\,\delta v^1.$$

Let $\mathbf{w}_o = w_o^k\,\mathbf{e}_k$, and $\delta\mathbf{w}_{||} = \delta w_{||}^l\,\mathbf{e}_l$. Then

$$\begin{aligned} -\delta\mathbf{w}_{||} &= -\delta w_{||}^l\,\mathbf{e}_l \\ &= \mathbf{R}(\mathbf{u}\,\delta u, \mathbf{v}\,\delta v;\, \mathbf{w}_o) \\ &= \mathbf{R}(\delta u^i\,\mathbf{e}_i, \delta v^j\,\mathbf{e}_j;\, w_o^k\,\mathbf{e}_k) \\ &= \delta u^i\,\delta v^j\,w_o^k\,R_{ijk}{}^l\,\mathbf{e}_l. \end{aligned}$$

But, by virtue of (29.7),

$$R_{11k}{}^l = 0 = R_{22k}{}^l \qquad \text{and} \qquad R_{21k}{}^l = -R_{12k}{}^l.$$

Therefore, the components of the vector holonomy are given by

$$\boxed{-\delta\mathbf{w}_{||} = -\delta w_{||}^l\,\mathbf{e}_l \asymp \delta\mathcal{A}\,\left[R_{12k}{}^l\,w_o^k\right]\,\mathbf{e}_l,} \tag{29.12}$$

completing the proof of (29.11).

29.5.7 Symmetries of the Riemann Tensor

Naively, one would expect the number of independent components of the Riemann tensor R_{ijkm} in a 3-manifold, defined in (29.9), to be $3^4 = 81$. And yet we have already stated that the true number is only *six!* This dramatic reduction results from the fact that the Riemann tensor possesses *four* remarkable algebraic symmetries.

The first one we have already met in (29.7): $\mathbf{R}(\mathbf{u},\mathbf{v};\,\mathbf{w})$ is antisymmetric in its first two slots, and this implies that

$$\boxed{R_{jikm} = -R_{ijkm}.}$$

This immediately reduces the number of independent components [exercise] from 81 to 27.

We now prove a second, far less obvious symmetry, namely, that R_{ijkm} is also antisymmetric in its *last* two slots:

$$\boxed{R_{ijmk} = -R_{ijkm}} \tag{29.13}$$

This further reduces the number of independent components [exercise] from 27 to 9.

Since parallel transport preserves length, the vector $\mathbf{w}_o$ returns to o as the *slightly rotated vector* $\mathbf{w}_{||}(o)$, and therefore the connecting vector $\delta\mathbf{w}_{||}$ between the tips of these vectors is ultimately *orthogonal* to both of them:

$$\delta\mathbf{w}_{||} \bullet \mathbf{w}_o = 0 \quad \Longrightarrow \quad [\mathcal{R}(\mathbf{u},\mathbf{v})\,\mathbf{w}_o] \bullet \mathbf{w}_o = 0.$$

Now let $\mathbf{x}$ and $\mathbf{y}$ be arbitrary vectors, and set $\mathbf{w}_o = \mathbf{x}+\mathbf{y}$. Then,

$$\begin{aligned} 0 &= [\mathcal{R}(\mathbf{u},\mathbf{v})\,(\mathbf{x}+\mathbf{y})] \bullet (\mathbf{x}+\mathbf{y}) \\ &= [\mathcal{R}(\mathbf{u},\mathbf{v})\,\mathbf{x}] \bullet \mathbf{x} + [\mathcal{R}(\mathbf{u},\mathbf{v})\,\mathbf{x}] \bullet \mathbf{y} + [\mathcal{R}(\mathbf{u},\mathbf{v})\,\mathbf{y}] \bullet \mathbf{x} + [\mathcal{R}(\mathbf{u},\mathbf{v})\,\mathbf{y}] \bullet \mathbf{y} \\ &= 0 + [\mathcal{R}(\mathbf{u},\mathbf{v})\,\mathbf{x}] \bullet \mathbf{y} + [\mathcal{R}(\mathbf{u},\mathbf{v})\,\mathbf{y}] \bullet \mathbf{x} + 0. \end{aligned}$$

Therefore,

$$\boxed{[\mathcal{R}(\mathbf{u},\mathbf{v})\,\mathbf{x}] \bullet \mathbf{y} = -\,[\mathcal{R}(\mathbf{u},\mathbf{v})\,\mathbf{y}] \bullet \mathbf{x},} \tag{29.14}$$

and so (29.13) follows immediately from the definition (29.9).

For the sake of completeness, we now state the remaining symmetries of the Riemann tensor.

First, the ***Algebraic Bianchi Identity***[11] states that if the first three vectors are permuted cyclically, their sum vanishes:

$$\boxed{\mathcal{R}(\mathbf{u},\mathbf{v})\,\mathbf{w} + \mathcal{R}(\mathbf{v},\mathbf{w})\,\mathbf{u} + \mathcal{R}(\mathbf{w},\mathbf{u})\,\mathbf{v} = \mathbf{0} \iff R_{ijkm} + R_{jkim} + R_{kijm} = 0.} \tag{29.15}$$

Exercise 10 provides both a computational and a geometric proof of this result.

Next, the Riemann tensor is *symmetric under interchange of the first and second pairs of vectors*:

$$\boxed{[\mathcal{R}(\mathbf{u},\mathbf{v})\,\mathbf{x}] \bullet \mathbf{y} = [\mathcal{R}(\mathbf{x},\mathbf{y})\,\mathbf{u}] \bullet \mathbf{v} \iff R_{ijkm} = R_{kmij}.} \tag{29.16}$$

Although this is very useful, it is not a truly new symmetry, but rather a consequence of the previous symmetries. For a computational proof, see Exercise 10.

Using these symmetries, you may confirm [exercise][12] that in a 3-manifold the Riemann tensor has only six independent components. More generally, Exercise 11 shows that in an n-manifold the number of components is given by $\frac{1}{12}n^2(n^2-1)$, as previously claimed in (29.1).

Finally, in addition to the four algebraic symmetries above, there exists a fifth, *differential* symmetry, of fundamental importance to Einstein's theory of gravity—it is called the ***Differential Bianchi Identity:***[13]

$$\boxed{\nabla_{\mathbf{x}}\mathcal{R}(\mathbf{u},\mathbf{v})\,\mathbf{w} + \nabla_{\mathbf{u}}\mathcal{R}(\mathbf{v},\mathbf{x})\,\mathbf{w} + \nabla_{\mathbf{v}}\mathcal{R}(\mathbf{x},\mathbf{u})\,\mathbf{w} = \mathbf{0},} \tag{29.17}$$

in which the first three vectors are permuted cyclically, as they were in the Algebraic Bianchi Identity. The proof of this result will be deferred until we can derive it elegantly, using ***curvature 2-forms***, in Section 38.12.4.

---

[11] Also called the ***First Bianchi Identity*** or the ***Bianchi Symmetry***, but actually discovered by Ricci!

[12] Try simply listing six independent components, and then convince yourself that all other components can be found using the symmetries.

[13] Also called the ***Second Bianchi Identity***. According to Pais (1982, pp. 275–276), this was first discovered by Aurel Voss (1880), then Ricci (1889), and then Bianchi (1902). However, even Voss was not the first—see [29.10]!

29.5.8 Sectional Curvatures

In a 2-surface, the Gaussian curvature $\mathcal{K}$ emerged as the holonomy per unit area of a vector that *necessarily resided within the plane of the loop* around which it was parallel-transported. But in our 3-manifold (or n-manifold), $\mathbf{w}_o$ will typically stick *out* of this plane. Nevertheless, we are certainly free to *choose* it to lie within $\Pi(\mathbf{u}, \mathbf{v})$, in the hope of recovering a concept of curvature akin to the 2-dimensional case.

However, at first we seem to hit a road block:

> *Even if we insist that the initial vector $\mathbf{w}_o$ lie within $\Pi(\mathbf{u}, \mathbf{v})$, then parallel-transporting it around a loop within Π will* not *return it to o merely rotated within Π—the returning vector $\mathbf{w}_{||}(o)$ will typically stick out of Π.*

To overcome this, let us focus our attention on the orthogonal *projection* $\mathcal{P}[\mathbf{w}_{||}]$ of $\mathbf{w}_{||}$ into Π. More formally, if $\Pi = \Pi(\mathbf{e}_1, \mathbf{e}_2)$ then this ***orthogonal projection operator*** $\mathcal{P}$ is given by,

$$\mathcal{P}[a^1\mathbf{e}_1 + a^2\mathbf{e}_2 + a^3\mathbf{e}_3] = a^1\mathbf{e}_1 + a^2\mathbf{e}_2.$$

We will prove the following:

> *Let $\mathbf{w}_o$ be* any *vector in Π, and let $\mathbf{w}_{||}$ be parallel-transported around a loop in Π. Define $\mathcal{K}(\mathbf{u}, \mathbf{v})$ to be the rotation per unit area of the* projection $\mathcal{P}[\mathbf{w}_{||}]$ *of $\mathbf{w}_{||}$ into Π. Then $\mathcal{K}(\mathbf{u}, \mathbf{v})$* is independent *of the specific choice of $\mathbf{w}_o$.*

The curvature $\mathcal{K}(\mathbf{u}, \mathbf{v}) = \mathcal{K}(\Pi)$ therefore depends *only* on the plane Π, and is called the ***sectional curvature*** of $\Pi(\mathbf{u}, \mathbf{v})$.

To verify this, we simply take the general vector holonomy formula (29.12) and insist that $\mathbf{w}_o$ lie in $\Pi(\mathbf{u}, \mathbf{v}) = \Pi(\mathbf{e}_1, \mathbf{e}_2)$:

$$\mathbf{w}_o = w_o^1\,\mathbf{e}_1 + w_o^2\,\mathbf{e}_2,$$

and for simplicity's sake let us again assume it is a unit vector. Writing (29.12) in column-vector form, and using (29.10), we see that $\mathcal{P}[\delta\mathbf{w}_{||}]$ is

$$-\begin{bmatrix}\delta w_{||}^1\\ \delta w_{||}^2\end{bmatrix} = \begin{bmatrix}R_{12k}{}^1\,w_o^k\\ R_{12k}{}^2\,w_o^k\end{bmatrix}\delta\mathcal{A} = \begin{bmatrix}R_{121}{}^1\,w_o^1 + R_{122}{}^1\,w_o^2\\ R_{121}{}^2\,w_o^1 + R_{122}{}^2\,w_o^2\end{bmatrix}\delta\mathcal{A} = \begin{bmatrix}R_{1211}\,w_o^1 + R_{1221}\,w_o^2\\ R_{1212}\,w_o^1 + R_{1222}\,w_o^2\end{bmatrix}\delta\mathcal{A}.$$

But the antisymmetry (29.13) yields

$$R_{1211} = 0 = R_{1222} \qquad \text{and} \qquad R_{1221} = -R_{1212}.$$

Therefore, we now formally define the ***sectional curvature*** as

$$\mathcal{K}(\Pi) \equiv \mathcal{K}(\mathbf{e}_1,\, \mathbf{e}_2) \equiv [\mathcal{R}(\mathbf{e}_1,\, \mathbf{e}_2)\,\mathbf{e}_2] \bullet \mathbf{e}_1 = R_{1221}, \tag{29.18}$$

which is to say, the holonomy of $\mathbf{e}_2$, after parallel-transporting it around the $\{\mathbf{e}_1, \mathbf{e}_2\}$-loop, projected onto $\mathbf{e}_1$. Then

$$\mathcal{P}[\delta \mathbf{w}_{||}] = \begin{bmatrix} \delta w^1_{||} \\ \delta w^2_{||} \end{bmatrix} \asymp \begin{bmatrix} -w^2_o \\ w^1_o \end{bmatrix} \mathcal{K}(\Pi)\,\delta\mathcal{A} = \mathbf{w}_\perp\, \mathcal{K}(\Pi)\,\delta\mathcal{A}, \tag{29.19}$$

where $\mathbf{w}_\perp$ is simply $\mathbf{w}_o$ rotated through a right angle within Π, as illustrated in [29.9].

Thus, as the figure explains, the angle $\delta\Theta$ through which the *projection* $\mathcal{P}[\mathbf{w}_{||}]$ has been rotated by parallel-transport around the loop of area $\delta\mathcal{A}$ is $\delta\Theta \asymp \mathcal{K}(\Pi)\,\delta\mathcal{A}$, *independent* of $\mathbf{w}_o$. We have come full circle to our original view of curvature as the holonomy per unit area within a 2-surface:

$$\frac{\delta\Theta}{\delta\mathcal{A}} \asymp \mathcal{K}(\Pi) = R_{1221}. \tag{29.20}$$

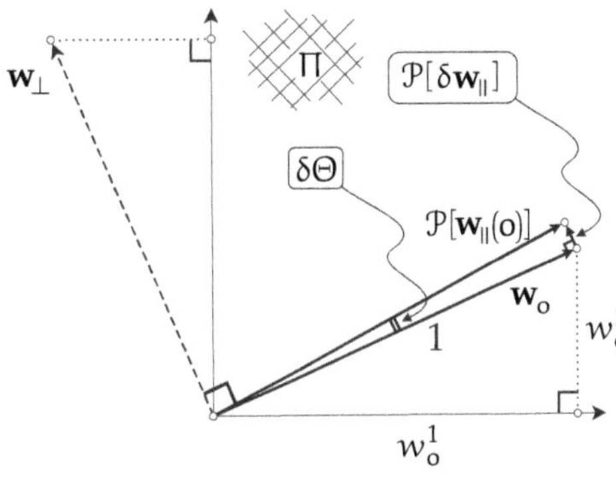

[29.9] *If* $\mathbf{w}_o$ *lies within the plane* Π *of the loop, then the projection into* Π *of the parallel-transported vector returns rotated by* $\delta\Theta \asymp |\delta\mathcal{P}[\mathbf{w}_{||}]| \asymp \mathcal{K}(\Pi)\,\delta\mathcal{A}$, *where* $\mathcal{K}(\Pi)$ *is the* **sectional curvature** *of* Π.

Only now there are *many* such sectional curvatures, depending upon *which* plane Π we parallel transport within.

As we mentioned earlier, knowing the sectional curvatures in just the orthogonal coordinate planes, for example, only gives a very incomplete[14] picture of the Riemann curvature of the space. But what if we are allowed to know $\mathcal{K}(\Pi)$ for *all possible* planes Π?

It turns out that this *is* sufficient to reconstruct the complete Riemann tensor! It is not easy to find this discussed in standard texts, but Kühnel (2015, p. 247) gives an *explicit formula* for $[\mathcal{R}(\mathbf{u},\mathbf{v})\,\mathbf{x}] \cdot \mathbf{y}$ in terms of a sum of *eighteen* sectional curvatures of eighteen planes constructed from linear combinations of the four input vectors of the Riemann curvature!

29.5.9 Historical Notes on the Origin of the Riemann Tensor

Riemann announced the discovery of his tensor, albeit obliquely, on the 10th of June, 1854, in a lecture at Göttingen University—as a mandatory part of his application to become an *unsalaried* faculty member!—delivered to a mainly non-mathematical audience.

The lecture made oracular pronouncements (without providing a shred of evidence) and it contained only a single formula: the metric for an n-manifold of constant curvature. But there was certainly *one* mathematician in the audience who grasped his meaning, the one who had *forced* Riemann to undertake this investigation in the first place: Carl Friedrich Gauss!

As was required by the rules, Riemann had submitted *three* possible topics for the mandatory lecture, the first of which was his recently completed, ground-breaking doctoral work on what are now called ***Riemann surfaces.*** Convention required that Gauss choose this one, which Riemann was well-prepared to address. Gauss did not! Instead, he picked Riemann's *third* choice: "The Foundations of Geometry."

Riemann was thereby forced into a brand new, major investigation, coming right on the heels of his doctoral masterpiece. The strain of this work, perhaps exacerbated by living in poverty, led

[14] In the final section of this chapter we shall see that (surprisingly!) the mere *sum* of these sectional curvatures nevertheless contains critical geometrical information!

Riemann to suffer another of his recurring mental breakdowns. But he recovered, and, following seven more weeks of intense effort, he was ready to deliver his June 10th lecture.

Whether Gauss knew of or cared about Riemann's plight in the run-up to the lecture is not known; what *is* known is how Gauss *reacted* to Riemann's lecture. Dedekind later recalled that Gauss sat at the lecture "which surpassed all his expectations, in the greatest astonishment, and on the way back from the faculty meeting he spoke to Wilhelm Weber, with the greatest appreciation, and with an excitement rare for him, about the depth of the ideas presented by Riemann."

In 1861 Riemann submitted a Prize Essay to the French Academy of Sciences that included a more mathematically explicit account of the Riemann tensor. But the essay—which did not win the prize!—concerned *heat conduction*, and Riemann gave only the briefest of hints regarding its significance to *geometry*.

Neither this essay nor the original 1854 lecture were published until after Riemann's death in 1866 (at the age of 39, from tuberculosis). Felix Klein (1928) summed up the reaction to their eventual publication in 1868: "I still have vivid memories of the extraordinary impression that Riemann's trains of thought made on young mathematicians. Much of it seemed obscure and hard to understand and yet of unfathomable depth."

Now we come face to face with the *same* mystery that opened Act IV, there regarding Einstein, but now regarding Riemann: Levi-Civita discovered parallel transport fully *63 years* after Riemann's lecture, so the modern geometrical interpretation of the Riemann tensor that we have provided would have been out of reach for Riemann himself. So how *did* Riemann discover his tensor?

The answer is that nobody knows! Spivak (1999) provides an English translation of the 1854 lecture, along with explanatory notes in which Spivak speculates on how Riemann *might* have done it. Spivak also includes an analysis of Riemann's Prize Essay of 1861.

What *is* clear from Riemann's own words is that his *initial* interpretation of his tensor was that *it measures the deviation of the* n*-manifold's metric from the Euclidean metric.* More precisely, Riemann discovered the following remarkably direct link between his tensor and the metric. If ds is the distance between the point p with Cartesian coordinates x_i and the neighbouring point with coordinates $(x_i + dx_i)$, then the difference between the actual metric and the Euclidean metric is given by this remarkable formula:

$$ds^2 - [dx_1^2 + \cdots + dx_n^2] \asymp \frac{1}{12} \sum_{i,j,k,m} R_{ijkm}(p)\,(x_i\,dx_j - x_j\,dx_i)\,(x_k\,dx_m - x_m\,dx_k).$$

For more on this, see Spivak (1999) and Berger (2003, §4.4).

The Mystery of Riemann's Curvature (Darrigol 2015) is a fascinating and thought-provoking investigation of these questions. Darrigol's analysis is based, in part, on photographs of previously unpublished, private notes by Riemann, which had been sitting in the archives of Göttingen University for more than 150 years.

One such scrap of paper is shown in [29.10]. Darrigol's painstaking decoding of Riemann's notation allows him to enter a fascinating correction into the annals of the history of mathematics. For that long, unintelligible final line turns out to be none other than the *Differential Bianchi Identity*, (29.17), which would be rediscovered decades later—first[15] by Aurel Voss (1880), then Ricci (1889), and then Bianchi (1902)—becoming one of the cornerstones[16] of Einstein's General Relativity!

For more on the life and science of Riemann, we recommend Monastyrskiĭ (1999), and Laugwitz (1999).

---

[15]See Pais (1982, pp. 275–276).

[16]See Exercise 14.

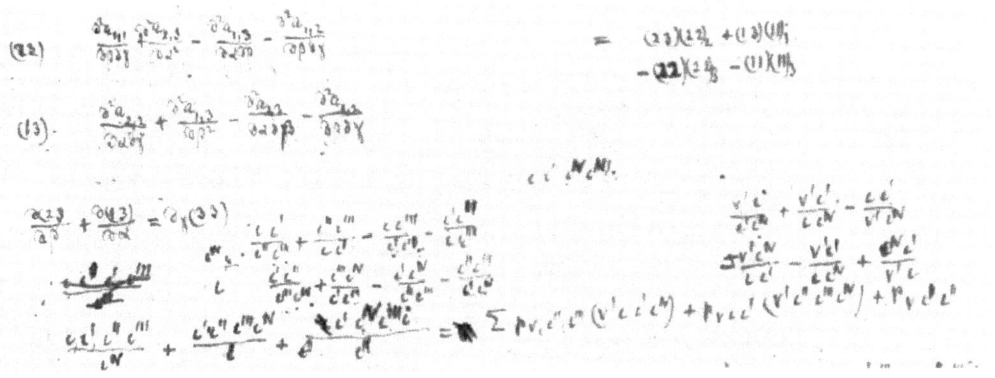

[29.10] *Extract of folio 4 of Cod. Ms. B. Riemann 9; courtesy of Göttingen University Archive. The final line is the* Differential Bianchi Identity, *(29.17), anticipating others by decades!*

29.6 The Jacobi Equation in an n-Manifold

29.6.1 Geometrical Proof of the Sectional Jacobi Equation

Within a 2-surface, the Jacobi Equation (the Equation of Geodesic Deviation) (28.2) describes the attraction (or repulsion) of neighbouring geodesics that pass through a region of positive (or negative) curvature, respectively.

In a 2-surface, this attraction or repulsion *only* depends on the curvature of the surface *at* the point through which the geodesics pass: it does *not* depend on the *direction* in which the geodesics pass through the point.

Within a 3-manifold, the attraction or repulsion again only depends on the (instantaneous) *plane* within which the neighbouring geodesics are travelling, not the direction of the geodesics within that plane. This plane $\Pi = \Pi(\mathbf{v}, \boldsymbol{\xi})$ is spanned by the velocity vector $\mathbf{v}$ and the connecting vector $\boldsymbol{\xi}$. But whereas there was only one possible "choice" for the plane for a 2-surface, now there are infinitely many such planes within our 3-manifold (or n-manifold), and, as already illustrated in [29.2], the sectional curvature in these planes can be positive in one and negative in another, causing attraction in one, and repulsion in the other.

It is therefore natural to guess that the generalization of the Jacobi Equation (28.2) to an n-manifold simply replaces the Gaussian curvature $\mathcal{K}$ with the *sectional curvature* $\mathcal{K}(\Pi)$ of the instantaneous plane within which the two neighbouring geodesics are travelling, and we shall now give two proofs that this is correct, but with a twist: *we must focus on the component of the relative acceleration that lies within* Π.

Whereas before we used a Newtonian superscript dot to denote the derivative along the geodesics, we now use the more standard $\nabla_{\mathbf{v}}$. Therefore, the relative velocity is now written $\nabla_{\mathbf{v}}\,\boldsymbol{\xi}$, and the relative acceleration, $\nabla_{\mathbf{v}}\nabla_{\mathbf{v}}\,\boldsymbol{\xi}$.

Reconsider our original proof of the Jacobi Equation in [28.7], but now imagine that the rectangle around which we parallel-transport the velocity lies within an arbitrary plane Π in our 3-manifold (or n-manifold). As before, parallel-transporting $\mathbf{v}_{||}$ around the loop with orthogonal edges $\boldsymbol{\xi}$ and $\mathbf{v}\delta t$, yields

$$\delta \mathbf{v}_{||} \asymp \delta t\, \nabla_{\mathbf{v}}\nabla_{\mathbf{v}}\,\boldsymbol{\xi}.$$

Since the speed of the particles is constant, $\delta \mathbf{v}_{||}$ is orthogonal to $\mathbf{v}$. In the 2-surface [28.7], this implied that it pointed along $\pm\boldsymbol{\xi}$, but in a 3-manifold *this is no longer true*: it can *also* have a component[17] *orthogonal to* Π. In summary, the relative acceleration $\nabla_{\mathbf{v}}\nabla_{\mathbf{v}}\,\boldsymbol{\xi}$ has a component

[17] In an n-manifold there will be $(n-2)$ components orthogonal to Π.

within Π that is responsible for pulling the geodesics together (or pushing them apart), *and* a component orthogonal to Π that does not affect the separation of the geodesics, but that causes them to *rotate* about one another.

We now focus attention on the attraction/repulsion component within Π, namely $\mathcal{P}[\nabla_{\mathbf{v}}\nabla_{\mathbf{v}}\,\boldsymbol{\xi}]$. As we have seen in (29.20), *if we project the parallel-transported vector into* Π then the holonomy is once again given by the curvature times the area of the rectangle, only now it is the *sectional curvature*, $\mathcal{K}(\Pi)$.

Thus, our original argument in [28.7] goes through essentially unchanged, except that now (29.19) *only* applies to the *projection* of $\nabla_{\mathbf{v}}\nabla_{\mathbf{v}}\,\boldsymbol{\xi}$ into Π, yielding,

$$\delta t\,\mathcal{P}[\nabla_{\mathbf{v}}\nabla_{\mathbf{v}}\,\boldsymbol{\xi}] \asymp \mathcal{P}[\delta \mathbf{v}_{||}] \asymp \mathbf{v}_{\perp}\,\mathcal{K}(\Pi)]\,\delta\mathcal{A} = \left[-\frac{\boldsymbol{\xi}}{|\boldsymbol{\xi}|}\right]\,\mathcal{K}(\Pi)\,|\boldsymbol{\xi}|\,\delta t,$$

and so we obtain what we shall christen[18] the

Sectional Jacobi Equation:

$$\boxed{\mathcal{P}\,[\nabla_{\mathbf{v}}\nabla_{\mathbf{v}}\,\boldsymbol{\xi}] = -\mathcal{K}(\Pi)\,\boldsymbol{\xi}.} \tag{29.21}$$

NOTE: This is *not* the standard "Jacobi Equation," which will be derived shortly (see (29.24)).

29.6.2 *Geometrical Implications of the Sectional Jacobi Equation*

To gain an intuitive grasp of the import of this equation, consider [29.11]. A set of particles are arranged around the perimeter of a small circle in a plane. A fiducial particle is added at the centre of the circle, the vector $\boldsymbol{\xi}$ connecting this central particle to a typical particle on the circle.

Now simultaneously fire these particles perpendicularly to the plane, with velocity $\mathbf{v}$. Initially, the particles will move in rigid unison, maintaining the same-size circular pattern. But then the curvature of the space begins to make itself felt, via the Sectional Jacobi Equation, (29.21). The motion of each individual particle on the circle, relative to the central one, is now governed by the *particular* $\mathcal{K}(\Pi)$ for the *particular* plane $\Pi(\mathbf{v}, \boldsymbol{\xi})$ for that *particular* particle's $\boldsymbol{\xi}$.

Thus as we move from a 2-surface to a 3-manifold (or n-manifold), a *fundamentally new feature* emerges: as the plane $\Pi(\mathbf{v}, \boldsymbol{\xi})$ rotates about the central axis (consisting of the central particle's trajectory) the sectional curvature $\mathcal{K}(\Pi)$ *varies*, both in magnitude and (possibly) sign.

Figure [29.11a] illustrates the case in which all sectional curvatures $\mathcal{K}(\mathbf{v}, \boldsymbol{\xi})$ are *negative*, regardless of the direction of $\boldsymbol{\xi}$, causing the particles to be repelled, accelerating apart on an expanding circle.[19] To use an optical analogy, negative curvature acts like a diverging lens.

Next, [29.11b] illustrates the case in which all curvatures are *positive*. In addition to the focusing attraction, governed by (29.21), here we have illustrated a case in which the relative acceleration has an appreciable component *orthogonal* to Π, causing the bundle of geodesics to rotate about the central one. This effect is intentionally ignored in our Sectional Jacobi Equation, (29.21), but it *is* taken into account by the (upcoming) full Jacobi Equation, (29.24). This is akin to the positive focussing of light by a magnifying lens.

Finally, [29.11c] depicts the most interesting case, and, as we shall see in the next chapter, the one of greatest importance to gravitational physics. Here the sectional curvature is positive in some planes (squeezing the circle of particles together) and negative in others (pulling the circle apart). The originally circular array of particles is therefore deformed into an ellipse, with minor

[18]Despite its twin virtues—(1) of providing a very direct interpretation of the sectional curvature, and (2) of being formally identical to the full Jacobi equation (28.2) in a 2-surface—we have not found this result written down in any standard text. We therefore felt the need to invent a logical name for this formula.

[19]If the negative curvatures were to vary in magnitude, then the circle would become oval as it grew.

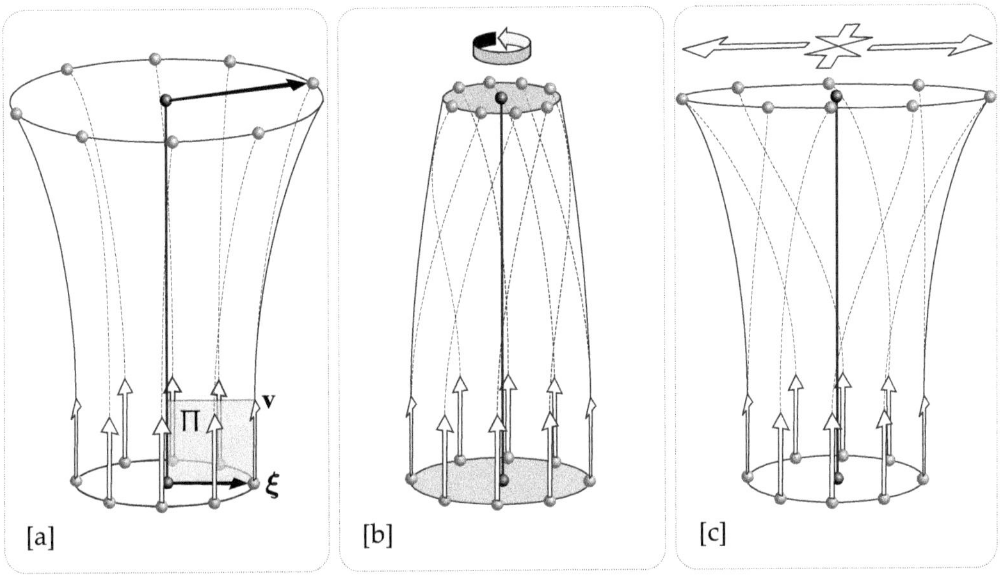

[29.11] *A small circle of particles are fired perpendicularly to the circle. Some planes $\Pi(\mathbf{v}, \boldsymbol{\xi})$ may have $\mathcal{K}(\Pi) > 0$, resulting in* attraction *towards the central particle, while other planes may have $\mathcal{K}(\Pi) < 0$, resulting in* repulsion *from the central particle.* **[a]** *$\mathcal{K}(\Pi) < 0$ in all planes.* **[b]** *$\mathcal{K}(\Pi) > 0$ in all planes.* **[c]** *Attraction in some planes squeezes the circle, and repulsion in others stretches it out, so that the circle is deformed into an ellipse.*

axis smaller than the original circle, and major axis larger than it. This is akin to the effects of an astigmatic lens.

In the next chapter we shall see that the optical analogy is not an analogy at all—it is reality! That is, the gravitational curvature of spacetime really *does* act like a lens as light passes through it!

29.6.3 Computational Proofs of the Jacobi Equation and the Sectional Jacobi Equation

For our orthonormal basis for Π, let us take $\mathbf{e}_1 = \widehat{\boldsymbol{\xi}} = \boldsymbol{\xi}/|\boldsymbol{\xi}|$ and $\mathbf{e}_2 = \mathbf{v}$. Then $\mathbf{e}_3 \equiv \mathbf{e}_1 \times \mathbf{e}_2$ will be the unit normal to Π. The definition (29.18) of the ***sectional curvature*** can now be written,

$$\mathcal{K}(\Pi) = R_{1221} = [\mathcal{R}(\mathbf{e}_1, \mathbf{e}_2)\,\mathbf{e}_2] \cdot \mathbf{e}_1 = [\mathcal{R}(\widehat{\boldsymbol{\xi}}, \mathbf{v})\,\mathbf{v})] \cdot \widehat{\boldsymbol{\xi}}. \tag{29.22}$$

By definition, $\boldsymbol{\xi}$ always connects the two neighbouring geodesics, so the parallelogram with edges $\mathbf{v}$ and $\boldsymbol{\xi}$ closes up:

$$[\mathbf{v}, \boldsymbol{\xi}] = 0 \quad \Longleftrightarrow \quad \nabla_{\mathbf{v}}\boldsymbol{\xi} = \nabla_{\boldsymbol{\xi}}\mathbf{v}. \tag{29.23}$$

Note (for use in just a moment) that this implies that the Riemann curvature operator (29.6) simplifies to

$$\mathcal{R}(\mathbf{v}, \boldsymbol{\xi}) = [\nabla_{\mathbf{v}}, \nabla_{\boldsymbol{\xi}}].$$

Taking the intrinsic derivative ($\nabla_{\mathbf{v}}$) of (29.23) along the geodesics, and recalling that $\mathbf{v}$ satisfies the geodesic equation, $\nabla_{\mathbf{v}}\mathbf{v} = \mathbf{0}$, we find that

$$\begin{aligned} \nabla_{\mathbf{v}}\nabla_{\mathbf{v}}\boldsymbol{\xi} &= \nabla_{\mathbf{v}}\nabla_{\boldsymbol{\xi}}\mathbf{v} \\ &= [\nabla_{\mathbf{v}}, \nabla_{\boldsymbol{\xi}}]\mathbf{v} + \nabla_{\boldsymbol{\xi}}(\nabla_{\mathbf{v}}\mathbf{v}) \\ &= \mathcal{R}(\mathbf{v}, \boldsymbol{\xi})\,\mathbf{v} + \nabla_{\boldsymbol{\xi}}\,(\mathbf{0}). \end{aligned}$$

Thus we have obtained the ***Equation of Geodesic Deviation***, or the

$$\textit{Jacobi Equation:} \qquad \boxed{\nabla_{\mathbf{v}}\nabla_{\mathbf{v}}\,\boldsymbol{\xi} = -\mathcal{R}(\boldsymbol{\xi},\,\mathbf{v})\,\mathbf{v}.} \tag{29.24}$$

This is the standard form of the Jacobi equation, which is found in all texts.

To instead recover our more intuitive "sectional" form, we decompose the vector on the right-hand side of the equation into its components, and find,

$$\begin{aligned}
\mathcal{R}(\boldsymbol{\xi},\,\mathbf{v})\,\mathbf{v} &= \left\{[\mathcal{R}(\boldsymbol{\xi},\,\mathbf{v})\,\mathbf{v}]\cdot\widehat{\boldsymbol{\xi}}\right\}\widehat{\boldsymbol{\xi}} + \{[\mathcal{R}(\boldsymbol{\xi},\,\mathbf{v})\,\mathbf{v}]\cdot\mathbf{v}\}\,\mathbf{v} + \left\{\left[\mathcal{R}(|\boldsymbol{\xi}|\widehat{\boldsymbol{\xi}},\,\mathbf{v})\,\mathbf{v}\right]\cdot\mathbf{e}_3\right\}\mathbf{e}_3 \\
&= \left\{\left[\mathcal{R}(\widehat{\boldsymbol{\xi}},\,\mathbf{v})\,\mathbf{v}\right]\cdot\widehat{\boldsymbol{\xi}}\right\}\boldsymbol{\xi} + 0 + \{[\mathcal{R}(\mathbf{e}_1,\,\mathbf{e}_2)\,\mathbf{e}_2]\cdot\mathbf{e}_3\}\,|\boldsymbol{\xi}|\,\mathbf{e}_3 \\
&= \mathcal{K}(\Pi)\,\boldsymbol{\xi} + R_{1223}|\boldsymbol{\xi}|\mathbf{e}_3,
\end{aligned}$$

by virtue of (29.22).

The second term, $R_{1223}|\boldsymbol{\xi}|\mathbf{e}_3$, is the component of the relative acceleration orthogonal to Π, which is responsible for the *rotation* of the bundle of geodesics, as illustrated in [29.11b].

The first term represents the *attraction* or *repulsion* within Π. Indeed, applying the projection operator $\mathcal{P}$ to both sides of the Jacobi Equation, (29.24), we recover the Sectional Jacobi Equation, (29.21):

$$\boxed{\mathcal{P}\,[\nabla_{\mathbf{v}}\nabla_{\mathbf{v}}\,\boldsymbol{\xi}] = -\mathcal{P}\,[\mathcal{R}(\boldsymbol{\xi},\,\mathbf{v})\,\mathbf{v}] = -\mathcal{K}(\Pi)\,\boldsymbol{\xi}.}$$

29.7 The Ricci Tensor

29.7.1 Acceleration of the Area Enclosed by a Bundle of Geodesics

Reconsider [29.11b]. The shaded area $\delta\mathcal{A}$ of the disc (inside the circle of particles) clearly shrinks as the particles are pulled together by the positive sectional curvatures. But according to what law?

For simplicity's sake, let us initially suppose that as Π rotates about the central particle's trajectory, all the sectional curvatures have the *same* positive value $\mathcal{K}$. Now let $r(t) = |\boldsymbol{\xi}(t)|$, so that $\delta\mathcal{A}(t) = \pi[r(t)]^2$ is the area at time t after launch. Reverting to Newtonian dots for time derivatives, it is clear that the initial rate of change of the area *vanishes*, because the particles head off in rigid unison, so that $\dot{r}(0) = 0$, and therefore

$$(\dot{\delta\mathcal{A}})(0) = 2\pi r(0)\,\dot{r}(0) = 0.$$

The sectional Jacobi Equation (29.21) tells us that the acceleration of the particles towards the centre is $\ddot{r}(0) = -\mathcal{K}r(0)$. Therefore the curvature manifests itself in the initial *acceleration* of the area:

$$(\ddot{\delta\mathcal{A}})(0) = 2\pi\,[\dot{r}^2(0) + r(0)\,\ddot{r}(0)] = -2\pi\,\mathcal{K}[r(0)]^2 = -2\mathcal{K}\,\delta\mathcal{A}(0).$$

Therefore, for small t, the Maclaurin expansion shows that the area decreases in proportion to the curvature and as the square of the time:

$$\text{change in the area after time } t = \delta\mathcal{A}(t) - \delta\mathcal{A}(0) \asymp -\mathcal{K}\,\delta\mathcal{A}(0)t^2.$$

Of course if $\mathcal{K}$ were *negative*, as in [29.11a], then the right-hand side would be positive, corresponding accelerating *growth* of the area.

Now that we have a rough idea of what is happening, let us move on to the general case, in which $\mathcal{K}(\Pi)$ varies in magnitude (and possibly sign) as Π rotates about the central axis.

Let θ denote the angle around this axis, starting from an arbitrary initial $\boldsymbol{\xi}$-direction. Consider the narrow (ultimately vanishing) sector of the disc located at angle θ and of angular width $\delta\theta$. The analysis above still applies to this sector, provided we replace $\mathcal{K}$ with the sectional curvature $\mathcal{K}(\theta)$ of the *particular* plane $\Pi(\theta)$ in this direction.

It follows that, in the general case, the previous equation becomes

$$\delta\mathcal{A}(t) - \delta\mathcal{A}(0) \asymp -\mathcal{K}_{\text{mean}}\, \delta\mathcal{A}(0) t^2,$$

where

$$\mathcal{K}_{\text{mean}} \equiv \frac{1}{2\pi}\int_0^{2\pi} \mathcal{K}(\theta)\, d\theta. \qquad (29.25)$$

Here, $\mathcal{K}_{\text{mean}}$ is the *average* (or "mean") of the sectional curvatures in all the planes containing the central particle's trajectory, in the direction $\mathbf{v}$.

[29.12] *Gregorio Ricci-Curbastro (1853–1925), who later in life became known simply as* Gregorio Ricci *(pronounced "reechee").*

Next, we shall find a formula for $\mathcal{K}(\theta)$ in terms of the Riemann tensor. To do so, let us choose $\mathbf{e}_3 = \mathbf{v}$ to be the direction of the central axis, in the direction that the particles are launched. See [29.13]. Then $\mathbf{e}_1$ and $\mathbf{e}_2$ are orthogonal unit vectors in the plane of the circle of particles. Having chosen the direction of $\mathbf{e}_1$ *arbitrarily*, let us agree to measure θ from this direction, as shown.

The plane $\Pi(\theta)$ is then spanned by $\mathbf{v}$ and the unit vector $\widehat{\boldsymbol{\xi}}(\theta)$ in the direction of the particle at angle θ:

$$\widehat{\boldsymbol{\xi}}(\theta) = \cos\theta\, \mathbf{e}_1 + \sin\theta\, \mathbf{e}_2 \equiv c\, \mathbf{e}_1 + s\, \mathbf{e}_2,$$

for short. Then the sectional curvature of $\Pi(\theta)$ is given by (29.22):

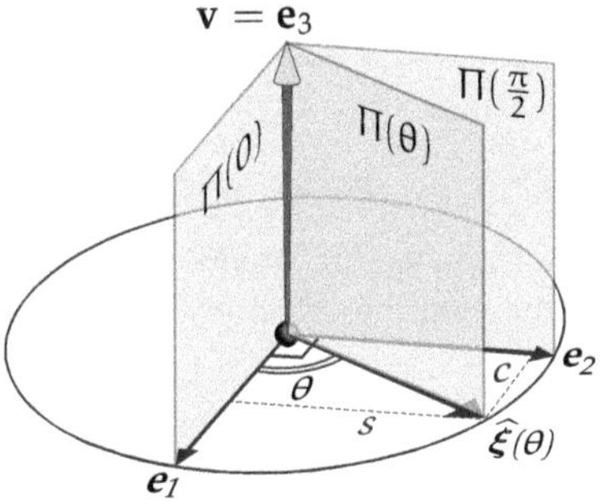

[29.13]

$$\begin{aligned}
\mathcal{K}(\theta) &= [\mathbf{R}(\widehat{\boldsymbol{\xi}}, \mathbf{v}; \mathbf{v})] \bullet \widehat{\boldsymbol{\xi}} \\
&= [\mathbf{R}(c\, \mathbf{e}_1 + s\, \mathbf{e}_2, \mathbf{v}; \mathbf{v})] \bullet (c\, \mathbf{e}_1 + s\, \mathbf{e}_2) \\
&= [c\, \mathbf{R}(\mathbf{e}_1, \mathbf{v}; \mathbf{v}) + s\, \mathbf{R}(\mathbf{e}_2, \mathbf{v}; \mathbf{v})] \bullet (c\, \mathbf{e}_1 + s\, \mathbf{e}_2) \\
&= c^2\, \mathbf{R}(\mathbf{e}_1, \mathbf{v}; \mathbf{v}) \bullet \mathbf{e}_1 + s^2\, \mathbf{R}(\mathbf{e}_2, \mathbf{v}; \mathbf{v}) \bullet \mathbf{e}_2 + sc\, [\mathbf{R}(\mathbf{e}_1, \mathbf{v}; \mathbf{v}) \bullet \mathbf{e}_2 + \mathbf{R}(\mathbf{e}_2, \mathbf{v}; \mathbf{v}) \bullet \mathbf{e}_1] \\
&= c^2\, \mathcal{K}(0) + s^2\, \mathcal{K}(\tfrac{\pi}{2}) + 2sc R_{1332},
\end{aligned}$$

where the last term follows from the symmetry [exercise], $R_{2331} = R_{1332}$.

To compute $\mathcal{K}_{\text{mean}}$, first verify that

$$\int_0^{2\pi} \cos^2\theta\, d\theta = \int_0^{2\pi} \sin^2\theta\, d\theta = \pi, \qquad \text{and} \qquad \int_0^{2\pi} \sin\theta\, \cos\theta\, d\theta = 0.$$

Then,

$$\boxed{\mathcal{K}_{\text{mean}} = \frac{\mathcal{K}(0) + \mathcal{K}(\frac{\pi}{2})}{2}.} \tag{29.26}$$

Recall that the direction $\theta = 0$ was chosen *arbitrarily*, so we have proved the remarkable fact that the average of the sectional curvatures over *all* directions is obtained by averaging it in *any two orthogonal* directions.

Combining (29.25) with (29.26), we may sum up our findings as follows:

If a small circle of particles are fired with velocity **v** *perpendicularly to the circle, the acceleration of the area* $\delta\mathcal{A}$ *of the circle is controlled by the average of the sectional curvatures in all planes containing* **v**, *but this is given by* the sum of the sectional curvatures in *any* two orthogonal planes containing **v**:

$$(\ddot{\delta\mathcal{A}})(0) = -[\mathcal{K}(0) + \mathcal{K}(\tfrac{\pi}{2})]\, \delta\mathcal{A}(0), \tag{29.27}$$

and so

$$\delta\mathcal{A}(t) - \delta\mathcal{A}(0) \asymp -\frac{1}{2}\, [\mathcal{K}(0) + \mathcal{K}(\tfrac{\pi}{2})]\, \delta\mathcal{A}(0)t^2.$$

29.7.2 *Definition and Geometrical Meaning of the Ricci Tensor*

Let us focus our attention on this sum of sectional curvatures in orthogonal coordinate planes containing the velocity $\mathbf{v} = \mathbf{e}_3$ of the particles. Recalling that $R_{3333} = 0$, by antisymmetry, we may write this as

$$\begin{aligned}
\mathcal{K}(0) + \mathcal{K}(\tfrac{\pi}{2}) &= \mathbf{R}(\mathbf{e}_1, \mathbf{e}_3; \mathbf{e}_3) \cdot \mathbf{e}_1 + \mathbf{R}(\mathbf{e}_2, \mathbf{e}_3; \mathbf{e}_3) \cdot \mathbf{e}_2 \\
&= R_{1331} + R_{2332} + R_{3333} \\
&= R_{m33}{}^{m}.
\end{aligned}$$

In a moment, this expression will guide us to the introduction of a brand new tensor, the *Ricci curvature tensor*. Whereas the Riemann tensor takes three vectors as input, and outputs a vector, the new Ricci tensor takes *two* vectors as input, and outputs a *scalar*.

NOTE ON NOTATION: It is an unfortunate accident of history that both the names Riemann and Ricci begin with the letter "R"! This has resulted in the universally accepted notation whereby the components of the Ricci tensor are denoted R_{jk}. We dare not defy this tradition—and do not even wish to!—but please be on high alert that an R with *four* indices is a component of the *Riemann* tensor, while an R with *two* indices is a component of the *Ricci* tensor. That said, it would be simply *too* confusing to represent these two (very different) geometrical tensors with the *same* symbol, **R**!

Therefore, we shall adopt (essentially[20]) the notation employed by Misner, Thorne, and Wheeler (1973), reserving **R** for the Riemann tensor, and writing the geometrical Ricci tensor itself as **Ricci**.

The definition of the ***Ricci curvature tensor*** is this:

$$\mathbf{Ricci}(\mathbf{v},\mathbf{w}) \equiv \sum_{m=1}^{n} \mathbf{R}(\mathbf{e}_m,\mathbf{v};\mathbf{w}) \cdot \mathbf{e}_m \quad \Longleftrightarrow \quad R_{jk} = \mathbf{Ricci}(\mathbf{e}_j,\mathbf{e}_k) = R_{mjk}{}^{m}.$$

CONVENTION WARNING: Many authors instead define $R_{jk} \equiv R_{jmk}{}^{m} = -R_{mjk}{}^{m}$, i.e., the negative of ours. Notably, this is true of Misner, Thorne, and Wheeler (1973). However, as we noted earlier, their definition of **R** is *also* the opposite of ours, and therefore the two sign-convention disagreements *cancel*: the sign of their ***Ricci*** is the *same* as our **Ricci**! As we shall see in a moment, *positive* Ricci curvature corresponds to *attraction* (both for us and for Misner, Thorne, and Wheeler). On the other hand, to cite just one other important example, Penrose (2005) shares our definition of **R** but uses the opposite definition of **Ricci**, so for Penrose it is *negative* Ricci curvature that causes attraction.

The fact that **Ricci** is indeed a tensor in its own right follows immediately [exercise] from the fact that **R** is a tensor. It also follows from the symmetries of **R** that **Ricci** is *symmetric*:

$$\mathbf{Ricci}(\mathbf{w},\mathbf{v}) = \mathbf{Ricci}(\mathbf{v},\mathbf{w}) \quad \Longleftrightarrow \quad R_{kj} = R_{jk}. \tag{29.28}$$

To see this, swap the *first pair* of vectors in its Riemann-tensor definition with the *second pair*, then swap the vectors within each pair.

We can now see that

$$\mathcal{K}(0) + \mathcal{K}(\tfrac{\pi}{2}) = R_{m33}{}^{m} = \mathbf{Ricci}(\mathbf{v},\mathbf{v}) = R_{jk}v^j v^k.$$

We have therefore arrived at the beautifully simple, Jacobi-type equation that governs the acceleration of the area of the bundle of geodesics launched with velocity **v**:

$$\ddot{\delta\mathcal{A}} = -\mathbf{Ricci}(\mathbf{v},\mathbf{v})\,\delta\mathcal{A}.$$

The role of the Gaussian curvature in the Jacobi equation of a 2-surface (or sectional curvature in an n-manifold) is here taken over by the Ricci curvature. We stress that *positive* Ricci curvature causes *attraction*, resulting in *shrinking* area as the geodesics converge.

Now imagine the generalization of the foregoing analysis to a 4-manifold. Again let **v** be the velocity with which we will launch a group of particles. This group is again chosen to be equidistant from the central particle, and located in the space orthogonal to **v**. But instead of yielding a *circle* in a *plane*, this construction now yields a small *sphere* of particles, enclosing a *volume* $\delta\mathcal{V}$, in the *3-dimensional space* orthogonal to **v**!

Although we shall not work out the details, the foregoing analysis goes through unchanged, and, specifically, the Ricci curvature now dictates the acceleration of the *volume* of the small sphere of particles that are simultaneously launched with velocity **v**:

---

[20] Verily, in the Bible it is written, ***Ricci*** (in all bold, and italics). By instead only putting the "**R**" in (nonitalic) bold, we hope to mentally connect **Ricci** to its components, R_{jk}.

$$\ddot{\delta V} = -\mathbf{Ricci}(\mathbf{v}, \mathbf{v})\, \delta V. \tag{29.29}$$

We stress for future use that *positive* Ricci curvature causes *attraction*, resulting in *shrinking* volume as the geodesics converge.

As before, this implies that (initially) the *volume changes in proportion to the Ricci curvature, and as the square of the time*:

$$\delta V(t) - \delta V(0) \asymp -\tfrac{1}{2}\, \mathbf{Ricci}(\mathbf{v}, \mathbf{v})\, \delta V(0) t^2 \tag{29.30}$$

29.8 Coda

This chapter has introduced a wealth of new ideas and results, but one formula is conspicuous by its absence: an analogue of (27.1) for expressing the Riemann curvature of an n-manifold in terms of its metric. The principal reason for this omission was a disciplined focus on providing the *minimal* set of concepts and results that will be needed to understand Einstein's theory of gravity. A second reason, however, is that the formula for **R** in terms of the metric coefficients turns out to be extremely complicated, at least when expressed in standard tensor formalism. In Act V, we shall see how the calculation of **R** can be greatly simplified—at both the theoretical level and the pragmatic, computational level—by the use of Forms.

As you began this chapter, you may have been preemptively flummoxed by the very notion that space could be *curved*—and who could blame you! But we very much hope that the multiple geometrical interpretations we have provided (especially via the Sectional Jacobi Equation and the Ricci tensor) have not only put you at your ease regarding Riemann's curvature within a 3-manifold, but have actually given you a very concrete, *tangible* grasp of exactly what this curvature *does*.

But 3-manifolds were merely the warm-up act. For it is a psychological *illusion* that you are sitting in a 3-dimensional room. In *reality*, you and your room are hurtling into the future, within Einstein's 4-dimensional spacetime. It is to the curvature of *that* manifold—what we call "reality"—that we now turn.

Study formula (29.29) well, for we shall see that it is *the key to the Universe!*

Chapter 30

Einstein's Curved Spacetime

30.1 Introduction: *"The Happiest Thought of My Life."*

This chapter lowers the curtain on Act IV—*it is the end of our self-contained introduction to (Visual) Differential Geometry.* We mark this milestone by returning to the subject that *began* Act IV: Einstein's extraordinarily beautiful *General Theory of Relativity*, which declares that gravity *is* the curvature of the 4-dimensional spacetime that comprises the framework of what we call "reality."

The great American physicist John Archibald Wheeler (1911–2008) famously distilled the theory down to a single sentence: *"Space tells matter how to move and matter tells space how to curve."* In greater detail, free-falling matter moves along the *geodesics* of the curved spacetime: these are still the *straightest* paths, but they now *maximize* (instead of minimize) the "distance" as measured by the spacetime metric. The second half of Wheeler's aphorism is the grand finale of Act IV: the precise law that Einstein discovered in 1915 that describes *how* matter and energy curve spacetime—the *Gravitational Field Equation* of General Relativity.

We begin at the historical beginning, with the (probably oversimplified[1]) tale of Newton's discovery of the inverse-square law of gravity, which would have us picture him sitting in his garden, in 1666, watching an apple fall from a tree. Suddenly, so the story goes, Newton realized that the *same* force that had pulled the apple to Earth might reach out to the Moon, pulling it towards the Earth, holding it in its orbit.

The genesis story of Einstein's geometrical theory of gravity is remarkably similar, but Einstein's falling object was not an apple, it was a *man!*

In 1907, fully two years after his discovery of Special Relativity, Einstein still held no academic position. He had spent the previous *five years* as a patent clerk (third class, promoted to second class) at the Federal Office for Intellectual Property, Bern, Switzerland.

As we noted at the outset of Act IV, Einstein knew that despite the superb accuracy of Newton's gravitational law, its instantaneous action-at-a-distance was fundamentally incompatible with the finite speed of light, and hence with his 1905 discovery of Special Relativity. Thus began Einstein's long struggle to understand gravitation, which would end almost exactly eight years later, on the 25th of November, 1915, when he finally wrote down the *Gravitational Field Equation.*

Einstein took the first, crucial step of that long journey in November of 1907, when he had what he later described as *"the happiest thought of my life"*:[2]

> I was sitting in a chair in the patent office at Bern when all of a sudden a thought occurred to me: If a person falls freely he will not feel his own weight . . . I was startled. This simple thought made a deep impression on me. It impelled me towards a theory of gravitation.

When we watch astronauts on TV, floating about within their orbiting space station, it appears that gravity has been totally eliminated—this is precisely the import of Einstein's insight. This *seemingly* total elimination of gravity arises from the fact that *the astronauts and their spacecraft are*

[1]Newton himself did, however, tell this story as related here to four different people on four different occasions. See Westfall (1980, pp. 154–155).

[2]See (Pais 1982 §9).

falling freely together in the Earth's gravitational field—it does *not* arise from the spacecraft being so far from Earth that it has escaped the force of gravity!

Yes, Newton's inverse-square law tells us that gravity is weaker up there than down here, but a quick calculation shows that the strength of the Newtonian gravitational field pulling on the International Space Station (orbiting at an altitude of 250 miles) is only about 12% weaker than it is for us down here. No, the seemingly *complete* elimination of gravity has nothing to do with this modest effect.

Instead, the *total* elimination of gravity is the result of the remarkable empirical fact—first discovered by Galileo around 1590, by (supposedly) dropping objects from the Leaning Tower of Pisa—that *all objects are accelerated equally by gravity, regardless of their mass or composition.*[3] Thus the astronauts and their spacecraft all follow the *same*, free-falling orbit *together*.

This empirical fact is simultaneously enshrined in and explained by Newton's Law of Gravitation. If a particle of very small mass m lies at distance r from a particle with a very large mass M, then the gravitational effect of m on M can be neglected, and we may think of m as simply being pulled towards M (with force F) by the latter's gravitational field. Then the acceleration a of m towards M is given by Newton's Second Law of Motion:

$$F = ma.$$

But if r is the distance of m from M, then ***Newton's Inverse-Square Law of Gravitation*** states that

$$\boxed{F = \frac{GmM}{r^2},}$$

where G is the ***gravitational constant***. Combining these facts,

$$m\ddot{r} = -\frac{GmM}{r^2} \quad \Longrightarrow \quad \boxed{\ddot{r} = -\frac{GM}{r^2},} \tag{30.1}$$

which is independent of m.

So Newton's gravitational law *explains* Galileo's otherwise deeply mysterious empirical fact that all objects (launched from the same place in the same manner) follow the *same* trajectory, regardless of their composition.

But is it possible to go even deeper? Can we in turn *explain* this aspect of Newton's gravitational law? Einstein's geometrical theory does exactly that! If a particle is destined to travel along the *geometrically* determined geodesics of spacetime—*which have nothing whatsoever to do with the particle!*—this *explains* Galileo's empirical fact, and likewise provides the *reason* that the gravitational force F *must* be proportional to the mass, m!

30.2 Gravitational Tidal Forces

But what exactly *remains* of gravity if it seemingly vanishes for the astronauts within their spacecraft, all free-falling together? Surely it must leave *some* trace of its existence?! This is the fundamentally new question to which we are led by Einstein's startling insight.

---

[3]This is only true if air resistance can be ignored. There is a wonderful film (from the 1971 Apollo 15 mission) of astronaut David Scott simultaneously dropping a feather and a hammer while standing on the surface of the Moon—they both hit the lunar surface at the same time!

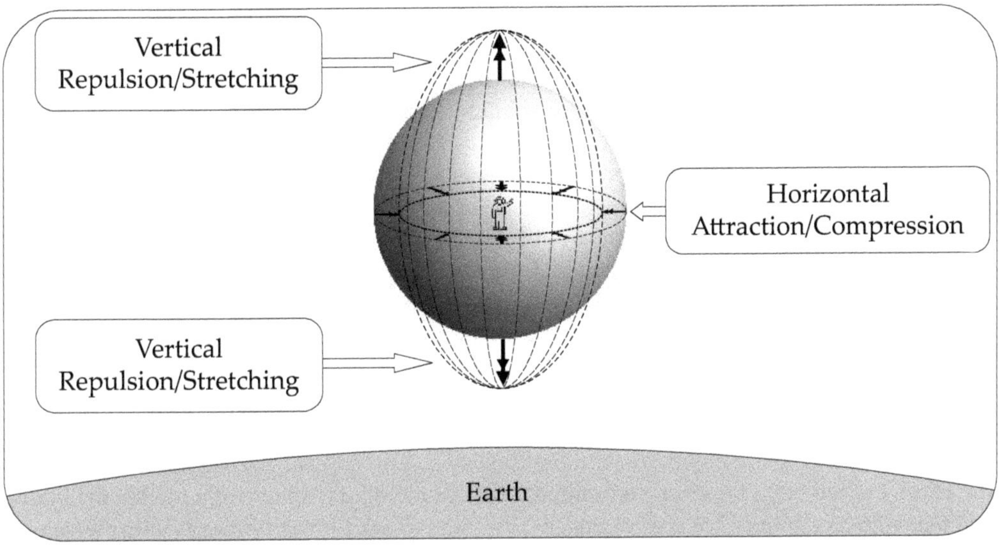

[30.1] The gravitational tidal forces of the Earth act on a free-falling sphere of particles, *resulting in* compression in the horizontal "equatorial" plane, parallel to the Earth's surface, and stretching in the vertical direction of the gravitational field itself. *As the sphere falls, it therefore starts to deform into an ellipsoidal egg. For the inverse-square law, the vertical stretching turns out to be exactly* twice *as powerful as the equatorial compression.*

Well, imagine that you yourself are that astronaut, high above the Earth, but rather than being safely in orbit inside the space station, you are outside, in your spacesuit, using your jetpack to hover at a fixed location. Further imagine that you are oriented straight up and down, with your head furthest from Earth, and your feet closest to it. Surrounding you, evenly spread over the surface of a sphere centred on you, are hundreds of shiny ball bearings, momentarily at rest, like you. See [30.1].

Now you turn off your jetpack. You and the sphere of ball bearings surrounding you begin to free-fall *together*, from rest, all accelerating downward towards the Earth. What *changes*, if any, will you observe in the sphere of particles as you fall?

If the gravitational field were *perfectly uniform*, with constant magnitude and fixed direction, your answer would be, "I see no changes at all!" Gravity would be truly invisible to you. And since the actual gravitational field of the Earth is *approximately* uniform over the small sphere, you won't see any changes, either ... at least not *at first*.

However, slowly, you will start to notice that the parts of the sphere above your head and below your feet are accelerating *away* from you, while the "equatorial" particles are accelerating *towards* to you. The net result is that the sphere starts turning into an egg!

To understand why this is happening, see [30.1], and let us immediately clarify that this diagram shows the *forces* that are going to *drive* the deformation of the sphere into the egg, not the egg itself, which is the new shape of the sphere after it has fallen down towards the Earth for a period of time.

First, realize that the equatorial particles, at the same altitude as you, are not accelerating in precisely the same *direction* that you are. Yes, you and the particles are accelerating downward, and at the same rate, but you are all *accelerating towards the centre of the Earth*,[4] and therefore your

---

[4]In reality, you are being pulled towards every little part of the Earth, with a force that is inversely proportional to the square of the distance of that particular part from you. But it turns out that the *net* effect of this horrendously complex sum of forces is *precisely* as though all the mass of the Earth were concentrated at its centre! Newton himself was shocked to discover this genuinely miraculous property of his inverse-square law, and he provided an elegant geometrical proof of it in the *Principia* (Newton 1687, Theorem 31). Note that this is another *signature* of the inverse-square law—it is *not* shared by other force laws.

trajectories are slowly *converging*. Since you only see the *relative acceleration*, the equatorial particles appear to be accelerating directly towards you, as if *attracted* to you, as illustrated in [30.1].

Now look up at the particles above your head at the top of the sphere, and below your feet at the bottom of the sphere. Unlike the previous case, you and the particles *are* now all heading in the same direction, straight down towards the centre of the Earth. However, the acceleration of the particles below your feet is *greater* than your acceleration, because they are closer to the centre of attraction than you are. Thus you see them start to accelerate *away* from you downward towards the Earth. Likewise, the acceleration of the particles above your head is *smaller* than your acceleration, because they are further away from the centre of attraction, so you are accelerating downward faster than they are, so they too seem to accelerate away from you, *upward*. Thus these particles appear to be *repelled* from you, as illustrated in [30.1].

This phenomenon, of compression in one plane and stretching in the orthogonal direction, deforming a sphere into an egg, is called the gravitational ***tidal force***. It is the *answer* to our Einstein-inspired question, "What remains of gravity in free-fall?"

Let us connect these observations more directly to our original astronauts in orbit around the Earth. Instead of falling from rest, suppose that you and your sphere of particles all blast off together with an arbitrary but shared velocity; the egg-inducing forces will be exactly the same as before! In particular, suppose this velocity is chosen to point horizontally, parallel to the Earth's surface, and suppose the launch speed is chosen so as to set you and your sphere into a *circular orbit* around the Earth. Finally, imagine that you and your sphere are inside the orbiting space station, and the ball bearings are replaced with fellow astronauts! If you look very closely, we will see that gravity has *not* been completely eliminated: a vertically (i.e., radially) separated astronaut will gradually accelerate *away* from you, while a horizontally separated astronaut will gradually accelerate *towards* you. However, as we shall see in the next section, these tidal forces are *proportional to the separation* of the particles, and over the small distances within the space station they are undetectable. But over larger scales, the impact of these tidal forces becomes obvious, which brings us to our next topic. . . .

Why is this gravitational distortion force field called "tidal"? What is the connection between this phenomenon and the *tides* of our oceans?

Imagine in [30.1] that the Earth is instead the *Sun*, and our small sphere is instead the *Earth*, or rather the thin skin of the Earth—its *oceans*. The tidal force of the Sun on the Earth now causes a bulging of the oceans towards the Sun on the side of the Earth facing the Sun, and a bulging of the oceans away from the Sun on the opposite side, representing a rise in sea level in these two opposite directions. This is the fixed, *dotted* ellipsoidal egg shown in all four panels of [30.2].

The directions of the bulges remain fixed as the Earth rotates once every 24 hours, so each coastal area encounters one of these bulges every 12 hours—*these are the high tides!* Well, as illustrated in [30.2], it's actually much more complicated than that, because the Moon exerts exactly the same kind of gravitational distortion on our oceans, and, despite its minuscule mass compared to the Sun, its proximity to Earth induces tidal forces that are roughly *twice*[5] as great as the Sun's! The actual tides are the complex[6] result of the superposition of these twin tidal forces of the Sun and Moon; their interaction over one lunar orbit is illustrated and explained by [30.2]. This was first understood by Newton, and was another triumph of the *Principia* (Newton, 1687).

The geometry is simplest when we see a full moon (top left of [30.2]), or a new moon (bottom left of [30.2]), for this means that the Moon, Earth, and Sun are lined up, in which case the tidal bulges induced by the Moon and Sun on Earth's oceans are lined up, too, reinforcing each other, producing the greatest high tides, and also the lowest low tides; these are called ***Spring tides***. (NOTE: These operate year-round, and have nothing to do with the season we call "Spring.")

[5]See Exercise 13.

[6]See Schutz (2003, §5).

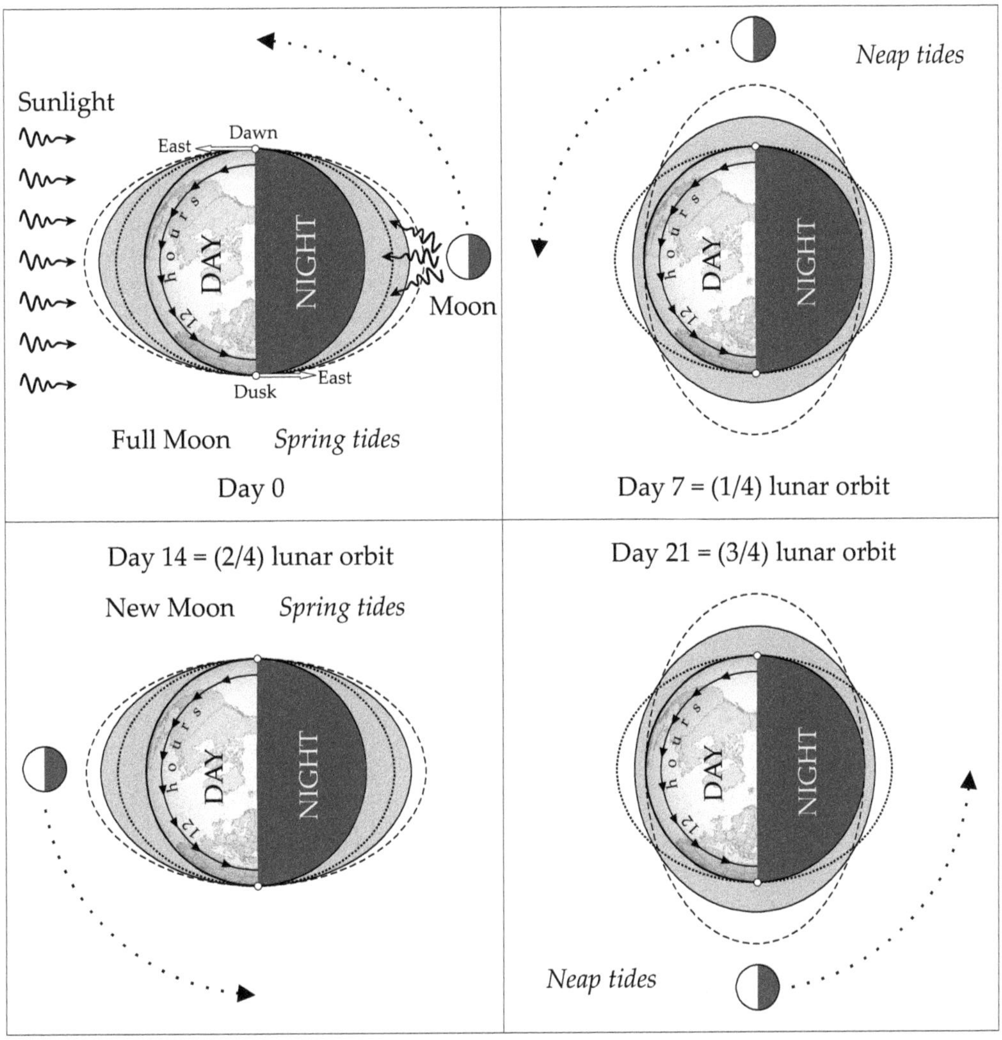

[30.2] Explanation of the Ocean Tides. *The tidal forces of the Moon (dashed line) and of the Sun (dotted oval) combine to distort the oceans (shaded oval areas), creating the tides.* (NOTE: *Nothing is to scale! The oceans and the tides are massively exaggerated, and the distance to the Moon is massively shrunken!) Top left, the illuminated half of the Moon faces the Earth, the reflected sunlight being seen by us as a Full Moon. Since the tidal forces of the Sun and Moon are now aligned, they reinforce each other to produce the greatest high tides, and the lowest low tides; these are the* Spring tides. *Top right, the more powerful lunar tidal forces are partially cancelled out by the orthogonal solar tidal forces, resulting in the smallest difference between high and low tides; these are the* Neap tides.

When the Moon and Sun form a right angle at the Earth (top right and bottom right of [30.2]), they produce bulges in orthogonal directions, so the smaller solar bulge partially cancels out the larger lunar bulge, resulting in the smallest difference between high and low tide; these are called ***Neap tides***.

Since—by definition!—the Moon executes one orbit of Earth each month, it is lined up with the Earth and the Sun *twice* per month: the New Moon corresponds to the Moon lying between the Sun and the Earth, while the Full Moon corresponds to the Earth lying between the Sun and the Moon. Thus the Spring tides occur every two weeks, and the same is true of the Neap tides, the alternation between the two being weekly—the time it takes for the Moon to execute one-quarter of an orbit around the Earth, as shown in [30.2].

To learn much more about the physics of gravity (and the tides in particular) we strongly recommend *Gravity From The Ground Up* (Schutz 2003).

30.3 Newton's Gravitational Law in Geometrical Form

(NOTE: The following exposition of Einstein's geometrical theory of gravity is inspired by Penrose's beautiful essay, *The Geometry of the Universe* (Penrose 1978) and also by the three relevant chapters of his remarkable work, *The Road to Reality* (Penrose 2005, §§17–19). However, we shall explicitly prove several crucial results that Penrose states but leaves as exercises for the reader, and we shall also include some additional geometrical observations that Penrose does not mention.)

The tidal forces we have just described would be *qualitatively* the same for *any* conceivable gravitational law of attraction that decreased with distance. But the *actual* law of gravitation, discovered by Newton and published in his *Principia* of 1687, *decreases as the square of the distance.* As we will now show, this law, and this law *alone*, results in tidal forces that have a very specific and very beautiful geometrical signature.

Let r be your distance from the centre of the Earth, let ξ be the radius of the horizontal, "equatorial" circle of the sphere of particles that surround you, and let $\delta\varphi$ be the small angle subtended by ξ at the centre of the Earth. Then $\xi \asymp r\,\delta\varphi$. Since $\delta\varphi$ remains fixed as you and the equatorial particles fall radially towards the centre of the Earth, the inward acceleration of this circle towards you is given by

$$\ddot{\xi} \asymp \ddot{r}\,\delta\varphi = -\frac{GM}{r^2}\,\delta\varphi = -\frac{GM}{r^3}\,r\,\delta\varphi \asymp -\frac{GM}{r^3}\,\xi,$$

by virtue of (30.1).

If the previous two chapters have done their job, your immediate, Pavlovian reaction to the mere sight[7] of this formula is to exclaim, *"That's a Jacobi equation!"* We shall make the connection explicit before long, but this attraction of horizontally separated particles is indeed due to *positive sectional curvature*, $\mathcal{K}_+$, where

$$\ddot{\xi} = -\mathcal{K}_+\,\xi \quad \Longrightarrow \quad \mathcal{K}_+ = +\frac{GM}{r^3}. \tag{30.2}$$

Let Ξ denote the distance from you to the particles above your head and below your feet—we chose a Greek letter that depicts this! As illustrated in [30.1], and as previously discussed, the relative acceleration of these particles *away* from you is given by the difference $\delta\ddot{r}$ in the acceleration $\ddot{r}$—given by (30.1)—at your height r and at height $r+\delta r = r+\Xi$. Thus,

$$\ddot{\Xi} = \delta\ddot{r} \asymp [\partial_r \ddot{r}]\,\delta r = \partial_r\left[-\frac{GM}{r^2}\right]\Xi = \left[+\frac{2GM}{r^3}\right]\Xi.$$

This repulsion of vertically separated particles is due to *negative sectional curvature*, $\mathcal{K}_-$, where

$$\ddot{\Xi} = -\mathcal{K}_-\,\Xi \quad \Longrightarrow \quad \mathcal{K}_- = -\frac{2GM}{r^3}. \tag{30.3}$$

[7] See (28.2) and (29.21).

Thus *the vertical repulsion is exactly double the horizontal attraction*, as illustrated in [30.1]. We are now poised to recognize the true geometrical significance of this fact—the geometrical *signature* of Newton's inverse-square tidal forces. The key is to follow the evolution of the *volume* δV of the sphere as it falls to Earth, the tidal forces deforming it into an ellipsoidal egg.

Imagine that the initial sphere of particles is enclosed in a cube. As the sphere falls and deforms into an egg, the enclosing cube and its contents undergo a linear transformation that deforms it into a box with edges of lengths 2ξ, 2ξ, and 2Ξ, and hence of volume $8\xi^2\,\Xi$. The volume occupied by the egg is a fixed proportion of this box, so

$$\boxed{\delta V = \tfrac{4\pi}{3}\,\xi^2\,\Xi.}$$

Since you and the sphere of particles start from rest, clearly the rate of change of this volume must initially vanish: $\dot{\delta V} = 0$. This can be confirmed by calculation:

$$\dot{\delta V} = \frac{4\pi}{3}\left[2\xi\dot{\xi}\,\Xi + \xi^2\,\dot{\Xi}\right] = 0,$$

since $\dot{\xi} = 0 = \dot{\Xi}$.

But the tidal forces do instantly *accelerate* the particles of the sphere, so let us now calculate the *acceleration* of the volume of the egg:

$$\ddot{\delta V} = \frac{4\pi}{3}\left[2(\dot{\xi})^2\,\Xi + 2\xi\,\ddot{\xi}\,\Xi + 2\xi\dot{\xi}\dot{\Xi} + 2\xi\dot{\xi}\dot{\Xi} + \xi^2\,\ddot{\Xi}\right] = \frac{4\pi}{3}\left[2\xi\,\ddot{\xi}\,\Xi + \xi^2\,\ddot{\Xi}\right].$$

But, initially, $\xi = \delta r = \Xi$, and we know that the horizontal attraction and vertical repulsion are governed by (30.2) and (30.3), respectively, so,

$$\ddot{\delta V} = \frac{4\pi}{3}\left[2\,\ddot{\xi} + \ddot{\Xi}\right](\delta r)^2 = -\frac{4\pi}{3}\,(2\mathcal{K}_+ + \mathcal{K}_-)\,(\delta r)^3 = 0!$$

We have arrived at the beautiful

Geometrical Signature of the Inverse-Square Tidal Forces:
The tidal forces generated by the Inverse-Square Law—and only *the Inverse-Square Law—are* **volume-preserving**, *in the precise sense that the acceleration of the volume vanishes, so that the volume remains constant to order* t^2*:*

$$\ddot{\delta V} = 0. \qquad (30.4)$$

We have not yet proved the "only" part of this proposition. To do so, suppose that $\ddot{r} = f(r)$, where f is an unknown function. Substituting this in place of (30.1) in the above analysis, we find [exercise],

$$\ddot{\delta V} = 0 \quad\Longrightarrow\quad \frac{df}{dr} + \frac{2f}{r} = 0 \quad\Longrightarrow\quad f(r) \propto \frac{1}{r^2},$$

as was to be shown.

In the foregoing analysis we have, understandably, neglected *your* microscopic gravitational field and its influence on the particles that surround you. But suppose we remove you from the interior of the sphere, and instead fill it with extremely dense matter of density ρ. Then the sphere of particles of radius $\xi = \Xi$ will be accelerated inward towards the centre, just as if there were a point mass $\rho\,\delta\mathcal{V}$ at the centre, pulling them in with acceleration,

$$\ddot{\xi} = -\frac{G\rho\,\delta\mathcal{V}}{\xi^2}.$$

Now let us compute the acceleration of the volume in this new circumstance. Since now $\delta\mathcal{V} = \frac{4\pi}{3}\xi^3$,

$$\delta\dot{\mathcal{V}} = 4\pi\xi^2\,\dot{\xi} \quad \Longrightarrow \quad \delta\ddot{\mathcal{V}} = 8\pi\xi(\dot{\xi})^2 + 4\pi\xi^2\,\ddot{\xi} = 8\pi\xi(0)^2 + 4\pi\xi^2\left(-\frac{G\rho\,\delta\mathcal{V}}{\xi^2}\right).$$

Thus we have discovered this

Geometrical Signature of the Inverse-Square Attraction:
Consider a sphere of volume $\delta\mathcal{V}$ that is filled with matter of density ρ, and suppose that just above its surface are tiny test particles, released from rest. Instantly they begin to accelerate towards the centre, and the inverse-square law causes the volume they enclose to implode with an acceleration governed by this geometrical law: (30.5)

$$\delta\ddot{\mathcal{V}} = -4\pi G\,\rho\,\delta\mathcal{V}.$$

If we now imagine this material sphere (and its skin of test particles) to be launched with arbitrary velocity in the Earth's gravitational field, then we obtain a superposition of these two effects: the Earth's tidal forces start to deform the sphere into an egg of equal volume, but the attraction of the matter within the sphere also exerts its volume-reducing effect. The net result is that the initial sphere evolves into an egg, but with its volume shrinking in proportion to the mass within, and in proportion to the square of the time.

30.4 The Spacetime Metric

Soon we will be able to recast the foregoing analysis in terms of Einstein's 4-dimensional spacetime, and we will then be led very naturally to Einstein's Gravitational Field Equation. But before we can do that, we must first discuss the metric structure of spacetime, and we must also learn how to draw *pictures* of spacetime.

The first thing to understand is that the local tangent-space structure is *not* $\mathbb{R}^4$ with the standard Euclidean metric; it is instead *Minkowski* spacetime with the ***Minkowski metric*** given by the spacetime interval we introduced in (6.15), on page 75. (We remind the reader that we have chosen units such that light travels one unit of distance in one unit of time, so that its speed is $c=1$. In these units, Einstein's famous equation, $E=mc^2$, becomes, simply, $E=m$, i.e., "energy is mass"!)

Thus,

$$ds^2 = dt^2 - (dx^2 + dy^2 + dz^2) = \left[dx^0\right]^2 - \left(\left[dx^1\right]^2 + \left[dx^2\right]^2 + \left[dx^3\right]^2\right). \qquad (30.6)$$

This is the analogue of distance within the tangent plane to a surface at a point, and just as the tangent plane is flat, so too is Minkowski spacetime—its Riemann tensor vanishes identically.

The analogue of the true distance between neighbouring points of the surface itself is given by the ***metric tensor*** of the curved spacetime, which is usually denoted **g**—it takes two vectors as input, and outputs a scalar. It generalized the ***dot product*** (or ***scalar product***), and it is symmetrical:

$$\mathbf{g}(\mathbf{u},\mathbf{v}) \equiv \mathbf{u}\cdot\mathbf{v} = \mathbf{v}\cdot\mathbf{u} = \mathbf{g}(\mathbf{v},\mathbf{u}).$$

Just as in a 2-surface, the metric constitutes the most *fundamental* information about spacetime: it defines *distance*, and once we know distance, we know *everything*: geodesics, parallel transport, and Riemann curvature. If ϵ is a small (ultimately vanishing) connecting vector between two neighbouring events in spacetime, then the metric tells us the Einsteinian distance between them:

$$ds^2 = \mathbf{g}(\epsilon, \epsilon)$$

But recall that $ds^2 > 0$ only when ϵ connects two points on the spacetime trajectory (aka ***world-line***) of a material particle with nonzero mass, which is constrained to travel at less than the speed of light; in this case we say that the separation is ***timelike***. As we stated in (6.16) on page 76, in this case ds simply measures wristwatch time elapsed for the traveller of the world-line. If ϵ is instead along a light ray (the world-line of a photon) then $ds^2 = 0$; in this case we say the separation is ***null***. Finally, if $ds^2 < 0$ we say that the separation is ***spacelike***.

If $\{\mathbf{e}_i\}$ is a set of four basis vectors—called a ***tetrad***—which need *not* be orthonormal, then we can obtain the *components* of the metric tensor in exactly the same way as we obtained the components of the Ricci tensor—simply apply the tensor to pairs of basis vectors:

$$g_{ij} \equiv \mathbf{g}(\mathbf{e}_i, \mathbf{e}_j) = \mathbf{g}(\mathbf{e}_j, \mathbf{e}_i) = g_{ji}.$$

Thus,

$$\boxed{ds^2 = \mathbf{g}(\epsilon, \epsilon) = \mathbf{g}(dx^i\, \mathbf{e}_i,\ dx^j\, \mathbf{e}_j) = \mathbf{g}(\mathbf{e}_i, \mathbf{e}_j)\, dx^i\, dx^j = g_{ij}\, dx^i\, dx^j.} \tag{30.7}$$

For example, in Minkowski spacetime, as expressed in (30.6), $g_{00} = +1$, while $g_{11} = g_{22} = g_{33} = -1$, and $g_{ij} = 0$ if $i \neq j$. In general, however, these g_{ij} are *functions* in spacetime, and $g_{ij} \neq 0$ if $i \neq j$. It is important to realize that, even in flat Minkowski spacetime, these components can look drastically different, depending on the choice of coordinates. For example, if we use standard spherical polar coordinates for the spatial coordinates, then

$$ds^2 = dt^2 - dr^2 - r^2(d\varphi^2 + \sin^2\varphi\, d\theta^2), \tag{30.8}$$

so, $g_{tt} = 1$, $g_{rr} = -1$, $g_{\varphi\varphi} = -r^2$, and $g_{\theta\theta} = -r^2 \sin^2\varphi$, and yet these functions describe exactly the same, flat geometry as before.

NOTE ON NOTATION: This g_{ij}-notation is *universally* accepted: it can be found in *all* modern texts, whether they be on mathematics or on physics. So we too shall employ this notation, which is the n-dimensional generalization of Gauss's original (E, F, G)-notation for a 2-surface; see (4.8), page 37. On the other hand, at least in the 2-dimensional case, we hope we have convinced you—several times over!—of the conceptual and computational advantages of our *alternative* (A^2, B^2)-notation. That said, into n dimensions our notation dare not tread....

30.5 Spacetime Diagrams

The vast majority of humans (including the author) cannot directly visualize 4-dimensional spacetime. How, then, can we draw diagrams that will enable us to reason geometrically about spacetime?! The most common and useful answer to this conundrum is to simply *suppress one of the three spatial directions*. More specifically, we shall draw time going *up* the page, and we shall represent two (out of the three) spatial directions as forming a horizontal plane, perpendicular to the vertical time direction.

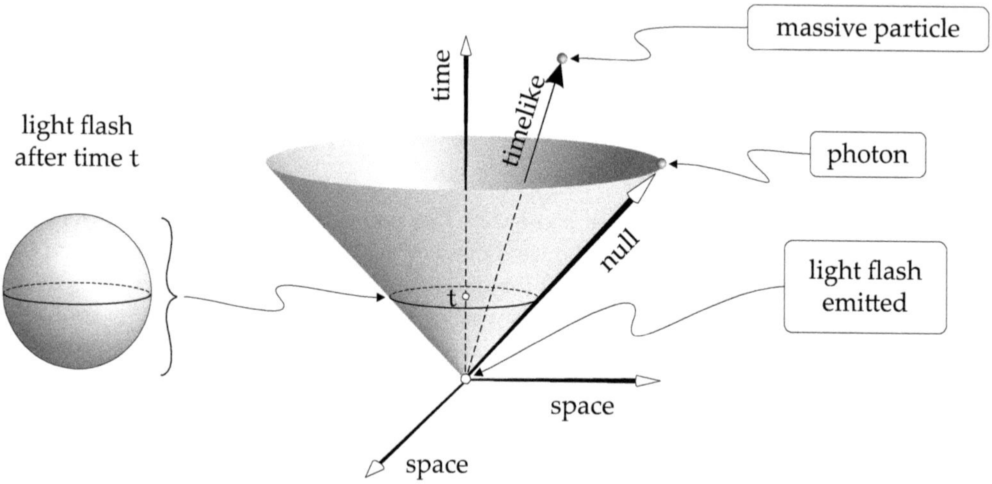

[30.3] The Null Cone (Light Cone) in Spacetime. *Time is represented by the vertical direction, and two (out of three) spatial directions are represented by the orthogonal, horizontal directions. A flash of light is emitted at an event in spacetime, and the expanding sphere of light is represented at each time* t *by its circular cross section. The entire future of the flash is therefore represented by a cone, the world-lines of the photons being the null generators of the cone. Material particles are constrained to travel at less than the speed of light, along timelike world-lines,* inside *the cone.*

As drastic as this loss of information may seem, symmetry often comes to our rescue: if two of the spatial directions are physically equivalent, then discarding one of them costs us nothing. For example, and of great upcoming importance, in [30.1] all horizontal directions—in the plane parallel to the Earth's surface—are indistinguishable, so if we picture this plane spanned by two orthonormal vectors, then we can safely suppress one of them, with no loss of information. In stark contrast to this, the vertical direction is the direction of the gravitational field, and it is physically quite distinct from the horizontal directions, so in a spacetime diagram we must not throw *that* direction away!

But let us begin with an even more symmetrical situation: empty, flat, Minkowski spacetime. Here *all* spatial directions are physically indistinguishable from one another. One of the most useful objects to depict in a spacetime diagram is a flash of light emitted from a particular point at a particular time—an ***event***. This results in a sphere of light that expands one unit of distance for every unit of time. Since Einstein tells us that material particles cannot travel faster than light, any massive particles ejected from the same event as the light flash must remain *inside* this expanding sphere. How does all this look in a spacetime diagram?

Suppressing one spatial dimension, the expanding sphere of light is now represented by the expanding *circle* shown in [30.3], which generates a cone. For this reason, the spacetime depiction of the light flash is called the ***light cone*** or ***null cone***. The world-lines of material particles stay inside this cone, as illustrated. The fundamental importance of the null cone is that its interior therefore represents the set of events that can be influenced by the original event—it tells us about the so-called ***causal structure*** of spacetime, i.e., can this event, here and now, cause something to happen at that event, there and then?

The tangent vector to the world-line of a particle is called the ***4-velocity*** of the particle. A particle at rest has a *nonzero* 4-velocity: it points straight up the time axis! Note that for a massive particle it is possible to normalize the 4-velocity in the same way that we have usually insisted that our particles travel over a 2-surface at unit speed, but this is *not* possible for the 4-velocity of a photon, because the "length" of a photon's 4-velocity is always *zero*.

30.6 Einstein's Vacuum Field Equation in Geometrical Form

In order to "derive"[8] Einstein's Vacuum Field Equation, let us return to the vacuum of space above the Earth's surface, and to the Newtonian inverse-square tidal forces that operate there, as depicted in [30.1]. But now let us analyze this from the perspective of *spacetime*.

We have already noted in [30.1] that all horizontal directions are indistinguishable, so there is no loss of information if we draw a spacetime diagram that only retains one such direction. Therefore, consider any vertical great circle of the sphere, spanned by one such horizontal direction, and the vertical direction of Earth's gravity. As the sphere of particles falls, the tidal forces distort the sphere into an ellipsoidal egg, but this evolution is *completely and faithfully captured* by looking at the evolution of merely this single vertical, great-circle cross section of the original sphere. If we follow just these particles as they fall, the tidal forces will distort their circle in to an ellipse, but then the *full*, physical ellipsoid can be recovered simply by rotating the ellipse about its vertical axis of symmetry.

Imagine, once again, that you are the astronaut in the centre of this sphere. Let us draw a spacetime diagram from *your* perspective. Having turned off your jetpack, you are in free-fall, as are the particles surrounding you. Therefore, so long as you don't look down at the Earth, you feel that you are simply *floating, motionless*, and the sphere of particles centred on you is likewise almost motionless... except that you gradually witness the tidal evolution of the sphere into an egg. If you *do* look down at the Earth, you see it moving towards you, but you are still at liberty to consider yourself motionless—surely, it is the *Earth* that is moving towards *you!*

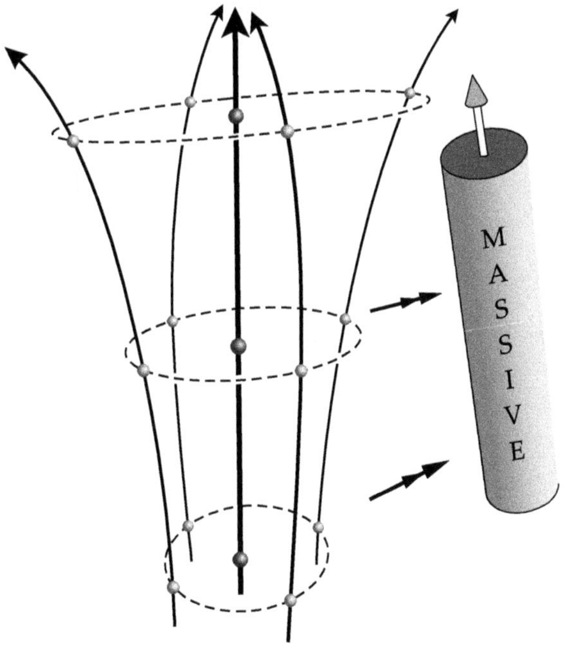

[30.4] Spacetime Depiction of the Tidal Distortion of a Free-Falling Sphere into an Egg. *As usual, time runs vertically.*

It is from this perspective that we have drawn [30.4], with you being the darker world-line in the centre. It depicts the spacetime evolution of the sphere (represented by a vertical circle) into an ellipsoidal egg, as it falls to the Earth. Here the Earth is schematically represented by the object labelled MASSIVE.

This diagram should ring a bell! It looks almost exactly like [29.11c], depicting the evolution of a circular bundle of geodesics in the presence of sectional curvature that is positive in one direction (causing attraction), and negative in the perpendicular direction (causing repulsion). This analogy becomes precise if we consider the sectional curvature in spacetime, and compare (30.2) and (30.3) with the (now 4-dimensional) Sectional Jacobi Equation, (29.21).

The 4-velocity **v** of the circle of particles (starting from rest) is vertical, along the time axis. As the vertical spacetime plane containing this 4-velocity rotates around it, it passes through the direction parallel to the Earth's surface, with the positive, attractive sectional curvature (30.2):

---

[8]In our present state of understanding, the physical laws are irreducible: they cannot be logically derived from anything more primitive. However, from Newton onward, consistency with other physical laws, combined with mathematical *beauty*, have served as remarkably effective guiding principles that have allowed us to correctly *guess* some of Nature's deepest secrets.

$$\mathcal{K}_{+} = +(GM/r^3).$$

Rotating through another right angle, the spacetime plane contains the direction of the Earth's gravitational field, yielding negative, repulsive sectional curvature (30.3):

$$\mathcal{K}_{-} = -(2GM/r^3).$$

But, of course, there is a *third* spatial direction that is orthogonal to these two directions, but it is suppressed in our spacetime diagram—it is the third direction orthogonal to the time axis, lying in the plane parallel to the Earth's surface.

Let us spell this out in terms of an orthonormal tetrad adapted to this situation. Let $\mathbf{e}_0 = \mathbf{v}$ point along the (vertical) time axis, let $\mathbf{e}_1$ and $\mathbf{e}_2$ span the plane parallel to the Earth's surface, and let $\mathbf{e}_3$ point radially away from the Earth, in the direction of the gravitational field.

The generalization of (29.26) from 3-manifolds to 4-manifolds is that the average $\mathcal{K}_{\text{mean}}$ of the sectional curvatures in spacetime over *all* planes containing $\mathbf{e}_0 = \mathbf{v}$ can be obtained by averaging over just these three orthogonal planes containing $\mathbf{e}_0 = \mathbf{v}$:

$$\Pi_1 \equiv \Pi(\mathbf{e}_0, \mathbf{e}_1), \quad \Pi_2 \equiv \Pi(\mathbf{e}_0, \mathbf{e}_2), \quad \text{and} \quad \Pi_3 \equiv \Pi(\mathbf{e}_0, \mathbf{e}_3), \quad .$$

So,

$$\mathcal{K}_{\text{mean}} = \frac{\mathcal{K}(\Pi_1) + \mathcal{K}(\Pi_2) + \mathcal{K}(\Pi_3)}{3} = \frac{1}{3}\mathbf{Ricci}(\mathbf{e}_0, \mathbf{e}_0) = \frac{1}{3} R_{00}.$$

Very remarkably, when viewed in this geometrical way, *the passage from Newton's inverse-square tidal forces to Einstein's curved spacetime is completely seamless!*

We saw in (30.4) that the signature of Newton's inverse-square tidal forces is that it is *volume-preserving*, and we saw in (29.29) that the acceleration of the volume is governed by the *Ricci curvature*. Combining these two results, we find that

$$\boxed{\mathbf{Ricci}(\mathbf{v}, \mathbf{v})\, \delta V = -\ddot{\delta V} = 0.} \tag{30.9}$$

More explicity, from (30.2) and (30.3),

$$R_{00} = \mathcal{K}(\Pi_1) + \mathcal{K}(\Pi_2) + \mathcal{K}(\Pi_3) = \mathcal{K}_{+} + \mathcal{K}_{+} + \mathcal{K}_{-} = \frac{GM}{r^3} + \frac{GM}{r^3} - \frac{2GM}{r^3},$$

so

$$\boxed{R_{00} = 0.} \tag{30.10}$$

Here we have imagined that the sphere of particles starts from rest, so that $\mathbf{v} = \mathbf{e}_0$ is a purely vertical 4-velocity, along the time axis. But we know that the volume-preserving signature of Newton's tidal forces continues to hold if we launch the sphere with any velocity, and therefore (30.9) holds for *any* timelike 4-velocity $\mathbf{v}$.

The final step in the derivation of Einstein's Vacuum Field Equation depends on the *symmetry*, (29.28), of **Ricci**:

$$\mathbf{Ricci}(\mathbf{x}, \mathbf{y}) = \mathbf{Ricci}(\mathbf{y}, \mathbf{x}).$$

Now let $\mathbf{v} = \mathbf{x} + \mathbf{y}$, for *arbitrary* (timelike) $\mathbf{x}$ and $\mathbf{y}$. Then,

$$\begin{aligned} 0 &= \mathbf{Ricci}(\mathbf{v}, \mathbf{v}) \\ &= \mathbf{Ricci}([\mathbf{x}+\mathbf{y}], [\mathbf{x}+\mathbf{y}]) \\ &= \mathbf{Ricci}(\mathbf{x}, \mathbf{x}) + \mathbf{Ricci}(\mathbf{x}, \mathbf{y}) + \mathbf{Ricci}(\mathbf{y}, \mathbf{x}) + \mathbf{Ricci}(\mathbf{y}, \mathbf{y}) \\ &= 0 + \mathbf{Ricci}(\mathbf{x}, \mathbf{y}) + \mathbf{Ricci}(\mathbf{y}, \mathbf{x}) + 0 \\ &= 2\,\mathbf{Ricci}(\mathbf{x}, \mathbf{y}). \end{aligned}$$

We have thus arrived at the

Einstein Vacuum Field Equation: $\mathbf{Ricci} = \mathbf{0} \iff R_{ik} = 0.$ (30.11)

Let us not forget where this came from and what it means:

> *In vacuum, in order for the tidal forces to preserve volume, positive sectional curvatures must be perfectly balanced by negative sectional curvatures, so as to exactly cancel each other out on average.*

While the Ricci curvature vanishes identically in vacuum, this is only the volume-reducing *average* of the positive and negative sectional curvatures. The Riemann tensor itself certainly does not vanish in general; witness the positive and negative tidal curvatures in (30.2) and (30.3).

In general, it is possible to split the Riemann curvature into a volume-reducing Ricci part, plus a purely tidal, volume-preserving part, called the ***Weyl curvature***; see Exercise 15. This split is most elegantly and naturally accomplished via Penrose's 2-spinor formalism; see Penrose and Rindler (1984, §4.6) or Wald (1984, §13.2).

Lest you doubt that (30.11) is in fact Einstein's equation, [30.5] shows Einstein himself proudly writing it down, and while he has modestly added a question mark, his equation now encompasses a remarkable array of startling, testable predictions, *all of which have been confirmed experimentally*, often with extraordinary precision.

We shall outline several of these experimental triumphs, but their mathematical derivations must be left to the technical references we will provide, as well as to those in the General Relativity section of *Further Reading*, at the end of this book.

30.7 The Schwarzschild Solution and the First Tests of the Theory

When we speak of a *solution* to Einstein's equation, we mean a *geometry* of spacetime (defined by its *metric*) that satisfies the equation.

The single most important such solution (at least historically) describes the spacetime in the vacuum region outside a spherically symmetric (nonspinning) mass, M. This solution, the *first* exact solution ever found, was discovered by Karl Schwarzschild[9] (1873–1916)—shown in [30.6]—almost *instantly* after Einstein announced his theory. Schwarzschild discovered it while on active duty on the Russian front, as a 42-year-old artillery leftenant in the German army during World War I. He sent the solution to Einstein in a letter dated the 22nd of December, 1915:

[9]Schwarzschild was a child prodigy, publishing papers on celestial mechanics when he was 16 years old. Later, from 1901 to 1909, he was a professor at Göttingen, where his colleagues included Hilbert, Klein, and Minkowski.

[30.5] *Albert Einstein (1879–1955) circa 1931, delivering a talk at the Mount Wilson Observatory's Hale Library, Pasadena, California. On the blackboard, Einstein has just written his vacuum field equation:* $R_{ik} = 0$ (plus a question mark). *Image courtesy of the Observatories of the Carnegie Institution for Science Collection at the Huntington Library, San Marino, California.*

> As you see, the war treated me kindly enough, in spite of the heavy gunfire, to allow me to get away from it all and take this walk in the land of your ideas.

Decades later, physicists would slowly come to realize that Schwarzschild's solution not only describes the geometry surrounding a spherical mass, such as the Sun or the Earth, it *also* describes the pure vacuum gravitational field of a black hole!

Schwarzschild's solution builds on the spherical polar metric formula of Minkowski spacetime, (30.8):

Schwarzschild Solution

$$ds^2 = \left(1 - \frac{2GM}{r}\right) dt^2 - \frac{dr^2}{\left(1 - \frac{2GM}{r}\right)} - r^2(d\varphi^2 + \sin^2 \varphi \, d\theta^2). \tag{30.12}$$

NOTES:

- Here the radial coordinate r is defined in such a way as to maintain its Euclidean relation to the area $\mathcal{A}(r)$ of a sphere of radius r, namely, $r = \sqrt{\mathcal{A}(r)/4\pi}$. Thus dr does *not* measure radial distance; radial distance is instead equal to $dr/\sqrt{\left(1 - \frac{2GM}{r}\right)}$. Only as we recede from the massive body (or black hole) does the radial coordinate asymptotically regain its Euclidean interpretation as radial distance.
- If we do not choose units in which the speed of light $c = 1$, then $(2GM/r)$ must be replaced by $(2GM/c^2 r)$.
- At the conclusion of our drama, we will use ***curvature 2-forms*** to confirm that this geometry does indeed satisfy the Einstein Vacuum Equation, (30.11).

[30.6] *Karl Schwarzschild (1873–1916).*

- The g_{rr} metric component blows up when

$$r = r_s \equiv \textbf{\textit{Schwarzschild radius}} \equiv \frac{2GM}{c^2}. \tag{30.13}$$

Einstein and others were initially confused by this, and thought this meant that there was something singular about spacetime itself when $r = r_s$. However, this ultimately turned out to be a nonphysical artifact of the particular *choice of coordinates*. If an object of length l is located at the Schwarzschild radius, it experiences ordinary, nonsingular tidal forces of order $(GM/r_s^3)\,l$, in other words, of order $(c^6 l/G^2)\,(1/M^2)$.
- If the mass M is extremely *large*, then the tidal forces at the Schwarzschild radius are extremely *small*.
- If the spherical star or planet has radius R, then the solution only applies to the vacuum region *outside* it, i.e., $r > R$. In the interior, $r < R$, a quite different metric applies, which satisfies the *full* Einstein Field Equation (with matter), which we shall derive shortly. In the case of a sphere of uniform density, the same Karl Schwarzschild found the exact solution to this nonvacuum equation, almost immediately after finding the vacuum solution; it is called, logically enough, the ***Interior Schwarzschild Solution***.
- For the Sun, which has a radius of 430,000 miles, the Schwarzschild radius is only 2 miles! The Schwarzschild radius of the Earth is only about a third of an inch! Thus we must switch from the vacuum Schwarzschild solution to a different interior solution long before we get anywhere near the Schwarzschild radius.

Although the Earth and the Sun *do* spin, Schwarzschild's solution is nevertheless an excellent approximation[10] to the spacetime geometry of both. Indeed this solution suffices to give a complete analysis of our solar system, and to make three crucial predictions about phenomena in our Solar System that *deviate* from Newton's theory.

In the interest of historical accuracy, we must point out that all three of these initial predictions were made by Einstein *before* he knew Schwarzschild's exact solution. Einstein did so by means of an *approximate* solution to his equation, which—bizarrely and inelegantly—he obtained using *rectangular* coordinates!

Here were Einstein's three initial predictions/tests of his theory:

1. When light travels upward through a gravitational field, its frequency should be reduced—we say it is ***redshifted***. Einstein made this prediction in 1907, long before he had discovered the field equation, and thus it is not a direct test of that equation. Nevertheless, had experiments *denied* this prediction, the entire framework of Einstein's theory would have been invalidated. Sadly, the first definitive, terrestrial confirmation of this prediction was not carried out by Pound and Rebka until 1959, four years after Einstein's death.
2. The major axis of the elliptical orbit of Mercury rotates a minuscule amount with each orbit. Newtonian theory predicted that the rotation per *century* should be 532 arcseconds—less than one degree. However, Urbain Le Verrier and later Simon Newcomb conducted extremely accurate analyses of observations collected since 1697, and by 1882 they had established that the *actual* rotation of the orbit was 575 arcseconds per century—a discrepancy of 43 arcseconds per century: a fantastically small (but also indelible and mysterious) blemish on Newton's theory.

   When Einstein calculated the correction according to his own law of gravitation, it made a definite, unambiguous prediction—whatever the verdict, there could be no appeal.

   Einstein's formula yielded 43 arcseconds per century! In that moment, Einstein realized that Nature had spoken to him—his law was built into the eternal structure of the world. Mercury's orbit had been rotating at this rate 4 billion years before mankind existed, and it would continue to do so billions of years into the future.

   Einstein told one friend that, in that moment, he had experienced physical heart palpitations; he told another friend that he felt that something actually "*snapped*" inside him. See Pais (1982, p. 253).
3. Einstein also calculated that light passing close to the rim of the Sun should be bent by 1.75 arcseconds. But starlight grazing the Sun can only be seen during a total solar eclipse, so the prediction could not be tested until one occurred, which happened on the 29th of May, 1919. In anticipation of the eclipse, Sir Arthur Eddington organized expeditions to Brazil and west Africa, in order to take the critical photographs during the brief eclipse, and to perform the measurement of the deflection, indeed, to see if there *were* any deflection.

   The dramatic *confirmation* of Einstein's prediction of the bending of light, and the precise amount by which it was bent, was an international sensation, splashed across the front pages of newspapers around the world. Overnight, the unknown German scientist became a household name, and "Einstein" became a synonym for genius.

After these early triumphs, there followed several fallow decades, in which physicists shifted their attention to all things quantum. However, beginning in the 1960s, General Relativity underwent something of a renaissance, which has continued to the present day, attracting the

[10] Rapidly rotating stars and black holes must be described by a different, vitally important solution, discovered in 1963 by the New Zealand mathematician, Roy Kerr (born 1934); accordingly, it is called the ***Kerr solution***. For Kerr's personal account of his remarkable discovery, see Kerr (2008) and Wiltshire et al. (2009).

finest theoreticians and experimentalists, with many new theoretical predictions, and many new experimental confirmations.

Some of these newer tests still depend on the simple Schwarzschild solution of 1915. Our successful daily use of GPS navigation—see Taylor and Wheeler (2000, §A-1)—is one of these newer confirmations of both Einstein's theory and Schwarzschild's solution!

Tragically, Schwarzschild himself did not live to see any of these triumphs: he died of a rare autoimmune disease on the 11th of May, 1916, mere months after discovering his two remarkable solutions of Einstein's equation. In remembering him, Sir Arthur Eddington wrote, "... his joy was to range unrestricted over the pastures of knowledge, and, like a guerrilla leader, his attacks fell where they were least expected."

30.8 Gravitational Waves

Other confirmations of Einstein's equation have arisen from completely different kinds of solutions to the vacuum field equation, (30.11). As we first outlined on page 231, the existence of gravitational waves was predicted by Einstein in 1916, and the very first experimental detection of a gravitational wave occurred almost exactly a century later, on the 14th of September, 2015.

We are accustomed to the idea that electromagnetic waves—such as the light by means of which we see the world—are photons travelling *through* space. Gravitational waves are something astonishingly different: they are ripples of curvature in the fabric *of* spacetime itself! But not just any old ripples are possible: like the Schwarzschild solution, they must satisfy the Einstein vacuum equation. They are oscillating waves of pure tidal force that have broken free from the violent events that created them, travelling out across space (*as* space!) at the speed of light.

As a gravitational wave passes through a sphere of particles at a fixed location, it has no effect on the particles whose separation lies in the travel direction of the wave. But in the plane *orthogonal* to the direction of the wave, it causes *oscillating*, tidal, egg deformation, stretching the sphere in one direction, and compressing it in the orthogonal direction. See [30.7a], which shows the ***field lines*** (i.e., streamlines of the tidal force field) at one particular moment.

The tidal force field of the gravitational wave is *different* in character from the tidal force field in the vacuum above the Earth, in two important respects:

- First, whereas the gravitational wave only compresses in *one* direction orthogonal to the stretching direction, the Earth's tidal force field compresses in *both* orthogonal directions.
- Second, the Einstein vacuum equation tells us that the positive curvatures in any two orthogonal directions lying within the plane parallel to the Earth's surface must exactly *cancel* the negative curvature in the radial direction of the Earth's gravitational field. Therefore these two positive curvatures must each be *half* as strong as the radial negative curvature, as illustrated in [30.1]. But in the case of the gravitational wave, the curvature associated with the direction of travel *vanishes*. Therefore the Einstein equation tells us that the remaining two curvatures must cancel, and must therefore be of *equal* strength. In other words, in [30.7a] the stretching force in the horizontal direction must *equal* the compressing force in the vertical direction, as illustrated.

Returning to [30.7a], the dashed circle of particles is initially deformed into the solid ellipse. But, half a wavelength later, the tidal forces have completely flipped, as illustrated in [30.7b]: now there is compression in the direction where the wave previously caused stretching, and visa versa. As the wave passes, the force field oscillates back and forth between these two opposite patterns of force.

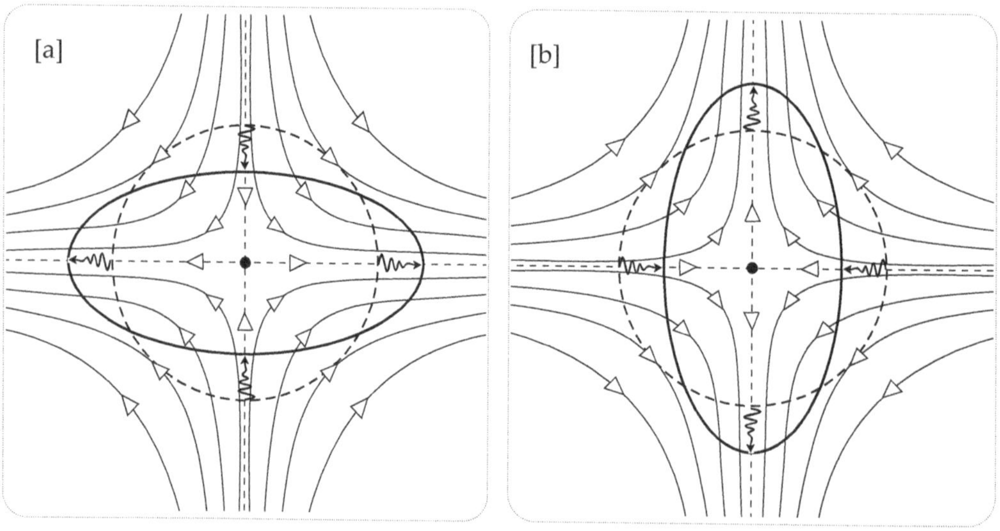

[30.7] The Oscillating Tidal Force Field of a Gravitational Wave. *As the gravitational wave passes through a sphere of particles, it has no effect on the particles whose separation lies in the travel direction of the wave, which is here taken to be perpendicular to the page. In the plane orthogonal to this direction of travel, the wave causes oscillating expansion and contraction in two perpendicular directions, here shown as horizontal and vertical. Figure* **[a]** *shows the deformation of the (dashed) circle of test particles into the (solid) ellipse. Figure* **[b]** *shows the reversal of the tidal force field of the wave, half a wavelength later, now with the stretching orthogonal to the original stretching. As the wave passes, the force field oscillates back and forth between these two opposite patterns.*

As we discussed in Section 19.2, in general a phase portrait cannot represent the *magnitude* of the underlying vector field. In [30.7a] we have merely drawn a few random field lines, and there is no way to tell how strong the tidal forces are by looking at this. However, suppose the vector field is *divergence-free*, like an electric field in vacuum. If we draw the field lines in such a way that their *density/crowding* is proportional to the magnitude of the electric field in one place, then, as we follow the field lines away from that place, the density of field lines in the new place will automatically *continue*[11] to faithfully represent the magnitude of the electric field in the new place. So, although we have not done so in [30.7a], the streamlines for the gravitational wave's tidal force field can *also* be drawn in this way, because (as we now explain) it too is divergence-free.

If we take the plane [30.7] to be the *complex* plane, then the tidal force field in [30.7a] looks like the saddle point (aka crosspoint) first encountered in [19.4h] (p. 200), representing the complex function $(1/z)$. However, the Jacobi Equation tells us that the tidal force of the gravitational wave actually *increases linearly* with distance from the origin, and so it is represented by $\overline{z}=x-iy$, i.e., by the "Pólya vector field of z"; see page 201. Thus,

$$\nabla\cdot\begin{bmatrix} x \\ -y \end{bmatrix}=0.$$

NOTE: One can also understand this from a more advanced point of view: the Pólya vector fields of *all* complex-analytic functions are *automatically* divergence-free (*and* curl-free); see VCA, Section 11.2.2.

But the above interaction between the wave and the sphere of test particles is *unbelievably weak*. Einstein himself doubted that we would *ever* be able to detect his waves. While Einstein did indeed underestimate our future technological prowess, more significantly, he could not have imagined the *extraordinary violence* of cosmic events capable of generating gravitational waves. We

---

[11] See (VCA, Section 11.3) and (Thorne and Blandford, 2017, §27.3.2)

will now describe one such event, which converted *astonishing* quantities of energy (in a fraction of a second!) into hugely powerful gravitational waves—veritable "tidal waves" [*sic*] in spacetime, that were capable of triggering detectors on Earth after spreading out and travelling for more than a billion years!

This 2015 detection of the first gravitational wave (at two independent detectors) was epoch-making in and of itself, but the scientists were able to go much further: they used Einstein's equation to work backwards from the *details* of the detected signal to determine, in remarkable detail, the cataclysmic event that had *created* the gravitational waves.

We will let one of these Nobel Prize–winning scientists, Kip Thorne, speak for himself. (Note that $M_\odot$ denotes the mass of our Sun.)

> On September 14, 2015, the advanced LIGO gravitational wave detectors made their first detection: a wave burst named GW150914 with amplitude 1.0×10^{-21}, duration ~ 150ms, and frequency chirping up from ~ 50Hz (when entering the LIGO band) to 240Hz. By comparing the observed waveform with those from numerical relativity simulations, the LIGO-VIRGO scientists deduced that the waves came from the merger of a $29M_\odot$ Black Hole with a $36M_\odot$ Black Hole, 1.2 billion light years from Earth, to form a $62M_\odot$ Black Hole, with a release of $3M_\odot c^2$ of energy in gravitational waves.—(Thorne and Blandford 2017, p. 1346)

We cannot allow the enormity of the energy release to pass without comment. Einstein's $E = mc^2$ translates this statement as saying that *during the last fraction of a second of the black hole collision, the equivalent of three times the mass of our Sun was converted into gravitational wave energy.* What does that mean? Let us begin with a grim but familiar instantiation of $E = mc^2$: the atomic bomb that destroyed Hiroshima in 1945, killing 75,000 people. In that atomic blast, the total amount of matter that was converted into energy weighed *less than a single raisin*. Let that sink in.

How, then, can we even *conceive* of the entire mass of our *Sun* being converted into gravitational wave energy?! Thorne and Blandford (2017, §27.5.5) offer this interpretation: in the last tenth of a second of the observed black hole collision, *the power output generating the GW150914 gravitational wave was 100 times the luminosity of all the stars in the observable Universe combined!*

For the fullest, most up-to-date—as of this writing!—technical treatment of the generation, propagation, and detection of gravitational waves, see Thorne and Blandford (2017, §27), and Schutz (2022).

Several more gravitational waves have been detected since GW150914.[12] In 2017 the same scientists achieved another first: they detected a burst, GW170817, which Einstein's equation allowed them to decode as arising from the merger of two *neutron stars*! Using the prescient theoretical work of Schutz (1986),[13] from 30 years earlier, these gravitational wave scientists were then able to use this data (in combination with optical data) to calculate the Hubble constant in a totally novel way, shining new light on the expansion of the Universe! Clearly, we have entered a new era of ***gravitational wave astronomy***.

Likewise, the gravitational bending and focusing of light, which Sir Arthur Eddington was the first to witness in 1919, has now led to ***gravitational lensing***: a technique that harnesses the gravitational focusing of light (and radio waves) as a *tool* for the study of very faint and distant objects.

Thus, in hindsight, our testing of Einstein's theory has been akin to looking down the wrong end of a telescope, checking to see if its lenses have been correctly configured. Scientists are now looking through the theory from the *correct* end, gazing out into spacetime, and using Einstein's

[12]The naming of these events is simple: GW stands for "gravitational wave"; 15 denotes 2015; 09 denotes the month, September; and 14 denotes the day of the month.

[13]For this work, the Royal Astronomical Society awarded Bernard Schutz the 2019 Eddington Medal.

equation to witness and decipher phenomena that would otherwise have remained invisible and inexplicable.

Everything we have described thus far has depended only on Einstein's *vacuum* field equation, (30.11). Even the black holes that generated GW150914 were describable by solutions of the vacuum equation, as were the gravitational waves themselves. But without the *matter* within the stars that collapsed billions of years ago, there would have been no black holes, no black hole collision, and no emitted gravitational waves!

We therefore now turn to the *full* Einstein Field Equation, with matter.

30.9 The Einstein Field Equation (with Matter) in Geometrical Form

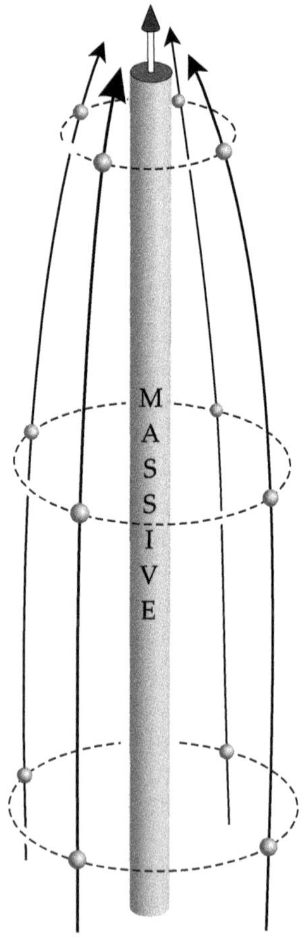

[30.8] Spacetime Depiction of the Volume Reduction of a Free-Falling Sphere Containing Matter. *As usual, time runs vertically.*

Figure [30.4] illustrated the volume-preserving effect of the tidal forces when our sphere of particles enclosed only a vacuum. Figure [30.8] is instead a spacetime depiction of the *volume-reducing effect* when matter is present within the sphere. Again, this picture should ring a bell! It is almost exactly like [29.11b]! And, once again, this is no accident: [30.8] shows the effect—via the Sectional Jacobi Equation, (29.21)—of purely *positive* sectional curvatures in the presence of matter.

If the volume $\delta\mathcal{V}$ is filled with matter of density ρ, the Newtonian, volume-reducing acceleration is given by (30.5), but, in spacetime terms, this is again described by the average of the sectional curvatures, i.e, by the Ricci curvature formula, (29.29). Therefore, combining these results, we find that

$$\mathbf{Ricci}(\mathbf{v},\mathbf{v})\,\delta\mathcal{V} = -\ddot{\delta\mathcal{V}} = 4\pi G\rho\delta\mathcal{V},$$

and therefore

$$\mathbf{Ricci}(\mathbf{v},\mathbf{v}) = 4\pi G\rho. \qquad (30.14)$$

To make further progress, we must express the right-hand side of this equation in terms of the tensor that describes matter and energy in both Special and General Relativity. In common with the Ricci tensor, this new tensor, denoted $\mathbf{T}$, takes two vectors as input, and outputs a scalar. It is called the ***energy–momentum tensor*** or, equally commonly, the ***stress–energy tensor***. Like the Ricci tensor, it is *symmetric*:

$$\boxed{\mathbf{T}(\mathbf{w},\mathbf{v}) = \mathbf{T}(\mathbf{v},\mathbf{w}) \quad \Longleftrightarrow \quad T_{ki} = T_{ik}.} \qquad (30.15)$$

Despite appearances to the contrary, this is not (at least primarily!) a physics book, so for a full description of $\mathbf{T}$ we must refer the reader to Misner, Thorne, and Wheeler (1973) or Thorne and Blandford (2017). For our purposes, the key feature of $\mathbf{T}$ is that for an observer with (normalized) 4-velocity $\mathbf{v}$, the total density of matter *and* energy is given by

$$\mathbf{T}(\mathbf{v},\mathbf{v}) = \rho_{\text{matter}} + \rho_{\text{energy}} \equiv \rho_{\text{total}}.$$

Recall that Einstein discovered that matter and energy are equivalent, so both of them curve spacetime. For example, the electromagnetic field has an energy density that Maxwell discovered to be proportional to $(|\mathbf{E}|^2 + |\mathbf{B}|^2)$, where $\mathbf{E}$ is the electric field, and $\mathbf{B}$ is the magnetic field. This electromagnetic energy curves spacetime just as surely as a planet does. Other fields, too, must be added in to obtain the complete energy–momentum tensor.

Thus, taking these relativistic effects into account, (30.14) becomes

$$\mathbf{Ricci}(\mathbf{v}, \mathbf{v}) = 4\pi G \rho_{\text{total}} = 4\pi G\, \mathbf{T}(\mathbf{v}, \mathbf{v}). \tag{30.16}$$

Note, for future comparison, that if we use the same tetrad as before, with $\mathbf{v} = \mathbf{e}_0$, then

$$\boxed{R_{00} = 4\pi G \rho_{\text{total}}.} \tag{30.17}$$

But (30.16) is true if $\mathbf{v}$ is an *arbitrary* timelike vector. So, using the same trick as before, writing $\mathbf{v} = \mathbf{x} + \mathbf{y}$ and appealing to the symmetry of $\mathbf{T}$, we deduce [exercise] that

$$\boxed{\mathbf{Ricci} = 4\pi G\, \mathbf{T} \quad \Longleftrightarrow \quad R_{ik} = 4\pi G\, T_{ik}.} \tag{30.18}$$

This was in fact one of Einstein's original proposals for his field equation, beginning in 1913. (Of course if $\mathbf{T} = \mathbf{0}$ then it reduces to the (physically correct) vacuum equation, (30.11).) Indeed, Einstein still believed in (30.18) as late as the *11th* of November, 1915, exactly two weeks prior to his discovery of the *correct*, final field equation of General Relativity, on the 25th of November. For more on Einstein's decade-long struggle towards the correct equation, see Misner, Thorne, and Wheeler (1973, §17.7), or, better yet, Pais (1982, §14), which chronicles (essentially day by day) Einstein's mental thrashings throughout that fateful November.

The *modern* way to recognize the fatal flaw inherent in (30.18) is to observe that if this equation *were* true, the Differential Bianchi Identity, (29.17), would imply that *energy is not conserved!* To fully explain this would take us too far afield, but the essential *mathematical* fact is derived in Exercise 14. To properly understand the *physics*, we refer you to Misner, Thorne, and Wheeler (1973, §17), or Penrose (2005, §19.6), or Thorne and Blandford (2017, §25.8), or Wald (1984, §4.3). As we noted earlier, we should also point out that this is *not* how Einstein finally arrived at the truth, because *in 1915 he did not know the Differential Bianchi Identity!* See Pais (1982, p. 256).

The upshot is that the Differential Bianchi Identity almost uniquely determines the correction that *must* be made to (30.18) in order for conservation of energy to be restored.

To make this correction, we introduce the ***trace*** of the energy–momentum tensor:

$$T \equiv T^m{}_m = \rho_{\text{total}} - (P_1 + P_2 + P_3),$$

where P_1, P_2, P_3 are the *pressures* within the matter in our three orthogonal spatial directions, $\mathbf{e}_1$, $\mathbf{e}_2$, $\mathbf{e}_3$. Note that while pressure and density may seem very different concepts, a connection can be seen as follows:

$$\text{pressure} = \frac{\text{force}}{\text{area}} = \frac{\text{force} \times \text{distance}}{\text{volume}} = \frac{\text{energy}}{\text{volume}} = \frac{\text{mass}}{\text{volume}} = \text{density}.$$

We can now state the *correct*, final form of the ***Einstein Field Equation:***

$$\boxed{\mathbf{Ricci} = 8\pi G \left[\mathbf{T} - \tfrac{1}{2} T \mathbf{g}\right] \quad \Longleftrightarrow \quad R_{ik} = 8\pi G \left[T_{ik} - \tfrac{1}{2} T g_{ik}\right].} \tag{30.19}$$

We note that this is *not* how you will find the equation written in almost all texts on General Relativity. Nevertheless, this is how *Einstein himself* originally wrote his equation, on the 25th of November, 1915! See Pais (1982, p. 256).

Let us state the more standard, mathematically equivalent form, though we leave the short proof to the technical references above:

$$\mathbf{Ricci} - \tfrac{1}{2}R\mathbf{g} = 8\pi G\,\mathbf{T} \quad \Longleftrightarrow \quad R_{ik} - \tfrac{1}{2}R g_{ik} = 8\pi G\,T_{ik}, \tag{30.20}$$

where $R \equiv R^m{}_m$ is the trace of **Ricci**, called the ***curvature scalar***. The tensor on the left-hand side,

$$\mathbf{G} \equiv \mathbf{Ricci} - \tfrac{1}{2}R\,\mathbf{g}, \tag{30.21}$$

is called the ***Einstein tensor***.[14] This notation leads to (by far) the most common way of writing Einstein's equation:[15,16]

$$\mathbf{G} = 8\pi G\,\mathbf{T}. \tag{30.22}$$

The final Einstein Field Equation, (30.19), appears very different from the original equation, (30.18), to which we were naturally led by our geometrical reasoning regarding Newton's Inverse-Square Law. In reality, they differ very little. Indeed, we now show that, physically, the Einstein equation only adds a correction to the original Newtonian equation, a correction that is extremely small under normal circumstances.

To see this, again choose the time axis along the observer's 4-velocity: $\mathbf{e}_0 = \mathbf{v}$. Then, since $g_{00} = 1$, we find that

$$R_{00} = 8\pi G\left[T_{00} - \frac{1}{2}T g_{00}\right] = 8\pi G\left[\rho_{\text{total}} - \frac{1}{2}\left\{\rho_{\text{total}} - (P_1 + P_2 + P_3)\right\}\right],$$

so

$$R_{00} = 4\pi G\left(\rho_{\text{total}} + P_1 + P_2 + P_3\right). \tag{30.23}$$

Comparing this to our original, Newtonian-based equation, (30.17), we see that Einstein's equation *only* differs by the addition of the three pressure terms, and for matter under normal circumstances these terms are tiny compared to the mass/energy term, ρ_{total}.

It is important to remember the *significance* of the Ricci tensor, and hence of (30.23)—it tells us the *volume-compressing effect of the spacetime curvature*, via (29.29). We have finally arrived at the ***Einstein Field Equation in geometrical form***:

$$\ddot{\delta V} = -R_{00}\,\delta V = -4\pi G\left(\rho_{\text{total}} + P_1 + P_2 + P_3\right)\delta V. \tag{30.24}$$

---

[14]For more on the geometry of the Einstein tensor, see Frankel (2011, §4).

[15]We remind the reader that the definition of the Ricci curvature varies, affecting its sign; thus, for example, Penrose (2005, §19.6) instead writes $\mathbf{G} = -8\pi G\,\mathbf{T}$.

[16]It is also common to use ***geometrized units***, in which both $c = 1$ *and* $G = 1$, in which case Einstein's equation becomes, simply, $\mathbf{G} = 8\pi\,\mathbf{T}$.

30.10 Gravitational Collapse to a Black Hole

During the normal life of a star, it is the continuous hydrogen bomb–like explosion within it that makes the star shine and that keeps the material of the star from collapsing inward under its own weight. But the star eventually exhausts its supply of its primary nuclear fuel, hydrogen, and as the nuclear fire goes out, the star starts to lose its power to resist gravity. Let us briefly consider the fate of normal star as its endgame begins.

In the course of gravitational collapse, the temperature and pressure can rise to the point that *new* nuclear reactions begin, renewing the fight against gravity by burning helium, and heavier elements. Also, the extreme circumstances of gravitational collapse can even cause the atoms themselves to be crushed, at which point, new *quantum-mechanical* forces resist the collapse, by virtue of the Pauli Exclusion Principle.

Precisely which nuclear reactions and which quantum-mechanical forces become dominant, and in what sequence, is the complicated dance[17] that governs the death throes of a star. But the *key* piece of information that largely determines this ultimate path is the *initial mass* M of the star.

Small stars (like our Sun) with $M < 8\,M_{\odot}$ may end up as stable white dwarfs, in which quantum-mechanical electron degeneracy pressure prevents further collapse. But much larger stars, with $M > 10\,M_{\odot}$, may ultimately go supernova, with much of the material of the star being exploded out into space, leaving only a core behind that then undergoes gravitational collapse. If the original star has $M < 30\,M_{\odot}$, then it is possible that the collapse of the postsupernova core will be halted by neutron degeneracy pressure (and other forces), leading to the formation of a stable, rapidly spinning neutron star.

But Einstein's equation predicts that if the collapsing core is sufficiently massive, something counterintuitive and almost paradoxical-seeming can happen, resulting in the creation of a *black hole*.

In the extreme circumstances of such a gravitational collapse, the pressure terms in Einstein's equation, (30.24), can become very significant as the matter is compressed and the speed of the atoms approaches the speed of light. Ordinary physical intuition tells us that these internal pressures will fight the collapse and perhaps even halt it. But, on the contrary, the Einstein equation tells us that the mounting pressures only *increase* the volume-crushing power of gravity! The harder the star fights its collapse, the more gravity tightens its stranglehold.

If, ultimately, the postsupernova core that undergoes gravitational collapse has a *mass that is greater than about*[18] $2\,M_{\odot}$—the ***Tolman–Oppenheimer–Volkoff limit***, TOV for short—this purely Einsteinian effect leads to *a point of no return*: now *no* force in the Universe can halt the collapse. See [30.9]. Gravity will relentlessly crush the entire core down to a point of infinite density and infinite tidal forces at $r = 0$, called a ***spacetime singularity***. After the collapse has finished, what remains is a pure, vacuum gravitational field.[19]

Imagine that flashes of light are emitted from the centre of the core during the collapse, and further imagine that these flashes can pass through the matter of the core, as though they were neutrinos. The fate of an individual flash depends on *when* it is emitted. If the flash is emitted early enough, then, as illustrated, it is able to escape the gravitational field.

But there comes a critical moment when the flash from the centre expands at first but then slows and ultimately *hovers* at the Schwarzschild radius, $r_s = (2GM/c^2)$. This hovering sphere of light is the ***event horizon***, and its interior is a ***black hole***: no matter or information can escape this

[17] For the fascinating details of this dance, see Schutz (2003, §12).

[18] The precise value of the TOV limit is a work in progress; $2.2\,M_{\odot}$ was the best estimate when this was written, in 2019.

[19] ***Birkhoff's Theorem*** tells us that as the spherical mass M shrinks and remains spherical, the geometry of the vacuum outside it *must* be the Schwarzschild solution, and that the M-value in the metric does not change as the object's radius R shrinks; the only change is the size of the vacuum region $r > R$ to which the solution applies.

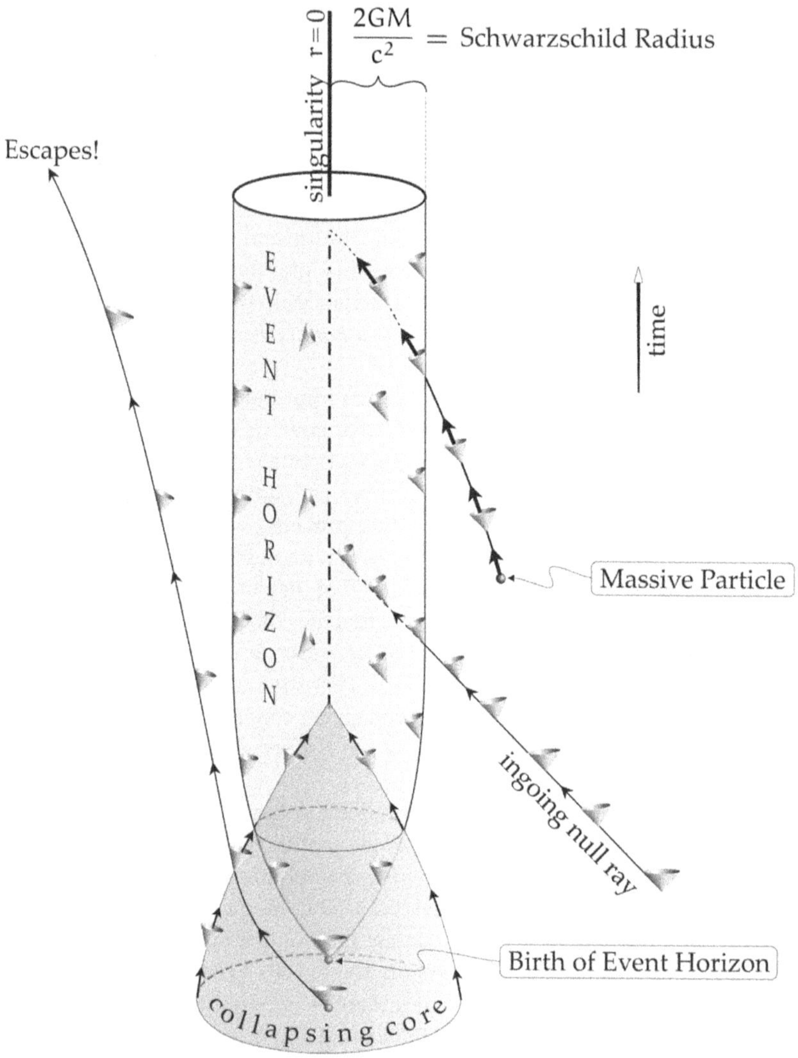

[30.9] The Birth of a Black Hole: the Gravitational Collapse of a Postsupernova Core. *Einstein's field equation tells us that the collapse of a core of sufficient mass will reach a point of no return: gravity will relentlessly crush the entire core down to a point of infinite density and infinite tidal forces at* $r = 0$, *called a* **spacetime singularity**; *what remains is a pure, vacuum gravitational field. If a flash is emitted from the centre of the collapsing core early enough, it can escape the gravitational field. But there comes a critical moment when the flash from the centre expands at first but then slows and ultimately* hovers *at the Schwarzschild radius,* $r_s = (2GM/c^2)$. *This hovering sphere of light is the* **event horizon**, *and its interior is a* **black hole**: *no matter or information can escape this region once it has been formed. The null cones are tangent to the horizon and so they allow matter and light to pass inwards, but never outwards, for matter must always travel* inside *the null cones.*

region once it has been formed. The null cones are tangent to the horizon and so they allow matter and light to pass inwards, but never outwards, for matter must always travel *inside* the null cones.

Therefore, once an object has fallen into the black hole, it is relentlessly dragged towards the central singularity, and the tidal forces exerted on it grow in proportion to $(1/r^3)$. Any solid object must ultimately succumb to these rapidly growing tidal forces, being stretched out in the radial direction, and compressed in the perpendicular directions. As the tidal forces tends to infinity, this process will stretch/compress the object into a long thin shape, like a piece of spaghetti. For this reason, physicists actually call this process ***spaghettification***!

Black holes can also form in other ways. We have mentioned that a gravitational wave (GW170817) detected in 2017 originated from two neutron stars spiralling around and into each other, finally colliding and merging. After the merger, the resulting object's mass exceeded the TOV limit. Remarkably, the gravitational wave scientists were able to spatially *localize* the source to a sufficiently small patch of the southern sky that traditional X-ray astronomers were then able to look there, successfully identifying and studying the aftermath of the neutron-star collision. This X-ray data now seems to confirm that the merged neutron stars immediately underwent full collapse to form a black hole!

We now know that the Universe is littered with black holes of these types, a few solar masses each. However, in addition, astronomers have now confirmed that almost every galaxy (including our own Milky Way) has, at its centre, a totally different kind of black hole—a ***supermassive black hole*** that is *millions* or even *thousands of millions* times more massive than our Sun! Although astrophysicists have hypothesized different mechanisms by which they might have been formed, the true origin of the supermassive black holes remains a mystery.

As we noted earlier, if a body of linear dimensions l passes through the event horizon of a black hole of mass M, the tidal forces exerted on it there are of order $(c^6 l/G^2)(1/M^2)$, which gets *smaller* as M gets bigger. For a supermassive black hole with $M = 10^8\, M_\odot$, this force is much less than the tidal force we experience on the surface of the Earth! Thus, if you were to fly your spaceship into such black hole, you would not even realize that you had done so—*you would feel nothing* as you crossed the event horizon.

Nevertheless, regardless of the power of your spaceship's engines (and how you might attempt to deploy them), after you cross the horizon, the *maximum* time you have till you hit the singularity is proportional to the mass of the black hole, and is given by[20]

$$t_{max} = \frac{M}{M_\odot} \times 15.5 \times 10^{-6} \text{ seconds.}$$

If you were "lucky enough" to fly your spaceship into a black hole of sufficient mass—a few galaxies' worth of mass will do!—then you could even *live* inside the black hole for *years* thereafter; remember, your friends outside the hole can still send care packages *into* the hole.

However, for more realistic supermassive black holes, the outlook is more grim. For example, the black hole at the centre of our own Milky Way galaxy (called Sagittarius A*) has an estimated mass[21] $M = 3.6 \times 10^6\, M_\odot$, so if you were to cross its horizon, your maximum survival time would be *less than one minute*!

Regardless of t_{max}, once you have entered a black hole, there can be no escape—sadly, your ultimate and completely inevitable demise is spaghettification!

30.11 The Cosmological Constant: *"The Greatest Blunder of My Life."*

We began this chapter with Einstein's happiest thought, and we shall end it with his *least* happy thought.

But this seemingly sad movie has a sudden, bizarre, *superhappy* plot twist at the very end! Indeed, after the curtain falls, and you finally walk through the exit door of Act IV, you will probably be muttering to yourself, "That could *never* happen in real life!"

[20]See Taylor and Wheeler (2000, §3–21).

[21]Measurements in 2018 revised upward the earlier estimate of $M = 2.6 \times 10^6\, M_\odot$.

In 1916 the widespread belief (based more on philosophy than science) was that the Universe was *static*, unchanging. But Einstein's field equation was quickly discovered to be in conflict with this idea—*it automatically led to a Universe that was expanding or contracting.*

Mind you, at this time, the "Universe" was thought to comprise *only* of our Milky Way galaxy—the existence of galaxies beyond our own was first discovered by Edwin Hubble in 1924.

In order to salvage a static Universe,[22] Einstein employed a daring and desperate gambit—despite the fact that it was the fruit of ten years of intense labour, he *changed* his field equation, (30.22)!

We have noted that the Differential Bianchi Identity, (29.17), *almost* uniquely determines the change that must be made to Einstein's original equation (30.18) in order to restore conservation of energy. However, Einstein realized that there remained one, and *only* one, additional freedom in his equation: he could add to the left-hand side a constant multiple Λ of the metric tensor, **g**.

Einstein made this change to his equation in 1917, and to distinguish this new equation from the original 1915 equation, we shall call it the

Cosmological Einstein Field Equation: $$\boxed{\mathbf{G}+\Lambda\mathbf{g}=8\pi G\,\mathbf{T}.} \qquad (30.25)$$

If true, this constant Λ—which is called the ***cosmological constant***—would need to be extraordinarily small to avoid conflict with observational evidence. At the scale of the solar system, or even the galaxy, the new Cosmological Field Equation would then make the same predictions as the original field equation. Only at the cosmological scale would its predictions be starkly different.

In 1929, twelve years after Einstein had made this change to his equation, Edwin Hubble made a *second* extraordinary discovery. Building on his earlier discovery of galaxies beyond our own, he sought to measure their distances from us and then to correlate those distances with the radial speed of the galaxies (measured using their redshifts). He found that the galaxies all had velocities directed *away* from us, and that these *velocities of the galaxies were proportional to their distances from us*. This is now known as ***Hubble's Law***.

Thus the Universe was *not* static, after all—it was *expanding!* Einstein realized that if he had only remained faithful to his original equation of 1915, he could have made the most extraordinary scientific prediction in human history! He told George Gamow that his introduction of the cosmological constant had been, *"The greatest blunder of my life."* Thereafter, Einstein retracted the Cosmological Field Equation in favour of the original.

In subsequent decades, many (most?) experts tended to side with Einstein's retraction, believing that Λ was identically zero, and that the more elegant, original Einstein equation was correct. Nevertheless, once the Λ-genie was out of the bottle, it was impossible to put it back in. Thus, research papers always tended to hedge their bets, at least examining how their conclusions might be affected if $\Lambda \neq 0$.

We now leap forward in time to 1998, and to a cosmological discovery[23] as great as Hubble's discovery, 70 years earlier. Although Hubble had found that the Universe was expanding, the expectation had been that the positive, attracting gravitational pull of the matter within would *slow* the expansion of the Universe over time. However, by observing ***type Ia supernovas***—which are exploding white dwarfs that have exceeded their stability limit, and which therefore have similar masses and therefore standardized intrinsic luminosity—the scientists made the astonishing discovery that *the expansion of the Universe is accelerating!*

But how can we *explain* this gobsmacking discovery?! Einstein's answer had been sitting there, patiently waiting, since 1917!

---

[22]The full story of Einstein's motivations is far more complex and subtle. See Pais (1982, §15c).

[23]The discovery was made by two independent research teams, and the 2011 Nobel Prize in Physics was divided, one-half awarded to Saul Perlmutter, the other half jointly to Brian P. Schmidt and Adam G. Riess.

If $\Lambda > 0$, then (moving $\Lambda\mathbf{g}$ to the right-hand side of the equation) its effect is *mathematically equivalent* to a fictitious[24] *negative energy density* $\rho = -(\Lambda/8\pi G)$, *resulting in* ***gravitational repulsion***. While [30.8] illustrates the attractive nature of ordinary matter, and positive sectional curvatures, the *opposite* behaviour in the presence of *negative* energy and *negative* sectional curvatures is illustrated in [29.11a], page 301. When viewed as a spacetime picture of the expansion of the Universe, this shows the *repulsive* effect of *negative* energy arising from a *positive* value of Λ, evidently resulting in *accelerating* expansion of the Universe.

In short, the observed accelerating expansion of the Universe agrees (in detail) with the Cosmological Einstein Field equation, with $\Lambda > 0$. This, then, is the happy, scarcely believable, final plot twist: Einstein's "greatest blunder" turned out to be, instead, one of his greatest strokes of genius!

30.12 The End

We hope that these final two chapters of Act IV have inspired you to grapple in greater depth with the connected wonders of Riemann's curvature and Einstein's curved spacetime. However, *we* must stop here—a *full* explanation would require Volumes 2 and 3!

If we have done our job, you cannot wait to learn more—let the *Further Reading* section at the end of this book be your guide. There we recommend *many* excellent works, each with its own, distinctive *spécialité de la maison*.

That said, let us impatiently assert that we can think of no smoother or faster road to enlightenment than Penrose's *Road to Reality* (2005), for it contains his countless original and revelatory insights into both the mathematics and the physics, all expressed with astonishing clarity by means of his singularly beautiful hand drawings.

For even greater technical detail, we also *strongly* recommend—see Prologue—*Gravitation* by Misner, Thorne, and Wheeler (1973). And for more up-to-date developments, especially on gravitational waves and modern cosmology, we highly recommend the massive and authoritative work, *Modern Classical Physics*, by Thorne and Blandford (2017).

Last, after reading Act V, we recommend Dray (2015), which uses Forms to clarify the geometry of General Relativity. Indeed, it is to *Forms* that we now turn, for they are the subject of our final Act.

But, as for our treatment of purely *Visual* Differential Geometry, this is, finally, the end.

---

[24]While I am personally content to accept the Cosmological Field Equation with $\Lambda > 0$, many famous scientists *instead* believe that there exists an actual, *nonfictitious thing*, permeating the entire Universe, that causes this repulsion—they call it ***dark energy***.

Chapter 31

Exercises for Act IV

Extrinsic Constructions

1. **Geodesic Curvature on the Sphere.** On the sphere of radius R, suppose a particle travels at unit speed along the circle at latitude ϕ.

   (i) What is the magnitude and direction of the acceleration of the particle?

   (ii) Sketch the projection of this acceleration onto the tangent plane of the sphere, and deduce that the geodesic curvature is $\kappa_g = \frac{\cot\phi}{R}$.

   (iii) Verify that this formula for κ_g yields the geometrically correct answer as $\phi \to 0$ and as $\phi \to (\pi/2)$.

   (iv) Show that the component of the acceleration directed towards the centre of the sphere has magnitude $(1/R)$, *independent* of ϕ. Explain this geometrically.

2. **Geodesic Curvature on the Cone.** Reconsider the cone of semivertical angle α shown in [14.3], page 145. Suppose a particle travels at unit speed along a horizontal circle of radius r on this cone.

   (i) What is the magnitude and direction of the acceleration of the particle?

   (ii) Sketch the projection of this acceleration onto the tangent plane of the cone, and deduce that the geodesic curvature is $\kappa_g = \frac{\sin\alpha}{r}$.

   (iii) Let s be the distance within the cone (along a geodesic generator) from the vertex to the circle. Cut the cone along a generator and press it flat onto the plane, as in [14.3], page 145, so that the circle becomes an arc of a circle in the plane of radius s and therefore of curvature $(1/s)$ within the plane. Verify that this curvature is the *same* as κ_g. (NOTE: What is the radius of this circle from the *intrinsic* point of view? Well, if two ants at neighbouring points of the circle start walking inward along radii—along geodesic generators orthogonal to the circle—they will meet at the vertex, so the intrinsic radius is simply the distance s from the vertex to the circle. The natural *intrinsic* definition of the curvature of the circle is therefore, once again, $(1/s)$.)

   (iv) Let ρ be the distance along the normal from the particle to the axis of symmetry. Show that the component of the acceleration along the normal is $(1/\rho)$, and explain this geometrically.

3. **Geodesic Curvature along Touching Surfaces.** Suppose that two surfaces touch along a common curve $\mathcal{C}$.

   (i) If $\mathcal{C}$ is a geodesic of one surface, explain geometrically why it must also be a geodesic of the other surface.

   (ii) If $\mathcal{C}$ has geodesic curvature κ_g for one surface, explain geometrically why it must also have geodesic curvature κ_g for the other surface.

(iii) Deduce (i) as a special case of (ii).

(iv) Use the *formulas* for κ_g in the previous two exercises to confirm (ii) in the case that one surface is a sphere, and the other surface is a cone, and $\mathcal{C}$ is a circle of latitude on the sphere.

Intrinsic Constructions

4. Geodesic Curvature via Intrinsic Differentiation. Let $\mathbf{v}$ be the velocity of a particle travelling at unit speed along a trajectory K over the surface $\mathcal{S}$ of some peelable fruit. If o is the launch point, let $(\mathbf{e}_1, \mathbf{e}_2)$ be an orthonormal basis for T_o. Now parallel transport $\mathbf{e}_1$ along K, and let $\theta_{||}$ denote the angle between this parallel-transported $\mathbf{e}_1$ and $\mathbf{v}$.

(i) Explain why $\mathbf{e}_2$ is automatically parallel-transported along K, too:

$$D_{\mathbf{v}}\mathbf{e}_1 = 0 \qquad \Longrightarrow \qquad D_{\mathbf{v}}\mathbf{e}_2 = 0.$$

(ii) Let $\mathbf{v}^{\perp}$ be the unit vector obtained by rotating $\mathbf{v}$ by $(\pi/2)$ within T_p. Write $\mathbf{v}$ and $\mathbf{v}^{\perp}$ in terms of $(\mathbf{e}_1, \mathbf{e}_2)$. Now use *calculation* to prove that

$$\boldsymbol{\kappa}_g = D_{\mathbf{v}}\mathbf{v} = \left[D_{\mathbf{v}}\theta_{||}\right]\mathbf{v}^{\perp},$$

thereby proving (23.4): $|\boldsymbol{\kappa}_g| = |D_{\mathbf{v}}\theta_{||}|$.

(iii) Draw a sketch illustrating $\mathbf{v}$, $(\mathbf{e}_1, \mathbf{e}_2)$, and $\theta_{||}$ at several points along K, *after* the narrow strip of peel surrounding K has been removed from $\mathcal{S}$ and pressed flat onto the plane.

(iv) Deduce that the geodesic curvature of K is indeed the ordinary curvature of the plane curve obtained by flattening the narrow strip surrounding K.

Holonomy

5. Holonomy on Cone and Sphere.

(i) Reconsider the cone of the semivertical angle α shown in [14.3], page 145. There we showed that if the tip of the cone is blunted, its spherical image allows us to assign a definite curvature $\mathcal{K}$ to the tip of the otherwise intrinsically flat cone, given by (14.2):

$$\mathcal{K}(\text{spike}) = \beta = \text{split angle of flattened spike} = 2\pi(1 - \sin\alpha).$$

By carrying out parallel transport within the flattened cone in [14.3], show that holonomy assigns the *same* total curvature to the spike.

(ii) On $\mathbb{S}^2$, let us use extrinsic, potato-peeler parallel transport to find the holonomy of a circle of latitude of fixed angle ϕ. Imagine a cone of semivertical angle α resting on the sphere so that it touches the sphere along this circle. The strip of peel of the sphere along this circle is therefore (ultimately) the same as the strip of the touching cone. Show that $\alpha = \frac{\pi}{2} - \phi$, and use (i) to confirm that the total curvature within the polar cap is indeed the holonomy of the circle of latitude that bounds it.

(iii) On the equator of $\mathbb{S}^2$, imagine a vector pointing due east. Now parallel transport it due east along the geodesic equator, so that it returns home seemingly unchanged, i.e., with *vanishing* holonomy. But this loop encloses half the sphere, with total curvature 2π! Use (ii) to reconcile these facts, by gradually increasing ϕ from 0 to $(\pi/2)$.

6. General Local Gauss–Bonnet Theorem.

(i) By approximating a smooth, closed loop L with a geodesic m-gon, then letting $m \to \infty$, deduce that the holonomy formula (24.6) becomes,

$$\mathcal{R}(L) = 2\pi - \oint_L \kappa_g \, ds,$$

where κ_g is the geodesic curvature along L, and s is distance along L.

(ii) If P is a closed "polygon" with external angles φ_i, but with edges that are *not* geodesic (i.e., $\kappa_g \neq 0$), then show that the generalization of (24.6) is,

$$\boxed{\mathcal{R}(P) = 2\pi - \left[\oint_P \kappa_g \, ds + \sum_i \varphi_i\right].}$$

(iii) If R denotes the interior of P, deduce the ***General Local Gauss–Bonnet Theorem***:

$$\boxed{\iint_R \mathcal{K} \, d\mathcal{A} = 2\pi - \left[\oint_P \kappa_g \, ds + \sum_i \varphi_i\right].}$$

Curvature as a Force between Neighbouring Geodesics

7. Minding's Theorem. Using the same geodesic polar coordinates as in Section 28.3, the metric takes the form

$$ds^2 = dr^2 + g^2(r) \, d\theta^2,$$

so that

$$g'' = -\mathcal{K} g.$$

Using the fact that $g(r) \asymp r$ (as r vanishes), solve this differential equation in each of the three cases of *constant* $\mathcal{K}$:

(i) If $\mathcal{K} = 0$ everywhere, deduce that the surface is locally isometric to the Euclidean plane.

(ii) If $\mathcal{K} = (1/R^2)$ is constant throughout the surface, deduce that the space is locally isometric to the sphere of radius R.

(iii) If $\mathcal{K} = -(1/R^2)$ is constant throughout the surface, deduce that the space is locally isometric to the pseudosphere of radius R.

8. Jacobi Equation on a General Surface of Revolution. As in Exercise 22, page 89, imagine a particle travelling along a curve in the (x, y)-plane at unit speed, and let its position at time t be $[x(t), y(t)]$. Now imagine rotating this plane through angle θ about the x-axis. As θ varies from 0 to 2π, the curve sweeps out a surface of revolution.

(i) Explain why $\dot{x}^2 + \dot{y}^2 = 1$, where the dot represents the time derivative.

(ii) Show geometrically that the metric of the surface is $ds^2 = dt^2 + y^2 \, d\theta^2$.

(iii) By considering the relative acceleration of two neighbouring meridian geodesics, deduce from the Jacobi Equation (28.2) that $\mathcal{K} = -\ddot{y}/y$. (Recall that this is the same formula (13.4) that we previously obtained in the course of proving the *Theorema Egregium* for surfaces of revolution.)

9. Gauss's Lemma via Computation. As in [28.6], consider two neighbouring, unit-speed geodesics launched from o with angular separation $\delta\theta$, and let $\mathbf{v}$ denote their unit velocity vectors. Let $\boldsymbol{\xi}$ be the connecting vector between the geodesics, connecting two points at the *same distance* $\sigma = t$ from o, so that both lie on the geodesic circle $K(\sigma)$. Then, as $\delta\theta \to 0$, $\boldsymbol{\xi}$ is tangent to $K(\sigma)$. To establish Gauss's Lemma, (28.4), we must therefore show that $\mathbf{v} \cdot \boldsymbol{\xi} = 0$. (NOTE: In the following, $\nabla_{\mathbf{v}}$ is the ordinary $\mathbb{R}^3$ derivative, not the intrinsic surface derivative, $D_{\mathbf{v}} = \boldsymbol{\nabla}_{\mathbf{v}}$.)

(i) Explain why $\lim_{\sigma \to 0} \mathbf{v} \cdot \boldsymbol{\xi} = 0$.

(ii) Deduce that to prove $\mathbf{v} \cdot \boldsymbol{\xi} = 0$, it suffices to show that $\nabla_{\mathbf{v}} [\mathbf{v} \cdot \boldsymbol{\xi}] = 0$.

(iii) Explain why $\mathbf{v} \cdot \nabla_{\boldsymbol{\xi}} \mathbf{v} = 0$.

(iv) Explain why $[\mathbf{v}, \boldsymbol{\xi}] = 0$.

(v) Using the fact that $\mathbf{v}$ is the velocity of a *geodesic*, deduce that $\boldsymbol{\xi} \cdot \nabla_{\mathbf{v}} \mathbf{v} = 0$.

(vi) Combine the three previous results to prove that

$$\nabla_{\mathbf{v}} [\mathbf{v} \cdot \boldsymbol{\xi}] = 0,$$

thereby completing the computational proof of Gauss's Lemma.

Riemann's Curvature

10. Two Symmetries of the Riemann Tensor.

(i) Let us confirm the First (Algebraic) Bianchi Identity, (29.15), which is also called the ***Bianchi Symmetry***:

$$\mathcal{R}(\mathbf{u}, \mathbf{v})\, \mathbf{w} + \mathcal{R}(\mathbf{v}, \mathbf{w})\, \mathbf{u} + \mathcal{R}(\mathbf{w}, \mathbf{u})\, \mathbf{v} = \mathbf{0} \quad \Longleftrightarrow \quad R_{ijkm} + R_{jkim} + R_{kijm} = 0.$$

To prove the general result it actually suffices (by virtue of linearity) to prove it in the case that all three vector fields are ***coordinate vector fields***, in which case their commutators all vanish. Prove that in this case,

$$\mathcal{R}(\mathbf{u}, \mathbf{v})\, \mathbf{w} + \mathcal{R}(\mathbf{v}, \mathbf{w})\, \mathbf{u} + \mathcal{R}(\mathbf{w}, \mathbf{u})\, \mathbf{v} = \boldsymbol{\nabla}_{\mathbf{u}} [\mathbf{v}, \mathbf{w}] + \boldsymbol{\nabla}_{\mathbf{v}} [\mathbf{w}, \mathbf{u}] + \boldsymbol{\nabla}_{\mathbf{w}} [\mathbf{u}, \mathbf{v}] = 0.$$

A more elegant proof using curvature 2-forms can be found in Section 38.12.4.

(ii) With the same commuting vector fields as in (i), we can give a *geometrical* explanation of the Bianchi Symmetry, as follows. (NOTE: I certainly do not believe I am the first to discover this proof, but I have not been able to find it in print, and therefore do not know to whom credit should be assigned.) Starting at some point, draw $\epsilon\mathbf{u}$, $\epsilon\mathbf{v}$, $\epsilon\mathbf{w}$, where ϵ is small and ultimately vanishing. Begin to construct a polyhedral "box" by completing each pair of these edges, creating three parallelogram faces, which are *closed* by virtue of their vanishing commutators. Now construct (and sketch) two new vector edges $\mathbf{A}$ and $\mathbf{B}$ of this box, obtained by parallel transporting $\epsilon\mathbf{w}$ first along $\epsilon\mathbf{v}$ and then along $\epsilon\mathbf{u}$ (yielding $\mathbf{A}$), and, in the reverse order, first along $\epsilon\mathbf{u}$ and then along $\epsilon\mathbf{v}$ (yielding $\mathbf{B}$). Next connect the end of $\mathbf{B}$ to the end of $\mathbf{A}$, thereby creating a new vector edge of a new triangular face of the box. Deduce that this new edge is given by,

$$\mathbf{A} - \mathbf{B} \asymp \epsilon^3 \mathcal{R}(\mathbf{u}, \mathbf{v})\, \mathbf{w}.$$

Repeat this construction using the other two faces, thereby constructing two more triangular faces of the box. Your figure should now reveal that $\epsilon^3\,\mathcal{R}(\mathbf{u},\mathbf{v})\,\mathbf{w}$, $\epsilon^3\,\mathcal{R}(\mathbf{v},\mathbf{w})\,\mathbf{u}$, and $\epsilon^3\,\mathcal{R}(\mathbf{w},\mathbf{u})\,\mathbf{v}$ form the vector edges of a fourth *triangular face* that closes the box, thereby proving the Bianchi Symmetry.

(iii) Let us confirm that *the Riemann tensor is symmetric under interchange of the first and second pairs of vectors*, (29.16):

$$[\mathcal{R}(\mathbf{u},\mathbf{v})\,\mathbf{x}] \bullet \mathbf{y} = [\mathcal{R}(\mathbf{x},\mathbf{y})\,\mathbf{u}] \bullet \mathbf{v} \quad \Longleftrightarrow \quad R_{ijkl} = R_{klij}.$$

As of this writing, I have failed to make geometrical sense of this result, so we must now resort to proving it by means of Satanic (see Prologue) Algebra. If we define

$$\mathbf{B}(\mathbf{u},\mathbf{v},\mathbf{x},\mathbf{y}) \equiv [\mathcal{R}(\mathbf{u},\mathbf{v})\,\mathbf{x}] \bullet \mathbf{y} + [\mathcal{R}(\mathbf{v},\mathbf{x})\,\mathbf{u}] \bullet \mathbf{y} + [\mathcal{R}(\mathbf{x},\mathbf{u})\,\mathbf{v}] \bullet \mathbf{y},$$

then

$$\mathbf{B}(\mathbf{u},\mathbf{v},\mathbf{x},\mathbf{y}) = 0,$$

by virtue of the Bianchi Symmetry in (i). Recalling that the Riemann tensor is antisymmetric in both the first two slots and in the second two slots, prove the result by cancelling like terms from the following (manifestly trivial!) identity:

$$\mathbf{B}(\mathbf{u},\mathbf{v},\mathbf{x},\mathbf{y}) + \mathbf{B}(\mathbf{v},\mathbf{x},\mathbf{y},\mathbf{u}) = 0 = \mathbf{B}(\mathbf{x},\mathbf{y},\mathbf{u},\mathbf{v}) + \mathbf{B}(\mathbf{y},\mathbf{u},\mathbf{v},\mathbf{x}).$$

(Can you smell the sulphur?)

11. Counting the Components of the Riemann Tensor. (The following proof, which we found in Lightman et al. (1975), is shorter and simpler than other standard proofs we have seen.)

(i) Given that the components R_{ijkl} of the Riemann tensor in an n-manifold are antisymmetric in ij and kl, deduce that there are $P = \frac{1}{2}n(n-1)$ nontrivial ways of choosing *pairs* ij, and likewise P ways of choosing pairs kl.

(ii) Given the symmetry (29.16) under interchange of the first and second pair of indices, $R_{ijkl} = R_{klij}$ (see previous exercise), deduce that if we also take these pair symmetries into account, there are $\frac{1}{2}P(P+1)$ independent ways of choosing $ijkl$.

(iii) Defining $B_{ijkl} \equiv R_{ijkl} + R_{jkil} + R_{kijl}$, as in the previous exercise, the *Bianchi Symmetry* (29.15) states that $B_{ijkl} = 0$ (see previous exercise). Verify that the pair symmetries now ensure that B_{ijkl} is *totally* antisymmetric on all four indices, and that the constraint $B_{ijkl} = 0$ is therefore trivially satisfied *unless all four indices are distinct*.

(iv) If $n < 4$, the Bianchi Symmetry therefore does not impose any new constraints. Use (ii) to deduce that if $n = 2$ the Riemann tensor has one component (the Gaussian curvature), and if $n = 3$ it has six components.

(v) If $n \geqslant 4$, deduce that the number of additional constraints resulting from the Bianchi symmetry is equal to the number of ways of choosing four objects from n objects.

(vi) Deduce that the number of independent components of the Riemann tensor is

$$\frac{1}{2}P(P+1) - \frac{n!}{(n-4)!\,4!},$$

in which the second term correctly disappears if $n < 4$.

(vii) Verify that this does indeed yield formula (29.1): *The number of independent components of the Riemann tensor is*

$$\frac{1}{12}n^2(n^2-1).$$

12. The Exponential Operator and Curvature. Recall the exponential series:

$$\exp(x) = e^x = 1 + x + \frac{1}{2!}x^2 + \frac{1}{3!}x^3 + \frac{1}{4!}x^4 + \cdots.$$

Now compare this to the Taylor series of a general function $f(x)$:

$$\begin{aligned} f(a+\delta x) &= f + \delta x \frac{df}{dx} + \frac{1}{2!}(\delta x)^2 \frac{d^2 f}{dx^2} + \frac{1}{3!}(\delta x)^3 \frac{d^3 f}{dx^3} + \cdots \Big|_a \\ &= \left[1 + \delta x \frac{d}{dx} + \frac{1}{2!}(\delta x)^2 \left(\frac{d}{dx}\right)^2 + \frac{1}{3!}(\delta x)^3 \left(\frac{d}{dx}\right)^3 + \cdots\right] f \Big|_a \\ &\equiv \exp\left[\delta x \frac{d}{dx}\right] f \Big|_a, \end{aligned}$$

in which the last line serves as the definition of the ***exponential operator***. The extension to vector fields follows naturally. If **w** is vector field defined in the vicinity of a, then, as we move distance δu in the direction of the unit vector **u**,

$$\mathbf{w}(a + \mathbf{u}\,\delta u) = \exp\left[\delta u \nabla_{\mathbf{u}}\right] \mathbf{w}\big|_a .$$

(i) Reconsider the derivation of the Riemann tensor in [29.8], but suppose for simplicity's sake that $[\mathbf{u}, \mathbf{v}] = 0$, so that the parallelogram closes up, in which case the curvature operator simplifies to

$$\mathcal{R}(\mathbf{u}, \mathbf{v}) = [\nabla_{\mathbf{u}}, \nabla_{\mathbf{v}}].$$

Explain why the vector holonomy of the parallelogram is given by

$$-\delta \mathbf{w}_{||} = [\exp(\delta u \nabla_{\mathbf{u}}), \exp(\delta v \nabla_{\mathbf{v}})]\, \mathbf{w}.$$

(ii) Deduce that

$$-\delta \mathbf{w}_{||} = (\delta u\, \delta v)\, \mathcal{R}(\mathbf{u}, \mathbf{v})\, \mathbf{w} + \text{(third-order error)},$$

where the "third-order error" is made up of terms involving $(\delta u)^p (\delta v)^q$, where $(p+q) \geqslant 3$.

Einstein's Curved Spacetime

13. Eclipses and the Tides. The tidal influence of the Moon on the oceans is more than twice as great as the Sun's, despite the fact the Sun's gravitational pull on the Earth is about 200 times more powerful than the Moon's. Let us try to understand this seemingly paradoxical reversal.

(i) It is a remarkable fact that total solar eclipses occur. This is only possible because of an empirical coincidence: the Moon and Sun look almost exactly the same (angular) size as seen from Earth. If the radii of the Moon and Sun are r_m and r_s, and their distances from Earth are R_m and R_s, respectively, deduce that $(R_m/R_s) \approx (r_m/r_s)$.

(ii) We have seen that the tidal force exerted on the oceans by a body of mass M at distance R from Earth is proportional to $\frac{M}{R^3}$. By appealing to (i), deduce that *the ratio of the lunar to solar tidal forces is the ratio of the* densities *of the Moon to the Sun.*

(iii) The average density of the Moon is approximately 3300 kilograms per cubic meter, while that of the Sun is approximately 1400 kilograms per cubic meter. Use (ii) to *explain* the strange opening fact!

14. Conservation of the Einstein Tensor. (NOTE: Unless you are already familiar with tensor ***contractions***, and with raising/lowering tensor indices, it would be best to defer this exercise until you have studied the relevant sections of Act V, namely, 33.7 and 33.8.) Recall from (30.21) that the *Einstein tensor* **G** is

$$\mathbf{G} \equiv \mathbf{Ricci} - \frac{1}{2} R\, \mathbf{g}.$$

This exercise demonstrates that the Einstein tensor is "conserved," in the same way that energy–momentum is ***conserved***:

$$\boxed{\nabla^a G_{ab} = 0.}$$

This crucial, *purely mathematical* fact was unknown to Einstein when he wrote down his field equation, (30.19), on the 25th of November, 1915 (Einstein's original form being mathematically equivalent of the modern form, $\mathbf{G} = 8\pi \mathbf{T}$). *Subsequently*, it was recognized that the link that Einstein had discovered between geometry and matter actually *implies* that energy–momentum *must* be conserved: $\nabla^a T_{ab} = 0$!

(i) Check that the *Second (Differential) Bianchi Identity*, (29.17), can be written

$$\nabla_a R_{bcd}{}^e + \nabla_b R_{cad}{}^e + \nabla_c R_{abd}{}^e = 0.$$

(ii) Now perform two *contractions* of this equation, by (1) raising a and renaming d to a, so that the two a's are summed over; (2) renaming e to c, so that the two c's are summed over. Confirm that this yields

$$\nabla^a R_{bca}{}^c + \nabla_b R_c{}^a{}_a{}^c + \nabla_c R^a{}_{ba}{}^c = 0.$$

(iii) Show that the previous equation can be rewritten as

$$-\nabla^a R_{ba} + \nabla_b R - \nabla_c R_b{}^c = 0.$$

(iv) Show that this may in turn be rewritten as

$$\nabla^a R_{ba} - \frac{1}{2} \nabla_b R = 0.$$

(v) Deduce that $\nabla^a G_{ab} = 0$, as was to be shown.

15. Weyl Curvature. The 20 components of the Riemann tensor in spacetime can be split between the 10 components of the Ricci tensor (generated by matter and energy, via the Einstein field equation) and the 10 components that represent pure gravitational degrees of freedom, present in vacuum. These gravitational degrees of freedom are completely encoded by the ***Weyl curvature tensor***:

$$\boxed{C_{ij}{}^{kl} \equiv R_{ij}{}^{kl} + 2R_{[i}{}^{[k}\, g_{j]}{}^{l]} - \tfrac{1}{3} R\, g_{[i}{}^{k}\, g_{j]}{}^{l}.}$$

Here the square brackets enclosing pairs of indices denote the operation of ***antisymmetrization***, as defined later, in (33.9), on page 369.

(i) Explicitly perform these antisymmetrizations, and thereby deduce the following unwieldy (and hard-to-remember) formula:

$$C_{ijkl} = R_{ijkl} + \frac{1}{2}(R_{ik}\, g_{jl} + R_{jl}\, g_{ik} - R_{il}\, g_{jk} - R_{jk}\, g_{il}) + \frac{1}{6}R\,(g_{jk}\, g_{il} - g_{ik}\, g_{jl}).$$

(ii) Deduce from (i) that the Weyl tensor has the same symmetries as the full Riemann tensor.

(iii) Verify that the Einstein Field Equation immediately implies that in vacuum, the Weyl tensor *is* the Riemann tensor.

(iv) Show that all the *traces* of the Weyl tensor vanish: in particular,

$$\boxed{C_{ij}{}^{ki} = 0.}$$

Thus the equivalent of the Ricci tensor—the matter–energy part of the Riemann tensor—vanishes for the Weyl tensor.

(v) (NOTE: The remainder of this exercise is more advanced, and really requires more than we have explained thus far. The solution can be found in the references cited at the conclusion of this exercise.) Suppose we subject the spacetime metric to a ***conformal transformation***:

$$\mathbf{g} \longrightarrow \Omega^2\, \mathbf{g},$$

where Ω is a function that *varies* from point to point throughout spacetime. But, at each point, all local distances are stretched by the *same* factor Ω, evaluated at that point. Such a transformation *preserves angles*, and also preserves the shapes of small objects. The Riemann tensor transforms in an extremely complex manner under such a conformal transformation. *However*, prove that the Weyl tensor—the purely gravitational/vacuum part—enjoys the remarkable and vitally important property that it merely *scales*:

$$\boxed{C_{ijkl} \longrightarrow \Omega^2\, C_{ijkl}.}$$

(vi) If we raise one index, show that we may rephrase this as saying that $C_{ijk}{}^{l}$ is ***conformally invariant***:

$$\boxed{C_{ijk}{}^{l} \longrightarrow C_{ijk}{}^{l}.}$$

NOTE: The most natural and elegant expression of the Weyl tensor is in terms of Penrose's 2-spinor formalism, where it takes the form of the ***Weyl conformal spinor*** Ψ_{ABCD}, which is *totally symmetric* and *conformally invariant*:

$$\boxed{\Psi_{ABCD} = \Psi_{(ABCD)} \longrightarrow \Psi_{ABCD}.}$$

Ψ_{ABCD} is also called, simply, the ***gravitational spinor***. See Penrose and Rindler (1984, §4.6, §6.8) or Wald (1984, §13.2). For an intuitive discussion of Weyl curvature, and Penrose's conjecture on its vital importance in characterizing the extraordinarily special nature of the Big Bang, see Penrose (2005, §19.7, §28.8).

ACT V

Forms

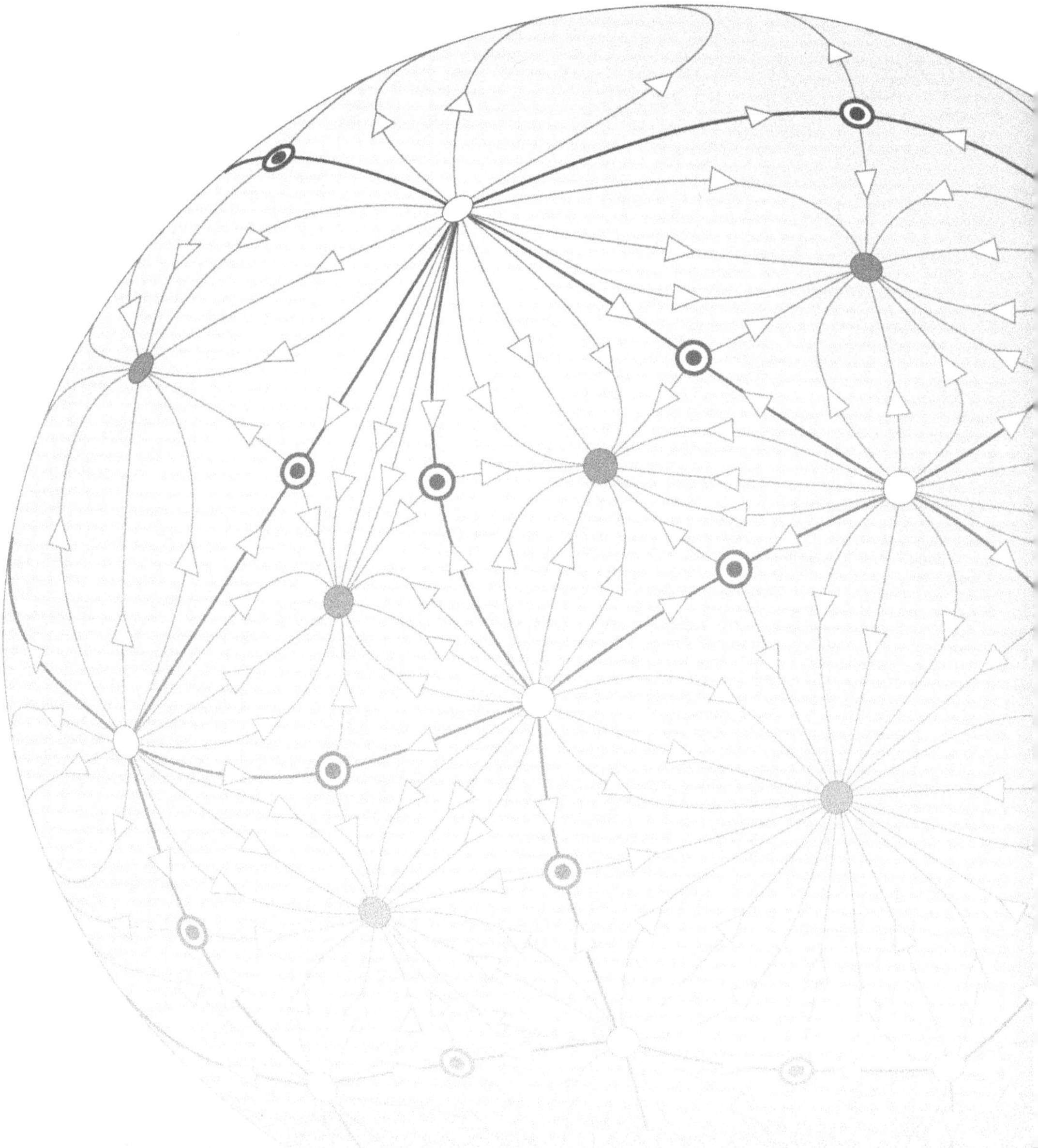

Chapter 32

1-Forms

32.1 Introduction

Act V represents cathartic release from four successive Acts of relentlessly strict geometrical rule.

As the Prologue foretold, our purpose now shall be to build the "Devil's machine," and to make it available to *undergraduate* students. We are speaking of a seductively powerful and elegant method of computation whose full name is ***The Exterior Calculus of Differential Forms***, which we have abbreviated here to ***Forms***.

Forms were discovered by the extraordinarily profound, original, and wide-ranging Élie Cartan (pictured in [32.1]) around 1900, more than a century ago, but it took even Cartan another 40 years to harness the *full* power of his discovery.

[32.1] *Élie Cartan (1869–1951).*

Our aim in Act V is to confront—succinctly, and in the plainest of Anglo-Saxon—a century-long scandal, namely, that the vast majority of undergraduates (in both mathematics and physics) will obtain their degrees without ever having glimpsed Cartan's Forms.

Although our principal purpose is to develop a novel method of *computation*, we desire to have our cake and eat it, too. That is, we shall endeavour to shed as much *geometrical* light as possible upon Cartan's Forms—much more so than in standard[1] treatments, which frequently careen into a perversely formal, shockingly abstract parallel Universe, in which Forms have been stripped of all vestiges of *meaning!*

Instead, our approach will be brutally concrete and vivid—occasionally downright lurid. We shall hold up Cartan's (Platonic!) Forms directly before your eyes, light gleaming upon their facets; we shall place them in your hands, so that you may feel their intricate shape and satisfying heft; finally, you will bear witness to their fearsome power.

[1]Misner, Thorne, and Wheeler (1973) and Schutz (1980) are two superb exceptions, and we *strongly* recommend both of them as companion (or follow-up) reading for Act V. Other noteworthy efforts can found in the *Further Reading* section at the end of this book.

But heed the warning that heralded the Prologue: do not be blinded by this power! Put this diabolical machine to work in the *service* of geometry, not in place of it!

—Here endeth the lesson. Amen!—

While the applications of Cartan's Forms are legion, *our* primary application (in the final chapter of Act V) will naturally be to Differential Geometry, enabling us to *reprove by symbolic means results that were proved geometrically in the first four Acts.*

First, however, *we shall fully develop Cartan's ideas in their own right, providing a self-contained introduction to Forms that is completely independent of the first four Acts.* We have done this because Forms find fruitful applications across diverse areas of mathematics, physics, and other disciplines. In short, *our aim is to make Forms accessible to the widest possible range of readers, even if their primary interest is* not *Differential Geometry.*

32.2 Definition of a 1-Form

The starting point, and the fundamental building block of Cartan's diabolically ingenious machine, is the concept of a ***1-form***. Let us waste no time in stating its definition:

A **1-form** *is a linear, real-valued function of a single vector input.*

NOTES: The "1-" refers to the *single* vector input; later we will meet ***2-forms*** that take *two* vectors as their input, ***3-forms*** that take *three* vectors as their input, and so on. A 1-form is therefore an especially simple kind of tensor. Older works instead call this concept a ***covariant vector***, or a ***covector***.[2] We will denote 1-forms by lowercase bold Greek letters, while continuing to denote vectors by lowercase bold Roman letters.

More explicitly, if k_1 and k_2 are arbitrary constants, and $\mathbf{v}_1$ and $\mathbf{v}_2$ are arbitrary vectors, then

$$\boldsymbol{\omega} \text{ is a 1-form} \quad \Longleftrightarrow \quad \boldsymbol{\omega}(k_1\mathbf{v}_1 + k_2\mathbf{v}_2) = k_1\,\boldsymbol{\omega}(\mathbf{v}_1) + k_2\,\boldsymbol{\omega}(\mathbf{v}_2). \tag{32.1}$$

When checking that a particular $\boldsymbol{\omega}$ is a 1-form, it can be conceptually helpful to break down this single requirement into two simpler ones:

$$\boldsymbol{\omega}(\mathbf{v}_1 + \mathbf{v}_2) = \boldsymbol{\omega}(\mathbf{v}_1) + \boldsymbol{\omega}(\mathbf{v}_2), \tag{32.2}$$

and

$$\boldsymbol{\omega}(k\mathbf{v}) = k\,\boldsymbol{\omega}(\mathbf{v}). \tag{32.3}$$

Check for yourself that (32.2) and (32.3) together imply (32.1), and vice versa: the two definitions of linearity are equivalent.

[2]This terminology is explained in Exercise 3.

A 1-form is *defined* by its action on vectors: two 1-forms are equal if and only if they have the same effect on all vectors. Given two distinct 1-forms, $\boldsymbol{\omega}$ and $\boldsymbol{\varphi}$, there is therefore a natural way to define their sum, $(\boldsymbol{\omega}+\boldsymbol{\varphi})$, by its action on a general vector $\mathbf{v}$:

$$(\boldsymbol{\omega}+\boldsymbol{\varphi})(\mathbf{v}) \equiv \boldsymbol{\omega}(\mathbf{v})+\boldsymbol{\varphi}(\mathbf{v}),$$

and it is easy to check [exercise] using (32.2) and (32.3) that this sum is itself a 1-form. Likewise, we may multiply a 1-form $\boldsymbol{\omega}$ by a constant k to obtain a new 1-form, $[k\boldsymbol{\omega}]$, defined by

$$[k\boldsymbol{\omega}](\mathbf{v}) \equiv k[\boldsymbol{\omega}(\mathbf{v})].$$

Thus the set of 1-forms is closed under addition and multiplication by constants, and therefore constitutes what is called a ***vector space***. This vector space of 1-forms is said to be ***dual*** to the space of vectors upon which it acts. The reason for this terminology is that there is a symmetrical relationship between these two spaces: we can also think of the space of vectors as "dual" to the space of 1-forms.

To see this symmetry, let us *think of a vector* $\mathbf{v}$ *as a function that acts on 1-forms* $\boldsymbol{\omega}$, this action being defined by

$$\boxed{\mathbf{v}(\boldsymbol{\omega}) \equiv \boldsymbol{\omega}(\mathbf{v}).}$$

This symmetrical action of a vector and a 1-form upon each other is often called the ***contraction*** of one with the other, and is sometimes denoted[3] $\langle\boldsymbol{\omega},\mathbf{v}\rangle$, to emphasize the equal footing of the two kinds of object.

It follows from this that a vector $\mathbf{v}$ is a *linear* function of 1-forms:

$$\begin{aligned}\mathbf{v}(\boldsymbol{\omega}+\boldsymbol{\varphi}) &= (\boldsymbol{\omega}+\boldsymbol{\varphi})(\mathbf{v}) \\ &= \boldsymbol{\omega}(\mathbf{v})+\boldsymbol{\varphi}(\mathbf{v}) \\ &= \mathbf{v}(\boldsymbol{\omega})+\mathbf{v}(\boldsymbol{\varphi}),\end{aligned}$$

and,

$$\mathbf{v}(k\boldsymbol{\omega}) = k\boldsymbol{\omega}(\mathbf{v}) = k\mathbf{v}(\boldsymbol{\omega}).$$

Just as T_p denotes the space of vectors at the point p, so T_p^* denotes the dual vector space of 1-forms at p. And just as a vector field assigns a vector $\mathbf{v}_p$ to each point p, so a ***field of 1-forms*** assigns a 1-form $\boldsymbol{\omega}_p$ to each point p.

The next section justifies the introduction of 1-forms (albeit after the fact!) by revealing a secret: you have spent your adult life surrounded by 1-forms—you just didn't know it.

32.3 Examples of 1-Forms

32.3.1 Gravitational Work

Let F be the magnitude of the gravitational force exerted on a unit mass near the surface of the Earth. We shall make the approximation that F is constant if we only move the mass short distances. If we lift the mass through a vertical distance h, then the ***work*** done—i.e., the energy we must expend to make this happen—is $\omega = Fh$.

[3] The similarity to the Dirac bra–ket notation is not coincidental, as we shall see in a moment.

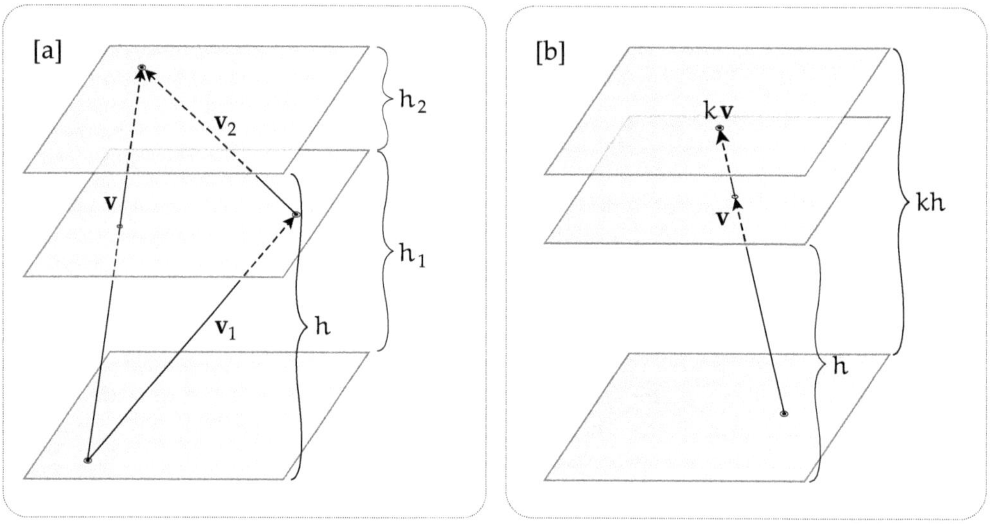

[32.2] **[a]** *When the mass is moved along* $\mathbf{v} = (\mathbf{v}_1 + \mathbf{v}_2)$*, the increase* h *in its height is simply the sum* $(h_1 + h_2)$ *of the separate increases in height resulting from moving along* $\mathbf{v}_1$ *and* $\mathbf{v}_2$*, separately.* **[b]** *Stretching the movement* $\mathbf{v}$ *by a factor of* k *also stretches the vertical movement by the same factor.*

Now suppose that we move the mass along the vector **v**, and define

$$\omega(\mathbf{v}) \equiv \text{work done moving the mass along } \mathbf{v}.$$

Then ω is a 1-form!

To verify this, we must check (32.2) and (32.3). As illustrated in [32.2a], when the mass is moved along $\mathbf{v} = (\mathbf{v}_1 + \mathbf{v}_2)$, the increase h in its height is simply the sum $(h_1 + h_2)$ of the separate increases in height resulting from moving along $\mathbf{v}_1$ and $\mathbf{v}_2$, separately. Likewise, as illustrated in [32.2b], stretching the movement **v** by a factor of k also stretches the vertical movement (and hence the work) by the same factor. Done!

32.3.2 Visualizing the Gravitational Work 1-Form

This example naturally suggests a means of *visualizing* a 1-form. Imagine a family of equally spaced surfaces of constant height. These are spheres centred at the centre of the Earth, but, in the local vicinity of the mass, these spheres will look like a stack of equally spaced horizontal planes. See the right-hand side of [32.3].

As we move the mass along **v**, the change in height, and hence the work, will be proportional to the number of planes pierced by **v**. By adjusting the spacing of the surfaces/planes to be $(1/F)$, we can therefore make

$$\omega(\mathbf{v}) = \text{number of surfaces of } \omega \text{ pierced by } \mathbf{v}.$$

Of course, in general, the tip of **v** will lie between two planes, but we can easily imagine filling the space between two planes with interpolating planes, so that if, for example, the tip of **v** lies half way between plane 17 and plane 18, we can say that it pierces 17.5 planes.

Thus the 1-form ω itself may be visualized as this stack of equally spaced surfaces of constant gravitational potential energy, and *the greater the force* F *represented by* ω*, the more densely packed the surfaces of* ω*.*

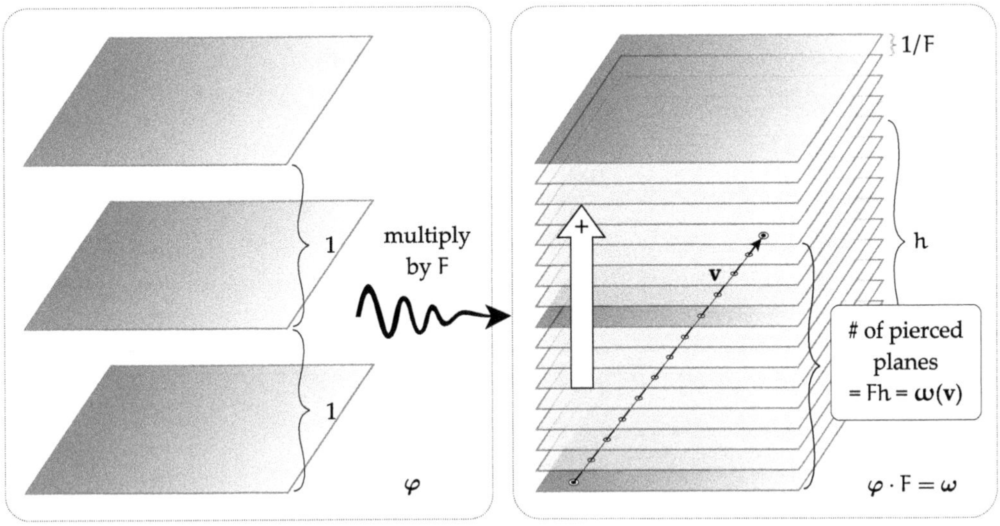

[32.3] Visualization of a 1-Form. *On the right is the stack representing the gravitational work 1-form ω, its planes being $(1/F)$ apart, and its direction upward. Thus $\omega(\mathbf{v})$ can be visualized as the (signed) number of planes pierced by* $\mathbf{v}$. *On the left is the* unit-spaced *1-form φ. When this is multiplied by F, the density of its planes is increased by F,* shrinking *their spacing to $(1/F)$, yielding $\varphi \cdot F = \omega$.*

To drive home this point, [32.3] shows a 1-form φ (on the left) with *unit* spacing being multiplied by F to yield our work 1-form, $\varphi \cdot F = \omega$, on the right.

Note that this visualization of the contraction of a vector with a 1-form has the virtue of very clearly putting the two kinds of objects on an equal footing. We call the set of equally spaced parallel planes a ***stack***. Note that in order to complete the interpretation in [32.3], we must attach a ***direction*** (variously known as a *sense* or an *orientation*) to the stack, upward in our case, as illustrated. If the direction of $\mathbf{v}$ *agrees* with the direction of ω, then the piercings are counted as *positive*, but if the vector goes *against* the direction of the stack, then the piercings are counted as *negative*.

This visualization can be applied more generally to an arbitrary 1-form. At the point p, the representative surface $\mathcal{S}$ of ω is defined by the fact that vectors emanating from p tangent to $\mathcal{S}$ satisfy $\omega(\mathbf{v}) = 0$, corresponding to zero surfaces being pierced. To put this into the language of Linear Algebra, the set of tangent vectors to $\mathcal{S}$ satisfy $\omega(\mathbf{v}) = 0$, so they comprise the ***kernel*** of ω.

In general, it is only possible to represent a 1-form field in this way as a stack of planes at a point—typically the small pieces of planes at *neighbouring* points will fail to mesh together to create a smooth family of space-filling surfaces, called a ***foliation***, although they *do* happen to in our gravitational example. For a discussion of this point, see Penrose (2005, §12.3), Bachman (2012, §5.7), or Dray (2015, §13.8).

In $\mathbb{R}^3$, the set of vectors satisfying $\omega(\mathbf{v}) = 0$ spans a 2-dimensional plane, but in n-dimensional space ω will instead be represented by $(n-1)$-dimensional spaces.

32.3.3 Topographic Maps and the Gradient 1-Form

If you are planning a long hike through rugged, mountainous terrain, you will be well-advised to plan your route using a *topographic map*, so as to avoid having to climb too hard, or to scamble down a slope that is dangerously steep. Recall from the example, [19.6], on page 203 that such a map shows contours of constant height $h(x, y)$: as the point $p = (x, y)$ in the map travels along such a contour, $h(p) = \text{constant}$.

Consider the two hills and the pass depicted in [32.4], and its topographic map below. If you hike along one of these contours, the ground is flat: you neither climb nor descend. If, on the other hand, you set off in the direction orthogonal to the contour, you will rise (or descend) as rapidly as possible: *the direction orthogonal to a contour is the direction of steepest ascent/descent*. The contours

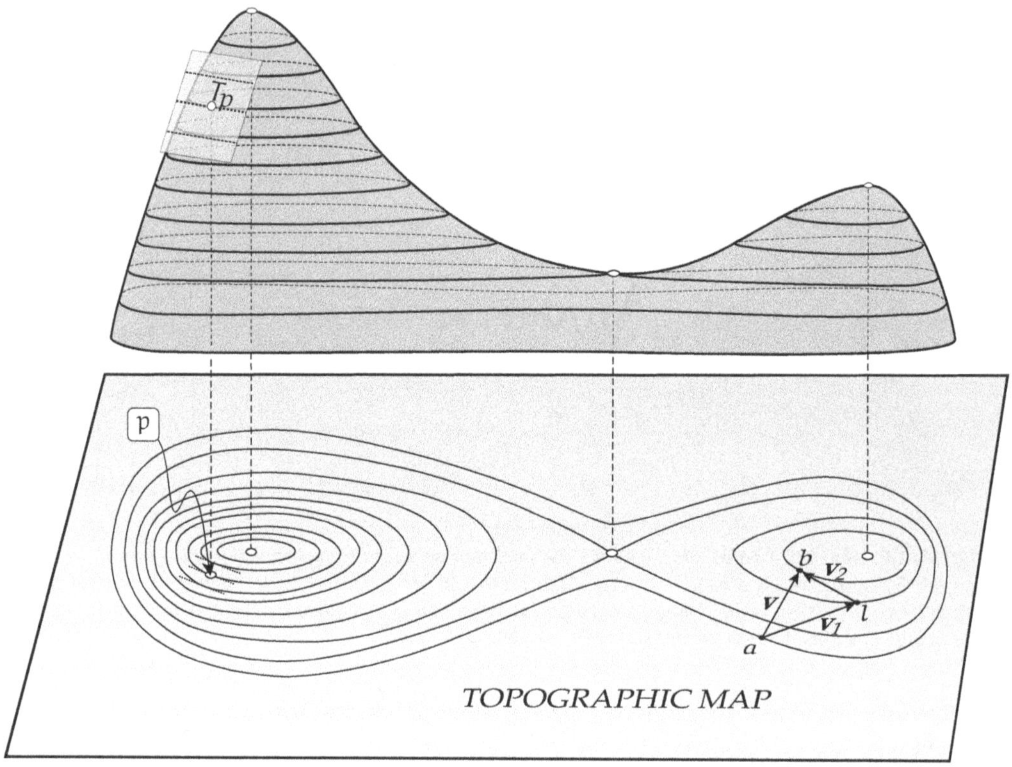

[32.4] Topographic Map and the Gradient 1-Form. *As we zoom in on a small region surrounding a point* p *in the topographic map of a surface, the contours look straighter and straighter, and more and more evenly spaced. Ultimately they become the representatives of the topographic map of the tangent plane* T_p *to the surface at the point directly above* p, *the latter map being a depiction of the* gradient 1-form $\boldsymbol{\zeta}$.

are drawn with equal increments δh in the height h, so if you walk in the direction orthogonal to a contour, *the more crowded together the contours in the map, the steeper the terrain.*

In greater detail, suppose we draw lots of contours with a fixed, small difference δh between one and the next. In the immediate vicinity of the point p, the density of the contours is roughly constant; let's call this roughly constant horizontal distance between neighbouring contours in the map δr. Then the slope of terrain in that small region is

$$\text{slope} \asymp \frac{\delta h}{\delta r}.$$

In other words, *the steepness of the terrain is inversely proportional to the gap between neighbouring contours in the map.*

This is very reminiscent of our visualization of the gravitational work 1-form: the stronger the gravitational force, the denser the packing of the surfaces representing the corresponding work 1-form. With this in mind, let us attempt to explicitly identify a steepness 1-form associated with our topographic map.

In the map plane, let $\mathbf{v} = \overrightarrow{ab}$ be a vector emanating from a, with its tip at b, as illustrated. Suppose we *attempt* to define a steepness 1-form as follows:

$$\begin{aligned} \eta(\mathbf{v}) &\equiv (\text{number of contours pierced by } \mathbf{v}) \cdot (\delta h) \\ &= h(b) - h(a) \\ &= \text{change in elevation from tail to tip of } \mathbf{v}. \end{aligned}$$

Does this satisfy the linearity conditions, (32.2) and (32.3)?

Let $\mathbf{v}_1 = \overrightarrow{al}$, $\mathbf{v}_2 = \overrightarrow{lb}$, and $\mathbf{v} = \mathbf{v}_1 + \mathbf{v}_2$, as illustrated; then, at first glance, it might appear that all goes well:

$$\eta(\mathbf{v}_1) + \eta(\mathbf{v}_2) = [h(l) - h(a)] + [h(b) - h(l)] = h(b) - h(a) = \eta(\mathbf{v}).$$

However, unlike a true 1-form, the value of $\eta(\mathbf{v})$ depends on *where* the vector $\mathbf{v}$ is drawn: if we imagine $\mathbf{v}_2$ drawn at a instead of at l, then clearly (32.2) is *no longer* satisfied! Thus our putative 1-form η at p has *not* actually passed the first test. Furthermore, even if we restrict ourselves to vectors emanating from a, it fails the second test, too, for it is clear that if we double the length of the illustrated vector $\mathbf{v}$ then we end up on the far side of the hill, and the altitude actually *decreases,* instead of doubling, as it should. Thus η does *not* satisfy (32.3).

Nevertheless, this discussion points the way to a genuine 1-form. The smaller the region surrounding a point p that we examine, the more precisely uniform the pattern of contours: *they look like evenly spaced, parallel lines.* Thus if we only apply η to very short (ultimately vanishing) vectors emanating from p, then η *does* (ultimately) satisfy the linearity requirements!

Now let us try to extend this local action of η at p from tiny vectors emanating from p to large vectors, drawn anywhere in the map plane. All we need to do is to take this local pattern of parallel, equally spaced lines in the immediate vicinity of p (all parallel to the tangent line to the contour through p) and extend it out to cover the entire plane. If we look thoughtfully at [32.4], we see that what we have just constructed is in fact *the topographic map of the tangent plane* T_p *to the surface at the point directly above* p!

In order to obtain a genuine 1-form field ζ, *we need only apply our old definition of* η *to this topographic map of* T_p. See [32.5]. We repeat that this topographic map consists of *equally spaced parallel lines*, all running in the same direction as the original contour through p, and *the gap between one line and the next is* $\delta r \asymp (\delta h)/(\textit{slope of } T_p)$: the bigger the slope of T_p, the more densely packed the parallel lines in its topographic map. The direction orthogonal to these lines is the direction of steepest ascent up the surface at the point p.

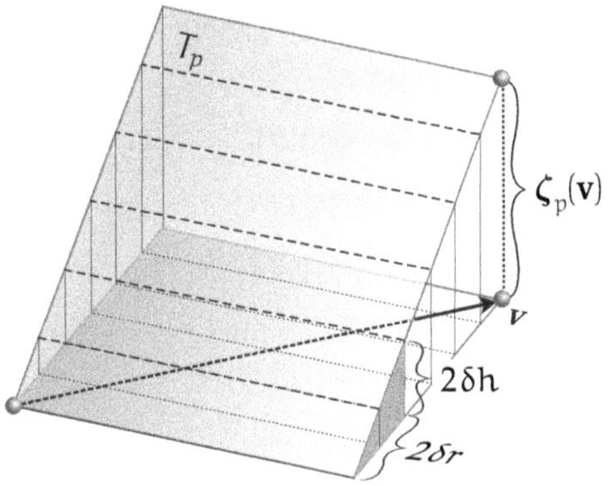

[32.5] The Gradient 1-Form ζ_p *of the height of a surface, when applied to a vector* $\mathbf{v}$ *in the plane, yields the change in the height of* T_p *as we travel along* $\mathbf{v}$*, where* T_p *is the tangent plane to the surface at the point directly above the point* p.

The ***value*** ζ_p of the 1-form field ζ at p is the 1-form that is defined by the tangent plane T_p *at* the point of the surface directly above p. As illustrated in [32.5],

$$\begin{aligned} \zeta_p(\mathbf{v}) &\equiv (\textit{number of lines pierced by } \mathbf{v}) \cdot (\delta h) \\ &= \textit{change in height of } T_p \textit{ from tail to tip of } \mathbf{v}. \end{aligned} \tag{32.4}$$

As p varies, the direction and spacing of the original contour lines around p both vary, reflecting the fact that the tangent plane T_p to the surface is varying, and so we have a *different* topographic map of T_p at each point, and hence a different value ζ_p of the 1-form field ζ.

If we apply ζ_p to very short (ultimately vanishing) vectors emanating from p, then its effect is ultimately equal to that of η; that is, $\zeta_p(\mathbf{v}) \asymp \eta_p(\mathbf{v})$. But as the vector gets bigger, $\eta(\mathbf{v})$ tells us the nonlinear variation in the height of the actual terrain/surface, while the 1-form $\zeta_p(\mathbf{v})$ tells us the perfectly linear variation in the height of the tangent plane to the surface at that point.

This 1-form field ζ is called the ***gradient*** of h. You are probably asking yourself if this is connected with the familiar, eponymous concept from Vector Calculus. Indeed it is! The explanation of the connection is the subject of Section 32.6. This idea of the gradient 1-form field ζ is extremely important, and we shall see how it can be generalized to functions h that depend on more than two variables, which therefore cannot be so simply visualized as the height $z = h(x, y)$ of a surface about the (x, y)-plane.

32.3.4 *Row Vectors*

Consider a specific 2-dimensional row vector, such as $\boldsymbol{\omega} = [-3, 2]$. We can define the action of $\boldsymbol{\omega}$ on 2-dimensional column vectors $\mathbf{v} = \begin{bmatrix} x \\ y \end{bmatrix}$ by means of standard matrix multiplication on the left:

$$\boldsymbol{\omega}(\mathbf{v}) = [-3, 2] \begin{bmatrix} x \\ y \end{bmatrix} = -3x + 2y.$$

We leave it to you to check that this does indeed satisfy (32.2) and (32.3). Thus, viewed in this way, *row vectors are 1-forms.*

We can relate this example to the gravitational work example. To do so, let us choose Cartesian coordinates (x, y, z) with the z-axis vertical, in the direction of the Earth's gravitational field. Then we can represent the gravitational work as the 1-form (i.e., row vector),

$$\boldsymbol{\omega} = [0, 0, F], \qquad \text{because then} \qquad \boldsymbol{\omega}(\mathbf{v}) = [0, 0, F] \begin{bmatrix} x \\ y \\ h \end{bmatrix} = Fh = \text{work}.$$

32.3.5 *Dirac's Bras*

NOTE: This example assumes that you are already familiar with quantum mechanics; feel free to skip it if you are not.

Although we have defined 1-forms to be *real*-valued, and shall continue to do so—with one exception—for the remainder of the book, it is often useful to broaden the definition to include *complex*-valued 1-forms. Indeed, this generalization arises naturally in quantum mechanics, and it is *essential* in that context.

If we think of ***Dirac kets*** $|\mathbf{v}\rangle$ (i.e., the quantum states) as the "vectors," then *a **Dirac bra*** $\langle\boldsymbol{\omega}|$ *is a 1-form*, provided we define the contraction of a 1-form and a vector to be the standard (complex) bra–ket inner product:

$$\boldsymbol{\omega}(\mathbf{v}) \equiv \langle\boldsymbol{\omega}|\mathbf{v}\rangle.$$

To confirm that a bra $\langle\boldsymbol{\omega}|$ is a 1-form, we simply check the definition, (32.1):

$$\begin{aligned} \boldsymbol{\omega}(k_1\mathbf{v}_1 + k_2\mathbf{v}_2) &= \langle\boldsymbol{\omega}|k_1\mathbf{v}_1 + k_2\mathbf{v}_2\rangle \\ &= k_1\,\langle\boldsymbol{\omega}|\mathbf{v}_1\rangle + k_2\,\langle\boldsymbol{\omega}|\mathbf{v}_2\rangle \\ &= k_1\,\boldsymbol{\omega}(\mathbf{v}_1) + k_2\,\boldsymbol{\omega}(\mathbf{v}_2). \end{aligned}$$

32.4 Basis 1-Forms

At a point p in an n-manifold, suppose that we have chosen a basis $\{\mathbf{e}_j\}$ for T_p, so that a general vector can be written, using the Einstein summation convention, as

$$\mathbf{v} = v^j\,\mathbf{e}_j.$$

We shall *not* assume that this basis is orthonormal. Given this basis, there exists a natural way to associate with it a ***dual basis*** $\{\boldsymbol{\omega}^i\}$ for the space T_p^* of 1-forms at p:

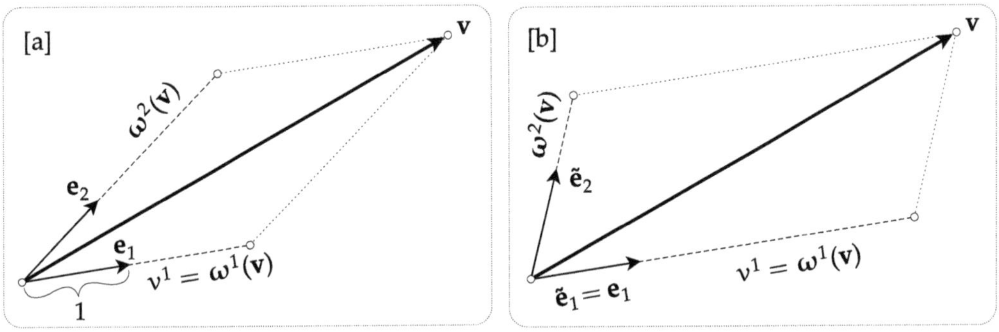

[32.6] [a] *The first basis 1-form ω^1 picks out the 1st component v^1 of $\mathbf{v}$; likewise, the second basis 1-form ω^2 picks out the second component.* **[b]** *With the same $\mathbf{e}_1$ as before, merely changing $\mathbf{e}_2$ causes both of the basis 1-forms to change.*

$$\omega^i \text{ picks out the i-th component of } \mathbf{v} \quad \Longleftrightarrow \quad \omega^i(\mathbf{v}) = v^i. \tag{32.5}$$

These $\{\omega^i\}$ are indeed 1-forms, which can be confirmed by checking (32.2) and (32.3): the i-th component of the sum of two vectors is the sum of their i-th components, and the i-th component of $k\mathbf{v}$ is just kv^i. Done!

Figure [32.6a] illustrates this definition in the case of a 2-manifold (i.e., a surface) for which the tangent plane $T_p = \mathbb{R}^2$. Here, as illustrated, we have chosen a basis $\{\mathbf{e}_1, \mathbf{e}_2\}$ of nonorthogonal but ***unit vectors***, and

$$\mathbf{v} = v^1\,\mathbf{e}_1 + v^2\,\mathbf{e}_2 = \omega^1(\mathbf{v})\,\mathbf{e}_1 + \omega^2(\mathbf{v})\,\mathbf{e}_2.$$

Thus, in this case, $\omega^1(\mathbf{v})$ is simply the length of the projection (parallel to $\mathbf{e}_2$) of $\mathbf{v}$ onto $\mathbf{e}_1$, and correspondingly for $\omega^2(\mathbf{v})$, as illustrated.

Let us nip the following common misconception in the bud. While the *set* of basis 1-forms $\{\omega^1, \omega^2\}$ is dual to the *set* of basis vectors $\{\mathbf{e}_1, \mathbf{e}_2\}$, it is simply wrong to think that ω^1 is dual to $\mathbf{e}_1$, and that ω^2 is dual to $\mathbf{e}_2$.

Figure [32.6b] reveals the error of this thinking. Here we have illustrated the same vector $\mathbf{v}$ broken down into components relative to a new basis $\{\tilde{\mathbf{e}}_1, \tilde{\mathbf{e}}_2\}$, in which $\tilde{\mathbf{e}}_1 = \mathbf{e}_1$, but $\tilde{\mathbf{e}}_2 \neq \mathbf{e}_2$. As you can see, even though we have only changed one basis vector, *both* basis 1-forms have changed! Likewise, in the general n-dimensional case, changing a single basis vector can change the *entire* dual basis of 1-forms.

We should note that our preferred definition (32.5) of the dual basis is not the conventional one, though it is the definition used by Schutz (1980), and it is mathematically equivalent to the standard one, as we now explain.

First recall the definition of the

$$\textit{\textbf{Kronecker delta}}: \quad \delta^i_j \equiv \left\{ \begin{array}{ll} 1 & \text{if } i = j \\ 0 & \text{if } i \neq j \end{array} \right\}.$$

It is easy to prove that our definition (32.5) is equivalent to the following, standard definition:

$$\text{Equivalent definitions of a } \textit{\textbf{basis}}\ \{\omega^i\}: \quad \omega^i(\mathbf{e}_j) = \delta^i_j \quad \Longleftrightarrow \quad \omega^i(\mathbf{v}) = v^i.$$

The implication from right to left is immediate, and the implication from left to right goes like this:

$$\omega^i(\mathbf{v}) = \omega^i(v^j\,\mathbf{e}_j) = v^j\,\omega^i(\mathbf{e}_j) = v^j\,\delta^i_j = v^i.$$

32.5 Components of a 1-Form

The ever-vigilant reader will have noticed that we have not actually proved that the set $\{\omega^i\}$ is a *basis* for the 1-forms. Rather than proving directly that they are linearly independent, we shall follow the more elegant and enlightening approach employed by Schutz (1980, §2.20).

Let the general 1-form φ act on the general vector $\mathbf{v}$:

$$\begin{aligned}\varphi(\mathbf{v}) &= \varphi(v^j\,\mathbf{e}_j)\\ &= v^j\,\varphi(\mathbf{e}_j)\\ &= \omega^j(\mathbf{v})\,\varphi(\mathbf{e}_j).\end{aligned}$$

We now define the

> ***Components*** ϕ_j of φ: $\quad \phi_j \equiv \varphi(\mathbf{e}_j)$.

Thus,

$$\varphi(\mathbf{v}) = \phi_j\,\omega^j(\mathbf{v}).$$

But since a 1-form is *defined* by its action on a general vector, we may "abstract away" the vector $\mathbf{v}$ on both sides of this equation, and thereby equate the 1-forms *themselves*, and, in so doing, decompose the arbitrary 1-form φ into its unique components in the 1-form basis $\{\omega^i\}$ dual to the vector basis $\{\mathbf{e}_j\}$:

$$\varphi = \phi_j\,\omega^j = \varphi(\mathbf{e}_j)\,\omega^j. \tag{32.6}$$

32.6 The Gradient as a 1-Form: $\mathbf{d}f$

32.6.1 Review of the Gradient as a Vector: ∇f

Recall from Vector Calculus that in $\mathbb{R}^2$ the ***gradient*** of a function f is defined to be the *vector*

$$\nabla f \equiv \begin{bmatrix}\partial_x f\\ \partial_y f\end{bmatrix}.$$

The *significance* of this vector is that,

> ∇f *points in the direction of most rapid increase of* f*, and its magnitude* $|\nabla f|$ *equals that maximum rate of increase of* f *with distance as we move in that direction.* (32.7)

This interpretation springs from a more primitive fact, which we now derive. Let $\{\mathbf{e}_1, \mathbf{e}_2\}$ be the standard orthonormal basis along the $(x^1, x^2) = (x, y)$ axes, and let δf be the small change in f resulting from moving along the short (ultimately vanishing) vector

$$\mathbf{v} = \delta x^1\, \mathbf{e}_1 + \delta x^2\, \mathbf{e}_2 = \begin{bmatrix} \delta x \\ \delta y \end{bmatrix}.$$

By definition, $\partial_x f$ is the rate of change of f as we move in the x-direction, so the change in f resulting from moving δx is ultimately equal to $(\partial_x f)\, \delta x$; and likewise for y. Therefore,

$$\delta f \asymp (\partial_x f)\, \delta x + (\partial_y f)\, \delta y = (\nabla f) \cdot \mathbf{v}. \tag{32.8}$$

In the classical notation of the eighteenth century, this was instead expressed as

$$df = (\partial_x f)\, dx + (\partial_y f)\, dy, \tag{32.9}$$

in which df, dx, and dy were understood to be *infinitesimals*—a concept we have eschewed in this work. *However*, as we shall see shortly, this ancient formula takes on a new life, with a precise and rigorous meaning, when viewed through the modern prism of 1-forms.

In order to derive (32.7), consider [32.7], which interprets (32.8) geometrically. If we keep the length $|\mathbf{v}| \equiv \delta s$ fixed, then as **v** rotates around the circle,

$$\delta f \asymp |\nabla f|\, (\delta s\ \cos\theta),$$

where θ is the angle that **v** makes with the gradient ∇f, so that $(\delta s\ \cos\theta)$ is simply the projection of **v** onto the direction of ∇f, as illustrated.

Let $\hat{\mathbf{v}} = \mathbf{v}/|\mathbf{v}|$ denote the ***unit vector*** in the direction of **v**. Then the above result can instead be expressed as

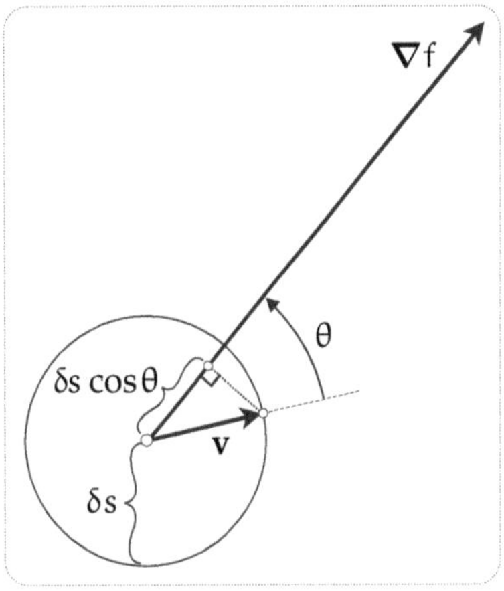

[32.7] The Gradient Vector ∇f. *The small change δf resulting from moving along the short vector* **v** *is ultimately equal to* $|\nabla f|\, (\delta s\ \cos\theta)$.

$$|\nabla_{\hat{\mathbf{v}}} f| = \frac{df}{ds} = |\nabla f|\ \cos\theta.$$

This immediately confirms the interpretation of ∇f given in (32.7). It also shows that if we move in the direction *orthogonal* to ∇f, then the rate of change of f vanishes. This is the direction of the tangent to the curve defined by f = *constant*.

In $\mathbb{R}^3$ the analysis is exactly as before, but now there is an entire *plane* of directions orthogonal to ∇f: this is the tangent plane to the *surface*, f = *constant*.

32.6.2 The Gradient as a 1-Form: **d**f

Although we are accustomed to thinking of the gradient as a vector, it is in fact more naturally a 1-form. Let us immediately define this 1-form, and then proceed to make sense of it.

The ***gradient 1-form*** **d**f of a function f is defined by its action on vectors:

$$\mathbf{d}f(\mathbf{v}) \equiv \nabla_{\mathbf{v}} f. \tag{32.10}$$

The bold **d** operator is called the ***exterior derivative***; it will play a central role in what follows.

Let us check that $\mathbf{d}f$ really is a 1-form, by verifying (32.2) and (32.3):

$$(\mathbf{d}f)(\mathbf{v}_1+\mathbf{v}_2)=\nabla_{\mathbf{v}_1+\mathbf{v}_2}f=\nabla_{\mathbf{v}_1}f+\nabla_{\mathbf{v}_2}f=\mathbf{d}f(\mathbf{v}_1)+\mathbf{d}f(\mathbf{v}_2)$$

and,

$$(\mathbf{d}f)(k\mathbf{v})=\nabla_{k\mathbf{v}}f=k\,\nabla_{\mathbf{v}}f=k\,\mathbf{d}f(\mathbf{v}).$$

Observe that since $\nabla_{\mathbf{v}}$ obeys the Leibniz Rule (aka Product Rule), so does the exterior derivative:

$$\mathbf{d}(fg)=f\,\mathbf{d}g+g\,\mathbf{d}f. \tag{32.11}$$

32.6.3 *The Cartesian 1-Form Basis:* $\{\mathbf{d}x^j\}$

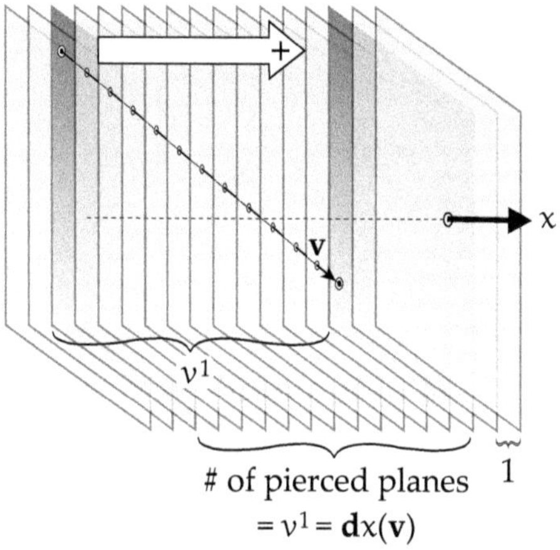

[32.8] The Cartesian Basis 1-form $\mathbf{d}x$. *When applied to a vector $\mathbf{v}$, the 1-form $\mathbf{d}x$ picks out the x-component: $\mathbf{d}x(\mathbf{v})=v^1$.*

What is the meaning of 1-forms $\mathbf{d}x=\mathbf{d}x^1$ and $\mathbf{d}y=\mathbf{d}x^2$ in $\mathbb{R}^2$? To answer this question, we must determine their effect on a general vector, $\mathbf{v}=\begin{bmatrix}v^1\\v^2\end{bmatrix}$. The definition (32.10) tells us that

$$\begin{aligned}(\mathbf{d}x)\begin{bmatrix}v^1\\v^2\end{bmatrix} &= (\mathbf{d}x)\mathbf{v}\\ &= \nabla_{\mathbf{v}}x\\ &= \nabla_{v^1\mathbf{e}_1+v^2\mathbf{e}_2}x\\ &= (v^1\nabla_{\mathbf{e}_1}+v^2\nabla_{\mathbf{e}_2})x\\ &= (v^1\partial_x+v^2\partial_y)x\\ &= v^1.\end{aligned}$$

In other words, $\mathbf{d}x$ *is the 1-form that picks out the x-component of a vector*. Likewise, we see that $(\mathbf{d}y)\mathbf{v}=v^2$. But according to the definition (32.5), this means that $\{\mathbf{d}x,\mathbf{d}y\}$ is the *dual basis* of $\{\mathbf{e}_1,\mathbf{e}_2\}$. Equivalently, $(\mathbf{d}x^i)\mathbf{e}_j=\delta^i_j$.

Geometrically, $\mathbf{d}x$ may be pictured as the family of unit-spaced lines perpendicular to the x-axis, and $(\mathbf{d}x)\mathbf{v}$ is the number of these lines that $\mathbf{v}$ pierces. The above reasoning applies equally well to $\mathbb{R}^3$, in which case $\mathbf{d}x$ is the family of unit-spaced *planes* perpendicular to the x-axis, as illustrated in [32.8].

Clearly, all of this immediately generalizes to any number of dimensions. Let $\{\mathbf{e}_i\}$ be an orthonormal basis for $\mathbb{R}^n$, with Cartesian coordinates $\{x^j\}$, so that $\mathbf{v}=v^i\,\mathbf{e}_i$ is a general vector. Then $\mathbf{d}x^j$ picks out the j-th component of the vector, i.e., $(\mathbf{d}x^j)\,\mathbf{v}=v^j$. In particular,

$$(\mathbf{d}x^i)\mathbf{e}_j=\delta^i_j. \tag{32.12}$$

We call this the ***Cartesian basis***:

$\{\mathbf{d}x^j\}=\{\omega^j\}$ is the Cartesian basis of 1-forms dual to $\{\mathbf{e}_j\}$.

It follows from (32.6) that a general 1-form φ can be decomposed into components in this dual basis as

$$\boxed{\varphi = \varphi_j \, \omega^j = \varphi(\mathbf{e}_j) \, \mathbf{d}x^j.} \tag{32.13}$$

32.6.4 *The 1-Form Interpretation of* $\mathrm{d}f = (\partial_x f)\,\mathrm{d}x + (\partial_y f)\,\mathrm{d}y$

Taking $\varphi = \mathbf{d}f$ in (32.13), we have come full circle, decomposing the gradient 1-form $\mathbf{d}f$ of a general function f into its Cartesian basis 1-form components, as follows:

$$\boxed{\mathbf{d}f = \left[(\mathbf{d}f)\,\mathbf{e}_j\right] \mathbf{d}x^j = \left[\partial_{x^j} f\right] \mathbf{d}x^j.} \tag{32.14}$$

On a formal level, this is *identical* to the classical formula (32.9), but now it has a precise, rigorous meaning that does not call upon infinitesimals. And yet it is very directly and intuitively connected to the geometrically meaningful *ultimate equality*, (32.8), illustrated in [32.7].

To see this, once again think of $\mathbf{v}$ as a short, ultimately vanishing vector,

$$\mathbf{v} = \delta x^1 \, \mathbf{e}_1 + \delta x^2 \, \mathbf{e}_2 = \begin{bmatrix} \delta x \\ \delta y \end{bmatrix}.$$

Then,

$$\mathbf{d}f(\mathbf{v}) = \left\{\left[\partial_x f\right] \mathbf{d}x + \left[\partial_y f\right] \mathbf{d}y\right\} \begin{bmatrix} \delta x \\ \delta y \end{bmatrix} = (\partial_x f)\,\delta x + (\partial_y f)\,\delta y \asymp \delta f,$$

which is *exactly* (32.8). Thus the 1-form $\mathbf{d}f$ gives us the best of both worlds!

The above analysis sought to connect the new 1-form $\mathbf{d}f$ to your prior knowledge of the vector gradient $\boldsymbol{\nabla} f$. However, if you look back at our discussion of topographic maps, you will see that we had *already* arrived there at this new 1-form by geometrical reasoning alone.

Indeed, if you look at our definition (32.4) (illustrated in [32.5]) of the 1-form ζ associated with a topographic map, which describes how the height of the tangent plane varies as we move off in an arbitrary direction, you will see that it is indeed the gradient of the height function h:

$$\boxed{\zeta = \mathbf{d}h.}$$

In particular, if $\mathbf{v}$ is the direction at p of the contour of constant height passing through that point, then $\zeta(\mathbf{v}) = \mathbf{d}h(\mathbf{v}) = 0$, as it should.

32.7 Adding 1-Forms Geometrically

We already know the geometric meaning of multiplying a 1-form by a constant k: it *compresses* the stack by k. But what does *addition* of 1-forms mean?

Figure [32.9] superimposes the stacks representing the 1-forms $2\mathbf{d}x$ and $\mathbf{d}y$, then joins the resulting intersection points to create a new stack. *This* is the geometrical construction of their sum, $\varphi = 2\mathbf{d}x + \mathbf{d}y$!

If $\widetilde{\varphi}$ denotes the 1-form corresponding to the constructed stack, then we must prove that $\widetilde{\varphi} = \varphi$.

We will offer *three* proofs, not because the result is especially important, but simply as valuable practice with thinking about and manipulating 1-forms.

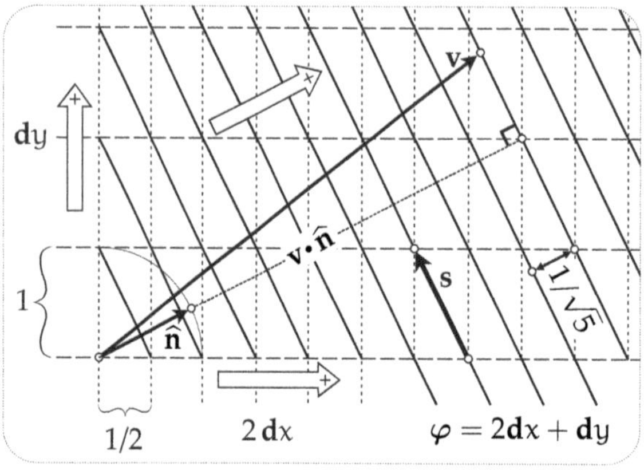

[32.9] Geometric Addition of 1-Forms. *To add the 1-forms* $2\,\mathbf{d}x$ *and* $\mathbf{d}y$, *superimpose their stacks, then join their intersection points to create the stack of* $\varphi = 2\mathbf{d}x + \mathbf{d}y$.

First note [exercise] that the unit vector normal to $\widetilde{\varphi}$ is

$$\widehat{\mathbf{n}} = \frac{1}{\sqrt{5}}\begin{bmatrix} 2 \\ 1 \end{bmatrix}.$$

Also note that this serves to specify the *direction* of the stack, which [exercise] must have positive components along both of the stacks that are being added.

If $\mathbf{v} = \begin{bmatrix} a \\ b \end{bmatrix}$ denotes a general vector, then its projection onto the direction of $\widehat{\mathbf{n}}$ is $\mathbf{v} \cdot \widehat{\mathbf{n}}$, as illustrated. But [exercise] the spacing of $\widetilde{\varphi}$ is $1/\sqrt{5}$, and so the number of lines of the stack $\widetilde{\varphi}$ pierced by $\mathbf{v}$ is

$$\widetilde{\varphi}(\mathbf{v}) = \text{number of lines pierced} = \frac{\mathbf{v} \cdot \widehat{\mathbf{n}}}{(1/\sqrt{5})} = 2a + b = (2\mathbf{d}x + \mathbf{d}y)\begin{bmatrix} a \\ b \end{bmatrix} = \varphi(\mathbf{v}).$$

Since $\mathbf{v}$ is a general vector, we may abstract it away, confirming our claim that $\widetilde{\varphi} = \varphi$.

Here is a second proof, which is simpler, but less direct. We know that for any 1-form, such as our φ, the direction(s) $\mathbf{s}$ lying within the stack are the kernel of the 1-form: $\mathbf{s}$ does not pierce any lines (or planes, in $\mathbb{R}^3$) of the stack, so

$$0 = \varphi(\mathbf{s}) = (2\mathbf{d}x + \mathbf{d}y)\begin{bmatrix} p \\ q \end{bmatrix} = 2p + q \qquad \Longrightarrow \qquad \mathbf{s} \propto \begin{bmatrix} -1 \\ 2 \end{bmatrix},$$

but, as illustrated, this is *also* the direction of the $\widetilde{\varphi}$ stack, connecting grid points. Thus the two stacks are parallel and can only differ as regards their spacing. Put algebraically, the two 1-forms must be *proportional*: $\widetilde{\varphi} = K\,\varphi$, for some constant K. But if we apply this equation to a specific vector for which the number of pierced lines is evident—try $\begin{bmatrix} 0 \\ 1 \end{bmatrix}$—then we instantly deduce that $K = 1$. Done!

The third proof is perhaps the simplest of all. In [32.9], think of $\mathbf{v} = \begin{bmatrix} x \\ y \end{bmatrix}$ as being a *position vector*, emanating from a fixed origin. What is the equation of a line in the φ-stack? Well, as the tip of $\mathbf{v}$ moves along such a line, $\mathbf{v}$ pierces a *fixed* number of stack lines, so

$$\textit{constant} = \varphi(\mathbf{v}) = (2\mathbf{d}x + \mathbf{d}y)\begin{bmatrix} x \\ y \end{bmatrix} = 2x + y \qquad \Longrightarrow \qquad y = -2x + \textit{constant}.$$

But these lines, with slope −2, are *parallel* to the $\widetilde{\varphi}$-stack. The rest of the proof is the same as before.

It is easy [exercise] to generalize these arguments to confirm that this geometrical construction works for $p\,\mathbf{dx} + q\,\mathbf{dy}$, yielding a stack with

$$\text{normal vector } \begin{bmatrix} p \\ q \end{bmatrix} \text{ and spacing } \frac{1}{\sqrt{p^2 + q^2}}.$$

Indeed, in $\mathbb{R}^3$, we may [exercise] construct $p\,\mathbf{dx} + q\,\mathbf{dy} + r\,\mathbf{dz}$ by superimposing the three orthogonal stacks of planes, constructing a new stack of planes passing through the intersection points, with

$$\text{normal vector } \begin{bmatrix} p \\ q \\ r \end{bmatrix} \text{ and spacing } \frac{1}{\sqrt{p^2 + q^2 + r^2}}.$$

Chapter 33

Tensors

33.1 Definition of a Tensor: Valence

In our discussion of the Riemann tensor, we tentatively defined a tensor as a *multilinear function of multiple vector inputs*. This was the best we could do at the time, because we lacked the concept of a 1-form. But a fully general tensor is, in fact, a multilinear function of vectors *and* 1-forms, and the ***valence*** tells us how many of each:

> *A* **tensor H of valence** $\left\{ \begin{smallmatrix} \mathfrak{f} \\ \mathfrak{v} \end{smallmatrix} \right\}$ *at a point* p *is a multilinear, real-valued function of* $\mathfrak{f}$ *1-forms, and* $\mathfrak{v}$ *vectors, such that the value of* **H** *at* p *only depends on the values of the 1-forms and vectors* at p.

Thus a 1-form is a tensor of valence $\left\{ \begin{smallmatrix} 0 \\ 1 \end{smallmatrix} \right\}$, because it has one "slot" into which a single vector can be fed, the output being the contraction of that vector with the 1-form. Likewise, a vector is a tensor of valence $\left\{ \begin{smallmatrix} 1 \\ 0 \end{smallmatrix} \right\}$, because it has one slot into which a single 1-form can be fed, the output being the contraction of that 1-form with the vector.

In the general case, we may divide the input slots of **H** into two groups: $\mathfrak{f}$ slots to input the $\mathfrak{f}$ 1-forms, $\varphi_1, \ldots, \varphi_{\mathfrak{f}}$, and a second group of $\mathfrak{v}$ slots to input the $\mathfrak{v}$ vectors, $\mathbf{v}_1, \ldots, \mathbf{v}_{\mathfrak{v}}$. Using $\|$ to denote the boundary between these two groups, we have

$$\mathbf{H}(\varphi_1, \ldots, \varphi_{\mathfrak{f}} \,\|\, \mathbf{v}_1, \ldots, \mathbf{v}_{\mathfrak{v}}).$$

In general, the *order* in which we feed 1-forms and vectors into these slots *matters*: if we swap a pair of 1-forms or a pair of vectors in these slots, the output will in general be completely unrelated to the original value.

Let us re-examine the Riemann tensor **R** from this new, general point of view. In (29.8) we defined this as a *vector*-valued, multilinear function of three vector inputs: **w** is the vector that is parallel-transported around the parallelogram with edges defined by **u** and **v**, which we now imagine to be very short, and the output is the vector holonomy $\delta\mathbf{w}$:

$$\delta\mathbf{w} = \mathbf{R}(\mathbf{u}, \mathbf{v}; \mathbf{w}) = \left\{ [\nabla_{\mathbf{u}}, \nabla_{\mathbf{v}}] - \nabla_{[\mathbf{u},\mathbf{v}]} \right\} \mathbf{w}.$$

In order to make this into a tensor in the new sense, it must output a real *number*, instead of the vector $\delta\mathbf{w}$. To extract a real number from $\delta\mathbf{w}$, we must contract it with a 1-form. For this reason, in place of our original definition, the *standard* definition of the Riemann tensor is instead a tensor of valence $\left\{ \begin{smallmatrix} 1 \\ 3 \end{smallmatrix} \right\}$:

$$\mathbf{R}(\varphi \,\|\, \mathbf{u}, \mathbf{v}, \mathbf{w}) \equiv \langle \varphi, \delta\mathbf{w} \rangle = \varphi(\delta\mathbf{w}).$$

NOTE ON NOTATION: We shall reserve uppercase bold Roman letters for tensors with multiple inputs, such as the Riemann tensor, $\mathbf{R}$, the energy-momentum tensor, $\mathbf{T}$, and the Einstein tensor, $\mathbf{G}$. On the other hand, some such higher-valence tensors are traditionally represented by *lowercase* bold Roman letters, the most notable example being the metric tensor, $\mathbf{g}$, which must not be confused with a vector!

33.2 Example: Linear Algebra

It should already be clear that an enormous range of mathematical and physical objects fall into our new, generalized category of tensors. To drive home this point, consider the vast and vastly important subject of *Linear Algebra*. From our new vantage point, this is "merely" the study of tensors $\mathbf{L}(\varphi \,\|\, \mathbf{v})$ of valence $\left\{ {1 \atop 1} \right\}$!

To see why this is, consider $\mathbf{L}(\ldots \,\|\, \mathbf{v})$, in which we have left the first slot empty. In order to yield a number, we must complete this by re-inserting the 1-form φ. Therefore, $\mathrm{L}(\mathbf{v}) \equiv \mathbf{L}(\ldots \,\|\, \mathbf{v})$ must be a *vector*, ready to be contracted with a 1-form φ to yield a number. This is perfectly analogous to our original definition of the Riemann tensor: there we had a vector-valued function of three vectors; now we have a vector-valued function L of a single vector, which only yields a number once we contract it with a 1-form.

But, by the definition of a tensor, $\mathbf{L}(\varphi \,\|\, \mathbf{v})$ is a *linear* function of both φ and $\mathbf{v}$. Thus,

$$\begin{aligned} \mathrm{L}(k_1\mathbf{v}_1 + k_2\mathbf{v}_2) &= \mathbf{L}(\ldots \,\|\, k_1\mathbf{v}_1 + k_2\mathbf{v}_2) \\ &= k_1\,\mathbf{L}(\ldots \,\|\, \mathbf{v}_1) + k_2\,\mathbf{L}(\ldots \,\|\, \mathbf{v}_2) \\ &= k_1\,\mathrm{L}(\mathbf{v}_1) + k_2\,\mathrm{L}(\mathbf{v}_2). \end{aligned}$$

Therefore L is a *linear transformation*, sending vectors to vectors—the fundamental object of study in Linear Algebra! Of course, by the same token, if we leave the *vector* slot open, we may instead interpret $\mathbf{L}(\varphi \,\|\, \ldots)$ as a linear, 1-form-valued function of 1-forms.

33.3 New Tensors from Old

33.3.1 Addition

Clearly it makes no sense to try to add tensors with different valences, which require entirely different kinds of inputs. However, if the two tensors have the *same* valence, the definition of addition is obvious. For example, if $\mathbf{H}$ and $\mathbf{J}$ are both of valence $\left\{ {2 \atop 1} \right\}$, then

$$(\mathbf{H}+\mathbf{J})(\varphi, \psi \,\|\, \mathbf{v}) \equiv \mathbf{H}(\varphi, \psi \,\|\, \mathbf{v}) + \mathbf{J}(\varphi, \psi \,\|\, \mathbf{v}).$$

It is easy to check [exercise] that $(\mathbf{H}+\mathbf{J})$ is indeed another *tensor* of valence $\left\{ {2 \atop 1} \right\}$.

33.3.2 Multiplication: The Tensor Product

Given two 1-forms, φ and ψ, each of which acts on a single vector, it is natural to define their ***tensor product***, acting on *two* vectors, and therefore of valence $\left\{ {0 \atop 2} \right\}$, as follows:

$$\boxed{(\varphi \otimes \psi)(\mathbf{v}, \mathbf{w}) \equiv \varphi(\mathbf{v})\, \psi(\mathbf{w}).} \tag{33.1}$$

Note that *order matters:* $\varphi \otimes \psi \neq \psi \otimes \varphi$.

TERMINOLOGY: The tensor product is also commonly called the ***direct product*** or the ***outer product***.

The tensor product of higher valence tensors goes the same way. For example, we may multiply a tensor $\mathbf{J}(\varphi, \psi \,\|\, \mathbf{u})$ of valence $\left\{ {2 \atop 1} \right\}$ and a tensor $\mathbf{T}(\mathbf{v}, \mathbf{w})$ of valence $\left\{ {0 \atop 2} \right\}$ to obtain the tensor $\mathbf{J} \otimes \mathbf{T}$ of valence $\left\{ {2 \atop 3} \right\}$:

$$(\mathbf{J} \otimes \mathbf{T})(\varphi, \psi \,\|\, \mathbf{u}, \mathbf{v}, \mathbf{w}) \equiv \mathbf{J}(\varphi, \psi \,\|\, \mathbf{u}) \cdot \mathbf{T}(\mathbf{v}, \mathbf{w}).$$

Note that under tensor multiplication, valences add like vectors: in our example, $\left\{ {2 \atop 1} \right\} + \left\{ {0 \atop 2} \right\} = \left\{ {2 \atop 3} \right\}$. Since multiplication by a scalar does not change the valence, the valence of a scalar must be taken to be $\left\{ {0 \atop 0} \right\}$.

33.4 Components

As usual, let $\{\mathbf{e}_i\}$ be an orthonormal vector basis, and let $\{\mathbf{d}x^j\}$ be the dual Cartesian basis of 1-forms. Just as we did to obtain the components of vectors and 1-forms, we obtain the ***components*** of a more general tensor by filling its slots with basis 1-forms and vectors. For example, the components of $\mathbf{T}(\mathbf{v}, \mathbf{w})$ are

$$T_{ij} = \mathbf{T}(\mathbf{e}_i, \mathbf{e}_j)$$

We can decompose the full tensor $\mathbf{T}$ into tensor components, as follows:

$$\begin{aligned} \mathbf{T}(\mathbf{v}, \mathbf{w}) &= \mathbf{T}(v^i \mathbf{e}_i, w^j \mathbf{e}_j) \\ &= \mathbf{T}(\mathbf{e}_i, \mathbf{e}_j)\, v^i w^j \\ &= T_{ij}\, v^i w^j \\ &= T_{ij} \left[\mathbf{d}x^i(\mathbf{v})\right] \left[\mathbf{d}x^j(\mathbf{w})\right] \\ &= T_{ij}\, (\mathbf{d}x^i \otimes \mathbf{d}x^j)(\mathbf{v}, \mathbf{w}) \end{aligned}$$

But since both $\mathbf{v}$ and $\mathbf{w}$ are general vectors, we may abstract them away, to find that the tensor itself is

$$\mathbf{T} = T_{ij}\, (\mathbf{d}x^i \otimes \mathbf{d}x^j). \tag{33.2}$$

Thus we see that

The set of tensors $\{\mathbf{d}x^i \otimes \mathbf{d}x^j\}$ *forms a* **basis** *for tensors of valence* $\left\{ {0 \atop 2} \right\}$.

Note that if we apply this idea to (33.1), then we find that

$$\varphi \otimes \psi = \varphi_i \psi_j\, (\mathbf{d}x^i \otimes \mathbf{d}x^j).$$

However, such tensors are special: a general $\mathbf{T}$ *cannot* be factorized in this way.

We can see this from the fact that while $\mathbf{T}$ has n^2 components T_{ij} in n dimensions, $(\varphi \otimes \psi)$ is uniquely determined by just $2n$ numbers: the n components φ_i of φ, and the n components ψ_i of ψ.

In exactly the same way, we can take the tensor product of two vectors:

$$(\mathbf{v} \otimes \mathbf{w})(\varphi, \psi) = \mathbf{v}(\varphi)\, \mathbf{w}(\psi) = v^i\, w^j\, (\mathbf{e}_i \otimes \mathbf{e}_j)(\varphi, \psi).$$

And, likewise, a general tensor $\mathbf{K}(\varphi, \psi)$ of valence $\left\{ {2 \atop 0} \right\}$ can be decomposed into basis tensors as

$$\boxed{\mathbf{K} = K^{ij}\, (\mathbf{e}_i \otimes \mathbf{e}_j).}$$

Clearly the above technique can be applied to tensors of *arbitrary* valence $\left\{ {\mathfrak{f} \atop \mathfrak{v}} \right\}$, decomposing them into components using the basis tensors,

$$(\mathbf{e}_{i_1} \otimes \mathbf{e}_{i_2} \otimes \cdots \otimes \mathbf{e}_{i_\mathfrak{f}}) \otimes (\mathbf{dx}^{j_1} \otimes \mathbf{dx}^{j_2} \otimes \cdots \otimes \mathbf{dx}^{j_\mathfrak{v}}).$$

Note that the components of a tensor of valence $\left\{ {\mathfrak{f} \atop \mathfrak{v}} \right\}$ have $\mathfrak{f}$ *upper* indices, and $\mathfrak{v}$ *lower* indices.[1]

33.5 Relation of the Metric Tensor to the Classical Line Element

Earlier we discussed the fact that the modern gradient 1-form $\mathbf{d}f$ is closely connected to the classical, infinitesimal-based formula (32.9) for the differential df. Similarly, let us now consider how the modern metric tensor is related to the classical, infinitesimal-based ***line element*** ds of Gauss (1827).

Gauss considered infinitesimal changes du and dv in the coordinates of a point on the surface, resulting in an infinitesimal line element ds within the surface. Then, in Gauss's original[2] notation,

$$ds^2 = E\, du^2 + 2F\, du\, dv + G\, dv^2.$$

In place of infinitesimals, we instead employ ultimate equalities, expressing this as

$$\delta s^2 \asymp E\, \delta u^2 + 2F\, \delta u\, \delta v + G\, \delta v^2.$$

Now let

$$x^1 = u, \quad x^2 = v \qquad \text{and} \qquad g_{11} = E, \quad g_{12} = g_{21} = F, \quad g_{22} = G.$$

Then, using (33.2), the metric tensor can be expressed as

$$\mathbf{g} = g_{ij}\, (\mathbf{dx}^i \otimes \mathbf{dx}^j) = E\,(\mathbf{d}u \otimes \mathbf{d}u) + F\,(\mathbf{d}u \otimes \mathbf{d}v) + F\,(\mathbf{d}v \otimes \mathbf{d}u) + G\,(\mathbf{d}v \otimes \mathbf{d}v),$$

which looks tantalizingly similar to Gauss's formula for ds^2.

[1] In older literature, upper indices were called ***contravariant***, and lower indices were called ***covariant***.

[2] Actually, as we remarked earlier, Gauss (1827) used p and q instead of u and v.

To make the connection explicit, let $\delta\mathbf{r} = \begin{bmatrix} \delta u \\ \delta v \end{bmatrix}$ be a short (ultimately vanishing) vector in the coordinate map plane, resulting in a small movement δs within the surface. Then, since

$$\mathbf{du}\begin{bmatrix} \delta u \\ \delta v \end{bmatrix} = \delta u, \qquad \text{and} \qquad \mathbf{dv}\begin{bmatrix} \delta u \\ \delta v \end{bmatrix} = \delta v,$$

we do indeed recover the line-element formula from the metric tensor:

$$\begin{aligned}
\delta s^2 &\asymp \mathbf{g}(\delta\mathbf{r}, \delta\mathbf{r}) \\
&= [E\,(\mathbf{du} \otimes \mathbf{du}) + F\,(\mathbf{du} \otimes \mathbf{dv}) + F\,(\mathbf{dv} \otimes \mathbf{du}) + G\,(\mathbf{dv} \otimes \mathbf{dv})]\left(\begin{bmatrix} \delta u \\ \delta v \end{bmatrix}, \begin{bmatrix} \delta u \\ \delta v \end{bmatrix}\right) \\
&= E\,\delta u^2 + F\,\delta u\,\delta v + F\,\delta v\,\delta u + G\,\delta v^2.
\end{aligned}$$

33.6 Example: Linear Algebra (Again)

What is the connection between the familiar matrices of a traditional Linear Algebra course, and the seemingly abstract tensor view of the subject, as the study of tensors of valence $\left\{ \begin{smallmatrix} 1 \\ 1 \end{smallmatrix} \right\}$? This section uses the concepts of components, and of tensor bases, to explain the connection.

First, our tensor $\mathbf{L}(\varphi \,\|\, \mathbf{v})$ has components

$$L^i{}_j = \mathbf{L}(\mathbf{dx}^i \,\|\, \mathbf{e}_j).$$

In order to keep things as simple as possible and as concrete as possible, we restrict ourselves to two dimensions.

If the vector basis is represented by the column vectors

$$\{\mathbf{e}_1, \mathbf{e}_2\} = \left\{ \begin{bmatrix} 1 \\ 0 \end{bmatrix}, \begin{bmatrix} 0 \\ 1 \end{bmatrix} \right\},$$

then, as discussed in Section 32.3.4, the dual 1-form basis corresponds to row vectors:

$$\left\{\mathbf{dx}^1, \mathbf{dx}^2\right\} = \left\{ \begin{bmatrix} 1 & 0 \end{bmatrix}, \begin{bmatrix} 0 & 1 \end{bmatrix} \right\}.$$

Thus,

$$\mathbf{dx}^1(\mathbf{v}) = \begin{bmatrix} 1 & 0 \end{bmatrix} \begin{bmatrix} v^1 \\ v^2 \end{bmatrix} = v^1,$$

as it should. Likewise for $\mathbf{dx}^2$.

It follows that $\mathbf{L}$ is indeed represented by a familiar 2×2 matrix, whose entries are the components of the tensor:

$$\begin{aligned}
\mathbf{L} &= L^i{}_j\, \mathbf{e}_i \otimes \mathbf{dx}^j \\
&= L^1{}_1 \begin{bmatrix} 1 \\ 0 \end{bmatrix} \begin{bmatrix} 1 & 0 \end{bmatrix} + L^1{}_2 \begin{bmatrix} 1 \\ 0 \end{bmatrix} \begin{bmatrix} 0 & 1 \end{bmatrix} + L^2{}_1 \begin{bmatrix} 0 \\ 1 \end{bmatrix} \begin{bmatrix} 1 & 0 \end{bmatrix} + L^2{}_2 \begin{bmatrix} 0 \\ 1 \end{bmatrix} \begin{bmatrix} 0 & 1 \end{bmatrix}
\end{aligned}$$

$$= L^1{}_1\begin{bmatrix}1 & 0\\0 & 0\end{bmatrix} + L^1{}_2\begin{bmatrix}0 & 1\\0 & 0\end{bmatrix} + L^2{}_1\begin{bmatrix}0 & 0\\1 & 0\end{bmatrix} + L^2{}_2\begin{bmatrix}0 & 0\\0 & 1\end{bmatrix}$$

$$= \begin{bmatrix}L^1{}_1 & L^1{}_2\\L^2{}_1 & L^2{}_2\end{bmatrix}.$$

In Linear Algebra, this matrix describes a linear transformation, by means of standard matrix multiplication:

$$\begin{bmatrix}v^1\\v^2\end{bmatrix} \to \begin{bmatrix}L^1{}_1 & L^1{}_2\\L^2{}_1 & L^2{}_2\end{bmatrix}\begin{bmatrix}v^1\\v^2\end{bmatrix} = \begin{bmatrix}L^1{}_1 v^1 + L^1{}_2 v^2\\L^2{}_1 v^1 + L^2{}_2 v^2\end{bmatrix} = \begin{bmatrix}L^1{}_j v^j\\L^2{}_j v^j\end{bmatrix}.$$

In other words,

$$v^i \longrightarrow L^i{}_j v^j \quad \Longleftrightarrow \quad \mathbf{v} = v^i \mathbf{e}_i \longrightarrow (L^i{}_j v^j)\, \mathbf{e}_i.$$

Expressed in this form, we can now see how this is equivalent to our discussion in Section 33.2, in which we used the tensor to define the linear transformation $L(\mathbf{v})$:

$$\mathbf{v} \longrightarrow L(\mathbf{v}) = \mathbf{L}(\ldots \,\|\, \mathbf{v}) = L^i{}_j\,(\mathbf{e}_i \otimes \mathbf{dx}^j)\,(\ldots, \mathbf{v}) = L^i{}_j\, \mathbf{e}_i(\ldots)\, \mathbf{dx}^j(\mathbf{v}) = (L^i{}_j v^j)\, \mathbf{e}_i(\ldots).$$

33.7 Contraction

Consider the contraction of a 1-form and a vector from the point of view of components:

$$\varphi(\mathbf{v}) = (\varphi_i\, \mathbf{dx}^i)(v^j\, \mathbf{e}_j) = (\varphi_i\, v^j)\, \mathbf{dx}^i(\mathbf{e}_j) = (\varphi_i\, v^j)\, \delta^i_j = \varphi_j\, v^j.$$

As we know, this contraction is a *geometrical* operation, and the answer is independent of the specific components of φ and $\mathbf{v}$, *despite* the fact that these components *do* depend on the choice of the dual bases $\{\mathbf{e}_j\}$ and $\{\mathbf{dx}^i\}$.

Next, consider the components $L^i{}_j$ of the tensor in the previous section. We now define, analogously, the ***contraction*** of this tensor to be

$$L^j{}_j = \mathbf{L}(\mathbf{dx}^j \,\|\, \mathbf{e}_j) = L^1{}_1 + L^2{}_2 = \mathrm{Tr}\begin{bmatrix}L^1{}_1 & L^1{}_2\\L^2{}_1 & L^2{}_2\end{bmatrix}.$$

As we know from Linear Algebra, this trace is *also* basis-independent.[3]

We will prove a greatly generalized version of this basis-independence result—using purely tensor methods, and without restricting ourselves to two dimensions—but first we must explain what *contraction* means in the general case.

The idea of contraction can be applied to *any* tensor whose inputs include at least one 1-form and at least one vector: *we sum over one upper index and one lower index.* We have already encountered one very important and natural example of such a contraction, namely, the contraction of the Riemann tensor that yields the Ricci tensor:

$$R_{mij}{}^m = R_{ij}.$$

Note that this contraction process changes the valence, in this example, from $\left\{ {1 \atop 3} \right\}$ to $\left\{ {0 \atop 2} \right\}$. In general, the input slots of the new, contracted tensor accept one fewer 1-form and one fewer vector, thereby eliminating one upper and one lower index.

[3]There is a *geometrical* reason for this, although it is hard to find in Linear Algebra texts. See Arnol'd (1973, §16.3).

In fact contraction has a still broader meaning in the context of tensor products. Suppose we form $\mathbf{A} \otimes \mathbf{B}$, and then perform a summation over an upper index of $\mathbf{A}$ and a lower index of $\mathbf{B}$. Let us prove that the result of this contraction is another *tensor*, independent of the specific $\{\mathbf{e}_j\}$ and $\{\mathbf{dx}^i\}$ that gave rise to the components being summed.

Following the example of Schutz (1980, §2.25), we shall communicate the idea of the proof by means of a concrete example. Let $\mathbf{A}$ be of valence $\left\{ {2 \atop 0} \right\}$, and let $\mathbf{B}$ be of valence $\left\{ {0 \atop 2} \right\}$. Let us prove that the contraction

$$A^{ij} B_{jk} \equiv C^i{}_k$$

yields the components of a new *tensor* $\mathbf{C}$ of valence $\left\{ {1 \atop 1} \right\}$, such that

$$\mathbf{C}(\varphi \| \mathbf{v}) = C^i{}_k \varphi_i v^k.$$

To prove this, first observe that

$$\begin{aligned} C^i{}_k \varphi_i v^k &= A^{ij} B_{jk} \varphi_i v^k \\ &= [\varphi_i \mathbf{A}(\mathbf{dx}^i, \mathbf{dx}^j)] \, [\mathbf{B}(\mathbf{e}_j, \mathbf{e}_k) \, v^k] \\ &= \mathbf{A}(\varphi_i \, \mathbf{dx}^i, \mathbf{dx}^j) \, \mathbf{B}(\mathbf{e}_j, v^k \, \mathbf{e}_k) \\ &= \mathbf{A}(\varphi, \mathbf{dx}^j) \, \mathbf{B}(\mathbf{e}_j, \mathbf{v}). \end{aligned}$$

But the $\mathbf{B}(\mathbf{e}_j, \mathbf{v})$ are numbers, and therefore the linearity of $\mathbf{A}$ in the second slot implies that

$$C^i{}_k \varphi_i v^k = \mathbf{A}(\varphi, \mathbf{B}(\mathbf{e}_j, \mathbf{v}) \, \mathbf{dx}^j).$$

But, for fixed $\mathbf{v}$,

$$\mathbf{B}(\mathbf{e}_j, \mathbf{v}) \, \mathbf{dx}^j = \mathbf{B}(\ldots, \mathbf{v})$$

is a 1-form, requiring the insertion of a vector to produce a number. Therefore,

$$\mathbf{C}(\varphi \| \mathbf{v}) = \mathbf{A}(\varphi, \mathbf{B}(\ldots, \mathbf{v}))$$

is indeed a tensor.

33.8 Changing Valence with the Metric Tensor

We already know that the metric tensor $\mathbf{g}$ of valence $\left\{ {0 \atop 2} \right\}$ is *the* fundamental structure of a manifold: it gives us the geodesics, parallel transport, and the curvature. But $\mathbf{g}$ also plays another crucial role: *it allows us to change the valence of a tensor.* The first step is to see how the metric allows us to associate a particular 1-form with a particular vector, and vice versa.

If we leave one slot of $\mathbf{g}$ open, and insert the vector $\mathbf{n}$ into the other, we obtain a unique 1-form $\boldsymbol{\nu}$ corresponding to $\mathbf{n}$:

$$\text{vector } \mathbf{n} \longrightarrow \text{1-form } \boldsymbol{\nu}, \quad \text{where } \boldsymbol{\nu}(\mathbf{w}) \equiv \mathbf{g}(\mathbf{w}, \mathbf{n}). \tag{33.3}$$

How are the *components* of the 1-form $\boldsymbol{\nu}$ related to the components of the original vector $\mathbf{n}$? We need only apply $\boldsymbol{\nu}$ to the basis vectors. But first ...

NOTATIONAL WARNING: *the universal convention is that the components of the 1-form corresponding to the vector* $\mathbf{n}$ *are denoted* n_i, violating our Greek/Roman dichotomy between 1-forms and vectors. Henceforth, the reader must therefore be more vigilant than before: if the index is up, then it belongs to a vector, and if the index is down, it belongs to a 1-form.

With this convention in place,

$$\boldsymbol{\nu} = n_i\, \mathbf{d}x^i.$$

Thus,

$$n_i = \boldsymbol{\nu}(\mathbf{e}_i) = \mathbf{g}(\mathbf{e}_i, \mathbf{n}) = \mathbf{g}(\mathbf{e}_i, n^j\, \mathbf{e}_j) = \mathbf{g}(\mathbf{e}_i, \mathbf{e}_j) n^j,$$

so,

$$\boxed{n_i = g_{ij}\, n^j.} \tag{33.4}$$

For example, if $\mathbf{n} = n^j\, \mathbf{e}_j$ is a vector in Minkowski spacetime, with metric (30.6), then [exercise] the corresponding 1-form is

$$\boldsymbol{\nu} = n_i\, \mathbf{d}x^i = n^0\, \mathbf{d}t - n^1\, \mathbf{d}x - n^2\, \mathbf{d}y - n^3\, \mathbf{d}z.$$

The metric's ability to transform a vector into a 1-form allows us to change the valence of any tensor. For example, consider the Riemann tensor $\mathbf{R}(\boldsymbol{\psi} \parallel \mathbf{u}, \mathbf{v}, \mathbf{w})$, of valence $\left\{ {1 \atop 3} \right\}$, with components $R_{ijk}{}^{m}$. Let us demonstrate how we may change this into a tensor of valence $\left\{ {0 \atop 4} \right\}$, by changing the single 1-form input into a vector input. By convention, this new tensor is *still* denoted $\mathbf{R}$, and its components are written R_{ijkl}.

To evaluate the new tensor with four vector inputs, we simply replace the extra vector input $\mathbf{n}$ with its corresponding 1-form $\boldsymbol{\nu}$ in the *original* definition of the Riemann tensor:

$$\mathbf{R}(\mathbf{u}, \mathbf{v}, \mathbf{w}, \mathbf{n}) \equiv \mathbf{R}(\boldsymbol{\nu} \parallel \mathbf{u}, \mathbf{v}, \mathbf{w}).$$

In component form, this equation becomes

$$R_{ijkl} u^i v^j w^k n^l = R_{ijk}{}^{m} u^i v^j w^k n_m = R_{ijk}{}^{m} u^i v^j w^k (g_{ml}\, n^l).$$

Therefore,

$$\boxed{R_{ijkl} = R_{ijk}{}^{m}\, g_{ml}.}$$

This process is called (logically enough) ***index lowering***.

It is also possible to go in the opposite direction, changing a vector input into a 1-form input; this results in ***index raising***. In order to follow the same path as before, we need a mapping from 1-forms $\boldsymbol{\nu}$ to vectors $\mathbf{n}$. What is needed is an analogue $\widetilde{\mathbf{g}}$ of the metric tensor, now of valence $\left\{ {2 \atop 0} \right\}$, taking two 1-forms as input.

The mapping is then obtained exactly as before, by inserting $\boldsymbol{\nu}$ into one slot, leaving the other slot open, resulting in a vector:

$$\boxed{\text{1-form } \boldsymbol{\nu} \longrightarrow \text{vector } \mathbf{n}, \qquad \text{where } \ \mathbf{n}(\boldsymbol{\varphi}) \equiv \widetilde{\mathbf{g}}(\boldsymbol{\varphi}, \boldsymbol{\nu}).} \tag{33.5}$$

In terms of components, we will then have,

$$n^i = \tilde{g}^{ik}\, n_k. \tag{33.6}$$

And, by the same token, we can now raise any tensor index we wish, e.g.,

$$R_{ijkl}\, \tilde{g}^{km} = R_{ij}{}^{m}{}_{l}$$

The final step is to realize that $\tilde{\mathbf{g}}$ is uniquely determined by the metric $\mathbf{g}$, as follows. Lowering and raising of indices must be *inverse* operations: if we lower an index and then raise it again, we should obtain the *same* tensor that we started with. Put differently, changing a vector into a 1-form with (33.3), then changing it back again with (33.5), should result in the *identity*. In other words, $\tilde{\mathbf{g}}$ is the *inverse* of $\mathbf{g}$:

$$\mathbf{n} \xrightarrow{\mathbf{g}} \boldsymbol{\nu} \xrightarrow{\tilde{\mathbf{g}}} \mathbf{n}.$$

Inserting (33.4) into (33.6), this implies that

$$n^i = \tilde{g}^{ik}\, n_k = \tilde{g}^{ik}\, g_{kj}\, n^j.$$

Therefore, the components of $\tilde{\mathbf{g}}$ must be related to the components of the metric by

$$\tilde{g}^{ik}\, g_{kj} = \delta^i_j. \tag{33.7}$$

Just as we always use the same symbol $\mathbf{R}$ to denote the Riemann tensor, regardless of the number of 1-forms and vectors it accepts as inputs, so it is traditional (if confusing) to simply write $\mathbf{g}$ instead of $\tilde{\mathbf{g}}$, and to correspondingly write its components as g^{ik}. Thus, (33.7) is conventionally written, $g^{ik}\, g_{kj} = \delta^i_j$.

33.9 Symmetry and Antisymmetry

Recall the defining properties of an ***even function***, $f^+(x)$, and of an ***odd function***, $f^-(x)$:

$$f^+(-x) = +f^+(x) \qquad \text{and} \qquad f^-(-x) = -f^-(x).$$

We may instead describe this by saying that $f^+(x)$ is ***symmetric***, and $f^-(x)$ is ***antisymmetric***. Geometrically, the graph $y = f^+(x)$ has mirror symmetry in the y-axis, e.g., $y = x^2$ or $y = \cos x$, whereas $y = f^-(x)$ does the *opposite* at x and its mirror image $-x$, e.g., $y = x^3$ or $y = \sin x$.

Suppose we try to split an *arbitrary* function $f(x)$ (lacking any special symmetry) into the sum of a symmetric function and an antisymmetric function:

$$f(x) = f^+(x) + f^-(x).$$

If this is possible, then, by definition,

$$f(-x) = f^+(x) - f^-(x).$$

By adding and then subtracting the previous two equations, we deduce that the split *is* always possible, and that it is explicitly given by

$$f^{+}(x)=\left[\frac{f(x)+f(-x)}{2}\right] \qquad \text{and} \qquad f^{-}(x)=\left[\frac{f(x)-f(-x)}{2}\right].$$

For example, if we take $f(x)=e^{x}$ then this split naturally leads us to the hyperbolic functions: the symmetric part of e^{x} is $f^{+}(x)=\cosh x$, and the antisymmetric part is $f^{-}(x)=\sinh x$.

Now let us try to do something analogous with a general tensor $\mathbf{E}(\mathbf{v},\mathbf{w})$ of valence $\left\{\begin{smallmatrix}0\\2\end{smallmatrix}\right\}$. By analogy, let the defining properties of a ***symmetric tensor***, $\mathbf{E}^{+}$, and of an ***antisymmetric***[4] ***tensor***, $\mathbf{E}^{-}$, be

$$\mathbf{E}^{+}(\mathbf{w},\mathbf{v})=+\mathbf{E}^{+}(\mathbf{v},\mathbf{w}) \qquad \text{and} \qquad \mathbf{E}^{-}(\mathbf{w},\mathbf{v})=-\mathbf{E}^{-}(\mathbf{v},\mathbf{w}).$$

By following the same line of reasoning as before, we discover that it is always possible to split a general tensor of valence $\left\{\begin{smallmatrix}0\\2\end{smallmatrix}\right\}$ into a symmetric part and an antisymmetric part:

$$\mathbf{E}(\mathbf{v},\mathbf{w})=\mathbf{E}^{+}(\mathbf{v},\mathbf{w})+\mathbf{E}^{-}(\mathbf{v},\mathbf{w}), \tag{33.8}$$

where

$$\mathbf{E}^{+}(\mathbf{v},\mathbf{w})=\left[\frac{\mathbf{E}(\mathbf{v},\mathbf{w})+\mathbf{E}(\mathbf{w},\mathbf{v})}{2}\right] \qquad \text{and} \qquad \mathbf{E}^{-}(\mathbf{v},\mathbf{w})=\left[\frac{\mathbf{E}(\mathbf{v},\mathbf{w})-\mathbf{E}(\mathbf{w},\mathbf{v})}{2}\right].$$

If we insert basis vectors into the slots, we obtain the component forms of these equations. While we are at it, let us also introduce the following standard notation: ***round brackets denote symmetrization*** and ***square brackets denote antisymmetrization***:

$$\boxed{E_{(ij)}\equiv E^{+}_{ij}=\tfrac{1}{2}\left[E_{ij}+E_{ji}\right] \qquad \text{and} \qquad E_{[ij]}\equiv E^{-}_{ij}=\tfrac{1}{2}\left[E_{ij}-E_{ji}\right]} \tag{33.9}$$

Therefore,

$$\boxed{E_{ij}=E_{(ij)}+E_{[ij]}.}$$

[4]Also often called ***skew symmetric***.

Chapter 34

2-Forms

34.1 Definition of a 2-Form and of a p-Form

We have already encountered several vitally important tensors of valence $\left\{\begin{smallmatrix}0\\2\end{smallmatrix}\right\}$ that are *symmetric*: the metric tensor, the Ricci tensor, the energy–momentum tensor, and the Einstein tensor.

It may therefore come as a surprise to learn that the *secret power source* of Cartan's Forms, which will drive our drama to its conclusion, is *anti*symmetry!

Let us waste no time in introducing you to the second-born of Cartan's Forms, the first to appear after the 1-forms:

> *A* **2-form** $\boldsymbol{\Psi}$ *is an antisymmetric tensor of valence* $\left\{\begin{smallmatrix}0\\2\end{smallmatrix}\right\}$. *That is,*
>
> $$\boldsymbol{\Psi}(\mathbf{v},\mathbf{u}) = -\boldsymbol{\Psi}(\mathbf{u},\mathbf{v}). \tag{34.1}$$

As a matter of simple housekeeping, let us immediately state the general definition of a p-form, knowing full well that this will be essentially *meaningless* to you in the absence of concrete examples:

> *A* **p-form** *(also called a differential form of* **degree** p*) is a* **completely antisymmetric** *tensor of valence* $\left\{\begin{smallmatrix}0\\p\end{smallmatrix}\right\}$, *meaning that swapping any two of the vector inputs reverses its sign.*

Naturally, if we take $p = 2$ then we recover our definition (34.1) of a 2-form.

Our plan now is to build up an intuitive, *geometrical* understanding of p-forms by gradually increasing p, one step at a time. Act V opened with an entire chapter devoted to 1-forms, and this chapter is likewise completely devoted to understanding 2-forms. After that, we shall turn to 3-forms, ...

This sounds like a recipe for a book of infinite length! Fortunately, by the time we arrive at 3-forms, all the essential ideas needed to understand general p-forms will be in play. Also, as we shall explain in the next chapter, the highest p-form that can exist in an n-manifold is an n-form, and since spacetime has only four dimensions, 4-forms will therefore suffice for the purposes of this book. (NOTE: Higher values of p are, however, required in some important applications of Forms, such as the ***symplectic manifolds*** that naturally arise in Hamiltonian mechanics; see Arnol'd (1989, Ch. 8).)

But why devote precious time and energy to such *antisymmetric* tensors in the first place? After all, with the one notable exception of the Riemann tensor, *every* important tensor we have encountered thus far has been *symmetric*!

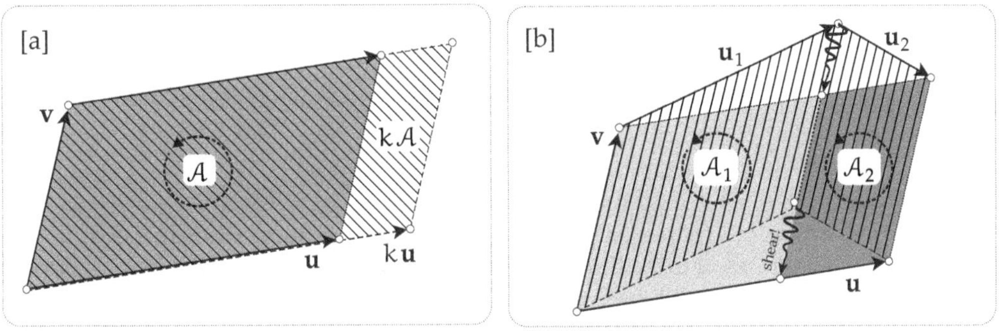

[34.1] Geometrical Proof That Oriented Area Is a 2-Form. **[a]** *Expanding an edge by* k *expands the area by* k*:* $\boldsymbol{\mathcal{A}}(\mathrm{k}\mathbf{u},\mathbf{v}) = \mathrm{k}\,\boldsymbol{\mathcal{A}}(\mathbf{u},\mathbf{v})$. **[b]** *Shearing* $\mathcal{A}_1 = \boldsymbol{\mathcal{A}}(\mathbf{u}_1,\mathbf{v})$ *and* $\mathcal{A}_2 = \boldsymbol{\mathcal{A}}(\mathbf{u}_2,\mathbf{v})$ *in the direction of* $\mathbf{v}$ *preserves their areas, so* $\mathcal{A} = \boldsymbol{\mathcal{A}}(\mathbf{u},\mathbf{v}) = \mathcal{A}_1 + \mathcal{A}_2$.

Only gradually will it become apparent the extent to which Élie Cartan was crazy like a fox. As was the case with 1-forms, so too with 2-forms: Cartan recognized that they had been hiding in plain sight!

34.2 Example: The Area 2-Form

In $\mathbb{R}^2$, let us define

$$\boldsymbol{\mathcal{A}}(\mathbf{u},\mathbf{v}) = \text{oriented area of the parallelogram with edges } \mathbf{u} \text{ and } \mathbf{v}.$$

Then $\boldsymbol{\mathcal{A}}$ is a 2-form!

First, it is immediately clear that $\boldsymbol{\mathcal{A}}(\mathbf{u},\mathbf{v})$ is antisymmetric: if we swap $\mathbf{u}$ and $\mathbf{v}$, then the magnitude of the area is unaltered, but the orientation of the parallelogram is reversed. It remains to verify that $\boldsymbol{\mathcal{A}}$ is a tensor, i.e., that it is *linear* in each slot. As we did with 1-forms, we may break down the linearity requirement into two parts: (32.2) and (32.3), applied to each slot.

The truth of (32.3) is explained by [34.1a], which illustrates the fact that if we expand either edge of a parallelogram by k, then its area is expanded by k, too:

$$\boldsymbol{\mathcal{A}}(\mathrm{k}\mathbf{u},\mathbf{v}) = \boldsymbol{\mathcal{A}}(\mathbf{u},\mathrm{k}\mathbf{v}) = \mathrm{k}\,\boldsymbol{\mathcal{A}}(\mathbf{u},\mathbf{v}).$$

Note that if $\boldsymbol{\Psi}$ is an arbitrary 2-form, then if (32.3) is true for one slot, it *must* be true for the other slot, too:

$$\boldsymbol{\Psi}(\mathrm{k}\mathbf{u},\mathbf{v}) = \mathrm{k}\,\boldsymbol{\Psi}(\mathbf{u},\mathbf{v}) \quad \Longrightarrow \quad \boldsymbol{\Psi}(\mathbf{u},\mathrm{k}\mathbf{v}) = -\boldsymbol{\Psi}(\mathrm{k}\mathbf{v},\mathbf{u}) = -\mathrm{k}\,\boldsymbol{\Psi}(\mathbf{v},\mathbf{u}) = \mathrm{k}\,\boldsymbol{\Psi}(\mathbf{u},\mathbf{v}).$$

The truth of (32.2) is less obvious, but the argument just given means that we need only prove it for the first slot—the truth for the second slot then follows from antisymmetry.

Let $\mathbf{u} = \mathbf{u}_1 + \mathbf{u}_2$, and define

$$\mathcal{A} = \boldsymbol{\mathcal{A}}(\mathbf{u},\mathbf{v}), \qquad \mathcal{A}_1 = \boldsymbol{\mathcal{A}}(\mathbf{u}_1,\mathbf{v}), \qquad \mathcal{A}_2 = \boldsymbol{\mathcal{A}}(\mathbf{u}_2,\mathbf{v}).$$

Then (32.2) requires that

$$\mathcal{A} = \mathcal{A}_1 + \mathcal{A}_2.$$

That this is indeed true is demonstrated geometrically in [34.1b]. Thus $\boldsymbol{\mathcal{A}}$ is indeed a 2-form.

Note that if $\boldsymbol{\Psi}$ is a general 2-form, then the antisymmetry under swapping of the two vector inputs implies that $\boldsymbol{\Psi}(\mathbf{u},\mathbf{u}) = -\boldsymbol{\Psi}(\mathbf{u},\mathbf{u})$, and so for arbitrary $\mathbf{u}$,

$$\boxed{\boldsymbol{\Psi}(\mathbf{u},\mathbf{u}) = 0.} \tag{34.2}$$

In the case of the area 2-form, this makes direct geometrical sense: if in $\mathcal{A}(\mathbf{u},\mathbf{v})$ we let $\mathbf{v}$ become $\mathbf{u}$, then the parallelogram collapses into a line segment of vanishing area, so $\mathcal{A}(\mathbf{u},\mathbf{u}) = 0$.

In fact, (34.2) is *equivalent* to our original antisymmetry-based definition of a 2-form, (34.1). To see this, we recycle an argument we previously employed in the context of the Riemann tensor. Let $\mathbf{x}$ and $\mathbf{y}$ be arbitrary vectors, and let $\mathbf{u} = \mathbf{x} + \mathbf{y}$. Then,

$$\begin{aligned} 0 &= \boldsymbol{\Psi}(\mathbf{u},\mathbf{u}) \\ &= \boldsymbol{\Psi}([\mathbf{x}+\mathbf{y}],\, [\mathbf{x}+\mathbf{y}]) \\ &= \boldsymbol{\Psi}(\mathbf{x},\mathbf{x}) + \boldsymbol{\Psi}(\mathbf{x},\mathbf{y}) + \boldsymbol{\Psi}(\mathbf{y},\mathbf{x}) + \boldsymbol{\Psi}(\mathbf{y},\mathbf{y}) \\ &= 0 + \boldsymbol{\Psi}(\mathbf{x},\mathbf{y}) + \boldsymbol{\Psi}(\mathbf{y},\mathbf{x}) + 0. \end{aligned}$$

Therefore, $\boldsymbol{\Psi}(\mathbf{y},\mathbf{x}) = -\boldsymbol{\Psi}(\mathbf{x},\mathbf{y})$. Done.

34.3 The Wedge Product of Two 1-Forms

Recall from (33.8) that any tensor of valence $\left\{ \begin{smallmatrix} 0 \\ 2 \end{smallmatrix} \right\}$ can be split into the sum of a symmetric tensor and an antisymmetric tensor. If we apply this procedure to the tensor product of two arbitrary 1-forms, then by swapping the 1-forms in the tensor products (instead of swapping the vectors) we find [exercise] that,

$$\boldsymbol{\varphi} \otimes \boldsymbol{\psi} = \frac{1}{2}[\boldsymbol{\varphi} \otimes \boldsymbol{\psi} + \boldsymbol{\psi} \otimes \boldsymbol{\varphi}] + \frac{1}{2}[\boldsymbol{\varphi} \otimes \boldsymbol{\psi} - \boldsymbol{\psi} \otimes \boldsymbol{\varphi}].$$

The antisymmetric part of this is, by definition, a *2-form* created by multiplying two 1-forms. This is a new kind of multiplication, called the ***wedge product***,[1] denoted $\wedge$:

$$\boxed{\boldsymbol{\varphi} \wedge \boldsymbol{\psi} \equiv \boldsymbol{\varphi} \otimes \boldsymbol{\psi} - \boldsymbol{\psi} \otimes \boldsymbol{\varphi}.} \tag{34.3}$$

Just as the tensor product allows us to systematically build tensors of higher valence from those of lower valence, so the wedge product allows us to build forms of higher degree from those of lower degree.

Let us stress that what makes $(\boldsymbol{\varphi} \wedge \boldsymbol{\psi})$ into a *2-form* is the fact that when it acts on a pair of vectors, it is linear in both slots, and *swapping the two vector inputs simply reverses the sign of the output, while keeping its magnitude fixed:*

$$(\boldsymbol{\varphi} \wedge \boldsymbol{\psi})(\mathbf{v}_1,\mathbf{v}_2) = \boldsymbol{\varphi}(\mathbf{v}_1)\boldsymbol{\psi}(\mathbf{v}_2) - \boldsymbol{\psi}(\mathbf{v}_1)\boldsymbol{\varphi}(\mathbf{v}_2) = -(\boldsymbol{\varphi} \wedge \boldsymbol{\psi})(\mathbf{v}_2,\mathbf{v}_1).$$

But observe that there is a *second* antisymmetry present, namely, that of the wedge product itself. Keeping the order of the vector inputs the same, but *swapping the 1-forms of the wedge product,* we see that

$$(\boldsymbol{\varphi} \wedge \boldsymbol{\psi})(\mathbf{v}_1,\mathbf{v}_2) = -(\boldsymbol{\psi} \wedge \boldsymbol{\varphi})(\mathbf{v}_1,\mathbf{v}_2).$$

[1] Also called the ***exterior product***.

Abstracting the vectors away, we have the antisymmetry of the wedge product itself, clearly evident in the definition in (34.3):

$$(\varphi \wedge \psi) = -(\psi \wedge \varphi).$$

Note that this implies, for arbitrary ψ,

$$\psi \wedge \psi = 0.$$

Also note that the wedge product distributes over addition:

$$\varphi \wedge (\psi + \sigma) = \varphi \wedge \psi + \varphi \wedge \sigma.$$

In $\mathbb{R}^2$, consider all possible wedge products of the Cartesian basis 1-forms. Since $\mathbf{d}x \wedge \mathbf{d}x = 0$ and $\mathbf{d}y \wedge \mathbf{d}y = 0$, the only nonvanishing wedge product is $\mathbf{d}x \wedge \mathbf{d}y = -\mathbf{d}y \wedge \mathbf{d}x$. And, in fact, this is none other than the *area 2-form*!

$$\mathcal{A} = \mathbf{d}x \wedge \mathbf{d}y. \tag{34.4}$$

Note that this again bears a striking resemblance to the classical expression for area: when we do a double integral, we write the element of area as $dx\, dy$. The explicit connection follows the same pattern as our earlier examples of the differential df and of the line element ds. If we take $\mathbf{u} = \delta x\, \mathbf{e}_1$ and $\mathbf{v} = \delta y\, \mathbf{e}_2$ to be the sides of a small rectangle, then

$$(\mathbf{d}x \wedge \mathbf{d}y)(\mathbf{u}, \mathbf{v}) = (\mathbf{d}x \wedge \mathbf{d}y)(\delta x\, \mathbf{e}_1, \delta y\, \mathbf{e}_2) = \delta x\, \delta y.$$

The general proof of (34.4) is easily achieved using components:

$$\begin{aligned}
(\mathbf{d}x \wedge \mathbf{d}y)(\mathbf{u}, \mathbf{v}) &= (\mathbf{d}x \otimes \mathbf{d}y - \mathbf{d}y \otimes \mathbf{d}x)\left(\begin{bmatrix} u^1 \\ u^2 \end{bmatrix}, \begin{bmatrix} v^1 \\ v^2 \end{bmatrix}\right) \\
&= u^1 v^2 - u^2 v^1 \\
&= \det \begin{bmatrix} u^1 & v^1 \\ u^2 & v^2 \end{bmatrix} \\
&= \mathcal{A}(\mathbf{u}, \mathbf{v}).
\end{aligned}$$

Let us now describe the geometrical meaning of the wedge product $(\varphi \wedge \psi)$ of two general 1-forms acting on vectors in $\mathbb{R}^n$. The argument will apply to any n, but to keep the idea vivid, let us illustrate it in the case of $\mathbb{R}^3$. In this case, both φ and ψ are represented by stacks of planes, and their action on a vector $\mathbf{v}$ in $\mathbb{R}^3$ is to count how many planes of each stack are pierced, as illustrated in [34.2].

Now comes the crucial idea: *let us use these two numbers $\varphi(\mathbf{v})$ and $\psi(\mathbf{v})$ as the x and y coordinates of a point in $\mathbb{R}^2$*. The two 1-forms thereby define a mapping F of vectors in $\mathbb{R}^3$ (or any $\mathbb{R}^n$) to vectors in $\mathbb{R}^2$:

$$\mathbf{v} \longrightarrow F(\mathbf{v}) \equiv \begin{bmatrix} \varphi(\mathbf{v}) \\ \psi(\mathbf{v}) \end{bmatrix} \tag{34.5}$$

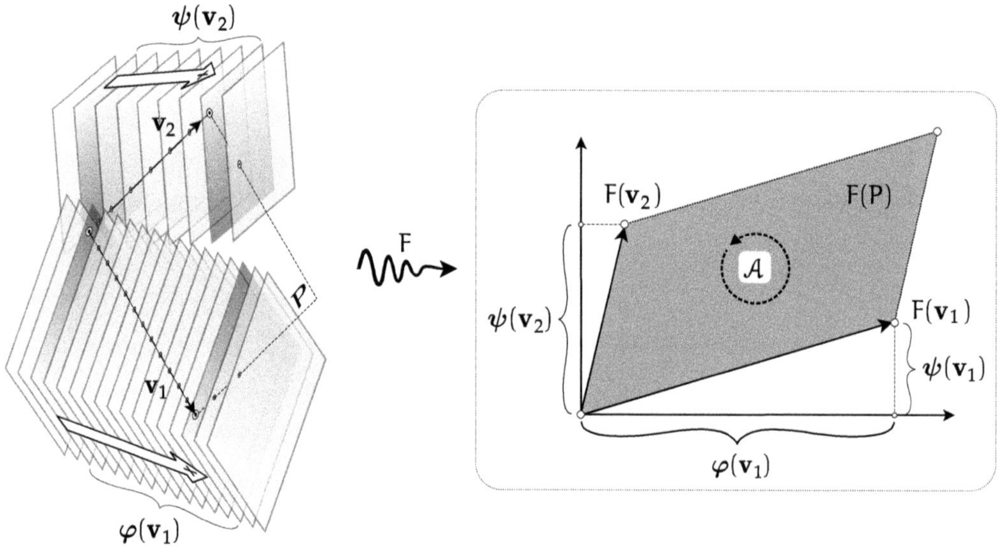

[34.2] Geometrical Meaning of the Wedge Product. *A parallelogram* P *in* $\mathbb{R}^3$ *has edges* $\mathbf{v}_1$ *and* $\mathbf{v}_2$, *and* $\varphi(\mathbf{v}_k)$ *and* $\psi(\mathbf{v}_k)$ *count how many planes of each stack are pierced by the edge* $\mathbf{v}_k$. *The mapping* F *fuses these pairs of piercing numbers* $(\varphi(\mathbf{v}_k), \psi(\mathbf{v}_k))$ *into Cartesian coordinates in* $\mathbb{R}^2$, *mapping* P *to the parallelogram* F(P). *Then* $(\varphi \wedge \psi)(\mathbf{v}_1, \mathbf{v}_2)$ *is the oriented area* $\mathcal{A}$ *of* F(P).

Under this mapping F, a parallelogram in $\mathbb{R}^n$ with edges $\mathbf{v}_1$ and $\mathbf{v}_2$ is mapped to a parallelogram in $\mathbb{R}^2$ with edges $F(\mathbf{v}_1)$ and $F(\mathbf{v}_2)$, as illustrated in [34.2]. We can now state the beautifully simple *meaning* of the wedge product:[2]

> *When the wedge product* $(\varphi \wedge \psi)$ *is applied to an arbitarary parallelogram in* $\mathbb{R}^n$, *it outputs the* oriented area of the image parallelogram in $\mathbb{R}^2$ *under the mapping* F:
>
> $$(\varphi \wedge \psi)(\mathbf{v}_1, \mathbf{v}_2) = \mathcal{A}\big[F(\mathbf{v}_1), F(\mathbf{v}_2)\big]. \tag{34.6}$$

The proof is very short:

$$\begin{aligned}(\varphi \wedge \psi)(\mathbf{v}_1, \mathbf{v}_2) &= \varphi(\mathbf{v}_1)\psi(\mathbf{v}_2) - \psi(\mathbf{v}_1)\varphi(\mathbf{v}_2) \\ &= \det\begin{bmatrix} \varphi(\mathbf{v}_1) & \varphi(\mathbf{v}_2) \\ \psi(\mathbf{v}_1) & \psi(\mathbf{v}_2) \end{bmatrix} \\ &= \mathcal{A}\big[F(\mathbf{v}_1), F(\mathbf{v}_2)\big].\end{aligned}$$

34.4 The Area 2-Form in Polar Coordinates

When we do a double integral in $\mathbb{R}^2$, we write the element of area as $d\mathcal{A} = dx\,dy$, and we have previously explained how this is connected to the area 2-form, (34.4): $\mathcal{A} = \mathbf{d}x \wedge \mathbf{d}y$. But when we switch to polar coordinates we instead write the element of area as $d\mathcal{A} = r\,dr\,d\theta$. In this case,

[2] Arnol'd (1989, §32) goes a step further, taking (34.6) to be the *definition* of the wedge product.

there is a simple geometric derivation of this formula, shown in [34.3]. The corresponding 2-form is therefore $\mathcal{A} = r\,\mathbf{d}r \wedge \mathbf{d}\theta$.

However, when we do a more complicated change of coordinates, the traditional method (which you undoubtedly learned in multivariable calculus) requires us to use the *Jacobian* to find the expression for the element of area in the new coordinate system. As we now demonstrate, using polar coordinates as our example, the use of the area 2-form allows us to jettison the Jacobian, and turns the entire affair into (mindless) algebraic child's play!

Applying the Leibnitz Rule (32.11), and remembering that $\mathbf{d}r \wedge \mathbf{d}r = 0$, and $\mathbf{d}\theta \wedge \mathbf{d}\theta = 0$, and $\mathbf{d}\theta \wedge \mathbf{d}r = -\mathbf{d}r \wedge \mathbf{d}\theta$, we find that

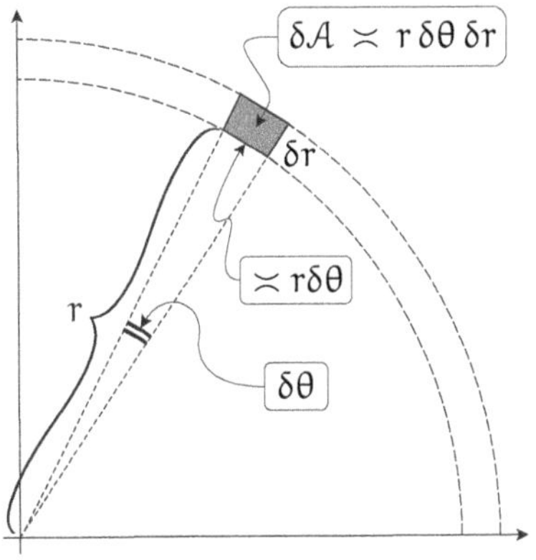

[34.3] *Geometric proof that* $\delta\mathcal{A} \asymp r\,\delta\theta\,\delta r$.

$$
\begin{aligned}
\mathbf{d}x \wedge \mathbf{d}y &= \mathbf{d}(r\cos\theta) \wedge \mathbf{d}(r\sin\theta) \\
&= [(\mathbf{d}r)\cos\theta - r\,\sin\theta\,\mathbf{d}\theta] \wedge [(\mathbf{d}r)\sin\theta + r\,\cos\theta\,\mathbf{d}\theta] \\
&= \cos^2\theta\, r\,\mathbf{d}r \wedge \mathbf{d}\theta - \sin^2\theta\, r\,\mathbf{d}\theta \wedge \mathbf{d}r \\
&= r\,\mathbf{d}r \wedge \mathbf{d}\theta.
\end{aligned}
$$

34.5 Basis 2-Forms and Projections

It is clear that the sum of two 2-forms is another 2-form, and likewise multiplication by a constant also yields another 2-form. Therefore, the 2-forms constitute a vector space, and so it is natural to seek a *basis* for this space.

Recall from (33.2) that the set of tensors $\{\mathbf{d}x^i \otimes \mathbf{d}x^j\}$ forms a basis for tensors of valence $\left\{ {0 \atop 2} \right\}$. This applies to *all* such tensors, including the 2-forms, but in the case of 2-forms we go a step further:

> *The set of 2-forms* $\{\mathbf{d}x^i \wedge \mathbf{d}x^j\}$, *with* $i<j$, *is a* **basis** *for the 2-forms.* (34.7)

The condition $i<j$ is imposed simply to avoid listing duplicate 2-forms: e.g., $(\mathbf{d}x^3 \wedge \mathbf{d}x^2) = -(\mathbf{d}x^2 \wedge \mathbf{d}x^3)$.

Since $\mathbf{d}x^i \wedge \mathbf{d}x^i = 0$, the nonvanishing (nonredundant) basis 2-forms result from picking an unordered pair of distinct $\mathbf{d}x^i$'s, (out of n) then forming their wedge product. So, if we assume (34.7) is true, then,

> *In* $\mathbb{R}^n$, *the set of 2-forms is a vector space of dimension* $\frac{1}{2}n(n-1)$. (34.8)

To confirm (34.7), let us begin in $\mathbb{R}^2$. If $\boldsymbol{\Psi}$ were merely a generic tensor of valence $\left\{ {0 \atop 2} \right\}$, then we know from (33.2) that it can be expanded as

$$\boldsymbol{\Psi} = \Psi_{11}(\mathbf{d}x \otimes \mathbf{d}x) + \Psi_{12}(\mathbf{d}x \otimes \mathbf{d}y) + \Psi_{21}(\mathbf{d}y \otimes \mathbf{d}x) + \Psi_{22}(\mathbf{d}y \otimes \mathbf{d}y).$$

But, because $\boldsymbol{\Psi}$ is a *2-form*

$$\Psi_{11}=0=\Psi_{22} \quad \text{and} \quad \Psi_{21}=-\Psi_{12}.$$

Therefore,

$$\boxed{\boldsymbol{\Psi}=\Psi_{12}\,(\mathbf{d}x\wedge\mathbf{d}y)=\Psi_{12}\,\boldsymbol{\mathcal{A}}.}$$

In other words, $(\mathbf{d}x\wedge\mathbf{d}y)$ is a *basis* for 2-forms in $\mathbb{R}^2$, and every 2-form is proportional to the area 2-form.

If we now go up one dimension to $\mathbb{R}^3$, then [exercise] exactly the same reasoning yields

$$\boxed{\boldsymbol{\Psi}=\Psi_{23}\,(\mathbf{d}y\wedge\mathbf{d}z)+\Psi_{31}\,(\mathbf{d}z\wedge\mathbf{d}x)+\Psi_{12}\,(\mathbf{d}x\wedge\mathbf{d}y),} \tag{34.9}$$

confirming (34.7) for this case. (NOTE: The reason for writing these three terms in the way that we have, and in the strange order that we have, will be explained in the next section.)

The proof in $\mathbb{R}^n$ is a straightforward extension of these examples. The nonvanishing terms in the expansion (33.2) always come in opposite pairs:

$$\Psi_{ij}\,(\mathbf{d}x^i\otimes\mathbf{d}x^j)+\Psi_{ji}\,(\mathbf{d}x^j\otimes\mathbf{d}x^i)=\Psi_{ij}\,(\mathbf{d}x^i\otimes\mathbf{d}x^j)-\Psi_{ij}\,(\mathbf{d}x^j\otimes\mathbf{d}x^i)=\Psi_{ij}\,(\mathbf{d}x^i\wedge\mathbf{d}x^j).$$

Let us return to $\mathbb{R}^3$ and seek out the *meaning* of the basis 2-forms that arise in (34.9), beginning with $(\mathbf{d}x\wedge\mathbf{d}y)$. In $\mathbb{R}^2$ this was simply the area 2-form in the (x, y)-plane, and the meaning in $\mathbb{R}^3$ is closely related, as illustrated in [34.4]:

> *If* $(\mathbf{d}x\wedge\mathbf{d}y)$ *is applied to a parallelogram* P *in* $\mathbb{R}^3$, *the output is the* oriented area $\mathcal{A}_z$ of the orthogonal projection *(in the z-direction)* of P onto the (x, y)-plane.

This is a direct consequence of the geometric interpretation (34.6) of the wedge product of two arbitrary 1-forms. The key is to realize that in this case the mapping F simply reduces to orthogonal projection in the z-direction onto the (x, y)-plane. Simply compare with the general construction of the wedge product in [34.2]: here $\varphi=\mathbf{d}x$ is a unit-spaced stack perpendicular to the x-axis, and likewise $\psi=\mathbf{d}y$ is a unit-spaced stack perpendicular to the y-axis. (Please take a moment to make sure that you can actually *see* all this.)

As illustrated in [34.4], the meaning of the other two basis 2-forms is completely analogous: $(\mathbf{d}y\wedge\mathbf{d}z)$ is the area $\mathcal{A}_x$ of the projection (in the x-direction) onto the (y, z)-plane; and $(\mathbf{d}z\wedge\mathbf{d}x)$ is the area $\mathcal{A}_y$ of the projection (in the y-direction) onto the (z, x)-plane.

In $\mathbb{R}^4$, the generalization of (34.9) instead contains *six* basis 2-forms, $(\mathbf{d}x^i\wedge\mathbf{d}x^j)$, but each one has the same meaning as above: for example, $(\mathbf{d}x^1\wedge\mathbf{d}x^3)$ yields the oriented area in the (x^1, x^3)-plane after projection along the x^2 and x^4 axes.

34.6 Associating 2-Forms with Vectors in $\mathbb{R}^3$: Flux

As we shall see, it is possible to do calculus with Forms in any number of dimensions. But the *Vector Calculus* we learn as undergraduates in college is *only* possible in *three* dimensions, and in this section and the next we begin to understand why.

Geometrically, there is certainly no confusion possible between 2-forms and vectors. Even algebraically the distinction is clear: in n dimensions, a vector has n components, whereas a 2-form has $\frac{1}{2}n(n-1)$ components, as we saw in (34.8).

But this implies that something remarkable happens when $n=3$: *In three dimensions, and* only *in three dimensions, 2-forms have the same number of components as vectors.*

Vector Calculus, so successfully pioneered by physicists in the 1880s, and still considered an indispensable tool of modern science in the twenty-first century, can now be seen to rest upon this singular *numerical coincidence*. Though the pioneers of Vector Calculus could not have had any inkling as to why their mathematical engine purred with such power, the *reason* was that it actually had *2-forms* hidden under the bonnet, but *masquerading* as vectors!

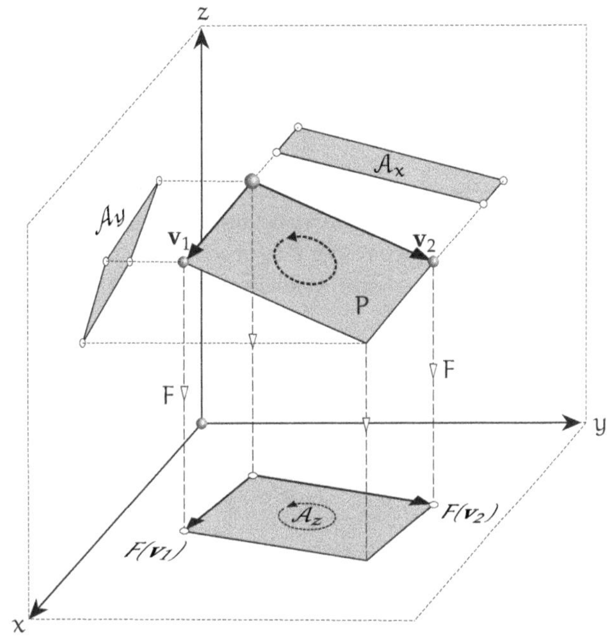

[34.4] Geometric Meaning of Basis 2-Forms. *Each basis 2-form yields the area of the projection of P onto the relevant coordinate plane. For example, $(dx \wedge dy)(v_1, v_2) = \mathcal{A}_z$, the area of the projection of P in the z-direction onto the (x, y)-plane.*

To begin to look under the bonnet, let us rewrite the components (34.9) of the general 2-form $\boldsymbol{\Psi}$ with a single raised index, and immediately identity these with components of a corresponding *vector*, which we shall denote by the same symbol as the 2-form, but ***underlined***: $\underline{\boldsymbol{\Psi}}$, i.e.,

$$\boldsymbol{\Psi} = \Psi^1\,(dx^2 \wedge dx^3) + \Psi^2\,(dx^3 \wedge dx^1) + \Psi^3\,(dx^1 \wedge dx^2) \quad \rightleftarrows \quad \underline{\boldsymbol{\Psi}} = \begin{bmatrix} \Psi^1 \\ \Psi^2 \\ \Psi^3 \end{bmatrix}. \tag{34.10}$$

The most natural mathematical way to understand this correspondence is via the ***Hodge star duality operator*** ($\star$), which is the subject of Exercise 15. It is a purely mathematical operation that (in n dimensions) maps a p-form to an $(n-p)$-form, called its ***Hodge dual*** (usually abbreviated simply to ***dual***), named after the important British mathematician, Sir W.V.D. Hodge (1903–1975).[3]

Here, however, we shall instead pursue a logical and compelling *physical* reason for associating a 2-form with a vector in this way, and to understand it we must now introduce the physicist's powerful concept of ***flux***.[4]

---

[3]Hodge was my mathematical grandfather: I studied under Penrose, who studied under Hodge.

[4]We shall explain the concept of flux using fluid flow, but it was originally introduced intuitively by Faraday to describe his ingenious experiments (beginning in the 1820s) on the flow of electric and magnetic fields through a surface in space; in the decades that followed, Maxwell gave the concept its current mathematical form. See Forbes and Mahon (2014, §10).

Reconsider [34.4], and imagine a uniform flow of fluid through space with velocity $\underline{\Psi}$. Now define,

$$\boldsymbol{\Phi}(\mathbf{v}_1, \mathbf{v}_2) \equiv \text{Amount of fluid crossing P per unit time} = \textbf{\textit{flux}} \text{ of } \underline{\Psi} \text{ through P.} \qquad (34.11)$$

We count the flux as *positive* if the orientation of P is *counterclockwise* around $\underline{\Psi}$. If we picture all the fluid that passes through P in unit time, it will fill a parallelepiped with edges $\mathbf{v}_1$, $\mathbf{v}_2$, $\underline{\Psi}$, and the flux will be its volume, as illustrated in [34.5]. It follows [exercise] that the flux $\boldsymbol{\Phi}$ is actually a *2-form.*

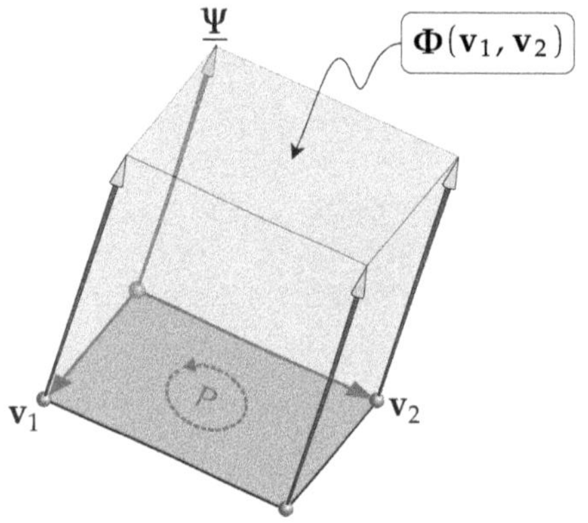

[34.5] Flux *of* $\underline{\Psi}$ *through* P = *volume* $\boldsymbol{\Phi}(\mathbf{v}_1, \mathbf{v}_2)$

It is natural to now ask the following question: *How is this flux 2-form* $\boldsymbol{\Phi}$ *related to the 2-form* $\boldsymbol{\Psi}$ *associated with the flow velocity* $\underline{\Psi}$*?*

We begin with a simple example. Suppose in [34.4] that $\underline{\Psi} = \Psi^3\,\mathbf{e}_3$ is purely upward, parallel to the z-axis. Clearly the flux through P is the same as the flux through its projection $\mathcal{A}_z = \mathcal{A}_3$. But if we picture the fluid flowing upward with speed Ψ^3 through $\mathcal{A}_3$ for one unit of time, then the fluid will fill a solid whose base area is $\mathcal{A}_3$ and whose height is Ψ^3, so the flux is $\Psi^3 \mathcal{A}_3$.

In the general case, the same reasoning applies to each component of $\underline{\Psi}$, so the total flux is simply the sum:

$$\begin{aligned}
\boldsymbol{\Phi}(\mathbf{v}_1, \mathbf{v}_2) &= \Psi^1 \mathcal{A}_1 + \Psi^2 \mathcal{A}_2 + \Psi^3 \mathcal{A}_3 \\
&= \Psi^1\,(\mathbf{dx}^2 \wedge \mathbf{dx}^3)(\mathbf{v}_1, \mathbf{v}_2) + \Psi^2\,(\mathbf{dx}^3 \wedge \mathbf{dx}^1)(\mathbf{v}_1, \mathbf{v}_2) + \Psi^3\,(\mathbf{dx}^1 \wedge \mathbf{dx}^2)(\mathbf{v}_1, \mathbf{v}_2) \\
&= \boldsymbol{\Psi}(\mathbf{v}_1, \mathbf{v}_2).
\end{aligned}$$

Therefore,

$$\boldsymbol{\Phi} = \boldsymbol{\Psi}\,!$$

Thus we have obtained a satisfying physical justification for associating a 2-form with a vector via (34.10):

> *If fluid flows through 3-dimensional space with velocity* $\underline{\Psi}$*, its flux 2-form is* $\boldsymbol{\Psi}$*, and vice versa.* (34.12)

In addition to using components to connect the 2-form $\boldsymbol{\Psi}$ to its associated vector $\underline{\Psi}$, as we initially did in (34.10), we may also identify the *direction* of the associated vector *geometrically*, instead. The flux of $\underline{\Psi}$ across the parallelogram P *vanishes* if and only if it contains the direction $\underline{\Psi}$, so *we can uniquely characterize the flow direction by means of vanishing flux:*

$$\boldsymbol{\Psi}(\underline{\Psi}, \ldots) = 0. \qquad (34.13)$$

34.7 Relation of the Vector and Wedge Products in $\mathbb{R}^3$

The conjuring trick that pulls Vector Calculus out of the top hat labelled "Forms" has a second component: we also associate *1-forms* with vectors, just as we previously did in Section 33.8. Given a 1-form φ, we will again denote the corresponding vector with same symbol as the 1-form, but ***underlined***:

$$\varphi = \varphi_1 \, \mathbf{dx}^1 + \varphi_2 \, \mathbf{dx}^2 + \varphi_3 \, \mathbf{dx}^3 \quad \rightleftarrows \quad \underline{\varphi} = \begin{bmatrix} \varphi_1 \\ \varphi_2 \\ \varphi_3 \end{bmatrix}. \tag{34.14}$$

The two fundamental kinds of multiplication in Vector Calculus are the ***scalar product*** (aka ***dot product***) and the ***vector product*** (aka ***cross product***). The former readily generalizes to any number of dimensions, and can be expressed as

$$\varphi(\mathbf{v}) = \underline{\varphi} \cdot \mathbf{v}. \tag{34.15}$$

But the vector product, illustrated in [34.6], is peculiar to three dimensions. Given two vectors $\underline{\varphi}$ and $\underline{\sigma}$, with the angle from the first to the second being θ, recall that we define the ***vector product*** $\underline{\varphi} \times \underline{\sigma}$ to be the vector orthogonal to both, its direction given by the right-hand rule, and its length being the area of the parallelogram they span:

$$|\underline{\varphi} \times \underline{\sigma}| \equiv \mathcal{A}(\underline{\varphi}, \underline{\sigma}) = |\underline{\varphi}| \, |\underline{\sigma}| \, \sin\theta.$$

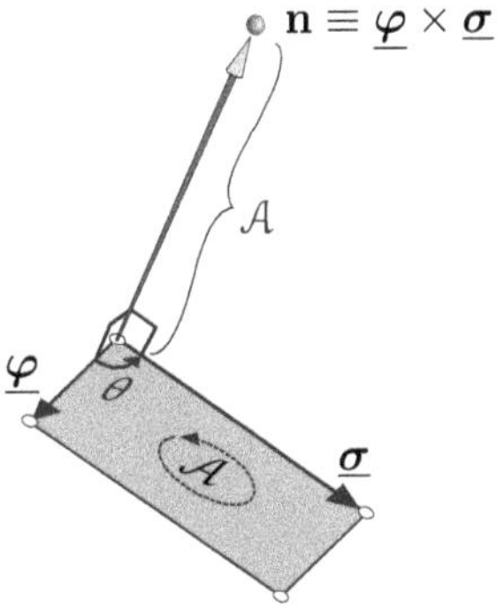

[34.6] The Vector Product.

We should immediately be suspicious: the *length* of a vector is equal to an *area*?! The reality is that we are dealing with a second sleight of hand: the vector product is not what it appears to be—it is actually the *wedge* product of the corresponding *1-forms*! The full explanation for this lies with the previously mentioned *Hodge duality*; see Exercise 15.

$$\underline{\varphi} \times \underline{\sigma} \quad \rightleftarrows \quad \varphi \wedge \sigma. \tag{34.16}$$

Before proving this correspondence, let us pause to recognize how special it is: in *any* number of dimensions, we are free to associate two vectors with two 1-forms, and we may then form their wedge product, but the resulting 2-form will have $\frac{1}{2}n(n-1)$ components, and so in general it *cannot* be reinterpreted as a vector. *Only* in three dimensions can a 2-form be identified as the flux 2-form $\boldsymbol{\Psi}$ of a vector flow $\underline{\boldsymbol{\Psi}}$.

In order to prove (34.16), let us define the 2-form $\boldsymbol{\Psi} \equiv \varphi \wedge \sigma$. Then we must show that $\underline{\boldsymbol{\Psi}} = \underline{\varphi} \times \underline{\sigma}$. We will prove this in two stages. First we will show that the vectors $\underline{\boldsymbol{\Psi}}$ and $\mathbf{n} \equiv \underline{\varphi} \times \underline{\sigma}$ point in the same direction, and then we will show that they have the same length.

Since $\mathbf{n} \equiv \underline{\varphi} \times \underline{\sigma}$ is, by definition, orthogonal to $\underline{\varphi}$ and $\underline{\sigma}$,

$$\varphi(\mathbf{n}) = \underline{\varphi} \bullet \mathbf{n} = 0 \qquad \text{and} \qquad \sigma(\mathbf{n}) = \underline{\sigma} \bullet \mathbf{n} = 0.$$

This then implies that the flux vanishes for any parallelogram containing $\mathbf{n}$:

$$\boldsymbol{\Psi}(\mathbf{n}, \ldots) = (\varphi \wedge \sigma)(\mathbf{n}, \ldots) = \varphi(\mathbf{n})\, \sigma(\ldots) - \sigma(\mathbf{n})\, \varphi(\ldots) = 0.$$

Thus, as discussed in (34.13), this implies that $\mathbf{n}$ is aligned with the flow: $\underline{\boldsymbol{\Psi}} \propto \mathbf{n}$.

It remains to show that the *speeds* of the two flows match: $|\underline{\boldsymbol{\Psi}}| = |\mathbf{n}| = \mathcal{A}(\underline{\varphi}, \underline{\sigma}) = |\underline{\varphi}|\, |\underline{\sigma}| \sin\theta$. We begin by noting that since we now know that $\underline{\boldsymbol{\Psi}}$ is orthogonal to the parallelogram P with edges $\underline{\varphi}$ and $\underline{\sigma}$, its flux across P is given by

$$\text{flux} = |\underline{\boldsymbol{\Psi}}|\, \mathcal{A}(\underline{\varphi}, \underline{\sigma}).$$

But,

$$\begin{aligned}
\text{flux} &= \boldsymbol{\Psi}(\underline{\varphi}, \underline{\sigma}) \\
&= (\varphi \wedge \sigma)(\underline{\varphi}, \underline{\sigma}) \\
&= \varphi(\underline{\varphi})\, \sigma(\underline{\sigma}) - \sigma(\underline{\varphi})\, \varphi(\underline{\sigma}) \\
&= |\underline{\varphi}|^2\, |\underline{\sigma}|^2 - |\underline{\varphi}|^2\, |\underline{\sigma}|^2 \cos^2\theta \\
&= [\mathcal{A}(\underline{\varphi}, \underline{\sigma})]^2.
\end{aligned}$$

Combining the previous two results,

$$|\underline{\boldsymbol{\Psi}}|\, \mathcal{A}(\underline{\varphi}, \underline{\sigma}) = \text{flux} = [\mathcal{A}(\underline{\varphi}, \underline{\sigma})]^2.$$

Therefore,

$$|\underline{\boldsymbol{\Psi}}| = \mathcal{A}(\underline{\varphi}, \underline{\sigma}) = |\underline{\varphi} \times \underline{\sigma}| \qquad \Longrightarrow \qquad \underline{\boldsymbol{\Psi}} = \underline{\varphi} \times \underline{\sigma},$$

proving (34.16).

Having established (34.16) geometrically, we may now use it to derive the standard algebraic expression for the vector product from the underlying formula for the wedge product:

$$\begin{aligned}
\boldsymbol{\Psi} &= \varphi \wedge \sigma \\
&= (\varphi_1\, \mathbf{d}x^1 + \varphi_2\, \mathbf{d}x^2 + \varphi_3\, \mathbf{d}x^3) \wedge (\sigma_1\, \mathbf{d}x^1 + \sigma_2\, \mathbf{d}x^2 + \sigma_3\, \mathbf{d}x^3) \\
&= (\varphi_2\, \sigma_3 - \varphi_3\, \sigma_2)(\mathbf{d}x^2 \wedge \mathbf{d}x^3) + (\varphi_3\, \sigma_1 - \varphi_1\, \sigma_3)(\mathbf{d}x^3 \wedge \mathbf{d}x^1) + (\varphi_1\, \sigma_2 - \varphi_2\, \sigma_1)(\mathbf{d}x^1 \wedge \mathbf{d}x^2).
\end{aligned}$$

The associated vector $\underline{\boldsymbol{\Psi}} = \underline{\varphi} \times \underline{\sigma}$ is therefore given by the familiar formula from Vector Calculus:

$$\begin{bmatrix} \varphi_1 \\ \varphi_2 \\ \varphi_3 \end{bmatrix} \times \begin{bmatrix} \sigma_1 \\ \sigma_2 \\ \sigma_3 \end{bmatrix} = \begin{bmatrix} \varphi_2\, \sigma_3 - \varphi_3\, \sigma_2 \\ \varphi_3\, \sigma_1 - \varphi_1\, \sigma_3 \\ \varphi_1\, \sigma_2 - \varphi_2\, \sigma_1 \end{bmatrix}.$$

Last, referring back to [34.5] and (34.12), we also deduce the standard formula for the volume of a parallelepiped with edges **u**, **v**, and $\underline{\mathbf{\Omega}}$:

$$\text{Volume} = \text{flux} = \mathbf{\Omega}(\mathbf{u},\mathbf{v}) = \begin{bmatrix} \Omega^1 \\ \Omega^2 \\ \Omega^3 \end{bmatrix} \cdot \begin{bmatrix} u^2 v^3 - u^3 v^2 \\ u^3 v^1 - u^1 v^3 \\ u^1 v^2 - u^2 v^1 \end{bmatrix} = \det \begin{bmatrix} \Omega^1 & u^1 & v^1 \\ \Omega^2 & u^2 & v^2 \\ \Omega^3 & u^3 & v^3 \end{bmatrix}. \tag{34.17}$$

34.8 The Faraday and Maxwell Electromagnetic 2-Forms

INTRODUCTORY NOTES: This section is entirely optional; it is not required to understand the mathematics that will follow (though it is a prerequisite for Section 36.6). That said, this section *is* very important, for it is our first demonstration that Forms are not only inhabitants of a Platonic, *mathematical* universe—they also reach out and weave their way into the very fabric of the laws of the *physical* Universe! One more thing: in order to ease the transition of the reader in passing from our treatment to more advanced works—such as Misner, Thorne, and Wheeler (1973), Schutz (1980), Baez and Muniain (1994), and Frankel (2012)—we have (reluctantly) adopted their convention, in which *the metric coefficients of spacetime are the negatives of those we have employed up to this point.* So, now and hereafter, the Minkowski metric is

$$ds^2 = -dt^2 + dx^2 + dy^2 + dz^2.$$

[34.7] *Michael Faraday (1791–1867).*

There are several 2-forms in 4-dimensional spacetime that play fundamental roles in the laws of physics. In this section we shall describe two of these that together lead to an extraordinarily compact and elegant description of the laws of electromagnetism (detailed in Section 36.6). These 2-forms are named after Michael Faraday (1791–1867) and James Clerk Maxwell (1831–1879), who are pictured in [34.7] and [34.8], respectively.

The laws of electromagnetism were first discovered empirically in a long, symphonic sequence of utterly ingenious experiments conducted by the self-educated Michael Faraday in his basement laboratory within the Royal Institution of London.[5] These experiments began in the 1820s, and reached their climax in 1830s. See Forbes and Mahon (2014, §§4–5).

[5]We *strongly* encourage you to visit the remains of Faraday's laboratory, where you may kneel down (reverentially!) to study the *first* electric generator—carefully preserved, mere inches away from you behind glass—ever constructed by the human race!

[34.8] *James Clerk Maxwell (1831–1879).*

Maxwell was quick to recognize the explanatory value of Faraday's ideas of lines of force and of flux, and, unlike Faraday, Maxwell possessed the mathematical skills to give them precise form. Indeed, not since Newton had the world witnessed such an alchemical fusion of profound physical insight and raw mathematical power.

By 1873 the job was done: Maxwell announced four[6] equations that completely described electricity, magnetism, and the *interaction* between the two. In particular, Maxwell used his equations to calculate the speed with which purely theoretical *waves* of electromagnetic energy would travel through empty space. His numerical answer coincided with the experimentally measured speed of light—*Maxwell was the first human to understand what light* is—*an electromagnetic wave!*

It is moving to read the letters[7] that passed between these two giants of science. One is struck by Maxwell's deep admiration of the profundity of Faraday's physical insights and his experimental ingenuity, and, symmetrically, Faraday's awe of Maxwell's extraordinary prowess in crystalizing physical intuitions into precise mathematical laws.

Einstein's 1905 discovery of Special Relativity was directly connected to these electromagnetic revelations of Faraday and Maxwell. Indeed, the title of Einstein's epoch-making paper *makes no mention of either space or time*—it reads, "On the Electrodynamics of Moving Bodies."[8]

The impetus for Einstein's breakthrough was his realization that electric and magnetic fields do not *separately* have an absolute existence—they are merely two, observer-dependent aspects of a single *electromagnetic* field, and waves of this field (light!) travel at the same speed for all observers. This *led* him (via Minkowski) to the realization that space and time cannot separately have an absolute existence—they are observer-dependent aspects of the geometry of *spacetime*, which does have absolute existence.

Let us simply quote the opening paragraph of Einstein's paper of 1905:

> It is known that Maxwell's electrodynamics—as usually understood at the present time—when applied to moving bodies, leads to asymmetries which do not appear to be inherent in the phenomena. Take, for example, the reciprocal electrodynamic action of a magnet and a conductor. The observable phenomenon here depends only on the relative motion of the conductor and the magnet, whereas the customary view draws a sharp distinction between the two cases in which either the one or the other of these bodies is in motion. For if the magnet is in motion and the conductor at rest, there arises in the neighbourhood of the magnet an electric field with a certain definite energy, producing a current at the places where parts of the conductor are situated. But if the magnet is stationary

---

[6]It was actually Oliver Heaviside who in 1885 reduced Maxwell's original set of 20 equations to the four we now know as "Maxwell's Equations"! See Forbes and Mahon (2014, §16) and Mahon (2017).

[7]See Forbes and Mahon (2014), and Jones (1870).

[8]The entire paper (translated into English) is freely available on the Internet: simply search for its title.

and the conductor in motion, no electric field arises in the neighbourhood of the magnet. In the conductor, however, we find an electromotive force, to which in itself there is no corresponding energy, but which gives rise—assuming equality of relative motion in the two cases discussed—to electric currents of the same path and intensity as those produced by the electric forces in the former case.

The classical description of these fields is as two 3-dimensional vector fields, with six components in all, each one being a function of space and of time:

$$\text{Electric field} = \underline{E} = \begin{bmatrix} E_x \\ E_y \\ E_z \end{bmatrix} \qquad \text{and} \qquad \text{Magnetic field} = \underline{\mathbf{B}} = \begin{bmatrix} B_x \\ B_y \\ B_z \end{bmatrix}.$$

Each of these may be associated both with a flux 2-form—via (34.10)—*and* with a 1-form, via (34.14). For the electric field we shall call the flux 2-form **E**, and the 1-form ε:

$$\boxed{\mathbf{E} = E_x\,(\mathbf{d}y \wedge \mathbf{d}z) + E_y\,(\mathbf{d}z \wedge \mathbf{d}x) + E_z\,(\mathbf{d}x \wedge \mathbf{d}y),} \tag{34.18}$$

and

$$\boxed{\varepsilon = E_x\,\mathbf{d}x + E_y\,\mathbf{d}y + E_z\,\mathbf{d}z.} \tag{34.19}$$

We do the same for the magnetic field, writing the flux 2-form as **B**, and the 1-form as β:

$$\boxed{\mathbf{B} = B_x\,(\mathbf{d}y \wedge \mathbf{d}z) + B_y\,(\mathbf{d}z \wedge \mathbf{d}x) + B_z\,(\mathbf{d}x \wedge \mathbf{d}y)} \tag{34.20}$$

and

$$\boxed{\beta = B_x\,\mathbf{d}x + B_y\,\mathbf{d}y + B_z\,\mathbf{d}z.} \tag{34.21}$$

As Einstein explained (see above), *none* of these 2-forms and 1-forms has an absolute, observer-independent existence. That is, observers in relative motion will disagree about the electric and magnetic fields, but they will *also* disagree about space and time in precisely such a way as to make physical phenomena independent of the observer's motion: there can be only *one* physical reality!

Very remarkably—and very beautifully!—*Nature informs us that the electric and magnetic 2-forms and 1-forms combine into a single* electromagnetic *2-form that* does *have absolute meaning in spacetime!*

It is called[9] the ***Faraday 2-form***, and is denoted **F**:

$$\boxed{\mathbf{F} = \textbf{Faraday} = \varepsilon \wedge \mathbf{d}t + \mathbf{B}.} \tag{34.22}$$

Let us immediately describe the physical (observer-independent) meaning of **F**.

[9]This is the terminology employed by Misner, Thorne, and Wheeler (1973), but it is not universal: for example, Penrose (2005) instead calls **F** the "Maxwell 2-form." As we shall explain in a moment, we reserve the name ***Maxwell 2-form*** for a *different* 2-form.

If we insert a vector $\mathbf{u}$ into the second slot of F then we obtain $\mathsf{F}(\ldots, \mathbf{u})$. This will yield a number when a vector is inserted into the first slot; therefore, as it stands, it must be a 1-form.

Suppose we place a charge q in the electromagnetic field described by F. Let the particle's 4-velocity vector in spacetime be $\mathbf{u}$, let its 4-momentum be described by the *1-form* $\boldsymbol{\pi}$, and let its proper time (aka "wrist-watch time") be τ. *Then* $q\,\mathsf{F}(\ldots, \mathbf{u})$ *describes the electromagnetic force exerted on the particle*

$$\boxed{\frac{d\pi}{d\tau} = q\,\mathsf{F}(\ldots, \mathbf{u}).} \tag{34.23}$$

This is a 1-form, awaiting the insertion of a vector, and (physically) the resulting number is the magnitude of the force acting in the direction of the inserted vector.

Contrast this with the more complicated formula—the ***Lorentz Force Law***—that one learns in introductory electrodynamics for the rate of change of the spatial vector momentum $\mathbf{p}$ of a particle that has spatial vector velocity $\mathbf{v}$:

$$\frac{d\mathbf{p}}{dt} = q\,(\underline{\mathsf{E}} + \mathbf{v} \times \underline{\mathbf{B}}).$$

In this classical formulation, *different observers in relative motion will all disagree about the values of* $\mathbf{p}$, $\underline{\mathsf{E}}$, $\mathbf{v}$, *and* $\underline{\mathbf{B}}$! (For a fuller discussion of this point, see Misner, Thorne, and Wheeler (1973, §3.3).) Contrast this with the elegance of the geometrical, observer-independent, spacetime law, (34.23)!

Before we introduce the Maxwell 2-form, let us state ***Maxwell's Equations*** of electromagnetism, which we shall divide into two pairs of equations.

The first pair describes the electromagnetic field in the absence of *sources* (meaning the (scalar) electric charge density ρ, and the (vector) current density $\underline{\mathbf{j}}$):

$$\boxed{\textit{Maxwell's Source-Free Equations:} \qquad \begin{aligned} \nabla \cdot \underline{\mathbf{B}} &= 0, \\ \nabla \times \underline{\mathsf{E}} + \partial_t \underline{\mathbf{B}} &= 0. \end{aligned}} \tag{34.24}$$

As we shall see in Section 36.6, when expressed in terms of the Faraday 2-form, these two equations reduce to a *single*, elegant equation.

The second pair of Maxwell's equations describes the electromagnetic field generated by sources:

$$\boxed{\textit{Maxwell's Source Equations:} \qquad \begin{aligned} \nabla \cdot \underline{\mathsf{E}} &= 4\pi\rho, \\ \nabla \times \underline{\mathbf{B}} - \partial_t \underline{\mathsf{E}} &= 4\pi\underline{\mathbf{j}}. \end{aligned}} \tag{34.25}$$

This second pair *also* reduces to a single, elegant equation—to be derived in Section 36.6—but this time in terms of a different 2-form that is closely to related to F, and is denoted $\star\mathsf{F}$. This is the ***Maxwell 2-form***[10], which is *almost* (but not quite) the result of interchanging the roles of the electric and magnetic fields in the Faraday 2-form:

[10] As noted in footnote 9, this terminology is not universal: Penrose (2005) instead calls F the ***Maxwell 2-form***.

$$\star F = \textbf{Maxwell} = \beta \wedge dt - E. \tag{34.26}$$

The star operator ($\star$) is the previously mentioned *Hodge duality operator*; see Exercise 15. It is a purely mathematical operation that (in n dimensions) maps a p-form to an $(n-p)$-form. Since we are in 4 dimensions, $\star$ maps the 2-form F to another 2-form $\star F$, the ***(Hodge) dual of*** F. If we apply the star operator a second time, in other words take the dual of Maxwell, then we recover Faraday (only with a minus):

$$\star\star F = -F. \tag{34.27}$$

Thus we may say that Maxwell and Faraday are the duals of one another—a nomenclature that we suspect both scientists would have relished, if only they had lived to see it.

Chapter 35

3-Forms

35.1 A 3-Form Requires Three Dimensions

Let us begin by recognizing that 3-forms require at least three dimensions in order to exist.

In $\mathbb{R}^2$, it is easy enough to create a tensor $\mathbf{H}$ of valence $\left\{ \begin{smallmatrix} 0 \\ 3 \end{smallmatrix} \right\}$. For example, take any 1-form φ, and construct,

$$\mathbf{H} = \varphi \otimes \varphi \otimes \varphi \qquad \Longrightarrow \qquad \mathbf{H}(\mathbf{v}_1, \mathbf{v}_2, \mathbf{v}_3) = \varphi(\mathbf{v}_1)\, \varphi(\mathbf{v}_2)\, \varphi(\mathbf{v}_3).$$

As we see, this tensor is *totally symmetric*, meaning that swapping any pair of input vectors returns the same output.

But to create a *3-form*, Ξ, we require (by definition) that our tensor be *totally antisymmetric*, and this is impossible! Consider the components of Ξ:

$$\Xi_{ijk} = \Xi\,(\mathbf{e}_i, \mathbf{e}_j, \mathbf{e}_k).$$

In $\mathbb{R}^2$ there are only two basis vectors, $\{\mathbf{e}_1, \mathbf{e}_2\}$, so that means two of the slots of Ξ contain the *same* vector, but then antisymmetry demands that the output vanishes.

In order for Ξ_{ijk} not to vanish identically, it is therefore essential that all three indices be distinct: in other words, we must be in *at least* three dimensions. Clearly, exactly the same reasoning shows, quite generally, that the existence of a p-form requires at least p dimensions.

We can also turn this around and observe that *the highest degree form that can exist in n dimensions is an n-form.*

35.2 The Wedge Product of a 2-Form and 1-Form

The wedge product of two 1-forms yielded a 2-form, and we now seek to extend the definition of the wedge product so as to ensure that multiplication of a 2-form Ψ and a 1-form σ automatically yields a *3-form*, $\Psi \wedge \sigma$.

Let us begin with the simplest possible (ill-fated) guess: how about

$$(\Psi \wedge \sigma)(\mathbf{v}_1, \mathbf{v}_2, \mathbf{v}_3) \overset{???}{=} \Psi(\mathbf{v}_1, \mathbf{v}_2)\, \sigma(\mathbf{v}_3)\,?$$

Well, this is clearly antisymmetric under $\mathbf{v}_1 \leftrightarrow \mathbf{v}_2$, but not under $\mathbf{v}_2 \leftrightarrow \mathbf{v}_3$, so let us *impose* antisymmetry, using the same trick we used in (33.8). That is, let us subtract the same expression with $\mathbf{v}_2 \leftrightarrow \mathbf{v}_3$:

$$(\Psi \wedge \sigma)(\mathbf{v}_1, \mathbf{v}_2, \mathbf{v}_3) \overset{???}{=} \Psi(\mathbf{v}_1, \mathbf{v}_2)\, \sigma(\mathbf{v}_3) - \Psi(\mathbf{v}_1, \mathbf{v}_3)\, \sigma(\mathbf{v}_2)\,?$$

But now the second term has ruined antisymmetry under $\mathbf{v}_1 \leftrightarrow \mathbf{v}_2$, so repeat the trick and subtract that same term, but with $\mathbf{v}_1 \leftrightarrow \mathbf{v}_2$:

$$(\Psi \wedge \sigma)(\mathbf{v}_1, \mathbf{v}_2, \mathbf{v}_3) \overset{???}{=} \Psi(\mathbf{v}_1, \mathbf{v}_2)\, \sigma(\mathbf{v}_3) - \left[\Psi(\mathbf{v}_1, \mathbf{v}_3)\, \sigma(\mathbf{v}_2) - \Psi(\mathbf{v}_2, \mathbf{v}_3)\, \sigma(\mathbf{v}_1)\right]\,?$$

Success!

$$(\boldsymbol{\Psi}\wedge\boldsymbol{\sigma})(\mathbf{v}_1,\mathbf{v}_2,\mathbf{v}_3)=\boldsymbol{\Psi}(\mathbf{v}_1,\mathbf{v}_2)\,\boldsymbol{\sigma}(\mathbf{v}_3)+\boldsymbol{\Psi}(\mathbf{v}_3,\mathbf{v}_1)\,\boldsymbol{\sigma}(\mathbf{v}_2)+\boldsymbol{\Psi}(\mathbf{v}_2,\mathbf{v}_3)\,\boldsymbol{\sigma}(\mathbf{v}_1). \tag{35.1}$$

Note the cyclic permutation of the vectors, which makes the formula easy to remember.

We can play the same game to define a seemingly different 3-form, $(\boldsymbol{\sigma}\wedge\boldsymbol{\Psi})$, but this turns out to be the *same* 3-form, $(\boldsymbol{\Psi}\wedge\boldsymbol{\sigma})$! To see why, simply move each $\boldsymbol{\sigma}(\mathbf{v}_i)$ to the front of each term on the right-hand side of (35.1), and observe that this preserves the cyclic permutation:

$$(\boldsymbol{\sigma}\wedge\boldsymbol{\Psi})(\mathbf{v}_1,\mathbf{v}_2,\mathbf{v}_3)=\boldsymbol{\sigma}(\mathbf{v}_1)\,\boldsymbol{\Psi}(\mathbf{v}_2,\mathbf{v}_3)+\boldsymbol{\sigma}(\mathbf{v}_2)\,\boldsymbol{\Psi}(\mathbf{v}_3,\mathbf{v}_1)+\boldsymbol{\sigma}(\mathbf{v}_3)\,\boldsymbol{\Psi}(\mathbf{v}_1,\mathbf{v}_2). \tag{35.2}$$

Therefore, although the wedge product is *antisymmetric when it combines two 1-forms*, i.e.,

$$(\boldsymbol{\varphi}\wedge\boldsymbol{\psi})=-(\boldsymbol{\psi}\wedge\boldsymbol{\varphi}),$$

it is *symmetric when it combines a 2-form* $\boldsymbol{\Psi}$ *and a 1-form* $\boldsymbol{\sigma}$:

$$\boldsymbol{\Psi}\wedge\boldsymbol{\sigma}=\boldsymbol{\sigma}\wedge\boldsymbol{\Psi}.$$

Here is the general rule, which we will justify shortly:

$$\textit{If } \boldsymbol{\Psi} \textit{ is a p-form and } \boldsymbol{\Omega} \textit{ is a q-form, then } \boldsymbol{\Psi}\wedge\boldsymbol{\Omega}=(-1)^{pq}\,\boldsymbol{\Omega}\wedge\boldsymbol{\Psi}. \tag{35.3}$$

You may easily check that the previous two equations conform to this rule.

Assuming (35.3) to be true, we also deduce that if the degree of $\boldsymbol{\Omega}$ is *odd*, then [exercise],

$$\boldsymbol{\Omega}\wedge\boldsymbol{\Omega}=0. \tag{35.4}$$

Of course the wedge product of a 1-form with itself was our first instance of this phenomenon: $\boldsymbol{\psi}\wedge\boldsymbol{\psi}=0$.

35.3 The Volume 3-Form

Now let us take $\boldsymbol{\Psi}=\mathbf{d}x^1\wedge\mathbf{d}x^2$ and $\boldsymbol{\sigma}=\mathbf{d}x^3$, and define the ***volume 3-form*** to be

$$\boldsymbol{\mathcal{V}}\equiv(\mathbf{d}x^1\wedge\mathbf{d}x^2)\wedge\mathbf{d}x^3. \tag{35.5}$$

We now justify the naming of $\boldsymbol{\mathcal{V}}$.

Look again at the standard formula (34.17) for the volume of a parallelepiped with edges **u**, **v**, and $\underline{\boldsymbol{\Omega}}$. If we apply our 3-form $\mathcal{V}$ to these three vectors, then (35.1) yields

$$\begin{aligned}\mathcal{V}(\mathbf{u},\mathbf{v},\underline{\boldsymbol{\Omega}}) &= [(\mathbf{dx}^1 \wedge \mathbf{dx}^2) \wedge \mathbf{dx}^3](\mathbf{u},\mathbf{v},\underline{\boldsymbol{\Omega}}) \\ &= (\mathbf{dx}^1 \wedge \mathbf{dx}^2)(\mathbf{u},\mathbf{v})\,\mathbf{dx}^3(\underline{\boldsymbol{\Omega}}) + (\mathbf{dx}^1 \wedge \mathbf{dx}^2)(\mathbf{v},\underline{\boldsymbol{\Omega}})\,\mathbf{dx}^3(\mathbf{u}) + (\mathbf{dx}^1 \wedge \mathbf{dx}^2)(\underline{\boldsymbol{\Omega}},\mathbf{u})\,\mathbf{dx}^3(\mathbf{v}) \\ &= (u^1 v^2 - u^2 v^1)\,\Omega^3 + (v^1 \Omega^2 - v^2 \Omega^1)\,u^3 + (\Omega^1 u^2 - \Omega^2 u^1)\,v^3 \\ &= \Omega^1(u^2 v^3 - u^3 v^2) + \Omega^2(u^3 v^1 - u^1 v^3) + \Omega^3(u^1 v^2 - u^2 v^1) \\ &= \underline{\boldsymbol{\Omega}}(\mathbf{u},\mathbf{v}) \\ &= \text{Volume of the parallelepiped with edges } \mathbf{u}, \mathbf{v}, \text{ and } \underline{\boldsymbol{\Omega}}.\end{aligned}$$

We shall see that the wedge product of these three 1-forms is *associative*, so the volume 3-form can be expressed as

$$(\mathbf{dx}^1 \wedge \mathbf{dx}^2) \wedge \mathbf{dx}^3 = \mathbf{dx}^1 \wedge \mathbf{dx}^2 \wedge \mathbf{dx}^3 = \mathbf{dx}^1 \wedge (\mathbf{dx}^2 \wedge \mathbf{dx}^3).$$

35.4 The Volume 3-Form in Spherical Polar Coordinates

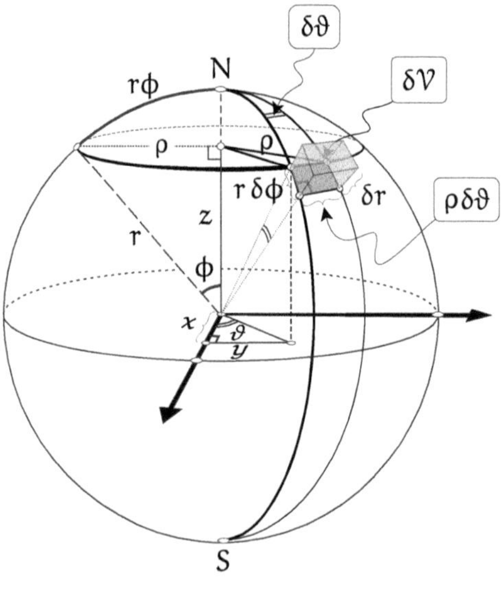

[35.1] *Geometric proof that* $\delta\mathcal{V} \asymp \delta r\ (r\,\delta\phi)\ (\rho\,\delta\vartheta) = r^2 \sin\phi\ \delta r\ \delta\phi\ \delta\vartheta$*, and derivation of formulas for* x*,* y*, and* z*.*

Once again, (35.5) reminds us of a classical expression: when doing a triple integral in space, we write the element of volume as $d\mathcal{V} = dx\, dy\, dz$. This connection can easily be made explicit, in the same way as before: applying $\mathbf{dx} \wedge \mathbf{dy} \wedge \mathbf{dz}$ to the vector edges of a small box with sides δx, δy, δz, we obtain its volume, $\delta\mathcal{V} = \delta x\, \delta y\, \delta z$.

If we switch to spherical polar coordinates, (r, ϕ, ϑ), then the classical element of volume becomes,

$$d\mathcal{V} = r^2 \sin\phi\ dr\ d\phi\ d\vartheta.$$

Figure [35.1] contains a direct, geometrical proof, enabling us to *understand* this result. Note that this figure also provides the *formulas* for the coordinate change: since the distance of a point on the sphere from the NS, z-axis is $\rho = r \sin\phi$,

$$\begin{aligned} x &= \rho\cos\vartheta = (r\sin\phi)\cos\vartheta, \\ y &= \rho\sin\vartheta = (r\sin\phi)\sin\vartheta, \\ z &= r\cos\phi. \end{aligned}$$

Finally, note that the *order* (r, ϕ, ϑ) of the coordinates[1] has been chosen so that the movements resulting from small increases in the coordinates are *right handed*. Make sure you can see this.

[1] We are using the (logical!) convention of American mathematicians, whereby ϑ denotes the angle within the (x, y)-plane, *just as it does in* $\mathbb{R}^2$. However, British mathematicians, and essentially *all* scientists, interchange the meanings of ϑ and ϕ, so *their* order of the coordinates, and their volume 3-form, will appear to be different from ours. There are actually compelling historical reasons for adopting the physicists' choice—spherical harmonics immediately come to mind—and the conflict could be resolved if only we could *all* agree to change the angle in $\mathbb{R}^2$ to ϕ! Dray (2015) is a rare, brave adopter of this solution.

In the case of a more complicated change of coordinates, we would traditionally be *forced* to calculate the Jacobian. Well, no longer!

Let us use the example of spherical polar coordinates to demonstrate how (in more difficult cases) the volume 3-form saves us time, memorization, and even the need to *think!*

Well, a *little* thinking is certainly helpful: when multiplying out the wedge products below, we are *only* looking for ways to obtain $\mathbf{d}r \wedge \mathbf{d}\phi \wedge \mathbf{d}\vartheta$, because all other terms vanish. Towards the end of the calculation, we write $c_{\vartheta} \equiv \cos\vartheta$, and so forth, to save space, but we also recommend this as a clarifying notation in private, pencil-and-paper calculations.

$$\begin{aligned}
\mathbf{d}x \wedge \mathbf{d}y \wedge \mathbf{d}z &= \mathbf{d}(r \sin\phi \cos\vartheta) \wedge \mathbf{d}(r \sin\phi \sin\vartheta) \wedge \mathbf{d}(r \cos\phi) \\
&= [(\mathbf{d}r) \sin\phi \cos\vartheta + r \cos\phi \, \mathbf{d}\phi \cos\vartheta - r \sin\phi \sin\vartheta \, \mathbf{d}\vartheta] \\
&\wedge \; [(\mathbf{d}r) \sin\phi \sin\vartheta + r \cos\phi \, \mathbf{d}\phi \sin\vartheta + r \sin\phi \cos\vartheta \, \mathbf{d}\vartheta] \\
&\wedge \; [(\mathbf{d}r) \cos\phi - r \sin\phi \, \mathbf{d}\phi] \\
&= r^2 \sin\phi \left[s_{\phi}^2 c_{\vartheta}^2 + s_{\phi}^2 s_{\vartheta}^2 + c_{\phi}^2 s_{\vartheta}^2 + c_{\phi}^2 c_{\vartheta}^2 \right] \mathbf{d}r \wedge \mathbf{d}\phi \wedge \mathbf{d}\vartheta \\
&= r^2 \sin\phi \, \mathbf{d}r \wedge \mathbf{d}\phi \wedge \mathbf{d}\vartheta.
\end{aligned}$$

Thinking geometrically enabled us to *understand* this result, while the "Devil's machine" only gave us the *answer*, and without any need for thought. Heed the warning that opened the Prologue: "...the danger to our soul is there, because when you pass over into algebraic calculation, essentially you stop thinking: you stop thinking geometrically, you stop thinking about the meaning."

These are but the first glimpses of the power of the "Devil's machine." You must therefore remain resolute, and on constant guard, lest you succumb to its temptations and thereby lose your soul!

35.5 The Wedge Product of Three 1-Forms and of p 1-Forms

To confirm the associativity we have just claimed, let us generalize the discussion, and define the wedge product of three *arbitrary* 1-forms. We wish to label the three 1-forms systematically, but we also wish to avoid any possible confusion with *components*, so let us call them $\underset{1}{\sigma}, \underset{2}{\sigma}, \underset{3}{\sigma}$.

The formula for $\underset{1}{\sigma} \wedge \underset{2}{\sigma} \wedge \underset{3}{\sigma}$ can be found by taking $\Psi = \underset{1}{\sigma} \wedge \underset{2}{\sigma}$ in (35.1), but it is simpler to generalize the geometrical mapping (34.5) and the resulting geometrical interpretation given in [34.2].

That is, let us use the three 1-forms to define a mapping F of a vector $\mathbf{v}$ in $\mathbb{R}^n$ to a vector $F(\mathbf{v})$ in $\mathbb{R}^3$:

$$\boxed{\mathbf{v} \longrightarrow F(\mathbf{v}) \equiv \begin{bmatrix} \underset{1}{\sigma}(\mathbf{v}) \\ \underset{2}{\sigma}(\mathbf{v}) \\ \underset{3}{\sigma}(\mathbf{v}) \end{bmatrix}} \tag{35.6}$$

Thus a parallelepiped in $\mathbb{R}^n$ with edges $\mathbf{v}_1$, $\mathbf{v}_2$, and $\mathbf{v}_3$, is mapped by F to a parallelepiped in $\mathbb{R}^3$ with edges $F(\mathbf{v}_1)$, $F(\mathbf{v}_2)$, and $F(\mathbf{v}_3)$, and, by analogy with [34.2], we can define $\underset{1}{\sigma} \wedge \underset{2}{\sigma} \wedge \underset{3}{\sigma}$ as the operator that yields the *volume* of this image parallelepiped:

$$\left[\underset{1}{\sigma}\wedge\underset{2}{\sigma}\wedge\underset{3}{\sigma}\right](\mathbf{v}_1,\mathbf{v}_2,\mathbf{v}_3)=\det\begin{bmatrix} | & | & | \\ F(\mathbf{v}_1) & F(\mathbf{v}_2) & F(\mathbf{v}_3) \\ | & | & | \end{bmatrix}. \tag{35.7}$$

Clearly, this can be immediately generalized to define the wedge product of p 1-forms:

$$\left[\underset{1}{\sigma}\wedge\cdots\wedge\underset{p}{\sigma}\right](\mathbf{v}_1,\cdots,\mathbf{v}_p)=\det\begin{bmatrix} | & \cdots & | \\ F(\mathbf{v}_1) & \cdots & F(\mathbf{v}_p) \\ | & \cdots & | \end{bmatrix}.$$

Equation (35.7) describes the *action* of a 3-form on three vectors, but can we abstract away the vectors to reveal the 3-form itself? Yes, but to do so, we need to deal directly with tensor products, just as we did in the original definition (34.3) of the wedge product of two 1-forms.

To achieve our goal, we must try to write down an expression that is antisymmetric under exchange of any pair of 1-forms. It is possible to do this [exercise] by working backwards from (35.7), but it is actually simpler to attack the problem head on. Try this yourself, *before* studying the answer, which is *here*:

$$\begin{aligned}\underset{1}{\sigma}\wedge\underset{2}{\sigma}\wedge\underset{3}{\sigma} &= \underset{1}{\sigma}\otimes\underset{2}{\sigma}\otimes\underset{3}{\sigma}+\underset{2}{\sigma}\otimes\underset{3}{\sigma}\otimes\underset{1}{\sigma}+\underset{3}{\sigma}\otimes\underset{1}{\sigma}\otimes\underset{2}{\sigma}\\ &\quad -\underset{2}{\sigma}\otimes\underset{1}{\sigma}\otimes\underset{3}{\sigma}-\underset{3}{\sigma}\otimes\underset{2}{\sigma}\otimes\underset{1}{\sigma}-\underset{1}{\sigma}\otimes\underset{3}{\sigma}\otimes\underset{2}{\sigma}.\end{aligned} \tag{35.8}$$

Note that the positive terms on the top line are the cyclic permutations of 123, whereas the negative terms on the second line are the noncyclic permutations. Each term on the second line is obtained by swapping the first two 1-forms in the cyclic term directly above it.

We leave it you to check that (35.8) is indeed equivalent to (35.7).

35.6 Basis 3-Forms

Just as $\{\mathbf{dx}^i\wedge\mathbf{dx}^j\}$ formed a basis for 2-forms, so $\{\mathbf{dx}^i\wedge\mathbf{dx}^j\wedge\mathbf{dx}^k\}$ forms a basis for 3-forms. In $\mathbb{R}^3$ there is only one such basis 3-form, namely, the volume 3-form, $\boldsymbol{\mathcal{V}}=\mathbf{dx}^1\wedge\mathbf{dx}^2\wedge\mathbf{dx}^3$; all 3-forms in $\mathbb{R}^3$ are simply multiples of $\boldsymbol{\mathcal{V}}$.

In $\mathbb{R}^4$, however, there are four distinct basis 3-forms, and a general 3-form $\boldsymbol{\Omega}$ can be decomposed into components as follows:

$$\begin{aligned}\boldsymbol{\Omega} &= \Omega_{123}\,(\mathbf{dx}^1\wedge\mathbf{dx}^2\wedge\mathbf{dx}^3)+\Omega_{234}\,(\mathbf{dx}^2\wedge\mathbf{dx}^3\wedge\mathbf{dx}^4)+\Omega_{341}\,(\mathbf{dx}^3\wedge\mathbf{dx}^4\wedge\mathbf{dx}^1)\\ &\quad+\Omega_{412}\,(\mathbf{dx}^4\wedge\mathbf{dx}^1\wedge\mathbf{dx}^2).\end{aligned}$$

Let us now use such components to explain the rule (35.3). To make the argument vivid, let us take $p=2$ and $q=3$, so that $\boldsymbol{\Psi}$ is a 2-form and $\boldsymbol{\Omega}$ is a 3-form. When we form $\boldsymbol{\Psi}\wedge\boldsymbol{\Omega}$, a typical component $(\mathbf{dx}^{j_1}\wedge\mathbf{dx}^{j_2})$ of $\boldsymbol{\Psi}$ is multiplied by a typical component $(\mathbf{dx}^{k_1}\wedge\mathbf{dx}^{k_2}\wedge\mathbf{dx}^{k_3})$

of $\boldsymbol{\Omega}$. (We assume that we are in at least five dimensions, otherwise this product vanishes identically!)

Now we wish to transform this into a term of the *reversed* product, $\boldsymbol{\Omega} \wedge \boldsymbol{\Psi}$. To do so, we must begin by moving $\mathbf{d}x^{j_2}$ through the three 1-forms to its right, with each successive swap producing a change of sign:

$$\begin{aligned}
(\mathbf{d}x^{j_1} \wedge \mathbf{d}x^{j_2}) \wedge (\mathbf{d}x^{k_1} \wedge \mathbf{d}x^{k_2} \wedge \mathbf{d}x^{k_3}) &= -\mathbf{d}x^{j_1} \wedge \mathbf{d}x^{k_1} \wedge \mathbf{d}x^{j_2} \wedge \mathbf{d}x^{k_2} \wedge \mathbf{d}x^{k_3} \\
&= +\mathbf{d}x^{j_1} \wedge \mathbf{d}x^{k_1} \wedge \mathbf{d}x^{k_2} \wedge \mathbf{d}x^{j_2} \wedge \mathbf{d}x^{k_3} \\
&= -\mathbf{d}x^{j_1} \wedge (\mathbf{d}x^{k_1} \wedge \mathbf{d}x^{k_2} \wedge \mathbf{d}x^{k_2}) \wedge \mathbf{d}x^{j_2}.
\end{aligned}$$

With $q = 3$, this has resulted in a factor of $(-1)^3 = -1$, and in the general case the factor would be $(-1)^q$. To complete the reversal of $\boldsymbol{\Psi} \wedge \boldsymbol{\Omega}$ to create $\boldsymbol{\Omega} \wedge \boldsymbol{\Psi}$, we must now move $\mathbf{d}x^{j_1}$ through the 1-forms of $\boldsymbol{\Omega}$ so as to reform $(\mathbf{d}x^{j_1} \wedge \mathbf{d}x^{j_2})$ on the right. In so doing, we pick up a second factor of $(-1)^q$. In the general case, we must repeat this for each of the p 1-forms of $\boldsymbol{\Psi}$, picking up p factors of $(-1)^q$. Therefore, the net factor is $(-1)^{pq}$, which completes the proof of (35.3).

35.7 Is $\Psi \wedge \Psi \neq 0$ Possible?

One consequence of the rule (35.3), which we have now proved, is (35.4): if $\boldsymbol{\Psi}$ is of *odd* degree, then $\boldsymbol{\Psi} \wedge \boldsymbol{\Psi} = 0$.

Thus, if it were possible for $\boldsymbol{\Psi} \wedge \boldsymbol{\Psi} \neq 0$, then $\boldsymbol{\Psi}$ would have to be of *even* degree. The simplest possibility is therefore that $\boldsymbol{\Psi}$ is a 2-form, in which case $\boldsymbol{\Psi} \wedge \boldsymbol{\Psi}$ would be a 4-form. The smallest number of dimensions in which a 4-form can exist is 4, so let us take $\boldsymbol{\Psi}$ to be a 2-form in spacetime, with coordinates (t, x, y, z).

Consider,

$$\boldsymbol{\Psi} = \mathbf{d}t \wedge \mathbf{d}x + \mathbf{d}y \wedge \mathbf{d}z.$$

If $\boldsymbol{\mathcal{V}}$ denotes the *volume 4-form* of spacetime, then [exercise],

$$\boldsymbol{\Psi} \wedge \boldsymbol{\Psi} = 2\,\mathbf{d}t \wedge \mathbf{d}x \wedge \mathbf{d}y \wedge \mathbf{d}z = 2\boldsymbol{\mathcal{V}}.$$

Chapter 36

Differentiation

36.1 The Exterior Derivative of a 1-Form

In the context of Forms, a function f is taken to be a 0-form. When the ***exterior derivative*** $\mathbf{d}$ is applied to f, it yields the gradient 1-form $\mathbf{d}\mathsf{f}$, which encodes *how fast* f *changes in all possible directions*.

If we wish to know the rate of change of f in a *specific* direction $\mathbf{u}$, we must submit it as a question to the oracle 1-form $\mathbf{d}\mathsf{f}$, which then declares the answer to be (32.10):

$$\mathbf{d}\mathsf{f}(\mathbf{u}) = \nabla_{\mathbf{u}}\mathsf{f}.$$

In summary, $\mathbf{d}$ measures the rate of change in all possible directions, and it increases the degree of the form by one, to allow for the input of an additional vector, along which to measure the rate of change of the form.

Expressed in this manner, our task is clear: we must extend the definition of $\mathbf{d}$ so that when it is applied to a p-form $\boldsymbol{\Psi}$ it yields a $(p+1)$-form $\mathbf{d}\boldsymbol{\Psi}$ that measures the rate of change of $\boldsymbol{\Psi}$ in all possible directions.

As is our custom, we shall begin with the simplest possible case, then gradually work our way up. So, let us try to differentiate a 1-form φ to obtain a 2-form $\mathbf{d}\varphi$, acting on pairs of vectors, $\mathbf{u}$ and $\mathbf{v}$.

In order for a rate of change to even make sense, we must imagine that the 1-form, and the vectors upon which it acts, are defined in at least a neighborhood of the point at which we seek the rate of change. Thus we take φ to be a 1-form *field*, and likewise take $\mathbf{u}$ and $\mathbf{v}$ to be vector *fields*.

Let us begin by imagining that $\mathbf{u}$ is the direction in which we want to know the rate of change of φ, the same role it played in $\mathbf{d}\mathsf{f}(\mathbf{u}) = \nabla_{\mathbf{u}}\mathsf{f}$. Since $\mathbf{v}$ is the only other vector in play, the function whose rate of change will be sought can only be $\varphi(\mathbf{v})$. In other words, we are essentially forced to consider the rate of change along $\mathbf{u}$ of $\varphi(\mathbf{v})$:

$$\nabla_{\mathbf{u}}\varphi(\mathbf{v}).$$

We are off to a good start, but we require that $\mathbf{d}\varphi$ be a *2-form*, so its action $\mathbf{d}\varphi(\mathbf{u}, \mathbf{v})$ on pairs of vectors must be *antisymmetric* in those vectors, which this expression is not. We therefore employ a trick that should be very familiar by now: we *impose* antisymmetry, by subtracting the same expression, but with $\mathbf{u} \leftrightarrow \mathbf{v}$:

$$\nabla_{\mathbf{u}}\varphi(\mathbf{v}) - \nabla_{\mathbf{v}}\varphi(\mathbf{u}). \tag{36.1}$$

This expression is now antisymmetric in the two vector inputs, so we are certainly getting closer, but this *still* cannot be the correct answer.

The problem is that $\mathbf{d}\varphi$ is supposed to measure the rate of change (in all directions) of the 1-form φ—the vector fields $\mathbf{u}$ and $\mathbf{v}$ should have nothing to do with this. And yet, even before we look into the matter in detail, it is clear that the variation of $\varphi(\mathbf{v})$, for example, will depend on both the variation of φ (which is the thing we care about) *and* on the variation of $\mathbf{v}$ (which we do *not* care about).

In order to define the exterior derivative $\mathbf{d}\varphi$, we must therefore discover how the variations of $\mathbf{u}$ and $\mathbf{v}$ contribute to (36.1), and then we must surgically *remove* their unwanted contributions!

In order to accomplish this, let us express the action of φ on a vector in terms of its vector counterpart, $\underline{\varphi}$, as we did in (34.15):

$$\varphi(\mathbf{u}) = \underline{\varphi} \cdot \mathbf{u}.$$

Then,

$$\begin{aligned}
\nabla_{\mathbf{u}}\varphi(\mathbf{v}) - \nabla_{\mathbf{v}}\varphi(\mathbf{u}) &= \nabla_{\mathbf{u}}(\underline{\varphi} \cdot \mathbf{v}) - \nabla_{\mathbf{v}}(\underline{\varphi} \cdot \mathbf{u}) \\
&= \left\{\mathbf{v} \cdot \nabla_{\mathbf{u}}\underline{\varphi} - \mathbf{u} \cdot \nabla_{\mathbf{v}}\underline{\varphi}\right\} + \left\{\underline{\varphi} \cdot \nabla_{\mathbf{u}}\mathbf{v} - \underline{\varphi} \cdot \nabla_{\mathbf{v}}\mathbf{u}\right\} \\
&= \left\{\mathbf{v} \cdot \nabla_{\mathbf{u}}\underline{\varphi} - \mathbf{u} \cdot \nabla_{\mathbf{v}}\underline{\varphi}\right\} + \left\{\underline{\varphi} \cdot [\mathbf{u}, \mathbf{v}]\right\} \\
&= \left\{\mathbf{v} \cdot \nabla_{\mathbf{u}}\underline{\varphi} - \mathbf{u} \cdot \nabla_{\mathbf{v}}\underline{\varphi}\right\} + \varphi([\mathbf{u}, \mathbf{v}]),
\end{aligned}$$

where $[\mathbf{u}, \mathbf{v}]$ is the ***commutator***, pictured in [29.7], page 288.

The last term, $\varphi([\mathbf{u}, \mathbf{v}])$, contains the variation in the vector fields—*this* is what must be surgically removed, leaving only the *first*, bracketed term as our definition of the exterior derivative $\mathbf{d}\varphi$ of φ.

Having used the vector correspondence as a stepping stone, we can now dispense with it, and express the final result purely in terms of the 1-form φ:

$$\boxed{\mathbf{d}\varphi(\mathbf{u}, \mathbf{v}) = \nabla_{\mathbf{u}}\varphi(\mathbf{v}) - \nabla_{\mathbf{v}}\varphi(\mathbf{u}) - \varphi([\mathbf{u}, \mathbf{v}]).} \tag{36.2}$$

One advantage of this formula is that it is geometrically defined, independently of any choice of basis and coordinates. However, we also perceive two *problems* with this formula, one aesthetic, and one philosophical: (i) it is disappointingly complicated, and (ii) it is very unclear what it *means*. We shall address the first problem immediately, and the second problem later in this chapter, and, at a deeper level, in the next chapter.

The key to simplifying (36.2) is to abstract the vectors away entirely, leaving behind the 2-form $\mathbf{d}\varphi$ itself. Since we have gone to so much trouble to make (36.2) independent of the variations of the vector fields, let us simply *choose* the vector fields to be *constant*, in which case their commutator vanishes. Next, imagine that a basis has been chosen, enabling us to express the formula in terms of the Cartesian basis 1-forms, $\{\mathbf{d}x^k\}$.

In this case, writing $\varphi = \varphi_i \, \mathbf{d}x^i$, and using the abbreviation $\partial_k \equiv \partial_{x^k}$, we find that (36.2) reduces to,

$$\begin{aligned}
\mathbf{d}\varphi(\mathbf{u}, \mathbf{v}) &= \nabla_{\mathbf{u}}\varphi(\mathbf{v}) - \nabla_{\mathbf{v}}\varphi(\mathbf{u}) \\
&= u^i \partial_i (v^j \, \varphi_j) - v^i \partial_i (u^j \, \varphi_j) \\
&= u^i v^j \, \partial_i \varphi_j - u^j v^i \, \partial_i \varphi_j \\
&= \partial_i \varphi_j \left\{\mathbf{d}x^i(\mathbf{u})\mathbf{d}x^j(\mathbf{v}) - \mathbf{d}x^j(\mathbf{u})\mathbf{d}x^i(\mathbf{v})\right\} \\
&= \partial_i \varphi_j \, (\mathbf{d}x^i \wedge \mathbf{d}x^j)(\mathbf{u}, \mathbf{v}).
\end{aligned}$$

Therefore, abstracting away the arbitrary vector fields,

$$\boxed{\mathbf{d}\varphi = \partial_i \varphi_j \, (\mathbf{d}x^i \wedge \mathbf{d}x^j).} \tag{36.3}$$

Much better!

But we can do better still! Since $\mathbf{d}f = \partial_i f \, \mathbf{d}x^i$, we now deduce the following elegant, natural-looking, and highly *practical* expression for the action of $\mathbf{d}$:

$$\boxed{\mathbf{d}\varphi = \mathbf{d}(\varphi_j \, \mathbf{d}x^j) = \mathbf{d}\varphi_j \wedge \mathbf{d}x^j.} \tag{36.4}$$

36.2 The Exterior Derivative of a 2-Form and of a p-Form

The obvious next step is to try to discover the action of $\mathbf{d}$ on a *2-form*, $\boldsymbol{\Psi}$. This should be a 3-form, taking three vector fields as inputs: $\mathbf{v}_1$, $\mathbf{v}_2$, $\mathbf{v}_3$. Clearly, $\mathbf{d}\boldsymbol{\Psi}(\mathbf{v}_1, \mathbf{v}_2, \mathbf{v}_3)$ should be built out of elements such as $\boldsymbol{\nabla}_{\mathbf{v}_1}\boldsymbol{\Psi}(\mathbf{v}_2, \mathbf{v}_3)$, but how?

Since $\mathbf{d}$ increases the degree of a form by 1, it is analogous to a 1-form. The answer to the construction of $\mathbf{d}\boldsymbol{\Psi}(\mathbf{v}_1, \mathbf{v}_2, \mathbf{v}_3)$ can therefore be found in our earlier work on the construction of the wedge product of a 1-form and a 2-form, as given in (35.2). Assuming constant vector fields from the outset, we therefore find that

$$\boxed{\mathbf{d}\boldsymbol{\Psi}(\mathbf{v}_1, \mathbf{v}_2, \mathbf{v}_3) = \boldsymbol{\nabla}_{\mathbf{v}_1}\boldsymbol{\Psi}(\mathbf{v}_2, \mathbf{v}_3) + \boldsymbol{\nabla}_{\mathbf{v}_2}\boldsymbol{\Psi}(\mathbf{v}_3, \mathbf{v}_1) + \boldsymbol{\nabla}_{\mathbf{v}_3}\boldsymbol{\Psi}(\mathbf{v}_1, \mathbf{v}_2).}$$

Again, we can simplify this formula in a gratifying manner, by abstracting the vectors away to reveal the 3-form itself. We find [exercise] that (36.4) generalizes in a way that is clearly part of a larger pattern:

$$\boxed{\mathbf{d}\boldsymbol{\Psi} = \mathbf{d}\left(\Psi_{ij} \, \mathbf{d}x^i \wedge \mathbf{d}x^j\right) = \mathbf{d}\Psi_{ij} \wedge \mathbf{d}x^i \wedge \mathbf{d}x^j.} \tag{36.5}$$

It is now clear how $\mathbf{d}$ acts upon a p-form $\boldsymbol{\Phi}$ to create a $(p+1)$-form $\mathbf{d}\boldsymbol{\Phi}$:

$$\boxed{\mathbf{d}\boldsymbol{\Phi} = \mathbf{d}\left(\Phi_{i_1 \ldots i_p} \, \mathbf{d}x^{i_1} \wedge \cdots \wedge \mathbf{d}x^{i_p}\right) = \mathbf{d}\Phi_{i_1 \ldots i_p} \wedge \mathbf{d}x^{i_1} \wedge \cdots \wedge \mathbf{d}x^{i_p}.} \tag{36.6}$$

36.3 The Leibniz Rule for Forms

It is simple to verify [exercise] that if f is a function (0-form) then

$$\boxed{\mathbf{d}(f\boldsymbol{\Psi}) = (\mathbf{d}f) \wedge \boldsymbol{\Psi} + f \, \mathbf{d}\boldsymbol{\Psi}.} \tag{36.7}$$

However, the general Leibniz Rule (aka Product Rule) for Forms is much less obvious:

$$\boxed{\mathbf{d}(\boldsymbol{\Phi} \wedge \boldsymbol{\Psi}) = (\mathbf{d}\boldsymbol{\Phi}) \wedge \boldsymbol{\Psi} + (-1)^{\deg \boldsymbol{\Phi}} \, \boldsymbol{\Phi} \wedge (\mathbf{d}\boldsymbol{\Psi}),} \tag{36.8}$$

where $\deg \boldsymbol{\Phi}$ denotes the degree of $\boldsymbol{\Phi}$.

For example, if $\deg \boldsymbol{\Phi} = 1$, so that $\boldsymbol{\Phi} = \boldsymbol{\phi}$ is a 1-form, then

$$\mathbf{d}(\boldsymbol{\phi} \wedge \boldsymbol{\Psi}) = (\mathbf{d}\boldsymbol{\phi}) \wedge \boldsymbol{\Psi} - \boldsymbol{\phi} \wedge (\mathbf{d}\boldsymbol{\Psi}). \tag{36.9}$$

We first prove this simple example, for its proof will immediately generalize to provide an explanation of the full rule, (36.8).

To be concrete, suppose that $\boldsymbol{\Psi}$ is a 2-form. Then, using the Leibniz Rule for functions, (32.11), we find that,

$$\begin{aligned}
\mathbf{d}(\boldsymbol{\phi} \wedge \boldsymbol{\Psi}) &= \mathbf{d}([\phi_i \, \mathbf{d}x^i] \wedge [\Psi_{jk} \, \mathbf{d}x^j \wedge \mathbf{d}x^k]) \\
&= \mathbf{d}(\phi_i \, \Psi_{jk}) \wedge \mathbf{d}x^i \wedge \mathbf{d}x^j \wedge \mathbf{d}x^k \\
&= (\Psi_{jk} \, \mathbf{d}\phi_i + \phi_i \, \mathbf{d}\Psi_{jk}) \wedge \mathbf{d}x^i \wedge \mathbf{d}x^j \wedge \mathbf{d}x^k \\
&= (\mathbf{d}\phi_i \wedge \mathbf{d}x^i) \wedge [\Psi_{jk} \, \mathbf{d}x^j \wedge \mathbf{d}x^k] + (\phi_i \, \mathbf{d}\Psi_{jk}) \wedge \mathbf{d}x^i \wedge \mathbf{d}x^j \wedge \mathbf{d}x^k \\
&= (\mathbf{d}\boldsymbol{\phi}) \wedge \boldsymbol{\Psi} - [\phi_i \mathbf{d}x^i] \wedge [\mathbf{d}\Psi_{jk} \wedge \mathbf{d}x^j \wedge \mathbf{d}x^k] \\
&= (\mathbf{d}\boldsymbol{\phi}) \wedge \boldsymbol{\Psi} - \boldsymbol{\phi} \wedge (\mathbf{d}\boldsymbol{\Psi}).
\end{aligned}$$

The key to understanding the general formula (36.8) is the passage from the fourth to the fifth line, in which we pushed $\mathbf{d}\Psi_{jk}$ past $\mathbf{d}x^i$ (belonging to $\boldsymbol{\phi}$) in order to return it to its rightful place, next to its own $\mathbf{d}x^j \wedge \mathbf{d}x^k$, obtaining

$$\mathbf{d}x^i \wedge \mathbf{d}\Psi_{jk} = -\mathbf{d}\Psi_{jk} \wedge \mathbf{d}x^i.$$

In the general case, we must push $\mathbf{d}\Psi_{jk}$ past all the $\mathbf{d}x^i$'s belonging to $\boldsymbol{\Phi}$—numbering $\deg \boldsymbol{\Phi}$—resulting in $\deg \boldsymbol{\Phi}$ changes of sign. In other words, the sign of $\boldsymbol{\Phi} \wedge (\mathbf{d}\boldsymbol{\Psi})$ is $(-1)^{\deg \boldsymbol{\Phi}}$, which completes the proof.

A helpful way of remembering the general formula is to think of $\mathbf{d}$ as a 1-form. Then, before $\mathbf{d}$ can *reach* $\boldsymbol{\Psi}$, on the right of $\mathbf{d}(\boldsymbol{\Phi} \wedge \boldsymbol{\Psi})$, *it must push past all the 1-forms of* $\boldsymbol{\Phi}$, resulting in $\deg \boldsymbol{\Phi}$ changes of sign. As we have just seen, this mnemonic is actually quite close to the truth of the matter!

36.4 Closed and Exact Forms

36.4.1 *A Fundamental Result:* $\mathbf{d}^2 = 0$

Reconsider the formula (36.3):

$$\mathbf{d}\boldsymbol{\varphi} = \partial_i \varphi_j \, (\mathbf{d}x^i \wedge \mathbf{d}x^j).$$

Now let us take this 2-form and apply $\mathbf{d}$ for a *second* time, using (36.5):

$$\begin{aligned}
\mathbf{d}^2\boldsymbol{\varphi} &= \mathbf{d}\Big[\partial_i \varphi_j \, (\mathbf{d}x^i \wedge \mathbf{d}x^j)\Big] \\
&= \mathbf{d}\Big\{\partial_i \varphi_j\Big\} \wedge \mathbf{d}x^i \wedge \mathbf{d}x^j \\
&= \Big\{[\partial_k \partial_i \varphi_j] \, \mathbf{d}x^k \wedge \mathbf{d}x^i\Big\} \wedge \mathbf{d}x^j \\
&= 0,
\end{aligned}$$

because $\partial_k \partial_i \varphi_j = \partial_i \partial_k \varphi_j$, but $(\mathbf{d}x^k \wedge \mathbf{d}x^i) = -(\mathbf{d}x^i \wedge \mathbf{d}x^k)$.

By virtue of (36.6), it is clear that the above argument generalizes to arbitrary p-forms: $\mathbf{d}^2\boldsymbol{\Phi}=0$. We may therefore abstract the p-form away, obtaining this absolutely fundamental result:

$$\textit{Every form is annihilated by two applications of } \mathbf{d}\textit{:} \qquad \boxed{\mathbf{d}^2=0.} \tag{36.10}$$

The computational *proof* is certainly short and clear, but what does the result itself actually *mean*? The resolution of this mystery must await our discussion of integration. (If you insist on ruining the suspense, see Section 37.4.)

36.4.2 Closed and Exact Forms

A form is called ***closed*** if its exterior derivative vanishes:

$$\boxed{\boldsymbol{\Upsilon} \text{ is } \textit{closed} \qquad \Longleftrightarrow \qquad \mathbf{d}\boldsymbol{\Upsilon}=0.}$$

We shall eventually see that a closed form is a higher-dimensional analogue of the flux 2-form of an incompressible fluid flow.

A p-form is called ***exact*** if it is the exterior derivative of a $(p-1)$-form:

$$\boxed{\boldsymbol{\Upsilon} \text{ is } \textit{exact} \qquad \Longleftrightarrow \qquad \boldsymbol{\Upsilon}=\mathbf{d}\boldsymbol{\Psi} \qquad (\text{for some } \boldsymbol{\Psi}).}$$

In the language of physics, we call $\boldsymbol{\Psi}$ a ***potential*** for $\boldsymbol{\Upsilon}$. If such a potential exists, it is far from unique, for if $\boldsymbol{\Theta}$ is an *arbitrary* $(p-2)$-form, and we change

$$\boldsymbol{\Psi} \rightsquigarrow \widetilde{\boldsymbol{\Psi}}=\boldsymbol{\Psi}+\mathbf{d}\boldsymbol{\Theta},$$

then [exercise] $\widetilde{\boldsymbol{\Psi}}$ is *also* a potential: $\boldsymbol{\Upsilon}=\mathbf{d}\widetilde{\boldsymbol{\Psi}}$. This freedom in choosing the potential is called ***gauge freedom***, and the transformation $\boldsymbol{\Psi} \rightsquigarrow \widetilde{\boldsymbol{\Psi}}$ is called a ***gauge transformation***.

It follows immediately from (36.10) that

$$\boxed{\textit{Every exact form is closed:} \qquad \boldsymbol{\Upsilon}=\mathbf{d}\boldsymbol{\Psi} \;\Longrightarrow\; \mathbf{d}\boldsymbol{\Upsilon}=0}$$

In light of this, it is very natural to ask about a possible converse: *Is a closed form always exact?* This turns out to be a very interesting, deep question.

The short answer is that it depends on the *topology* of the region in which the form is defined.

Poincaré Lemma. *If* $\mathbf{d}\boldsymbol{\Upsilon}=0$ *throughout a simply connected region, then* $\boldsymbol{\Upsilon}=\mathbf{d}\boldsymbol{\Psi}$, *for some* $\boldsymbol{\Psi}$. (36.11)

Thus a potential *does* always exist *locally*. But global problems arise if the region is *not* simply connected, in which case it is possible that no such $\boldsymbol{\Psi}$ exists. The study of the closed Forms that are *not* exact then encodes detailed information about the topology of the space, called the ***de Rham cohomology***. This is the subject of Section 37.9.

Let us end this subsection with some practice with these concepts. We ask that you prove to yourself each of the following interesting facts:

If $\boldsymbol{\Upsilon}$ and $\boldsymbol{\Phi}$ are closed, then $\boldsymbol{\Upsilon} \wedge \boldsymbol{\Phi}$ is closed, too.

If $\boldsymbol{\Upsilon}$ is closed, then $\boldsymbol{\Upsilon} \wedge \mathbf{d}\boldsymbol{\Phi}$ is closed for all $\boldsymbol{\Phi}$.

If $\deg \boldsymbol{\Phi}$ *is* even, *then* $\boldsymbol{\Phi} \wedge \mathbf{d}\boldsymbol{\Phi}$ *is closed.*

Hint: (35.4).

36.4.3 *Complex Analysis: Cauchy–Riemann Equations*

In VCA (Section 5.1.2) we showed geometrically that a complex function $f(z) = u + iv$ is locally an amplitwist—as defined in Section 4.6—if and only if

$$i\,\partial_x f = \partial_y f. \tag{36.12}$$

Equating real and imaginary parts, we recover the more common form of the celebrated ***Cauchy–Riemann equations***, characterizing a conformal (aka complex analytic) mapping:

$$\partial_x u = \partial_y v \qquad \text{and} \qquad \partial_x v = -\partial_y u.$$

We now show that these equations may be elegantly (and consequentially) reformulated in the language of Forms.

Consider the complex-valued 1-form,

$$f\,\mathbf{dz} = (u + iv)(\mathbf{dx} + i\mathbf{dy}) = [u\,\mathbf{dx} - v\,\mathbf{dy}] + i\,[v\,\mathbf{dx} + u\,\mathbf{dy}].$$

Then,

$$\begin{aligned}
\mathbf{d}(f\,\mathbf{dz}) &= \mathbf{d}f \wedge \mathbf{dz} \\
&= (\partial_x f\,\mathbf{dx} + \partial_y f\,\mathbf{dy}) \wedge (\mathbf{dx} + i\,\mathbf{dy}) \\
&= (i\,\partial_x f - \partial_y f)\,\mathcal{A},
\end{aligned}$$

where, as usual, $\mathcal{A} = \mathbf{dx} \wedge \mathbf{dy}$ is the area 2-form.

Therefore, by virtue of the compact form of the Cauchy–Riemann equations, (36.12), we conclude that

The complex function f is locally an amplitwist if and only if the 1-form $f\,\mathbf{dz}$ is closed: $\mathbf{d}(f\,\mathbf{dz}) = 0$. (36.13)

The central result regarding integration of Forms is called (in this book[1]) *The Fundamental Theorem of Exterior Calculus*; it is the main result of the next chapter. Armed with that theorem, we shall see (in Section 37.7) that (36.13) *instantly* implies the pivotal result of Complex Analysis—*Cauchy's Theorem!*

The above formulation is valuable even if we are interested in complex functions that are *not* locally an amplitwist (i.e., that are ***nonanalytic***). In particular, we note for future use that reflection across the real axis is *anticonformal*, so the mapping $f(z) = \overline{z} = x - iy$ is *not* subject to $\mathbf{d}(f\,\mathbf{d}z) = 0$. Indeed, we find [exercise] that

$$\boxed{\mathbf{d}(\overline{z}\,\mathbf{d}z) = 2i\,\mathcal{A}.} \tag{36.14}$$

36.5 Vector Calculus via Forms

In this section we simply apply the exterior derivative to form fields in $\mathbb{R}^3$, and discover that the fundamental operations and identities of Vector Calculus emerge automatically, as if by magic—automagically!

As we explained in Sections 34.6 and 34.7, the exterior calculus works perfectly in *all* dimensions and on *all* p-forms, but *only* in $\mathbb{R}^3$ can a 2-form masquerade as a vector, and *only* in $\mathbb{R}^3$ can the wedge product of two 1-forms masquerade as the vector product of two vectors—these are the twin spells that conjure Vector Calculus into existence!

First, we remind the reader of the notation: $\underline{\varphi}$ is the vector corresponding ($\rightleftarrows$) to the 1-form φ, and $\underline{\Psi}$ is the vector corresponding to the flux 2-form Ψ, and $\mathcal{V} = \mathbf{d}x^1 \wedge \mathbf{d}x^2 \wedge \mathbf{d}x^3$ is the volume 3-form.

We have seen that the Vector Calculus concept of the gradient vector ∇f corresponds to the gradient 1-form $\mathbf{d}f$ of the 0-form f. Now let us step up one level, and take the exterior derivative of a 1-form, writing out explicitly the three resulting components of (36.3) that occur in $\mathbb{R}^3$; finally, let us *interpret* the answer as the flux 2-form of a flow in space:

$$\begin{aligned}
\mathbf{d}\varphi &= \partial_i\varphi_j\,(\mathbf{d}x^i \wedge \mathbf{d}x^j) \\
&= (\partial_2\varphi_3 - \partial_3\varphi_2)\,(\mathbf{d}x^2 \wedge \mathbf{d}x^3) + (\partial_3\varphi_1 - \partial_1\varphi_3)\,(\mathbf{d}x^3 \wedge \mathbf{d}x^1) + (\partial_1\varphi_2 - \partial_2\varphi_1)\,(\mathbf{d}x^1 \wedge \mathbf{d}x^2) \\
&\rightleftarrows \begin{bmatrix} \partial_2\varphi_3 - \partial_3\varphi_2 \\ \partial_3\varphi_1 - \partial_1\varphi_3 \\ \partial_1\varphi_2 - \partial_2\varphi_1 \end{bmatrix},
\end{aligned}$$

by virtue of (34.10).

Lo and behold, out has popped the ***curl*** of the vector field corresponding to the 1-form! To summarize,

$$\boxed{\mathbf{d}\varphi \quad \rightleftarrows \quad \nabla \times \underline{\varphi} = \mathbf{curl}\,\underline{\varphi} = \begin{bmatrix} \partial_1 \\ \partial_2 \\ \partial_3 \end{bmatrix} \times \begin{bmatrix} \varphi_1 \\ \varphi_2 \\ \varphi_3 \end{bmatrix}.} \tag{36.15}$$

[1]The more common name is, instead, the Generalized Stokes's Theorem, the twisted origin of which is described in Section 37.3.2.

Thus if φ is *closed*, meaning $\mathbf{d}\varphi = 0$, then $\underline{\varphi}$ has vanishing curl: if $\underline{\varphi}$ is pictured as the velocity of a fluid flow in space, then if we insert a tiny ball—*not* a point particle—into the flow, it will be carried off with velocity $\underline{\varphi}$, but *it will not spin*. In this case the flow is called ***irrotational***. If $\mathbf{d}\varphi \neq 0$, then the ball will spin, and **curl** $\underline{\varphi}$ points along the spin axis, and it magnitude gives double[2] the rate of spin. In this context, **curl** $\underline{\varphi}$ is called the ***vorticity*** vector.

Alternatively, we may picture the closed form as corresponding to a ***conservative force field***, with the attendant advantage that the 1-form φ itself can immediately be given meaning, as the *work*, exactly as in the gravitational example of Section 32.3.1. There, φ was closed by virtue of the rather uninteresting reason that it was *constant*. In that gravitational case, the path-independent work done transporting a particle along a path from p to q could therefore be obtained simply by applying φ to the direct path along $\overrightarrow{pq}$.

In accord with the Poincaré lemma, in this gravitational example, the closed 1-form φ is *exact*: if h again represents height, so that gh is the potential energy of a unit mass, then $\varphi = \mathbf{d}(gh) = g\,\mathbf{d}h$. This connects naturally with the classical expression $g\, dh$, for if the particle is raised δh by moving it along the short vector $\mathbf{v}$, then the work done is

$$g\,\mathbf{d}h(\mathbf{v}) = g\,\delta h.$$

As we will discuss in the next chapter, for a *nonconstant* φ, the work done along a path must be evaluated by breaking the path down into lots of small vectors, applying φ to each one, and summing. As we shall discuss, if φ is *closed*, then the work done in transporting a particle around a closed loop vanishes.

Next, let us take the exterior derivative of a 2-form Ψ, but labelling its component with the corresponding vector $\underline{\Psi}$, as we did in (34.10):

$$\Psi = \Psi^1\,(\mathbf{d}x^2 \wedge \mathbf{d}x^3) + \Psi^2\,(\mathbf{d}x^3 \wedge \mathbf{d}x^1) + \Psi^3\,(\mathbf{d}x^1 \wedge \mathbf{d}x^2) \quad \rightleftarrows \quad \underline{\Psi} = \Psi^j\,\mathbf{e}_j = \begin{bmatrix} \Psi^1 \\ \Psi^2 \\ \Psi^3 \end{bmatrix}.$$

Then,

$$\begin{aligned}
\mathbf{d}\Psi &= \mathbf{d}\Psi^1 \wedge \mathbf{d}x^2 \wedge \mathbf{d}x^3 + \mathbf{d}\Psi^2 \wedge \mathbf{d}x^3 \wedge \mathbf{d}x^1 + \mathbf{d}\Psi^3 \wedge \mathbf{d}x^1 \wedge \mathbf{d}x^2 \\
&= \partial_1\Psi^1\,\mathbf{d}x^1 \wedge \mathbf{d}x^2 \wedge \mathbf{d}x^3 + \partial_2\Psi^2\,\mathbf{d}x^2 \wedge \mathbf{d}x^3 \wedge \mathbf{d}x^1 + \partial_3\Psi^3\,\mathbf{d}x^3 \wedge \mathbf{d}x^1 \wedge \mathbf{d}x^2 \\
&= \left(\partial_1\Psi^1 + \partial_2\Psi^2 + \partial_3\Psi^3\right)\mathbf{d}x^1 \wedge \mathbf{d}x^2 \wedge \mathbf{d}x^3 \\
&= \left\{ \begin{bmatrix} \partial_1 \\ \partial_2 \\ \partial_3 \end{bmatrix} \bullet \begin{bmatrix} \Psi^1 \\ \Psi^2 \\ \Psi^3 \end{bmatrix} \right\} \mathcal{V}.
\end{aligned}$$

Lo and behold, out has popped the ***divergence*** of the vector field flow corresponding to the flux 2-form! To summarize,

$$\boxed{\mathbf{d}\Psi = (\mathbf{div}\,\underline{\Psi})\,\mathcal{V} = (\boldsymbol{\nabla} \bullet \underline{\Psi})\,\mathcal{V}.} \tag{36.16}$$

Thus if Ψ is *closed*, meaning $\mathbf{d}\Psi = 0$, then $\underline{\Psi}$ has vanishing divergence, corresponding to the flow of an incompressible fluid in a region where no fluid is being pumped in or sucked out, so that exactly as much fluid flows out of a closed surface as flows into it. In general, applying $\mathbf{d}\Psi$ to

[2]See Feynman et al. (1963, Vol. 1, §40).

a very small parallelepiped ultimately yields the *net flux of fluid out of it*. This will be explained in the next chapter.

The fundamental identity of exterior differentiation is (36.10):

$$\mathbf{d}^2 = 0,$$

and this has important implications for vector fields in $\mathbb{R}^3$. If in (36.15) we take $\varphi = \mathbf{d}f$, then we instantly obtain a classical result of Vector Calculus:

$$\boxed{\mathbf{d}\varphi = \mathbf{d}^2 f = 0 \qquad \Longleftrightarrow \qquad \text{curl}\left(\text{grad}\, f\right) = \boldsymbol{\nabla} \times \boldsymbol{\nabla} f = 0.}$$

If in (36.16) we take $\Psi = \mathbf{d}\varphi$, then we instantly obtain a second classical result of Vector Calculus:

$$\boxed{\mathbf{d}\Psi = \mathbf{d}^2 \varphi = 0 \qquad \Longleftrightarrow \qquad \text{div}\left(\text{curl}\, \underline{\varphi}\right) = \boldsymbol{\nabla} \boldsymbol{\cdot} \left(\boldsymbol{\nabla} \times \underline{\varphi}\right) = 0.}$$

Working with Forms obviates the need to remember these identities, as well as the many more complicated ones that are routinely employed by working scientists when using Vector Calculus. Furthermore, if we *do* wish to employ such an identity, it may be *derived* using Forms, and often much more quickly and elegantly than by direct vector methods. We illustrate this point with two examples of differential identities, leaving others to Exercise 14.

First, however, we shall state a very useful *algebraic* link between the world of Forms and the world of ($\mathbb{R}^3$) vectors, leaving the simple proof to you:

$$\boxed{\varphi \wedge \Psi = (\underline{\varphi} \boldsymbol{\cdot} \underline{\Psi})\, \mathcal{V}} \tag{36.17}$$

As our first example, suppose we wish to find an identity for $\boldsymbol{\nabla} \boldsymbol{\cdot} \left[f \underline{\Psi}\right]$. According to (36.16), we should therefore evaluate $\mathbf{d}\left[f \Psi\right]$, using (36.7), like this:

$$\left(\boldsymbol{\nabla} \boldsymbol{\cdot} \left[f \underline{\Psi}\right]\right) \mathcal{V} = \mathbf{d}\left[f \Psi\right] = (\mathbf{d}f) \wedge \Psi + f\, \mathbf{d}\Psi = \left[(\boldsymbol{\nabla} f) \boldsymbol{\cdot} \underline{\Psi} + f\, \boldsymbol{\nabla} \boldsymbol{\cdot} \underline{\Psi}\right] \mathcal{V},$$

yielding the familiar identity,

$$\boxed{\boldsymbol{\nabla} \boldsymbol{\cdot} \left[f \underline{\Psi}\right] = (\boldsymbol{\nabla} f) \boldsymbol{\cdot} \underline{\Psi} + f\, \boldsymbol{\nabla} \boldsymbol{\cdot} \underline{\Psi}.}$$

In this case, it is almost as easy [exercise] to derive the identity directly. So let us move on to an example in which the advantages of Forms start to emerge clearly.

Suppose we wish to find an identity for $\boldsymbol{\nabla} \times \left[f \underline{\varphi}\right]$. According to (36.15), this corresponds to the flux 2-form

$$\mathbf{d}\left[f \varphi\right] = \mathbf{d}f \wedge \varphi + f\, \mathbf{d}\varphi,$$

so we *effortlessly* deduce another familiar identity of Vector Calculus:

$$\boxed{\boldsymbol{\nabla} \times \left[f \underline{\varphi}\right] = \boldsymbol{\nabla} f \times \underline{\varphi} + f\, \boldsymbol{\nabla} \times \underline{\varphi}.}$$

As the complexity of the vector identity increases, so does the simplifying power of Forms; see Exercise 14.

36.6 Maxwell's Equations

We end this chapter with a beautiful and *profound* application of the above ideas: the encapsulation of Maxwell's complete laws of electromagnetism in just *two*, supremely beautiful equations.

We begin with the source-free equations, (34.24), which we restate here, for your convenience:

$$\begin{aligned} \nabla \cdot \underline{\mathbf{B}} &= 0, \\ \nabla \times \underline{\mathbf{E}} + \partial_t \underline{\mathbf{B}} &= 0. \end{aligned}$$

Following the example of Baez and Muniain (1994, §5), let us use $\mathbf{d}_S$ to denote the *spatial* part of the spacetime exterior derivative, so that

$$\mathbf{d}f = \mathbf{d}_S f + \partial_t f \, \mathbf{dt}.$$

Now consider the exterior derivative of the Faraday 2-form, (34.22):

$$\mathbf{dF} = \mathbf{d}(\varepsilon \wedge \mathbf{dt}) + \mathbf{dB}.$$

Let us evaluate these two terms separately.

First,

$$\begin{aligned} \mathbf{d}(\varepsilon \wedge \mathbf{dt}) &= \mathbf{d}(\varepsilon) \wedge \mathbf{dt} \\ &= \mathbf{d}_S(\varepsilon) \wedge \mathbf{dt} \\ &= \left[\text{flux 2-form of } \boldsymbol{\nabla} \times \underline{\mathbf{E}}\right] \wedge \mathbf{dt}, \end{aligned}$$

by virtue of (36.15).

Second,

$$\begin{aligned} \mathbf{dB} &= \mathbf{d}_S \mathbf{B} + \left[\partial_t B_x \, \mathbf{dt} \wedge (\mathbf{dy} \wedge \mathbf{dz}) + \cdots\right] \\ &= (\boldsymbol{\nabla} \cdot \underline{\mathbf{B}}) \, \mathcal{V} + \mathbf{dt} \wedge \left[\partial_t B_x \, (\mathbf{dy} \wedge \mathbf{dz}) + \cdots\right] \\ &= (\boldsymbol{\nabla} \cdot \underline{\mathbf{B}}) \, \mathcal{V} + \mathbf{dt} \wedge \partial_t \mathbf{B} \end{aligned}$$

by virtue of (36.16).

But $\mathbf{dt} \wedge (\partial_t \mathbf{B}) = (\partial_t \mathbf{B}) \wedge \mathbf{dt}$, by virtue of (35.3), because $\partial_t \mathbf{B}$ is a 2-form. Therefore, since

$$\partial_t \mathbf{B} = \text{flux 2-form of } \partial_t \underline{\mathbf{B}},$$

combining the previous two results yields

$$\mathbf{dF} = (\boldsymbol{\nabla} \cdot \underline{\mathbf{B}}) \, \mathcal{V} + \left\{\text{flux 2-form of } \left[\boldsymbol{\nabla} \times \underline{\mathbf{E}} + \partial_t \underline{\mathbf{B}}\right]\right\} \wedge \mathbf{dt}.$$

Therefore, reaching upward, we may now touch the face of God:

> *Maxwell's Source-Free Equations state that the Faraday 2-form is* closed:
>
> $$\mathbf{dF} = 0.$$

As we have stated, and will discuss further in the next chapter, *locally*, every closed form is exact. In other words, there exists a 1-form[3] potential, $\mathbf{A}$, such that

$$\boxed{\mathsf{F} = \mathbf{d}\mathbf{A}.}$$

NOTE: The classical literature on electrodynamics instead employs the corresponding *vector field* $\underline{\mathbf{A}}$, which is called the ***vector potential***.

Next, we turn our attention to the second pair of Maxwell's Equations, (34.25), which describe the fields generated by *sources*—an electrical charge density ρ, and a current density $\underline{\mathbf{j}}$:

$$\begin{aligned} \nabla \cdot \underline{\mathsf{E}} &= 4\pi\rho, \\ \nabla \times \underline{\mathbf{B}} - \partial_t \underline{\mathsf{E}} &= 4\pi \underline{\mathbf{j}}. \end{aligned}$$

In order to see what these equations are truly trying to show us, we must first introduce the *spacetime* version of the sources, combined into a single 1-form:

$$\mathbf{J} = -\rho\, \mathbf{d}t + \mathbf{j},$$

where the (spatial) 1-form[4] $\mathbf{j}$ corresponds to the (spatial) vector current density $\underline{\mathbf{j}}$:

$$\underline{\mathbf{j}} = \begin{bmatrix} j^1 \\ j^2 \\ j^3 \end{bmatrix} \qquad \rightleftarrows \qquad \mathbf{j} = j^1\, \mathbf{d}x^1 + j^2\, \mathbf{d}x^2 + j^3\, \mathbf{d}x^3.$$

Thus when $\mathbf{J}$ acts on a purely spatial 4-vector, it yields the amount of current flowing in that spatial direction, but when it acts on the time-axis 4-velocity of a stationary observer, it describes the charge flowing with the observer in time, i.e., the charge density ρ at the observer's location. More generally, if $\mathbf{u}$ is the 4-velocity of an observer, then she measures the charge density to be $\mathbf{J}(\mathbf{u})$.

In Exercise 15 you will see that in 4-dimensional spacetime, the Hodge star operator ($\star$) maps a p-form to a ***dual*** $(4-p)$-form. In particular, the dual of the source density 1-form $\mathbf{J}$ is the 3-form,

$$\star\mathbf{J} = -\rho\, \mathcal{V} + \left[\text{flux 2-form of } \underline{\mathbf{j}}\right] \wedge \mathbf{d}t, \tag{36.18}$$

and the dual of the Faraday 2-form is the Maxwell 2-form, (34.26):

$$\star\mathsf{F} = \mathbf{Maxwell} = \boldsymbol{\beta} \wedge \mathbf{d}t - \mathsf{E}.$$

Since Maxwell can be obtained from Faraday by interchanging the electric and magnetic fields, *and* changing the sign of the first term, the entire calculation above of $\mathbf{d}\mathsf{F}$ may be coopted to simply write down the answer for $\mathbf{d}\star\mathsf{F}$:

$$\begin{aligned} \mathbf{d}\star\mathsf{F} &= -(\boldsymbol{\nabla} \cdot \underline{\mathsf{E}})\, \mathcal{V} + \left\{\text{flux 2-form of } \left[\boldsymbol{\nabla} \times \underline{\mathbf{B}} - \partial_t \underline{\mathsf{E}}\right]\right\} \wedge \mathbf{d}t \\ &= 4\pi \left[-\rho\, \mathcal{V} + \left\{\text{flux 2-form of } \underline{\mathbf{j}}\right\} \wedge \mathbf{d}t\right], \end{aligned}$$

by virtue of the two Maxwell source equations above.

---

[3]This notation is universal, so in this case we must abandon our convention of using lowercase Greek letters to represent 1-forms.

[4]This notation is universal, so, once again, we must abandon our convention of using lowercase Greek letters to represent 1-forms.

We have arrived at a very remarkable—and a very *beautiful*—conclusion:

The Maxwell 2-form is subject to this Law of Nature:

$$\mathbf{d} \star \mathbf{F} = 4\pi \star \mathbf{J}.$$

For further discussion of Maxwell's equations in the language of Forms, see Hubbard and Hubbard (2009, §6.11), Baez and Muniain (1994, §§1,5), and Misner, Thorne, and Wheeler (1973, § 4).

Chapter 37

Integration

37.1 The Line Integral of a 1-Form

37.1.1 Circulation and Work

The *circulation* $\mathcal{C}_{\mathsf{K}}(\underline{\varphi})$ along a directed curve K of a fluid flow with velocity $\underline{\varphi}$ was previously defined ((27.2), p. 263) as the integral of the component of the fluid flow *along* K:

$$\mathcal{C}_{\mathsf{K}}(\underline{\varphi}) \equiv \int_{\mathsf{K}} \underline{\varphi} \cdot d\mathbf{r},$$

as illustrated in [37.1]. Alternatively, and equally importantly, we may think of $\underline{\varphi}$ as a *force field*, in which case exactly the same integral is now interpreted as the *work* done by the force field in moving a particle along K.

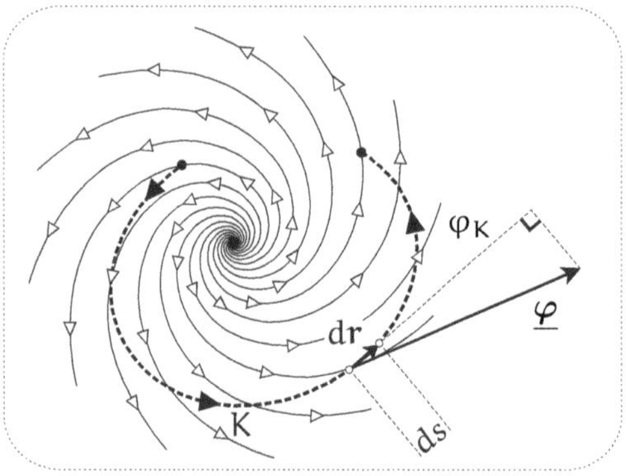

[37.1] *The work done in moving a particle a distance* ds *along the path* K *is* $\varphi_{\mathsf{K}}\,ds$*, where* φ_{K} *is the component of the force field* $\underline{\varphi}$ *in the direction of* K.

We have chosen to illustrate this idea in $\mathbb{R}^2$ in order to keep things simple, but we should stress that there is nothing intrinsically 2-dimensional about this concept: imagine the fluid to be the ocean, or the force field to be gravity. Indeed, everything we are about to discuss applies (without change) to 1-forms acting on vectors in $\mathbb{R}^n$.

Translating this into the language of Forms, we may write,

$$\varphi(d\mathbf{r}) = \underline{\varphi} \cdot d\mathbf{r} = \varphi_{\mathsf{K}}\, ds.$$

Here, as illustrated in [37.1], φ_{K} is the projection of $\underline{\varphi}$ onto the direction of K, and $ds = |d\mathbf{r}|$. In other words, the 1-form φ acts on a small piece[1] $d\mathbf{r}$ of the directed curve, yielding the component of the force in the direction of the movement, times the length of that small movement, which is the work done in carrying out that movement. The complete integral may then be thought of as the total work done, obtained by summing the increments of work, as the 1-form φ eats its way along the directed curve K, one small $d\mathbf{r}$ bite at a time.

We stress that *only the component of the force in the direction of movement does any work*. Thus, for example, if K were, instead, an orthogonal trajectory of the field lines, then no work would be done, i.e., the integral would vanish.

---

[1]Here we are being lazy: we *should* write $\delta\mathbf{r}$, and use ultimate equalities.

Thought of in this way, we have arrived at our definition of the integral of the 1-form φ along K, written

$$\int_{K} \varphi \equiv \mathcal{C}_{K}(\underline{\varphi}) \equiv \int_{K} \varphi_{K}\, ds. \tag{37.1}$$

In this definition, the absence (on the left) of d (*something*) under the integral sign is unfamiliar and perhaps disconcerting at first: just remember that this is not needed, since it is *understood* that φ is going to eat its way along K, one small $d\mathbf{r}$ bite at a time.

In the illustrated case, it is clear that K is always moving in roughly the same direction as the flow or force field, so the integral is clearly positive. But suppose we traverse the *same curve, but in the opposite direction* denoted $-K$. Now we are swimming upstream against the current, or, using the other interpretation, we are pushing the particle against the resistance of the force field. Thus the integral is now negative. More precisely, since $\varphi(-d\mathbf{r}) = -\varphi(d\mathbf{r})$, it follows that

$$\int_{-K} \varphi = -\int_{K} \varphi. \tag{37.2}$$

37.1.2 Path-Independence $\Longleftrightarrow$ Vanishing Loop Integrals

If K connects two fixed points a and b, then, in general, the value of $\mathcal{C}_{K}(\underline{\varphi})$ will depend on the specific path K from a to b that is chosen. However, for many important physical fluid flows and force fields, we have ***path independence***, which means what it says, and says what it means: all paths K from a to b yield the *same* value of the integral.

This concept can be rephrased in an important and useful way. As illustrated in [37.2], consider any two paths K_1 and K_2 connecting a to b. Then we can create a *closed loop* L by first travelling along K_1 from a to b, and then travelling back from b to a along $-K_2$, so that $L = K_1 - K_2$.

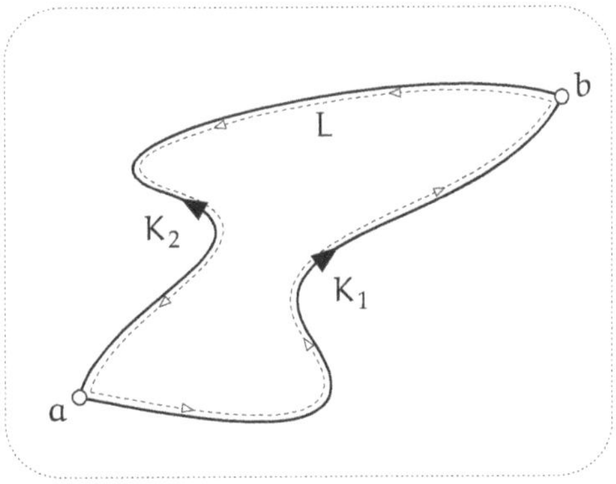

[37.2] *If the integral is independent of the path K between a and b, then the integral around the closed loop L vanishes. Conversely, if all loop integrals vanish, then the integral is path-independent.*

But then path-independence implies that the loop integral *vanishes*:

$$\int_{L} \varphi = \int_{K_1 - K_2} \varphi = \int_{K_1} \varphi - \int_{K_2} \varphi = 0.$$

Conversely, suppose that we know that all loop integrals vanish. It follows that the integral is path-independent. For if K_1 and K_2 are any two paths connecting a to b, then we may form the closed loop $L = K_1 - K_2$, and the fact that $\mathcal{C}_{L}(\underline{\varphi}) = 0$ then implies

$$\int_{K_1} \varphi = \int_{K_2} \varphi.$$

In brief,

$$\textit{Path-independence} \iff \textit{vanishing loop integrals.} \tag{37.3}$$

Look again at [37.1], and imagine that we extend K so that it closes up to become an elliptical loop encircling the vortex. Clearly the integral around this loop does *not* vanish, and is in fact positive. Thus here we have an example in which the integral of φ *does* depend on the path connecting two points.

37.1.3 The Integral of an Exact Form: $\varphi = \mathbf{d}f$

If φ is *exact* then, by definition, $\varphi = \mathbf{d}f$, for some function f. If $\delta\mathbf{r}$ is a short, ultimately vanishing, movement along the integration path K, then its contribution to the integral is

$$\varphi(\delta\mathbf{r}) = \mathbf{d}f(\delta\mathbf{r}) \asymp \delta f \equiv \text{change in f from the tail to the tip of } \delta\mathbf{r}.$$

Thus the complete integral is simply the net change in f from the start of K to the end of K. To summarize,

The integral of an exact 1-form $\varphi = \mathbf{d}f$ *is path-independent, and is given by the change in* f *along* K*:*

$$\int_K \mathbf{d}f = f(b) - f(a). \tag{37.4}$$

In the language of physics, $\varphi = \mathbf{d}f$ corresponds to a ***conservative*** force field, $\underline{\varphi} = \nabla f$, and f is the ***potential energy***. Then the change in potential energy $[f(b) - f(a)]$ is the path-independent amount of work done by the force field in carrying the particle from a to b, so (37.4) is simply another way of writing the familiar fact that

$$\int_K (\nabla f) \cdot \mathbf{dr} = f(b) - f(a). \tag{37.5}$$

As discussed by Feynman et al. (1963, Vol. 1, §14-4), at the most fundamental level *all* the forces in Nature are conservative, in accordance with the fundamental *principle of conservation of energy*.

37.2 The Exterior Derivative as an Integral

37.2.1 $\mathbf{d}$*(1-Form)*

From our very first exposure to calculus, we are taught that differentiation and integration are *inverse operations*—this is the Fundamental Theorem of Calculus. Indeed, we have just seen (or so it would seem) a fresh instance of this idea in (37.4).

It may therefore come as disorienting shock to learn that, when it comes to Forms, exterior differentiation *is* integration—around a small closed loop, or over a small closed surface—*not* its inverse!

Let $\Pi(\epsilon\mathbf{u}, \epsilon\mathbf{v})$ be the oriented boundary of a small—ultimately vanishing, as ϵ vanishes—parallelogram with first edge $\epsilon\mathbf{u}$ and second edge $\epsilon\mathbf{v}$, as illustrated in [37.3]. Let us define $\mathbf{\Omega}(\epsilon\mathbf{u}, \epsilon\mathbf{v})$ to be the integral of the 1-form φ around Π:

$$\mathbf{\Omega}(\epsilon\mathbf{u}, \epsilon\mathbf{v}) \equiv \oint_{\Pi(\epsilon\mathbf{u}, \epsilon\mathbf{v})} \varphi.$$

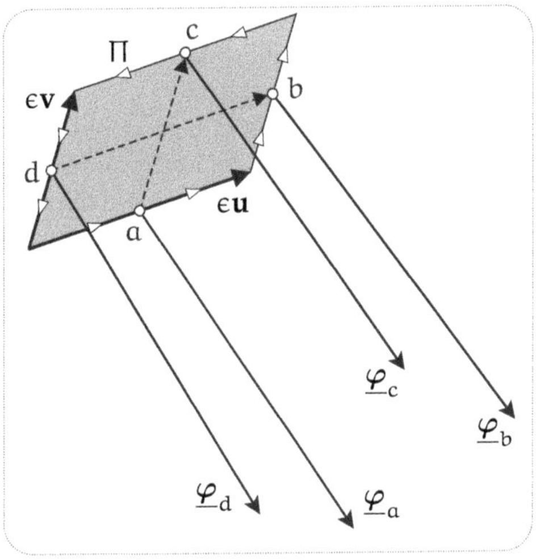

[37.3] *The integral of the 1-form φ around the small parallelogram Π is ultimately equal to its exterior derivative, $\mathbf{d}\varphi$, applied to the edges of Π.*

We will now show that this integral is a 2-*form*, and not just any 2-form—it is the *exterior derivative* of φ! That is,

$$\mathbf{\Omega}(\epsilon\mathbf{u}, \epsilon\mathbf{v}) \asymp \mathbf{d}\varphi(\epsilon\mathbf{u}, \epsilon\mathbf{v}).$$

Abstracting the vectors away,

$$\mathbf{\Omega} \asymp \mathbf{d}\varphi. \qquad (37.6)$$

That $\mathbf{\Omega}$ does indeed possess the antisymmetry required of a 2-form follows immediately from the fact that $\Pi(\epsilon\mathbf{v}, \epsilon\mathbf{u}) = -\Pi(\epsilon\mathbf{u}, \epsilon\mathbf{v})$, for this implies that

$$\mathbf{\Omega}(\epsilon\mathbf{v}, \epsilon\mathbf{u}) = \oint_{\Pi(\epsilon\mathbf{v}, \epsilon\mathbf{u})} \varphi = \oint_{-\Pi(\epsilon\mathbf{u}, \epsilon\mathbf{v})} \varphi = -\oint_{\Pi(\epsilon\mathbf{u}, \epsilon\mathbf{v})} \varphi = -\mathbf{\Omega}(\epsilon\mathbf{u}, \epsilon\mathbf{v}).$$

It is also possible to confirm geometrically that the linearity requirements are met, directly from the definition, but we shall instead cut directly to the chase, proving the *full* result, (37.6), thereby confirming linearity, and (re)confirming antisymmetry, in the process. To do so, we need only generalize the argument we previously gave in [27.2], page 263.

First, however, let us try to picture a flag flying over our ultimate destination, so that we have some idea in which direction we should march. In order for $\Pi(\epsilon\mathbf{u}, \epsilon\mathbf{v})$ to be a *closed* parallelogram loop, it is necessary—as explained in [29.7], page 288, and again in the upcoming [37.4]—that the *commutator* of its sides vanishes:

$$[\epsilon\mathbf{u}, \epsilon\mathbf{v}] = 0.$$

This is readily achieved by choosing the vector fields to be constant. In this event, our very first expression for the exterior derivative, (36.2), reduces to

$$\mathbf{d}\varphi(\epsilon\mathbf{u}, \epsilon\mathbf{v}) = \nabla_{\epsilon\mathbf{u}}\, \varphi(\epsilon\mathbf{v}) - \nabla_{\epsilon\mathbf{v}}\, \varphi(\epsilon\mathbf{u}), \qquad (37.7)$$

and *this* is our flag.

With our objective clear, let us return to the evaluation of the integral $\mathbf{\Omega}(\epsilon\mathbf{u}, \epsilon\mathbf{v})$. Since we will ultimately shrink the parallelogram to nothing by letting ϵ go to zero, we may use a Riemann sum with just a single term per side of the parallelogram. Usually a Riemann sum is used to evaluate an integral along some fixed interval, breaking it down into a larger and larger number of smaller and smaller subintervals, so that the precise point at which we evaluate φ on each subinterval becomes irrelevant in this limit.

However, that is *not* what we are doing here: we seek the *limiting form*—pun intended—of the integral as ϵ vanishes, *and the integral vanishes with it*. If we simply pick a *random* point on each side, then the error dies away as ϵ^2, and that is simply not good enough. For it is intuitively clear that the contributions to Ω from opposite sides of Π will *cancel* to order ϵ, so that Ω itself will be of order ϵ^2—the *same* order as the error incurred in a random Riemann sum.

The answer is to instead choose the *midpoints* of the sides, for then *the error incurred dies away much faster—in fact as* ϵ^3. (This is proved geometrically in VCA, Section 8.2.3.) By using midpoints, we therefore immediately attain the accuracy we require[2] to find a net result for Ω that is of order ϵ^2.

Therefore, as illustrated in [37.3], let us approximate $\mathbf{\Omega}(\epsilon\mathbf{u}, \epsilon\mathbf{v})$ using these midpoints, a, b, c, d, and let the corresponding vector at a be $\underline{\varphi}_a$, and so on, as illustrated. Then, taking all derivatives to be evaluated at the centre of the parallelogram,

$$
\begin{aligned}
\mathbf{\Omega}(\epsilon\mathbf{u}, \epsilon\mathbf{v}) &\asymp \varphi_a(\epsilon\mathbf{u}) + \varphi_b(\epsilon\mathbf{v}) + \varphi_c(-\epsilon\mathbf{u}) + \varphi_d(-\epsilon\mathbf{v}) \\
&= \left[\varphi_b(\epsilon\mathbf{v}) - \varphi_d(\epsilon\mathbf{v})\right] - \left[\varphi_c(\epsilon\mathbf{u}) - \varphi_a(\epsilon\mathbf{u})\right] \\
&\asymp \nabla_{\epsilon\mathbf{u}}\, \varphi(\epsilon\mathbf{v}) - \nabla_{\epsilon\mathbf{v}}\, \varphi(\epsilon\mathbf{u}) \\
&= \mathbf{d}\varphi(\epsilon\mathbf{u}, \epsilon\mathbf{v}),
\end{aligned}
$$

by virtue of (37.7). Abstracting away the vectors, this completes the proof of (37.6).

We stress that there is nothing intrinsically 2-dimensional about the preceding argument: picture the $\underline{\varphi}$ vectors in [37.3] as sticking out of the page of the book, and pick up and tilt the book so that Π takes up any position in space!

If we now think of the vector field $\underline{\varphi}$ as a fluid flow in space, then the result (37.6) attaches a wonderfully concrete and vivid *meaning* to the exterior derivative:

> *If fluid flows with velocity $\underline{\varphi}$, then its circulation Ω around a small (ultimately vanishing) parallelogram is ultimately equal to the result of applying the flux 2-form* $\mathbf{d}\varphi$ *to its edges.* (37.8)

For future use, let us express this crucial result as an explicit formula in the language of classical Vector Calculus. Let $\widehat{\mathbf{n}}$ be the unit normal to the parallelogram, its direction determined by the right-hand rule, curling our fingers from $\mathbf{u}$ to $\mathbf{v}$. Also, let $\delta\mathcal{A}$ be the area of the parallelogram:

$$\delta\mathcal{A} = |\epsilon\,\mathbf{u} \times \epsilon\,\mathbf{v}|,$$

so that

$$\epsilon\,\mathbf{u} \times \epsilon\,\mathbf{v} = \widehat{\mathbf{n}}\,\delta\mathcal{A}.$$

The flux of $\nabla \times \underline{\varphi}$ through the parallelogram is its component perpendicular to the parallelogram (i.e., its component in the direction of $\widehat{\mathbf{n}}$) multiplied by the area of the parallelogram:

$$\text{flux} \asymp (\nabla \times \underline{\varphi}) \cdot \widehat{\mathbf{n}}\,\delta\mathcal{A}. \tag{37.9}$$

Then, combining (36.15), (37.8), and (37.9),

$$\oint_{\Pi(\epsilon\mathbf{u}, \epsilon\mathbf{v})} \varphi = \mathbf{\Omega}(\epsilon\mathbf{u}, \epsilon\mathbf{v}) \asymp \mathbf{d}\varphi(\epsilon\mathbf{u}, \epsilon\mathbf{v}) = (\nabla \times \underline{\varphi}) \cdot \widehat{\mathbf{n}}\,\delta\mathcal{A}. \tag{37.10}$$

---

[2]We first used this reasoning in VCA (Section 8.10.2) to provide a direct, geometrical proof of Cauchy's Theorem.

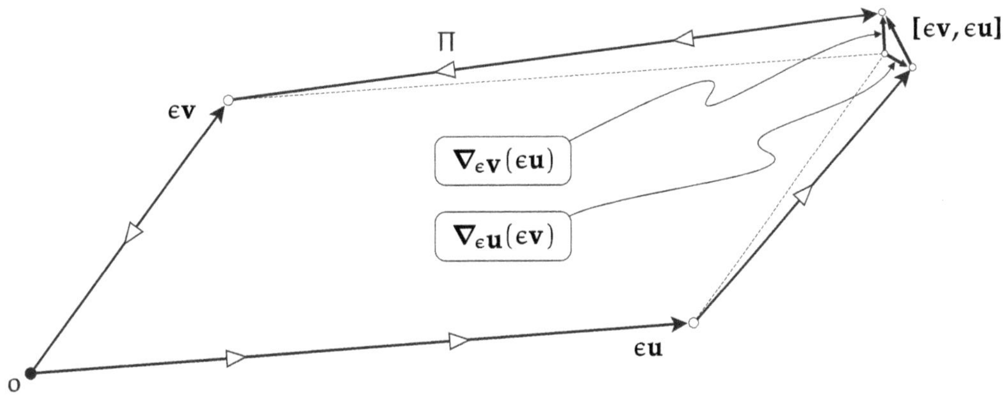

[37.4] *General vector fields do not create a closed parallelogram. However, we may create a closed loop* Π *by bridging the gap with the commutator:* $[\epsilon\mathbf{v}, \epsilon\mathbf{u}] = \nabla_{\epsilon\mathbf{v}}\,\epsilon\mathbf{u} - \nabla_{\epsilon\mathbf{u}}\epsilon\mathbf{v}$.

We end with another bonus: a *geometrical* explanation of the original, full formula for the exterior derivative, (36.2), *including* its formerly mysterious commutator term. Recall that we first arrived at the following formula for $\mathbf{d}\varphi$ by means of a purely formal calculation (albeit an enlightened calculation):

$$\mathbf{d}\varphi(\mathbf{u}, \mathbf{v}) = \nabla_{\mathbf{u}}\varphi(\mathbf{v}) - \nabla_{\mathbf{v}}\varphi(\mathbf{u}) - \varphi([\mathbf{u}, \mathbf{v}]).$$

We chose to simplify the foregoing exposition by insisting on constant vector fields, but what if $\mathbf{u}$ and $\mathbf{v}$ are *not* constant? Then, as illustrated in [37.4], the vector fields do not generally create a closed loop. However, we may *create a closed loop* Π *by bridging the gap with the commutator* $[\epsilon\mathbf{v}, \epsilon\mathbf{u}]$. The bulk of the foregoing analysis goes through unchanged, but now *the integral* $\boldsymbol{\Omega}(\epsilon\mathbf{u}, \epsilon\mathbf{v})$ *gains one extra contribution from the gap-closing commutator edge:*

$$\varphi([\epsilon\mathbf{v}, \epsilon\mathbf{u}]) = -\varphi([\epsilon\mathbf{u}, \epsilon\mathbf{v}]),$$

thereby explaining the formula for $\mathbf{d}\varphi$!

37.2.2 **d***(2-Form)*

Let us repeat the above analysis for the integral of a 2-form $\boldsymbol{\Psi}$ over the oriented 2-dimensional boundary $\Pi(\epsilon\mathbf{u}, \epsilon\mathbf{v}, \epsilon\mathbf{w})$ of a small—ultimately vanishing, as ϵ vanishes—right-handed parallelepiped with first edge $\epsilon\mathbf{u}$, second edge $\epsilon\mathbf{v}$, and third edge $\epsilon\mathbf{w}$.

Let us define $\boldsymbol{\Omega}(\epsilon\mathbf{u}, \epsilon\mathbf{v}, \epsilon\mathbf{w})$ to be the integral of $\boldsymbol{\Psi}$ over Π:

$$\boldsymbol{\Omega}(\epsilon\mathbf{u}, \epsilon\mathbf{v}, \epsilon\mathbf{w}) \equiv \iint\limits_{\Pi(\epsilon\mathbf{u}, \epsilon\mathbf{v}, \epsilon\mathbf{w})} \boldsymbol{\Psi}.$$

In the language of classical Vector Calculus, this is the net *outward flux* of $\underline{\boldsymbol{\Psi}}$ through Π:

$$\Omega(\Pi) = \iint\limits_{\Pi} \underline{\boldsymbol{\Psi}} \cdot \hat{\mathbf{n}}\, dA,$$

where $\hat{\mathbf{n}}$ is the unit normal pointing *out* of the parallelepiped.

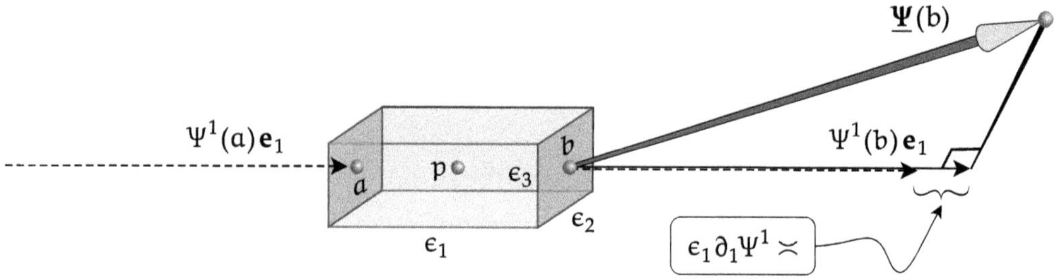

[37.5] *Flux out of right face* $\asymp \epsilon_2\epsilon_3\,\Psi^1(b)$. *Flux out of left face* $\asymp -\epsilon_2\epsilon_3\,\Psi^1(a)$. *Therefore, the net outflux through this pair of faces is* $\asymp [\partial_1\Psi^1\epsilon_1]\,\epsilon_2\epsilon_3 = [\partial_1\Psi^1]\,\mathcal{V}$.

In light of the 1-form result (37.6), we anticipate that this integral is a *3-form*, and not just any 3-form—it should be the *exterior derivative* of $\boldsymbol{\Psi}$:

$$\iint_{\Pi(\epsilon\mathbf{u},\epsilon\mathbf{v},\epsilon\mathbf{w})} \boldsymbol{\Psi} = \boldsymbol{\Omega}(\epsilon\mathbf{u},\epsilon\mathbf{v},\epsilon\mathbf{w}) \asymp \mathbf{d}\boldsymbol{\Psi}(\epsilon\mathbf{u},\epsilon\mathbf{v},\epsilon\mathbf{w}) = (\boldsymbol{\nabla}\boldsymbol{\cdot}\underline{\boldsymbol{\Psi}})\,\mathcal{V}(\epsilon\mathbf{u},\epsilon\mathbf{v},\epsilon\mathbf{w}),$$

by virtue of (36.16). So, abstracting the vectors away, we are led to believe that

$$\boldsymbol{\Omega} \asymp \mathbf{d}\boldsymbol{\Psi} = (\boldsymbol{\nabla}\boldsymbol{\cdot}\underline{\boldsymbol{\Psi}})\,\mathcal{V}. \tag{37.11}$$

This is indeed true, and the proof is simplest and most transparent in the case that the edges of the parallelepiped (which we take to be centred at p) are orthogonal, directed along the orthonormal basis vectors: $\{\epsilon_1\,\mathbf{e}_1, \epsilon_2\,\mathbf{e}_2, \epsilon_3\,\mathbf{e}_3\}$, so that its volume is $\mathcal{V} = \epsilon_1\epsilon_2\epsilon_3$.

As illustrated in [37.5], consider the flux of $\underline{\boldsymbol{\Psi}}$ through the two faces that are orthogonal to $\mathbf{e}_1$, that are separated by distance ϵ_1, and whose centres are a and b. Only the component $\Psi^1\mathbf{e}_1$ carries fluid across these these faces; the other two components flow parallel to the faces. (For the sake of definiteness, in this figure we have assumed that $\Psi^1 > 0$.)

To evaluate the flux out of the right-hand face, we multiply its area $\epsilon_2\epsilon_3$ by the orthogonal component $\Psi^1(b)$ of the velocity at the centre of the face, so that

$$\text{flux out of right face} \asymp \epsilon_2\epsilon_3\,\Psi^1(b).$$

For the left-hand face, the outward-pointing normal $\widehat{\mathbf{n}} = -\mathbf{e}_1$ is *opposite* to the outward normal of the right face, so

$$\textit{outward}\text{ velocity of fluid on left face} = \underline{\boldsymbol{\Psi}}\boldsymbol{\cdot}\widehat{\mathbf{n}} = -\Psi^1(a).$$

Therefore,

$$\text{flux } \textit{out} \text{ of left face} = -(\text{flux } \textit{into} \text{ left face}) \asymp -\epsilon_2\epsilon_3\,\Psi^1(a).$$

Note that in [37.5] we have drawn $\Psi^1(a)\mathbf{e}_1$ (dashed) with its arrowhead *at* a, both to declutter the picture and to make it clear that this represents flow *into* the box. Note that we have also drawn a *copy* of $\Psi^1(a)\mathbf{e}_1$ at b, to facilitate an upcoming comparsion to $\Psi^1(b)\mathbf{e}_1$.

Adding these outfluxes,

$$\begin{aligned}\text{NET out-flux through this pair of faces} &\asymp [\Psi^1(b) - \Psi^1(a)]\,\epsilon_2\epsilon_3\\ &\asymp [\partial_1\Psi^1\epsilon_1]\,\epsilon_2\epsilon_3\\ &= [\partial_1\Psi^1]\,\mathcal{V},\end{aligned}$$

in which the derivative is evaluated at the centre p of the box.

Of course exactly the same reasoning applies to the pair of faces orthogonal to $\mathbf{e}_2$, yielding a net outflux of $[\partial_2\Psi^2]\mathcal{V}$ through that pair. The final pair likewise contributes $[\partial_3\Psi^3]\mathcal{V}$. Thus,

$$\Omega(\Pi) = \text{Total flux out of } \Pi \asymp [\partial_1\Psi^1 + \partial_2\Psi^2 + \partial_3\Psi^3]\mathcal{V} = (\boldsymbol{\nabla}\cdot\underline{\Psi})\mathcal{V},$$

thereby proving (37.11).

The results (37.6) and (37.11) are part of a pattern that continues on into higher dimensions. Though visualization fails us, it is nevertheless true that **d**(3-form) applied to a small, compact, ultimately vanishing region of a 4-dimensional space yields the integral of the 3-form over its 3-dimensional boundary, and so on. For more on these higher-dimensional results, see the *Further Reading* section at the end of this book.

37.3 Fundamental Theorem of Exterior Calculus (Generalized Stokes's Theorem)

37.3.1 *Fundamental Theorem of Exterior Calculus*

All of the integral theorems you learned in Vector Calculus—(37.5), Green's Theorem, Stokes's Theorem, and Gauss's Theorem—are merely special cases of *one* elegant theorem about Forms.

The Fundamental Theorem of Exterior Calculus:

$$\int_R \mathbf{d}\boldsymbol{\Phi} = \int_{\partial R} \boldsymbol{\Phi} \tag{37.12}$$

On the left, the $(p+1)$-form $\mathbf{d}\boldsymbol{\Phi}$ is integrated over a compact (oriented) $(p+1)$-dimensional region R, and, on the right, the p-form $\boldsymbol{\Phi}$ is integrated over the (oriented) p-dimensional *boundary* ∂R of R. Note that we must dispense with multiple integral signs, instead using a single integral sign on both sides of this equation, because the degree p is general and unknown.

WARNING: What we have called the ***Fundamental Theorem of Exterior Calculus*** (hereinafter, FTEC) is instead called, in essentially all other books, the ***Generalized Stokes's Theorem***, often abbreviated as GST.

37.3.2 *Historical Aside*

Unquestionably, Stokes made great contributions to science, but it is singularly inappropriate that *this* theorem should bear his name, and his name alone. Vladimir Arnol'd (1989, p. 192) refers to (37.12)—tongue-in-cheek—as the "*Newton-Leibniz-Gauss-Green-Ostrogradskii-Stokes-Poincaré formula*"! But, out of this list, Stokes actually had the *least* to do with the theorem.

In 1854, Cambridge University's Smith's Prize examination was devised by Stokes. Question number 8 asked the students to prove the original, pregeneralized result—the one we now call *Stokes's Theorem*; this was the very first appearance of the result in print. One of the candidates who sat for that examination was none other than James Clerk Maxwell—who tied for first place, by the way—and the initial naming of the theorem after Stokes likely stemmed from this event!

But Stokes only knew of the result because William Thomson (Lord Kelvin) had sent it to him in a letter, on the 2nd of July, 1850! But Kelvin seems to have been led to the result by studying Green ...! You begin to see what a pickle we are in; for extra pickles, see Katz (1979) and Crowe (1985).

An Alexandrian solution to this Gordian Knot of a naming conundrum was proposed by N.M.J. Woodhouse to Roger Penrose: jettison *all* these names, and call it instead the *Fundamental Theorem of Exterior Calculus*. That is the solution adopted in Penrose and Rindler (1984), and subsequently in Penrose (2005), and we hereby follow those precedents.

As for the history of the general result, (37.12), we shall only say this: the first complete statement occurred in a lecture course delivered by Cartan in Paris in 1936–1937, in which he explicitly pointed out that all the results of Vector Calculus were special cases; these lectures first appeared in print in Cartan (1945).

37.3.3 Example: Area

Returning to the result itself, let us begin with an elementary geometrical example of the theorem in action.

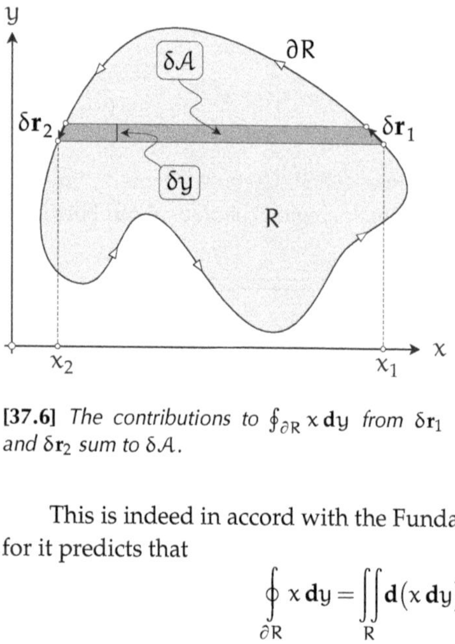

[37.6] *The contributions to* $\oint_{\partial R} x\,\mathbf{d}y$ *from* $\delta\mathbf{r}_1$ *and* $\delta\mathbf{r}_2$ *sum to* $\delta\mathcal{A}$.

As illustrated in [37.6], in $\mathbb{R}^2$ consider the integral of the 1-form $x\,\mathbf{d}y$ around the counterclockwise boundary ∂R of the region R of area $\mathcal{A}(R)$.

Let us evaluate the contributions to the integral from the illustrated matching pair of movements $\delta\mathbf{r}_1$ and $\delta\mathbf{r}_2$ sandwiched between the horizontal lines at heights y and $y+\delta y$:

$$(x\,\mathbf{d}y)(\delta\mathbf{r}_1)+(x\,\mathbf{d}y)(\delta\mathbf{r}_2)=(x_1-x_2)\,\delta y \asymp \delta\mathcal{A}.$$

Thus,

$$\oint_{\partial R} x\,\mathbf{d}y=\mathcal{A}(R).$$

(Check that this argument continues to work properly even if we slide the shaded horizonal strip down until it splits into two strips.)

This is indeed in accord with the Fundamental Theorem of Exterior Calculus (FTEC), (37.12), for it predicts that

$$\oint_{\partial R} x\,\mathbf{d}y=\iint_R \mathbf{d}(x\,\mathbf{d}y)=\iint_R \mathbf{d}x\wedge\mathbf{d}y=\iint_R \mathcal{A}=\mathcal{A}(R).$$

This also makes physical sense. According to the correspondence (34.10), $\mathbf{d}x\wedge\mathbf{d}y$ corresponds to a flow in the z-direction, orthogonal to the (x,y)-plane, with unit speed. Thus its flux through a region of area $\mathcal{A}$ is equal to $\mathcal{A}$.

37.4 The Boundary of a Boundary Is Zero!

Before we prove the Fundamental Theorem of Exterior Calculus—or FTEC—let us see how it resolves a mystery we have been living with for some time: *why* is $\mathbf{d}^2=0$? The computational *proof* of this fundamental result was short and easy (see Section 36.4.1), but what does the result itself actually *mean*?!

If we apply FTEC *twice*, we find that

$$0=\int_R \mathbf{d}^2\Phi=\int_{\partial R}\mathbf{d}\Phi=\int_{\partial(\partial R)}\Phi.$$

But since Φ and R are arbitrary, this can only be true if

$\partial^2 = 0 \iff$ *The boundary of a boundary is zero!*

Conversely, if we can understand this geometrical statement, we will have explained the geometrical *meaning* of $\mathbf{d}^2 = 0$:

$$\mathbf{d}^2 = 0 \iff \partial^2 = 0.$$

Let us explain this in 3-2-1 dimensions. Consider [37.7], which shows a right-handed, 3-dimensional parallelepiped with edges $\{\mathbf{v}_1, \mathbf{v}_2, \mathbf{v}_3\}$. Its boundary consists of six 2-dimensional faces, with indicated orientations. Finally, the boundary of the boundary consists of 12 edges, *each of which is traversed once in one direction, and once in the opposite direction.*

Thus, as illustrated, we see that the 1-dimensional boundary of the 2-dimensional boundary of the 3-dimensional solid is indeed *zero!*

For an explanation in 4-3-2 dimensions and beyond, see Misner, Thorne, and Wheeler (1973, §15). As they explain, but we shall not, this is also the key to understanding the Differential Bianchi Identity, (29.17).

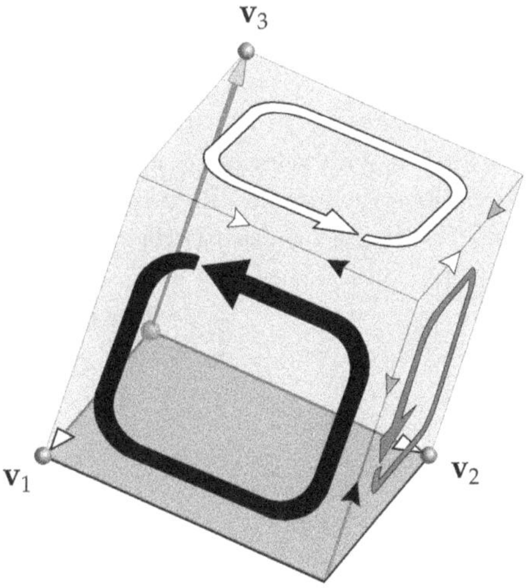

[37.7] *The boundary of a boundary is zero:* $\partial^2 = 0$.

37.5 The Classical Integral Theorems of Vector Calculus

Of course the above geometrical explanation of $\mathbf{d}^2 = 0$ hinges on FTEC, and we have yet to explain why *that* is true! But before we offer a proof of FTEC, we shall ramp up the suspense still further, in order to convince you that the Fundamental Theorem of Exterior Calculus truly is *"Fundamental"!* To that end, we now verify our earlier claim that this single theorem contains *all* of the classical theorems of Vector Calculus as special cases.

37.5.1 $\boldsymbol{\Phi} = 0$-Form

Taking $\boldsymbol{\Phi} = f$ to be a function (0-form) in FTEC, $\mathbf{d}\boldsymbol{\Phi} = \mathbf{d}f$ is a 1-form, so R must be taken to be a 1-dimensional curve K, and its boundary ∂K must be taken to be its end points, say, a and b. Since K points at b and points away from a, we count the arrival point as $+b$, and the departure point as $-a$. Then FTEC yields the previously established (37.4), and its classical vector equivalent (37.5):

$$\int_K \mathbf{d}f = f(b) - f(a) \iff \int_K (\boldsymbol{\nabla} f) \cdot \mathbf{d}\mathbf{r} = f(b) - f(a).$$

37.5.2 $\boldsymbol{\Phi} = 1$-Form

Taking $\boldsymbol{\Phi} = \varphi$ to be a 1-form in FTEC, $\mathbf{d}\boldsymbol{\Phi} = \mathbf{d}\varphi$ is the flux 2-form of the vector field $\boldsymbol{\nabla} \times \underline{\varphi}$, so R must be taken to be an oriented 2-dimensional surface $\mathcal{S}$, and its boundary $\partial\mathcal{S}$ must be taken to be its 1-dimensional boundary curve.

Let us begin with the simplest case, and historically first case, in which the surface $\mathcal{S}$ is simply a planar region R of the (x, y)-plane, exactly as in our area example above, [37.6]. Here the vector field is also planar, so that the 1-form is given by

$$\varphi = \varphi_x \, \mathbf{d}x + \varphi_y \, \mathbf{d}y.$$

Then the corresponding flux 2-form is

$$\begin{aligned} \mathbf{d}\varphi &= \mathbf{d}\varphi_x \wedge \mathbf{d}x + \mathbf{d}\varphi_y \wedge \mathbf{d}y \\ &= \partial_y \varphi_x \, \mathbf{d}y \wedge \mathbf{d}x + \partial_x \varphi_y \, \mathbf{d}x \wedge \mathbf{d}y \\ &= (\partial_x \varphi_y - \partial_y \varphi_x) \, \mathcal{A}. \end{aligned}$$

According to (34.10), this $\mathbf{d}\varphi$ corresponds to a flow in the z-direction, orthogonal to the (x, y)-plane, with velocity $(\partial_x \varphi_y - \partial_y \varphi_x)$. According to FTEC, its flux through R is equal to the circulation of $\underline{\varphi}$ around ∂R. In other words, if we begin and end with classical notation, we see that FTEC implies ***Green's Theorem***:[3]

$$\begin{aligned} \oint_{\partial R} \varphi_x \, dx + \varphi_y \, dy &= \oint_{\partial R} \varphi \\ &= \iint_R \mathbf{d}\varphi \\ &= \iint_R (\partial_x \varphi_y - \partial_y \varphi_x) \, \mathcal{A} \\ &= \iint_R (\partial_x \varphi_y - \partial_y \varphi_x) \, dx \, dy. \end{aligned}$$

This is a generalized version of the area example in [37.6].

Turning now to the case of a more general, curved surface $\mathcal{S}$ in space, let $\widehat{\mathbf{n}}$ be the unit normal to surface, the choice of direction being determined by the orientation of the surface via the right-hand rule. Then, by virtue of (37.9), FTEC yields ***Stokes's Theorem***:

$$\begin{aligned} \oint_{\partial\mathcal{S}} \underline{\varphi} \cdot d\mathbf{r} &= \oint_{\partial\mathcal{S}} \varphi \\ &= \iint_{\mathcal{S}} \mathbf{d}\varphi \\ &= \iint_{\mathcal{S}} (\boldsymbol{\nabla} \times \underline{\varphi}) \cdot \widehat{\mathbf{n}} \, d\mathcal{A}. \end{aligned}$$

This theorem provides the crucial bridge between the macroscopic electromagnetic phenomena that Faraday observed in his laboratory, and the microscopic differential equations to which Maxwell ultimately reduced them mathematically.

[3]Green never actually wrote down the 2-dimensional formula that now bears his name, but it *is* a logical consequence of a 3-dimensional formula that he did discover. The first known publication of Green's Theorem (but without any proof) was by Cauchy, in 1846. See Katz (1979).

Take ***Faraday's Law of Electromagnetic Induction***, which Faraday discovered in 1831:

> *If the surface $\mathcal{S}$ spans a loop of wire $\partial\mathcal{S}$, and the field lines of a varying magnetic field pass through it, then an electromotive force is induced in the loop that is equal to the negative of the rate of change of the magnetic flux through $\mathcal{S}$.*

Expressed in mathematical form,

$$\oint_{\partial\mathcal{S}} \underline{\mathbf{E}} \bullet \mathrm{d}\mathbf{r} = -\partial_t \iint_{\mathcal{S}} \underline{\mathbf{B}} \bullet \hat{\mathbf{n}}\, \mathrm{d}\mathcal{A}.$$

But then Stokes's Theorem allows us to re-express this as

$$\iint_{\mathcal{S}} \left[\nabla \times \underline{\mathbf{E}} + \partial_t \underline{\mathbf{B}}\right] \bullet \hat{\mathbf{n}}\, \mathrm{d}\mathcal{A} = 0.$$

But if this is to be true for all surfaces $\mathcal{S}$, then the integrand itself must vanish, and we thereby find that Faraday's Law of 1831 is equivalent to one of Maxwell's equations of 1873:

$$\nabla \times \underline{\mathbf{E}} + \partial_t \underline{\mathbf{B}} = 0. \tag{37.13}$$

37.5.3 $\boldsymbol{\Phi} = 2$-Form

If we take $\boldsymbol{\Phi} = \boldsymbol{\Psi}$ to be a flux 2-form in FTEC, then, by virtue of (37.11), $\mathbf{d}\boldsymbol{\Psi} = (\nabla \bullet \underline{\boldsymbol{\Psi}})\, \mathcal{V}$ is the 3-form describing the source density. Now R must be taken to be a an oriented volume V, and its boundary $\mathcal{S} = \partial V$ must be taken to be its 2-dimensional boundary surface. So, if we begin and end with classical notation, we see that FTEC implies ***Gauss's Theorem***, also known as the ***Divergence Theorem:***

$$\iiint_V (\nabla \bullet \underline{\boldsymbol{\Psi}})\, \mathrm{d}\mathcal{V} = \iiint_V \mathbf{d}\boldsymbol{\Psi} = \oiint_{\partial V} \boldsymbol{\Psi} = \oiint_{\mathcal{S}} \underline{\boldsymbol{\Psi}} \bullet \hat{\mathbf{n}}\, \mathrm{d}\mathcal{A}. \tag{37.14}$$

37.6 Proof of the Fundamental Theorem of Exterior Calculus

We shall prove the 2-dimensional form (Stokes's Theorem) of FTEC, and we shall also extend the proof (essentially unchanged) to the 3-dimensional case (Gauss's Theorem). But we shall not develop the concepts, terminology, and notation required to explicitly extend the proof, in a uniform manner, to all dimensions. For that, see the *Further Reading* section at the end of this book.

Consider [37.8], which shows a region R of the (x, y)-plane divided up into small (ultimately vanishing) parallelograms, each of which has first edge $\epsilon\mathbf{u}$ and second edge $\epsilon\mathbf{v}$. The orientation associated with this ordering is indicated by the swirl within each cell. Note that the parallelograms *abutting* the boundary ∂R are cut off into irregular shapes.

Now let us sum the integrals of the 1-form φ around all of the parallelograms and truncated parallelograms into which R has been divided. We shall collectively refer to the parallelograms and the cut-off parallelograms as the ***cells*** of R.

Consider the illustrated, black, interior parallelogram, and the four white copies that abut it. Every black edge is matched with an oppositely directed white edge of a neighbouring cell, and so the integrals along these two edges *cancel*. The only edges that are *not* cancelled are the small pieces of the boundary, ∂R, such as the illustrated $\delta\mathbf{r}$. Therefore,

$$\oint_{\partial R} \varphi = \sum_{\text{cells in R}} \oint_{\partial(\text{cell})} \varphi \qquad (37.15)$$

One might naively guess that the integral around each cell should be of order ϵ, since each edge is of that order. But that cannot be correct, as one can see from the following rough argument.

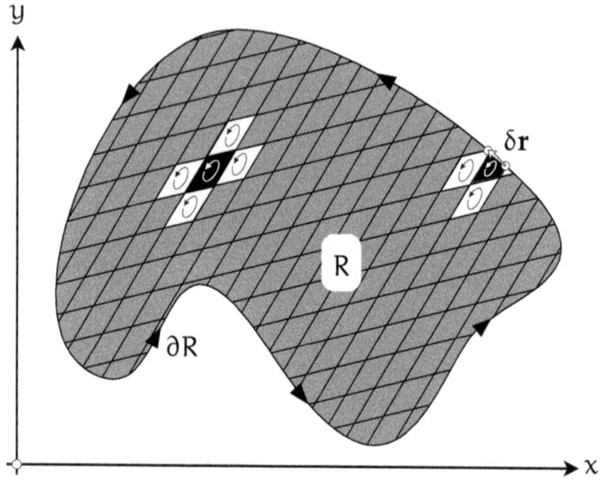

[37.8] *The circulations along the edges of the interior, black parallelogram are all cancelled by the oppositely directed edges of the abutting white parallelograms. Summing over all cells, the only surviving contributions are from the uncancelled boundary edges, such as the illustrated* $\delta\mathbf{r}$.

We know that for a general 1-form, the integral round ∂R—hence the sum (37.15)—will be nonzero and finite. This leads us to believe that each term must die away with the reciprocal dependence on ϵ as governs the growth of the number of terms in the series. But the number of terms grows as *(fixed area of R, divided by the area of each cell)*—that is, as $(1/\epsilon^2)$. Thus the magnitude of each term is expected to die away as ϵ^2. If our original guess had been correct, the order of the sum in (37.15) would have been $\epsilon(1/\epsilon^2)$, yielding an *infinite* result as the parallelograms shrunk. Conversely, any contributions to the terms involving powers of ϵ *greater* than two cannot have any influence on the final result.

In fact, there is no need to appeal to such rough reasoning, because we previously derived the precise result (37.8) governing the integral of φ around such a shrinking parallelogram, and this is the key to understanding and proving FTEC. We see that, as anticipated, the contribution from each parallelogram is indeed of order ϵ^2, and is given by the application of the 2-form $\mathbf{d}\varphi$ to the edges of the parallelogram. Thus, (37.15) becomes

$$\oint_{\partial R} \varphi = \sum_{\text{cells in R}} \mathbf{d}\varphi(\epsilon\mathbf{u}, \epsilon\mathbf{v}).$$

But, in the limit that ϵ vanishes, this sum is precisely the definition of the integral of $\mathbf{d}\varphi$ over R, thereby completing our proof of the Fundamental Theorem of Exterior Calculus:

$$\oint_{\partial R} \varphi = \iint_{R} \mathbf{d}\varphi.$$

Now let us generalize this argument to prove the Stokes's Theorem form of FTEC. To do so, we need only suppose that the region R upon which our parallelograms are drawn is actually a rubber membrane, spanning a stiff but bendable wire, ∂R. We may now deform this wire ∂R out of $\mathbb{R}^2$ into space, so that R takes up the form of some curved surface in $\mathbb{R}^3$. We may then further deform the surface R as we please, while leaving the boundary wire in place.

The two critical ingredients of the above proof were the cancellation of all interior edges, and (37.8). *Both* of these remain true in this new 3-dimensional context, the latter taking the form (37.10):

$$\oint_{\Pi(\epsilon\mathbf{u},\epsilon\mathbf{v})} \varphi \asymp \mathbf{d}\varphi(\epsilon\mathbf{u}, \epsilon\mathbf{v}) = (\boldsymbol{\nabla} \times \underline{\varphi}) \boldsymbol{\cdot} \widehat{\mathbf{n}}\, \delta\mathcal{A}.$$

Thus we have proved the Fundamental Theorem of Exterior Calculus in this case, thereby also proving Stokes's Theorem in the process:

$$\oint_{\partial \mathcal{S}} \underline{\varphi} \bullet d\mathbf{r} = \oint_{\partial \mathcal{S}} \varphi = \iint_{\mathcal{S}} \mathbf{d}\varphi = \iint_{\mathcal{S}} (\nabla \times \underline{\varphi}) \bullet \hat{\mathbf{n}}\, d\mathcal{A}.$$

Finally, let us step up a dimension to see why the above proof of FTEC goes through essentially unchanged, and yields, in the process, a proof of Gauss's Theorem. Let V be a compact region of $\mathbb{R}^3$ with boundary surface $\mathcal{S} = \partial V$. Now fill V into parallelepipeds with edges $\{\epsilon \mathbf{u}, \epsilon \mathbf{v}, \epsilon \mathbf{w}\}$, and volume $\mathcal{V}(\epsilon \mathbf{u}, \epsilon \mathbf{v}, \epsilon \mathbf{w})$. The ones at the surface are truncated, just as happened with the parallelograms of [37.8]. Again, let us collectively refer the interior parallelepipeds and truncated ones at the surface as the cells of V.

Now consider the flux of the 2-form Ψ out of an interior cell. Each of its faces is also the face of a neighbouring cell, of *opposite orientation*: the outward normal of our cell points *into* the neighbouring cell. Thus if we sum the fluxes out of all the cells, the fluxes of all the interior faces *cancel*: physically, the fluid that flows out of a face flows into its neighbour. Thus the only faces that are not cancelled are those comprising part of the boundary surface, and therefore,

$$\oiint_{\partial V} \Psi = \sum_{\text{cells in } V} \oiint_{\partial(\text{cell})} \Psi.$$

But we know from (37.11) that the flux out of a small, ultimately vanishing parallelepiped is ultimately of order ϵ^3 and is obtained by applying the 3-form $\mathbf{d}\Psi$ to its edges:

$$\oiint_{\Pi(\epsilon \mathbf{u}, \epsilon \mathbf{v}, \epsilon \mathbf{w})} \Psi \asymp \mathbf{d}\Psi(\epsilon \mathbf{u}, \epsilon \mathbf{v}, \epsilon \mathbf{w}) = (\nabla \bullet \underline{\Psi})\, \mathcal{V}(\epsilon \mathbf{u}, \epsilon \mathbf{v}, \epsilon \mathbf{w}),$$

Thus,

$$\oiint_{\partial V} \Psi \asymp \sum_{\text{cells in } V} \mathbf{d}\Psi(\epsilon \mathbf{u}, \epsilon \mathbf{v}, \epsilon \mathbf{w}).$$

Finally, as ϵ vanishes, the right-hand side becomes the integral of $\mathbf{d}\Psi$ over V. Thus we have proved the Fundamental Theorem of Exterior Calculus in this case.

Beginning and ending with the notation of classical Vector Calculus, we have also proved Gauss's Theorem in the process:

$$\oiint_{\mathcal{S}} \underline{\Psi} \bullet \hat{\mathbf{n}}\, d\mathcal{A} = \oiint_{\partial V} \Psi = \iiint_{V} \mathbf{d}\Psi = \iiint_{V} (\nabla \bullet \underline{\Psi})\, d\mathcal{V}.$$

37.7 Cauchy's Theorem

Suppose that the region R depicted in [37.8] resides not in $\mathbb{R}^2$ but in the *complex* plane, and that a complex function $f(z)$ is analytic throughout it, so that its local geometric effect is an amplitwist. Then, as we saw in (36.13), page 397, the fact that $f(z)$ is locally an amplitwist is equivalent to the statement that the 1-form $f\,\mathbf{d}z$ is *closed*: $\mathbf{d}(f\,\mathbf{d}z) = 0$.

Therefore, FTEC immediately yields ***Cauchy's Theorem***:

$$\oint_{\partial R} f\,\mathbf{d}z = \iint_{R} \mathbf{d}(f\,\mathbf{d}z) = 0.$$

FTEC can also be used to obtain interesting results when $f(z)$ is *not* locally an amplitwist. For example, in the case of the anticonformal mapping $f(z) = \bar{z}$, we saw in (36.14) that $\mathbf{d}(\bar{z}\,\mathbf{d}z) = 2i\,\mathcal{A}$, so now FTEC yields

$$\oint_{\partial R} \bar{z}\,\mathbf{d}z = \iint_R \mathbf{d}(\bar{z}\,\mathbf{d}z) = 2i \iint_R \mathcal{A} = 2i\,\mathcal{A}(R).$$

For geometrical and physical explanations of this result, see VCA.

37.8 The Poincaré Lemma for 1-Forms

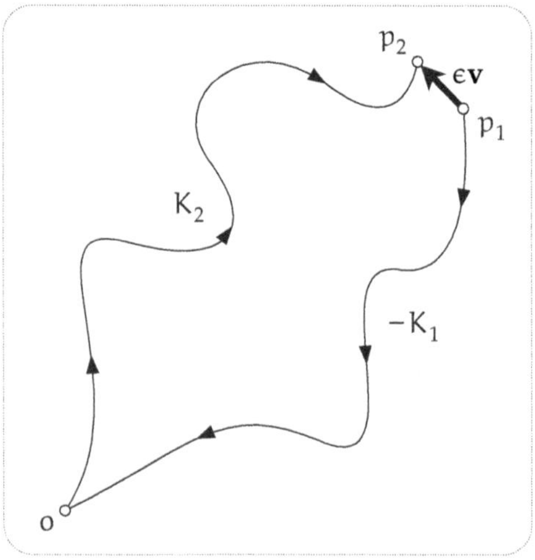

[37.9] *If* $\mathbf{d}\varphi = 0$ *then* $f(p) \equiv \int_o^p \varphi$ *is path-independent, and* $\varphi = \mathbf{d}f$.

Suppose that φ is *closed* throughout a simply connected region R of $\mathbb{R}^n$, i.e., $\mathbf{d}\varphi = 0$. Then, by FTEC, if L is a closed loop in R and we span it with a surface S, so that $L = \partial S$, then

$$\oint_L \varphi = \int_S \mathbf{d}\varphi = 0.$$

This implies, by virtue of (37.3), that $\int \varphi$ is *path-independent*.

Now let us pick an arbitrary fixed point o and treat it as our origin. Thinking of $\underline{\varphi}$ as a force field, we may then define a potential energy function $f(p)$ as the path-independent work done in carrying a particle from o to p:

$$f(p) \equiv \int_o^p \varphi.$$

We will now show that φ is *exact*, and is given by

$$\varphi = \mathbf{d}f, \tag{37.16}$$

thereby proving the Poincaré Lemma in the case of a 1-form.

Let p_1 and p_2 be two neighbouring points, as illustrated in [37.9], and let the short (ultimately vanishing) connecting vector between them be $\epsilon\mathbf{v} = \overrightarrow{p_1 p_2}$. Then the energy difference δf between the two points is

$$\delta f = f(p_2) - f(p_1) \asymp \mathbf{d}f(\epsilon\mathbf{v}).$$

But, as illustrated in [37.9], if K_1 is a path from o to p_1, and K_2 is a path from o to p_2, then $-K_1 + K_2$ is a path from p_1 to p_2. And since $\int \varphi$ is path-independent, we may replace this long, indirect route with the short, direct route along $\epsilon\mathbf{v}$. Therefore,

$$\mathbf{d}f(\epsilon\mathbf{v}) \asymp \int_{-K_1+K_2} \varphi = \int_{\epsilon\mathbf{v}} \varphi \asymp \varphi(\epsilon\mathbf{v}).$$

Since $\mathbf{v}$ is arbitrary, we may abstract it away, completing the proof of (37.16), and with it the proof of the Poincaré Lemma, (36.11), for 1-forms.

37.9 A Primer on de Rham Cohomology

37.9.1 Introduction

What if the region R is *not* simply connected? Then it turns out that a closed Form need *not* be exact, and the study of the Forms that are closed but not exact yields detailed information on the topology of R.

This information is encoded in what are called the ***de Rham cohomology groups***, $H^k(R)$, named after Georges de Rham (1903–1990), who discovered them in 1931. Here k refers to k-forms defined on R that are closed but not exact.

We will only provide a small taste of the subject, beginning with the 1-forms that are closed but not exact, yielding the *first* de Rham cohomology group, $H^1(R)$.

37.9.2 A Special 2-Dimensional Vortex Vector Field

In $\mathbb{R}^2$, with polar coordinates (r, θ), consider the special circular, counterclockwise vortex[4] centred at the origin, shown in [37.10], in which the speed of the flow is chosen to be $|\underline{\varphi}| = (q/2\pi r)$, where q is a constant that measures the strength of the vortex. The reason for this special choice will be revealed in just a moment.

As we approach the origin, the speed goes to infinity, so the origin is a ***singularity***, and we exclude this one point: the flow is defined in the ***punctured plane***, $\mathbb{R}^2 -$ origin.

Since the speed along the circle K_r of radius r is $(q/2\pi r)$, and the length of K_r is $2\pi r$, the circulation around it is

$$\mathcal{C} = \oint_{K_r} \varphi = q,$$

independent of the radius r. Indeed, it was precisely in order to make the circulation independent of the radius that we needed to choose the speed of the flow to be inversely proportional to the radius.

The mere fact that $\mathcal{C} \neq 0$ implies that

The 1-form φ of the vortex cannot be exact: $\varphi \neq \mathbf{d}f$, for any f.

Nevertheless, in a moment we will show that

The 1-form φ of the vortex is closed: $\mathbf{d}\varphi = 0$.

The existence of such a 1-form, that is closed but not exact, is made possible by the *puncturing* of the plane. A loop that encircles the vortex's singularity at the origin cannot be shrunk to a point without crossing the origin, so the punctured plane is *not* simply connected, and therefore there is no conflict with the Poincaré Lemma.

[4]Here we prefer the more evocative word ***vortex*** over the more technically precise term ***centre***. See [19.4], page 200.

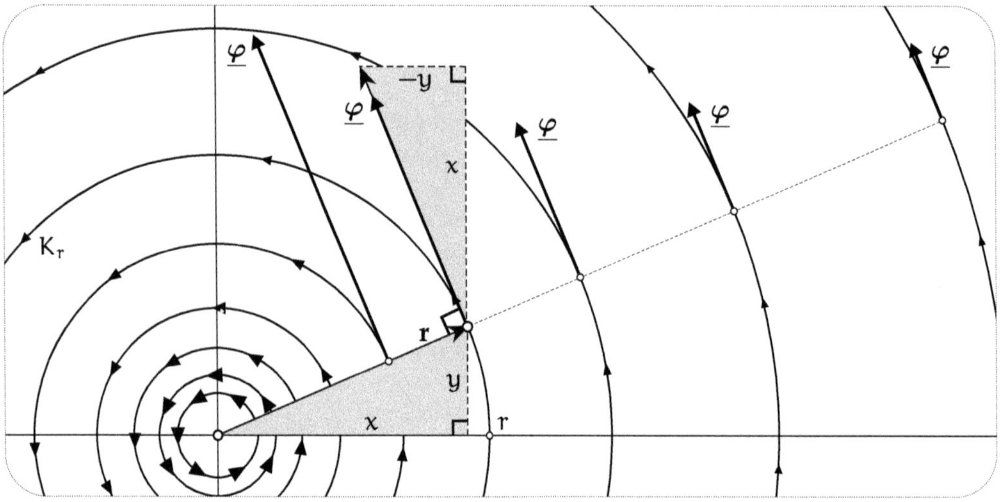

[37.10] *A special circular vortex vector field* $\underline{\varphi}$ *with speed* $|\underline{\varphi}| = (q/2\pi r)$. *Note the decreasing speed as we move away from the origin. The direction of* $\underline{\varphi}$ *is obtained by rotating the radius vector* $\mathbf{r}$ *through a right angle, but since this has length* $r = |\mathbf{r}|$, *we must multiply it by* $(q/2\pi r^2)$. *Thus the 1-form of the vortex is* $\varphi = (q/2\pi r^2)$ $[-y\,\mathbf{d}x + x\,\mathbf{d}y]$. *The origin is a singularity of infinite speed, and must be excluded.*

37.9.3 The Vortex 1-Form Is Closed

Figure [37.10] shows a typical radius vector $\mathbf{r}$ of length r, which is then rotated by $(\pi/2)$ to obtain a vector in the direction of the vortex flow. But this vector has length r, whereas we wish to construct a vortex that has speed $(q/2\pi r)$, so we must multiply this vector by $(q/2\pi r^2)$.

Expressed as a 1-form, the vortex vector field is therefore

$$\boxed{\varphi = \frac{q}{2\pi r^2}\,[-y\,\mathbf{d}x + x\,\mathbf{d}y],} \tag{37.17}$$

where $r^2 = x^2 + y^2$.

Therefore,

$$\begin{aligned}
\left[\frac{2\pi}{q}\right] \mathbf{d}\varphi &= -\frac{1}{r^4}(\mathbf{d}r^2)\wedge\Big[-y\,\mathbf{d}x + x\,\mathbf{d}y\Big] + 2\,\frac{1}{r^2}\,\mathbf{d}x\wedge\mathbf{d}y \\
&= -\frac{1}{r^4}\,2\,(x\,\mathbf{d}x + y\,\mathbf{d}y)\wedge\Big[-y\,\mathbf{d}x + x\,\mathbf{d}y\Big] + 2\,\frac{1}{r^2}\,\mathbf{d}x\wedge\mathbf{d}y \\
&= -\frac{1}{r^4}\,2\,(x^2 + y^2)\,\mathbf{d}x\wedge\mathbf{d}y + 2\,\frac{1}{r^2}\,\mathbf{d}x\wedge\mathbf{d}y \\
&= 0,
\end{aligned}$$

as claimed.

37.9.4 Geometrical Meaning of the Vortex 1-Form

We presented the above calculation in order to provide useful practice with the mechanics of exterior differentiation. However, this calculation is actually unnecessary, for it is possible to instead give a simple geometrical *reason* why φ is closed.

Consider [37.11], which shows a short, ultimately vanishing vector $\begin{bmatrix} \delta x \\ \delta y \end{bmatrix}$ emanating from the tip of $\mathbf{r} = \begin{bmatrix} x \\ y \end{bmatrix}$, subtending an angle of $\delta\theta$ at the origin. Let the shaded area of the triangle with these two edges be $\delta\mathcal{A}$.

On the one hand, we may evaluate $\delta\mathcal{A}$ as one-half of the determinant of the triangle's edges, and on the other hand it is ultimately equal to the area of the triangle with base r and height $r\,\delta\theta$, so,

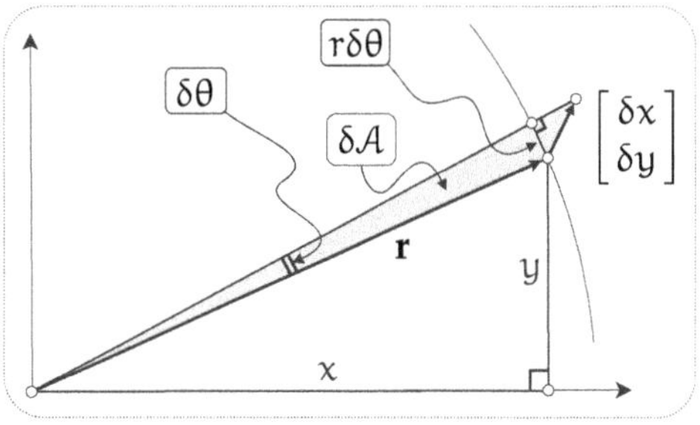

[37.11] *The area $\delta\mathcal{A}$ of the shaded triangle can be evaluated via the determinant of its edges, yielding $\delta\mathcal{A} = \frac{1}{2}(-y\,\delta x + x\,\delta y)$, or geometrically from the figure, yielding $\delta\mathcal{A} \asymp \frac{1}{2}r^2\,\delta\theta$. Equating these expressions, we find that the vortex 1-form is $\varphi = \frac{q}{2\pi}\,\mathbf{d}\theta$.*

$$\frac{1}{2}(-y\,\delta x + x\,\delta y) = \frac{1}{2}\det\begin{bmatrix} x & \delta x \\ y & \delta y \end{bmatrix}$$
$$= \delta\mathcal{A} \asymp \frac{1}{2}r(r\,\delta\theta).$$

The vortex 1-form (37.17) therefore has a beautifully simple geometrical meaning:

$$\boxed{\varphi = \tfrac{q}{2\pi}\,\mathbf{d}\theta,}$$

making it crystal clear *why* φ is closed: $\mathbf{d}\varphi = \frac{q}{2\pi}\,\mathbf{d}^2\theta = 0$.

But *wait*—wasn't the entire point of this example that φ is *not* exact? Sweet baby Newton, what the dickens is going on here?!

The resolution of this appararent paradox is somewhat subtle. Although $\mathbf{d}\theta$ has a perfectly well-defined geometrical meaning, it *cannot* be interpreted as ($\mathbf{d}$ of θ), because θ is not even a *function!* Which is to say, any given point must be assigned infinitely many angles.

If we restrict our attention to a simply connected region S that does not contain the singularity at the origin, such as the one shown in [37.12], then we may (nonuniquely) define a genuine, single-valued angle function θ on S, with $\theta_1 \leqslant \theta \leqslant \theta_2$, as illustrated. Now the Poincaré Lemma applies, so the closed 1-form φ must be exact, and we already know the explicit formula: $\varphi = \frac{q}{2\pi}\,\mathbf{d}\theta$.

Furthermore, the circulation of the vortex around any loop L contained within S does indeed vanish. For, as the illustrated particle traverses L, the position vector $\mathbf{r}$ rocks back and forth, but when it returns to its starting point, the *net* change in its angle θ is zero, so $\mathcal{C} = \oint_L \frac{q}{2\pi}\,\mathbf{d}\theta = 0$.

But if we attempt to do this in the entire punctured plane, then we immediately run into trouble. Yes, we can *try* to define θ by insisting that $0 \leqslant \theta < 2\pi$, for example, but then θ is not even continuous, let alone differentiable. Thus it is the global topology of the region on which φ is defined that prevents it from being exact throughout that region.

37.9.5 The Topological Stability of the Circulation of a Closed 1-Form

We previously saw that the circulation $\mathcal{C} = \oint_{K_r} \varphi$ of our special vortex was independent of the size of the circle K_r, and the new geometric interpretation of φ makes it obvious why: as we traverse any K_r, the angle increases by 2π, and therefore $\int_0^{2\pi} \frac{q}{2\pi}\,d\theta = q$.

Indeed, this geometrical interpretation makes it clear that the size-independence of $\mathcal{C}$ is in fact a very special case of a far more general phenomenon: if we gradually deform such a circle into *any* shape whatsoever, but being careful not to make it cross the singularity at the origin as it evolves, then the value of the integral does not change!

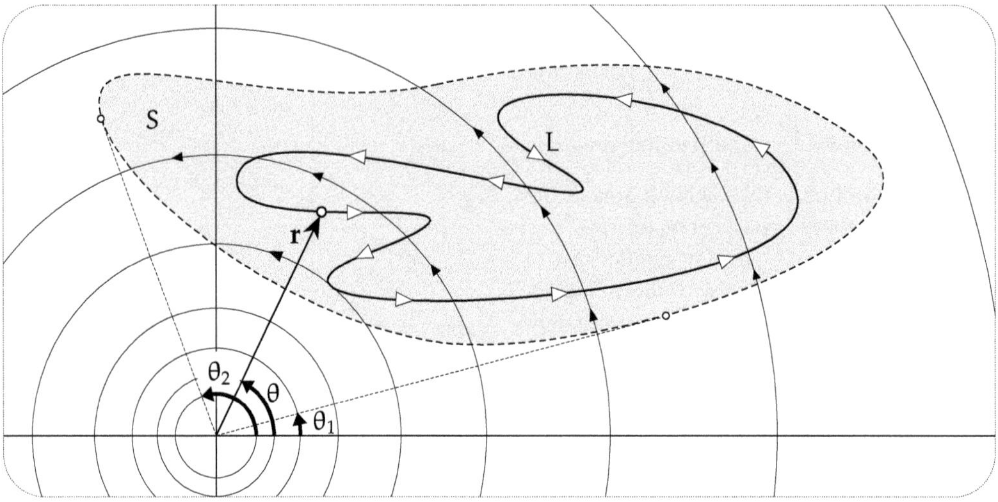

[37.12] *The simply connected region* S *does not contain the origin, so we we may define a single-valued angle function* θ *on it, where* $\theta_1 \leqslant \theta \leqslant \theta_2$. *Then* $\varphi = \frac{q}{2\pi}\,\mathbf{d}\theta$ *is exact on* S. *As* $\mathbf{r}$ *traces the loop* L, *it rocks back and forth, but when it returns to its starting point, the net change in the angle* θ *is zero, so* $\mathcal{C} = \oint_L \frac{q}{2\pi}\,\mathbf{d}\theta = 0$.

In other words,

> *All simple loops that enclose the vortex's singularity at the origin have the same circulation,* $\mathcal{C} = q$. *On the other hand, any loop that does not enclose the singularity has vanishing circulation:* $\mathcal{C} = 0$. (37.18)

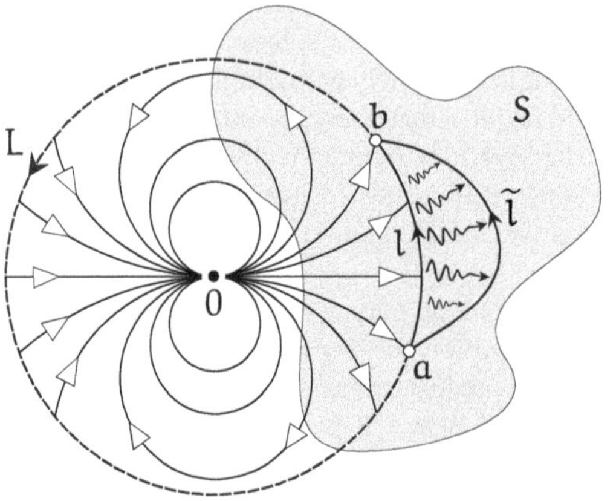

[37.13] *The simply connected region* S *does not contain the singularity of the closed dipole field* χ. *The Poincaré Lemma then tells us that* $\oint_L \chi$ *is invariant under the deformation of* l *into* $\widetilde{l}$.

Observe that the distinction between these two kinds of loops is purely *topological*.

NOTE: If a loop goes around the origin m times, then one says that it has ***winding number*** m, in which case $\mathcal{C} = mq$. For much more on the concept of the winding number, and its many applications, see VCA.

In fact, the topological stability of the integral, as expressed in (37.18), is true of *all* closed 1-forms, not just our specific vortex. To understand this broader result, consider the dipole field χ shown in [37.13], assumed to be closed: $\mathbf{d}\chi = 0$. There is no special significance to this particular field; we simply wish to make it clear that we may apply the following reasoning to *any* closed 1-form.

Now let S be any simply connected region that does *not* contain the singularity at the origin, such as the shaded region shown in [37.13]. Since χ is closed within S, the Poincaré Lemma implies that $\chi = \mathbf{d}f$ is exact within S, and hence its integral is path-independent within $\mathcal{S}$.

Now, as illustrated, consider an origin-centred circle L, part of which is contained in S. If we take the illustrated arc l (connecting a to b) and deform it into the alternative route $\widetilde{l}$, then the path-independence tells us that

$$\int_{\widetilde{l}} \chi = \int_{l} \chi = f(b) - f(a).$$

Thus if $\widetilde{L}$ denotes the new deformed loop, in which l has been deformed into $\widetilde{l}$, then

$$\boxed{\oint_{\widetilde{L}} \chi = \oint_{L} \chi.}$$

It is intuitively clear (and we prove it in VCA) that we may now continue deforming L into *any* simple loop around the singularity, and since the circulation does not change throughout this deformation, we have proved the generalized version of (37.18):

Deformation Theorem for Closed 1-Forms. *The circulation (called the* **(de Rham) period***) of a closed 1-form along a loop is invariant under a continuous deformation of the loop.* (37.19)

Given this independence of the circulation from the specific shape of the loop encircling the singularity, we are entitled to think of it as being *a property of the singularity itself*. This is a generalization of the concept in Complex Analysis of the ***residue*** of a complex analytic function $f(z)$ at each of its singularities. Indeed, in VCA (Ch. 11) we explicitly describe residues in terms of the circulation and flux of the Pólya vector field, $\overline{f(z)}$.

37.9.6 The First de Rham Cohomology Group

Now let us return to the definition of the ***first de Rham cohomology group***, for the current case, $R = (\mathbb{R}^2 - \text{origin})$, written $H^1(\mathbb{R}^2 - \text{origin})$. An element of this group is not an individual 1-form, rather it is an *equivalence class* of 1-forms, in which

Two 1-forms are considered fundamentally the same, "equivalent," if they have the same circulation around all loops.

(NOTE: The technical terminology (which we shall not employ) is to say that the two 1-forms are ***cohomologous***.)

For example, suppose we add to the vortex 1-form φ a constant 1-form, corresponding to a steady flow in some direction, which itself clearly has no circulation around any loop. Then the circulation of the new flow is the sum of the circulation of the old flow and the circulation of the steady flow (which vanishes). Therefore, the circulation is unaltered around all loops, and therefore this new 1-form belongs to the same equivalence class as φ.

More generally, we may add *any* flow that has vanishing circulation around all loops. In other words, we may add any 1-form that is *exact*. To sum up,

Let $\widetilde{\varphi} \sim \varphi$ denote the equivalence of the 1-forms $\widetilde{\varphi}$ and φ. Then,

$$\widetilde{\varphi} \sim \varphi \iff \widetilde{\varphi} - \varphi = \mathbf{d}f, \quad \textit{for some } f.$$

Each equivalence class is defined by its value of the circulation, $\mathcal{C}$. (NOTE: In standard texts on de Rham cohomology, the circulation $\mathcal{C}$ is called the ***period*** of the equivalence class, and the class itself is called a ***cohomology class***.)

TERMINOLOGY WARNING: Historically, the use of the term ***period*** in this context stems from the theory of *elliptic integrals*. But in physics we are accustomed to the period referring to the time it takes for a pendulum to swing back and forth, or the time it takes for a planet to go once around the Sun. In looking at [37.10], we should therefore warn you that the de Rham period *has nothing to do with* the time it takes for a particle of fluid to perform one orbit around the singularity. Indeed, the time T_r it takes to go once around K_r *depends on* r, and is given by $T_r = 4\pi^2(r^2/q)$, whereas the fixed de Rham ***period*** is $\mathcal{C} = q$.

All of the above reasoning only depended on FTEC, not on the specific form of the vortex with which we chose to illustrate the idea. Thus it all carries through to more general 1-forms φ that are closed but not exact.

If we add two such 1-forms, φ_1 and φ_2, from two different equivalence classes, then we obtain a new 1-form from the class that has period

$$\mathcal{C}(\varphi_1 + \varphi_2) = \mathcal{C}(\varphi_1) + \mathcal{C}(\varphi_2).$$

Thus, the group $H^1(\mathbb{R}^2 - \text{origin})$ is isomorphic to the set of real periods under addition, and we may simply write

$$\boxed{H^1(\mathbb{R}^2 - \text{origin}) = \mathbb{R}.}$$

Next, suppose we step up a dimension and consider 1-forms in $\mathbb{R}^3$. If we remove the origin from $\mathbb{R}^3$ then we *can* contract every loop to a point, because now we have enough room to slip a contracting loop *around* the origin, without having to pass through it—so now the space *is* simply connected again. Thus there are no closed 1-forms that are not exact, and we record this fact as

$$\boxed{H^1(\mathbb{R}^3 - \text{origin}) = 0.} \tag{37.20}$$

On the other hand, if we remove the entire z-axis (for example) from $\mathbb{R}^3$ then we have the analogue of the punctured $\mathbb{R}^2$: it is *not* simply connected. Indeed, to deepen the analogy, imagine the flow [37.10] extending upward out of the page, creating a 3-dimensional vortex swirling around the z-axis, with the fluid rotating in concentric cylinders centred on the singular z-axis, where the velocity goes to infinity. Indeed, the analogy with the punctured plane is perfect:

$$\boxed{H^1(\mathbb{R}^3 - \text{z-axis}) = \mathbb{R}.}$$

37.9.7 The Inverse-Square Point Source in $\mathbb{R}^3$

As a prelude to our discussion of the second de Rham cohomology group, $H^2(R)$, we now define a special vector field with a singularity at the origin of $\mathbb{R}^3$. Let $\widehat{\mathbf{r}}$ be the unit radial vector:

$$\widehat{\mathbf{r}} = \frac{1}{r}\mathbf{r} = \frac{1}{r}\begin{bmatrix} x \\ y \\ z \end{bmatrix}, \qquad \text{where} \quad r = \sqrt{x^2 + y^2 + z^2}.$$

We now present to you perhaps the single most important vector field in all of physics, the ***inverse-square point source***:

$$\underline{\Psi} = \frac{q}{4\pi}\,\frac{\widehat{\mathbf{r}}}{r^2},$$

where q is a constant that determines the strength of the field. Since the field has a singularity at the origin, this point must be excluded: the field is defined on $\mathbb{R}^3$ – origin.

This field has (at least!) three very important, physically distinct interpretations:

- If fluid is pumped into 0 at rate q, and flows radially and symmetrically outward, then $\underline{\Psi}$ is the velocity of the fluid flow.
- If a point electric charge q sits at 0, Coulomb's Law says that $\underline{\Psi}$ is its electric field.
- If we reverse the direction of the field, then Newton's Law of Universal Gravitation says that $-\underline{\Psi}$ is the attractive gravitational field of the point mass q located at 0.

  As Newton proved in the *Principia*, this field also represents (very remarkably!) the external gravitational field of *any* spherically symmetric body of mass q, such as the Earth or the Sun.

In $\mathbb{R}^3$ we may interpret any vector field as either a 1-form or a 2-form. Our focus in the next section will be on the 2-form, but first let us discuss the 1-form ψ corresponding to the vector field $\underline{\Psi}$:

$$\psi = \frac{q}{4\pi r^3}\left(x\,\mathbf{d}x + y\,\mathbf{d}y + z\,\mathbf{d}z\right).$$

Let us view this in terms of the third physical interpretation: gravity. Then instead of thinking of the line integral along an open path J as representing circulation, we should now think of it representing *work*:

$$\mathcal{W}_J = \text{work} = \int_J \psi.$$

We know that the gravitational force is *conservative*: carrying a mass around a loop within this gravitational field *must* result in zero net work, or else we could construct a perpetual motion machine! Thus we anticipate that ψ is closed. We leave it as an exercise to confirm by direct calculation that, indeed, $\mathbf{d}\psi = 0$.

However, we can instead demonstrate this by showing that ψ is exact: $\psi = \mathbf{d}f$. More explicitly, let us show that the potential energy is

$$f = \text{potential energy} = -\frac{q}{4\pi r}.$$

As a lemma, first observe [exercise] that

$$\mathbf{d}r = \frac{1}{r}\left(x\,\mathbf{d}x + y\,\mathbf{d}y + z\,\mathbf{d}z\right).$$

Therefore,

$$\begin{aligned} df &= -\frac{q}{4\pi}\,d\left[\frac{1}{r}\right] \\ &= \frac{q}{4\pi r^2}\,dr \\ &= \frac{1}{4\pi r^3}\,(x\,dx + y\,dy + z\,dz) \\ &= \psi. \end{aligned}$$

Thus if J is any path connecting **a** to **b**, then

$$\mathcal{W}_J = \int_J \psi = \int_J df = f(\mathbf{b}) - f(\mathbf{a}) = \frac{q}{4\pi}\left[\frac{1}{|\mathbf{a}|} - \frac{1}{|\mathbf{b}|}\right].$$

Note that, unlike our previous vortex example, here the potential f is a well-defined function throughout $(\mathbb{R}^3 - \text{origin})$. Thus this result is in keeping with the claim, (37.20), that the first de Rham cohomology group is trivial in this case.

37.9.8 The Second de Rham Cohomology Group

Next, let us calculate a specific example of the *second* de Rham cohomology group, $H^2(R)$, which measures the 2-forms on R that are closed but not exact. Let us stay in $\mathbb{R}^3$, and consider the flux of a closed 2-form $\boldsymbol{\Psi}$ out of a closed surface $\mathcal{S}$.

Let us continue thinking about the inverse-square point source, defined on $R = (\mathbb{R}^3 - \text{origin})$, but let us now switch to thinking of it as a *fluid flow*, in which case it is natural to represent the field as a flux 2-form, via (34.10) and (34.12):

$$\boldsymbol{\Psi} = \frac{q}{4\pi}\frac{1}{r^3}(x\,dy \wedge dz + y\,dz \wedge dx + z\,dx \wedge dy).$$

That this field cannot be exact can be seen by looking at the flux out of an origin-centred sphere $\mathcal{S}_r$ of radius r. The sphere has area $4\pi r^2$, and the field is orthogonal to it, so the flux Ω is simply the product of the area and the field strength:

$$\Omega = \oiint_{\mathcal{S}} \boldsymbol{\Psi} = q,$$

independent of r.

Despite the fact that $\boldsymbol{\Psi}$ is not exact, we will now prove that it is *closed*:

$$\begin{aligned} \left[\frac{4\pi}{q}\right] d\boldsymbol{\Psi} &= \frac{3}{r^3}\mathcal{V} - \frac{3}{r^4}\,dr \wedge (x\,dy \wedge dz + y\,dz \wedge dx + z\,dx \wedge dy) \\ &= \frac{3}{r^3}\left[\mathcal{V} - \frac{1}{r^2}\,(x\,dx + y\,dy + z\,dz) \wedge (x\,dy \wedge dz + y\,dz \wedge dx + z\,dx \wedge dy)\right] \\ &= \frac{3}{r^3}\left[\mathcal{V} - \frac{1}{r^2}\,(x^2 + y^2 + z^2)\mathcal{V}\right] \\ &= 0. \end{aligned}$$

Next, let us show that (37.19) generalizes:

Deformation Theorem for Closed 2-Forms. *The flux* (**period**) *of a closed 2-form out of a closed 2-surface is invariant under a continuous deformation of the surface.* (37.21)

Figure [37.14] shows fluid flowing radially outward through the sphere $\mathcal{S}$ from the inverse-square point source at the origin. A circular patch of $\mathcal{S}$ is deformed outward—indicated by squiggly arrows—to create the illustrated bump or blister on the surface of $\mathcal{S}$.

All the fluid that flows into the base of the blister also flows out of its outer surface, so if $\widetilde{\mathcal{S}}$ denotes the new deformed surface, then,

$$\oiint_{\widetilde{\mathcal{S}}} \boldsymbol{\Psi} = \oiint_{\mathcal{S}} \boldsymbol{\Psi}.$$

This argument easily generalizes to apply to *any* closed 2-form $\boldsymbol{\Psi}$. Since $\mathbf{d}\boldsymbol{\Psi} = 0$ within the blister, FTEC implies that the net *outflux* from it vanishes. But this can be rephrased as we just did: what flows into the blister must flow out. Thus we have proved (37.21): the flux out of $\mathcal{S}$ is invariant under deformations of $\mathcal{S}$ that do not force it to pass through a singularity of $\boldsymbol{\Psi}$.

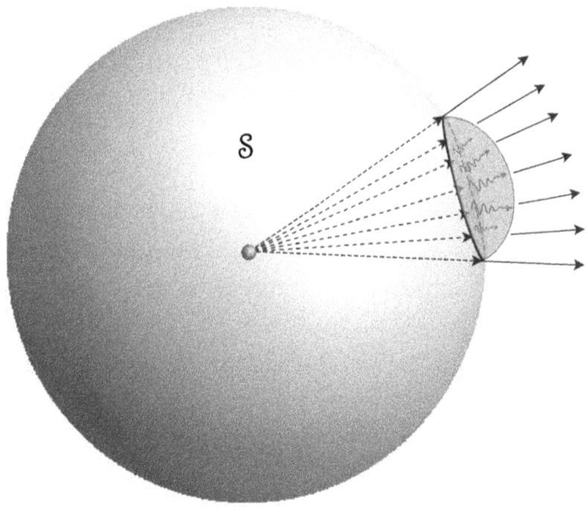

[37.14] *Because the 2-form is closed, what flows into the blister must flow out. So the flux out of $\mathcal{S}$ is invariant under deformations.*

Returning to the specific inverse-square point source, a closed surface enclosing the singularity at the origin cannot be contracted to a point without crossing the origin, so all surfaces containing the singularity have the same flux q. This topologically stable flux measures the strength of the source at the origin, which has the three physically distinct interpretations listed above. (Again, we repeat that more advanced texts on de Rham cohomology theory call this source strength the *period*.)

The definition of equivalence classes of 2-forms is completely analogous to that for 1-forms; that is, *two 2-forms* $\widetilde{\boldsymbol{\Psi}}$ *and* $\boldsymbol{\Psi}$ *are* equivalent *if they have the same flux out of every closed surface.* Expressed in more conventional terms,

$$\widetilde{\boldsymbol{\Psi}} \sim \boldsymbol{\Psi} \quad \Longleftrightarrow \quad \widetilde{\boldsymbol{\Psi}} - \boldsymbol{\Psi} = \mathbf{d}\phi, \quad \text{for some 1-form } \phi.$$

If we add two 2-forms then their fluxes add, too. So,

$$\mathrm{H}^2(\mathbb{R}^3 - \text{origin}) = \mathbb{R}.$$

37.9.9 The First de Rham Cohomology Group of the Torus

What if the nonexact, closed Forms are defined on a manifold other than $\mathbb{R}^n$, such as closed 2-surface of genus g?

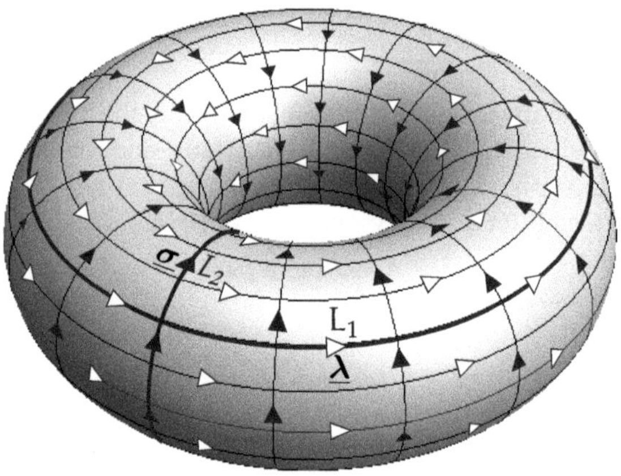

[37.15] *The loop L_1 around the equator of the torus, and the loop L_2 through the hole of the torus, are topologically distinct: we cannot deform one into the other. The equivalence class of a closed 1-form flow on the torus therefore requires the specification of* two *independent periods.*

Consider the torus shown in [37.15]. Here we see two different possible flows on the surface: the white flow $\underline{\boldsymbol{\lambda}}$ goes around the axis of symmetry of the torus, in the direction the equator(s), while the black flow $\underline{\boldsymbol{\sigma}}$ goes through the hole. We shall assume that both flows are *closed*:

$$\mathbf{d}\boldsymbol{\lambda} = 0 = \mathbf{d}\boldsymbol{\sigma}.$$

Now let L_1 be a closed loop streamline of the white flow $\underline{\boldsymbol{\lambda}}$, and let L_2 be a closed-loop streamline of the black flow $\underline{\boldsymbol{\sigma}}$, as illustrated. As we discussed in connection with [19.9], L_1 and L_2 are topologically *distinct*: we cannot continuously deform one into the other. However, *any* loop that goes once around the axis of symmetry (and zero times through the hole) is topologically equivalent to L_1, and *any* loop that goes once through the hole (and zero times around the axis of symmetry) is topologically equivalent to L_2.

Next, observe that neither of these flows can be exact. Since the white vector field $\underline{\boldsymbol{\lambda}}$ flows directly along L_1, it has a nonzero circulation along it. Let us call this period $\omega_1(\boldsymbol{\lambda})$, the subscript identifying which of the topologically distinct loops is being considered:

$$\omega_1(\boldsymbol{\lambda}) = \oint_{L_1} \boldsymbol{\lambda} \neq 0.$$

The Deformation Theorem (37.19) then tells us that this period will be the same for any loop into which we may deform L_1. That is, the period $\omega_1(\boldsymbol{\lambda})$ is a topological property of the white vector field, associated with any loop that goes once around the axis of symmetry (and zero times through the hole).

On the other hand, $\underline{\boldsymbol{\lambda}}$ is everywhere orthogonal to L_2, so it has zero circulation along it:

$$\omega_2(\boldsymbol{\lambda}) = \oint_{L_2} \boldsymbol{\lambda} = 0.$$

Furthermore, by the Deformation Theorem, this will be true of any loop that goes once through the hole (and zero times around the axis of symmetry).

A completely symmetrical analysis applies to the black vector field $\underline{\boldsymbol{\sigma}}$ through the hole:

$$\omega_1(\boldsymbol{\sigma}) = \oint_{L_1} \boldsymbol{\sigma} = 0, \qquad \text{and} \qquad \omega_2(\boldsymbol{\sigma}) = \oint_{L_2} \boldsymbol{\sigma} \neq 0,$$

and again these are true of any loops into which L_1 and L_2 may be deformed.

Now consider a more general vector field $\underline{\phi}$ on the torus, obtained by taking a linear combination of $\underline{\boldsymbol{\lambda}}$ and $\underline{\sigma}$:

$$\phi = a\boldsymbol{\lambda} + b\sigma, \qquad (37.22)$$

where a and b are constants. This both flows through the hole *and* around the axis of symmetry, as illustrated in [37.16]. But it is still closed, because

$$\mathbf{d}\phi = a\,\mathbf{d}\boldsymbol{\lambda} + b\,\mathbf{d}\sigma = 0,$$

and thus its periods are still topologically defined.

But unlike either of the fields from which it was built, $\underline{\phi}$ has *two* nonvanishing, independent periods:

$$\omega_1(\phi) = a\,\omega_1(\boldsymbol{\lambda}) \neq 0,$$
$$\omega_2(\phi) = b\,\omega_2(\sigma) \neq 0.$$

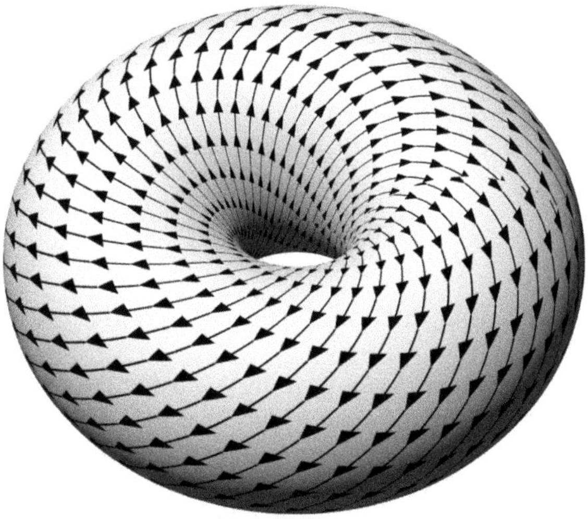

[37.16] *The equivalence class of a closed 1-form ϕ on the torus is determined by two independent, topological periods: the circulation $\omega_1(\phi)$ around the axis of symmetry, and the circulation $\omega_2(\phi)$ through the hole. Courtesy of RokerHRO, CC by SA 3.0.*

This characterization of ϕ by two independent periods also applies to a completely general closed 1-form on the torus, even if ϕ doesn't happen to have the special form (37.22). To sum up,

The equivalence class of a closed 1-form ϕ on the torus is determined by two topologically defined periods: $\omega_1(\phi)$ measures the circulation around the axis of symmetry, and $\omega_2(\phi)$ measures the circulation through the hole.

We may therefore neatly identify the equivalence class of ϕ with a 2-dimensional vector, which we shall call the ***period vector***, $\omega(\phi)$, given by

$$\omega(\phi) \equiv \begin{bmatrix} \omega_1(\phi) \\ \omega_2(\phi) \end{bmatrix}.$$

Finally, observe that *if we add two flows from two different equivalence classes, their period vectors add, as follows:*

$$\omega(\widetilde{\phi}+\phi) = \begin{bmatrix} \omega_1(\widetilde{\phi}+\phi) \\ \omega_2(\widetilde{\phi}+\phi) \end{bmatrix} = \begin{bmatrix} \omega_1(\widetilde{\phi}) \\ \omega_2(\widetilde{\phi}) \end{bmatrix} + \begin{bmatrix} \omega_1(\phi) \\ \omega_2(\phi) \end{bmatrix} = \omega(\widetilde{\phi}) + \omega(\phi).$$

We have thus found the first de Rham cohomology group of the torus:

$$H^1(\text{torus}) = \mathbb{R}^2.$$

Entire books are devoted to de Rham's theory, but we must come to a full stop here, lest *this* book become one of them! We have glimpsed the tip of a massive, beautiful iceberg, and we must reluctantly turn hard about. But *you* should feel free to go exploring deep beneath the surface, with the *Further Reading* section as your guide.

Chapter 38

Differential Geometry via Forms

38.1 Introduction: Cartan's Method of Moving Frames

At long last, we are finally ready to put the "Devil's machine"[1] to work in the service of geometry!

In this final chapter of this final Act, we shall apply the machinery of Cartan's Forms to Differential Geometry, just as Cartan himself did, reproving by computation many of the same fundamental results that were proved geometrically in the first four Acts. Furthermore, we shall see that Forms naturally lead us to many *new* results. We begin by outlining our battle plans, and their historical origins.

Recall that Frenet in 1847, and (independently) Serret in 1851, both analyzed curves in $\mathbb{R}^3$ by attaching an orthonormal frame $(\mathbf{T}, \mathbf{N}, \mathbf{B})$ to the curve, finding that its rate of change along the curve is given by the *Frenet–Serret Equations*, (9.3), which we repeat here:

$$\begin{bmatrix} \mathbf{T} \\ \mathbf{N} \\ \mathbf{B} \end{bmatrix}' = \begin{bmatrix} 0 & \kappa & 0 \\ -\kappa & 0 & \tau \\ 0 & -\tau & 0 \end{bmatrix} \begin{bmatrix} \mathbf{T} \\ \mathbf{N} \\ \mathbf{B} \end{bmatrix} = [\Omega] \begin{bmatrix} \mathbf{T} \\ \mathbf{N} \\ \mathbf{B} \end{bmatrix}. \tag{38.1}$$

Thus, as this frame moves along the curve, its rate of rotation is fully encoded as the matrix $[\Omega]$, which is skew-symmetric (i.e., $[\Omega]^{\mathrm{T}} = -[\Omega]$).

The success of this enterprise hinged upon two ideas:

IDEA 1: Adapt the frame to the curve in a geometrically meaningful way, called the ***Frenet Frame***: $\mathbf{T}$ is tangent to the curve, and $\mathbf{N}$ is the *principal normal*, lying within the instantaneous plane of motion, and pointing at the instantaneous centre of curvature. $\mathbf{B}$ is then determined by the right hand rule.

IDEA 2: Express the rate of change of the frame in terms of the frame *itself*.

Around 1880, Jean-Gaston Darboux took the next step, extending these ideas to *surfaces*:

IDEA 3: Adapt the frame to a *surface*, by choosing one of the vectors to be the *normal* to the surface; the remaining two vectors are then automatically tangent vectors to the surface.

Starting around 1900, Élie Cartan developed Forms and brought them to bear on Differential Geometry, his efforts reaching substantial completion in Cartan (1928), though the final *perfection* of his ideas was perhaps Cartan (1945):

IDEA 4: Free the frame completely, and allow it to vary in *any (differentiable) way whatsoever* throughout $\mathbb{R}^3$. Instead of calling it the *Frenet Frame*, we now call it ***Cartan's moving frame field***, referring to this approach as a whole as ***Cartan's method of moving frames.*** As we shall see shortly, this again yields a skew-symmetric matrix, but it is now a matrix of *1-forms*.

---

[1] See Prologue.

IDEA 5: Replace the moving frame with its *dual frame of 1-forms*, and look at its rate of change in an arbitrary direction in space. Recapitulating **IDEA 2**, express this rate of change of the dual basis in terms of the dual basis *itself*.

Let us pause in order to comment on **IDEA 4**, which would appear to be an ill-fated step *backwards*. Frenet, Serret, and Darboux had all adapted their frames to a geometrical object in a geometrically meaningful way, so it made sense that their examination of its rate of change would yield geometrically meaningful information about that object.

But (diabolical) *Cartan* instead instructs us to do the following. Take your right hand, and arrange your thumb (imagined Satanically elongated!), forefinger, and middle finger into an orthonormal frame. Now wave your hand about in space, twisting your hand as you go![2]

Even if fleetingly, you have created an arbitrary *moving frame field*, at least at each point of the region of space that your hand passed through. Now imagine an entire region of space *filled* (differentiably) with such an arbitrary moving frame field. How on earth can anything mathematically interesting arise from looking at the arbitrary gyrations of your three fingers?!

Very remarkably, Cartan found that such an *arbitrary* moving frame field is subject to *two* extremely elegant laws, called *Cartan's First* and *Second Structural Equations*.[3] These are the subject of Section 38.4.

Most authors pass over the existence of these laws in silence, as if to say, "Keep moving! Nothing to see here." But when *I* first saw these two equations as a student, I was *shocked*—perceiving them to be magical, and possibly *black*-magical, at that! Even now, I cannot offer you a truly compelling explanation for the existence of these beautiful laws, but consider this....

To specify a unit vector in $\mathbb{R}^3$, or, equivalently, a point on the unit sphere, we need *two* numbers—say, longitude and latitude. Thus three arbitrary unit vectors require *six* numbers. But now consider Cartan's *orthonormal* moving frame field, and suppose we are told just *one* of his vectors (using two numbers). We then know that the tips of the remaining two, orthogonal vectors of the frame lie somewhere on the unit circle orthogonal to the given vector. Thus the remaining two vectors can be pinned down with the specification of a *single* angle (of either one).

In other words, the extra structure provided by the *orthogonality* of Cartan's moving frame field has reduced the amount of data by *half*. To phrase this the other way around, Cartan's *orthonormal* moving frame field has *twice* the structure of a nonorthogonal field.

IDEA 6: Having risen above the fray to obtain his completely general Structural Equations for an arbitrary moving frame field, Cartan then descended to Earth, and adapted them to specific geometrical objects.

When Cartan adapted them to a curve, he simply recovered (38.1). But when he adapted his moving frame field to a surface—Darboux's **IDEA 3**—Cartan's Structural Equations immediately yielded all the *Fundamental Equations of a Surface*, but in a much more compact and elegant form than had been obtained by Gauss, Darboux, Codazzi, and other early masters of Differential Geometry—all thanks to their expression in Cartan's language of *Forms*. These Form equations are the subject of Section 38.5.

In this chapter we shall follow closely in the footsteps of the trail-blazing text of UCLA Professor Barrett O'Neill (1924–2011), *Elementary Differential Geometry* (second revised edition), the first edition of which appeared in 1966.

Today, more than a half-century later, O'Neill's work remains, in our view, the single most clear-eyed, elegant, and (ironically) *modern* treatment of the subject available—present company

[2]We humbly suggest that you do not attempt this in public!

[3]These are also commonly referred to as *Cartan's Structure Equations*.

excepted!—at the undergraduate level. Indeed, as of this writing, it still stands essentially[4] *alone* in its pioneering embrace of Forms as the principal means of communicating Differential Geometry to an undergraduate audience.

We shall carry our investigations substantially further than O'Neill's text, calculating the Riemann tensor of an n-manifold using curvature 2-forms, and also providing many *geometrical* explanations that he does not. Nevertheless, *O'Neill investigates many very interesting topics that we shall not.* Therefore, we *strongly* recommend O'Neill's pioneering work as a companion or follow-up text to this final chapter.

38.2 Connection 1-Forms

38.2.1 Notational Conventions and Two Definitions

Before we get going, let us remind you of some of our (ongoing) notational conventions:

- Vectors are denoted by lowercase bold Roman letters, e.g., $\mathbf{v}$.
- 1-forms are denoted by lowercase bold Greek letters, e.g., $\boldsymbol{\theta}$.
- 2-forms are denoted by uppercase bold Greek letters, e.g., $\boldsymbol{\Psi}$.
- Matrices are denoted by square brackets, e.g., $[A]=[a_{ij}]$, where a_{ij} denotes the matrix entry in *row* i, and *column* j.

To avoid any possibility of confusion while you become acclimatized to Cartan's ideas, we shall depart from usual practice and dispense with the Einstein summation convention, instead writing out summation signs, or even writing out the sums explicitly. This is actually a very small sacrifice because we shall see that once we specialize to 2-surfaces many of the "sums" reduce to a *single* term!

Finally, here are two definitions, one old and one new:

- As usual, let $\{\mathbf{e}_j\}$ be the fixed Euclidean orthonormal basis, with dual 1-form basis $\{\mathbf{dx}^i\}$. Thus, as we saw in (32.12), $\mathbf{dx}^i(\mathbf{e}_j)=\delta^i_j$.
- Let $\{\mathbf{m}_j\}$ be Cartan's (arbitrary, but differentiable) orthonormal ***moving frame field,*** and let its dual 1-form basis be $\{\boldsymbol{\theta}^i\}$. Then, by definition, $\boldsymbol{\theta}^i(\mathbf{m}_j)=\delta^i_j$.

  Note that while [32.6] showed that a general, nonorthogonal basis does not allow us to pair an *individual* basis vector with a matching 1-form, this is no longer true of our *orthonormal* moving frame field: we *can* think of $\boldsymbol{\theta}^i$ as the dual of $\mathbf{m}_i$.

38.2.2 Connection 1-Forms

Let us apply **IDEA 2** to Cartan's moving frame field. That is, let us *express the rate of change of the moving frame field in terms of the frame field itself*.

In the case of the *Frenet–Serret Equations*, the Frenet frame was only defined along the curve, and we only needed to look at its rate of change along that curve, but *now* we must examine how our moving frame field rotates as we move off in some *general* direction $\mathbf{v}$ in space:

$$\begin{aligned}\nabla_{\mathbf{v}}\mathbf{m}_1 &= c_{11}\,\mathbf{m}_1 + c_{12}\,\mathbf{m}_2 + c_{13}\,\mathbf{m}_3\\ \nabla_{\mathbf{v}}\mathbf{m}_2 &= c_{21}\,\mathbf{m}_1 + c_{22}\,\mathbf{m}_2 + c_{23}\,\mathbf{m}_3\\ \nabla_{\mathbf{v}}\mathbf{m}_3 &= c_{31}\,\mathbf{m}_1 + c_{32}\,\mathbf{m}_2 + c_{33}\,\mathbf{m}_3.\end{aligned}$$

[4] Darling (1994) is the closest thing to an exception that we know of, but it smacks more of a graduate text.

If we collect these coefficients into the matrix $[C] \equiv [c_{ij}]$, and make the abbreviation,

$$[\mathbf{m}] \equiv \begin{bmatrix} \mathbf{m}_1 \\ \mathbf{m}_2 \\ \mathbf{m}_3 \end{bmatrix},$$

then we can rewrite these equations more neatly and compactly as

$$\nabla_{\mathbf{v}}[\mathbf{m}] = [C][\mathbf{m}].$$

Since the moving frame field is (by definition) orthonormal, it follows that these coefficients—making up the entries of the matrix $[C]$—are given by

$$c_{ij} = (\nabla_{\mathbf{v}}\mathbf{m}_i) \bullet \mathbf{m}_j.$$

But these coefficients depend on the choice of $\mathbf{v}$, so we change our notation[5] and write them as *functions* of $\mathbf{v}$:

$\omega_{ij}(\mathbf{v}) \equiv (\nabla_{\mathbf{v}}\mathbf{m}_i) \bullet \mathbf{m}_j$ *is the initial rate at which* $\mathbf{m}_i$ *tips towards* $\mathbf{m}_j$ *as the frame moves along* $\mathbf{v}$. (38.2)

Perhaps you will not be surprised to learn that these ω_{ij} are *1-forms!*

The ω_{ij} *are called the* **connection 1-forms**, *and* $\omega_{ji} = -\omega_{ij}$.

WARNING: the two subscripts attached to ω_{ij} immediately (but *wrongly!*) make one think of the components of tensor of valence $\left\{ {0 \atop 2} \right\}$. Let us therefore *stress* that these are all *1-forms*, housed within a matrix of 1-forms, and the subscripts serve to identify the *row and column* of each 1-form entry within this matrix.

We begin by verifying that they are indeed 1-forms:

$$\begin{aligned} \omega_{ij}(a\,\mathbf{v} + b\,\mathbf{w}) &= (\nabla_{a\mathbf{v}+b\mathbf{w}}\,\mathbf{m}_i) \bullet \mathbf{m}_j \\ &= (a\,\nabla_{\mathbf{v}}\mathbf{m}_i + b\,\nabla_{\mathbf{w}}\mathbf{m}_i) \bullet \mathbf{m}_j \\ &= a\,(\nabla_{\mathbf{v}}\mathbf{m}_i) \bullet \mathbf{m}_j + b\,(\nabla_{\mathbf{w}}\mathbf{m}_i) \bullet \mathbf{m}_j \\ &= a\,\omega_{ij}(\mathbf{v}) + b\,\omega_{ij}(\mathbf{w}). \end{aligned}$$

Next, to prove $\omega_{ji} = -\omega_{ij}$, we must show that $\omega_{ji}(\mathbf{v}) = -\omega_{ij}(\mathbf{v})$ for every vector $\mathbf{v}$:

$$0 = \nabla_{\mathbf{v}}\delta^i_j = \nabla_{\mathbf{v}}(\mathbf{m}_i \bullet \mathbf{m}_j) = \nabla_{\mathbf{v}}(\mathbf{m}_i) \bullet \mathbf{m}_j + (\nabla_{\mathbf{v}}\mathbf{m}_j) \bullet \mathbf{m}_i = \omega_{ij}(\mathbf{v}) + \omega_{ji}(\mathbf{v}).$$

A *geometrical* explanation of this result is implicit in the upcoming diagram, [38.2].

---

[5] Frustratingly, the *order* of the indices varies by author. We shall employ the notation of O'Neill (2006), whereas Dray (2015), for example, has i and j reversed: his ω_{ji} is our ω_{ij}.

It follows that

$$\omega_{11} = \omega_{22} = \omega_{33} = 0.$$

Therefore, the matrix [C] of coefficients is *a skew-symmetric matrix of 1-forms, with only* three *independent entries,* $\omega_{12}, \omega_{13}, \omega_{23}$:

$$[\omega] \equiv \begin{bmatrix} \omega_{11} & \omega_{12} & \omega_{13} \\ \omega_{21} & \omega_{22} & \omega_{23} \\ \omega_{31} & \omega_{32} & \omega_{33} \end{bmatrix} = \begin{bmatrix} 0 & \omega_{12} & \omega_{13} \\ -\omega_{12} & 0 & \omega_{23} \\ -\omega_{13} & -\omega_{23} & 0 \end{bmatrix}.$$

If we take it as understood that $[\omega](\mathbf{v}) = [\omega(\mathbf{v})]$ means that each 1-form entry within the matrix acts on $\mathbf{v}$, then the rotation of the moving frame field as it moves off in a general direction $\mathbf{v}$ can be elegantly described by the so-called ***Connection Equations***:

$$\boxed{\nabla_{\mathbf{v}}[\mathbf{m}] = [\omega(\mathbf{v})][\mathbf{m}].} \tag{38.3}$$

In component form, this states

$$\boxed{\nabla_{\mathbf{v}}\mathbf{m}_i = \textstyle\sum_j \omega_{ij}(\mathbf{v})\,\mathbf{m}_j.} \tag{38.4}$$

For fear that elegance become the enemy of clarity, let us simply spell out these equations *in full*:

$$\begin{aligned} \nabla_{\mathbf{v}}\mathbf{m}_1 &= \omega_{12}(\mathbf{v})\,\mathbf{m}_2 + \omega_{13}(\mathbf{v})\,\mathbf{m}_3 \\ \nabla_{\mathbf{v}}\mathbf{m}_2 &= -\omega_{12}(\mathbf{v})\,\mathbf{m}_1 + \omega_{23}(\mathbf{v})\,\mathbf{m}_3 \\ \nabla_{\mathbf{v}}\mathbf{m}_3 &= -\omega_{13}(\mathbf{v})\,\mathbf{m}_1 - \omega_{23}(\mathbf{v})\,\mathbf{m}_2. \end{aligned}$$

We can use **IDEA 6** to recover the Frenet–Serret equations from these general equations, as follows: we simply ***adapt*** the moving frame to the curve, by choosing

$$\mathbf{m}_1 = \mathbf{T}, \qquad \mathbf{m}_2 = \mathbf{N}, \qquad \mathbf{m}_3 = \mathbf{B}, \qquad \text{and} \qquad \mathbf{v} = \mathbf{T}.$$

In comparing these equations with (38.1), the only mystery is the absence of a term corresponding to $\omega_{13}(\mathbf{v})$.

To understand this, recall that **B** is the normal to the instantaneous plane of motion of the curve (*osculating plane*); the orthogonal pair $(\mathbf{T}, \mathbf{N})$ rotate together within this plane, about the axis **B**. This means that $\mathbf{T} = \mathbf{m}_1$ does not tip in the direction of $\mathbf{B} = \mathbf{m}_3$, so (38.2) implies $\omega_{13}(\mathbf{v}) = 0$, thereby explaining the mystery.

38.2.3 WARNING: *Notational Hazing Rituals Ahead!*

First the *good* news: the connection 1-forms are (*almost*) universally written ω_{ij}, just as we have written them, though often an index is raised $(\omega^i{}_j)$ in order to be able to employ the Einstein summation convention—we too shall adopt this notation in due course.

Our use of $\boldsymbol{\theta}^i$ to denote the dual basis of 1-forms follows O'Neill (2006). Happily, this notation is widespread, but other perfectly reasonable alternatives exist e.g., Dray (2015) writes $\boldsymbol{\theta}^i$ as σ^i.

Now the *bad* news: several highly respected mathematicians[6]—including more than one Fields Medalist—have elected to employ a notation that seems to us to have been expressly and perversely *designed* to induce maximum befuddlement of the hapless student—instead of $\boldsymbol{\theta}^i$, these mathematicians write $\boldsymbol{\omega}^i$! Yes, they use the *same* Greek letter to denote the connection 1-forms $\boldsymbol{\omega}^i{}_j$ and the basis 1-forms $\boldsymbol{\omega}^i$, with only the number of indices as clue that the symbols represent *entirely different concepts!*

Although rarer, there exists an equally perverse alternative notation—of which Chern et al. (1999) provide an exemplar—in which the dual basis is written $\boldsymbol{\theta}^i$, just as we have written it, but in which the *same* Greek letter is used to denote the *connection* 1-forms, written $\boldsymbol{\theta}^i{}_j$!

It is very hard to see these extraordinary notational follies as anything other than *hazing rituals*, designed to test the determination of impudent neophytes who would dare to master the learned treatises of their elders and betters!

We feel compelled to quote one of our scientific heroes, Cornelius Lanczos (1893–1974). In the preface to his (wonderful) *Variational Principles of Mechanics*, Lanczos (1970) writes,

> Many of the scientific treatises of today are formulated in a half-mystical language, as though to impress the reader with the uncomfortable feeling that he is in the permanent presence of a superman. The present book is conceived in a humble spirit and is written for humble people.

Well, tiptoeing in the giant footsteps of Lanczos, *we too* shall speak to you humbly, as plainly and as clearly as possible, *welcoming* you into the glorious garden of Cartan's discoveries. And, for that reason, we shall denote different concepts with different letters.

38.3 The Attitude Matrix

38.3.1 The Connection Forms via the Attitude Matrix

The ***attitude*** in space of the USS Enterprise[7] (or, less interestingly, *any* rigid body) means its *orientation* within $\mathbb{R}^3$.

To describe the attitude of Cartan's moving frame field $\{\mathbf{m}_j\}$, we need only specify the rotation A of the fixed Euclidean frame $\{\mathbf{e}_j\}$ that brings it into coincidence with $\{\mathbf{m}_j\}$.

This rotation is described by the ***attitude matrix*** $[A] = [a_{ij}]$, in which the entries are *functions* of position, since the whole idea is that the attitude of the **m**-frame *varies* in space. Thus, if we write

$$[\mathbf{e}] \equiv \begin{bmatrix} \mathbf{e}_1 \\ \mathbf{e}_2 \\ \mathbf{e}_3 \end{bmatrix},$$

then

$$\boxed{[\mathbf{m}] = [A]\,[\mathbf{e}].}$$

It follows from our geometrical interpretation of the transpose of a matrix (given in Ex. 12, p. 221) that the matrix of a rotation must satisfy

---

[6]Sadly, my beloved Misner, Thorne, and Wheeler (1973) must *also* be counted amongst these sinners! Saddest of all, the *origin* of this notation was none other than Cartan (1927) himself! Cartan's works have *always* "enjoyed" the reputation of being very hard to understand; I would argue that his choice of notation did not help matters!

[7]To avoid any ambiguity, we refer here to NCC-1701.

$$\boxed{[A]^T = [A]^{-1}.} \tag{38.5}$$

Recall from Linear Algebra that such a matrix is called ***orthogonal***.

It is intuitively clear that the rate of change of $[A]$ determines the rate of change of the $\mathbf{m}$-frame, and thus it must determine the connection 1-forms. To derive an explicit formula, let us first agree that the exterior derivative of a matrix is the matrix of the exterior derivatives of its entries:

$$\mathbf{d}[A] = \mathbf{d}[a_{ij}] = [\mathbf{d}a_{ij}].$$

We will now show that the matrix of connection 1-forms is given by

$$\boxed{[\omega] = (\mathbf{d}[A])\,[A]^T.} \tag{38.6}$$

Since $[a_{ij}]^T = [a_{ji}]$, the matrix equation (38.6) is equivalent to this component equation:

$$\boxed{\omega_{ij} = \sum_k (\mathbf{d}a_{ik})\,a_{jk}.} \tag{38.7}$$

To prove this, we first compute the rate of rotation of the moving frame itself:

$$\nabla_{\mathbf{v}}\mathbf{m}_i = \nabla_{\mathbf{v}} \sum_k a_{ik}\,\mathbf{e}_k = \sum_k (\mathbf{d}a_{ik})(\mathbf{v})\,\mathbf{e}_k,$$

because $\nabla_{\mathbf{v}}\mathbf{e}_k = 0$.

According to the definition (38.2) of the connection 1-forms, we must now determine how much of this tipping of $\mathbf{m}_i$ is in the direction of $\mathbf{m}_j$. Since

$$\mathbf{m}_j = \sum_l a_{jl}\,\mathbf{e}_l,$$

we find that

$$\mathbf{m}_j \cdot \mathbf{e}_k = \sum_l a_{jl}\,\mathbf{e}_l \cdot \mathbf{e}_k = a_{jk},$$

and therefore,

$$\omega_{ij}(\mathbf{v}) \equiv (\nabla_{\mathbf{v}}\mathbf{m}_i)\cdot\mathbf{m}_j = \sum_k (\mathbf{d}a_{ik})(\mathbf{v})\,\mathbf{e}_k \cdot \mathbf{m}_j = \sum_k (\mathbf{d}a_{ik}(\mathbf{v}))\,a_{jk}.$$

Abstracting away the arbitrary vector $\mathbf{v}$, we have therefore proved (38.7), and, with it, the matrix equation (38.6).

38.3.2 Example: The Cylindrical Frame Field

The left panel of [38.1] shows the definition of the ***cylindrical frame field***, derived from the illustrated cylindrical polar coordinate system,[8] (r, ϑ, z). From this figure, one easily deduces [exercise] that

$$\begin{aligned}
\mathbf{m}_1 &= \cos\vartheta\,\mathbf{e}_1 + \sin\vartheta\,\mathbf{e}_2 \\
\mathbf{m}_2 &= -\sin\vartheta\,\mathbf{e}_1 + \cos\vartheta\,\mathbf{e}_2 \\
\mathbf{m}_3 &= \mathbf{e}_3.
\end{aligned}$$

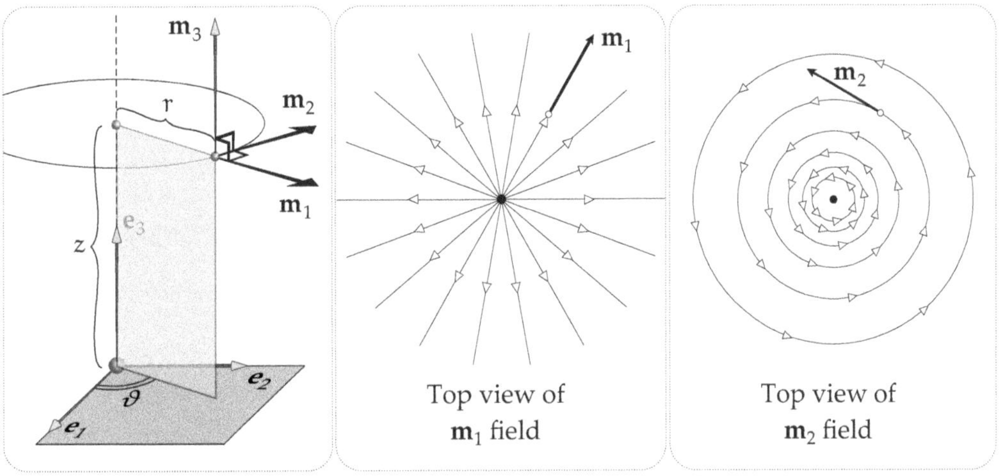

[38.1] *On the left, the definition of the* cylindrical frame field, *based on the illustrated cylindrical polar coordinates,* (r, ϑ, z). *Looking down the z-axis from above, the middle panel shows the* $\mathbf{m}_1$ *field, and the right panel shows the* $\mathbf{m}_2$ *field.*

It is desirable to get a feel for the cylindrical frame field throughout space, rather than at the single illustrated point. To that end, the middle panel of [38.1] shows a top view (looking down the z-axis) of the entire $\mathbf{m}_1$ vector field, which radiates outward (horizontally), away from the z-axis. Likewise, the panel on the right shows the top view of the $\mathbf{m}_2$ vector field, which swirls around the z-axis. We have not drawn it, but the $\mathbf{m}_3$ vector field is simply a uniform upward flow, of unit speed, parallel to the z-axis.

From the formulas above, we easily read off the attitude matrix:

$$[A] = \begin{bmatrix} \cos\vartheta & \sin\vartheta & 0 \\ -\sin\vartheta & \cos\vartheta & 0 \\ 0 & 0 & 1 \end{bmatrix}.$$

Therefore, (38.6) yields

$$\begin{aligned} [\boldsymbol{\omega}] = (\mathbf{d}[A])\,[A]^T &= \begin{bmatrix} -\sin\vartheta\,\mathbf{d}\vartheta & \cos\vartheta\,\mathbf{d}\vartheta & 0 \\ -\cos\vartheta\,\mathbf{d}\vartheta & -\sin\vartheta\,\mathbf{d}\vartheta & 0 \\ 0 & 0 & 0 \end{bmatrix} \begin{bmatrix} \cos\vartheta & -\sin\vartheta & 0 \\ \sin\vartheta & \cos\vartheta & 0 \\ 0 & 0 & 1 \end{bmatrix} \\ &= \begin{bmatrix} 0 & \mathbf{d}\vartheta & 0 \\ -\mathbf{d}\vartheta & 0 & 0 \\ 0 & 0 & 0 \end{bmatrix}. \end{aligned}$$

Thus the connection equations (38.3) become

$$\begin{aligned} \nabla_{\mathbf{v}}\mathbf{m}_1 &= \mathbf{d}\vartheta(\mathbf{v})\,\mathbf{m}_2 = (\nabla_{\mathbf{v}}\vartheta)\,\mathbf{m}_2 \\ \nabla_{\mathbf{v}}\mathbf{m}_2 &= -\mathbf{d}\vartheta(\mathbf{v})\,\mathbf{m}_1 = -(\nabla_{\mathbf{v}}\vartheta)\,\mathbf{m}_1 \\ \nabla_{\mathbf{v}}\mathbf{m}_3 &= 0. \end{aligned}$$

We presented this example as very useful practice with the mechanics of the attitude matrix, the connection 1-forms, and the connection equations. However, we can in fact obtain these final three equations directly and geometrically, without any computation.

---

[8]Here we have switched from the traditional θ to ϑ, in order to avoid any possible confusion with the dual-basis 1-forms, $\boldsymbol{\theta}^j$.

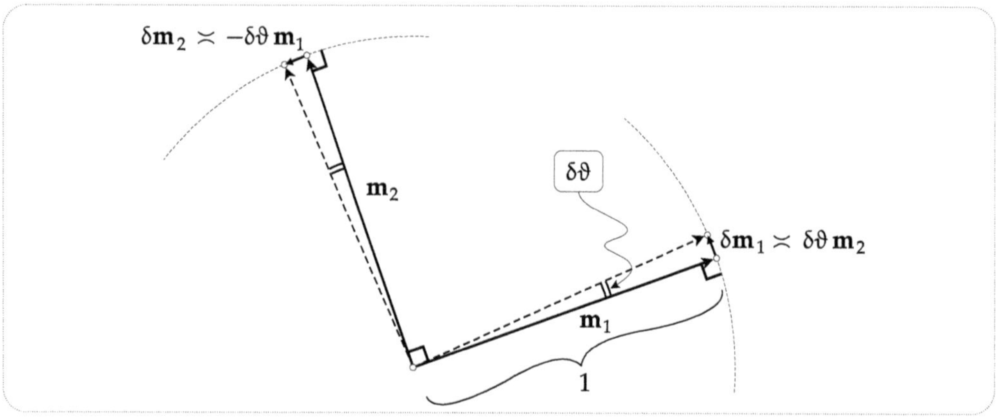

[38.2] *Moving along a short (ultimately vanishing) vector* $\mathbf{v}$ *in space results in a small (ultimately vanishing) rotation* $\delta\vartheta$ *of the orthogonal pair* $(\mathbf{m}_1, \mathbf{m}_2)$. *The tip of* $\mathbf{m}_1$ *moves through distance* $\delta\vartheta$ *along the illustrated arc of the unit circle. Since the tip initially moves in the direction orthogonal to itself,* $\delta\mathbf{m}_1 \asymp \delta\vartheta\,\mathbf{m}_2$. *Likewise,* $\delta\mathbf{m}_2 \asymp -\delta\vartheta\,\mathbf{m}_1$.

The last equation is trivial: $\mathbf{m}_3 = \mathbf{e}_3$ is constant, so of course its derivative vanishes. The first two equations are much more interesting, and we now derive them geometrically.

First, we observe that if $\mathbf{v}$ is very short, and ultimately vanishing, then

$$\mathbf{d}\vartheta(\mathbf{v}) \asymp \delta\vartheta \equiv \text{change in } \vartheta \text{ from tail to tip of } \mathbf{v}.$$

Next, by the same token,

$$\begin{aligned} \nabla_{\mathbf{v}}\mathbf{m}_j &\asymp \delta\mathbf{m}_j \\ &\equiv \text{change in } \mathbf{m}_j \text{ from tail to tip of } \mathbf{v} \\ &= \text{change in } \mathbf{m}_j \text{ brought about by the change } \delta\vartheta, \\ &\quad\ \text{resulting from the movement along } \mathbf{v}. \end{aligned}$$

Now consider [38.2], which shows the orthogonal pair $(\mathbf{m}_1, \mathbf{m}_2)$ rotating rigidly together through $\delta\vartheta$. Focus on the change $\delta\mathbf{m}_1$. The tip of $\mathbf{m}_1$ moves through distance $\delta\vartheta$ along the illustrated arc of the unit circle. Initially, the tip moves in the direction orthogonal to itself, in other words, in the direction of $\mathbf{m}_2$, as illustrated. Therefore, we have proved the first of the two connection equations, in the form

$$\delta\mathbf{m}_1 \asymp \delta\vartheta\,\mathbf{m}_2.$$

The same diagram simultaneously proves the second connection equation:

$$\delta\mathbf{m}_2 \asymp -\delta\vartheta\,\mathbf{m}_1.$$

38.4 Cartan's Two Structural Equations

38.4.1 The Duals $\boldsymbol{\theta}^i$ of $\mathbf{m}_i$ in Terms of the Duals $\mathbf{d}x^j$ of $\mathbf{e}_j$

Recall that (Cartan's) IDEA 5 is to replace the moving frame field $\mathbf{m}_i$ with its dual 1-form basis $\boldsymbol{\theta}^i$.

We know that (by definition) the attitude matrix $[a_{ij}]$ transforms the fixed Euclidean basis $\{\mathbf{e}_j\}$ into Cartan's moving frame field, according to

$$\mathbf{m}_i = \sum_k a_{ik}\, \mathbf{e}_k.$$

We now show that the *same* rotation matrix transforms the 1-form basis $\{\mathbf{d}x^j\}$ dual to $\{\mathbf{e}_j\}$ into the 1-form basis $\{\boldsymbol{\theta}^i\}$ dual to $\{\mathbf{m}_i\}$:

$$[\boldsymbol{\theta}] = [A]\, [\mathbf{d}x]. \tag{38.8}$$

Expressed in terms of components,

$$\boldsymbol{\theta}^i = \textstyle\sum_j a_{ij}\, \mathbf{d}x^j. \tag{38.9}$$

To prove this, first recall from (32.13), page 357, that *any* 1-form φ can be decomposed into components in the $\{\mathbf{d}x^j\}$ basis by applying φ to the Euclidean basis vectors. Thus, in particular,

$$\boldsymbol{\theta}^i = \sum_j \boldsymbol{\theta}^i(\mathbf{e}_j)\, \mathbf{d}x^j.$$

But because $\boldsymbol{\theta}^i$ is dual to $\mathbf{m}_i$,

$$\boldsymbol{\theta}^i(\mathbf{e}_j) = \mathbf{m}_i \bullet \mathbf{e}_j = \left[\sum_k a_{ik}\, \mathbf{e}_k\right] \bullet \mathbf{e}_j = a_{ij},$$

completing the proof of (38.9).

38.4.2 Cartan's First Structural Equation

The connection equations (38.3) tell us the rotation of the moving frame field $\mathbf{m}_i$ as we move in an arbitrary direction $\mathbf{v}$. It is therefore natural to ask how the dual 1-forms $\boldsymbol{\theta}^i$ vary in space. Since the exterior derivative measures the rate of change in *all* directions (simultaneously!), it is $\mathbf{d}[\boldsymbol{\theta}]$ that we seek.

The remarkably elegant answer is ***Cartan's First Structural Equation***:

$$\mathbf{d}[\boldsymbol{\theta}] = [\boldsymbol{\omega}] \wedge [\boldsymbol{\theta}], \tag{38.10}$$

where

$$[\boldsymbol{\theta}] \equiv \begin{bmatrix} \boldsymbol{\theta}^1 \\ \boldsymbol{\theta}^2 \\ \boldsymbol{\theta}^3 \end{bmatrix}.$$

The proof is a simple computation. Writing

$$[\mathbf{d}x] \equiv \begin{bmatrix} \mathbf{d}x_1 \\ \mathbf{d}x_2 \\ \mathbf{d}x_3 \end{bmatrix},$$

(38.9) becomes

$$[\boldsymbol{\theta}] = [A][\mathbf{d}x].$$

Differentiating, and using first (38.5) and then (38.6),

$$\mathbf{d}[\boldsymbol{\theta}] = \mathbf{d}[A] \wedge [\mathbf{d}x] = \mathbf{d}[A][A]^T \wedge [A][\mathbf{d}x] = [\omega] \wedge [\boldsymbol{\theta}].$$

Done!

When performing concrete calculations, it is sometimes useful to rephrase (38.10) in terms of its components:

$$\boxed{\mathbf{d}\theta^i = \textstyle\sum_j \omega_{ij} \wedge \theta^j} \tag{38.11}$$

38.4.3 Cartan's Second Structural Equation

Now let us ask how the connection 1-forms vary in space. Again, the exterior derivative simultaneously encodes the variation in all possible directions, so it is $\mathbf{d}[\omega]$ that we seek.

The remarkably elegant answer is ***Cartan's Second Structural Equation***:

$$\boxed{\mathbf{d}[\omega] = [\omega] \wedge [\omega].} \tag{38.12}$$

As we shall see in Section 38.12, the deeper geometric meaning of this equation is that it characterizes the lack of curvature, the *flatness* of the Euclidean space $\mathbb{R}^3$ within which we have been operating up till now. When we move on to *curved* manifolds, such as Einstein's gravitationally warped spacetime, the two sides of this equation are no longer equal, and in fact the *difference* between the two sides will be found to encode the curvature of the manifold.

To prove the equation, we require three simple lemmas. First, note that if f and g are functions, then

$$\mathbf{d}(\mathbf{d}f\, g) = \mathbf{d}(g\, \mathbf{d}f) = \mathbf{d}g \wedge \mathbf{d}f = -\mathbf{d}f \wedge \mathbf{d}g.$$

Next, recall that the geometrical interpretation of the transpose (given in Ex. 12, p. 221) implies that $(AB)^T = B^T A^T$. Finally, recall that $[\omega]$ is skew-symmetric: $[\omega]^T = -[\omega]$.

Therefore, differentiating (38.6),

$$\begin{aligned}
\mathbf{d}[\omega] &= \mathbf{d}\big[(\mathbf{d}[A])\,[A]^T\big] \\
&= -(\mathbf{d}[A]) \wedge \mathbf{d}[A]^T \\
&= -(\mathbf{d}[A])[A]^T \wedge [A]\,\mathbf{d}[A]^T \\
&= -(\mathbf{d}[A])[A]^T \wedge \big[(\mathbf{d}[A])[A]^T\big]^T \\
&= -[\omega] \wedge [\omega]^T \\
&= [\omega] \wedge [\omega].
\end{aligned}$$

Done!

When performing concrete calculations, it is sometimes useful to rephrase (38.12) in terms of its components:

$$\boxed{\mathbf{d}\omega_{ij} = \textstyle\sum_k \omega_{ik} \wedge \omega_{kj}} \tag{38.13}$$

Finally, note that it is easy to remember *both* Structural Equations at once, by remembering just this:

$$\mathbf{d}[?] = [\omega] \wedge [?].$$

38.4.4 Example: The Spherical Frame Field

In this section we shall find the dual basis and the connection forms associated with the spherical polar coordinates[9] (r, ϕ, ϑ) of Section 35.4, and we shall use them to test/illustrate Cartan's Structural Equations.

In the following, we shall write $c_\vartheta \equiv \cos\vartheta$, $s_\phi \equiv \sin\phi$, etc., in part to save space, but also because we find this to be a clarifying notation in private, pencil-and-paper calculations.

From [38.3], we see that,

$$[x] = \begin{bmatrix} x_1 \\ x_2 \\ x_3 \end{bmatrix} = \begin{bmatrix} \rho\, c_\vartheta \\ \rho\, s_\vartheta \\ r\, c_\phi \end{bmatrix} = \begin{bmatrix} r\, s_\phi\, c_\vartheta \\ r\, s_\phi\, s_\vartheta \\ r\, c_\phi \end{bmatrix}$$

From this same figure we also deduce [exercise] that the illustrated ***spherical frame field*** is given by

$$[\mathbf{m}] = \begin{bmatrix} \mathbf{m}_1 \\ \mathbf{m}_2 \\ \mathbf{m}_3 \end{bmatrix} = \begin{bmatrix} s_\phi\left[c_\vartheta \mathbf{e}_1 + s_\vartheta \mathbf{e}_2\right] + c_\phi \mathbf{e}_3 \\ c_\phi\left[c_\vartheta \mathbf{e}_1 + s_\vartheta \mathbf{e}_2\right] - s_\phi \mathbf{e}_3 \\ -s_\vartheta \mathbf{e}_1 + c_\vartheta \mathbf{e}_2 \end{bmatrix} = [A] \begin{bmatrix} \mathbf{e}_1 \\ \mathbf{e}_2 \\ \mathbf{e}_3 \end{bmatrix},$$

from which we immediately deduce the attitude matrix:

$$[A] = \begin{pmatrix} s_\phi\, c_\vartheta & s_\phi\, s_\vartheta & c_\phi \\ c_\phi\, c_\vartheta & c_\phi\, s_\vartheta & -s_\phi \\ -s_\vartheta & c_\vartheta & 0 \end{pmatrix}.$$

At this point we can calculate the connection forms using (38.6):

$$[\omega] = (\mathbf{d}[A])\, [A]^T.$$

Clearly, this will require *lengthy* and arduous computations, which we hereby assign as a (cruel)[10] exercise! However, let us at least offer some concrete guidance on *how* to carry out such computations.

First, we certainly *don't* need to multiply the *entire* matrix $\mathbf{d}[A]$ by the *entire* matrix $[A]$. For example, calculating a diagonal entry is a waste of time: we know it must be zero! In fact we *only* need to calculate the three entries in the top-right corner of the product: $\omega_{12}, \omega_{13}, \omega_{23}$.

In order to target just these three entries, it is simpler to use the component form of the matrix product, namely, (38.7):

$$\omega_{ij} = \sum_k (\mathbf{d}a_{ik})\, a_{jk}.$$

Thus we never need to actually form the transpose of $[A]$. Instead, we interpret this formula as telling us that

> *To obtain ω_{ij} we take* $\mathbf{d}$ *of the* i*-th row of* $[A]$*, then multiply each element by the corresponding one in the* j*-th row of* $[A]$.

[9] O'Neill (2006, pp. 86, 97) instead measures ϕ from the equator, so our ϕ is obtained by subtracting his from $(\pi/2)$. This has the effect of interchanging $\sin\phi$ and $\cos\phi$. Also, in order to make the frame right-handed, the *order* of O'Neill's coordinates is (r, ϑ, ϕ), and his spherical frame field, written $(\mathbf{F}_1, \mathbf{F}_2, \mathbf{F}_3)$, is therefore related to ours by $(\mathbf{m}_1 = \mathbf{F}_1, \mathbf{m}_2 = -\mathbf{F}_3, \mathbf{m}_3 = \mathbf{F}_2)$.

[10] When forced to assign such a cruel computation to my USF students, I simultaneously assign them a viewing of Tarentino's *Kill Bill, Vol. 2, Chapter 8: The Cruel Tutelage of Pai Mei*, for it demonstrates how such "cruelty" ultimately *saved* Beatrix Kiddo from the grave!

For example, to find $\boldsymbol{\omega}_{12}$, we must take $\mathbf{d}$ of the first row, and multiply it by the second row:

$$\boldsymbol{\omega}_{12} = \left[\mathbf{d}(s_\phi\, c_\vartheta)\right] c_\phi\, c_\vartheta + \left[\mathbf{d}(s_\phi\, s_\vartheta)\right] c_\phi\, s_\vartheta - \left[\mathbf{d}(c_\phi)\right] s_\phi.$$

Even if you don't feel up to calculating all three connection forms, we urge you to at least do this one. After taking the derivatives, you will be faced with five terms, each of which is the product of four trigonometric functions. But then, "miraculously," everything *cancels or simplifies*, and we are left with a beautifully simple answer: $\boldsymbol{\omega}_{12} = \mathbf{d}\phi$!

Equally "miraculous" cancellations and simplifications occur in the calculation of the other two connection forms, and we obtain the following simple formulas:

$$\begin{aligned} \boldsymbol{\omega}_{12} &= \mathbf{d}\phi \\ \boldsymbol{\omega}_{13} &= \sin\phi\, \mathbf{d}\vartheta \\ \boldsymbol{\omega}_{23} &= \cos\phi\, \mathbf{d}\vartheta. \end{aligned} \tag{38.14}$$

Recalling what was said in the Prologue, we have here three examples of what we call "false miracles": if all those terms cancelled, they should never have been there in the first place—we must be looking at the mathematics in the *wrong way!*

By this point in our drama it can surely come as no surprise to learn that the *correct* way to look at this is *geometrically*. *All* of the painful calculations you just performed are thereby avoided, and the results are obtained directly and intuitively! We shall provide this Newtonian, geometrical explanation at the end of this section, but for now we press on in our quest to test Cartan's Structural Equations.

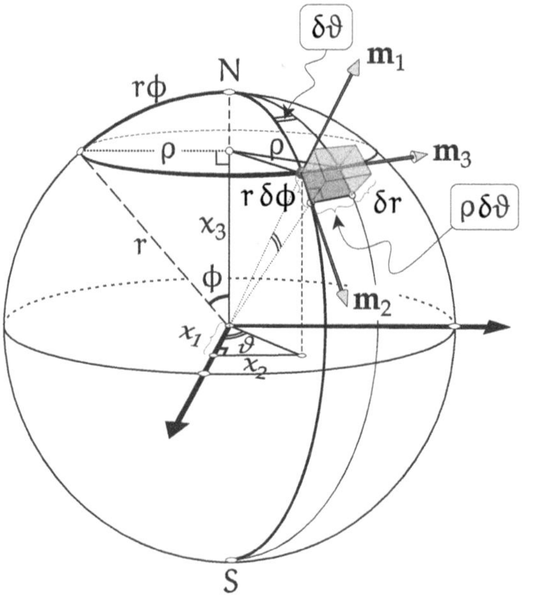

[38.3] *Geometric proof that* $\boldsymbol{\theta}^1 = \mathbf{d}r$, $\boldsymbol{\theta}^2 = r\,\mathbf{d}\phi$, *and* $\boldsymbol{\theta}^3 = r\sin\phi\,\mathbf{d}\vartheta$.

To that end, let us find the dual-basis 1-forms, $\boldsymbol{\theta}^i$. Again, we can certainly obtain these by blind computation: we know [A] and we know [x], so we can calculate $\mathbf{d}[x] = [\mathbf{d}x]$, and hence we can use (38.8):

$$[\boldsymbol{\theta}] = [A]\,[\mathbf{d}x].$$

But, *again*, this will clearly require a *lengthy* computation, which we hereby assign as *another* cruel exercise! (You will appreciate what comes next so much more if you actually *carry out* these calculations, which are filled with many cancellations and simplifications. Having successfully hacked your way through this dense thicket of symbols, you should emerge into a clearing, within which you will find these simple formulas: (38.16).)

Here, too, Newtonian geometry allows us leap through the air, flying over the lengthy and opaque calculations beneath us, landing directly at the answers!

Consider the shaded box in [38.3] in the limit that it shrinks and ultimately vanishes. Its edges are ultimately equal to

$$\delta r\, \mathbf{m}_1, \qquad r\,\delta\phi\, \mathbf{m}_2, \qquad \rho\,\delta\vartheta\, \mathbf{m}_3 = r\, s_\phi\, \delta\vartheta\, \mathbf{m}_3,$$

as illustrated. Therefore, the diagonal of the box, which points in an *arbitrary* direction, is the short (ultimately vanishing) vector **v**, given by

$$\boxed{\mathbf{v} \asymp \delta r\, \mathbf{m}_1 + r\, \delta\phi\, \mathbf{m}_2 + r\, s_\phi\, \delta\vartheta\, \mathbf{m}_3.} \tag{38.15}$$

By definition of the dual basis, $\boldsymbol{\theta}^i(\mathbf{m}_j) = \delta^i_j$, so,

$$\boldsymbol{\theta}^1(\mathbf{v}) \asymp \delta r \asymp \mathbf{d}r(\mathbf{v}) \qquad \Longrightarrow \qquad \boldsymbol{\theta}^1 = \mathbf{d}r.$$

This follows from the fact that **v** is *arbitrary*, and therefore can be abstracted away.

We find $\boldsymbol{\theta}^2$ and $\boldsymbol{\theta}^3$ in exactly the same way. Therefore, the geometry of [38.3] immediately yields

$$\boxed{[\boldsymbol{\theta}] = \begin{bmatrix} \boldsymbol{\theta}^1 \\ \boldsymbol{\theta}^2 \\ \boldsymbol{\theta}^3 \end{bmatrix} = \begin{bmatrix} \mathbf{d}r \\ r\,\mathbf{d}\phi \\ r\,s_\phi\,\mathbf{d}\vartheta \end{bmatrix}.} \tag{38.16}$$

Contrast this geometry with the calculations you just performed!

Now that we know the dual basis and the connection forms, let us check Cartan's *First* Structural Equation, (38.10),

$$\mathbf{d}[\boldsymbol{\theta}] = [\boldsymbol{\omega}] \wedge [\boldsymbol{\theta}].$$

On the left-hand side, we find

$$\mathbf{d}[\boldsymbol{\theta}] = \mathbf{d}\begin{bmatrix} \mathbf{d}r \\ r\,\mathbf{d}\phi \\ r\,s_\phi\,\mathbf{d}\vartheta \end{bmatrix} = \begin{bmatrix} 0 \\ \mathbf{d}r \wedge \mathbf{d}\phi \\ s_\phi\,\mathbf{d}r \wedge \mathbf{d}\vartheta + r\,c_\phi\,\mathbf{d}\phi \wedge \mathbf{d}\vartheta \end{bmatrix}.$$

On the right-hand side, we find

$$\begin{aligned}
[\boldsymbol{\omega}] \wedge [\boldsymbol{\theta}] &= \begin{bmatrix} 0 & \boldsymbol{\omega}_{12} & \boldsymbol{\omega}_{13} \\ -\boldsymbol{\omega}_{12} & 0 & \boldsymbol{\omega}_{23} \\ -\boldsymbol{\omega}_{13} & -\boldsymbol{\omega}_{23} & 0 \end{bmatrix} \wedge \begin{bmatrix} \boldsymbol{\theta}^1 \\ \boldsymbol{\theta}^2 \\ \boldsymbol{\theta}^3 \end{bmatrix} \\
&= \begin{bmatrix} 0 & \mathbf{d}\phi & s_\phi\,\mathbf{d}\vartheta \\ -\mathbf{d}\phi & 0 & c_\phi\,\mathbf{d}\vartheta \\ -s_\phi\,\mathbf{d}\vartheta & -c_\phi\,\mathbf{d}\vartheta & 0 \end{bmatrix} \wedge \begin{bmatrix} \mathbf{d}r \\ r\,\mathbf{d}\phi \\ r\,s_\phi\,\mathbf{d}\vartheta \end{bmatrix} \\
&= \begin{bmatrix} 0 \\ \mathbf{d}r \wedge \mathbf{d}\phi \\ s_\phi\,\mathbf{d}r \wedge \mathbf{d}\vartheta + r\,c_\phi\,\mathbf{d}\phi \wedge \mathbf{d}\vartheta \end{bmatrix},
\end{aligned}$$

thereby verifying Cartan's First Structural Equation for the spherical frame field.

Now let us check Cartan's *Second* Structural Equation, (38.12),

$$\mathbf{d}[\boldsymbol{\omega}] = [\boldsymbol{\omega}] \wedge [\boldsymbol{\omega}].$$

On the left-hand side, we find

$$\mathbf{d}\begin{bmatrix} 0 & \mathbf{d}\phi & s_\phi\,\mathbf{d}\vartheta \\ -\mathbf{d}\phi & 0 & c_\phi\,\mathbf{d}\vartheta \\ -s_\phi\,\mathbf{d}\vartheta & -c_\phi\,\mathbf{d}\vartheta & 0 \end{bmatrix} = \begin{bmatrix} 0 & 0 & c_\phi\,\mathbf{d}\phi \wedge \mathbf{d}\vartheta \\ 0 & 0 & -s_\phi\,\mathbf{d}\phi \wedge \mathbf{d}\vartheta \\ -c_\phi\,\mathbf{d}\phi \wedge \mathbf{d}\vartheta & s_\phi\,\mathbf{d}\phi \wedge \mathbf{d}\vartheta & 0 \end{bmatrix}.$$

On the right-hand side, we find

$$[\omega]\wedge[\omega] = \begin{bmatrix} 0 & \mathbf{d}\phi & s_\phi\,\mathbf{d}\vartheta \\ -\mathbf{d}\phi & 0 & c_\phi\,\mathbf{d}\vartheta \\ -s_\phi\,\mathbf{d}\vartheta & -c_\phi\,\mathbf{d}\vartheta & 0 \end{bmatrix} \wedge \begin{bmatrix} 0 & \mathbf{d}\phi & s_\phi\,\mathbf{d}\vartheta \\ -\mathbf{d}\phi & 0 & c_\phi\,\mathbf{d}\vartheta \\ -s_\phi\,\mathbf{d}\vartheta & -c_\phi\,\mathbf{d}\vartheta & 0 \end{bmatrix}$$

$$= \begin{bmatrix} 0 & 0 & c_\phi\,\mathbf{d}\phi\wedge\mathbf{d}\vartheta \\ 0 & 0 & -s_\phi\,\mathbf{d}\phi\wedge\mathbf{d}\vartheta \\ -c_\phi\,\mathbf{d}\phi\wedge\mathbf{d}\vartheta & s_\phi\,\mathbf{d}\phi\wedge\mathbf{d}\vartheta & 0 \end{bmatrix},$$

thereby verifying Cartan's Second Structural Equation for the spherical frame field.

As promised, we end this section by returning to the connection forms, showing that the lengthy calculations that led to (38.14) can be replaced with simple, clear, Newtonian geometry.

To do so, we shall apply Newtonian ultimate equalities *directly* to the definition (38.2):

> $\omega_{ij}(\mathbf{v}) \equiv (\nabla_{\mathbf{v}}\mathbf{m}_i)\bullet\mathbf{m}_j$ *is the initial rate at which* $\mathbf{m}_i$ *tips towards* $\mathbf{m}_j$ *as the frame moves along* $\mathbf{v}$.

Here we shall take $\mathbf{v}$ to be the arbitrary, short (ultimately vanishing) vector given by (38.15):

$$\mathbf{v} \asymp \delta r\,\mathbf{m}_1 + r\,\delta\phi\,\mathbf{m}_2 + r\,s_\phi\,\delta\vartheta\,\mathbf{m}_3.$$

The key tool in the following analysis is the fact that, since $\mathbf{v}$ is short, and ultimately vanishing,

$$\nabla_{\mathbf{v}}\mathbf{m}_i \asymp \text{change in } \mathbf{m}_i \text{ from tail to tip of } \mathbf{v} \equiv \delta\mathbf{m}_i.$$

Note that in order to easily visualize the changes $\delta\mathbf{m}_i$, *we shall draw all the vectors* $\mathbf{m}_i$ *as emanating from a common point; then* $\delta\mathbf{m}_i$ *is simply the movement of the tip of* $\mathbf{m}_i$.

We begin with $\omega_{12}(\mathbf{v}) \equiv (\nabla_{\mathbf{v}}\mathbf{m}_1)\bullet\mathbf{m}_2$. Consider the changes $\delta\mathbf{m}_1$ in $\mathbf{m}_1$ that result from moving along *each* of the three components of $\mathbf{v}$, separately:

• Moving radially outward along $\delta r\,\mathbf{m}_1$ keeps $\mathbf{m}_1$ *constant*: it does not tip towards $\mathbf{m}_2$ (or anything else!).

• Moving along $r\,s_\phi\,\delta\vartheta\,\mathbf{m}_3$ (due East, along the circle of latitude) causes $\mathbf{m}_1$ to rotate around the cone shown in the top half of [38.5], of semivertical angle ϕ and slant height 1, and therefore base radius s_ϕ. Evidently, $\delta\mathbf{m}_1$ is orthogonal to $\mathbf{m}_2$, so this too makes zero contribution to ω_{12}.

In greater detail, we note for future use that the figure demonstrates that

$$\delta\mathbf{m}_1 = s_\phi\,\delta\vartheta\,\mathbf{m}_3. \tag{38.17}$$

• As an aside, for future use, we note that this rotation causes $\mathbf{m}_2$ to rotate around the cone shown in the bottom half of [38.5], of semivertical angle $(\frac{\pi}{2}-\phi)$ and slant height 1, and therefore base radius c_ϕ. Thus, as illustrated,

$$\delta\mathbf{m}_2 = c_\phi\,\delta\vartheta\,\mathbf{m}_3. \tag{38.18}$$

• Returning our focus to $\mathbf{m}_1$, the *only* part of $\mathbf{v}$ that *does* cause $\mathbf{m}_1$ to tip towards $\mathbf{m}_2$ is the southward movement $r\,\delta\phi\,\mathbf{m}_2$ towards the equator, along the meridian. Figure [38.4] is drawn in the vertical plane of this meridian, containing $\mathbf{m}_1$ and $\mathbf{m}_2$. As $\mathbf{m}_1$ rotates through angle $\delta\phi$, its tip ultimately moves along $\delta\mathbf{m}_1 \asymp \delta\phi\,\mathbf{m}_2$, as illustrated. Therefore,

$$\omega_{12}(\mathbf{v}) \asymp \delta\mathbf{m}_1\bullet\mathbf{m}_2 \asymp (\delta\phi\,\mathbf{m}_2)\bullet\mathbf{m}_2 = \delta\phi \asymp \mathbf{d}\phi(\mathbf{v}) \quad\Longrightarrow\quad \omega_{12} = \mathbf{d}\phi.$$

Next, consider

$$\omega_{13}(\mathbf{v}) \equiv (\nabla_{\mathbf{v}}\mathbf{m}_1) \bullet \mathbf{m}_3.$$

Here we are considering the *same* changes $\delta\mathbf{m}_1$ in $\mathbf{m}_1$ as above, but now we want to know how much $\delta\mathbf{m}_1$ points in the direction of $\mathbf{m}_3$ (instead of $\mathbf{m}_2$).

As we have seen, the only part of $\mathbf{v}$ that produces a $\delta\mathbf{m}_1$ along $\mathbf{m}_3$ is the movement along $r\,s_\phi\,\delta\vartheta\,\mathbf{m}_3$. From (38.17) we immediately deduce that

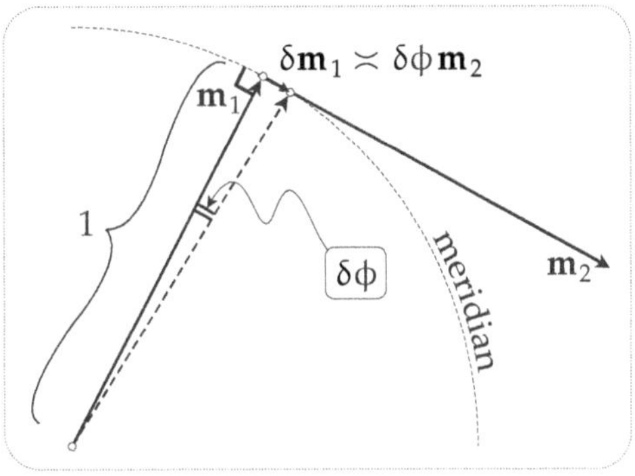

[38.4] *Geometric proof that* $\omega_{12} = \mathbf{d}\phi$.

$$\omega_{13}(\mathbf{v}) \asymp \delta\mathbf{m}_1 \bullet \mathbf{m}_3 \asymp (s_\phi\,\delta\vartheta\,\mathbf{m}_3) \bullet \mathbf{m}_3$$
$$= s_\phi\,\delta\vartheta \asymp s_\phi\,\mathbf{d}\vartheta(\mathbf{v}) \quad\Longrightarrow\quad \omega_{13} = s_\phi\,\mathbf{d}\vartheta.$$

Finally, consider $\omega_{23}(\mathbf{v}) \equiv (\nabla_{\mathbf{v}}\mathbf{m}_2) \bullet \mathbf{m}_3$. It is easy to see [exercise] that the only component of $\mathbf{v}$ that causes $\mathbf{m}_2$ to tip in the direction of $\mathbf{m}_3$ is the movement along $r\,s_\phi\,\delta\vartheta\,\mathbf{m}_3$ (due East, along the circle of latitude). Thus, using (38.18), illustrated in the bottom cone of [38.5], we deduce that

$$\omega_{23}(\mathbf{v}) \asymp \delta\mathbf{m}_2 \bullet \mathbf{m}_3 \asymp (c_\phi\,\delta\vartheta\,\mathbf{m}_3) \bullet \mathbf{m}_3 = c_\phi\,\delta\vartheta \asymp c_\phi\,\mathbf{d}\vartheta(\mathbf{v}) \quad\Longrightarrow\quad \omega_{23} = c_\phi\,\mathbf{d}\vartheta.$$

This completes our geometrical proof of (38.14). Again, contrast the illuminating simplicity and directness of this Newtonian geometrical reasoning with the three long, unilluminating calculations you were originally forced to perform in order to arrive at these three formulas!

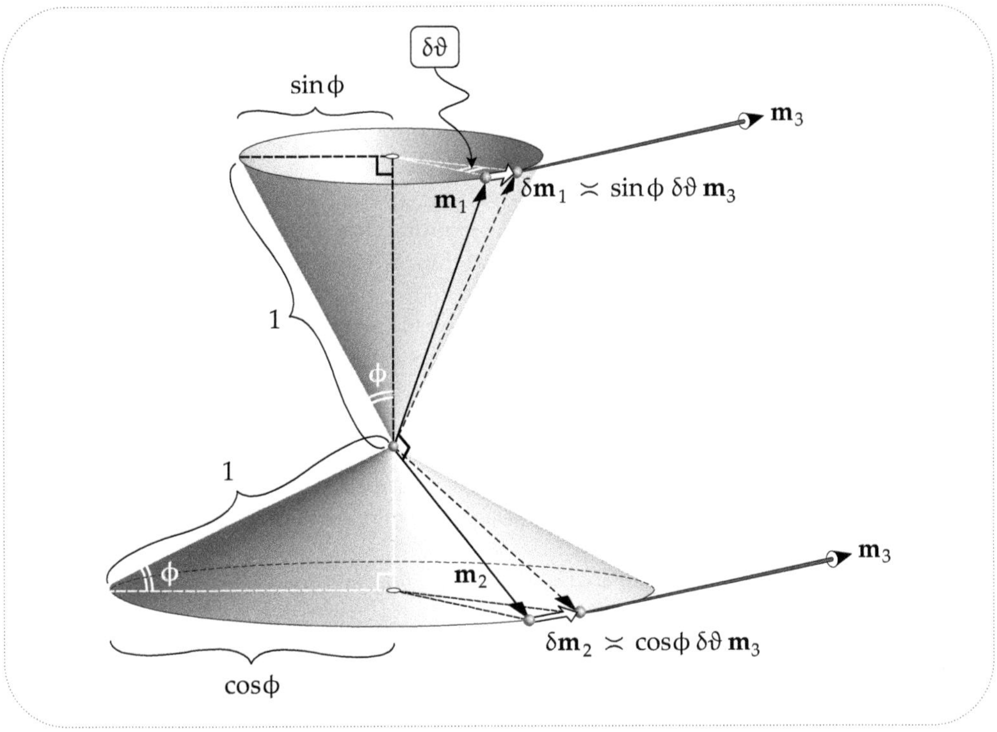

[38.5] *Geometric proof of* $\omega_{13} = \sin\phi\,\mathbf{d}\vartheta$ *(top cone), and of* $\omega_{23} = \cos\phi\,\mathbf{d}\vartheta$ *(bottom cone).*

38.5 The Six Fundamental Form Equations of a Surface

38.5.1 Adapting Cartan's Moving Frame to a Surface: The Shape Operator and the Extrinsic Curvature

We now implement Darboux's **IDEA 3** (adapt the frame to a surface) in combination with Cartan's **IDEA 6**: use Forms, specializing Cartan's two Structural Equations to this adapted frame field.

Given a surface $\mathcal{S}$ with normal $\mathbf{n}$, we *choose* $\mathbf{m}_3 = \mathbf{n}$; this is called the ***adapted frame field***. Then $(\mathbf{m}_1, \mathbf{m}_2)$ *becomes an intrinsic (i.e., tangent) basis for the vectors* $\mathbf{v}$ *"within"* $\mathcal{S}$.

Henceforth, we shall only *consider such vectors* $\mathbf{v}$ *that are tangent to* $\mathcal{S}$. When applied to such a tangent vector, the connection equations (38.4) yield the following simple expression for the Shape Operator S ((15.4), p. 151) in terms of the connection forms:

$$\mathsf{S}(\mathbf{v}) = \omega_{13}(\mathbf{v})\,\mathbf{m}_1 + \omega_{23}(\mathbf{v})\,\mathbf{m}_2. \tag{38.19}$$

The proof goes as follows:

$$\mathsf{S}(\mathbf{v}) = -\nabla_{\mathbf{v}}\mathbf{n} = -\nabla_{\mathbf{v}}\mathbf{m}_3 = -\left[\nabla_{\mathbf{v}}\mathbf{m}_3 \cdot \mathbf{m}_1\right]\mathbf{m}_1 - \left[\nabla_{\mathbf{v}}\mathbf{m}_3 \cdot \mathbf{m}_2\right]\mathbf{m}_2.$$

The last step is to insert the definitions of the connection 1-forms, (38.2), taking note of their skew-symmetry.

The first column of the *matrix* [S] representing the Shape Operator is the image under S of the first basis vector, namely, $\mathsf{S}(\mathbf{m}_1)$. Likewise, the second column is the image of the second basis vector. Therefore,

$$[\mathsf{S}] = \begin{bmatrix} \omega_{13}(\mathbf{m}_1) & \omega_{13}(\mathbf{m}_2) \\ \omega_{23}(\mathbf{m}_1) & \omega_{23}(\mathbf{m}_2) \end{bmatrix}. \tag{38.20}$$

Let us now use this to express the extrinsic curvature $\mathcal{K}_{\mathrm{ext}} = \kappa_1\kappa_2$ in terms of the connection forms. Since we intend to *reprove* fundamental results from the first four Acts, we must avoid making *assumptions* based on this prior knowledge. Thus we must revert to distinguishing between extrinsic and intrinsic measurements of curvature. Only then can we prove (and appreciate!) Gauss's *Theorema Egregium*, which reveals that these two measures of curvature are actually the *same*!

On the other hand, it is certainly *not* our intention to use Forms to rebuild the entire edifice of Differential Geometry from scratch! So long as we are explicit about which former results are being invoked, and circular reasoning is scrupulously avoided, no harm can arise from this hybrid approach.

Thus, now, recall from (15.8), page 153, that the *extrinsic* curvature $\mathcal{K}_{\mathrm{ext}}$ of $\mathcal{S}$ measures the *spreading* of the normal vectors $\mathbf{m}_3$ over a small patch of surface. More precisely, it is the local area expansion factor of S. In other words, it is the *determinant* of the matrix [S] representing S:

$$\mathcal{K}_{\mathrm{ext}} = \det[\mathsf{S}] = \omega_{13}(\mathbf{m}_1)\,\omega_{23}(\mathbf{m}_2) - \omega_{13}(\mathbf{m}_2)\,\omega_{23}(\mathbf{m}_1).$$

We have arrived at an expression for the extrinsic curvature that will soon prove its importance:

$$\mathcal{K}_{\mathrm{ext}} = (\omega_{13} \wedge \omega_{23})(\mathbf{m}_1, \mathbf{m}_2). \tag{38.21}$$

Finally, consider the 1-form basis $\boldsymbol{\theta}^i$ dual to the adapted frame field $(\mathbf{m}_1, \mathbf{m}_2, \mathbf{m}_3)$. These give the coordinates $\boldsymbol{\theta}^i(\mathbf{v}) = \mathbf{v} \cdot \mathbf{m}_i$ of any tangent vector $\mathbf{v}$ with respect to the adapted frame field. As with the connection forms, these dual-basis forms will be applied *only* to tangent vectors to $\mathcal{S}$, so that they become forms *on* $\mathcal{S}$. O'Neill (2006, p. 266) states the implication perfectly: "this restriction is fatal to $\boldsymbol{\theta}^3$, for if $\mathbf{v}$ is tangent to $\mathcal{S}$, it is orthogonal to $\mathbf{m}_3$, so $\boldsymbol{\theta}^3(\mathbf{v}) = \mathbf{v} \cdot \mathbf{m}_3 = 0$. Thus $\boldsymbol{\theta}^3$ is identically zero on $\mathcal{S}$."

38.5.2 Example: The Sphere

In order to adapt the spherical frame field to a particular sphere, $r = R$, we must relabel the vectors shown in [38.3]. The radial vector (which *was* $\mathbf{m}_1$) is the normal to the sphere, and therefore (according to our convention) it must be relabelled as $\mathbf{m}_3$. The remaining two members of the original frame field, tangent to the sphere, must likewise be relabelled, as shown in [38.6], subject to the ***right-hand rule***: $\mathbf{m}_1$ points South, and $\mathbf{m}_2$ points East.

It follows that the previously obtained 1-form basis $\boldsymbol{\theta}^i$ and the connection forms $\boldsymbol{\omega}_{ij}$ are *now* given by,[11]

$$\boldsymbol{\theta}^1 = R\,\mathbf{d}\phi \qquad \omega_{12} = c_\phi\,\mathbf{d}\vartheta$$
$$\boldsymbol{\theta}^2 = R\,s_\phi\,\mathbf{d}\vartheta \qquad \omega_{13} = -\mathbf{d}\phi$$
$$\omega_{23} = -s_\phi\,\mathbf{d}\vartheta$$

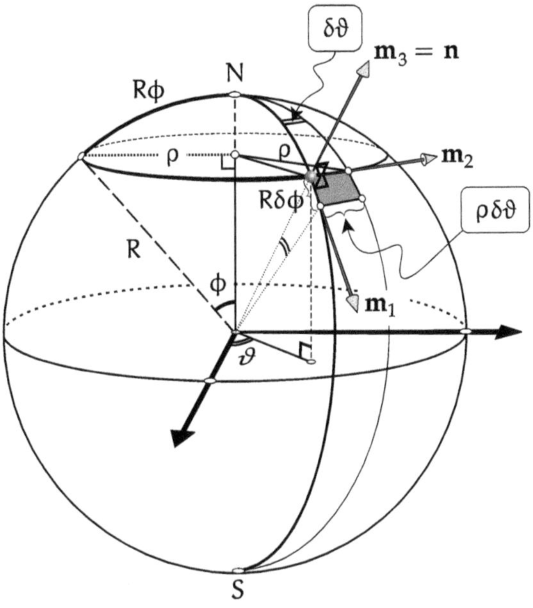

[38.6] *The* **adapted frame field** *has (by definition)* $\mathbf{m}_3 = \mathbf{n}$. *The resulting changes of names imply that, now,* $\boldsymbol{\theta}^1 = R\,\mathbf{d}\phi$, *and* $\boldsymbol{\theta}^2 = \rho\,\mathbf{d}\vartheta = R\,\sin\phi\,\mathbf{d}\vartheta$.

As valuable practice with the geometrical methods we have just explained, try checking each of these five results, simply by *inspecting* [38.6]—no calculations required!

Finally, note that the shaded element of area is $\mathcal{A} = (R\,\delta\phi)(\rho\,\delta\vartheta) = R^2\,\sin\vartheta\,\delta\phi\,\delta\vartheta$. Thus, the ***area 2-form of the sphere***—or, indeed, of *any* surface—is

$$\mathcal{A} = \boldsymbol{\theta}^1 \wedge \boldsymbol{\theta}^2. \tag{38.22}$$

38.5.3 Uniqueness of Basis Decompositions

Let us briefly review the uniqueness of the decomposition of 1-forms and 2-forms relative to a basis; both these facts will be needed in short order.

For 1-forms within $\mathcal{S}$, our basis is $(\boldsymbol{\theta}^1, \boldsymbol{\theta}^2)$. Recall that *two 1-forms* φ *and* ψ *are equal if and only if*

$$\varphi(\mathbf{m}_1) = \psi(\mathbf{m}_1) \qquad \text{and} \qquad \varphi(\mathbf{m}_2) = \psi(\mathbf{m}_2).$$

It follows immediately [exercise] that φ can be *uniquely* decomposed as

$$\varphi = \varphi(\mathbf{m}_1)\,\boldsymbol{\theta}^1 + \varphi(\mathbf{m}_2)\,\boldsymbol{\theta}^2. \tag{38.23}$$

[11]Again, these do *not* match the results you will find in O'Neill (2006, p. 267), because his different method of defining ϕ changes *everything*.

Likewise, *two 2-forms* $\mathbf{\Phi}$ *and* $\mathbf{\Psi}$ *are equal if and only if*

$$\mathbf{\Phi}(\mathbf{m}_1, \mathbf{m}_2) = \mathbf{\Psi}(\mathbf{m}_1, \mathbf{m}_2).$$

It follows immediately [exercise] that $\mathbf{\Phi}$ can be *uniquely* decomposed as

$$\boxed{\mathbf{\Phi} = \mathbf{\Phi}(\mathbf{m}_1, \mathbf{m}_2)\,\mathcal{A} = \mathbf{\Phi}(\mathbf{m}_1, \mathbf{m}_2)\,\boldsymbol{\theta}^1 \wedge \boldsymbol{\theta}^2.} \tag{38.24}$$

In particular, applying this result to (38.21), we find that

$$\boxed{\boldsymbol{\omega}_{13} \wedge \boldsymbol{\omega}_{23} = (\boldsymbol{\omega}_{13} \wedge \boldsymbol{\omega}_{23})(\mathbf{m}_1, \mathbf{m}_2)\,\boldsymbol{\theta}^1 \wedge \boldsymbol{\theta}^2 = \mathcal{K}_{\text{ext}}\,\boldsymbol{\theta}^1 \wedge \boldsymbol{\theta}^2.} \tag{38.25}$$

38.5.4 *The Six Fundamental Form Equations of a Surface*

Let us immediately state all six of these equations, and also state their names,[12] in one fell swoop. We will then prove them by calculation, and finally, gradually, extract their true geometrical meanings.

$\mathbf{d}\boldsymbol{\theta}^1 = \boldsymbol{\omega}_{12} \wedge \boldsymbol{\theta}^2$ $\mathbf{d}\boldsymbol{\theta}^2 = \boldsymbol{\omega}_{21} \wedge \boldsymbol{\theta}^1$	***First Structural Equations***	
$\boldsymbol{\omega}_{31} \wedge \boldsymbol{\theta}^1 + \boldsymbol{\omega}_{32} \wedge \boldsymbol{\theta}^2 = 0$	***Symmetry Equation***	
$\mathbf{d}\boldsymbol{\omega}_{12} = \boldsymbol{\omega}_{13} \wedge \boldsymbol{\omega}_{32}$	***Gauss Equation***	(38.26)
$\mathbf{d}\boldsymbol{\omega}_{13} = \boldsymbol{\omega}_{12} \wedge \boldsymbol{\omega}_{23}$ $\mathbf{d}\boldsymbol{\omega}_{23} = \boldsymbol{\omega}_{21} \wedge \boldsymbol{\omega}_{13}$	***Peterson–Mainardi–Codazzi Equations***	

In our description of the surface $\mathcal{S}$, only $\mathbf{m}_3$ is uniquely, geometrically determined by the surface itself, as its normal. Meanwhile, we may arbitrarily twirl $(\mathbf{m}_1, \mathbf{m}_2)$ within the surface, about the axis $\mathbf{m}_3$. All these arbitrarily different frame fields yield *different* descriptions of the surface—*different* dual-basis 1-forms, and *different* connection 1-forms. Nevertheless, despite these freedoms, the six equations above embody everything that can be known or said about the surface.

It is therefore remarkable how quickly and easily all six equations follow from Cartan's Structural Equations—all we need do is set $\boldsymbol{\theta}^3 = 0$.

Because $\boldsymbol{\omega}_{11} = 0$, the First Structural Equation, (38.11), immediately yields

$$\mathbf{d}\boldsymbol{\theta}^1 = \sum_j \boldsymbol{\omega}_{1j} \wedge \boldsymbol{\theta}^j = \boldsymbol{\omega}_{12} \wedge \boldsymbol{\theta}^2 + \boldsymbol{\omega}_{13} \wedge \boldsymbol{\theta}^3 = \boldsymbol{\omega}_{12} \wedge \boldsymbol{\theta}^2.$$

The calculation of $\mathbf{d}\boldsymbol{\theta}^2$ is essentially identical.

[12]Here we largely adopt the names employed by O'Neill (2006, p. 267), but the reader should be aware that, while widely used, these names are not *universal*. In particular, what O'Neill calls the Codazzi Equations are also commonly known as the *Mainardi–Codazzi Equations*, but we have attempted to give credit where credit is due, calling them instead the ***Peterson–Mainardi–Codazzi Equations***. For although Mainardi discovered the equations in 1856, and, independently, Codazzi rediscovered them in 1860, both were anticipated in 1853 by a young Latvian student, Karl M. Peterson. Peterson's work (and priority) remained unknown until his dissertation (supervised by Minding!) was finally translated into Russian, and published in 1952. See Phillips (1979).

The Symmetry Equation follows in the same way:

$$0 = \mathbf{d}\theta^3 = \sum_j \omega_{3j} \wedge \theta^j = \omega_{31} \wedge \theta^1 + \omega_{32} \wedge \theta^2.$$

Next, since $\omega_{11} = 0 = \omega_{22}$, the Gauss Equation follows immediately from the Second Structural Equation, (38.13):

$$\mathbf{d}\omega_{12} = \sum_k \omega_{1k} \wedge \omega_{k2} = \omega_{13} \wedge \omega_{32}.$$

The final two Peterson–Mainardi–Codazzi Equations follow from the Second Structural Equation in exactly the same way.

38.6 Geometrical Meanings of the Symmetry Equation and the Peterson–Mainardi–Codazzi Equations

But what do these fundamental form equations actually *mean*?

Let us begin with the Symmetry Equation. If we apply this vanishing 2-form to the basis vectors, we obtain

$$\begin{aligned} 0 &= [\omega_{31} \wedge \theta^1 + \omega_{32} \wedge \theta^2](\mathbf{m}_1, \mathbf{m}_2) \\ &= \omega_{31}(\mathbf{m}_1)\theta^1(\mathbf{m}_2) - \omega_{31}(\mathbf{m}_2)\theta^1(\mathbf{m}_1) + \omega_{32}(\mathbf{m}_1)\theta^2(\mathbf{m}_2) - \omega_{32}(\mathbf{m}_2)\theta^2(\mathbf{m}_1) \\ &= \omega_{32}(\mathbf{m}_1) - \omega_{31}(\mathbf{m}_2). \end{aligned}$$

Thus, invoking the skew-symmetry of the connection forms,

$$\boxed{\omega_{23}(\mathbf{m}_1) = \omega_{13}(\mathbf{m}_2).}$$

Therefore, referring back to (38.20), we deduce that

> *The Symmetry Equation is equivalent to the fact the Shape Operator is symmetric:* $S^T = S$.

The geometrical interpretation of this symmetry was previously derived in (15.13), on page 156: *The eigenvectors of* S *are orthogonal.*

If we allow ourselves the former knowledge that these eigenvectors are in fact the *principal directions* of maximum and minimum normal curvature, these curvatures being the associated eigenvalues, κ_1 and κ_2, then we have reproved a major part of Euler's discovery: these maximum and minimum curvatures occur in *orthogonal* directions.

Let us break new ground, and refine our adapted frame field so as to make $(\mathbf{m}_1, \mathbf{m}_2)$ align with these principal directions at each point of $\mathcal{S}$; this is called the ***principal frame field***. In this case, $S(\mathbf{m}_1) = \kappa_1 \mathbf{m}_1$ and $S(\mathbf{m}_2) = \kappa_2 \mathbf{m}_2$, so we deduce that (38.20) takes the form

$$[S] = \begin{bmatrix} \omega_{13}(\mathbf{m}_1) & \omega_{13}(\mathbf{m}_2) \\ \omega_{23}(\mathbf{m}_1) & \omega_{23}(\mathbf{m}_2) \end{bmatrix} = \begin{bmatrix} \kappa_1 & 0 \\ 0 & \kappa_2 \end{bmatrix}.$$

Thus, by virtue of (38.23),

$$\boxed{\omega_{13} = \kappa_1 \, \theta^1 \qquad \text{and} \qquad \omega_{23} = \kappa_2 \, \theta^2.} \tag{38.27}$$

We can now derive a geometrical interpretation of the Peterson–Mainardi–Codazzi Equations—this is genuinely new mathematics that we have not seen in the previous Acts. Inserting the equations (38.27) into the first Peterson–Mainardi–Codazzi Equation of (38.26), we find that

$$\mathbf{d}(\kappa_1\,\theta^1) = \omega_{12} \wedge \kappa_2\,\theta^2.$$

Thus,

$$\mathbf{d}\kappa_1 \wedge \theta^1 + \kappa_1\,\mathbf{d}\theta^1 = \kappa_2\,\omega_{12} \wedge \theta^2.$$

Using the First Structural Equation from (38.26), namely, $\mathbf{d}\theta^1 = \omega_{12} \wedge \theta^2$, we deduce that

$$\mathbf{d}\kappa_1 \wedge \theta^1 + \kappa_1\,\omega_{12} \wedge \theta^2 = \kappa_2\,\omega_{12} \wedge \theta^2,$$

and therefore,

$$\mathbf{d}\kappa_1 \wedge \theta^1 = (\kappa_2 - \kappa_1)\,\omega_{12} \wedge \theta^2.$$

Finally, applying these 2-forms (these are the left and right sides of this equation) to the basis vectors $(\mathbf{m}_1, \mathbf{m}_2)$, we obtain

$$0 - \mathbf{d}\kappa_1(\mathbf{m}_2) = (\kappa_2 - \kappa_1)\,\omega_{12}(\mathbf{m}_1) - 0.$$

Hence, the ***Peterson–Mainardi–Codazzi Equation*** now takes the geometrically meaningful form,

$$\boxed{\nabla_{\mathbf{m}_2} \kappa_1 = (\kappa_1 - \kappa_2)\,\omega_{12}(\mathbf{m}_1).} \tag{38.28}$$

Repeating this calculation for the other Peterson–Mainardi–Codazzi Equation, we obtain

$$\boxed{\nabla_{\mathbf{m}_1} \kappa_2 = (\kappa_1 - \kappa_2)\,\omega_{12}(\mathbf{m}_2).} \tag{38.29}$$

In both these equations, $\omega_{12}(\mathbf{v})$ tells us how fast the principal directions are rotating within $\mathcal{S}$ as we move along the surface in the direction $\mathbf{v}$. So, putting both equations into words, as we move off at right angles to a principal direction, the rate of change of that principal curvature is proportional to both the difference of the two principal curvatures, and to the rate at which the principal directions rotate.

38.7 Geometrical Form of the Gauss Equation

Of the six Form equations (38.26), the Gauss Equation is undoubtedly the most important, for it holds the key both to the intrinsic curvature, and to its relation to the extrinsic curvature. The connection to the latter is immediate from (38.25), for this allows us rewrite the Gauss Equation in a new, geometrically meaningful form.

$$\textit{Gauss Equation:} \quad \boxed{\mathbf{d}\omega_{12} = -\mathcal{K}_{\text{ext}}\,\theta^1 \wedge \theta^2.} \tag{38.30}$$

For example, consider the sphere of radius R, analyzed in Section 38.5.2:

$$\mathbf{d}\omega_{12} = \mathbf{d}(c_\phi\,\mathbf{d}\vartheta) = -s_\phi\,\mathbf{d}\phi \wedge \mathbf{d}\vartheta = -\frac{1}{R^2}[R\,\mathbf{d}\phi] \wedge [R\,s_\phi\,\mathbf{d}\vartheta] = -\frac{1}{R^2}\,\theta^1 \wedge \theta^2.$$

From this we immediately deduce (correctly!) that $\mathcal{K}_{\text{ext}} = +(1/R^2)$.

38.8 Proof of the Metric Curvature Formula and the *Theorema Egregium*

38.8.1 Lemma: Uniqueness of ω_{12}

The connection form $\omega_{12} = -\omega_{21}$ is the only *1-form that satisfies Cartan's First Structural Equations of (38.26):*

$$\mathbf{d}\theta^1 = \omega_{12} \wedge \theta^2 \qquad \textit{and} \qquad \mathbf{d}\theta^2 = -\omega_{12} \wedge \theta^1. \tag{38.31}$$

To prove this, we apply these 2-forms to the pair of basis vectors, $(\mathbf{m}_1, \mathbf{m}_2)$, and find that [exercise]

$$\omega_{12}(\mathbf{m}_1) = \mathbf{d}\theta^1(\mathbf{m}_1, \mathbf{m}_2) \qquad \text{and} \qquad \omega_{12}(\mathbf{m}_2) = \mathbf{d}\theta^2(\mathbf{m}_1, \mathbf{m}_2).$$

It follows from (38.23) that ω_{12} is uniquely and *explicitly* given by

$$\omega_{12} = \left[\mathbf{d}\theta^1(\mathbf{m}_1, \mathbf{m}_2)\right] \theta^1 + \left[\mathbf{d}\theta^2(\mathbf{m}_1, \mathbf{m}_2)\right] \theta^2, \tag{38.32}$$

thereby completing the proof of (38.31).

38.8.2 Proof of the Metric Curvature Formula

Mathematically speaking, we are now living in the twenty-third century—the future from which the "*Star Trek* phaser" formula (4.10) was delivered to us, back through time, long ago on page 38.

Cartan's Forms now make possible a computational proof of this Metric Curvature Formula that is startling in its brevity and simplicity, on a par with our geometrical demonstration in [27.4].

From

$$ds^2 = (A\,du)^2 + (B\,dv)^2,$$

we immediately deduce that the adapted 1-form basis for the surface is

$$\theta^1 = A\,\mathbf{d}u \qquad \text{and} \qquad \theta^2 = B\,\mathbf{d}v, \tag{38.33}$$

and that the area 2-form of the surface is

$$\mathcal{A} = \theta^1 \wedge \theta^2 = AB\,\mathbf{d}u \wedge \mathbf{d}v.$$

We now calculate the exterior derivatives of the dual-basis 1-forms:

$$\begin{aligned} \mathbf{d}\theta^1 &= \mathbf{d}A \wedge \mathbf{d}u = \partial_v A\,\mathbf{d}v \wedge \mathbf{d}u = -\frac{\partial_v A}{B}\,\mathbf{d}u \wedge \theta^2 \\ \mathbf{d}\theta^2 &= \mathbf{d}B \wedge \mathbf{d}v = \partial_u B\,\mathbf{d}u \wedge \mathbf{d}v = -\frac{\partial_u B}{A}\,\mathbf{d}v \wedge \theta^1 \end{aligned}$$

Comparing these equations with Cartan's First Structural Equations, (38.31), we see immediately that one possible solution for ω_{12} is

$$\omega_{12} = -\frac{\partial_v A}{B}\, du + \frac{\partial_u B}{A}\, dv. \tag{38.34}$$

But lemma (38.31) tells us that if this is *a* solution, then it is the *only* solution—it is *the* solution!

Taking the exterior derivative,

$$\begin{aligned} d\omega_{12} &= -d\left[\frac{\partial_v A}{B}\right] \wedge du + d\left[\frac{\partial_u B}{A}\right] \wedge dv \\ &= -\partial_v\left[\frac{\partial_v A}{B}\right] dv \wedge du + \partial_u\left[\frac{\partial_u B}{A}\right] du \wedge dv \\ &= \frac{1}{AB}\left(\partial_v\left[\frac{\partial_v A}{B}\right] + \partial_u\left[\frac{\partial_u B}{A}\right]\right) \theta^1 \wedge \theta^2. \end{aligned}$$

Comparing this with the Gauss Equation, (38.30), we see that, in a mere fraction of a page, our work is *done!* We have (re)proved the very remarkable formula, (4.10):

$$\mathcal{K}_{ext} = -\frac{1}{AB}\left(\partial_v\left[\frac{\partial_v A}{B}\right] + \partial_u\left[\frac{\partial_u B}{A}\right]\right). \tag{38.35}$$

The extrinsic curvature $\mathcal{K}_{ext} = \kappa_1\kappa_2$ on the left-hand side of this equation arose from (38.21)—it measures the spreading of the normal vectors to the surface in space. But the expression on the right-hand side of this equation depends *only* on the intrinsic metric geometry of the surface. Therefore, this formula is a very explicit and very pure expression of (and proof of!) Gauss's *Theorema Egregium.*

There is a *reason* that the simplicity of this computational proof rivals that of our geometrical proof in [27.4]—it is, fundamentally, the *same* proof!

To begin to understand this, first observe that the ***intrinsic*** connection form ω_{12} tells us about the rotation of $(\mathbf{m}_1, \mathbf{m}_2)$ within the surface, *relative to a parallel-transported vector.* Since this is the *opposite* of the rotation $\delta\mathcal{R}$ of a parallel-transported vector relative to the frame, this explains *why* (38.34) is simply the negative of our original, geometrically derived formula, (27.5), page 267.

The final step of the geometrical proof was to find the holonomy of a vector that was parallel-transported around a small closed loop. But, as we learned in (37.8), page 408, this integral of ω_{12} around a small loop is *precisely* what is measured by $d\omega_{12}$.

38.9 A New Curvature Formula

We now derive a new curvature formula that will enable us to reach new results. With the *Theorema Egregium* (re)established, we may revert to denoting *the* curvature of the surface as $\mathcal{K}$, since the extrinsic and intrinsic curvatures are in fact one and the same.

We will now show that

$$\mathcal{K} = \nabla_{\mathbf{m}_2}\left[\omega_{12}(\mathbf{m}_1)\right] - \nabla_{\mathbf{m}_1}\left[\omega_{12}(\mathbf{m}_2)\right] - \left[\omega_{12}(\mathbf{m}_1)\right]^2 - \left[\omega_{12}(\mathbf{m}_2)\right]^2. \tag{38.36}$$

To avoid clutter, let us write the components of $\boldsymbol{\omega}_{12}$ as f_1 and f_2, so that (38.32) takes the form

$$\boldsymbol{\omega}_{12} = f_1\,\boldsymbol{\theta}^1 + f_2\,\boldsymbol{\theta}^2,$$

where

$$f_1 = \boldsymbol{\omega}_{12}(\mathbf{m}_1) = \mathbf{d}\boldsymbol{\theta}^1(\mathbf{m}_1, \mathbf{m}_2) \qquad \text{and} \qquad f_2 = \boldsymbol{\omega}_{12}(\mathbf{m}_2) = \mathbf{d}\boldsymbol{\theta}^2(\mathbf{m}_1, \mathbf{m}_2).$$

If we apply the 2-form $\mathbf{d}\boldsymbol{\omega}_{12}$ to the pair of basis vectors, $(\mathbf{m}_1, \mathbf{m}_2)$, then the Gauss Equation (38.30) yields,

$$\begin{aligned}
\mathcal{K} &= -\mathbf{d}\boldsymbol{\omega}_{12}(\mathbf{m}_1, \mathbf{m}_2) \\
&= -\left[\mathbf{d}f_1 \wedge \boldsymbol{\theta}^1 + f_1\,\mathbf{d}\boldsymbol{\theta}^1 + \mathbf{d}f_2 \wedge \boldsymbol{\theta}^2 + f_2\,\mathbf{d}\boldsymbol{\theta}^2\right](\mathbf{m}_1, \mathbf{m}_2) \\
&= \mathbf{d}f_1(\mathbf{m}_2) - \mathbf{d}f_2(\mathbf{m}_1) - f_1^2 - f_2^2 \\
&= \boldsymbol{\nabla}_{\mathbf{m}_2} f_1 - \boldsymbol{\nabla}_{\mathbf{m}_1} f_2 - f_1^2 - f_2^2.
\end{aligned}$$

which is, indeed, the desired formula, (38.36).

38.10 Hilbert's Lemma

Hilbert showed that if a point p on a surface has the following three properties, then the curvature of the surface *cannot be positive* at p:

$$\left\{\begin{array}{l} \kappa_1 \text{ has local maximum at } p \\ \kappa_2 \text{ has local minimum at } p \\ \kappa_1(p) > \kappa_2(p) \end{array}\right\} \implies \mathcal{K}(p) \leqslant 0. \tag{38.37}$$

For example, these conditions are realized at each point on the inner equator of a torus, and the curvature at these points is indeed negative, in accord with Hilbert's Lemma.

Since p is a critical point for κ_1 and for κ_2, at p we have

$$\boldsymbol{\nabla}_{\mathbf{m}_2} \kappa_1 = 0 = \boldsymbol{\nabla}_{\mathbf{m}_1} \kappa_2. \tag{38.38}$$

Also, since κ_1 achieves a local *maximum* at p, and κ_2 achieves a local *minimum* at p,

$$\boldsymbol{\nabla}_{\mathbf{m}_2} \boldsymbol{\nabla}_{\mathbf{m}_2} \kappa_1 \leqslant 0 \qquad \text{and} \qquad \boldsymbol{\nabla}_{\mathbf{m}_1} \boldsymbol{\nabla}_{\mathbf{m}_1} \kappa_2 \geqslant 0. \tag{38.39}$$

Since $(\kappa_1 - \kappa_2) > 0$, inserting (38.38) into the Peterson–Mainardi–Codazzi Equations (38.28) and (38.29), we find that, at p,

$$\boldsymbol{\omega}_{12}(\mathbf{m}_1) = 0 = \boldsymbol{\omega}_{12}(\mathbf{m}_2). \tag{38.40}$$

Thus, the curvature formula (38.36) reduces to

$$\mathcal{K} = \boldsymbol{\nabla}_{\mathbf{m}_2}\left[\boldsymbol{\omega}_{12}(\mathbf{m}_1)\right] - \boldsymbol{\nabla}_{\mathbf{m}_1}\left[\boldsymbol{\omega}_{12}(\mathbf{m}_2)\right]. \tag{38.41}$$

Differentiating (38.28) and inserting (38.40),

$$\begin{aligned}
\boldsymbol{\nabla}_{\mathbf{m}_2} \boldsymbol{\nabla}_{\mathbf{m}_2} \kappa_1 &= \left[\boldsymbol{\nabla}_{\mathbf{m}_2}(\kappa_1 - \kappa_2)\right] \boldsymbol{\omega}_{12}(\mathbf{m}_1) + (\kappa_1 - \kappa_2)\, \boldsymbol{\nabla}_{\mathbf{m}_2}\left[\boldsymbol{\omega}_{12}(\mathbf{m}_1)\right] \\
&= (\kappa_1 - \kappa_2)\, \boldsymbol{\nabla}_{\mathbf{m}_2}\left[\boldsymbol{\omega}_{12}(\mathbf{m}_1)\right].
\end{aligned}$$

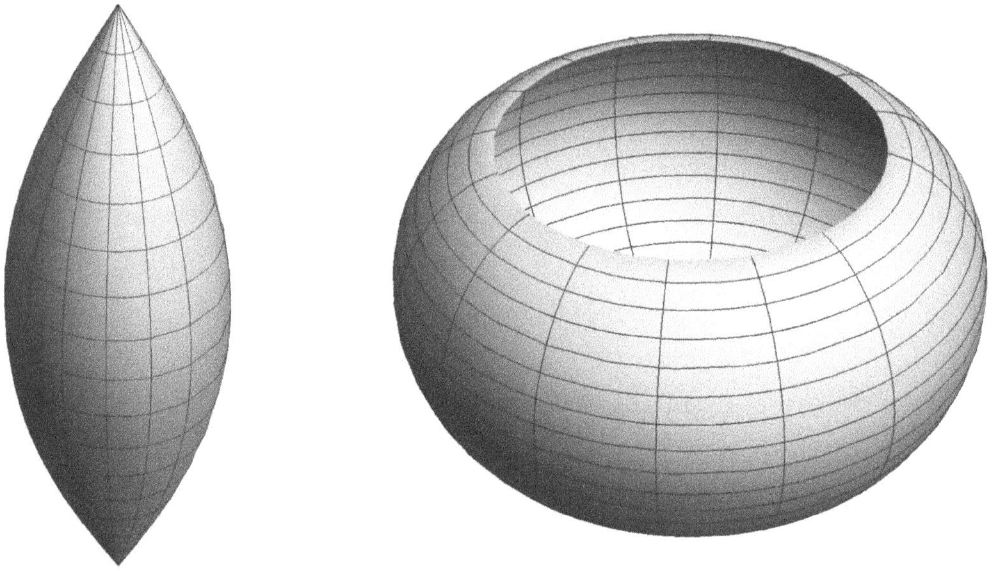

[38.7] *These two kinds of surfaces have constant positive curvature, and the same* intrinsic *geometry as a sphere, but they are clearly* not *spherical!*

Finally, since $(\kappa_1 - \kappa_2) > 0$, (38.39) implies

$$\nabla_{\mathbf{m}_2}\left[\omega_{12}(\mathbf{m}_1)\right] \leqslant 0.$$

Applying the same logic to the other Peterson–Mainardi–Codazzi Equation, (38.29), we find

$$\nabla_{\mathbf{m}_1}\left[\omega_{12}(\mathbf{m}_2)\right] \geqslant 0.$$

Inserting the previous two inequalities into the curvature formula (38.41), the lemma is proved.

38.11 Liebmann's Rigid Sphere Theorem

Recall that Minding's Theorem of 1839 (proved in Ex. 7, p. 336) implies that if a surface $\mathcal{S}$ has *constant* positive curvature $\mathcal{K}$, then its *intrinsic* metric geometry must be identical to that of a sphere of radius $1/\sqrt{\mathcal{K}}$, at least locally.

It is tempting to go further, speculating that $\mathcal{S}$ must actually *be* part of a sphere, manifesting the completely symmetrical *extrinsic* geometry of a sphere, in which every point is an *umbilic* point: $\kappa_1 = \kappa_2$. In fact, in the following, we shall take it as given that this condition does indeed *characterize* the sphere.

However, such a guess would be incorrect: [38.7] depicts two kinds of surfaces that are very definitely *not* spherical, and yet which *both have constant positive curvature, and the same intrinsic geometry as a sphere!* (For the equations of these surfaces, see Ex. 7, p. 336.)

We can also see this experimentally. Take a ping pong ball and cut it in half; now gently squeeze together antipodal points on the equator. The surface will flex without stretching, creating a new, nonspherical surface of constant positive curvature, and the same intrinsic geometry as the original sphere.

But what if we try to flex a *complete* ping pong ball? It would appear that we *cannot!* Minding himself had first speculated about this issue in 1839, but it took another 60 years before Heinrich

Liebmann was finally able to *prove* (in 1899) that a *closed* (bounded) surface of constant curvature embedded in $\mathbb{R}^3$ can *only* be a sphere.

Thus, if the geometry is intrinsically uniform (in the sense of constant $\mathcal{K}$), then the extrinsic geometry can also *only* be the extrinsically uniform geometry of a sphere, and therefore the sphere *cannot be deformed* into any other extrinsic shapes—a complete sphere is *rigid*.

The best known proof of Liebmann's Theorem was discovered, not long afterwards, by David Hilbert,[13] and the crucial ingredient of his proof was (38.37), hence its name—"Hilbert's Lemma."

We begin by observing that since $\mathcal{S}$ is bounded, its constant curvature must be *positive* curvature. By definition, being bounded means that $\mathcal{S}$ can be enclosed within a sufficiently large sphere. Now shrink the sphere till it first touches $\mathcal{S}$, at c, say. Then the little piece of $\mathcal{S}$ surrounding c all lies on one side of the common tangent plane to $\mathcal{S}$ and the sphere at c. Thus c must be an *elliptic* point of *positive* curvature, and since the curvature is assumed to be constant, it is positive everywhere on $\mathcal{S}$.

Let us assume that on a closed (bounded) surface the function κ_1 must achieve a maximum. (A proof can be found in O'Neill (2006, pp. 185–186).) Since $\mathcal{K} = \kappa_1 \kappa_2$ is assumed to be constant, κ_2 must achieve a minimum at a point at which κ_1 achieves a maximum. But this means that it is *impossible* that $\kappa_1 > \kappa_2$. For, if that were the case, Hilbert's Lemma would demand that $\mathcal{K} \leqslant 0$, contradicting the fact that the surface has (constant) *positive* curvature.

Thus $\kappa_1(p) = \kappa_2(p)$. And since $\kappa_1(p)$ is the *maximum* of κ_1 on $\mathcal{S}$, and $\kappa_2(p)$ is the *minimum* of κ_2 on $\mathcal{S}$, this means that $\kappa_1 = \kappa_2$ at *every* point of $\mathcal{S}$, and therefore $\mathcal{S}$ is a sphere.

38.12 The Curvature 2-Forms of an n-Manifold

38.12.1 Introduction and Summary

Recall the scene as the curtain rose on Act I of our drama. The *flatness* of the Euclidean plane $\mathbb{R}^2$ was *characterized* by the following property of an arbitrary triangle Δ constructed within it:

$$\boxed{\text{Angle sum of } \Delta = \pi.} \tag{38.42}$$

If this equation is *not* satisfied, it means that Δ lives within a surface that is *curved*.

The situation is analogous to the juxtaposition of Newton's First and Second Laws of Motion: if we observe an object that is *not* moving at constant speed in a straight line, there must be a force acting upon it. And the magnitude of the force manifests itself as the acceleration, that is, as the *difference* between the actual motion and force-free, constant-speed, straight-line motion.

Likewise, the amount of curvature within the surface is quantified by the *difference* between the two sides of the equation above. This is the *angular excess*, $\mathcal{E}$:

$$\boxed{\mathcal{E}(\Delta) \equiv (\text{Angle sum of } \Delta) - \pi.} \tag{38.43}$$

When applied to a small (ultimately vanishing) triangle, this allows us to *quantify* the curvature $\mathcal{K}$ *at* the triangle Δ, of area $\mathcal{A}(\Delta)$, by means of this fundamental formula:

$$\boxed{\boxed{\mathcal{E}(\Delta) \asymp \mathcal{K}\,\mathcal{A}(\Delta).}} \tag{38.44}$$

[13] Liebmann completed his doctoral work at Göttingen, but *not* as Hilbert's student. Yet one senses, perhaps, some jealousy on the part of Hilbert, for he writes that Liebmann proved the theorem "at my urging" (Hilbert 1902, p. 197). In any event, it is Hilbert's *proof* (given here) of Liebmann's theorem that lives on!

Now let us move up a dimension, to *3-dimensional* Euclidean space, $\mathbb{R}^3$. As we will prove, the analogue of (38.42) turns out to be none other than *Cartan's Second Structural Equation*, (38.12)—*this equation characterizes the flatness of the space:*

$$\mathbf{d}[\omega] = [\omega] \wedge [\omega].$$

In a curved n-manifold, this equation does *not* hold, and the *difference* between the two sides of the equation once again quantifies the curvature of the space. The analogue of (38.43) is now *a matrix* $[\Omega]$ *of 2-forms*, called the ***curvature matrix***:

$$[\Omega] \equiv \mathbf{d}[\omega] - [\omega] \wedge [\omega]. \tag{38.45}$$

The individual entries within this matrix are called[14] the

Curvature 2-forms:
$$\Omega_{ij} = \mathbf{d}\omega_{ij} - \textstyle\sum_k \omega_{ik} \wedge \omega_{kj}. \tag{38.46}$$

It is *this* (fully general) equation that is usually referred to (in more advanced works) as ***Cartan's Second Structural Equation***; in $\mathbb{R}^3$ it reduces to our original version, (38.13), for in that case $\Omega_{ij} = 0$.

IMPORTANT NOTATIONAL NOTE: Two different notations exist in the literature for both the connection 1-forms and the curvature 2-forms. In the more common notation, which we too shall adopt shortly, both these objects have one raised index and one lowered index. As we shall explain, the relation between the notations is, $\omega^i{}_j \equiv \omega_{ji}$ and $\Omega^i{}_j \equiv \Omega_{ji}$. This reversal of the order of the indices has the effect that *in both (38.45) and (38.46) the "−" becomes a "+".*

We will prove that *these curvature 2-forms contain exactly the same information as the Riemann tensor,* but packaged in a more compact and elegant form [*sic*].

For your convenience, we repeat here the definitions of the Riemann tensor $\mathbf{R}$ and of the Riemann curvature operator $\mathcal{R}$, both of which are stated in (29.8), page 290:

$$\mathbf{R}(\mathbf{u}, \mathbf{v}; \mathbf{w}) = \mathcal{R}(\mathbf{u}, \mathbf{v})\, \mathbf{w} = \left\{ [\nabla_{\mathbf{u}}, \nabla_{\mathbf{v}}] - \nabla_{[\mathbf{u}, \mathbf{v}]} \right\} \mathbf{w}. \tag{38.47}$$

Recall that in Chapter 29 we never actually worked out the generalization of the "*Star Trek* phaser" formula (27.1) for an n-manifold. That is, we never found a formula for the Riemann tensor explicitly expressed directly in terms of the metric coefficients. There *is* such a formula—it's ugly and it's complicated, involving things called ***Christoffel symbols***, and it can be found in *every* standard textbook on Riemannian Geometry—but we chose not to derive that formula for a good reason: *we can do better!*

---

[14]Multiple notations exist for the curvature 2-forms. Our Ω_{ij} is probably the most common, and it was the notation originally used by Cartan (1927) himself, but Flanders (1989) and Chern et al. (1999) instead use Θ_{ij}, while Misner, Thorne, and Wheeler (1973) use $\mathcal{R}_{ij}$.

The curvature 2-forms allow us to calculate the Riemann tensor much more efficiently and cleanly than any frontal assault could ever hope to achieve. Formula (38.46) allows us to calculate $\boldsymbol{\Omega}_{ij}$, and then we simply *read off* the components of the Riemann tensor, for *we shall prove that they are simply the components of the curvature 2-forms!* That is,

$$\boxed{\boldsymbol{\Omega}_{ij} = R_{ijkl}\, \boldsymbol{\theta}^k \wedge \boldsymbol{\theta}^l.} \tag{38.48}$$

NOTE: Here, and in the following, we have reverted to the Einstein summation convention: summation over the matching k-indices and l-indices is *understood*. However, we assume that the sum is over *distinct* basis 2-forms, or else we would require a factor of $(1/2)$.

38.12.2 The Generalized Exterior Derivative

Calculating the Riemann tensor via the curvature 2-forms, as defined in (38.46), only requires routine exterior differentiation of the connection 1-forms. However, in order to *reach* the formula (38.48), we shall need to generalize the action of **d** beyond Forms.

First, though, let us introduce an alternative notation for the connection 1-forms in which one index is raised and one is lowered: $\boldsymbol{\omega}^i{}_j$. This notation will not only allow us to reinstate the use of the Einstein summation convention, but, more importantly, it will also enable you to make an easy transition to studying such works as Misner, Thorne, and Wheeler (1973), Frankel (2012), Dray (2015), and Tu (2017), *all* of whom employ this notation.

These $\boldsymbol{\omega}^i{}_j$ are defined so that (38.4) now takes the form

$$\boxed{\nabla_{\mathbf{v}} \mathbf{m}_j = \textstyle\sum_i \boldsymbol{\omega}_{ji}(\mathbf{v})\, \mathbf{m}_i \equiv \sum_i \boldsymbol{\omega}^i{}_j(\mathbf{v})\, \mathbf{m}_i = \boldsymbol{\omega}^i{}_j(\mathbf{v})\, \mathbf{m}_i,} \tag{38.49}$$

in which the final equality signals the return to the Einstein summation convention—the sum over the matching pair or i-indices (one up and one down) is *understood*.

Note the *reversal* of the order of the indices![15]

$$\boxed{\boldsymbol{\omega}^i{}_j \equiv \boldsymbol{\omega}_{ji}.} \tag{38.50}$$

Now let us try to *generalize* the exterior derivative **d**, defining its action upon the moving frame field *vectors*, $\mathbf{m}_i$. By analogy with

$$\mathbf{d}f(\mathbf{v}) = \nabla_{\mathbf{v}} f,$$

let us define,

$$\mathbf{d}\mathbf{m}_j(\mathbf{v}) \equiv \nabla_{\mathbf{v}} \mathbf{m}_j = \boldsymbol{\omega}^i{}_j(\mathbf{v})\, \mathbf{m}_i.$$

Abstracting away the arbitrary vector **v**,

$$\boxed{\mathbf{d}\mathbf{m}_j = \boldsymbol{\omega}^i{}_j\, \mathbf{m}_i.}$$

[15] I do not understand the origin of this choice, and I do not welcome it, but it *is* the standard notation, so we shall abide by it!

This now allows us to apply $\mathbf{d}$ to a general vector field, $\mathbf{w} = w^j\,\mathbf{m}_j$. Renaming indices as we go, we find that

$$\mathbf{d}\mathbf{w} = \mathbf{d}\big[w^j\,\mathbf{m}_j\big] = \big[\mathbf{d}w^j\big]\mathbf{m}_j + w^j\,\mathbf{d}\mathbf{m}_j = \mathbf{m}_i\,\big(\mathbf{d}w^i + \boldsymbol{\omega}^i{}_j\,w^j\big).$$

Now let us differentiate *again*.

Wait!—isn't $\mathbf{d}^2 = 0$?! No, $\mathbf{d}^2$ vanishes when it is applied to *Forms*, but not now! Renaming indices as we go, we find,

$$\begin{aligned}
\mathbf{d}^2\mathbf{w} &= \mathbf{d}\mathbf{m}_k \wedge \big(\mathbf{d}w^k + \boldsymbol{\omega}^k{}_j\,w^j\big) + \mathbf{m}_i\,\big(\mathbf{d}^2 w^i + w^j\,\mathbf{d}\boldsymbol{\omega}^i{}_j - \boldsymbol{\omega}^i{}_j \wedge \mathbf{d}w^j\big) \\
&= \boldsymbol{\omega}^i{}_k\,\mathbf{m}_i \wedge \big(\mathbf{d}w^k + \boldsymbol{\omega}^k{}_i\,w^i\big) + \mathbf{m}_i\,\big(w^j\,\mathbf{d}\boldsymbol{\omega}^i{}_j - \boldsymbol{\omega}^i{}_k \wedge \mathbf{d}w^k\big) \\
&= \mathbf{m}_i\,\big(\mathbf{d}\boldsymbol{\omega}^i{}_j + \boldsymbol{\omega}^i{}_k \wedge \boldsymbol{\omega}^k{}_j\big)\,w^j.
\end{aligned}$$

Thus,

$$\boxed{\mathbf{d}^2\mathbf{w} = \mathbf{m}_i\,\boldsymbol{\Omega}^i{}_j\,w^j,} \tag{38.51}$$

where $\boldsymbol{\Omega}^i{}_j$ are, again, the

Curvature 2-Forms:
$$\boxed{\boldsymbol{\Omega}^i{}_j = \mathbf{d}\boldsymbol{\omega}^i{}_j + \boldsymbol{\omega}^i{}_k \wedge \boldsymbol{\omega}^k{}_j.} \tag{38.52}$$

IMPORTANT NOTATIONAL NOTE: Observe that there is an unfortunate sign discrepancy between this formula for the 2-forms and our original definition, (38.46). This discrepancy arises from (38.50):

$$\boxed{\boldsymbol{\Omega}_{ij} = \mathbf{d}\boldsymbol{\omega}_{ij} - \textstyle\sum_k \boldsymbol{\omega}_{ik} \wedge \boldsymbol{\omega}_{kj} = \mathbf{d}\boldsymbol{\omega}^j{}_i - \sum_k \boldsymbol{\omega}^k{}_i \wedge \boldsymbol{\omega}^j{}_k = \mathbf{d}\boldsymbol{\omega}^j{}_i + \sum_k \boldsymbol{\omega}^j{}_k \wedge \boldsymbol{\omega}^k{}_i = \boldsymbol{\Omega}^j{}_i.}$$

Likewise, note that the sign also changes in the formula (38.45) for the *curvature matrix*:

$$\boxed{[\boldsymbol{\Omega}^i{}_j] = \mathbf{d}[\boldsymbol{\omega}] + [\boldsymbol{\omega}] \wedge [\boldsymbol{\omega}].} \tag{38.53}$$

Likewise, the new notation flips the sign in ***Cartan's First Structural Equation***, so that (38.11) now becomes

$$\boxed{\mathbf{d}\boldsymbol{\theta}^i = -\boldsymbol{\omega}^i{}_j \wedge \boldsymbol{\theta}^j.} \tag{38.54}$$

Henceforth, we shall *only* employ $\boldsymbol{\Omega}^i{}_j$, given by (38.52), for this is the most common expression of the curvature 2-forms found in the literature.

38.12.3 Extracting the Riemann Tensor from the Curvature 2-Forms

Recall our very first formula (on p. 393) for the exterior derivative of a 1-form, (36.2):

$$\mathbf{d}\varphi(\mathbf{u},\mathbf{v}) = \nabla_{\mathbf{u}}\varphi(\mathbf{v}) - \nabla_{\mathbf{v}}\varphi(\mathbf{u}) - \varphi([\mathbf{u},\mathbf{v}]).$$

We now generalize this by substituting $\mathbf{dw}$ in place of φ, recalling that

$$\mathbf{dw}(\mathbf{v}) = \nabla_{\mathbf{v}}\mathbf{w}.$$

Thus,

$$\begin{aligned}\mathbf{d}^2\mathbf{w}(\mathbf{u},\mathbf{v}) &= \nabla_{\mathbf{u}}\mathbf{dw}(\mathbf{v}) - \nabla_{\mathbf{v}}\mathbf{dw}(\mathbf{u}) - \mathbf{dw}([\mathbf{u},\mathbf{v}]) \\ &= \nabla_{\mathbf{u}}\nabla_{\mathbf{v}}\mathbf{w} - \nabla_{\mathbf{v}}\nabla_{\mathbf{u}}\mathbf{w} - \nabla_{[\mathbf{u},\mathbf{v}]}\mathbf{w} \\ &= \mathcal{R}(\mathbf{u},\mathbf{v})\,\mathbf{w},\end{aligned}$$

by virtue of (38.47).

Combining this with (38.52), we have obtained a remarkable formula for the Riemann tensor in terms of the curvature 2-forms:

$$\boxed{-R^i{}_{jkl}\,\boldsymbol{\theta}^k \wedge \boldsymbol{\theta}^l = \boldsymbol{\Omega}^i{}_j = \mathbf{d}\boldsymbol{\omega}^i{}_j + \boldsymbol{\omega}^i{}_m \wedge \boldsymbol{\omega}^m{}_j.} \quad (38.55)$$

We obtained this formula by climbing the ladder of the *generalized* exterior derivative, but, having *arrived* in the promised land, we may now toss the ladder aside! For this final formula (38.55) uses nothing more than the *ordinary* exterior derivative applied to the connection 1-forms.

Thus we are now in the possession of an elegant and extraordinarily powerful method of computing the Riemann tensor using curvature 2-forms. We shall *end* our drama by applying this method to a specific example of enormous physical importance.

38.12.4 The Bianchi Identities Revisited

Both the First (Algebraic) Bianchi Identity (29.15) and the Second (Differential) Bianchi Identity (29.17) can be very elegantly derived, and compactly expressed, using our shiny new curvature 2-forms.

Since the basis 1-forms $\boldsymbol{\theta}^i$ are ordinary 1-forms, $\mathbf{d}^2$ annihilates them, yielding a vanishing 3-form. Thus, differentiating Cartan's First Structural Equation, (38.54),

$$\begin{aligned}0 &= -\mathbf{dd}\boldsymbol{\theta}^i \\ &= \mathbf{d}(\boldsymbol{\omega}^i{}_j \wedge \boldsymbol{\theta}^j) \\ &= \mathbf{d}\boldsymbol{\omega}^i{}_j \wedge \boldsymbol{\theta}^j - \boldsymbol{\omega}^i{}_k \wedge \mathbf{d}\boldsymbol{\theta}^k \\ &= \mathbf{d}\boldsymbol{\omega}^i{}_j \wedge \boldsymbol{\theta}^j + \boldsymbol{\omega}^i{}_k \wedge \boldsymbol{\omega}^k{}_j \wedge \boldsymbol{\theta}^j \\ &= (\mathbf{d}\boldsymbol{\omega}^i{}_j + \boldsymbol{\omega}^i{}_k \wedge \boldsymbol{\omega}^k{}_j) \wedge \boldsymbol{\theta}^j \\ &= \boldsymbol{\Omega}^i{}_j \wedge \boldsymbol{\theta}^j.\end{aligned}$$

Thus, employing the more compact matrix notation, we have obtained the

$$\textit{(First) Algebraic Bianchi Identity:} \qquad [\boldsymbol{\Omega}] \wedge [\boldsymbol{\theta}] = 0. \tag{38.56}$$

In its present form, this is certainly *unrecognizable* as the original First (Algebraic) Bianchi Identity (29.15)! To see that these two identities are, in fact, one and the same, we invoke (38.55):

$$0 = -\boldsymbol{\Omega}^{i}{}_{j} \wedge \boldsymbol{\theta}^{j} = R^{i}{}_{jkl}\, \boldsymbol{\theta}^{j} \wedge \boldsymbol{\theta}^{k} \wedge \boldsymbol{\theta}^{l},$$

and this does indeed imply [exercise] the equivalent [exercise] of (29.15):

$$R^{i}{}_{jkl} + R^{i}{}_{klj} + R^{i}{}_{ljk} = 0.$$

The Second (Differential) Bianchi Identity (29.17) can be obtained similarly, by applying $\mathbf{d}^2$ to the connection 1-forms. Thus, differentiating Cartan's Second Structural Equation, (38.52),

$$\begin{aligned}
0 &= \mathbf{d}\mathbf{d}\boldsymbol{\omega}^{i}{}_{j} \\
&= \mathbf{d}(\boldsymbol{\Omega}^{i}{}_{j} - \boldsymbol{\omega}^{i}{}_{k} \wedge \boldsymbol{\omega}^{k}{}_{j}) \\
&= \mathbf{d}\boldsymbol{\Omega}^{i}{}_{j} - \mathbf{d}\boldsymbol{\omega}^{i}{}_{k} \wedge \boldsymbol{\omega}^{k}{}_{j} + \boldsymbol{\omega}^{i}{}_{k} \wedge \mathbf{d}\boldsymbol{\omega}^{k}{}_{j} \\
&= \mathbf{d}\boldsymbol{\Omega}^{i}{}_{j} - (\boldsymbol{\Omega}^{i}{}_{k} - \boldsymbol{\omega}^{i}{}_{m} \wedge \boldsymbol{\omega}^{m}{}_{k}) \wedge \boldsymbol{\omega}^{k}{}_{j} + \boldsymbol{\omega}^{i}{}_{k} \wedge (\boldsymbol{\Omega}^{k}{}_{j} - \boldsymbol{\omega}^{k}{}_{m} \wedge \boldsymbol{\omega}^{m}{}_{j}) \\
&= \mathbf{d}\boldsymbol{\Omega}^{i}{}_{j} - \boldsymbol{\Omega}^{i}{}_{k} \wedge \boldsymbol{\omega}^{k}{}_{j} + \boldsymbol{\omega}^{i}{}_{k} \wedge \boldsymbol{\Omega}^{k}{}_{j}.
\end{aligned}$$

Thus, employing the more compact matrix notation, we have obtained the

$$\textit{(Second) Differential Bianchi Identity:} \qquad \mathbf{d}[\boldsymbol{\Omega}] = [\boldsymbol{\Omega}] \wedge [\boldsymbol{\omega}] - [\boldsymbol{\omega}] \wedge [\boldsymbol{\Omega}]. \tag{38.57}$$

Again, (38.55) provides the explicit link to the Riemann tensor that enables one to see that this is indeed equivalent [exercise] to our original version of the Differential Bianchi Identity, (29.17).

38.13 The Curvature of the Schwarzschild Black Hole

This section lowers the curtain on Act V, bringing our long drama to its end. It does so by applying the foregoing ideas to a mathematical and physical object of cosmic importance: the spacetime curvature of a black hole!

Specifically, we shall endeavour to verify that the Schwarzschild black hole geometry, given by[16] (30.12), which we repeat here,

---

[16] As we explained in the context of Maxwell's Equations, we have now reversed the sign of the metric coefficients, in order to bring our work into line with the majority of General Relativity texts that we recommend in the *Further Reading* section at the end of this book.

Schwarzschild Black Hole

$$ds^2 = -\left(1-\frac{2GM}{r}\right)dt^2 + \frac{dr^2}{\left(1-\frac{2GM}{r}\right)} + r^2(d\phi^2 + \sin^2\phi\, d\vartheta^2), \tag{38.58}$$

is *indeed* a solution of the Einstein Vacuum Field Equation, (30.11), which we also repeat here:

$$\textit{Einstein Vacuum Field Equation:} \qquad \mathbf{Ricci} = \mathbf{0} \quad \Longleftrightarrow \quad R_{ik} = 0. \tag{38.59}$$

Let us define[17]

$$f(r) = 1 - \frac{2GM}{r},$$

so that (for later use),

$$f' = \frac{2GM}{r^2}, \qquad \text{and} \qquad f'' = -\frac{4GM}{r^3}, \qquad \text{and} \qquad 1-f = \frac{2GM}{r}. \tag{38.60}$$

Let us label each basis 1-form with its associated spacetime coordinate. Then the Schwarzschild metric (38.58) may be written:

$$\mathbf{g} = g_{ij}(\boldsymbol{\theta}^i \otimes \boldsymbol{\theta}^j) = -\boldsymbol{\theta}^t \otimes \boldsymbol{\theta}^t + \boldsymbol{\theta}^r \otimes \boldsymbol{\theta}^r + \boldsymbol{\theta}^\phi \otimes \boldsymbol{\theta}^\phi + \boldsymbol{\theta}^\vartheta \otimes \boldsymbol{\theta}^\vartheta,$$

where

$$\boldsymbol{\theta}^t = \sqrt{f}\,\mathbf{d}t, \qquad \boldsymbol{\theta}^r = \frac{\mathbf{d}r}{\sqrt{f}}, \qquad \boldsymbol{\theta}^\phi = r\,\mathbf{d}\phi, \qquad \boldsymbol{\theta}^\vartheta = r\sin\phi\,\mathbf{d}\vartheta.$$

Writing the First Structural Equation on the left of each equation—with Einstein summation on m—then actually evaluating each derivative on the right, finally rewriting each result in terms of the basis 1-forms, we find that

$$\begin{aligned}
\boldsymbol{\omega}^t{}_m \wedge \boldsymbol{\theta}^m &= -\mathbf{d}\boldsymbol{\theta}^t = -\frac{f'}{2\sqrt{f}}\,\mathbf{d}r \wedge \mathbf{d}t = \frac{f'}{2\sqrt{f}}\sqrt{f}\,\mathbf{d}t \wedge \frac{\mathbf{d}r}{\sqrt{f}} = \frac{f'}{2\sqrt{f}}\,\boldsymbol{\theta}^t \wedge \boldsymbol{\theta}^r, \\
\boldsymbol{\omega}^r{}_m \wedge \boldsymbol{\theta}^m &= -\mathbf{d}\boldsymbol{\theta}^r = -\mathbf{d}\left[\frac{1}{\sqrt{f}}\right] \wedge \mathbf{d}r = -\partial_r\left[\frac{1}{\sqrt{f}}\right]\mathbf{d}r \wedge \mathbf{d}r = 0,
\end{aligned}$$

[17] In the following, we largely follow the notation of Dray (2015, pp. 259–261), although he writes the basis 1-forms as $\boldsymbol{\sigma}^i$, whereas we shall continue to write them as $\boldsymbol{\theta}^i$. Also note that our ϕ and ϑ are the reverse of his! See the footnote on page 388.

$$\omega^{\phi}{}_{m} \wedge \theta^{m} = -\mathbf{d}\theta^{\phi} = -\mathbf{d}r \wedge \mathbf{d}\phi = \frac{\sqrt{f}}{r} r\,\mathbf{d}\phi \wedge \frac{\mathbf{d}r}{\sqrt{f}} = \frac{\sqrt{f}}{r}\theta^{\phi} \wedge \theta^{r},$$

$$\omega^{\vartheta}{}_{m} \wedge \theta^{m} = -\mathbf{d}\theta^{\vartheta} = -s_{\phi}\,\mathbf{d}r \wedge \mathbf{d}\vartheta - r c_{\phi}\,\mathbf{d}\phi \wedge \mathbf{d}\vartheta = \frac{\sqrt{f}}{r}\theta^{\vartheta} \wedge \theta^{r} + \frac{\cot\phi}{r}\theta^{\vartheta} \wedge \theta^{\phi}.$$

In place of the more famous *Occam's Razor*, let us invoke (and simultaneously christen[18]) the logically indefensible, yet remarkably effective[19]

> ***Cartan's Razor:*** *If a term is not explicitly prohibited from vanishing, it does vanish.*

Inspecting the equations above, and writing down *only* the nonvanishing terms that are *explicitly* forced upon us, and *assuming that all other terms vanish* (as dictated by *Cartan's Razor!*), we find,

$$\omega^{t}{}_{r} = \frac{f'}{2\sqrt{f}}\theta^{t}, \qquad \omega^{\phi}{}_{r} = \frac{\sqrt{f}}{r}\theta^{\phi}, \qquad \omega^{\vartheta}{}_{r} = \frac{\sqrt{f}}{r}\theta^{\vartheta}, \qquad \omega^{\vartheta}{}_{\phi} = \frac{\cot\phi}{r}\theta^{\vartheta}.$$

The Uniqueness Lemma (38.31) easily generalizes, so these *are* the solutions!

Next, we must differentiate these connection 1-forms to obtain the curvature 2-forms, bearing in mind that their antisymmetry, and the metric coefficient $g_{tt} = -1$, together imply:

$$\omega^{r}{}_{t} = \omega^{t}{}_{r}, \qquad \omega^{r}{}_{\phi} = -\omega^{\phi}{}_{r}, \qquad \omega^{r}{}_{\vartheta} = -\omega^{\vartheta}{}_{r}, \qquad \omega^{\phi}{}_{\vartheta} = -\omega^{\vartheta}{}_{\phi},$$

and

$$\omega^{t}{}_{t} = \omega^{r}{}_{r} = \omega^{\phi}{}_{\phi} = \omega^{\vartheta}{}_{\vartheta} = 0.$$

There are six independent curvature 2-forms, and we now spell out (in detail) the calculation of the first of these, leaving it to you to verify the remaining five.

The first step in calculating $\mathbf{d}\omega^{i}{}_{j}$ is to revert to expressing $\omega^{i}{}_{j}$ in terms of the $\mathbf{d}x^{i}$; this makes it easy to calculate the exterior derivative. The final result is then expressed in terms of $\theta^{i} \wedge \theta^{j}$. From (38.52),

$$\begin{aligned}
\Omega^{t}{}_{r} &= \mathbf{d}\omega^{t}{}_{r} + \omega^{t}{}_{m} \wedge \omega^{m}{}_{r} \\
&= \mathbf{d}\left[\frac{f'}{2}\right] \wedge \mathbf{d}t + \omega^{t}{}_{t} \wedge \omega^{t}{}_{r} + \omega^{t}{}_{r} \wedge \omega^{r}{}_{r} + \omega^{t}{}_{\phi} \wedge \omega^{\phi}{}_{r} + \omega^{t}{}_{\vartheta} \wedge \omega^{\vartheta}{}_{r} \\
&= \frac{f''}{2}\,\mathbf{d}r \wedge \mathbf{d}t + 0 + 0 + 0 + 0 \\
&= -\frac{f''}{2}\theta^{t} \wedge \theta^{r}.
\end{aligned}$$

We now list all six curvature 2-forms, starting with the one we just calculated:

$$\begin{aligned}
\Omega^{t}{}_{r} &= -\frac{f''}{2}\theta^{t} \wedge \theta^{r}, \\
\Omega^{t}{}_{\phi} &= -\frac{f'}{2r}\theta^{t} \wedge \theta^{\phi},
\end{aligned}$$

[18] To be clear, Cartan himself never said any such thing (so far as I know)—nevertheless, I choose to *name* this principle after him!

[19] No doubt a counterexample exists, but I personally do not recall ever having seen one.

$$\begin{aligned}
\boldsymbol{\Omega}^{t}{}_{\vartheta} &= -\frac{f'}{2r}\,\boldsymbol{\theta}^{t}\wedge\boldsymbol{\theta}^{\vartheta},\\
\boldsymbol{\Omega}^{\phi}{}_{r} &= -\frac{f'}{2r}\,\boldsymbol{\theta}^{\phi}\wedge\boldsymbol{\theta}^{r},\\
\boldsymbol{\Omega}^{\vartheta}{}_{r} &= -\frac{f'}{2r}\,\boldsymbol{\theta}^{\vartheta}\wedge\boldsymbol{\theta}^{r},\\
\boldsymbol{\Omega}^{\vartheta}{}_{\phi} &= \left[\frac{1-f}{r^2}\right]\boldsymbol{\theta}^{\vartheta}\wedge\boldsymbol{\theta}^{\phi}.
\end{aligned}$$

Again, the curvature 2-forms that are purely spatial are antisymmetric, but those with a temporal component are symmetric, because $g_{tt}=-1$:

$$\boldsymbol{\Omega}^{\vartheta}{}_{r}=-\boldsymbol{\Omega}^{r}{}_{\vartheta} \qquad \boldsymbol{\Omega}^{\phi}{}_{r}=-\boldsymbol{\Omega}^{r}{}_{\phi} \qquad \boldsymbol{\Omega}^{\vartheta}{}_{\phi}=-\boldsymbol{\Omega}^{\phi}{}_{\vartheta},$$

and

$$\boldsymbol{\Omega}^{t}{}_{r}=\boldsymbol{\Omega}^{r}{}_{t} \qquad \boldsymbol{\Omega}^{\phi}{}_{t}=\boldsymbol{\Omega}^{t}{}_{\phi} \qquad \boldsymbol{\Omega}^{\vartheta}{}_{t}=\boldsymbol{\Omega}^{t}{}_{\vartheta}.$$

Comparing the formulas above for $\boldsymbol{\Omega}^{i}{}_{j}$ with (38.55), i.e., $\boldsymbol{\Omega}^{i}{}_{j}=R^{i}{}_{jlk}\,\boldsymbol{\theta}^{k}\wedge\boldsymbol{\theta}^{l}$, and appealing to (38.60), we can immediately read off the components of the Riemann tensor:

$$\begin{aligned}
R^{t}{}_{rrt}=-\frac{f''}{2} &= +\frac{2GM}{r^3}\\
R^{t}{}_{\phi\phi t}=R^{t}{}_{\vartheta\vartheta t}=R^{\phi}{}_{rr\phi}=R^{\vartheta}{}_{rr\vartheta}=-\frac{f'}{2r} &= -\frac{GM}{r^3}\\
R^{\vartheta}{}_{\phi\phi\vartheta}=\left[\frac{1-f}{r^2}\right] &= +\frac{2GM}{r^3}.
\end{aligned}$$

While the meaning of the r coordinate is no longer so simple as "radial distance," observe that these formulas are in perfect accord with the formulas (30.2) and (30.3) (p. 312) that we deduced from the tidal forces arising from Newton's Inverse-Square Law of Gravitation!

We have arrived at a moment of high drama: *Is this putative black hole spacetime geometry* physically possible—*does it satisfy the Einstein Vacuum Field Equation?* Let us find out!

Summing the relevant Riemann tensor components to find the Ricci tensor, we easily see that the off-diagonal components of the Ricci tensor vanish, and, using (38.60), we now calculate its diagonal components, as follows:

$$\begin{aligned}
R_{tt} &= R^{r}{}_{ttr}+R^{\phi}{}_{tt\phi}+R^{\vartheta}{}_{tt\vartheta}=+\frac{f''}{2}+\frac{f'}{2r}+\frac{f'}{2r}=-\frac{2GM}{r^3}+\frac{GM}{r^3}+\frac{GM}{r^3}=0,\\
R_{rr} &= R^{t}{}_{rrt}+R^{\phi}{}_{rr\phi}+R^{\vartheta}{}_{rr\vartheta}=-\frac{f''}{2}-\frac{f'}{2r}-\frac{f'}{2r}=+\frac{2GM}{r^3}-\frac{GM}{r^3}-\frac{GM}{r^3}=0,\\
R_{\phi\phi} &= R^{t}{}_{\phi\phi t}+R^{r}{}_{\phi\phi r}+R^{\vartheta}{}_{\phi\phi\vartheta}=-\frac{f'}{2r}-\frac{f'}{2r}+\frac{1-f}{r^2}=-\frac{GM}{r^3}-\frac{GM}{r^3}+\frac{2GM}{r^3}=0,\\
R_{\vartheta\vartheta} &= R^{t}{}_{\vartheta\vartheta t}+R^{r}{}_{\vartheta\vartheta r}+R^{\phi}{}_{\vartheta\vartheta\phi}=-\frac{f'}{2r}-\frac{f'}{2r}+\frac{1-f}{r^2}=-\frac{GM}{r^3}-\frac{GM}{r^3}+\frac{2GM}{r^3}=0.
\end{aligned}$$

Thus the black hole geometry discovered by Leftenant Schwarzschild in December of 1915, written down in his World War I trench on the Russian front, is *indeed* an exact solution of the equation that Einstein had discovered one month earlier!

To be clear, Schwarzschild had no idea at the time that his discovery represented a black hole. For him, and for Einstein, it *only* represented the spacetime geometry surrounding ordinary celestial bodies, such as the Earth or the Sun. It would take another *half-century* before physicists finally grasped that this solution, in its entirety, *also* presents the pure vacuum gravitational field of a black hole.

In January 1965, ten years after Einstein's death, Roger Penrose proved for the first time that Einstein's theory mathematically implies that the collapse of a sufficiently massive body must *necessarily* result in the formation of a black hole.[20] For this discovery, Penrose was awarded half of the 2020 Nobel Prize in Physics; the other half was divided equally between Reinhard Genzel and Andrea Ghez, for their independent experimental discovery of the supermassive black hole (Sagittarius A*) at the centre of our own Milky Way galaxy.

Before our long *Mathematical Drama* finally comes to its end, and the curtain falls, we have one last trick up our sleeve. Why did we insist on expressing the curvature in terms of f, f', and f'' till the bitter end, despite the fact that their values were known to us from the very beginning, on full display in (38.60)?

Instead of "merely" *verifying* that Schwarzschild's geometry solves Einstein's equation, we can instead *deduce* it *from* Einstein's equation!

Einstein's Field Equation implies

$$\boxed{R_{\phi\phi} = R_{\vartheta\vartheta} = -\frac{f'}{r} + \frac{1-f}{r^2} = 0 \quad \Longrightarrow \quad \frac{df}{dr} = \frac{1-f}{r} \quad \Longrightarrow \quad f(r) = 1 - \frac{C}{r},}$$

where C is a constant.

But then the remaining diagonal components of the Ricci tensor automatically vanish, too, thereby confirming that we have successfully solved Einstein's equation:

$$\boxed{R_{tt} = -R_{rr} = \frac{f''}{2} + \frac{f'}{r} = -\frac{C}{r^3} + \frac{C}{r^3} = 0.}$$

It remains to understand why the constant C is tied to the mass[21] of the black hole by $C = 2GM$. But to explain this would require that we delve much deeper into the fascinating world of gravitational physics than this *Mathematical Drama* will permit us to venture.

---

[20] In fact what Penrose proved was that the collapse must result in the formation of a spacetime singularity. But, in principle, this singularity could be something far more terrifying than a black hole—it could be a so-called ***naked singularity***, visible to the outside world! In 1969, Penrose conjectured that this could never happen, and that a "Cosmic Censor" would insist that every singularity be respectably ***clothed*** inside an event horizon, thereby forming a black hole! Many theoretical arguments lend credence to this ***Cosmic Censorship Hypothesis***, but it remains unproven in 2020.

[21] This connection can be established by a variety of arguments, and it is treated in *all* of the General Relativity texts we recommend in the *Further Reading* section at the end of this book. However, perhaps the single best, most *physical* treatment of the topic can be found in Schutz (2003, §§18, 21).

Chapter 39

Exercises for Act V

1-Forms

1. **Dirac Delta Function as 1-Form.** (This example is adapted from Schutz (1980).) Consider the vector space $C[-1,+1]$ of infinitely differentiable, real-valued functions $f(x)$ defined on the interval $-1 \leqslant x \leqslant +1$. The dual space of 1-forms is called the ***distributions***. An example of such a distribution is the ***Dirac Delta*** *Function* $\delta(x)$, which is the 1-form whose action on the "vector" $f(x)$ produces its value at $x=0$:

$$\langle \delta(x), f(x) \rangle = f(0).$$

As we now explain, this is not at all how Paul Dirac (1930) himself originally introduced his delta function into quantum mechanics.

(i) For any function $g(x)$ in $C[-1,+1]$, define a corresponding 1-form $\widetilde{g}(x)$ such that its action on the vector $f(x)$ is given by

$$\langle \widetilde{g}(x), f(x) \rangle \equiv \int_{-1}^{+1} g(x)\, f(x)\, dx.$$

Prove that $\widetilde{g}$ is indeed a 1-form.

(ii) Thus Dirac actually defined his delta function by insisting that

$$\int_{-1}^{+1} \delta(x)\, f(x)\, dx = f(0). \tag{39.1}$$

No ordinary function $\delta(x)$ can do this, but consider a smooth, bell-shaped curve centred at 0, and imagine it growing higher and higher as it becomes narrower and narrower, all the while obeying (39.1), with $f(x) = 1$:

$$\int_{-1}^{+1} \delta(x)\, 1\, dx = 1.$$

Explain why, in the limit that the width of the bell curve vanishes, we recover (39.1).

2. In $\mathbb{R}^3$, give an explicit *geometrical* interpretation of the action $\boldsymbol{\omega}(\mathbf{v})$ of the 1-form $\boldsymbol{\omega} = 2\,\mathbf{dx} + \mathbf{dy} + 2\,\mathbf{dz}$ on the vector $\mathbf{v}$.

3. **Explanation of *Covariant* and *Contravariant*.** Let R_θ denote a rotation of $\mathbb{R}^2$ by θ. Then, as explained in [15.3], page 153, its matrix is

$$[R_\theta] = \begin{bmatrix} c & -s \\ s & c \end{bmatrix}, \quad \text{with inverse} \quad [R_\theta]^{-1} = [R_{-\theta}] = [R_\theta]^T = \begin{bmatrix} c & s \\ -s & c \end{bmatrix},$$

where $c \equiv \cos\theta$ and $s \equiv \sin\theta$. Let $\{\mathbf{e}_1, \mathbf{e}_2\}$ be the standard Cartesian basis, and let $\{\widetilde{\mathbf{e}}_1, \widetilde{\mathbf{e}}_2\}$ be the new basis that results from rotating the old basis by θ, so that $\widetilde{\mathbf{e}}_j = R_\theta \mathbf{e}_j$.

(i) Draw both bases in the same picture, and also draw a general vector $\mathbf{v} = v^j \mathbf{e}_j$, and draw its components $\widetilde{v}^j$ in the rotated basis: $\mathbf{v} = \widetilde{v}^j \widetilde{\mathbf{e}}_j$. Use your picture to explain why the new components are obtained by applying the *opposite* rotation matrix $[R_\theta]^{-1}$ to the original components:

$$\begin{bmatrix} \widetilde{v}^1 \\ \widetilde{v}^2 \end{bmatrix} = [R_\theta]^{-1} \begin{bmatrix} v^1 \\ v^2 \end{bmatrix}.$$

Because the components vary in the *opposite* ("contra-") way to the basis vectors, in the older literature a vector is called ***contravariant***.

(ii) The dual basis of $\{\mathbf{e}_1, \mathbf{e}_2\}$ is $\{\mathbf{d}x^1, \mathbf{d}x^2\}$. Let $\varphi = \varphi_j\, \mathbf{d}x^j$ be a general 1-form, and let $\{\widetilde{\omega}^1, \widetilde{\omega}^2\}$ be the basis dual to $\{\widetilde{\mathbf{e}}_1, \widetilde{\mathbf{e}}_2\}$, so that the new components are given by $\varphi = \widetilde{\varphi}_j\, \widetilde{\omega}^j$. Show that the components of the 1-form transform in the *same* ("co-") way as the basis vectors:

$$[\widetilde{\varphi}_1, \widetilde{\varphi}_2] = [\varphi_1, \varphi_2]\, [R_\theta].$$

For this reason, 1-forms used to be called ***covariant vectors***, or ***covectors***. (NOTE: Modern Differential Geometry focuses on (coordinate-independent) geometric objects, which are unchanged by basis transformations such as this one. For this reason, the terms ***contravariant*** and ***covariant*** have largely disappeared from the modern literature. For an excellent, fully general discussion of this topic, see Schutz (1980, §2.26).)

Tensors

4. **Matrix multiplication.** (From Schutz (1980).) Show that in matrix algebra the action of one matrix on a second matrix, to produce their matrix product, is a tensor of valence $\left\{ \begin{smallmatrix} 2 \\ 2 \end{smallmatrix} \right\}$.

(*Hint:* Recall that each matrix is itself a tensor of valence $\left\{ \begin{smallmatrix} 1 \\ 1 \end{smallmatrix} \right\}$; see Section 33.2.)

5. **Tensor Contractions.** (From Schutz (1980).) How many different $\left\{ \begin{smallmatrix} 2 \\ 1 \end{smallmatrix} \right\}$ tensors can be made by contraction on pairs of indices of the $\left\{ \begin{smallmatrix} 3 \\ 2 \end{smallmatrix} \right\}$ tensor $Q^{ijk}{}_{lm}$? How many $\left\{ \begin{smallmatrix} 1 \\ 0 \end{smallmatrix} \right\}$ tensors (i.e., vectors) can be made by a second contraction?

6. **Tensor View of Matrix Multiplication as Contraction.** Referring back to Section 33.2 and Section 33.6, let $\mathbf{A}$ be a linear transformation, represented as a vector-valued tensor of valence $\left\{ \begin{smallmatrix} 1 \\ 1 \end{smallmatrix} \right\}$: $\mathbf{v} \to \mathbf{A}(\mathbf{v})$. Let $\mathbf{B}$ be a second linear transformation, and let $\mathbf{C}$ be their composition: $\mathbf{C}(\mathbf{v}) \equiv \mathbf{B}[\mathbf{A}(\mathbf{v})]$. Show that the components of the $\left\{ \begin{smallmatrix} 1 \\ 1 \end{smallmatrix} \right\}$ tensor $\mathbf{C}$ are given by contraction (aka matrix multiplication): $C^i{}_j = B^i{}_k A^k{}_j$.

2-Forms

7. **Factorizes in $\mathbb{R}^3$.** In $\mathbb{R}^3$, let Ψ be a general 2-form. Show that Ψ can always be factorized as the wedge product of two 1-forms: $\Psi = \alpha \wedge \beta$.

8. **Need not factorize in $\mathbb{R}^4$.** In $\mathbb{R}^4$, let Ψ be a general 2-form.

(i) Show that Ψ *cannot* always be factorized as the wedge product of two 1-forms. *Hint:* If it can be factorized, consider $\Psi \wedge \Psi$. But now look back at §35.7.

(ii) If $\Psi \wedge \Psi = 0$, prove that Ψ *can* be factorized as the wedge product of two 1-forms.

(iii) Show that Ψ can always be expressed as the *sum* of *two* wedge products:

$$\Psi = \alpha \wedge \beta + \gamma \wedge \delta.$$

9. Area formula for surface f = const. Let $\mathcal{S}$ be the surface with equation $f(x, y, z) = \text{const}$.

(i) Show that the unit normal $\mathbf{n}$ to $\mathcal{S}$ is

$$\mathbf{n} = \frac{1}{\sqrt{(\partial_x f)^2 + (\partial_y f)^2 + (\partial_z f)^2}} \begin{bmatrix} \partial_x f \\ \partial_y f \\ \partial_z f \end{bmatrix}.$$

(ii) By thinking of $\mathbf{n}$ as the velocity of fluid flowing at unit speed orthogonally across $\mathcal{S}$, deduce that the area 2-form on $\mathcal{S}$ is

$$\mathcal{A} = \frac{\partial_x f\, \mathbf{d}y \wedge \mathbf{d}z + \partial_y f\, \mathbf{d}z \wedge \mathbf{d}x + \partial_z f\, \mathbf{d}x \wedge \mathbf{d}y}{\sqrt{(\partial_x f)^2 + (\partial_y f)^2 + (\partial_z f)^2}}.$$

3-Forms

10. Factorizes in $\mathbb{R}^4$. Show that any 3-form in $\mathbb{R}^4$ can be expressed as the wedge product of three 1-forms.

Differentiation

11. Closed and Exact Forms. In $\mathbb{R}^3$, let

$$\begin{aligned} \varphi &= 2x\, \mathbf{d}x + 2y\, \mathbf{d}y + 2z\, \mathbf{d}z, \\ \chi &= xy\, \mathbf{d}z, \\ \Psi &= x\, \mathbf{d}y \wedge \mathbf{d}z + y\, \mathbf{d}x \wedge \mathbf{d}z. \end{aligned}$$

(i) Verify that φ is closed, and explain this by showing that it is *exact*: $\varphi = \mathbf{d}r^2$, where r is the distance from the origin.

(ii) Verify that Ψ is closed, and explain this by showing that it is *exact*: $\Psi = \mathbf{d}\chi$.

12. Closed and Exact Forms. Continuing with the Forms of Exercise 11, compute each of the following exterior derivatives, (A) directly; (B) using the Leibniz product rule, (36.8).

(i) $\mathbf{d}(\chi \wedge \chi)$.

(ii) $\mathbf{d}(\varphi \wedge \chi)$.

(iii) $\mathbf{d}(\varphi \wedge \Psi)$.

(iv) $\mathbf{d}(\chi \wedge \Psi)$.

13. Closed and Exact Forms. If you did not do so earlier, prove each of the following facts:

(i) If Υ and Φ are closed, then $\Upsilon \wedge \Phi$ is closed, too.

(ii) If Υ is closed, then $\Upsilon \wedge \mathbf{d}\Phi$ is closed for all Φ.

(iii) If deg Φ is *even*, then $\Phi \wedge \mathbf{d}\Phi$ is closed. *Hint:* See (35.4).

14. Vector Calculus Identities via Forms. Use Forms to derive the following identities of Vector Calculus:

(i)

$$\nabla \bullet \left[\underline{\varphi} \times \nabla f\right] = \left[\nabla \times \underline{\varphi}\right] \bullet \nabla f.$$

(ii)

$$\nabla \bullet \left[\underline{\varphi} \times \underline{\psi}\right] = \left[\nabla \times \underline{\varphi}\right] \bullet \underline{\psi} - \underline{\varphi} \bullet \left[\nabla \times \underline{\psi}\right].$$

(*Hint:* (36.9).) Note that (i) may now be derived as a special case of (ii).

15. Hodge Star Duality Operator ($\star$) (For a fully general discussion of this concept, see Schutz (1980), Baez and Muniain (1994), or Dray (2015). In particular, Baez and Muniain (1994, §§1, 5) contains an excellent discussion of the relation of Hodge duality to Maxwell's equations.)

(i) In n dimensions, show that the space of p-forms has the same dimension as the space of $(n-p)$-forms. These two spaces can therefore be put into 1-to-1 correspondence, and Hodge duality is a specific method of accomplishing this.

(ii) In $\mathbb{R}^3$, let $\mathcal{V}_3 = dx \wedge dy \wedge dz$ be the volume 3-form. Given a basis p-form σ, we define $\star$ to be the linear operator that fills in the "missing pieces" of $\mathcal{V}_3$:

$$\boxed{\sigma \wedge \star\sigma = \mathcal{V}_3.}$$

Deduce that

$$\begin{aligned} \star dx &= dy \wedge dz, \\ \star dy &= dz \wedge dx, \\ \star dz &= dx \wedge dy, \end{aligned}$$

and, symmetrically, deduce that

$$\begin{aligned} \star(dy \wedge dz) &= dx, \\ \star(dz \wedge dx) &= dy, \\ \star(dx \wedge dy) &= dz. \end{aligned}$$

Thus, in this case, two applications of the Hodge star operator yields the identity: $\star\star = 1$. Finally, in the interest of completeness, it is natural to define

$$\star 1 = \mathcal{V}_3 \quad \text{and} \quad \star\mathcal{V}_3 = 1.$$

(iii) Continuing in $\mathbb{R}^3$, we remind you of our earlier notation: $\underline{\varphi}$ is the vector corresponding ($\rightleftarrows$) to the 1-form φ. Show that Hodge duality yields two forms of the vector (cross) product:

$$\underline{\alpha} \times \underline{\beta} = \underline{\gamma} \iff \alpha \wedge \beta = \star\gamma \iff \star(\alpha \wedge \beta) = \gamma.$$

(iv) Still in $\mathbb{R}^3$, recall that every 2-form Ψ can be thought of as the flux associated with a flow of velocity $\underline{\Psi}$, via (34.10):

$$\Psi = \Psi^1\,(dx^2 \wedge dx^3) + \Psi^2\,(dx^3 \wedge dx^1) + \Psi^3\,(dx^1 \wedge dx^2) \rightleftarrows \underline{\Psi} = \begin{bmatrix} \Psi^1 \\ \Psi^2 \\ \Psi^3 \end{bmatrix}.$$

If ψ is the 1-form corresponding to $\underline{\Psi}$, deduce that the correspondence between Ψ and ψ is Hodge duality:

$$\star\Psi = \psi \quad \text{and} \quad \Psi = \star\psi.$$

(v) Now consider Minkowski spacetime, with metric

$$ds^2 = g_{ij}\, dx^i\, dx^j = -dt^2 + dx^2 + dy^2 + dz^2.$$

NOTE: Here, and henceforth, the metric coefficients g_{ij} are the *negatives* of those used earlier in Act II, for this is the convention favoured by the majority of more advanced physics texts we shall refer to, going forward.

Let the volume 4-form be $\mathcal{V}_4 = \mathbf{dx} \wedge \mathbf{dy} \wedge \mathbf{dz} \wedge \mathbf{dt}$, and define $\star\mathbf{dx}^i$ by

$$\boxed{\mathbf{dx}^i \wedge \star\mathbf{dx}^i = g_{ii}\, \mathcal{V}_4.}$$

Deduce that

$$\star\mathbf{dt} = \mathbf{dx} \wedge \mathbf{dy} \wedge \mathbf{dz} = \mathcal{V}_3,$$
$$\star\mathbf{dx} = \mathbf{dy} \wedge \mathbf{dz} \wedge \mathbf{dt},$$
$$\star\mathbf{dy} = \mathbf{dz} \wedge \mathbf{dx} \wedge \mathbf{dt},$$
$$\star\mathbf{dz} = \mathbf{dx} \wedge \mathbf{dy} \wedge \mathbf{dt}.$$

(vi) Still in Minkowski spacetime, we define the dual $\star\Psi$ of a basis 2-form Ψ by

$$\boxed{\Psi \wedge \star\Psi = \pm\mathcal{V}_4,}$$

in which we choose *minus* $(-)$ if Ψ contains $\mathbf{dt}$, and we choose *plus* $(+)$ otherwise. (Note that this rule encompasses the rule in the previous part as a special case.) Deduce that

$$\star(\mathbf{dx} \wedge \mathbf{dt}) = -\mathbf{dy} \wedge \mathbf{dz}, \qquad \star(\mathbf{dy} \wedge \mathbf{dz}) = \mathbf{dx} \wedge \mathbf{dt},$$
$$\star(\mathbf{dy} \wedge \mathbf{dt}) = -\mathbf{dz} \wedge \mathbf{dx}, \qquad \star(\mathbf{dz} \wedge \mathbf{dx}) = \mathbf{dy} \wedge \mathbf{dt},$$
$$\star(\mathbf{dz} \wedge \mathbf{dt}) = -\mathbf{dx} \wedge \mathbf{dy}, \qquad \star(\mathbf{dx} \wedge \mathbf{dy}) = \mathbf{dz} \wedge \mathbf{dt}.$$

(vii) Verify that for 2-forms in Minkowski spacetime, $\star\star = -1$. (This is in fact true of all p-forms.)

(viii) Recall that the Faraday 2-form is given by (34.22):

$$\mathbf{F} = \mathbf{Faraday} = \boldsymbol{\varepsilon} \wedge \mathbf{dt} + \mathbf{B}.$$

Use the previous parts to deduce that the Maxwell 2-form (34.26) is indeed the Hodge dual of the Faraday 2-form:

$$\star\mathbf{F} = \mathbf{Maxwell} = \boldsymbol{\beta} \wedge \mathbf{dt} - \mathbf{E}.$$

(ix) Likewise, verify that the dual of the spacetime current 1-form $\mathbf{J} = -\rho\, \mathbf{dt} + \mathbf{j}$, is indeed given by formula (36.18) for the current density 3-form:

$$\star\mathbf{J} = -\rho\, \mathcal{V}_3 + \left[\text{flux 2-form of } \underline{\mathbf{j}}\right] \wedge \mathbf{dt}$$

16. Conservation of Electric Charge

(i) Assuming that electric charge is ***conserved***, i.e., that it can neither be created nor destroyed, deduce that if V is the interior of a closed surface $\mathcal{S} = \partial V$, then

$$\frac{d}{dt}\iiint_V \rho \, d\mathcal{V} = -\oiint_{\mathcal{S}} \underline{\mathbf{j}} \cdot \mathbf{n} \, d\mathcal{A}.$$

(ii) Use Gauss's Theorem to deduce that the local expression of charge conservation is

$$\frac{d\rho}{dt} + \nabla \cdot \underline{\mathbf{j}} = 0.$$

(iii) By taking the exterior derivative of the second pair of Maxwell's Equations, in the form

$$\mathbf{d} \star \mathbf{F} = 4\pi \star \mathbf{J},$$

deduce that charge conservation is a logical *consequence* of Maxwell's Equations, taking the form

$$\boxed{\mathbf{d} \star \mathbf{J} = 0.}$$

HISTORICAL NOTE: Faraday's 1831 experimental discovery of electromagnetic induction was ultimately crystalized by Maxwell into mathematical form: a changing magnetic field generates curl in the electric field—see (37.13). Thirty years later, *without any experimental evidence*, the purely theoretical considerations of this exercise forced Maxwell to conclude, in 1861, that, symmetrically, a changing electric field *must* generate curl in the magnetic field. Without this symmetry, electromagnetic waves could not exist—*Let there be light!*

17. Spinning Photons and Self-Duality. In Minkowski spacetime, a *complex* 2-form $\boldsymbol{\Psi}$ is called ***self-dual*** if

$$\star\boldsymbol{\Psi} = i\,\boldsymbol{\Psi},$$

and ***anti-self-dual*** if

$$\star\boldsymbol{\Psi} = -i\,\boldsymbol{\Psi},$$

where $\star$ is the Hodge duality operator introduced in Exercise 15.

(i) Let us verify that the Faraday 2-form $\mathbf{F}$ (or any 2-form, for that matter) can always be split into a self-dual part ${}^{+}\mathbf{F}$ and an anti-self-dual part ${}^{-}\mathbf{F}$, so that

$$\mathbf{F} = {}^{+}\mathbf{F} + {}^{-}\mathbf{F}.$$

Starting from this formula, and using the fact that $\star{}^{\pm}\mathbf{F} = \pm i\,{}^{\pm}\mathbf{F}$, deduce that

$${}^{+}\mathbf{F} = \frac{1}{2}\left[\mathbf{F} - i\star\mathbf{F}\right] \quad \text{and} \quad {}^{-}\mathbf{F} = \frac{1}{2}\left[\mathbf{F} + i\star\mathbf{F}\right].$$

(ii) It turns out that in quantum mechanics, these complex conjugate components ${}^{+}\mathbf{F}$ and ${}^{-}\mathbf{F}$ describe, respectively, *right*-spinning and *left*-spinning photons (the quanta of the electromagnetic field). Verify that *all four of Maxwell's Equations can be encapsulated into a single complex equation:*

$$\boxed{\mathbf{d}\,{}^{+}\mathbf{F} = -2\pi i \star \mathbf{J}.}$$

Integration

18. Closed and Exact 1-Forms. In the language of physics, $\varphi = \mathbf{d}f$ corresponds to a ***conservative*** force field, $\underline{\varphi} = \nabla f$, and f is the ***potential energy***. Confirm that each of the following 1-forms is conservative by proving that each is *closed*: $\mathbf{d}\varphi = 0$. Then show that each is *exact*, by explicitly finding a matching potential energy function, f.

(i) $\varphi = x\,\mathbf{d}x + y^2\,\mathbf{d}y + z^3\,\mathbf{d}z$.

(ii) $\varphi = 3x^2y^2z\,\mathbf{d}x + 2x^3yz\,\mathbf{d}y + x^3y^2\,\mathbf{d}z$.

(iii) $\varphi = (2xy + z)\,\mathbf{d}x + (x^2 + 3y^2)\,\mathbf{d}y + x\,\mathbf{d}z$.

19. Closed and Exact 2-Forms. Verify that each of the following 2-forms is *closed*: $\mathbf{d}\Psi = 0$. Then show that each is *exact*, by explicitly finding a 1-form potential φ such that $\Psi = \mathbf{d}\varphi$.

(i) $\Psi = 3x^2y^4\,\mathbf{d}x \wedge \mathbf{d}y$.

(ii) $\Psi = 2xyz^3\,\mathbf{d}x \wedge \mathbf{d}y - 3x^2yz^2\,\mathbf{d}y \wedge \mathbf{d}z$.

20. Closed and Exact 3-Forms. In $\mathbb{R}^3$ every 3-form Υ is proportional to the volume 3-form, $\mathcal{V} = \mathbf{d}x \wedge \mathbf{d}y \wedge \mathbf{d}z$, and, since there are no 4-forms, every 3-form is *closed*: $\mathbf{d}\Upsilon = 0$. Show that each Υ is *exact*, by explicitly finding a 2-form potential Ψ such that $\Upsilon = \mathbf{d}\Psi$.

(i) $\Upsilon = yz\,\mathcal{V}$.

(ii) $\Upsilon = 2(x + y + z)\,\mathcal{V}$.

21. Homogeneous Functions. (From do Carmo (1994).) A function $g : \mathbb{R}^3 \to \mathbb{R}$ is called ***homogeneous*** and of ***degree*** k if $g(tx, ty, tz) = t^k\, g(x, y, z)$.

(i) Show that such a g satisfies ***Euler's Relation***:

$$x\,\partial_x g + y\,\partial_y g + z\,\partial_z g = kg.$$

Hint: Differentiate the defining relation with respect to t.

(ii) If the 1-form,

$$\psi = a\,\mathbf{d}x + b\,\mathbf{d}y + c\,\mathbf{d}z$$

is such that a, b, c are homogeneous of degree k, and $\mathbf{d}\psi = 0$, then show that $\psi = \mathbf{d}f$, where

$$f = \frac{xa + yb + zc}{k+1}.$$

Hint: Write out $\mathbf{d}\psi = 0$ in component form, and then apply Euler's Relation.

(iii) Consider the flux 2-form corresponding to ψ:

$$\Psi = a\,\mathbf{d}y \wedge \mathbf{d}z + b\,\mathbf{d}z \wedge \mathbf{d}x + c\,\mathbf{d}x \wedge \mathbf{d}y.$$

Show that if $\mathbf{d}\Psi = 0$, then $\Psi = \mathbf{d}\gamma$, where

$$\gamma = \frac{(zb - yc)\,\mathbf{d}x + (xc - za)\,\mathbf{d}y + (ya - xb)\,\mathbf{d}z}{k+2}.$$

(iv) If V is the interior of S^2, so that $\partial V = S^2$, use FTEC, in the form of (37.14), to show that

$$\iiint_V [\nabla^2 g]\; d\mathcal{V} = k \oiint_{S^2} g\; d\mathcal{A}.$$

Hint: If $\mathbf{n}$ is the unit normal to S^2, verify that Euler's Relation can be written $\mathbf{n} \bullet \nabla g = kg$.

Differential Geometry via Forms

22. Hairy Ball Theorem. (From do Carmo (1994).) As we discussed in connection with [19.8], the Poincaré–Hopf Theorem (19.6) implies that there cannot exist a singularity-free vector field on S^2. This is often called the *Hairy Ball Theorem*: we cannot comb the hair on a coconut! Here is a second proof, using FTEC. Suppose that such a nonvanishing field $\mathbf{v}$ exists, and use it to construct the first vector $\mathbf{m}_1 = \mathbf{v}/|\mathbf{v}|$ of an orthonormal basis $\{\mathbf{m}_1, \mathbf{m}_2\}$. Then the Gauss Equation (38.30) yields

$$\mathbf{d}\omega_{12} = -\boldsymbol{\theta}^1 \wedge \boldsymbol{\theta}^2 = -\mathcal{A}.$$

Now use FTEC to integrate this equation over S^2, thereby obtaining a contradiction.

23. Conformal Curvature Formula. If the conformal metric of a surface is $ds^2 = (du^2 + dv^2)/\Omega^2$, show that

$$\boxed{\mathcal{K} = \Omega\, \nabla^2\Omega - \left[(\partial_u\Omega)^2 + (\partial_v\Omega)^2\right].}$$

As a simple check, try this on the Beltrami–Poincaré half-plane model of the hyperbolic plane, for which $\Omega = v$.

24. Vanishing Curvature *Characterizes* Euclidean Geometry. (From do Carmo (1994).) This exercise will establish that

> *The Gaussian curvature vanishes if and only if a 2-surface is locally Euclidean.*

By "locally Euclidean" we mean that around each point we can find (u, v)-coordinates such the metric is $ds^2 = du^2 + dv^2$.

(i) Choose an orthonormal frame $\{\mathbf{m}_1, \mathbf{m}_2\}$ for tangent vectors to the surface. By the Gauss Equation, (38.30),

$$\mathbf{d}\omega_{12} = -\mathcal{K}\, \boldsymbol{\theta}^1 \wedge \boldsymbol{\theta}^2 = 0.$$

Deduce that there exists a function ϕ—which we shall interpret as an angle— such that $\omega_{12} = \mathbf{d}\phi$.

(ii) Consider a new orthonormal frame $\{\widetilde{\mathbf{m}}_1, \widetilde{\mathbf{m}}_2\}$, obtained by *rotating* $\{\mathbf{m}_1, \mathbf{m}_2\}$ *by angle* ϕ. Show that in the new frame, $\widetilde{\omega}_{12} = 0$.

(iii) Deduce that

$$\mathbf{d}\widetilde{\boldsymbol{\theta}}^1 = 0 = \mathbf{d}\widetilde{\boldsymbol{\theta}}^2.$$

(iv) Deduce the existence of coordinates in which the metric is Euclidean.

25. Connection Forms of the Beltrami–Poincaré Half-Plane. The metric (5.8) of the Beltrami–Poincaré half-plane immediately yields the basis 1-forms:

$$ds^2 = \frac{dx^2 + dy^2}{y^2} \quad \Longrightarrow \quad \boldsymbol{\theta}^1 = \frac{1}{y}\,dx \quad \text{and} \quad \boldsymbol{\theta}^2 = \frac{1}{y}\,dy.$$

(i) Calculate $\mathbf{d}\boldsymbol{\theta}^1$ and $\mathbf{d}\boldsymbol{\theta}^2$.

(ii) Use Cartan's First Structural Equation,

$$\mathbf{d}\boldsymbol{\theta}^1 = -\omega^1{}_2 \wedge \boldsymbol{\theta}^2, \qquad \mathbf{d}\boldsymbol{\theta}^2 = -\omega^2{}_1 \wedge \boldsymbol{\theta}^1 = \omega^1{}_2 \wedge \boldsymbol{\theta}^1,$$

to deduce that the (unique) solution is

$$\omega^1{}_2 = -\frac{1}{y}\,\mathbf{d}x.$$

(iii) Calculate $\mathbf{d}\omega^1{}_2$, and, by comparing the result with the Gauss Equation,

$$\mathbf{d}\omega^1{}_2 = \mathcal{K}\,\theta^1 \wedge \theta^2,$$

deduce that the curvature of the hyperbolic plane is indeed $\mathcal{K} = -1$.

26. Curvature 2-Forms of a 2-Surface. Let $\{\mathbf{m}_1, \mathbf{m}_2\}$ be an orthonormal basis field for tangent vectors $\mathbf{v}$ to a 2-surface, so that

$$\nabla_{\mathbf{v}}\mathbf{m}_1 = -\omega^1{}_2(\mathbf{v})\,\mathbf{m}_2$$
$$\nabla_{\mathbf{v}}\mathbf{m}_2 = \;\;\omega^1{}_2(\mathbf{v})\,\mathbf{m}_1.$$

(i) Write down the connection matrix $[\omega]$, and hence prove that

$$[\omega] \wedge [\omega] = 0.$$

(ii) Deduce that the curvature matrix (38.53) is

$$[\Omega] = \begin{bmatrix} 0 & 1 \\ -1 & 0 \end{bmatrix} \mathbf{d}\omega^1{}_2.$$

Thus, the curvature matrix is completely described by the single curvature 2-form $\Omega^1{}_2$, governed by the Gauss Equation:

$$\Omega^1{}_2 = \mathbf{d}\omega^1{}_2 = \mathcal{K}\mathcal{A}.$$

27. Curvature of Hyperbolic 3-Space, $\mathbb{H}^3$. Recall that the metric of $\mathbb{H}^3$ is given by (6.23), depicted in [6.6], page 80:

$$ds^2 = \frac{dx^2 + dy^2 + dz^2}{z^2} \qquad \Longrightarrow \qquad [\theta] = \begin{bmatrix} \theta^1 \\ \theta^2 \\ \theta^3 \end{bmatrix} = \frac{1}{z}\begin{bmatrix} \mathbf{d}x \\ \mathbf{d}y \\ \mathbf{d}z \end{bmatrix}.$$

(i) Recall that the switch from our original (O'Neill) notation, ω_{ij}, to our new, more standard notation, $\omega^i{}_j$, induces a change of sign in Cartan's First Structural Equation, so that it now takes the form (see (38.54)):

$$\mathbf{d}[\theta] = -[\omega] \wedge [\theta].$$

Deduce that the connection matrix is

$$[\omega] = \frac{1}{z}\begin{bmatrix} 0 & 0 & -\mathbf{d}x \\ 0 & 0 & -\mathbf{d}y \\ \mathbf{d}x & \mathbf{d}y & 0 \end{bmatrix}$$

(ii) Use (38.53) to deduce that the curvature matrix is

$$[\Omega] = \frac{1}{z^2}\begin{bmatrix} 0 & -\mathbf{d}x \wedge \mathbf{d}y & -\mathbf{d}x \wedge \mathbf{d}z \\ \mathbf{d}x \wedge \mathbf{d}y & 0 & -\mathbf{d}y \wedge \mathbf{d}z \\ \mathbf{d}x \wedge \mathbf{d}z & \mathbf{d}y \wedge \mathbf{d}z & 0 \end{bmatrix}.$$

(iii) Verify that $[\Omega]$ may be rewritten as

$$[\Omega] = \begin{bmatrix} 0 & -\theta^1 \wedge \theta^2 & -\theta^1 \wedge \theta^3 \\ \theta^1 \wedge \theta^2 & 0 & -\theta^2 \wedge \theta^3 \\ \theta^1 \wedge \theta^3 & \theta^2 \wedge \theta^3 & 0 \end{bmatrix},$$

and use (38.55) to deduce that the Riemann tensor has components

$$R^1{}_{212} = R^1{}_{313} = R^2{}_{323} = -1.$$

(iv) Interpret these components of the Riemann tensor as sectional curvatures. What does the Sectional Jacobi Equation (29.21) therefore tell us about the behaviour of neighbouring geodesics that are initially launched in the same direction within this space? Conclude that it is meaningful to say that "$\mathbb{H}^3$ has constant curvature -1."

28. Cosmic Curvature. (The detailed solution to this exercise can be found in Dray (2015, §A.9) and in Sternberg (2012, §6.7).) The ***Friedmann–Lemaître–Robertson–Walker*** (FLRW) solutions of Einstein's Field Equations are now considered the ***standard model*** for the large-scale geometry of our expanding Universe. They were first discovered by Alexander Friedmann in 1922, rediscovered by the Jesuit priest, Georges Lemaître in 1927, and in 1935 Robertson and Walker jointly proved their uniqueness as *the* spatially homogeneous and isotropic geometries that are possible for the Universe.

Friedmann sought and found solutions of the *original* Einstein Equation, whereas we now know that the Cosmological Einstein Equation (30.25) is in fact correct, with cosmological constant $\Lambda > 0$. Fortunately, it is possible to adapt Friedmann's discovery to this new reality.

The FLRW metrics are given by

$$ds^2 = -dt^2 + R^2(t)\left[\frac{dr^2}{\left(1 - Kr^2\right)} + r^2(d\phi^2 + \sin^2\phi\, d\vartheta^2)\right],$$

where $R(t)$ measures the size of the expanding Universe at time t after the big bang, and K is the constant spatial curvature of the Universe for all time. See Penrose (2005, §§27.11, 27.12) for an excellent discussion (and lovely drawings) of the three cases, $K > 0$, $K = 0$, $K < 0$, including the effect of $\Lambda > 0$.

By following in the footsteps of our calculation of the curvature of the Schwarzschild black hole, write down the basis 1-forms, $\boldsymbol{\theta}^i$, take their exterior derivatives to find the connection 1-forms, $\boldsymbol{\omega}^i{}_j$, then calculate $\mathbf{d}\boldsymbol{\omega}^i{}_j$, and finally deduce the curvature 2-forms $\boldsymbol{\Omega}^i{}_j$ of such a Universe.

Further Reading

NOTES: In the following, I merely list title and author; full details can be found in the bibliography. Some of the following works are included simply because they are highly relevant, and I therefore believe they deserve to be brought to your attention (even if they are not quite my cup of tea). Most, however, are included because I consider them to be gems, and I strongly recommend them to you. Many other excellent books sit on my bookshelves, and I seek their counsel often, and yet they are not included here simply in order to cut down this (already long) list to a manageable size—I apologize to the authors of all those excellent works for failing to highlight them here. Finally, I also apologize to the authors of the wonderful works I have yet to discover!

Global Recommendations

First, let me set the stage with six works that I hold to be *invaluable*, the content of each of which spans multiple Acts of this book.

- *The Road to Reality*, by Roger Penrose.
  An extraordinary panorama of almost all of physics, and most of mathematics, by a master of both. Many of the insights can only be found here, and they are brought to life by Penrose's remarkable (and beautiful!) hand drawings.

- *Gravitation*, by Misner, Thorne, and Wheeler.
  Almost 50 years after its original publication in 1973, this classic remains one of the very best introductions to Einstein's geometrical theory of gravity (General Relativity) and to the Differential Geometry upon which it rests. It also contains one of the best, most geometrical introductions to Forms, including the curvature 2-forms that allow one to calculate the Riemann tensor efficiently. The new (2017) edition from Princeton University Press is beautifully done, and contains a new introduction by Charles Misner and Kip Thorne, discussing the exciting developments in the field since the book's original publication.

- *Differential Geometry in the Large*, by Heinz Hopf.
  Hopf was not only one of the towering figures of twentieth-century mathematics, he was also a master of exposition. Here, ideas of Differential Geometry and Topology (many of which are due to Hopf himself) come together in a beautiful way, explained with remarkable clarity and simplicity. Every time I return to this *Meisterwerk*, I feel that some beneficent magician has inserted more wonderful ideas into its pages, for I swear that *this* beautiful idea wasn't on the page the last time I looked!

- *Elementary Differential Geometry* (revised 2nd edition), by Barrett O'Neill.
  First published in 1966, this trail-blazing text pioneered the use of Forms at the undergraduate level. Today, more than a half-century later, O'Neill's work remains, in my view, the single most clear-eyed, elegant, and (ironically) *modern* treatment of the subject available—present company excepted!—at the undergraduate level.

- *Geometrical Methods of Mathematical Physics*, by Bernard Schutz.

  This work—now 40 years old!—is a timeless treasure trove, covering manifolds, tensors, Lie derivatives and Lie groups, Forms, Riemannian Geometry, gauge theories, and a host of other applications to physics. To achieve this, Professor Schutz channels *Star Trek*'s Mr. Spock. His *Vulcan* half enables him to erect a logical structure of crystalline perfection, in which everything is concisely and rigorously proven, and—unlike my (I hope delicious) cheeseburger approach in this book—Kashrut is strictly observed: concepts that depend on the existence of a metric are scrupulously and explicitly separated from those that do not. But, in tandem with this, Schutz is able to harness his *human* half to provide a wealth of intuition that reveals the underlying *geometric* reality.

- *Mathematics and Its History* (3rd edition), by John Stillwell.

  A remarkable panorama of all of mathematics through the lens of history. But make no mistake, this is not primarily a book about history, rather it is fundamentally a work about the interconnectedness and meaning of mathematics itself, all explained in a rather concise style (relative to mine!), with deep insight and lucidity.

Geometry in General

The following works are concerned with geometry in general, but especially with Hyperbolic Geometry. (*Differential* Geometry has its own category.)

- *Geometry and the Imagination*, by David Hilbert and S. Cohn-Vossen.

  A magnificent, deeply insightful survey of geometry, focusing on intuitive understanding, by one of the greatest mathematicians of the 20th century. The diagrams (drawn by K. H. Naumann and H. Bödeker) are *astonishingly* beautiful, to the point of causing me envy!

- *Experiencing Geometry*, by David W. Henderson and Daina Taimina.

  A highly unusual approach, philosophically akin to mine (but using the Moore method), focused on intuitive, *experimental* investigations of geometry. It contains significant discussion of parallel transport and holonomy. The overlap of their approach with mine is made clear by this quotation from the preface: "This book is based on a view of proof as a *convincing communication that answers—Why?*" (Their italics.)

- *Introduction to Geometry* (2nd edition), by H.S.M. Coxeter.

  A wonderfully clear survey by a modern master.

- *Geometry*, by Brannan, Esplen, and Gray.

  An excellent modern survey of geometry, based on Klein's vision of groups of transformations.

- *Euclidean and Non-Euclidean Geometries: Development and History*, by Marvin J. Greenberg.

  A valuable, detailed history of the development of Hyperbolic Geometry, including lengthy quotations from critical, private letters of Gauss, Bolyai, and many others.

- *The Poincaré Half-Plane*, by Saul Stahl.

  The title says it all.

- *Geometry Revealed*, by Marcel Berger.
  A much more advanced survey of geometry, with a focus on conceptual proofs and unsolved problems, by one of the great geometers of the 20th century.

Topology

- *Intuitive Topology*, by V. V. Prasolov.
  Less than 100 pages long, and filled with diagrams, this super-friendly introduction lives up to its title!

- *Euler's Gem*, by David S. Richeson.
  A masterful, mathematically accurate, yet riveting account of Euler's polyhedral formula, its history and the connected mathematical ideas.

- *Surface Topology*, by P. A. Firby and C. F. Gardiner.
  A very gentle, nicely illustrated introduction to the fundamental *geometric* ideas of topology.

- *First Concepts of Topology*, by W. G. Chinn and N. E. Steenrod.
  Another very gentle, nicely illustrated introduction to the fundamental *geometric* ideas of topology.

- *Topology: A Very Short Introduction*, by Richard Earl
  This remarkable little book lives up to its title, covering a huge range of fundamental ideas in just 140 pages, and it does so in a very clear, elementary, informal style. This is my new favourite introduction to the subject.

- *The Shape of Space* (3rd edition), by Jeffrey R. Weeks.
  A wonderfully lucid, engaging, elementary treatment of the topology of two and three dimensional spaces. The last of the four parts of the book deals with the possibility of detecting the topology of the Universe! An appendix contains John Horton Conway's famous ZIP Proof of the Classification Theorem for surfaces, beautifully illustrated by George K. Francis.

- *Three-Dimensional Geometry and Topology*, by William P. Thurston.
  Thurston won the Fields Medal for discovering that 3-manifolds are fundamentally built out of Hyperbolic Geometry. In this book you will hear Thurston's discoveries in his own distinctive voice, and although the difficulty of the *topology* accelerates rapidly, the first 100 pages provide a relatively elementary, highly original introduction to Hyperbolic Geometry that should not be missed.

Hopf's Line Fields and the Poincaré–Hopf Theorem in Physics

In this book I have sought to draw attention to *line fields* and Hopf's beautiful result that the Poincaré–Hopf Theorem applies to them, too, ((19.9), p. 213). These ideas have all but disappeared from modern *mathematics* textbooks, and I strongly believe that it is past time for a revival. That said, *physicists* never lost sight of the value of these ideas, and they have sustained them with wonderful new discoveries.

Here I am forced to list research papers rather than expository textbooks. The *only* expository mathematical treatments I know of are the lectures by Hopf (1956) himself (see previously), and the book by his student, Stoker (1969). Although the latest contributions to optics explicitly cite Hopf's work—going so far as to call fractional indices "Hopf indices"—these ideas were pioneered by J. F. Nye, and later, Sir Michael Berry, neither of whom seem to have been aware of Hopf (1956). (NOTE: In reading these papers, it is important to understand that what mathematicians call *holonomy*, physicists sometimes call *anholonomy*!)

- *The Topology of Ridge Systems*, by Roger Penrose.

  A completely elementary introduction (for nonmathematicians) to this circle of ideas, using fingerprints and palm prints as exemplars, for these are indeed line fields!

- *The Fingerprint of the Weyl Tensor*, Spinors and Space-Time, Vol. 2, pp. 229–233, by Roger Penrose.

  Penrose's "fingerprint" description of the Weyl curvature tensor reveals line fields having singularities possessing fractional indices.

- *I. Liquid Crystals. On the Theory of Liquid Crystals*, by F. C. Frank.

  One of the earliest appearances of line fields in physics. Note that here they are called ***nematic fields*** and their singularities are called ***topological defects***.

- *The Poincaré–Hopf Theorem for Line Fields Revisited*, by Diarmuid Crowley and Mark Grant.

  A nice review of the history and applications of the theorem; it also includes a general proof, in all dimensions.

- *Generic Singularities of Line Fields on 2D Manifolds*, by Ugo Boscain, Ludovic Sacchelli, and Mario Sigalotti.

  A very nice review of the many appearances of line fields in nature, and their mathematical classification, related to umbilic points.

- *Natural Focusing and Fine Structure of Light: Caustics and Wave Dislocations*, by J. F. Nye

  Nye pioneered the introduction of line fields (polarization fields) into optics: the generic singularities are circularly polarized points in 2D or lines in 3D. Two other sorts of singularities are analyzed: those of intensity (caustics) and phase (vortices), which were pioneered by Nye and Sir Michael Berry.

- *The Quantum Phase, Five Years After*, by M. V. Berry, in Shapere and Wilczek (1989).

  A review by the discoverer of the quantum phase himself, explicitly relating the discovery to Differential Geometry, discussing the lines of curvature surrounding umbilic points, and making the crucial observation that a circuit around such a point results in a rotation of $\pm\pi$, corresponding to an index of $\pm\frac{1}{2}$, as illustrated in [19.14 a & A]. For a lovely illustration of this phenomenon, see Hilbert and S. Cohn-Vossen (1952, p. 189).

- *Geometry of Phase and Polarization Singularities, Illustrated by Edge Diffraction and the Tides*, by M. V. Berry.

  Edge diffraction and the tides also yield singular points with fractional indices.

- *Index Formulae for Singular Lines of Polarization*, by M. V. Berry.

  Formulas are obtained for the indices of lines in space on which the polarization of a monochromatic light field is purely circular (C) or purely linear (L). The indices ($\pm\frac{1}{2}$ for C lines and ± 1 for L lines) involve the electric or magnetic field and its derivatives on the line.

- *Polarization Singularities in the Clear Sky*, by M. V. Berry, M. R. Dennis, and R. L. Lee.

  An account of the pattern of polarization directions in daylight. The singularities (two near the Sun and two near the anti-Sun) are points in the sky where the polarization line field pattern has index $+\frac{1}{2}$ and the intensity of polarization is zero.

- *A Half-Century of Physical Asymptotics and Other Diversions: Selected Works by Michael Berry*, by M. V. Berry

  Sir Michael Berry's seminal contributions to optics, focusing on revealing ubiquitous manifestations of a variety of optical *singularities* in both everyday life and in cutting-edge, fundamental photonic research.

- *Singularities and Poincaré Indices of Electromagnetic Multipoles*, by Weijin Chen, Yuntian Chen, and Wei Liu.

  The authors map all the singularities of multipolar radiations of different orders, identify their indices, and show explicitly that the index sum over the entire momentum sphere is always 2, consistent with the Poincaré–Hopf Theorem.

- *Global Mie Scattering: Polarization Morphologies and the Underlying Topological Invariant*, by Weijin Chen, Qingdong Yang, Yuntian Chen, and Wei Liu.

  The Poincaré–Hopf Theorem (with fractional indices) is used to show that if you shine coherently polarized light of any form on any particle or particle clusters, there must be one direction where the scattering is either zero or circularly polarized.

- *Line Singularities and Hopf Indices of Electromagnetic Multipoles*, by Weijin Chen, Yuntian Chen, and Wei Liu

  The paper investigates the connections between half-integer Hopf indices, electromagnetic multipoles, and Bloch modes.

Differential Geometry

- *Differential Geometry: A Geometric Introduction*, by David W. Henderson.

  Of all the Differential Geometry books I will list, this one is the closest in spirit to my work, and that is why I have placed it first. I only discovered it late in the writing of VDGF, and was immediately convinced that Professor Henderson was a kindred spirit, a mathematical brother-in-arms. For the sad story of my attempt to make contact with him, see the footnote on page 252.

- *150 Years After Gauss' "Disquisitiones generales circa superficies curvas,"* by Peter Dombrowski.

  Using Gauss's letters to friends, private notebooks, and unpublished draft manuscripts, this uniquely valuable and extremely insightful work traces the evolution of Gauss's discoveries in Differential Geometry, culminating in his revolutionary, *General Investigations of Curved Surfaces*, which announced the *Theorema Egregium* to the world. As an added bonus, Dombrowski includes Gauss's 1827 masterpiece in its entirety, in the original Latin, with the English translation on facing pages.

- *A Comprehensive Introduction to Differential Geometry* (Vol. 2),
  by Michael Spivak

  Spivak's magnum opus actually comprises *five* volumes, but, for my purposes, Volume 2 is the most relevant, for it includes detailed, insightful analysis of the pioneering investigations of Gauss and of Riemann. This volume also includes Cartan's method of moving frames, and his two Structural Equations.

- *Differential Geometry of Curves and Surfaces* (3rd edition),
  by Thomas Banchoff and Stephen Lovett.

  While the mathematical machinery employed here is the standard one, featuring Christoffel symbols, and the like, it contains several highly original geometrical insights (due to Banchoff) that you will not find elsewhere. Furthermore, the accompanying website contains a wonderful array of Java applets for interactively exploring the concepts.

- *Differential Geometry of Curves and Surfaces*, by Kristopher Tapp.

  One of the best, most *geometrical* introductions of recent years. Although the mathematical machinery is the standard one—ugly equations filled with Christoffel symbols!—there are nevertheless many *conceptual* points of contact with my approach. It has many, nicely done colour diagrams, *many* excellent and interesting applications, and (like this book) it emphasizes the importance of parallel transport and holonomy. Highly recommended! (Note: It only treats 2-surfaces, so there is no discussion of the Riemann tensor or of General Relativity.)

- *Curved Spaces*, by P.M.H. Wilson.

  The first two-thirds of the book is a concise, modern, elegant treatment of Euclidean, Spherical, and Hyperbolic Geometry, employing the unifying power of Möbius transformations (exactly as I do). The last third is an introduction to Differential Geometry.

- *A First Course in Differential Geometry: Surfaces in Euclidean Space*,
  by L. M. Woodward and J. Bolton.

  The long final chapter of this work is the best introduction I know of (at the undergraduate level) to an important and beautiful topic that I have chosen to neglect entirely in this book: *minimal surfaces*. Recall that these are surfaces of vanishing mean curvature, realized by soap films: $\mathrm{H} \equiv \frac{\kappa_1+\kappa_2}{2} = 0$. The discussion includes the 1982 breakthrough discovery by Celso Costa of new minimal surfaces (beyond planes, catenoids, and helicoids) that are closed and without self-intersections. Furthermore, unusually, it also features a thorough discussion of the generalization to the case where $\mathrm{H} = \mathrm{const.} \neq 0$, called ***surfaces of constant mean curvature*** (CMC surfaces).

- *Elementary Differential Geomety* (2nd edition), by Andrew Pressley.

  An excellent, elementary introduction, and, like the previous book, it contains a very nice introduction to minimal surfaces.

- *A Course of Differential Geometry and Topology*,
  by A. Mishchenko and A. Fomenko.

  An excellent, somewhat advanced, wide-ranging introduction that includes Riemannian Geometry and Homology Theory. It is distinguished by its large number of remarkable illustrations in the unmistakable, manga-style of Fomenko.

- *Differential Geometry: Connections, Curvature, and Characteristic Classes*,
  by Loring W. Tu

  This is the best book I know of for making the transition from VDGF to the much more advanced works that deal with Chern's characteristic classes, which have become important both in pure mathematics and in physics. The book does assume that the reader is already familiar with Forms, but my Act V provides more than enough preparation. Tu also assumes the reader is familiar with de Rham cohomology, and here my "primer" in Act V will likely not be sufficient, though I hope my physical and geometrical treatment—which is hard to find elsewhere—will greatly ease further study of the subject.

- *Lectures on Differential Geometry*, by S. S. Chern, W. H. Chen, and K. S. Lam.

  This book is 356 pages long, and contains fewer than 10 diagrams. I include this advanced work for one reason, and one reason only: Chern (upon whose lectures this is based) was one of the greatest geometers of the 20th century, and therefore *anything* he has to say, we should all listen to intently. In particular, Chern includes a lengthy treatment of Finsler Geometry, a vast generalization of Riemannian Geometry, pioneered by Chern himself. The only point of contact with this book is that Chern employs Forms as his principal tool—no surprise, given that Chern actually studied under Cartan himself!

Riemann's Curvature

- *A Panoramic View of Riemannian Geometry*, by Marcel Berger.

  An invaluable, panoramic view of the subject from one of the great geometers of the 20th century. It focuses on results and concepts, clearly and intuitively explained, and it omits all but the most essential details.

- *Riemannian Geometry: A Beginner's Guide* (2nd edition), by Frank Morgan.

  This nicely written and well-illustrated introduction packs a lot into only 129 pages. Despite its subtitle, I suspect that a true beginner would have a much easier time tackling this work *after* reading mine!

- *Riemannian Geometry*, by Manfredo P. do Carmo.

  An excellent (but advanced) treatment of many deep theorems, some of which are hard to find discussed clearly anywhere else.

- *Semi-Riemannian Geometry: With Applications to Relativity*, by Barrett O'Neill

  Semi-Riemannian Geometry is the geometry of spacetime, and this is simply the best mathematical work on the subject that I am aware of.

Einstein's Curved Spacetime (General Relativity)

- *Gravity from the ground up*, by Bernard Schutz.

  A *second* masterpiece! Whereas Schutz channelled Mr. Spock in his first masterpiece (see above), here he channels Captain Kirk! He employs true, Feynmanesque, *physical*, intuitive reasoning—using only high-school mathematics—to treat *all* aspects of gravity. Simultaneously, he succeeds in painting a vivid picture of the historical, *human* context of our evolving understanding of this fundamental force of Nature.

- *A Journey into Gravity and Spacetime*, by John Archibald Wheeler.

  A deeply insightful, largely nonmathematical, *conceptual* explication of Einstein's vision, by one of the greatest physicists of the 20th century, and one of the key figures in the rebirth of General Relativity in the 1960s: for example, it was Wheeler who coined the term ***black hole***. I have a personal connection with this work: having read it, I discovered and sent to Wheeler a Newtonian, purely geometrical proof of his crucial *Law of Double Curvature* (p. 142). As explained in the Prologue, Wheeler (1911–2008), whom I was fortunate to meet several times as Penrose's student, was a pivotal figure in my life, so I was *thrilled* when he wrote back that my proof had delighted him!

- *A First Course in General Relativity* (3rd edition), by Bernard Schutz.

  As of this writing, the third edition does not exist, but thanks to my correspondence with its author, I know it is coming. The second edition was already excellent, so the third edition will only be better! In particular, I predict that the new edition will greatly expand upon the treatment of gravitational waves, for these had not yet even been detected at the time of the second edition, and Schutz has been a key player in the field for more than 30 years. In 2019 Schutz received the Eddington Medal for his prescient theoretical work (Schutz 1986) that has now provided an entirely new method of calculating the Hubble constant from gravitational wave data.

- *Gravity: An Introduction to Einstein's Gravity*, by James B. Hartle.

  One of the very best, most *physical* (less mathematical) introductions to Einstein's theory, written by a master practitioner of the theory.

- *General Relativity*, by Robert M. Wald.

  I view this superb work as a significantly updated successor to Misner, Thorne, and Wheeler's *Gravitation*. It gracefully balances physical intuition with advanced mathematics, including topics that are rarely covered in a first course, such as Penrose's pioneering work on 2-spinors, and the singularity theorems of Penrose and Hawking.

- *General Relativity: A Geometric Approach*, by Malcolm Ludvigsen.

  If you are in a hurry to master General Relativity, you can hardly do better than this. It is a remarkably clean, elegant, geometrical, and *concise* presentation of the theory, but this is achieved (miraculously) without feeling rushed.

- *Einstein Gravity in a Nutshell*, by A. Zee.

  An excellent introduction to General Relativity, covering many unusual topics, written in a lively, informal, engaging, and highly opinionated style. (Are we related?)

Forms (in Mathematics)

- *A Geometric Approach to Differential Forms*, by David Bachman.
  One of a handful of brave books that seek to bring Forms to an *undergraduate* audience. And while it does include a very brief application to Differential Geometry, it does not include any discussion of the Riemann tensor or of General Relativity.

- *A Visual Introduction to Differential Forms and Calculus on Manifolds*, by Jon Pierre Fortney.
  The most recent (and longest) book to attempt to bring Forms to an *undergraduate* audience. While it does include a discussion of electromagnetism, it does not include any Differential Geometry, and thus does not include any discussion of the Riemann tensor or of General Relativity.

- *Differential Forms: A Heuristic Introduction*, by M. Schreiber.
  One of the earliest attempts to bring Forms to the masses. It contains some nice insights, but hardly any pictures! It does not include any discussion of the Riemann tensor or of General Relativity.

- *Differential Forms: Theory and Practice*, by Steven H. Weintraub.
  As the title indicates, this is an attempt to bring Forms into the advanced undergraduate multivariable calculus classroom. While perfectly clear, it is not geometric in flavour, and has very few pictures.

- *Vector Calculus, Linear Algebra, and Differential Forms: A Unified Approach* (4th edition), by John H. Hubbard and Barbara Burke Hubbard
  As with the previous book, this is an attempt to bring Forms into the advanced undergraduate multivariable calculus classroom, and here the approach is much more geometrical, rubbing shoulders with mine. However, it must be noted that Forms only appear for the first time on page 558.

- *Differential Forms and Applications*, by Manfredo P. do Carmo.
  At only 118 pages, this slim volume (by a very well-respected student of Chern) packs in a wealth of ideas and information, and some lovely exercises, some of which I have appropriated here (with attribution!).

Forms (in Physics)

- *Mathematical Methods of Classical Mechanics* (2nd edition), by V. I. Arnol'd.
  A masterpiece by one of the most brilliant mathematicians of the twentieth century. Chapter 7 is a self-contained, elegant, and insightful introduction to Forms, which are then applied to Hamiltonian mechanics.

- *Gauge Fields, Knots and Gravity*, by John Baez and Javier P. Muniain.
  A wide-ranging, highly original, modern introduction to electromagnetism, gauge fields, and gravity, most of which is expressed in the language of Forms. Amongst its *many* excellent

features is an insightful discussion of the role of Hodge duality in Maxwell's Equations. Do not overlook the Notes that conclude each of the three parts of the book: they contain annotated recommendations for further study (very much like this section!), and they also contain fascinating historical vignettes and pithy quotations. It is immensely refreshing and helpful that the authors speak to the reader directly, in a friendly, informal way, as if addressing an intelligent friend sitting right next to them, instead of proclaiming dry theorems into the void. (Of course this is precisely what I have attempted to do in this book, too!)

- *Differential Forms and the Geometry of General Relativity*, by Tevian Dray.

  The only introduction to General Relativity to exclusively employ Forms as its mathematical engine. Of the many features that recommend it, I would note that it includes a much more complete discussion of Hodge duality than I have provided.

- *The Geometry of Physics* (3rd edition), by Theodore Frankel.

  An impressive tome that treats almost *every* topic covered by *all* of the books above, combined, *and* has more beyond that! Forms are the primary language of the book. Of particular note, however, is its unique treatment of the *geometry* of the Einstein tensor.

- *Curvature in Mathematics and Physics*, by Shlomo Sternberg.

  Forms are the primary language of this book. As the title suggests, it contains *many* interesting applications to both mathematics and physics. In particular, it treats the following physical topics in depth: Hodge duality and electromagnetism, explicit calculations of the geometry and orbits of the Schwarzschild solution, and the geometry of the extremely important Kerr solution (representing a spinning black hole), though he stops shy of actually computing the curvature 2-forms. But this list hardly does justice to the wealth of material that is covered. WARNING: the author describes the book as suitable for advanced undergraduates—it is not! That said, if you have mastered my Act V, then you stand to learn a great deal from this book.

Bibliography

Aleksandrov, A. D.
1969. *Mathematics: Its Content, Methods, Meaning*, Vol. 2. Chapter 7: *Curves and Surfaces*, 57–117. A. D. Aleksandrov, A. N. Kolmogorov, M. A. Lavrent'ev, eds. Translated from the Russian by S. H. Gould. 2nd ed., paperback. Cambridge, Mass.: MIT.

Arianrhod, R.
2019. *Thomas Harriot: A Life in Science*, Oxford University Press.

Arnol'd, V. and V. Vasil'ev
1991. Newton's Principia read 300 years later. *Current Science*, 61(2).

Arnol'd, V. I.
1973. *Ordinary Differential Equations*. Cambridge, Mass.: MIT.

Arnol'd, V. I.
1989. *Mathematical Methods of Classical Mechanics*, Vol 60 of *Graduate Texts in Mathematics*, 2nd ed. New York: Springer.

Arnol'd, V. I.
1990. *Huygens and Barrow, Newton and Hooke*. Basel: Birkhäuser Verlag.
Pioneers in mathematical analysis and catastrophe theory from evolvents to quasicrystals, Translated from the Russian by Eric J. F. Primrose.

Bachman, D.
2012. *A Geometric Approach to Differential Forms*, 2nd ed. New York: Birkhäuser.

Baez, J. C. and J. P. Muniain
1994. *Gauge Fields, Knots, and Gravity*, Vol. 4. Singapore: World Scientific.

Banchoff, T.
1970. Critical points and curvature for embedded polyhedral surfaces. *American Mathematical Monthly*, 77(5):475–485.

Banchoff, T. and S. Lovett
2023. *Differential Geometry of Curves and Surfaces*, 3rd ed. Natick, Mass.: AK Peters.

Berger, M.
2003. *A Panoramic View of Riemannian Geometry*. Berlin: Springer.

Berger, M.
2010. *Geometry Revealed: A Jacob's Ladder to Modern Higher Geometry*. Berlin: Springer.

Berry, M.
1991. Bristol anholonomy calendar. In *Sir Charles Frank, OBE, FRS: An eightieth birthday tribute*. Bristol: A. Hilger.

Berry, M.
2017. *A Half-Century of Physical Asymptotics and Other Diversions: Selected Works by Michael Berry*. Singapore: World Scientific.

Berry, M. V.
1989. The quantum phase, five years after. In *Geometric Phases in Physics*, Vol. 5., A. Shapere and F. Wilczek, eds. Singapore: World Scientific.

Berry, M. V.
1990. Anticipations of the geometric phase. *Physics Today*, 43(12):34–40.

Berry, M. V.
2001. Geometry of phase and polarization singularities illustrated by edge diffraction and the tides. In *Second International Conference on Singular Optics (Optical Vortices): Fundamentals and Applications*, M. S. Soskin and M. V. Vasnetsov, eds. 4403:(1–12). Bellingham, Wash.: International Society for Optics and Photonics.

Berry, M. V.
2004. Index formulae for singular lines of polarization. *Journal of Optics A: Pure and Applied Optics*, 6(7):675–678.

Berry, M. V., M. R. Dennis, and R. L. Lee
2004. Polarization singularities in the clear sky. *New Journal of Physics*, 6:162–162.

Blaschke, W.
1929. *Vorlesungen über Differentialgeometrie und geometrische Grundlagen von Einsteins Relativitätstheorie* [2 volumes in one]. Chelsea Publishing.

Bloye, N. and S. Huggett
2011. Newton, the Geometer. *Newsletter of the European Mathematical Society*, (82): 19–27.

Boscain, U., L. Sacchelli, and M. Sigalotti
2016. Generic singularities of line fields on 2D manifolds. *arxiv.org: 1605.06295.*

Bottazzini, U.
1999. Ricci and Levi-Civita: From Differential Invariants to General Relativity. In *The Symbolic Universe: Geometry and Physics 1890-1930*, J. Gray, ed. Oxford: Oxford University.

Boy, W.
1903. Über die Curvatura integra und die Topologie geschlossener Flächen. *Mathematische Annalen*, 57(2):151–184.

Brackenridge, J. B. and M. Nauenberg
2002. Curvature in Newton's Dynamics. In *The Cambridge Companion to Newton*, I. B. Cohen and G. E. Smith, eds., Cambridge Companions to Philosophy, Chapter 3, pp. 85–137. Cambridge: Cambridge University.

Brannan, D. A., M. F. Esplen, and J. Gray
2012. *Geometry*, 2nd ed. Cambridge: Cambridge University.

Carathéodory, C.
1937. The most general transformations of plane regions which transform circles into circles. *Bulletin of the American Mathematical Society*, 43(8):573–579.

Cartan, E.
1927. *Riemannian Geometry in an Orthogonal Frame: from lectures delivered by Élie Cartan at the Sorbonne in 1926-1927.* [Translated into Russian in 1960 by S. P. Finikov, and from Russian into English by V. V. Golberg in 2001.] Singapore: World Scientific.

Cartan, E.
1928. *Geometry of Riemannian Spaces*, Volume 13. Brookline, Mass.: Math Sci Press. Translation of 2nd ed., 1951, published 1983.

Cartan, E.
1945. *Les Systèmes Differentiels Extérieure et leur Applications Géométriques.* Paris: Hermann.

Casey, J.
1996. *Exploring Curvature.* Braunschweig: F. Vieweg und Sohn.

Chandrasekhar, S.
1995. *Newton's Principia for the Common Reader.* Oxford: The Clarendon Press.

Chen, W., Y. Chen, and W. Liu
2019. Singularities and Poincaré indices of electromagnetic multipoles. *Phys. Rev. Lett.*, 122:153907.

Chen, W., Y. Chen, and W. Liu
2020a. Line singularities and Hopf indices of electromagnetic multipoles. *Laser & Photonics Reviews*, 2000049.

Chen, W., Q. Yang, Y. Chen, and W. Liu
2020b. Global Mie scattering: Polarization morphologies and the underlying topological invariant. *ACS Omega*, 5(23):14157–14163.

Chern, S.-S., W.-H. Chen, and K. S. Lam
1999. *Lectures on Differential Geometry*, Volume 1. Singapore: World Scientific.

Chinn, W. G. and N. E. Steenrod
1975. *First Concepts of Topology: The Geometry of Mappings of Segments, Curves, Circles, and Disks.* Washington, D.C.: Mathematical Association of America.

Coxeter, H.S.M.
1967. The Lorentz group and the group of homographies. In *Proc. Internat. Conf. Theory of Groups*, L. G. Kovacs and B. H. Neumann, eds., New York: Gordon and Breach.

Coxeter, H.S.M.
1969. *Introduction to Geometry*, 2nd ed. New York: Wiley.

Crowe, M. J.
1985. *A History of Vector Analysis: the Evolution of the Idea of a Vectorial System.* New York: Dover.

Crowley, D. and M. Grant
2017. The Poincaré–Hopf theorem for line fields revisited. *Journal of Geometry and Physics*, 117:187–196.

Darling, R.W.R.
1994. *Differential Forms and Connections.* Cambridge: Cambridge University.

Darrigol, O.
2015. The Mystery of Riemann's Curvature. *Historia Mathematica*, 42(1):47–83.

de Gandt, F.
1995. *Force and Geometry in Newton's Principia.* Princeton, N.J.: Princeton University. [Translated from the French original and with an introduction by Curtis Wilson.]

Diacu, F. and P. Holmes
1996. *Celestial Encounters: the Origins of Chaos and Stability.* Princeton, N.J.: Princeton University.

Dirac, P.A.M.
1930. *The Principles of Quantum Mechanics* [The International Series Of Monographs On Physics]. Oxford: The Clarendon Press.

do Carmo, M. P.
1992. *Riemannian Geometry.* Basel: Birkhäuser.

do Carmo, M. P.
1994. *Differential Forms and Applications*, Universitext. Berlin: Springer.

Dombrowski, P.
1979. *150 Years after Gauss' "Disquisitiones generales circa superficies curvas,"* Vol. 62 of *Astérisque.* Paris: Société Mathématique de France.
With the original text of Gauss.

Dray, T.
2015. *Differential Forms and the Geometry of General Relativity.* Boca Raton: CRC Press.

Earl, R.
2019. *Topology: A Very Short Introduction.* Oxford: Oxford University.

Feynman, R. P.
1985. *QED: the Strange Theory of Light and Matter.* Princeton, N.J.: Princeton University.

Feynman, R. P., R. B. Leighton, and M. L. Sands
1963. *The Feynman Lectures on Physics.* Reading, Mass.: Addison-Wesley.

Firby, P. and C. F. Gardiner
2001. *Surface Topology*, 3rd ed.: Cambridge: Woodhead.

Flanders, H.
1989. *Differential Forms with Applications to the Physical Sciences.* Mineola, N.Y.: Dover.

Forbes, N. and B. Mahon
2014. *Faraday, Maxwell, and the Electromagnetic Field: How Two Men Revolutionized Physics.* Amherst, N.Y.: Prometheus.

Fortney, J. P.
2018. *A Visual Introduction to Differential Forms and Calculus on Manifolds.* Basel: Birkhäuser.

Frank, F. C.
1958. I. Liquid crystals: On the theory of liquid crystals. *Discuss. Faraday Soc.*, 25:19–28.

Frankel, T.
2011. *Gravitational Curvature: An Introduction to Einstein's Theory.* Mineola, N.Y.: Dover.

Frankel, T.
2012. *The Geometry of Physics: An Introduction*, 3rd ed. Cambridge: Cambridge University.
Fricke, R.
1926. *Lehrbuch der Algebra*, Vol. 2. Braunschweig: F. Vieweg und Sohn.
Gauss, C. F.
1827. *General Investigations of Curved Surfaces*. [Translated from the Latin and German by Adam Hiltebeitel and James Morehead, 1965 edition.] Hewlett, N.Y.: Raven Press.
Gauß, C. F.
1973. *Werke. Band VIII*. Hildesheim: Georg Olms Verlag. [Reprint of the 1900 original.]
Gindikin, S.
2007. *Tales of Mathematicians and Physicists*, 2nd English ed. Berlin: Springer.
Goodstein, D. L. and J. R. Goodstein
1996. *Feynman's Lost Lecture: The Motion of Planets Around the Sun*, 1st ed. New York: Norton.
Goodstein, J. R.
2018. *Einstein's Italian Mathematicians: Ricci, Levi-Civita, and the Birth of General Relativity*. Providence, R.I.: American Mathematical Society.
Gray, J.
1989. *Ideas of Space: Euclidean, Non-Euclidean, and Relativistic*, Oxford: Clarendon Press.
Greenberg, M. J.
2008. *Euclidean and Non-Euclidean Geometries*, 4th ed. New York: W.H. Freeman.
Guicciardini, N.
1999. *Reading the Principia*. Cambridge: Cambridge University.
The debate on Newton's mathematical methods for natural philosophy from 1687 to 1736.
Guicciardini, N.
2009. *Isaac Newton on Mathematical Certainty and Method* [Transformations: Studies in the History of Science and Technology]. Cambridge, Mass.: MIT.
Guillemin, V. and A. Pollack
1974. *Differential Topology*. Englewood Cliffs, N.J.: Prentice Hall.
Hankins, T. L.
1970. *Jean d'Alembert: Science and the Enlightenment*. Oxford: Clarendon Press.
Hartle, J. B.
2021. *Gravity: An Introduction to Einstein's Gravity*. Cambridge: Cambridge University. [Reissue of the 2002 original.]
Hayes, D. F., T. Shubin, G. L. Alexanderson, and P. Ross
2004. *Mathematical Adventures for Students and Amateurs*. Washington, D.C.: Mathematical Association of America.
Henderson, D. W.
2013. *Differential Geometry: A Geometric Introduction*, 3rd ed. Project Euclid.
Henderson, D. W. and D. Taimina
2005. *Experiencing Geometry: Euclidean and non-Euclidean with History*, 3rd ed. Upper Saddle River, N.J.: Pearson Prentice Hall.
Hilbert, D.
1902. *The Foundations of Geometry*. [2nd English ed. of 10th German ed. Translated by Leo Unger. Revised and enlarged by Dr. Paul Bernays, published 1971.] Chicago: The Open Court.
Hilbert, D. and S. Cohn-Vossen
1952. *Geometry and the Imagination*. New York: Chelsea.
Hirsch, M. W.
1976. *Differential Topology*, Vol. 33. Berlin: Springer.
Hopf, H.
1935. Über die drehung der tangenten und sehnen ebener kurven. *Compositio Mathematica*, 2:50–62.
Hopf, H.
1946–1956. *Differential Geometry In The Large: Seminar Lectures, New York University, 1946 And Stanford University*, 1956, 2nd ed., 1983, Vol. 1000 of *Lecture Notes In Mathematics*. Berlin: Springer.
Hubbard, J. and B. Hubbard
2009. *Vector Calculus, Linear Algebra, and Differential Forms*, 4th ed. Ithaca, N.Y.: Matrix Editions.
James, I. M.
1999. *History of Topology*, 1st ed. Amsterdam: Elsevier.
Jones, B.
1870. *The Life and Letters of Faraday*. London: Longmans, Green and Co.
Katz, V.
1979. The History of Stokes's Theorem. *Mathematics Magazine*, 52(3):146–156.
Kerr, R. P.
2008. Discovering the Kerr and Kerr-Schild metrics. *arXiv:0706.1109v2*.
Klein, F.
1928. *Development of Mathematics in the 19th Century* [English translation by M. Ackerman of the German original, with an extensive appendix by R. Hermann, published 1979]. Vol. 9 of *Lie groups*. Brookline, Mass.: Math Sci Press.
Knoebel, A.
2007. *Mathematical Masterpieces: Further Chronicles by the Explorers*, Undergraduate texts mathematics. Readings in mathematics. Berlin: Springer.
Koenderink, J. J.
1990. *Solid Shape*, Artificial intelligence. Cambridge, Mass.: MIT.
Kolmogorov, A. N. and A. P. Yushkevich
1996. *Mathematics of the 19th Century: Geometry, Analytic Function Theory*. Basel: Birkhäuser Verlag.
Kühnel, W.
2015. *Differential Geometry: Curves - Surfaces - Manifolds*, Vol. 77, 3rd ed. Providence, R.I.: American Mathematical Society.

Lanczos, C.
1970. *The Variational Principles of Mechanics*, 4th ed. Mineola, N.Y.: Dover.

Laugwitz, D.
1999. *Bernhard Riemann, 1826-1866: Turning Points In The Conception Of Mathematics*. Boston: Birkhäuser.

Levi-Civita, T.
1917. Nozione di parallelismo in una varietà qualunque e conseguente specificazione geometrica della curvatura riemanniana. In *Opere Matematiche: Memorie e Note, Pubblicate a cura dell'Accademia Nazionale dei Lincei*, Vol. IV (1917 to 1928), published 1960, pp. 1–39. Bologna: N. Zanichelli.

Levi-Civita, T.
1926. *The Absolute Differential Calculus: (Calculus of Tensors)*. Mineola, N.Y.: Dover, 1977.

Levi-Civita, T.
1931. Calcolo differenzale assoluto. In *Opere Matematiche: Memorie e Note, Pubblicate a cura dell'Accademia Nazionale dei Lincei*, Vol. V (1929 to 1937), published 1970, pp. 219–229. Bologna: N. Zanichelli.

Lightman, A. P., W. H. Press, R. H. Price, and S. A. Teukolsky
1975. *Problem Book in Relativity and Gravitation*. Princeton, N.J.: Princeton University.

Ludvigsen, M.
1999. *General Relativity: A Geometric Approach*. Cambridge: Cambridge University.

Mahon, B.
2017. *The Forgotten Genius of Oliver Heaviside: A Maverick of Electrical Science*. Buffalo, N.Y.: Prometheus.

Mahoney, M. S.
1994. *The Mathematical Career of Pierre de Fermat, 1601–1665*, 2nd ed. Princeton, N.J.: Princeton University.

Maxwell, J. C.
2002. *The Scientific Letters and Papers of James Clerk Maxwell*. Cambridge: Cambridge University.

Maxwell, J. C.
2003. *On the Transformation of Surfaces by Bending. The Scientific Papers Of James Clerk Maxwell*, Vol. 1, pp. 80–114. Mineola, N.Y.: Dover.

Mishchenko, A. S. and A. T. Fomenko
1988. *A Course of Differential Geometry and Topology*. Moscow: Mir Publishers.

Misner, C. W., K. S. Thorne, and J. A. Wheeler
1973. *Gravitation*. San Francisco: W. H. Freeman.
A marvellous new hardback edition was published by Princeton University Press in 2017. It contains a new introduction by Charles Misner and Kip Thorne, discussing exciting developments in the field since the book's original publication.

Monastyrskiĭ, M. I.
1999. *Riemann, Topology, and Physics*, 2nd ed. Boston: Birkhäuser.

Morgan, F.
1998. *Riemannian Geometry: A Beginner's Guide*, 2nd ed. Wellesley, Mass.: AK Peters.

Needham, T.
1993. Newton and the transmutation of force. *Amer. Math. Monthly*, 100(2):119–137.

Needham, T.
2014. Visual Differential Geometry and Beltrami's Hyperbolic Plane. In *The Art of Science: From Perspective Drawing to Quantum Randomness*, R. Lupacchini and A. Angelini, eds., pp. 71–99. Berlin: Springer.

Needham, T.
2023. *Visual Complex Analysis: 25th Anniversary Edition*. Oxford: The Clarendon Press.

Newton, I.
1687. *The Principia: Mathematical Principles of Natural Philosophy*, 1999 edition. Berkeley, Calif.: University of California.
A new translation by I. Bernard Cohen and Anne Whitman, assisted by Julia Budenz, Preceded by "A guide to Newton's *Principia*" by Cohen.

Nye, J.
1999. *Natural Focusing and Fine Structure of Light: Caustics and Wave Dislocations*. Boca Raton: Taylor & Francis.

O'Neill, B.
1983. *Semi-Riemannian Geometry With Applications to Relativity*. Cambridge, Mass.: Academic Press.

O'Neill, B.
2006. *Elementary Differential Geometry*, Revised 2nd ed. Amsterdam: Elsevier.

Pais, A.
1982. *"Subtle is the Lord...": The Science and the Life of Albert Einstein*. Oxford: Oxford University.

Penrose, R.
1978. The geometry of the universe. In *Mathematics Today: Twelve Informal Essays*, L. A. Steen, ed., pp. 83–127. New York: Vintage Books.

Penrose, R.
1979. The topology of ridge systems. *Annals of Human Genetics*, 42(4):435–444. [Also reprinted in his *Collected Works*, Oxford University, 2010: Vol. 3, pp. 347–357.]

Penrose, R.
2005. *The Road to Reality*. New York: Alfred A. Knopf. A complete guide to the laws of the universe.

Penrose, R. and W. Rindler
1984. *Spinors and Space-Time*. Cambridge: Cambridge University.

Phillips, E. R.
1979. Karl M. Peterson: The earliest derivation of the Mainardi–Codazzi equations and the fundamental theorem of surface theory. *Historia Mathematica*, 6(2):137–163.

Poincaré, H.
1892, 1893, 1899. *New Methods of Celestial Mechanics*, Vol. 13. American Institute of Physics, published 1993.
Pólya, G.
1954. *Mathematics and Plausible Reasoning*. Princeton, N.J.: Princeton University.
Prasolov, V. V.
1995. *Intuitive Topology*, Vol. 4. Providence: American Mathematical Society.
Pressley, A.
2010. *Elementary Differential Geometry* [Springer Undergraduate Mathematics Series], 2nd ed. Berlin: Springer.
Richeson, D. S.
2008. *Euler's Gem: The Polyhedron Formula and the Birth of Topology*. Princeton, N.J.: Princeton University.
Rosenfeld, B. A.
1988. *A History of Non-Euclidean Geometry: Evolution of the Concept of a Geometric Space*, Vol 12. Berlin: Springer.
Schreiber, M.
1977. *Differential Forms: A Heuristic Introduction*, Universitext. Berlin: Springer.
Schutz, B. F.
1980. *Geometrical Methods of Mathematical Physics*. Cambridge: Cambridge University.
Schutz, B. F.
1986. Determining the Hubble constant from gravitational wave observations. *Nature*, 323(6086):310–311.
Schutz, B. F.
2003. *Gravity from the Ground Up*. Cambridge: Cambridge University.
Schutz, B. F.
2022. *A First Course in General Relativity*, 3rd ed. Cambridge: Cambridge University.
Segerman, H.
2016. *Visualizing Mathematics with 3D Printing*. Baltimore: Johns Hopkins.
Shapere, A. and F. Wilczek
1989. *Geometric Phases in Physics*, Vol. 5. Singapore: World Scientific.
Shenitzer, A. and J. Stillwell
2002. *Mathematical Evolutions*. Washington, D.C.: Mathematical Association of America.
Spivak, M.
1999. *A Comprehensive Introduction to Differential Geometry*, Vol. 2, 3rd ed. Houston: Publish or Perish.
Stahl, S.
1993. *The Poincaré Half-Plane: A Gateway to Modern Geometry*. Boston: Jones and Bartlett.
Sternberg, S.
2012. *Curvature in Mathematics and Physics*, Dover books on mathematics. Dover.
Stewart, G. W.
1993. On the early history of the singular value decomposition. *SIAM Review*, 35(4):551–566.
Stillwell, J.
1995. *Classical Topology and Combinatorial Group Theory*, Vol. 72, 2nd ed. Berlin: Springer.
Stillwell, J.
1996. *Sources of Hyperbolic Geometry*, Vol. 10 of *History of Mathematics*. Providence: American Mathematical Society.
Stillwell, J.
2005. *The Four Pillars of Geometry*. Berlin: Springer.
Stillwell, J.
2010. *Mathematics and its History*, 3d ed. Berlin: Springer.
Stoker, J. J.
1969. *Differential Geometry*, Vol. 20 of *Pure and Applied Mathematics*. New York: Wiley-Interscience.
Tapp, K.
2016. *Differential Geometry of Curves and Surfaces*. Cham, Switzerland: Springer Cham.
Taylor, E. F. and J. A. Wheeler
1992. *Spacetime Physics: Introduction to Special Relativity*, 2nd ed. New York: W.H. Freeman.
Taylor, E. F. and J. A. Wheeler
2000. *Exploring Black Holes: Introduction to General Relativity*. San Francisco: Addison Wesley Longman.
Thorne, K. S. and R. D. Blandford
2017. *Modern Classical Physics: Optics, Fluids, Plasmas, Elasticity, Relativity, and Statistical Physics*. Princeton, N.J.: Princeton University.
Thurston, W. P.
1997. *Three-Dimensional Geometry and Topology*, Vol. 35. Princeton, N.J.: Princeton University.
Tu, L. W.
2017. *Differential Geometry: Connections, Curvature, and Characteristic Classes*. Berlin: Springer.
Wald, R. M.
1984. *General Relativity*. Chicago: University of Chicago.
Watson, G. N.
1917. A problem of analysis situs. *Proceedings of the London Mathematical Society*, 2(1): 227–242.
Weeks, J. R.
2020. *The Shape of Space*, 3rd ed. Boca Raton: CRC Press.
Weintraub, S. H.
2014. *Differential Forms: Theory and Practice*. San Diego: Academic Press.
Westfall, R. S.
1980. *Never at Rest: A biography of Isaac Newton*. Cambridge: Cambridge University.
Wheeler, J. A.
1990. *A Journey into Gravity and Spacetime*. New York: W.H. Freeman and Co. (Scientific American).

Wilson, P.M.H.
2008. *Curved Spaces: From Classical Geometries to Elementary Differential Geometry.* Cambridge: Cambridge University.

Wiltshire, D. L., M. Visser, and S. Scott
2009. *The Kerr Spacetime: Rotating Black Holes in General Relativity.* Cambridge: Cambridge University.

Woodward, L. M. and J. Bolton
2019. *A First Course in Differential Geometry: Surfaces in Euclidean Space.* Cambridge: Cambridge University.

Zee, A.
2013. *Einstein Gravity in a Nutshell.* Princeton, N.J.: Princeton University.

Index

aberration formula, 93
Absolute Differential Calculus, 231
amplification, 43, 156
amplitwist, 73, 85, 156, 417
  amplification of, 43
  Cauchy's Theorem, 417
  Cauchy–Riemann equations, 397
  definition of, 43
  of z^2, 43
  of z^m, 44, 90
  twist of, 43
analytic mapping, 42
angle of parallelism, 60
angular excess
  definition of, 8
  equivalence to holonomy, 247
  in a 3-manifold, 282
  is additive, 23
  of polygon, 19, 26, 83
  of quadrilateral, 26
  of spherical polygon, 147
  proportional to area, 15
angular momentum, 128
  and geodesics in $\mathbb{H}^2$, 129
anti-self-dual 2-form, 470
anticonformal
  definition of, 39
antipodal points, 49, 83
antipodal triangle, 8
Apollo 15, 308
Apollonius, xviii
Archimedes, 86
Archimedes–Lambert projection, 85
area of surface element, 38
Arnol'd
  and Newton's *Principia*, xviii
  as champion of Newton's geometrical method, xx
  Newtonian problem, 24
asymptotic, 60
asymptotic direction, 162
  at parabolic point, 164
  effect of Shape Operator on, 163
  generalized definition, 164
Atiyah
  The Faustian Offer, xvii
  Atiyah–Singer Index Theorem, 165
Attitude Matrix, 435–438

Bach, BWV 101, xxv
Baker, Michael, xxvii
banana skin
  spherical image under deformation, 139
  Theorema Egregium illustrated with, 140
banana, fried
  Honey-Flow on, 202
  topographic map of, 203
Banchoff, Thomas
  and *Theorema Egregium*, 252
  and Polyhedral *Theorema Egregium*, 148
  author's debt to, xxv
  Global Gauss–Bonnet Theorem via pancakes, 171
Barish, Barry C., 231
Beautiful Theorem, 139
  geometric proof of, 255
  predates *Theorema Egregium* by eleven years, 253
Beethoven
  death of, 138
  Grosse Fuge, xxv
Beltrami
  conformal metric of $\mathbb{H}^3$ and $\mathbb{H}^n$, 79
  curvature at hyperbolic point, 164
  discovered Beltrami–Poincaré disc model, 62
  discovered Beltrami–Poincaré half-plane, 57
  discovered SVD, 154
  discovery of concrete interpretation of Hyperbolic Geometry, 5, 51
  explanation of horosphere, 82
  portrait of, 51
  pseudospherical surfaces, 22
Beltrami–Laplace equation, 84
Beltrami–Poincaré disc, 63
  apparent size of objects within, 63
  conformal mapping to half-plane, 70
  curvature of, 90
  geodesics of, 62
  mapping to half-plane, 90
  metric of, 62, 90
Beltrami–Poincaré half-plane, 56
  apparent size of objects within, 57
  conformal mapping to disc, 70
  connection forms of, 472
  discovery by Beltrami, 57
  geodesics of, 60
  mapping to disc, 90
  semi-circular geodesics within, 57
Berger, Marcel, 131
  on known proofs of *Theorema Egregium*, 252
  on new proof of *Theorema Egregium*, 252
Bernoulli, Johann
  Brachistochrone Problem, 58
  geodesics, 119
Berry phase, 246
Berry, Sir Michael, 246, 478
Bianchi, 298
Bianchi Identities, 295
  Algebraic (First), 295, 337
  Differential (Second), 295, 298, 327, 332, 340, 413
  via Forms, 459–460
binormal, 107
  only spins, 219
black hole, 329
  as solution of Einstein's Vacuum Field Equation, 460–464
  Birkhoff's Theorem, 329
  Cosmic Censorship Hypothesis, 464
  curvature of, 460–464
  event horizon, 329, 330, 464
  formation of, 329–331
  Kerr solution, 322
  maximum survival time within, 331
  Sagittarius A*, 331, 464
  Schwarzschild radius, 321, 329
  Schwarzschild Solution, 320
  spacetime depiction of birth, 330
  spacetime singularity, 330
  spaghettification, 330
  supermassive, 331
  Tolman–Oppenheimer–Volkoff limit, 329
  versus naked singularity, 464
Blaschke
  metric notation, 37
Bolyai
  co-discovered horosphere, 82
  discovery of Hyperbolic Geometry, 4
  initally sought 3-dimensional non-Euclidean Geometry, 79
  treatment of by Gauss, 7
Bolyai–Lobachevsky Formula, 61, 62
Bonnet
  Local Gauss–Bonnet Theorem, 174
  unaware of Global Gauss–Bonnet Theorem, 174
Bowler, Michael G., xxvii
Boy, Werner, 174
Brachistochrone Problem, 58
Brahe, Tycho, 123
Bramson, Brian D., xxvii
branch points, 170
bridge (between bagels), 171
Brouwer degree, 173
Brouwer, L.E.J., 173
Burnett-Stuart, George, xxiii, xxvii

Carathéodory, 81
Cartan
  "debauch of indices", xvii
  Differential Geometry via Forms, 430–474
  discovery of Forms, xxii
  portrait of, 345

Cartan's Razor, 462
Cartan's Structural Equations, 438–445
  First Equation, 439
  Second Equation, 440, 456
catenoid, 480
Catmull, Edwin, xxvi
Cauchy
  polyhedra as hollow surfaces, 186
Cauchy's Theorem, 398, 417
Cauchy–Riemann equations, 85, 397
Celestial Mechanics, 210
celestial sphere, 44, 77
Central Cylindrical Projection, 86
central force field, 123
Central Projection, 32
centre of curvature, 98
Chandrasekhar
  and $\asymp$-notation, xviii
  and Newton's *Principia*, xviii
Chern, 221, 252, 435, 481
  Characteristic Classes, 481
Cicero, 86
circle of curvature, 98
circulation, 404
circulation of vector field, 263
Clairaut's Theorem
  angular momentum explanation of, 128
  dynamical proof of, 126–128
  experimental investigation of, 27
  on the sphere, 121
Clairaut's theorem, 122
Clairaut, Alexis Claude, 123
CMC surfaces, 480
Codazzi–Mainardi Equations, *see* Peterson–Mainardi–Codazzi Equations
commutator, 287–288, 407, 409
complex inversion, 67
  is rotation of Riemann sphere, 69
complex mapping, 41
  amplitwist of, 43
  analytic, 42
  anticonformality, 398
  Cauchy's Theorem, 398, 417
  Cauchy–Riemann equations, 398
  conformality, 41
  Pólya vector field of, 201
cone
  curvature of, 134, 142
  geodesics on, 27
  spike of, 143
conformal
  amplitwist, 43
  Cauchy–Riemann Equations, 397
  coordinates, 40, 84
  curvature formula, 41, 89
  definition of, 39
  map, 39
  map of hyperbolic plane, 56
  map of pseudosphere, 54–55
  mapping between hyperbolic models, 63
  mapping of $\mathbb{C}$, 41
  Mercator projection of sphere, 86
  metric, 39
  property of stereographic projection, 45
  stereographic metric of sphere, 47
conformal disc model, 63
conjugate point, 269, 272
connection 1-forms, 432–435
  definition of, 433
  differing conventions, 433
conservation of energy, 327, 340, 406
contraction:
  of 1-form and vector, 347
  of tensors, 365–366
convex, 186
coordinates
  breakdown of, 37
  conformal, 40, 84
  homogeneous, 70
  isothermal, 40, 84
  orthogonal, 37
  projective, 70
Cosmic Censorship Hypothesis, 464
cosmological constant, Λ, 332
Costa, Celso, 480
Coxeter, 63
crosspoint, 196
curl (of vector field), 264, 398
Curvatura Integra, 165
curvature
  and exponential operator, 339
  as angular excess per unit area, 18
  as attractive force, 272
  as rate of turning, 101
  as repulsive force, 272, 274
  as spreading of normals, 131, 132
  as stiffness of spring, 272
  Cartesian formula, 222
  centre of, 98
  circle of, 98
  conformal formula, 41, 89, 472
  controls area of small circle, 20, 278–279
  controls circumference of small circle, 20, 28, 278–279
  definition for plane curve, 98
  dependence of sign on shape, 19
  Euclidean Geometry *characterized* by vanishing of, 472
  Euler's Formula for surfaces, 110
  extrinsic equals intrinsic, 142
  extrinsic formula for, 134
  Gaussian, 17
  geometric formula for, 99
  Hilbert's Theorem on surfaces of constant negative, 22, 52
  holonomy as measure of, 246
  intrinsic, 17
  mean, 110, 130, 222
  metric formula, 38
  metric formula (forms proof of), 451–452
  metric formula (geometric proof of), 261–268
  Minding's Theorem on surfaces of constant, 21
  Newton's formula for, 100
  nonisometric surfaces with equal curvature, 224
  normal, 115
  of n-saddle, 224
  of black hole, 460–464
  of cone, 134, 142
  of conical spike, 143–145
  of cosine graph, 99
  of cylinder, 132, 134, 142
  of hyperbolic point, 164
  of monkey saddle, 224
  of parabola, 99
  of polyhedral spike, 145–147
  of polyhedral spike, defined, 147
  of pseudosphere, 53, 142
  of saddle, 224
  of sphere, 132, 134, 142
  of spike, defined, 144
  of surface of revolution, 114, 142
  of the Beltrami–Poincaré disc, 90
  of the Beltrami–Poincaré half-plane, 251
  of the Universe, 474
  of torus, 21, 90, 142
  of tractrix, 105
  parametric formula, 103
  polar formula, 221, 224
  principal, 109
  radius of, 98
  related to acceleration, 102
  related to force, 98
  Rodrigues's extrinsic formula for, 134
  sign defined by orientation, 135
  sign for plane curves, 99
  simplified formula for unit speed orbit, 104
  surface of revolution, 220, 221
  surfaces of constant, 89
  surfaces of constant negative, 22, 52
  surfaces of constant positive, 21
  Theorema Egregium, *see Theorema Egregium*
  total, 165
  umbilic point, 110
  via Shape Operator, 222
curvature vector, 107
  geodesic curvature component, 117
  normal curvature component, 117
cylinder
  curvature of, 134, 142
  extrinsic curvature, 132

Darboux vector, 219
Darboux, Jean-Gaston, 219, 430
dark energy, 333
de Rham cohomology, 397, 419–429
  Closed 1-Form Deformation Theorem, 423
  Closed 2-Form Deformation Theorem, 427
  cohomologous 1-forms, 423
  cohomology class, 423
  first de Rham cohomology group, 423
  first de Rham cohomology group of torus, 428–429
  inverse-square point source, 424–426
  period of 1-form, 423

period of 2-form, 427
second de Rham cohomology group, 426
topological circulation, 421–423
vortex 1-form, 419–421
Dedekind, 298
degree (of mapping), 173, 177–180, 182
Descartes
anticipation of Euler's Polyhedral Formula, 184
determinant, 153
coordinate-independence of, 153
of Shape Operator, 153
Differential Equations
integral curves, 211
Poincaré's Qualitative Theory of, 211
Differential Forms, *see* Forms
Differential Geometry via Forms
1-forms in terms of basis, 447
2-forms in terms of basis, 448
adapted frame field, 446
adapted frame field of sphere, 447
area 2-form, 447
attitude matrix, 435–438
Bianchi Identities, 459–460
black hole curvature, 460–464
Cartan's First Structural Equation, 439, 458
Cartan's Method of Moving Frames, 430–432
Cartan's Razor, 462
Cartan's Second Structural Equation, 440
Cartan's Second Structural Equation (generalized), 456
Cartan's Two Structural Equations, 438–445
Cartan, Élie, 430
Codazzi Equations in geometric form, 450
connection 1-forms, 432–435
differing conventions, 433
connection equations, 434
curvature 2-forms, 455–460
curvature 2-forms (defined), 456
curvature 2-forms (notation), 456
curvature 2-forms (of 2-surface), 473
curvature matrix (of curvature 2-forms), 456
cylindrical frame field, 436–438
Darboux, Jean-Gaston, 430
dual 1-form basis, 438, 439, 430–474
extra structure of orthonormal frame, 431
extrinsic curvature 2-form, 446, 448
First Structural Equations of 2-surface, 448
Frenet Frame, 430
Frenet–Serret Equations, 430
Gauss Equation in geometric form, 450
Gauss Equation of 2-surface, 448
generalized exterior derivative, 457
Hilbert's Lemma, 453–454
Hilbert's Lemma curvature formula, 452
Liebmann's Rigid Sphere Theorem, 454–455
moving frame field, 432
O'Neill championed Forms approach in 1966, 431
Peterson–Mainardi–Codazzi Equations of 2-surface, 448
principal frame field, 449
Riemann tensor via curvature 2-forms, 457, 459
Shape Operator, 446
Shape Operator symmetry, 449
six Fundamental Form equations of a 2-surface, 448
spherical frame field, 441–445
Symmetry Equation (meaning of), 449
Symmetry Equation of 2-surface, 448
dipole, 196, 199
streamlines, 227
Dirac bra-ket, 352
directional derivative, 149
meaning of, 151
discontinuity (of vector field), 195
dodecahedron, 185
tesselation of sphere, 26
Dombrowski, Peter, 138, 252
dual (of vectors), 347
dual polyhedra, 227
Dupin indicatrix, 111, 162, 204
elliptic, 111
hyperbolic, 111
parabolic, 111
Dupin, Charles, 110, 111
Durian, 143
Dyck, Walther, 174, 215

Eddington, Sir Arthur, 322, 325
Einstein
1905 discovery of Special Relativity, 382
1915 discovery of General Relativity, 327
1916 prediction of gravitational waves, 323
1917 introduction of cosmological constant, Λ, 332
aberration formula, 93
and precession of orbit of Mercury, 322
curved spacetime, 82
discovered General Relativity *without* parallel transport, 232
electromagnetic motivation of relativity, 382
final Gravitational Field Equation, 327
first attempt at Gravitational Field Equation, 327
first three predictions of General Relativity, 322
geometric form of field equation, 328
Greatest Blunder, 332
Happiest Thought, 307
ignorance of Bianchi Identity in 1915, 327
need for new theory of gravity, 231
portrait of, 320
prediction of bending of light, 322
Special Theory of Relativity, 74, 75
tensor, 328
Vacuum Field Equation, 319
Einstein summation convention, 292
Einstein tensor, 328
conservation of, 340
Electromagnetism
1-form potential, 402
conservation of electric charge, 470
electric 1-form, 383
electric 2-form, 383
electric field, 383
Faraday 2-form, 383, 402, 469
Faraday's Law of Electromagnetic Induction, 415
Heaviside's simplification of Maxwell's Equations, 382
Hodge duality, 385, 402
magnetic 1-form, 383
magnetic 2-form, 383
magnetic field, 383
Maxwell 2-form, 384, 402, 469
Maxwell's Equations (1873 publication), 382
Maxwell's Equations (source-free), 384
Maxwell's Equations (via Forms), 401–403
Maxwell's Equations (with sources), 384
photons (spinning), 470
self-duality, 470
spacetime source of, 402, 469
vector potential, 402
elliptic point, 111, 134, 136
energy–momentum tensor, 326
Enneper, 164
Equation of Geodesic Deviation, 302
Escher, M. C., 63
Euclid's *Elements*, 4, 185
Euler characteristic
adding handle reduces by two, 191
definition of, 167, 184
determines degree of spherical map, 174
Hopf's calculation of, 227
of polygonal net, 187
related to degree, 217
vanishes for hollow handle, 192
Euler's Polyhedral Formula, 184
Cauchy's proof, 186–188
Euler's own proof, 184
Legendre's proof, 188–190, 226
Platonic solids for the, 185
Euler's Relation (for homogeneous functions), 471

Euler, Leonhard
breakthrough on curvature of surfaces, 109
curvature formula, 110
curvature formula (original form), 110
description of polyhedra, 183
generosity to young Lagrange, 7
geodesics, 119
impossibility of perfect map, 31
portrait of, 183
rescued Lambert's career, 8
rigid motions of the sphere, 73
Euler–L'Huilier Formula, 190
proof of, 209
event (in spacetime), 74
event horizon, 329, 330, 464
extended complex plane, 44
exterior derivative, *see* Forms
exterior product, 372
extrinsic geometry
definition of, 11

faithfullness of map, 33
Faltings, Gerd, 24
Faraday
2-form, 383
admiration of Maxwell, 382
first electric generator, 381
introduced flux concept, 377
laboratory, 381
portrait of, 381
Fermat, 58
Fermat's Last Theorem, 6, 58
Fermat's Principle, 58
Feynman
Newtonian derivation of Snell's Law, 58
on Newton, 126
quantum-mechanical explanation of Fermat's Principle, 58
fiducial vector field, 205
field of line elements, 211
fixed points, 77
flux
as 2-form, 378, 408
associated vector, 377
definition of, 378
introduced by Faraday, 377
refined by Maxwell, 377
Focus, 196, 200
folding paper, 221
Forms
1-forms
"contravariant" explained, 465
"covariant" explained, 465
as duals of vectors, 347
basis, 352–354
Cartesian basis, 356–357
components of, 354
connection 1-forms, 432–435
contraction of, 347
definition of, 346–347
Dirac bras as, 352
Dirac delta function, 465
direction (sense, orientation), 349
dual basis, 352
electromagnetic potential, 402
examples of, 347–352
field of, 347
geometrical addition of, 357–359
gradient, 350, 351
gradient (defined), 355
gravitational work as, 347–349
interpretation of df, 357
kernel of, 349
notation, 346
row vectors as, 352
stack, 349
topographic map, 350
visualized, 349
2-forms
anti-self-dual, 470
area 2-form, 371, 373
area 2-form (polar), 374–375
area formula for surface, $f = const.$, 467
as flux, 378, 398
associated vector (definition of), 377
basis, 375–376
basis (as area projection), 376
basis (geometric meaning), 377
definition of, 370
factorizes in $\mathbb{R}^3$, 466
Faraday, 383, 402
flux, 408
flux as vector, 376–378
Maxwell, 402
need not factorize in $\mathbb{R}^4$, 466
self-dual, 470
symplectic manifolds, 370
vector product as wedge product, 379–381
via wedge product, 372–374
3-forms
basis, 390
factorizes in $\mathbb{R}^4$, 467
need at least three dimensions, 386
volume (Cartesian), 387
volume (spherical polar), 389
4-forms
volume, 391
Bianchi Identities, 459–460
Cauchy's Theorem, 398, 417
Cauchy–Riemann Equations, 397
closed, 419–429, 467, 471
closed (definition), 396
de Rham cohomology, 397, 419–429
discovery by Cartan, 345
exact, 419–429, 467, 471
exact (definition), 396
exterior derivative
generalized to vectors, 457–460
exterior derivative ($\mathbf{d}$)
as integral, 406–411
closed Forms (definition), 396
exact Forms (definition), 396
explanation of $\mathbf{d}^2 = 0$, 413
Fundamental result: $\mathbf{d}^2 = 0$, 395–396
Leibniz Rule (Product Rule) for 0-forms, 356
Leibniz Rule (Product Rule) for p-forms, 394–395
of 0-form, 355
of 1-form, 392–394
of 2-form, 394
of p-form, 394
Poincaré Lemma, 396, 418
exterior product, 372
gauge freedom, 396
gauge transformation, 396
Maxwell's Equations, 401–403
p-forms
definition of, 370
need at least p dimensions, 386
Poincaré Lemma, 396, 418
potential, 396
Vector Calculus
curl, 398
divergence, 399
flux 2-form, 398
Gibbs and Heaviside, 274
identities via Forms, 400, 468
integral theorems, 413–415
irrotational, 399
via Forms, 398–400
vector potential, 402
wedge product, 372–374
as vector product, 379–381
definition of, 372
geometry of, 374
of 2-form and 1-form, 387
of three 1-forms, 390
vector product formula, 380
volume formula, 381
Frenet Approximation, 220
Frenet frame, 106
Frenet–Serret Equations, 108, 220, 430, 434
variable speed, 219
Friedmann, Alexander, 474
Friedmann–Lemaître–Robertson–Walker Universe, 474
Fundamental Forms
are *not* Differential Forms, 164
first, 164
second, 164
third, 164
fundamental group, 207
Fundamental Theorem of Exterior Calculus, 411–412
boundary of a boundary is zero: $\partial^2 = 0$, 413
Cauchy's Theorem, 417
circulation, 404
Divergence Theorem, 415
explanation of $\mathbf{d}^2 = 0$, 413
exterior derivative as integral, 406–411
Faraday's Law of Electromagnetic Induction, 415
flux 2-form, 408
Gauss's Theorem, 415
Green's Theorem, 414
history of, 411–412

integral theorems of vector calculus, 413–415
line integral of 1-form, 404–406
line integral of 1-form (path-independence), 405–406
line integral of exact 1-form, 406
named by N.M.J. Woodhouse, 412
Penrose precedent for FTEC terminology, 412
proof of, 415–417
role of Stokes, 411
statement of, 411
Stokes's Theorem, 414
work, 404

Galileo, 308
Gamow, George, 332
gauge freedom, 396
gauge transformation, 396
Gauss
and possibility of absolute unit of length, 15
and the spherical map, 131
Beautiful Theorem, 138
curvature, *see* curvature
discovery of intrinsic geometry, 11, 31
Dombrowski's analysis of, 138
experimental test of the curvature of physical space, 14
first curvature formula, 40
General Investigations of Curved Surfaces, 3, 17
introduction of metric, 31
isometries versus bendings, 141
metric notation, 36, 37
motto, 138
portrait of, 17
reaction to Riemann's ideas, 298
rotations of sphere as Möbius transformations, 73
Theorema Egregium, 138, 140, 142
treatment of Abel, 7
treatment of Bolyai, 7
treatment of Riemann, 297
unaware of Global Gauss–Bonnet Theorem, 174
Gauss map, 131
Gauss's Integral Theorem, *see* Fundamental Theorem of Exterior Calculus
Gauss's Lemma, 274
via computation, 337
visualized, 275
Gauss–Bonnet Theorem
General Local, 336
Global, *see* Global Gauss–Bonnet Theorem
Local, 22, 174, 336
Gaussian curvature, *see* curvature
General Relativity, *see* Gravity
Generalized Stokes's Theorem (GST), *see* Fundamental Theorem of Exterior Calculus (FTEC)
genus, 166
increasing by adding handles, 191
of sphere, 166
of torus, 166
of two-holed doughnut, 166
Genzel, Reinhard, 464
geodesic
as equivalent of straight line, 9
as shortest route, 9
as straightest route, 11, 13, 118
as taut string, 9, 118
equation of, 244
of $\mathbb{H}^3$, 79
of Beltrami–Poincaré disc model, 62
of Beltrami–Poincaré half-plane, 60, 128–129
of cone, 27
of pseudosphere, 28, 89
parallel transport via, 240–241
possible nonuniqueness of, 10
relative acceleration, 270
sticky-tape construction of, 14, 239
vanishing geodesic curvature, 119
via parallel transport, 235–236, 238, 239
geodesic circle, 10
geodesic curvature, 115
extrinsic construction, 120
intrinsic formula for, 244
intrinsic measurement of, 119
on a cone, 334
on a sphere, 334
on touching surfaces, 334
used to construct geodesics, 120
vanishes for geodesics, 119
vector, 117, 243
via intrinsic differentiation, 335
geodesic equation, 244, 286
geodesic polar coordinates, 274–276
geodesic triangle, 10
geometric inversion, 68, 81, 82
geometrization conjecture, 6
geometrized units, 328
GGB, *see* Global Gauss–Bonnet Theorem
Ghez, Andrea, 464
Gibbs, Josiah Willard, 274
gimel (from the Hebrew alphabet), 75
Global Gauss–Bonnet Theorem, 167
discovered by Kronecker and Dyck, 174
for sphere, 168–169
for torus, 90, 169–170
Hopf's intrinsic proof, 258–260
intuitive interpretation of, 167
paradoxes?, 225
some predictions of, 224
via angular excess, 194
via Bagels and Bridges, 171–172
via folded membrane, 182
via thick pancake, 171
via topological degree, 173
via vector fields, 217
Goldbach's Conjecture, 183
Goldbach, Christian, 183
gradient 1-form, 351
gradient vector, 354–355
graph (topological), 186
gravitation, *see* Gravity
gravitational lensing, 325
Gravitational Spinor, *see* Weyl Curvature
gravitational wave astronomy, 325
gravitational waves, 323–326
curvature of, 323
depiction of oscillating tidal forces, 324
details of first detection (14th of September, 2015), 325
Einstein's 1916 prediction of, 231
enormous energy of, 325
field lines of tidal forces, 323
first detection of, 231
harnessed for gravitational wave astronomy, 325
naming convention for, 325
gravity
bending of light, 322
Big Bang, 474
Birkhoff's Theorem, 329
birth of a black hole, 330
black hole, 329
Cosmological Einstein Field Equation, 332, 474
cosmological constant, Λ, 332, 474
curvature of Friedmann–Lemaître–Robertson–Walker Universe, 474
dark energy, 333
eclipses and the tides, 339
Einstein (matter) Field Equation, 327
Einstein (matter) Field Equation (*geometrical* form), 328
Einstein (vacuum) Field Equation, 319
Einstein tensor, 328, 340
Einstein's initial Gravitational Field Equation, 327
eliminated in free fall, 307
energy-momentum tensor, 326
event horizon, 329
geometrical signature of inverse-square attraction, 314
geometrical signature of inverse-square law, 313
geometrized units, 328
gravitational waves, 323–326
gravitational work as 1-form, 347–349
neutron star, 329
neutron star collision, 331
Newton's apple, 307
Newton's explanation of the ocean tides, 311
Newton's Inverse-Square Law, 98, 307–309
precession of orbit of Mercury, 322
redshifting of light, 322
repulsion of negative energy, 333
spacetime singularity, 464
spacetime tidal forces, 317
spherical Schwarzschild field, 320
standard model (cosmological), 474
static Universe, 332
stellar evolution, 329
stress–energy tensor, 326

gravity (*continued*)
tidal force field around spherical mass, 309
tidal forces, 308–312
Tolman–Oppenheimer–Volkoff limit, 329
Universe's expansion is accelerating, 332
Weyl curvature, 340, *see* Weyl curvature
great circle, 6
Green's theorem, *see* Fundamental Theorem of Exterior Calculus
Green, George, 411
group, 65
GST, *see* Fundamental Theorem of Exterior Calculus

$\mathbb{H}^2$, *see* hyperbolic plane
Hairy-Ball Theorem, 472
handle (topological), 192
harmonic oscillator, 271
Harriot
angular excess theorem on sphere, 8
discovered conformality of stereographic projection, 46
discovered Snell's Law, 58
portrait of, 8
Heaviside, Oliver, 274, 382
helicoid, 224, 480
helix, 220
Henderson, David W., 252
Hilbert
and Polyhedral *Theorema Egregium*, 148
impossibility of hyperbolic plane within Euclidean space, 82
spherical map, 131
theorem on surfaces of constant negative curvature, 22, 52
worked with Schwarzschild, 319
Hipparchus, 44
Hodge duality, 377, 379, 385, 402, 468
Hodge, Sir W.V.D., 377
Hofstadter, Douglas, xxvi
holonomy
along open curve, 257–258
definition of, 245
equivalence to angular excess, 247
invariance under spherical map, 255
is additive, 248
measures relative acceleration of geodesics, 278
measures total curvature, 246
of geodesic polygon, 247
of geodesic triangle, 247
of the hyperbolic plane, 248–251
on cone and sphere, 335
on the sphere, 245–246
Riemann curvature and, 290
vector, 293
homeomorphic, 166
homeomorphism, 165
homogeneous coordinates, 70
homogeneous function (defined), 471
homogeneous functions, 471
honey-flow, 202
as othogonal trajectories of topographic map, 204
on surface of genus $\mathfrak{g}$, 209, 227
precise definition of, 215
reversing flow preserves indices, 217
sign of index determines orientation of covering by spherical map, 216, 217
singular points related to spherical map, 215, 216
Hooke's Law, 123, 271
Hopf
calculation of Euler characteristic, 227
indices (fractional), 213
intrinsic proof of Global Gauss–Bonnet Theorem, 258–260
line field, 211
metric notation, 37
Poincaré–Hopf Theorem, 206, 213
portrait of, 176
spherical map, 131
Umlaufsatz, 176, 225
horosphere, 82
curvature of, 92
metric of, 92
Hubble Constant, 325
Hubble's Law (1929), 332
constant of, 325
Hubble, Edwin
discovered expansion of Universe (1929), 332
discovered galaxies beyond our own (1924), 332
Huygens
finite area of pseudosphere, 88
influence on Newton, xviii
investigated pseudosphere, 22, 53
Hyperbolic Axiom, 5, 52, 60, 63
Hyperbolic Geometry
$\mathcal{K} = -1$, by convention, 56
3-dimensional, 74, 79–82
absolute unit of length, 16
angle of parallelism, 60
angular excess proportional to area, 15
as geometry of surface of constant negative curvature, 51
aymptotics, 60
Beltrami's critical discovery, 51
Bolyai–Lobachevsky Formula, 61, 62
conformal Beltrami–Poincaré disc model of, 63
conformal Beltrami–Poincaré half-plane model of, 56
conformal mapping between models, 63
curvature of $\mathbb{H}^3$, 473
definition of, 5
discovery of, 4
Euclidean and Spherical Geometries subordinate to it, 82
Euclidean illusion for small figures, 16
Euclidean plane within $\mathbb{H}^3$, 82
Euclidean sphere within $\mathbb{H}^3$, 82
horizon, 57, 79
horosphere, 82, 92
isometries of $\mathbb{H}^3$, 81
Lambert as pioneer of, 5, 15
limit rotation, 74
maximum area of a triangle, 15
metric of $\mathbb{H}^3$, 79
Poincaré as prophet of, 5
points at infinity, 57
some properties of, 15–16
sphere in $\mathbb{H}^3$, 92
standardized half-plane metric, 56
ultraparallels, 60
visualizing length within, 57
hyperbolic plane, 5, 52
$\mathcal{K} = -1$, by convention, 56
conformal Beltrami–Poincaré disc model of, 63
conformal Beltrami–Poincaré half-plane model of, 56
definition of, 57
generalized to $\mathbb{H}^3$, 79
geodesics via Clairaut's Theorem, 128–129
hemispherical $\mathbb{H}^2$ in $\mathbb{H}^3$, 80
isometries as Möbius transformations, 73
metric within $\mathbb{H}^3$, 91
pseudosphere as flawed model of, 52
within $\mathbb{H}^3$, 80
hyperbolic point, 111, 134, 136

Ibn Sahl, 58
icosahedron, 185
tesselation of sphere, 26
ideal points, 57
index, 196
as winding number, 198
determined by infinitesimal neighbourhood, 199
formal definition, 197, 205
fractional, 213
illustrated for complex powers, 200
invariance of, 197, 198
of complex powers, 198–201
of dipole, 197, 199
of honey-flow on fried banana, 204
of saddle point, 197
of vortex, 197
on a surface, 204
on a surface (illustrated), 205
index of refraction, 59
integral curve, 195
integration, *see* Fundamental Theorem of Exterior Calculus
intrinsic derivative, 241–244
definition of, 242
formal properties of, 243
geodesic curvature via, 244
of n-manifold, 284–286
Shape Operator formula for, 243
visualized, 242

intrinsic geometry
definition of, 11
inversion, 68
effect on Riemann sphere, 68
is anticonformal, 69
preserves circles, 69
isometries
definition of, 61
direct, 65
group structure of, 65–66
of $\mathbb{H}^3$, 81
of flat spacetime, 74–79
of hyperbolic plane as Möbius transformations, 73
of Riemann sphere, 73, 90
opposite, 65
symmetry groups of surfaces of constant curvature, 72–74
versus bendings, 141
isothermal coordinates, 40, 84

Jacobi Equation
computational proof of, 301–302
defined, 271
geodesic polar coordinates proof of, 274–276
holonomy proof of, 276–278
in n-manifold, 302
introduced, 269
negative curvature, 272–274
on general surface, 336
positive curvature, 270–272
Sectional Jacobi Equation, 299–302
zero curvature, 269–270
Jacobi, Carl Gustav Jacob, 269
Jobs, Steve, xxvi

Kepler, 58, 123
Kepler's Laws, 124
kernel (of 1-form), 349
Kerr solution, 322
Kerr, Roy, 322
Klein
classification of closed surfaces, 191
reaction to Riemann's ideas, 298
worked with Schwarzschild, 319
Kronecker delta, 353
Kronecker, Leopold, 174

Lambert
and first theorems of Hyperbolic Geometry, 5
career rescued by Euler, 8
discovery of hyperbolic equivalent of Harriot's Theorem, 15
portrait of, 5
projection of sphere, 85
Laplace Equation, 85
Laplacian, 40
in polar coordinates, 47
Le Verrier, Urbain, 322
Legendre, 189
Leibniz, xviii
Leibniz Rule, *see* Forms, exterior derivative
Lemaître, Georges, 474
level curves, 203
Levi-Civita connection, 242
Levi-Civita, Tullio, 231
discoverer of parallel transport, 232
portrait of, 232
Levy, Anthony, xxvii
L'Huilier, Simon-Antoine-Jean, 190, 226
Lie Bracket, 287
Lie, Sophus, 287
Liebmann, Heinrich
equivalence of Möbius and Lorentz transformations, 74
sphere rigidity theorem, 21, 454–455
light cone
picture of, 316
LIGO detector, 231
limit rotation, 74
line field, 211
applications to physics, 214
as flow around barriers, 213
fractional indices, 212
Poincaré–Hopf Theorem applies to, 213
singular points of, 212
Linear Algebra (Visual), 221
Listing, 166
Lobachevsky, 4
initally sought 3-dimensional non-Euclidean Geometry, 79
named horosphere, 82
loop, 175
Lorentz
contraction, 75
Special Theory of Relativity, 75
symmetries of spacetime as Möbius transformations, 74–79
Lorentz transformations, 74–79
as spin-matrices, 77
as spin-transformations, 77
boost, 78, 93
classification of, 78
fixed null rays of, 77
four archetypes, 77
four-screw, 78
as Möbius transformations, 77
null rotation, 79
preservation of spacetime interval, 77
loxodrome, 87
lune, 168

Mainardi–Codazzi Equations, *see* Peterson–Mainardi–Codazzi Equations
manifold, 280
n-dimensional, 280
3-manifold as exemplar of n-manifold, 281
angular excess in a 3-manifold, 282
pseudo-Riemannian, 280
Riemannian, 280
map
conformal, 39
definition of, 31
element of area, 38
faithfulness of, 33
of hyperbolic plane, *see* hyperbolic plane
of pseudosphere, 55
of sphere, *see* sphere
Maxwell
admiration of Faraday, 382
and Polyhedral *Theorema Egregium*, 148
portrait of, 382
refined flux concept, 377
Smith's Prize examination, 411
Maxwell's Equations, *see* electromagnetism
via Forms, 401–403
mean curvature, 110
Mercator projection, 86
loxodrome, 87
rhumb line, 87
meridian, 121
Merton College, Oxford, xviii, xxvii
metric
conformal, 39
conformal stereographic, 47
curvature formula, 38, 261
definition of, 31
formula in orthogonal coordinates, 37
Gauss's notation differs from ours, 37
general formula for 2-surface, 36
in orthogonal coordinates, 262
of $\mathbb{H}^2$ in $\mathbb{H}^3$, 91
of $\mathbb{H}^3$, 79
of Beltrami–Poincaré disc model, 62, 90
of Beltrami–Poincaré half-plane model, 56
of conformal map of pseudosphere, 55
of hemispherical $\mathbb{H}^2$ in $\mathbb{H}^3$, 81
of horosphere, 92
of pseudosphere, 53
of pseudosphere via parameterized tractrix, 88
of torus, 90
spacetime interval, 75
tensor, 363–364
used to change tensor valence, 366–368
Meusnier's Theorem, 117
Meusnier, Jean-Baptiste, 110
Minding's Theorem, 21, 276, 336
Minding, Ferdinand, 53
minimal surface, 130
Minkowski, 382
spacetime geometry, 74
spacetime interval, 75
worked with Schwarzschild, 319
miracles
false, xxiii
true, xxiii
Misner, Charles, 475

Möbius
band, 166
classification of closed surfaces, 166
strip, 166
Möbius transformations, 67–74
as isometries of $\mathbb{H}^3$, 81
as Lorentz transformations, 77
as rotations of the Riemann sphere, 73, 90
as symmetry group of flat spacetime, 74–79
characterized by circle-preserving property, 81
classification of, 78
decomposition into simpler transformations, 67
elliptic, 78
explanation of matrix correspondence, 71
fixed points of, 77
form a group, 71
four achetypal Lorentz transformations, 77
hyperbolic, 78
inversion in a circle, 67
isometries of Riemann sphere, 73
loxodromic, 78
matrix representation of, 70–71
nonsingular, 67
normalized, 71
normalized matrix defined uniquely up to sign, 71
parabolic, 79
rotations of sphere via antipodal points, 73
singular, 67
symmetry groups of surfaces of constant curvature, 66, 72–74
moment of force, 128
monkey saddle, 136
complex equation, 137
curvature of, 224
generalized, 137
Morgan, Frank, xxvi
Riemannian Geometry, 481

naked singularity, 464
Needham, Faith, xxviii
Needham, Guy, xxvi
Needham, Hope, xxviii
Needham, Mary, xxvii
Nel, Stanley, xxvi
nematic field, 478
network (topological), 186
neutron star, 329, 331
Newcomb, Simon, 322
Newton
and Celestial Mechanics, 210
and the falling apple, 307
elliptical orbits in linear field, 123
embraced geometrical methods in 1680s, xviii
explained Kepler's Laws, 124
explained the ocean tides, 311
general curvature formula, 100
geometric curvature formula, 99
geometric definition of tractrix, 52
investigated tractrix, 22
Law of Gravitation, 308
parametric curvature formula, 103
Principia, xviii–xx
Principia, area is the clock, 124
Principia, Lemma II, 99
Principia, Proposition 1, 124
Principia, Proposition 2, 126
Principia, Proposition 10, 123
Principia Proposition 31, 309
priority battle with Leibniz, xviii
proof of Kepler's Second Law, 124–126
Second Law of Motion, 97
shunned his 1665 calculus in the *Principia*, xviii
synthetic method of fluxions, xviii
Westfall's biography of, xviii
Ultimate Equality, *see* Ultimate Equality
Newtonmas, 98
Nobel Prize (2011), 332
Nobel Prize (2017), 231
Nobel Prize (2020), 464
non-Euclidean Geometry, *see* Hyperbolic Geometry
nonorientable, 166
normal curvature, 115
normal curvature vector, 243
normal map, 131
null cone
picture of, 316
Riemann sphere representation of, 76
null vector, 76
Nye, J. F., 478

O'Neill, Barrett
championed Forms approach to Differential Geometry, 431, 475
championed Shape Operator, 164
octahedron, 185, 227
optics, 58
orientable surface, 165
orthogonal coordinates, 37
orthogonal linear transformation, 222
osculating plane, 106

Pappus, xviii
centroid theorem, 90
parabolic point, 111, 134, 136
Parallel Axiom
definition of, 4
via angle sum of a triangle, 4
via similar triangles, 4, 15
parallel transport
discovery of, 232
geodesics and, 235–236
in n-manifold via constant-angle cone, 282–283
in n-manifold via parallel-transported plane, 283–284
in n-manifold via Schild's Ladder, 284
preserved by spherical map, 255
used to define intrinsic derivative, 241–244
via geodesics, 240–241
via potato-peeler, 236–239
via projection into surface, 233–235
parallelepiped volume, 381
Pauli Exclusion Principle, 329
Penrose
and *Theorema Egregium*, 252
author's debt to, xxv
Cosmic Censorship Hypothesis, 464
Escher diagram, 63
labelling of light rays with complex numbers, 92
Nobel Prize in Physics, 464
on Lambert, 5
pioneering use of 2-spinors, 70
studied under Hodge, 377
Perelman, 6
Perlmutter, Saul, 332
Peterson–Mainardi–Codazzi Equations, 448
geometric meaning of, 450
use in Hilbert's Lemma, 453
phase portrait, 195
photons (spinning), 470
Pixar, xxvi
planar point, 136
surrounded by negative curvature, 137
surrounded by positive curvature, 137
Plato, 185
Platonic Forms, xxiii
Platonic solids, 185, 227
topologically determined, 186, 226
uniqueness of, 226
Poincaré, *see* Beltrami–Poincaré half-plane
and celestial mechanics, 210
and Euler characteristic, 184
as father of topology, 165
as prophet of Hyperbolic Geometry, 5
discovery of Möbius isometries of $\mathbb{H}^3$, 81
rediscovery of hyperbolic disc model, 62
rediscovery of hyperbolic half-plane model, 57
Special Theory of Relativity, 75
Poincaré Conjecture, 6
Poincaré disc, *see* Beltrami–Poincaré disc
Poincaré Lemma, 396, 418
Poincaré–Hopf Theorem, 206
also applies to line fields, 213
Hairy Ball Theorem, 472
on sphere, 206, 213
on torus, 207
physical applications of, 477
proof of, 207–208
point at infinity, 44
polar decomposition, 222

Pólya, 201
mechanical proof of Snell's Law, 87
Pólya vector field, 201
divergence-free and curl-free, 324
physical examples of, 201
polygonal net, 186
polyhedral spike, 145
curvature of, 146, 147
spherical image of, 147
Polyhedral *Theorema Egregium*, 147
attributed to Hilbert, 148
discovered by Maxwell, 148
visualization of, 146
polyhedron
and Theorema Egregium, 147
curvature of, 145–147
dual, 227
positive definite, 222
positive semidefinite, 222
potential, 396
Pound, Robert, 322
Prime Meridian, 274
principal curvatures, 109
as eigenvalues of Shape Operator, 152
extrinsic curvature in terms of, 134
Unit Speed Formula, 113, 114
principal directions, 109
as eigenvectors of Shape Operator, 152
as symmetry directions, 111, 133
rate of rotation of normal along, 134
principal normal, 107
principal radii of curvature
of pseudosphere, 105
of surface of revolution, 112
of torus, 112
Principia, *see* Newton
projective coordinates, 70
projective geometry, 70
projective map, 32
projective model, 32
pseudosphere, 22
as flawed model of Hyperbolic Plane, 52
building your own, 54, 88
conformal map of, 54–55
conformal metric of, 55
constant negative curvature of, 53
curvature of, 142
definition of, 53
element of area, 55
finite area of, 88
geodesics of, 28, 89
geodesics via Clairaut's Theorem, 128–129
metric of, 53
principal radii of curvature, 105, 113
rim, 56
tractrix, as generator of, 22
via parameterized tractrix, 88
Ptolemy, 44, 58
punctured plane, 419
Pythagoras, 3
Pythagoras's Theorem
characterizes flatness, 3
Pythagorean triples
Babylonian examples, 3
definition of, 3
general formula for, 25

radius of curvature, 98
Rebka, Glen, 322
redshift
gravitational, 322
of galaxies, 332
refraction, 58
rhumb line, 87
Ricci Calculus, 231
Ricci tensor, 302–306
definition of, 305
effect on bundle of geodesics, 305
geometrical meaning of, 304–306
notation, 304
sign conventions, 305
symmetry of, 305
Ricci, Gregorio, 231, 298
portrait of, 303
Riemann
as father of topology, 165
Darrigol's analysis of, 298
definition of genus, 166
discovered Differential Bianchi Identity, 298
Foundations of Geometry lecture (1854), 297
French Academy Prize Essay (1861), 298
Gauss's reaction to lecture by, 298
Klein's reaction to, 298
portrait of, 281
Spivak's analysis of, 298
Riemann curvature operator, 287, 290
antisymmetry of, 290
Riemann sphere, 44
as representation of null cone in spacetime, 76
fixed points under Möbius transformation, 77
rotated by complex inversion, 69
rotations as Möbius transformations, 73, 90
used to label light rays, 92
Riemann tensor, 281
and exponential operator, 339
and vector commutator, 287–288
antisymmetry of, 290
changing valence of, 367
components of, 292–293
defined, 290
defined via parallel transport, 286–287
different notational conventions, 290
history of, 297–298
is a tensor, 291–292
number of components, 281, 338
Riemann curvature operator, 287
standard definition of, 360
symmetries of, 294–295, 337
vector holonomy, 293
visualization of, 289
Weyl curvature, *see* Weyl curvature
Riess, Adam G., 332
Robertson, 474
Rodrigues, Olinde, 17
discovered extrinsic curvature formula, 134
introduced spherical map, 131
Rodrigues–Gauss map, 131
Rose, Jed, xxv
rotation matrix, 152
Royal Institution of London, 381
Royal Observatory at Greenwich, 274

Saccheri, 5
Saddle Point, 196
Sagittarius A*, 331, 464
Schild's Ladder, 284
Schild, Alfred, 284
Schmidt, Brian P., 332
Schutz, Bernard, 325
Schwarzschild radius, 321, 329
Schwarzschild Solution, 320
interior, 321
Schwarzschild, Karl, 319, 321
portrait of, 321
Second Fundamental Form, 151
sectional curvature, 282, 296–297
Segerman, Henry, 191
self-adjoint matrix, 153
self-dual 2-form, 470
Shape Operator, 151
also called the Second Fundamental Form, 164
Cartesian formula, 222
curvature formula, 222
curvature interpretation of, 159
curvature interpretation visualized, 160
determinant of, 153
determined by three normal curvatures, 161
diagonalized matrix of, 153
effect on asymptotic directions, 163
eigenvectors and eigenvalues of, 152, 223
general matrix of, 158, 161
geometric meaning of components, 160
is linear, 152
is symmetric, 153
matrix representation of, 152
of saddle, 223
visualization of, 150
Singer
Atiyah–Singer Index Theorem, 165
spherical map, 131
singular point, 195
singular value decomposition
discovered by Beltrami, 154
geometric derivation of, 154–156
in $\mathbb{R}^3$, 222
matrix form of, 155
singular values of, 154
statement of, 154
twist of, 154
visualization of, 155
Sink, 200
skew symmetric, 222

Snell, 58
Snell's Law, 58
  Fermat's two proofs of, 58
  Generalized, 59
  mechanical proof of, 87
  Newtonian proof of, 58
  Ptolemy's experiments, 58
soap films, 130
Source, 200
  physics of, 201
spacetime
  4-velocity, 316
  4-vector, 74
  absolute, observer-independent structure of, 75
  causal structure of, 316
  diagrams, 315–316
  event, 74, 316
  event horizon, 330
  Gravitational Field Equation, 319
  interval, 75
  light cone, 316
  Lorentz transformation, 76
  metric, 314–315
  metric tensor, 314
  Minkowski, 74
  null cone, 316
  null vector, 315
  singularity of, 330, 464
  spacelike vector, 315
  spherical Schwarzschild geometry, 320
  symmetries as Möbius Transformations, 74–79
  tetrad, 315
  tidal forces, 317
  timelike vector, 315
  Weyl curvature, 319
  Weyl curvature formula, 340
  world-line, 315
spacetime interval, 75
  preservation by Lorentz transformation, 77
  vivid interpretation of, 75
spaghettification, 330
Special Theory of Relativity, 74, 382
  aberration formula, 93
  discrepancy between clocks in relative motion, 75
  Einstein, 75
  Einstein's 1905 paper, 382
  Lorentz, 75
  Lorentz transformation, 76
  Minkowski, 74
  Poincaré, 75
  spacetime interval, 75
Spectral Theorem, 157
sphere
  Archimedes–Lambert metric, 86
  Archimedes–Lambert projection, 85
  area in cylindrical polar coordinates, 88
  area of small circle on, 20
  central cylindrical projection, 86
  central projection, 32, 83
  central-cylindrical metric, 86
  circumference of small circle on, 20, 28
  Clairaut's theorem, 121
  curvature of, 134, 142
  extrinsic curvature, 132
  genus of, 166
  Global Gauss–Bonnet Theorem for, 168–169
  holonomy on, 245–246
  in $\mathbb{H}^3$, 92
  isometries as Möbius Transformations, 73
  Liebmann's rigidity theorem, 21
  longitude-latitude metric formula, 34
  lune of, 168
  Mercator projection, 86
  other surfaces with same intrinsic geometry, 21
  preservation of antipodal points used to derive rotations of, 73
  projective map, 32
  projective metric formula, 33
  projective model, 32
  Riemann, 44
  rigidity of, 454–455
  stereographic metric, 47, 83
  stereographic projection of, 44
  tessellation of, 26
  vector fields on, 206
Spherical Geometry
  absolute unit of length, 15
  Euclidean illusion for small figures, 16
  geodesics are great circles, 6
  Harriot's angular excess theorem, 8
  perpendicular bisector of, 25
  Spherical Axiom, 6
spherical map, 131
  and orientation, 135
  covers closed surface, 214
  curvature as local expansion of area by, 132
  degree determined by Euler characteristic, 174, 217
  degree of, 180
  folds of image, 182
  "Gauss map", 131
  generalized, 144
  index of, 178
  negative covering by, 173
  of bridged bagels, 174
  of thick pancake, 173, 174
  positive covering by, 173
  precedence for terminology, 131
  preserves holonomy, 255
  preserves parallel transport, 255
  related to honey-flow, 215
  "Rodrigues–Gauss map", 131
  topological degree of, 173
spin-matrix, 77
spin-transformation, 77
2-spinor, 70
spinorial objects, 71
stable node, 200
Star Trek
  Captain Kirk, 191, 482
  Dr. McCoy, 38
  forms proof of ("*Star Trek* phaser") metric curvature formula, 452
  geometric proof of ("*Star Trek* phaser") metric curvature formula, 266
  Mr. Spock, 191, 476, 482
  NCC-1701, 435
  ("*Star Trek* phaser") metric curvature formula, 38
  The City on the Edge of Forever, 38
  The Doomsday Machine, 191
stereographic projection, 44
  conformality, 46
  formulas, 47–49
  image of line, 45
  image of point, 44
  preservation of metric under Möbius transformation, 73
  preserves circles, 49
  used to identify light rays with complex numbers, 77
Stiefel vector field, 208, 227
Stiefel, Eduard, 208
stiffness, 271
Stillwell, xxvii
  Mathematics and Its History, xxi, xxvii
  transformation of Escher diagram, xxvii, 63, 64
Stillwell, Elaine, xxvii
Stokes's Theorem, *see* Fundamental Theorem of Exterior Calculus
streamline, 195
stress-energy tensor, 326
surface of revolution
  Clairaut's Theorem, *see* Clairaut's Theorem
  curvature of, 114, 142, 220, 221
  meridian of, 121
  normal of, 168
  of constant curvature, 89
  of constant positive curvature, 21
  of tractrix is pseudosphere, 22
  parallel of, 123
  principal radii of curvature, 112
  total curvature, 169
surfaces of constant mean curvature, 480
SVD, *see* singular value decomposition
symmetric matrix, 153
  geometric meaning of, 156

Tabachnikov, Sergei, 58
Tensor Calculus, 231
tensors, 360–369
  addition of, 361
  and Linear Algebra, 361–362, 364–365
  antisymmetric, 369, 370
  basis tensors, 362
  changing valence, 366–368
  components of, 362–363
  contraction, 365–366
  contravariant indices, 363
  covariant indices, 363
  definition of, 360

Einstein tensor, 328, 340
index lowering, 367
index raising, 367
matrix multiplication, 466
metric, 363–364
preliminary definition, 291
product of, 361
Riemann, *see* Riemann tensor
skew symmetric, 369
symmetric, 369
valence of, 360
tessellation
of the plane, 25
of the sphere, 26
tetrad, 315
Thales, 3
Theaetetus, 185
Theorema Egregium, 138, 140
Beautiful Theorem as precursor of, 138
bending vs isometry, 141
confirmed for specific surfaces, 142
equality of extrinsic and intrinsic curvature, 142
experimental test of, 141
geometric proof of, 255
illustrated with banana skin, 140
polyhedral version of, 147
proved using Forms, 451–452
simple geometric proof of, 252–256
statement of, 140
Thomson, William (Lord Kelvin), 148, 411
Thorne, Kip S., 231, 475
Thurston, 6, 477
tidal forces, 308–312
around spherical mass, 309
field lines of a gravitational wave, 323
spacetime depiction of, 317
Weyl curvature, 319
Weyl curvature formula, 340
Tolman–Oppenheimer–Volkoff limit, 329
topographic map, 203, 349, 350
preservation of orthogonality, 204
topological defect, 478
topological degree, 173, 215
topological joke, 191
topological mapping, 165
topological transformation, 165
topologically equivalent, 166
topology, 165
Toponogov, V. A.
spherical map, 131
torsion, 106, 107, 164
curvature in terms of, 164
torus, 18, 89
curvature of, 21, 90, 142
genus of, 166
Global Gauss–Bonnet Theorem for, 169–170
metric of, 90
principal radii of curvature, 112
total curvature of, 90
vector fields on, 207
with g handles, 191
total curvature, 165
TOV, *see* Tolman–Oppenheimer–Volkoff limit
trace (of energy-momentum), 327
trace of matrix, 161
invariance of, 162
tractrix
as generator of pseudosphere, 22
as geodesic of pseudosphere, 28
curvature of, 105
equation of, 52
mechanical definition of, 52
Newton's geometric definition of, 52
parametric description of, 88
tractroid, 22
transpose of matrix, 153
composed with original matrix, 158
geometric meaning of, 156
geometric visualization of, 157
has same determinant, 157
triangulation, 187
twist, 43, 154

Ultimate Equality
$\asymp$–notation, xviii
adoption of $\asymp$–notation by Chandrasekhar, xviii
basic properties of, xix
definition, xviii
Newton's use of in the *Principia*, xviii–xx
of vectors, 151
practice exercises using, 24
used to calculate $(\sin\theta)'$ and $(\cos\theta)'$, 24
used to calculate $(\tan\theta)'$, xix
used to calculate $(x^3)'$, 24
ultra-parallel, 60
umbilic, 110
Umlaufsatz, 176
proof of, 225
unitary, 72
universal cover, 56
Unstable Node, 200

valence (of tensor), 360
VCA, *see* Visual Complex Analysis, in bibliography
Vector Calculus, *see* Forms: Vector Calculus
vector field, 195
and differential equations, 211
archetypal singular points, 198–201
centre, 200
circulation of, 263
commutator, 287–288, 407, 409
conservative, 399
crosspoint, 196
curl of, 264, 398
dipole, 196, 199, 200
divergence, 399
divergence-free, 324
double saddle point, 200
fiducial, 205
focus, 196, 200
generalization to line field, 211
Honey-Flow, 202
index, 196–198
index on surface, 205
integral curve, 195
irrotational, 399
on $\mathcal{S}_g$ with $-\chi$ saddle points, 227
on $\mathcal{S}_g$ with one singular point, 227
on a surface, 202
on sphere, 206
on torus, 207
phase portrait, 195
quadrupole, 200
reference, 205
saddle point, 196, 200
singular point, 195
singularity, 419
sink, 200
source, 200
stable node, 200
Stiefel, 208, 227
streamline, 195
unstable node, 200
vortex, 196, 200
vorticity, 399
with barriers, 213
vector potential, 402
vector product (as wedge product), 380
vector space (defined), 347
vortex, 196, 200
Voss, Aurel, 298

Wachter, 82
Walker, 474
Wallis, 4, 15
Ward, Richard S., xxvii
Weber, Wilhelm, 298
Weingarten Map, 151
Weiss, Rainer, 231
Weyl curvature, 319
conformal invariance of, 341
explicit formula for, 340
Penrose's "fingerprint" of, 478
requires $n \geqslant 4$, 281
Wheeler
General Relativity in one sentence, 307
impact on the author's life, xviii, 482
Wiles, 6
winding number, 198, 422
Wolf, Robert, 221
Woodhouse, N.M.J., 412
work, 404
wormhole, 172, 225
puzzle, 172, 225

Zeitz, Paul, xxvi
"Zeno's Revenge", 57

Milton Keynes UK
Ingram Content Group UK Ltd.
UKHW050453240924
448772UK00005B/20

9 780691 203706